Enumerative Combinatorics, Volume 1

Richard Stanley's two-volume basic introduction to enumerative combinatorics has become the standard guide to the topic for students and experts alike. This thoroughly revised second edition of Volume 1 includes ten new sections and more than 350 new exercises, most with solutions, reflecting numerous new developments since the publication of the first edition in 1986.

The material in Volume 1 was chosen to cover those parts of enumerative combinatorics of greatest applicability and with the most important connections with other areas of mathematics. The four chapters are devoted to an introduction to enumeration (suitable for advanced undergraduates), sieve methods, partially ordered sets, and rational generating functions. Much of the material is related to generating functions, a fundamental tool in enumerative combinatorics.

In this new edition, the author brings the coverage up to date and includes a wide variety of additional applications and examples, as well as updated and expanded chapter bibliographies. Many of the less difficult new exercises have no solutions so that they can more easily be assigned to students. The material on P-partitions has been rearranged and generalized; the treatment of permutation statistics has been greatly enlarged; and there are also new sections on q-analogues of permutations, hyperplane arrangements, the cd-index, promotion and evacuation, and differential posets.

RICHARD P. STANLEY is a professor of applied mathematics at the Massachusetts Institute of Technology. He is universally recognized as a leading expert in the field of combinatorics and its applications to a variety of other mathematical disciplines. In addition to the seminal two-volume book *Enumerative Combinatorics*, he is the author of *Combinatorics and Commutative Algebra* (1983) and more than 100 research articles in mathematics. Among Stanley's many distinctions are membership in the National Academy of Sciences (elected in 1995), the 2001 Leroy P. Steele Prize for mathematical exposition, and the 2003 Schock Prize.

Enumerative Combinatorics, Volume 1

Second Edition

RICHARD P. STANLEY
Massachusetts Institute of Technology

CAMBRIDGE
UNIVERSITY PRESS

CAMBRIDGE
UNIVERSITY PRESS

32 Avenue of the Americas, New York NY 10013-2473, USA

Cambridge University Press is part of the University of Cambridge.

It furthers the University's mission by disseminating knowledge in the pursuit of education, learning and research at the highest international levels of excellence.

www.cambridge.org
Information on this title: www.cambridge.org/9781107602625

First edition published by Wadsworth 1986
First Cambridge edition 1997
Second edition published 2012

A catalogue record for this publication is available from the British Library

Library of Congress Cataloguing in Publication data
Stanley, Richard P., 1944–
Enumerative combinatorics. Volume 1 / Richard P. Stanley. – 2nd ed.
p.   cm. – (Cambridge studies in advanced mathematics ; 49)
Includes bibliographical references and index.
ISBN 978-1-107-01542-5 (hardback) – ISBN 978-1-107-60262-5 (paperback)
1. Combinatorial enumeration problems. I. Title.
QA164.8.S73   2011
511'.62–dc23   2011037989

ISBN 978-1-107-01542-5 Hardback
ISBN 978-1-107-60262-5 Paperback

to msp

"Yes, wonderful things."
—Howard Carter when asked if he saw anything, upon his first glimpse into the tomb of Tutankhamun

Contents

Contents   ix

Preface

Enumerative combinatorics has undergone enormous development since the publication of the first edition of this book in 1986. It has become more clear what the essential topics are, and many interesting new ancillary results have been discovered. This second edition is an attempt to bring the coverage of the first edition more up to date and to impart a wide variety of additional applications and examples.

The main difference between this edition and the first is the addition of ten new sections (six in Chapter 1 and four in Chapter 3) and more than 350 new exercises. In response to complaints about the difficulty of assigning homework problems whose solutions are included, I have added some relatively easy exercises without solutions, marked by an asterisk. There are also a few organizational changes, the most notable being the transfer of the section on P-partitions from Chapter 4 to Chapter 3, and extending this section to the theory of (P, ω)-partitions for any labeling ω. In addition, the old Section 4.6 has been split into Sections 4.5 and 4.6.

There will be no second edition of volume 2 nor a volume 3. Since the references in volume 2 to information in volume 1 are no longer valid for this second edition, I have included a table entitled "First Edition Numbering," which gives the conversion between the two editions for all numbered results (theorems, examples, exercises, etc., but not equations).

Exercise 4.12 has some sentimental meaning for me. This result, and related results connected to other linear recurrences with constant coefficients, is a product of my earliest research, done around the age of 17 when I was a student at Savannah High School.

I have written my work, not as an essay which is to win the applause of the moment, but as a possession for all time.

It is ridiculous to compare *Enumerative Combinatorics* with *History of the Peloponnesian War*, but I can appreciate the sentiment of Thucydides. I hope this book will bring enjoyment to many future generations of mathematicians and aspiring mathematicians as they are exposed to the beauties and pleasures of enumerative combinatorics.

Acknowledgments

It is impossible to acknowledge the innumerable people who have contributed to this new volume. A number of persons provided special help by proofreading large portions of the text, namely, (1) Donald Knuth; (2) the Five Eagles (五只鹰): Yun Ding (丁云), Rosena Ruo Xia Du (杜若霞), Susan Yi Jun Wu (吴宜均), Jin Xia Xie (谢瑾霞), and Dan Mei Yang (杨丹梅); (3) Henrique Pondé de Oliveira Pinto; and (4) Sam Wong (王昭为). To these persons I am especially grateful. Undoubtedly many errors remain, whose fault is my own.

1

What Is Enumerative Combinatorics?

1.1 How to Count

The basic problem of enumerative combinatorics is that of counting the number of elements of a finite set. Usually we are given an infinite collection of finite sets S_i where i ranges over some index set I (such as the nonnegative integers $\mathbb{N}$), and we wish to count the number $f(i)$ of elements in each S_i "simultaneously." Immediate philosophical difficulties arise. What does it mean to "count" the number of elements of S_i? There is no definitive answer to this question. Only through experience does one develop an idea of what is meant by a "determination" of a counting function $f(i)$. The counting function $f(i)$ can be given in several standard ways:

1. The most satisfactory form of $f(i)$ is a completely explicit closed formula involving only well-known functions, and free from summation symbols. Only in rare cases will such a formula exist. As formulas for $f(i)$ become more complicated, our willingness to accept them as "determinations" of $f(i)$ decreases. Consider the following examples.

1.1.1 Example. For each $n \in \mathbb{N}$, let $f(n)$ be the number of subsets of the set $[n] = \{1, 2, \ldots, n\}$. Then $f(n) = 2^n$, and no one will quarrel about this being a satisfactory formula for $f(n)$.

1.1.2 Example. Suppose n men give their n hats to a hat-check person. Let $f(n)$ be the number of ways that the hats can be given back to the men, each man receiving one hat, so that no man receives his own hat. For instance, $f(1) = 0$, $f(2) = 1$, $f(3) = 2$. We will see in Chapter 2 (Example 2.2.1) that

$$f(n) = n! \sum_{i=0}^{n} \frac{(-1)^i}{i!}. \tag{1.1}$$

This formula for $f(n)$ is not as elegant as the formula in Example 1.1.1, but for lack of a simpler answer we are willing to accept (1.1) as a satisfactory formula. It certainly has the virtue of making it easy (in a sense that can be made precise) to compute the values $f(n)$. Moreover, once the derivation of (1.1) is understood

1

(using the Principle of Inclusion–Exclusion), every term of (1.1) has an easily understood combinatorial meaning. This enables us to "understand" (1.1) intuitively, so our willingness to accept it is enhanced. We also remark that it follows easily from (1.1) that $f(n)$ is the nearest integer to $n!/e$. This is certainly a simple explicit formula, but it has the disadvantage of being "noncombinatorial"; that is, dividing by e and rounding off to the nearest integer has no direct combinatorial significance.

1.1.3 Example. Let $f(n)$ be the number of $n \times n$ matrices M of 0's and 1's such that every row and column of M has three 1's. For example, $f(0) = 1$, $f(1) = f(2) = 0$, $f(3) = 1$. The most explicit formula known at present for $f(n)$ is

$$f(n) = 6^{-n} n!^2 \sum \frac{(-1)^\beta (\beta + 3\gamma)! \, 2^\alpha \, 3^\beta}{\alpha! \, \beta! \, \gamma!^2 \, 6^\gamma}, \tag{1.2}$$

where the sum ranges over all $(n + 2)(n + 1)/2$ solutions to $\alpha + \beta + \gamma = n$ in nonnegative integers. This formula gives very little insight into the behavior of $f(n)$, but it does allow one to compute $f(n)$ much faster than if only the combinatorial definition of $f(n)$ were used. Hence with some reluctance we accept (1.2) as a "determination" of $f(n)$. Of course, if someone were later to prove that $f(n) = (n-1)(n-2)/2$ (rather unlikely), then our enthusiasm for (1.2) would be considerably diminished.

1.1.4 Example. There are actually formulas in the literature ("nameless here for evermore") for certain counting functions $f(n)$ whose evaluation requires listing all (or almost all) of the $f(n)$ objects being counted! Such a "formula" is completely worthless.

2. A recurrence for $f(i)$ may be given in terms of previously calculated $f(j)$'s, thereby giving a simple procedure for calculating $f(i)$ for any desired $i \in I$. For instance, let $f(n)$ be the number of subsets of $[n]$ that do not contain two consecutive integers. For example, for $n = 4$ we have the subsets $\emptyset$, $\{1\}$, $\{2\}$, $\{3\}$, $\{4\}$, $\{1,3\}$, $\{1,4\}$, $\{2,4\}$, so $f(4) = 8$. It is easily seen that $f(n) = f(n-1) + f(n-2)$ for $n \geq 2$. This makes it trivial, for example, to compute $f(20) = 17711$. On the other hand, it can be shown (see Section 4.1 for the underlying theory) that

$$f(n) = \frac{1}{\sqrt{5}} \left(\tau^{n+2} - \bar{\tau}^{n+2} \right),$$

where $\tau = \frac{1}{2}(1 + \sqrt{5})$, $\bar{\tau} = \frac{1}{2}(1 - \sqrt{5})$. This is an explicit answer, but because it involves irrational numbers, it is a matter of opinion (which may depend on the context) whether it is a better answer than the recurrence $f(n) = f(n-1) + f(n-2)$.

3. An algorithm may be given for computing $f(i)$. This method of determining f subsumes the previous two, as well as method **5**, which follows. Any counting function likely to arise in practice can be computed from an algorithm, so the acceptability of this method will depend on the elegance and performance of the algorithm. In general, we would like the time that it takes the algorithm to compute $f(i)$ to be "substantially less" than $f(i)$ itself. Otherwise, we are accomplishing little more than a brute force listing of the objects counted by $f(i)$. It would take us too far afield to discuss the profound contributions that computer science has made to the problem of analyzing, constructing, and evaluating algorithms. We will be concerned almost exclusively with enumerative problems that admit solutions that are more concrete than an algorithm.

4. An estimate may be given for $f(i)$. If $I = \mathbb{N}$, this estimate frequently takes the form of an *asymptotic formula* $f(n) \sim g(n)$, where $g(n)$ is a "familiar function." The notation $f(n) \sim g(n)$ means that $\lim_{n\to\infty} f(n)/g(n) = 1$. For instance, let $f(n)$ be the function of Example 1.1.3. It can be shown that

$$f(n) \sim e^{-2}36^{-n}(3n)!.$$

For many purposes this estimate is superior to the "explicit" formula (1.2).

5. The most useful but most difficult to understand method for evaluating $f(i)$ is to give its *generating function*. We will not develop in this chapter a rigorous abstract theory of generating functions, but will instead content ourselves with an informal discussion and some examples. Informally, a generating function is an "object" that represents a counting function $f(i)$. Usually this object is a *formal power series*. The two most common types of generating functions are *ordinary* generating functions and *exponential* generating functions. If $I = \mathbb{N}$, then the ordinary generating function of $f(n)$ is the formal power series

$$\sum_{n\geq 0} f(n)x^n,$$

while the exponential generating function of $f(n)$ is the formal power series

$$\sum_{n\geq 0} f(n)\frac{x^n}{n!}.$$

(If $I = \mathbb{P}$, the positive integers, then these sums begin at $n = 1$.) These power series are called "formal" because we are not concerned with letting x take on particular values, and we ignore questions of convergence and divergence. The term x^n or $x^n/n!$ merely marks the place where $f(n)$ is written.

If $F(x) = \sum_{n\geq 0} a_n x^n$, then we call a_n the *coefficient* of x^n in $F(x)$, and write

$$a_n = [x^n]F(x).$$

Similarly, if $F(x) = \sum_{n\geq 0} a_n x^n/n!$, then we write

$$a_n = n![x^n]F(x).$$

In the same way, we can deal with generating functions of several variables, such as

$$\sum_{l \geq 0} \sum_{m \geq 0} \sum_{n \geq 0} f(l, m, n) \frac{x^l y^m z^n}{n!}$$

(which may be considered as "ordinary" in the indices l, m and "exponential" in n), or even of infinitely many variables. In this latter case every term should involve only finitely many of the variables. A simple generating function in infinitely many variables is $x_1 + x_2 + x_3 + \cdots$.

Why bother with generating functions if they are merely another way of writing a counting function? The answer is that we can perform various natural operations on generating functions that have a combinatorial significance. For instance, we can add two generating functions, say in one variable with $I = \mathbb{N}$, by the rule

$$\left(\sum_{n \geq 0} a_n x^n \right) + \left(\sum_{n \geq 0} b_n x^n \right) = \sum_{n \geq 0} (a_n + b_n) x^n$$

or

$$\left(\sum_{n \geq 0} a_n \frac{x^n}{n!} \right) + \left(\sum_{n \geq 0} b_n \frac{x^n}{n!} \right) = \sum_{n \geq 0} (a_n + b_n) \frac{x^n}{n!}.$$

Similarly, we can multiply generating functions according to the rule

$$\left(\sum_{n \geq 0} a_n x^n \right) \left(\sum_{n \geq 0} b_n x^n \right) = \sum_{n \geq 0} c_n x^n,$$

where $c_n = \sum_{i=0}^{n} a_i b_{n-i}$, or

$$\left(\sum_{n \geq 0} a_n \frac{x^n}{n!} \right) \left(\sum_{n \geq 0} b_n \frac{x^n}{n!} \right) = \sum_{n \geq 0} d_n \frac{x^n}{n!},$$

where $d_n = \sum_{i=0}^{n} \binom{n}{i} a_i b_{n-i}$, with $\binom{n}{i} = n!/i!(n-i)!$. Note that these operations are just what we would obtain by treating generating functions as if they obeyed the ordinary laws of algebra, such as $x^i x^j = x^{i+j}$. These operations coincide with the addition and multiplication of functions when the power series converge for appropriate values of x, and they obey such familiar laws of algebra as associativity and commutativity of addition and multiplication, distributivity of multiplication over addition, and cancellation of multiplication (i.e., if $F(x)G(x) = F(x)H(x)$ and $F(x) \neq 0$, then $G(x) = H(x)$). In fact, the set of all formal power series $\sum_{n \geq 0} a_n x^n$ with complex coefficients a_n (or more generally, coefficients in any integral domain R, where integral domains are assumed to be commutative with a multiplicative identity 1) forms a (commutative) integral domain under the operations just defined. This integral domain is denoted $\mathbb{C}[[x]]$ (or more generally, $R[[x]]$). Actually, $\mathbb{C}[[x]]$, or more generally $K[[x]]$ when K is a field, is a very special type of integral domain. For readers with some familiarity with algebra, we remark that $\mathbb{C}[[x]]$ is a principal ideal domain and therefore a unique factorization domain. In fact, every ideal of $\mathbb{C}[[x]]$ has the form (x^n) for some $n \geq 0$. From the viewpoint of commutative algebra, $\mathbb{C}[[x]]$ is a one-dimensional complete regular local ring. Moreover, the operation $[x^n] : \mathbb{C}[[x]] \to \mathbb{C}$ of taking the coefficient of x^n (and similarly $[x^n/n!]$) is a linear functional on $\mathbb{C}[[x]]$. These general algebraic considerations will not concern us here; rather we will discuss from an elementary viewpoint the properties of $\mathbb{C}[[x]]$ that will be useful to us.

There is an obvious extension of the ring $\mathbb{C}[[x]]$ to formal power series in m variables $x_1, \ldots, x_m$. The set of all such power series with complex coefficients is denoted $\mathbb{C}[[x_1, \ldots, x_m]]$ and forms a unique factorization domain (though not a principal ideal domain for $m \geq 2$).

It is primarily through experience that the combinatorial significance of the algebraic operations of $\mathbb{C}[[x]]$ or $\mathbb{C}[[x_1, \ldots, x_m]]$ is understood, as well as the problems of whether to use ordinary or exponential generating functions (or various other kinds discussed in later chapters). In Section 3.18 we will explain to some extent the combinatorial significance of these operations, but even then experience is indispensable.

If $F(x)$ and $G(x)$ are elements of $\mathbb{C}[[x]]$ satisfying $F(x)G(x) = 1$, then we (naturally) write $G(x) = F(x)^{-1}$. (Here 1 is short for $1 + 0x + 0x^2 + \cdots$.) It is easy to see that $F(x)^{-1}$ exists (in which case it is unique) if and only if $a_0 \neq 0$, where $F(x) = \sum_{n \geq 0} a_n x^n$. One commonly writes "symbolically" $a_0 = F(0)$, even though $F(x)$ is not considered to be a function of x. If $F(0) \neq 0$ and $F(x)G(x) = H(x)$, then $G(x) = F(x)^{-1}H(x)$, which we also write as $G(x) = H(x)/F(x)$. More generally, the operation^{-1} satisfies all the familiar laws of algebra, provided it is only applied to power series $F(x)$ satisfying $F(0) \neq 0$. For instance, $(F(x)G(x))^{-1} = F(x)^{-1}G(x)^{-1}$, $(F(x)^{-1})^{-1} = F(x)$, and so on. Similar results hold for $\mathbb{C}[[x_1, \ldots, x_m]]$.

1.1.5 Example. Let $\left(\sum_{n \geq 0} \alpha^n x^n \right)(1 - \alpha x) = \sum_{n \geq 0} c_n x^n$, where α is nonzero complex number. (We could also take α to be an indeterminate, in which case we should extend the coefficient field to $\mathbb{C}(\alpha)$, the field of rational functions over $\mathbb{C}$ in the variable α.) Then by definition of power series multiplication,

$$c_n = \begin{cases} 1, & n = 0 \\ \alpha^n - \alpha(\alpha^{n-1}) = 0, & n \geq 1. \end{cases}$$

Hence, $\sum_{n \geq 0} \alpha^n x^n = (1 - \alpha x)^{-1}$, which can also be written

$$\sum_{n \geq 0} \alpha^n x^n = \frac{1}{1 - \alpha x}.$$

This formula comes as no surprise; it is simply the formula (in a formal setting) for summing a geometric series.

Example 1.1.5 provides a simple illustration of the general principle that, informally speaking, if we have an identity involving power series that is valid when the power series are regarded as functions (so that the variables are sufficiently small complex numbers), then this identity continues to remain valid when regarded as an identity among formal power series, *provided* the operations defined in the formulas are well defined for formal power series. It would be unnecessarily pedantic for us to state a precise form of this principle here, since the reader should have little trouble justifying in any particular case the formal validity of our manipulations with power series. We will give several examples throughout this section to illustrate this contention.

1.1.6 Example. The identity

$$\left(\sum_{n\geq 0}\frac{x^n}{n!}\right)\left(\sum_{n\geq 0}(-1)^n\frac{x^n}{n!}\right) = 1 \tag{1.3}$$

is valid at the function-theoretic level (it states that $e^x e^{-x} = 1$) and is well defined as a statement involving formal power series. Hence, (1.3) is a valid formal power series identity. In other words (equating coefficients of $x^n/n!$ on both sides of (1.3)), we have

$$\sum_{k=0}^{n}(-1)^k\binom{n}{k} = \delta_{0n}. \tag{1.4}$$

To justify this identity directly from (1.3), we may reason as follows. Both sides of (1.3) converge for all $x \in \mathbb{C}$, so we have

$$\sum_{n\geq 0}\left(\sum_{k=0}^{n}(-1)^k\binom{n}{k}\right)\frac{x^n}{n!} = 1, \qquad \text{for all } x \in \mathbb{C}.$$

But if two power series in x represent the same function $f(x)$ in a neighborhood of 0, then these two power series must agree term-by-term, by a standard elementary result concerning power series. Hence, (1.4) follows.

1.1.7 Example. The identity

$$\sum_{n\geq 0}\frac{(x+1)^n}{n!} = e\sum_{n\geq 0}\frac{x^n}{n!}$$

is valid at the function-theoretic level (it states that $e^{x+1} = e \cdot e^x$) but does not make sense as a statement involving formal power series. There is no *formal* procedure for writing $\sum_{n\geq 0}(x+1)^n/n!$ as a member of $\mathbb{C}[[x]]$. For instance, the constant term of $\sum_{n\geq 0}(x+1)^n/n!$ is $\sum_{n\geq 0}1/n!$, whose interpretation as a member of $\mathbb{C}[[x]]$ involves the consideration of convergence.

Although the expression $\sum_{n\geq 0}(x+1)^n/n!$ does not make sense *formally*, there are nevertheless certain infinite processes that can be carried out formally in $\mathbb{C}[[x]]$. (These concepts extend straightforwardly to $\mathbb{C}[[x_1,\ldots,x_m]]$, but for simplicity we consider only $\mathbb{C}[[x]]$.) To define these processes, we need to put some additional structure on $\mathbb{C}[[x]]$—namely, the notion of *convergence*. From an algebraic standpoint, the definition of convergence is inherent in the statement that $\mathbb{C}[[x]]$ is *complete* in a certain standard topology that can be put on $\mathbb{C}[[x]]$. However, we will assume no knowledge of topology on the part of the reader and will instead give a self-contained, elementary treatment of convergence.

If $F_1(x), F_2(x),\ldots$ is a sequence of formal power series, and if $F(x) = \sum_{n\geq 0} a_n x^n$ is another formal power series, we say by definition that $F_i(x)$ *converges* to $F(x)$ as $i \to \infty$, written $F_i(x) \to F(x)$ or $\lim_{i\to\infty} F_i(x) = F(x)$, provided that for all $n \geq 0$ there is a number $\delta(n)$ such that the coefficient of x^n in $F_i(x)$ is a_n whenever $i \geq \delta(n)$. In other words, for every n the sequence

$$[x^n]F_1(x), \ [x^n]F_2(x), \ \ldots$$

of complex numbers eventually becomes constant (or *stabilizes*) with value a_n. An equivalent definition of convergence is the following. Define the *degree* of a nonzero formal power series $F(x) = \sum_{n\geq 0} a_n x^n$, denoted $\deg F(x)$, to be the least integer n such that $a_n \neq 0$. Note that $\deg F(x)G(x) = \deg F(x) + \deg G(x)$. Then $F_i(x)$ converges if and only if $\lim_{i\to\infty} \deg(F_{i+1}(x) - F_i(x)) = \infty$, and $F_i(x)$ converges to $F(x)$ if and only if $\lim_{i\to\infty} \deg(F(x) - F_i(x)) = \infty$.

We now say that an infinite sum $\sum_{j\geq 0} F_j(x)$ has the value $F(x)$ provided that $\sum_{j=0}^{i} F_j(x) \to F(x)$. A similar definition is made for the infinite product $\prod_{j\geq 1} F_j(x)$. To avoid unimportant technicalities we assume that, in any infinite product $\prod_{j\geq 1} F_j(x)$, each factor $F_j(x)$ satisfies $F_j(0) = 1$.

For instance, let $F_j(x) = a_j x^j$. Then for $i \geq n$, the coefficient of x^n in $\sum_{j=0}^{i} F_j(x)$ is a_n. Hence $\sum_{j\geq 0} F_j(x)$ is just the power series $\sum_{n\geq 0} a_n x^n$. Thus, we can think of the formal power series $\sum_{n\geq 0} a_n x^n$ as actually being the "sum" of its individual terms. The proofs of the following two elementary results are left to the reader.

1.1.8 Proposition. The infinite series $\sum_{j\geq 0} F_j(x)$ converges if and only if

$$\lim_{j\to\infty} \deg F_j(x) = \infty.$$

1.1.9 Proposition. The infinite product $\prod_{j\geq 1}(1 + G_j(x))$, where $G_j(0) = 0$, converges if and only if $\lim_{j\to\infty} \deg G_j(x) = \infty$.

It is essential to realize that in evaluating a convergent series $\sum_{j\geq 0} F_j(x)$ (or similarly a product $\prod_{j\geq 1} F_j(x)$), the coefficient of x^n for any given n can be computed using only *finite* processes. For if j is sufficiently large, say $j > \delta(n)$, then $\deg F_j(x) > n$, so that

$$[x^n] \sum_{j \geq 0} F_j(x) = [x^n] \sum_{j=0}^{\delta(n)} F_j(x).$$

The latter expression involves only a *finite* sum.

The most important combinatorial application of the notion of convergence is to the idea of power series composition. If $F(x) = \sum_{n \geq 0} a_n x^n$ and $G(x)$ are formal power series with $G(0) = 0$, define the *composition* $F(G(x))$ to be the infinite sum $\sum_{n \geq 0} a_n G(x)^n$. Since $\deg G(x)^n = n \cdot \deg G(x) \geq n$, we see by Proposition 1.1.8 that $F(G(x))$ is well defined as a *formal* power series. We also see why an expression such as e^{1+x} does not make sense formally; namely, the infinite series $\sum_{n \geq 0} (1+x)^n / n!$ does not converge in accordance with the preceding definition. On the other hand, an expression like $e^{e^x - 1}$ makes good sense formally, since it has the form $F(G(x))$ where $F(x) = \sum_{n \geq 0} x^n / n!$ and $G(x) = \sum_{n \geq 1} x^n / n!$.

1.1.10 Example. If $F(x) \in \mathbb{C}[[x]]$ satisfies $F(0) = 0$, then we can *define* for any $\lambda \in \mathbb{C}$ the formal power series

$$(1 + F(x))^{\lambda} = \sum_{n \geq 0} \binom{\lambda}{n} F(x)^n, \tag{1.5}$$

where $\binom{\lambda}{n} = \lambda(\lambda - 1) \cdots (\lambda - n + 1)/n!$. In fact, we may regard λ as an indeterminate and take (1.5) as the definition of $(1 + F(x))^{\lambda}$ as an element of $\mathbb{C}[[x, \lambda]]$ (or of $\mathbb{C}[\lambda][[x]]$; that is, the coefficient of x^n in $(1 + F(x))^{\lambda}$ is a certain *polynomial* in λ). All the expected properties of exponentiation are indeed valid, such as

$$(1 + F(x))^{\lambda + \mu} = (1 + F(x))^{\lambda} (1 + F(x))^{\mu},$$

regarded as an identity in the ring $\mathbb{C}[[x, \lambda, \mu]]$, or in the ring $\mathbb{C}[[x]]$ where one takes $\lambda, \mu \in \mathbb{C}$.

If $F(x) = \sum_{n \geq 0} a_n x^n$, define the *formal derivative* $F'(x)$ (also denoted $\frac{dF}{dx}$ or $DF(x)$) to be the formal power series

$$F'(x) = \sum_{n \geq 0} n a_n x^{n-1} = \sum_{n \geq 0} (n+1) a_{n+1} x^n.$$

It is easy to check that all the familiar laws of differentiation that are well defined formally continue to be valid for formal power series, In particular,

$$(F + G)' = F' + G',$$
$$(FG)' = F'G + FG',$$
$$F(G(x))' = G'(x) F'(G(x)).$$

We thus have a theory of *formal calculus* for formal power series. The usefulness of this theory will become apparent in subsequent examples. We first give an example of the use of the formal calculus that should shed some additional light on the validity of manipulating formal power series $F(x)$ as if they were actual functions of x.

1.1.11 Example. Suppose $F(0) = 1$, and let $G(x)$ be the power series (easily seen to be unique) satisfying

$$G'(x) = F'(x)/F(x), \qquad G(0) = 0. \tag{1.6}$$

From the function-theoretic viewpoint we can "solve" (1.6) to obtain $F(x) = \exp G(x)$, where by definition

$$\exp G(x) = \sum_{n \geq 0} \frac{G(x)^n}{n!}.$$

Since $G(0) = 0$ everything is well defined formally, so (1.6) should remain equivalent to $F(x) = \exp G(x)$ even if the power series for $F(x)$ converges only at $x = 0$. How can this assertion be justified without actually proving a combinatorial identity? Let $F(x) = 1 + \sum_{n \geq 1} a_n x^n$. From (1.6) we can compute explicitly $G(x) = \sum_{n \geq 1} b_n x^n$, and it is quickly seen that each b_n is a *polynomial* in finitely many of the a_i's. It then follows that if $\exp G(x) = 1 + \sum_{n \geq 1} c_n x^n$, then each c_n will also be a polynomial in finitely many of the a_i's, say $c_n = p_n(a_1, a_2, \ldots, a_m)$, where m depends on n. Now we know that $F(x) = \exp G(x)$ provided $1 + \sum_{n \geq 1} a_n x^n$ converges. If two Taylor series convergent in some neighborhood of the origin represent the same function, then their coefficients coincide. Hence $a_n = p_n(a_1, a_2, \ldots, a_m)$ provided $1 + \sum_{n \geq 1} a_n x^n$ converges. Thus, the two polynomials a_n and $p_n(a_1, \ldots, a_m)$ agree in some neighborhood of the origin of $\mathbb{C}^m$, so they must be equal. (It is a simple result that if two complex polynomials in m variables agree in some open subset of $\mathbb{C}^m$, then they are identical.) Since $a_n = p_n(a_1, a_2, \ldots, a_m)$ as polynomials, the identity $F(x) = \exp G(x)$ continues to remain valid for *formal* power series.

There is an alternative method for justifying the formal solution $F(x) = \exp G(x)$ to (1.6), which may appeal to topologically inclined readers. Given $G(x)$ with $G(0) = 0$, define $F(x) = \exp G(x)$ and consider a map $\phi : \mathbb{C}[[x]] \to \mathbb{C}[[x]]$ defined by $\phi(G(x)) = G'(x) - \frac{F'(x)}{F(x)}$. One easily verifies the following: (a) if G converges in some neighborhood of 0, then $\phi(G(x)) = 0$; (b) the set $\mathcal{G}$ of all power series $G(x) \in \mathbb{C}[[x]]$ that converge in some neighborhood of 0 is dense in $\mathbb{C}[[x]]$, in the topology defined earlier (in fact, the set $\mathbb{C}[x]$ of polynomials is dense); and (c) the function ϕ is continuous in the topology defined earlier. From this it follows that $\phi(G(x)) = 0$ for all $G(x) \in \mathbb{C}[[x]]$ with $G(0) = 0$.

We now present various illustrations in the manipulation of generating functions. Throughout we will be making heavy use of the principle that formal power series can be treated as if they were functions.

1.1.12 Example. Find a simple expression for the generating function $F(x) = \sum_{n \geq 0} a_n x^n$, where $a_0 = a_1 = 1$, $a_n = a_{n-1} + a_{n-2}$ if $n \geq 2$. We have

$$F(x) = \sum_{n \geq 0} a_n x^n = 1 + x + \sum_{n \geq 2} a_n x^n$$

$$= 1 + x + \sum_{n \geq 2} (a_{n-1} + a_{n-2}) x^n$$

$$= 1 + x + x \sum_{n \geq 2} a_{n-1} x^{n-1} + x^2 \sum_{n \geq 2} a_{n-2} x^{n-2}$$

$$= 1 + x + x(F(x) - 1) + x^2 F(x).$$

Solving for $F(x)$ yields $F(x) = 1/(1 - x - x^2)$. The number a_n is just the *Fibonacci number* F_{n+1}. For some combinatorial properties of Fibonacci numbers, see Exercises 1.35–1.42. For the general theory of rational generating functions and linear recurrences with constant coefficients illustrated in the present example, see Section 4.1.

1.1.13 Example. Find a simple expression for the generating function $F(x) = \sum_{n \geq 0} a_n x^n / n!$, where $a_0 = 1$,

$$a_{n+1} = a_n + n a_{n-1}, \ n \geq 0. \tag{1.7}$$

(Note that if $n = 0$ we get $a_1 = a_0 + 0 \cdot a_{-1}$, so the value of a_{-1} is irrelevant.) Multiply the recurrence (1.7) by $x^n / n!$ and sum on $n \geq 0$. We get

$$\sum_{n \geq 0} a_{n+1} \frac{x^n}{n!} = \sum_{n \geq 0} a_n \frac{x^n}{n!} + \sum_{n \geq 0} n a_{n-1} \frac{x^n}{n!}$$

$$= \sum_{n \geq 0} a_n \frac{x^n}{n!} + \sum_{n \geq 1} a_{n-1} \frac{x^n}{(n-1)!}.$$

The left-hand side is just $F'(x)$, while the right-hand side is $F(x) + x F(x)$. Hence, $F'(x) = (1 + x) F(x)$. The unique solution to this differential equation satisfying $F(0) = 1$ is $F(x) = \exp\left(x + \frac{1}{2}x^2\right)$. (As shown in Example 1.1.11, solving this differential equation is a purely formal procedure.) For the combinatorial significance of the numbers a_n, see equation (5.32).

NOTE. With the benefit of hindsight we wrote the recurrence $a_{n+1} = a_n + n a_{n-1}$ with indexing that makes the computation simplest. If for instance we had written $a_n = a_{n-1} + (n-1) a_{n-2}$, then the computation would be more complicated (though still quite tractable). In converting recurrences to generating function identities, it can be worthwhile to consider how best to index the recurrence.

1.1.14 Example. Let $\mu(n)$ be the Möbius function of number theory; that is, $\mu(1) = 1$, $\mu(n) = 0$ if n is divisible by the square of an integer greater than one, and $\mu(n) = (-1)^r$ if n is the product of r distinct primes. Find a simple expression for the power series

$$F(x) = \prod_{n \geq 1} (1 - x^n)^{-\mu(n)/n}. \tag{1.8}$$

First let us make sure that $F(x)$ is well defined as a formal power series. We have by Example 1.1.10 that

$$(1 - x^n)^{-\mu(n)/n} = \sum_{i \geq 0} \binom{-\mu(n)/n}{i} (-1)^i x^{in}.$$

Note that $(1 - x^n)^{-\mu(n)/n} = 1 + H(x)$, where $\deg H(x) = n$. Hence, by Proposition 1.1.9 the infinite product (1.8) converges, so $F(x)$ is well defined. Now apply log to (1.8). In other words, form $\log F(x)$, where

$$\log(1 + x) = \sum_{n \geq 1} (-1)^{n-1} \frac{x^n}{n},$$

the power series expansion for the natural logarithm at $x = 0$. We obtain

$$\log F(x) = \log \prod_{n \geq 1} (1 - x^n)^{-\mu(n)/n}$$

$$= -\sum_{n \geq 1} \log(1 - x^n)^{\mu(n)/n}$$

$$= -\sum_{n \geq 1} \frac{\mu(n)}{n} \log(1 - x^n)$$

$$= -\sum_{n \geq 1} \frac{\mu(n)}{n} \sum_{i \geq 1} \left(-\frac{x^{in}}{i} \right).$$

The coefficient of x^m in the preceding power series is

$$\frac{1}{m} \sum_{d \mid m} \mu(d),$$

where the sum is over all positive integers d dividing m. It is well known that

$$\frac{1}{m} \sum_{d \mid m} \mu(d) = \begin{cases} 1, & m = 1 \\ 0, & \text{otherwise.} \end{cases}$$

Hence, $\log F(x) = x$, so $F(x) = e^x$. Note that the derivation of this miraculous formula involved only *formal* manipulations.

1.1.15 Example. Find the unique sequence $a_0 = 1, a_1, a_2, \ldots$ of real numbers satisfying

$$\sum_{k=0}^{n} a_k a_{n-k} = 1 \tag{1.9}$$

for all $n \in \mathbb{N}$. The trick is to recognize the left-hand side of (1.9) as the coefficient of x^n in $\left(\sum_{n \geq 0} a_n x^n \right)^2$. Letting $F(x) = \sum_{n \geq 0} a_n x^n$, we then have

$$F(x)^2 = \sum_{n \geq 0} x^n = \frac{1}{1-x}.$$

Hence,

$$F(x) = (1-x)^{-1/2} = \sum_{n \geq 0} \binom{-1/2}{n} (-1)^n x^n,$$

so

$$
\begin{aligned}
a_n &= (-1)^n \binom{-1/2}{n} \\
&= (-1)^n \frac{\left(-\frac{1}{2}\right) \left(-\frac{3}{2}\right) \left(-\frac{5}{2}\right) \cdots \left(-\frac{2n-1}{2}\right)}{n!} \\
&= \frac{1 \cdot 3 \cdot 5 \cdots (2n-1)}{2^n n!}.
\end{aligned}
$$

Note that a_n can also be rewritten as $4^{-n} \binom{2n}{n}$. The identity

$$\binom{2n}{n} = (-1)^n 4^n \binom{-1/2}{n} \tag{1.10}$$

can be useful for problems involving $\binom{2n}{n}$.

Now that we have discussed the manipulation of formal power series, the question arises as to the advantages of using generating functions to represent a counting function $f(n)$. Why, for instance, should a formula such as

$$\sum_{n \geq 0} f(n) \frac{x^n}{n!} = \exp \left(x + \frac{x^2}{2} \right) \tag{1.11}$$

be regarded as a "determination" of $f(n)$? Basically, the answer is that there are many standard, routine techniques for extracting information from generating functions. Generating functions are frequently the most concise and efficient way of presenting information about their coefficients. For instance, from (1.11) an experienced enumerative combinatorialist can tell at a glance the following:

1. A simple recurrence for $f(n)$ can be found by differentiation. Namely, we obtain

$$\sum_{n \geq 1} f(n) \frac{x^{n-1}}{(n-1)!} = (1+x) e^{x+x^2/2} = (1+x) \sum_{n \geq 0} f(n) \frac{x^n}{n!}.$$

Equating coefficients of $x^n/n!$ yields

$$f(n+1) = f(n) + nf(n-1), \quad n \geq 1.$$

Note that in Example 1.1.13 we went in the opposite direction (i.e., we obtained the generating function from the recurrence, a less straightforward procedure).

2. An explicit formula for $f(n)$ can be obtained from $e^{x+(x^2/2)} = e^x e^{x^2/2}$. Namely,

$$\sum_{n \geq 0} f(n) \frac{x^n}{n!} = e^x e^{x^2/2} = \left(\sum_{n \geq 0} \frac{x^n}{n!} \right) \left(\sum_{n \geq 0} \frac{x^{2n}}{2^n n!} \right)$$

$$= \left(\sum_{n \geq 0} \frac{x^n}{n!} \right) \left(\sum_{n \geq 0} \frac{(2n)!}{2^n n!} \frac{x^{2n}}{(2n)!} \right),$$

so that

$$f(n) = \sum_{\substack{i \geq 0 \\ i \text{ even}}} \binom{n}{i} \frac{i!}{2^{i/2}(i/2)!} = \sum_{j \geq 0} \binom{n}{2j} \frac{(2j)!}{2^j j!}.$$

3. Regarded as a function of a complex variable, $\exp(x + \frac{x^2}{2})$ is a nicely behaved entire function, so that standard techniques from the theory of asymptotic analysis can be used to estimate $f(n)$. As a first approximation, it is routine (for someone sufficiently versed in complex variable theory) to obtain the asymptotic formula

$$f(n) \sim \frac{1}{\sqrt{2}} n^{n/2} e^{-\frac{n}{2} + \sqrt{n} - \frac{1}{4}}. \tag{1.12}$$

No other method of describing $f(n)$ makes it so easy to determine these fundamental properties. Many other properties of $f(n)$ can also be easily obtained from the generating function; for instance, we leave to the reader the problem of evaluating, essentially by inspection of (1.11), the sum

$$\sum_{i=0}^{n} (-1)^{n-i} \binom{n}{i} f(i) \tag{1.13}$$

(see Exercise 1.7). Therefore, we are ready to accept the generating function $\exp(x + \frac{x^2}{2})$ as a satisfactory determination of $f(n)$.

This completes our discussion of generating functions and more generally the problem of giving a satisfactory description of a counting function $f(n)$. We now turn to the question of what is the best way to *prove* that a counting function has some given description. In accordance with the principle from other branches of mathematics that it is better to exhibit an explicit isomorphism between two objects than merely prove that they are isomorphic, we adopt the general principle that it is better to exhibit an explicit one-to-one correspondence (bijection) between two finite sets than merely to prove that they have the same number of elements. A proof that shows that a certain set S has a certain number m of elements by constructing an explicit bijection between S and some other set that is known to have m elements is called a *combinatorial proof* or *bijective proof*. The precise border between combinatorial and noncombinatorial proofs is rather hazy, and certain arguments that to an inexperienced enumerator will appear noncombinatorial will be recognized by a more facile counter as combinatorial, primarily because he or she is aware of certain standard techniques for converting apparently noncombinatorial arguments into combinatorial ones. Such subtleties will not concern us here, and we now give some clear-cut examples of the distinction between combinatorial and noncombinatorial proofs. We use the notation $\#S$ or $|S|$ for the cardinality (number of elements) of the finite set S.

1.1.16 Example. Let n and k be fixed positive integers. How many sequences $(X_1, X_2, \ldots, X_k)$ are there of subsets of the set $[n] = \{1, 2, \ldots, n\}$ such that $X_1 \cap X_2 \cap \cdots \cap X_k = \emptyset$? Let $f(k, n)$ be this number. If we were not particularly inspired we could perhaps argue as follows. Suppose $X_1 \cap X_2 \cap \cdots \cap X_{k-1} = T$, where $\#T = i$. If $Y_j = X_j - T$, then $Y_1 \cap \cdots \cap Y_{k-1} = \emptyset$ and $Y_j \subseteq [n] - T$. Hence, there are $f(k-1, n-i)$ sequences $(X_1, \ldots, X_{k-1})$ such that $X_1 \cap X_2 \cap \cdots \cap X_{k-1} = T$. For each such sequence, X_k can be any of the 2^{n-i} subsets of $[n] - T$. As is probably familiar to most readers and will be discussed later, there are $\binom{n}{i} = n!/i!(n-i)!$ i-element subsets T of $[n]$. Hence,

$$f(k, n) = \sum_{i=0}^{n} \binom{n}{i} 2^{n-i} f(k-1, n-i). \tag{1.14}$$

Let $F_k(x) = \sum_{n \geq 0} f(k, n) x^n / n!$. Then (1.14) is equivalent to

$$F_k(x) = e^x F_{k-1}(2x).$$

Clearly $F_1(x) = e^x$. It follows easily that

$$F_k(x) = \exp(x + 2x + 4x + \cdots + 2^{k-1} x)$$
$$= \exp((2^k - 1)x)$$
$$= \sum_{n \geq 0} (2^k - 1)^n \frac{x^n}{n!}.$$

Hence, $f(k, n) = (2^k - 1)^n$. This argument is a flagrant example of a noncombinatorial proof. The resulting answer is extremely simple despite the contortions involved to obtain it, and it cries out for a better understanding. In fact, $(2^k - 1)^n$ is clearly the number of n-tuples $(Z_1, Z_2, \ldots, Z_n)$, where each Z_i is a subset of $[k]$ not equal to $[k]$. Can we find a bijection θ between the set S_{kn} of all $(X_1, \ldots, X_k) \subseteq [n]^k$ such that $X_1 \cap \cdots \cap X_k = \emptyset$, and the set T_{kn} of all $(Z_1, \ldots, Z_n)$ where $[k] \neq Z_i \subseteq [k]$? Given an element $(Z_1, \ldots, Z_n)$ of T_{kn}, define $(X_1, \ldots, X_k)$ by the condition that $i \in X_j$ if and only if $j \in Z_i$. This rule is just a precise way of saying the following: The element 1 can appear in any of the X_i's except all of them, so there are $2^k - 1$ choices for which of the X_i's contain 1; similarly there are $2^k - 1$ choices for which of the X_i's contain $2, 3, \ldots, n$, so there are $(2^k - 1)^n$ choices in all. Thus, the crucial point of the problem is that the different elements of $[n]$ behave *independently*, so we end up with a simple product. We leave to the reader the (rather dull) task of rigorously verifiying that θ is a bijection, but this fact should be intu-itively clear. The usual way to show that θ is a bijection is to construct explicitly a map $\phi : T_{kn} \to S_{kn}$, and then to show that $\phi = \theta^{-1}$; for example, by showing that $\phi\theta(X) = X$ and that θ is surjective. CAVEAT. Any proof that θ is bijective must not use a priori the fact that $\#S_{kn} = \#T_{kn}$!

Not only is the preceding combinatorial proof much shorter than our previous proof, but it also makes the reason for the simple answer completely transparent. It is often the case, as occurred here, that the first proof to come to mind turns out to be laborious and inelegant, but that the final answer suggests a simpler combinatorial proof.

1.1.17 Example. Verify the identity

$$\sum_{i=0}^{n} \binom{a}{i} \binom{b}{n-i} = \binom{a+b}{n}, \tag{1.15}$$

where a, b, and n are nonnegative integers. A noncombinatorial proof would run as follows. The left-hand side of (1.15) is the coefficient of x^n in the power series (polynomial) $\left(\sum_{i\geq 0} \binom{a}{i} x^i\right) \left(\sum_{j\geq 0} \binom{b}{j} x^j\right)$. But by the binomial theorem,

$$\left(\sum_{i\geq 0} \binom{a}{i} x^i\right) \left(\sum_{j\geq 0} \binom{b}{j} x^j\right) = (1+x)^a (1+x)^b$$

$$= (1+x)^{a+b}$$

$$= \sum_{n\geq 0} \binom{a+b}{n} x^n,$$

so the proof follows. A combinatorial proof runs as follows. The right-hand side of (1.15) is the number of n-element subsets X of $[a+b]$. Suppose X intersects $[a]$ in i elements. There are $\binom{a}{i}$ choices for $X \cap [a]$, and $\binom{b}{n-i}$ choices for the remaining $n-i$ elements $X \cap \{a+1, a+2, \ldots, a+b\}$. Thus, there are $\binom{a}{i}\binom{b}{n-i}$ ways that $X \cap [a]$ can have i elements, and summing over i gives the total number $\binom{a+b}{n}$ of n-element subsets of $[a+b]$.

There are many examples in the literature of finite sets that are known to have the same number of elements but for which no combinatorial proof of this fact is known. Some of these will appear as exercises throughout this book.

1.2 Sets and Multisets

We have (finally!) completed our description of the solution of an enumerative problem, and we are now ready to delve into some actual problems. Let us begin with the basic problem of counting subsets of a set. Let $S = \{x_1, x_2, \ldots, x_n\}$ be an

n-element set, or *n-set* for short. Let 2^S denote the set of all subsets of S, and let $\{0,1\}^n = \{(\varepsilon_1, \varepsilon_2, \ldots, \varepsilon_n) : \varepsilon_i = 0 \text{ or } 1\}$. Since there are two possible values for each ε_i, we have $\#\{0,1\}^n = 2^n$. Define a map $\theta : 2^S \to \{0,1\}^n$ by $\theta(T) = (\varepsilon_1, \varepsilon_2, \ldots, \varepsilon_n)$, where

$$\varepsilon_i = \begin{cases} 1, & \text{if } x_i \in T \\ 0, & \text{if } x_i \notin T. \end{cases}$$

For example, if $n = 5$ and $T = \{x_2, x_4, x_5\}$, then $\theta(T) = (0,1,0,1,1)$. Most readers will realize that $\theta(T)$ is just the *characteristic vector* of T. It is easily seen that θ is a bijection, so that we have given a combinatorial proof that $\#2^S = 2^n$. Of course, there are many alternative proofs of this simple result, and many of these proofs could be regarded as combinatorial.

Now define $\binom{S}{k}$ (sometimes denoted $S^{(k)}$ or otherwise, and read "S choose k") to be the set of all k-element subsets (or *k-subsets*) of S, and *define* $\binom{n}{k} = \#\binom{S}{k}$, read "$n$ choose k" (ignore our previous use of the symbol $\binom{n}{k}$) and called a *binomial coefficient*. Our goal is to prove the formula

$$\binom{n}{k} = \frac{n(n-1)\cdots(n-k+1)}{k!}. \tag{1.16}$$

Note that if $0 \le k \le n$ then the right-hand side of equation (1.16) can be rewritten $n!/k!(n-k)!$. The right-hand side of (1.16) can be used to define $\binom{n}{k}$ for any complex number (or indeterminate) n, provided $k \in \mathbb{N}$. The numerator $n(n-1)\cdots(n-k+1)$ of (1.16) is read "n lower factorial k" and is denoted $(n)_k$. CAVEAT. Many mathematicians, especially those in the theory of special functions, use the notation $(n)_k = n(n+1)\cdots(n+k-1)$.

We would like to give a bijective proof of (1.16), but the factor $k!$ in the denominator makes it difficult to give a "simple" interpretation of the right-hand side. Therefore, we use the standard technique of clearing the denominator. To this end we count in two ways the number $N(n,k)$ of ways of choosing a k-subset T of S and then linearly ordering the elements of T. We can pick T in $\binom{n}{k}$ ways, then pick an element of T in k ways to be first in the ordering, then pick another element in $k-1$ ways to be second, and so on. Thus,

$$N(n,k) = \binom{n}{k} k!.$$

On the other hand, we could pick any element of S in n ways to be first in the ordering, then another element in $n-1$ ways to be second, on so on, down to any remaining element in $n-k+1$ ways to be kth. Thus,

$$N(n,k) = n(n-1)\cdots(n-k+1).$$

We have therefore given a combinatorial proof that

$$\binom{n}{k} k! = n(n-1)\cdots(n-k+1),$$

and hence of equation (1.16).

A generating function approach to binomial coefficients can be given as follows. Regard $x_1, \ldots, x_n$ as independent indeterminates. It is an immediate consequence of the process of multiplication (one could also give a rigorous proof by induction) that

$$(1+x_1)(1+x_2)\cdots(1+x_n) = \sum_{T \subseteq S} \prod_{x_i \in T} x_i. \qquad (1.17)$$

If we put each $x_i = x$, then we obtain

$$(1 + x)^n = \sum_{T \subseteq S} \prod_{x_i \in T} x = \sum_{T \subseteq S} x^{\#T} = \sum_{k \geq 0} \binom{n}{k} x^k, \qquad (1.18)$$

since the term x^k appears exactly $\binom{n}{k}$ times in the sum $\sum_{T \subseteq S} x^{\#T}$. This reasoning is an instance of the simple but useful observation that if $\mathcal{S}$ is a collection of finite sets such that $\mathcal{S}$ contains exactly $f(n)$ sets with n elements, then

$$\sum_{S \in \mathcal{S}} x^{\#S} = \sum_{n \geq 0} f(n) x^n.$$

Somewhat more generally, if $g : \mathbb{N} \to \mathbb{C}$ is any function, then

$$\sum_{S \in \mathcal{S}} g(\#S) x^{\#S} = \sum_{n \geq 0} g(n) f(n) x^n.$$

Equation (1.18) is such a simple result (the binomial theorem for the exponent $n \in \mathbb{N}$) that it is hardly necessary to obtain first the more refined (1.17). However, it is often easier in dealing with generating functions to work with the most number of variables (indeterminates) possible and then specialize. Often the more refined formula will be more transparent, and its various specializations will be automatically unified.

Various identities involving binomial coefficients follow easily from the identity $(1 + x)^n = \sum_{k \geq 0} \binom{n}{k} x^k$, and the reader will find it instructive to find combinatorial proofs of them. (See Exercise 1.3 for further examples of binomial coefficient identities.) For instance, put $x = 1$ to obtain $2^n = \sum_{k \geq 0} \binom{n}{k}$; put $x = -1$ to obtain $0 = \sum_{k \geq 0} (-1)^k \binom{n}{k}$ if $n > 0$; differentiate and put $x = 1$ to obtain $n2^{n-1} = \sum_{k \geq 0} k \binom{n}{k}$, and so on.

There is a close connection between subsets of a set and compositions of a nonnegative integer. A *composition* of n can be thought of as an expression of n as an *ordered* sum of integers. More precisely, a composition of n is a sequence $\alpha = (a_1, \ldots, a_k)$ of positive integers satisfying $\sum a_i = n$. For instance, there are eight compositions of 4; namely,

$$
\begin{array}{ll}
1+1+1+1 & 3+1 \\
2+1+1 & 1+3 \\
1+2+1 & 2+2 \\
1+1+2 & 4.
\end{array}
$$

If exactly k summands appear in a composition α, then we say that α has k *parts*, and we call α a k-*composition*. If $\alpha = (a_1, a_2, \ldots, a_k)$ is a k-composition of n, then define a $(k-1)$-subset S_α of $[n-1]$ by

$$S_\alpha = \{a_1, a_1 + a_2, \ldots, a_1 + a_2 + \cdots + a_{k-1}\}.$$

The correspondence $\alpha \mapsto S_\alpha$ gives a bijection between all k-compositions of n and $(k-1)$-subsets of $[n-1]$. Hence, there are $\binom{n-1}{k-1}$ k-compositions of n and 2^{n-1} compositions of $n > 0$. The inverse bijection $S_\alpha \mapsto \alpha$ is often represented schematically by drawing n dots in a row and drawing vertical bars between $k-1$ of the $n-1$ spaces separating the dots. This procedure divides the dots into k linearly ordered (from left-to-right) "compartments" whose number of elements is a k-composition of n. For instance, the compartments

$$\cdot | \cdot \cdot | \cdot | \cdot | \cdot \cdot \cdot | \cdot \cdot \qquad (1.19)$$

correspond to the 6-composition $(1, 2, 1, 1, 3, 2)$ of 10. The diagram (1.19) illustrates another very general principle related to bijective proofs—it is often efficacious to represent the objects being counted geometrically.

A problem closely related to compositions is that of counting the number $N(n, k)$ of solutions to $x_1 + x_2 + \cdots + x_k = n$ in *nonnegative* integers. Such a solution is called a *weak composition* of n into k parts, or a *weak* k-*composition* of n. (A solution in *positive* integers is simply a k-composition of n.) If we put $y_i = x_i + 1$, then $N(n, k)$ is the number of solutions in positive integers to $y_1 + y_2 + \cdots + y_k = n + k$, that is, the number of k-compositions of $n + k$. Hence, $N(n, k) = \binom{n+k-1}{k-1}$. A further variant is the enumeration of $\mathbb{N}$-solutions (that is, solutions where each variable lies in $\mathbb{N}$) to $x_1 + x_2 + \cdots + x_k \leq n$. Again we use a standard technique, namely, introducing a *slack variable* y to convert the inequality $x_1 + x_2 + \cdots + x_k \leq n$ to the equality $x_1 + x_2 + \cdots + x_k + y = n$. An $\mathbb{N}$-solution to this equation is a weak $(k+1)$-composition of n, so the number $N(n, k+1)$ of such solutions is $\binom{n+(k+1)-1}{k} = \binom{n+k}{k}$.

A k-subset T of an n-set S is sometimes called a k-*combination of S without repetitions*. This suggests the problem of counting the number of k-combinations of S *with repetitions*; that is, we choose k elements of S, disregarding order and allowing repeated elements. Denote this number by $\left(\!\binom{n}{k}\!\right)$, which could be read "$n$ multichoose k." For instance, if $S = \{1, 2, 3\}$, then the combinations counted by $\left(\!\binom{3}{2}\!\right)$ are 11, 22, 33, 12, 13, 23. Hence, $\left(\!\binom{3}{2}\!\right) = 6$. An equivalent but more precise treatment of combinations with repetitions can be made by introducing the concept of a *multiset*. Intuitively, a multiset is a set with repeated elements; for instance, $\{1, 1, 2, 5, 5, 5\}$. More precisely, a *finite multiset* M on a set S is a pair (S, ν), where ν is a function $\nu : S \to \mathbb{N}$ such that $\sum_{x \in S} \nu(x) < \infty$. One regards $\nu(x)$ as the number of repetitions of x. The integer $\sum_{x \in S} \nu(x)$ is called the *cardinality*, *size*, or *number of elements* of M and is denoted $|M|$, $\#M$, or card M. If $S = \{x_1, \ldots, x_n\}$ and $\nu(x_i) = a_i$, then we call a_i the *multiplicity* of x_i in M and write $M = \{x_1^{a_1}, \ldots, x_n^{a_n}\}$. If $\#M = k$, then we call M a k-*multiset*. The set of all k-multisets on S is denoted $\left(\!\binom{S}{k}\!\right)$. If $M' = (S, \nu')$ is another multiset on S, then we say that M' is a *submultiset* of M if $\nu'(x) \le \nu(x)$ for all $x \in S$. The number of submultisets of M is $\prod_{x \in S}(\nu(x) + 1)$, since for each $x \in S$ there are $\nu(x) + 1$ possible values of $\nu'(x)$. It is now clear that a k-combination of S with repetition is simply a multiset on S with k elements.

Although the reader may be unaware of it, we have already evaluated the number $\left(\!\binom{n}{k}\!\right)$. If $S = \{y_1, \ldots, y_n\}$ and we set $x_i = \nu(y_i)$, then we see that $\left(\!\binom{n}{k}\!\right)$ is the number of solutions in nonnegative integers to $x_1 + x_2 + \cdots + x_n = k$, which we have seen is $\binom{n+k-1}{n-1} = \binom{n+k-1}{k}$.

There are two elegant direct combinatorial proofs that $\left(\!\binom{n}{k}\!\right) = \binom{n+k-1}{k}$. For the first, let $1 \le a_1 < a_2 < \cdots < a_k \le n + k - 1$ be a k-subset of $[n+k-1]$. Let $b_i = a_i - i + 1$. Then, $\{b_1, b_2, \ldots, b_k\}$ is a k-multiset on $[n]$. Conversely, given a k-multiset $1 \le b_1 \le b_2 \le \cdots \le b_k \le n$ on $[n]$, then defining $a_i = b_i + i - 1$ we see that $\{a_1, a_2, \ldots, a_k\}$ is a k-subset of $[n+k-1]$. Hence, we have defined a bijection between $\left(\!\binom{[n]}{k}\!\right)$ and $\binom{n+k-1}{k}$, as desired. This proof illustrates the technique of *compression*, where we convert a strictly increasing sequence to a weakly increasing sequence.

Our second direct proof that $\left(\!\binom{n}{k}\!\right) = \binom{n+k-1}{k}$ is a "geometric" (or "balls into boxes" or "stars and bars") proof, analogous to the preceding proof that there are $\binom{n-1}{k-1}$ k-compositions of n. There are $\binom{n+k-1}{k}$ sequences consisting of k dots and $n - 1$ vertical bars. An example of such a sequence for $k = 5$ and $n = 7$ is given by

$$|| \cdot \cdot | \cdot | ||| \cdot \cdot$$

The $n - 1$ bars divide the k dots into n compartments. Let the number of dots in the ith compartment be $\nu(i)$. In this way the diagrams correspond to k-multisets on $[n]$, so $\left(\!\binom{n}{k}\!\right) = \binom{n+k-1}{k}$. For the preceding example, the multiset is $\{3, 3, 4, 7, 7\}$.

The generating function approach to multisets is instructive. In exact analogy to our treatment of subsets of a set $S = \{x_1, \ldots, x_n\}$, we have

$$(1 + x_1 + x_1^2 + \cdots)(1 + x_2 + x_2^2 + \cdots) \cdots (1 + x_n + x_n^2 + \cdots) = \sum_{M=(S,\nu)} \prod_{x_i \in S} x_i^{\nu(x_i)},$$

where the sum is over all finite multisets M on S. Put each $x_i = x$. We get

$$(1 + x + x^2 + \cdots)^n = \sum_{M=(S,\nu)} x^{\nu(x_1) + \cdots + \nu(x_n)}$$

$$= \sum_{M=(S,\nu)} x^{\#M}$$

$$= \sum_{k \ge 0} \left(\!\binom{n}{k}\!\right) x^k.$$

But

$$(1 + x + x^2 + \cdots)^n = (1 - x)^{-n} = \sum_{k \geq 0} \binom{-n}{k}(-1)^k x^k, \qquad (1.20)$$

so $\left(\binom{n}{k}\right) = (-1)^k \binom{-n}{k} = \binom{n+k-1}{k}$. The elegant formula

$$\left(\binom{n}{k}\right) = (-1)^k \binom{-n}{k} \qquad (1.21)$$

is no accident; it is the simplest instance of a *combinatorial reciprocity theorem*. A partially ordered set generalization appears in Section 3.15.3, while a more general theory of such results is given in Chapter 4.

The binomial coefficient $\binom{n}{k}$ may be interpreted in the following manner. Each element of an n-set S is placed into one of two categories, with k elements in Category 1 and $n - k$ elements in Category 2. (The elements of Category 1 form a k-subset T of S.) This suggests a generalization allowing more than two categories. Let $(a_1, a_2, \ldots, a_m)$ be a sequence of nonnegative integers summing to n, and suppose that we have m categories $C_1, \ldots, C_m$. Let $\binom{n}{a_1, a_2, \ldots, a_m}$ denote the number of ways of assigning each element of an n-set S to one of the categories $C_1, \ldots, C_m$ so that exactly a_i elements are assigned to C_i. The notation is somewhat at variance with the notation for binomial coefficients (the case $m = 2$), but no confusion should result when we write $\binom{n}{k}$ instead of $\binom{n}{k, n-k}$. The number $\binom{n}{a_1, a_2, \ldots, a_m}$ is called a *multinomial coefficient*. It is customary to regard the elements of S as being n distinguishable balls and the categories as being m distinguishable boxes. Then $\binom{n}{a_1, a_2, \ldots, a_m}$ is the number of ways to place the balls into the boxes so that the ith box contains a_i balls.

The multinomial coefficient can also be interpreted in terms of "permutations of a multiset." If S is an n-set, then a *permutation* w of S can be defined as a linear ordering $w_1, w_2, \ldots, w_n$ of the elements of S. Think of w as a *word* $w_1 w_2 \cdots w_n$ in the alphabet S. If $S = \{x_1, \ldots, x_n\}$, then such a word corresponds to the bijection $w : S \to S$ given by $w(x_i) = w_i$, so that a permutation of S may also be regarded as a bijection $S \to S$. Many interesting combinatorics are based on these two different ways of representing permutations; a good example is the second proof of Proposition 5.3.2.

We write $\mathfrak{S}_S$ for the set of permutations of S. If $S = [n]$, then we write $\mathfrak{S}_n$ for $\mathfrak{S}_{[n]}$. Since we choose w_1 in n ways, then w_2 in $n - 1$ ways, and so on, we clearly have $\#\mathfrak{S}_S = n!$. In an analogous manner, we can define a permutation w of a multiset M of cardinality n to be a linear ordering $w_1, w_2, \ldots, w_n$ of the "elements" of M; that is, if $M = (S, \nu)$ then the element $x \in S$ appears exactly $\nu(x)$ times in the permutation. Again, we think of w as a word $w_1 w_2 \cdots w_n$. For instance, there are 12 permutations of the multiset $\{1, 1, 2, 3\}$; namely, 1123, 1132, 1213, 1312, 1231, 1321, 2113, 2131, 2311, 3112, 3121, 3211. Let $\mathfrak{S}_M$ denote the set of permutations of M. If $M = \{x_1^{a_1}, \ldots, x_m^{a_m}\}$ and $\#M = n$, then it is clear that

$$\#\mathfrak{S}_M = \binom{n}{a_1, a_2, \ldots, a_m}. \tag{1.22}$$

Indeed, if x_i appears in position j of the permutation, then we put the element j of $[n]$ into Category i.

Our results on binomial coefficients extend straightforwardly to multinomial coefficients. In particular, we have

$$\binom{n}{a_1, a_2, \ldots, a_m} = \frac{n!}{a_1! a_2! \cdots a_m!}. \tag{1.23}$$

Among the many ways to prove this result, we can place a_1 elements of S into Category 1 in $\binom{n}{a_1}$ ways, then a_2 of the remaining $n - a_1$ elements of $[n]$ into Category 2 in $\binom{n-a_1}{a_2}$ ways, and so on, yielding

$$\binom{n}{a_1, a_2, \ldots, a_m} = \binom{n}{a_1}\binom{n - a_1}{a_2} \cdots \binom{n - a_1 - \cdots - a_{m-1}}{a_m} \tag{1.24}$$

$$= \frac{n!}{a_1! a_2! \cdots a_m!}.$$

Equation (1.24) is often a useful device for reducing problems on multinomial coefficients to binomial coefficients. We leave to the reader the (easy) multinomial analogue (known as the *multinomial theorem*) of equation (1.18), namely,

$$(x_1 + x_2 + \cdots + x_m)^n = \sum_{a_1 + \cdots + a_m = n} \binom{n}{a_1, a_2, \ldots, a_m} x_1^{a_1} \cdots x_m^{a_m},$$

where the sum ranges over all $(a_1, \ldots, a_m) \in \mathbb{N}^m$ satisfying $a_1 + \cdots + a_m = n$. Note that $\binom{n}{1,1,\ldots,1} = n!$, the number of permutations of an n-element set.

Binomials and multinomial coefficients have an important geometric interpretation in terms of lattice paths. Let S be a subset of $\mathbb{Z}^d$. More generally, we could replace $\mathbb{Z}^d$ by any lattice (discrete subgroup of full rank) in $\mathbb{R}^d$, but for simplicity we consider only $\mathbb{Z}^d$. A *lattice path* L in $\mathbb{Z}^d$ of length k with steps in S is a sequence $v_0, v_1, \ldots, v_k \in \mathbb{Z}^d$ such that each consecutive difference $v_i - v_{i-1}$ lies in S. We say that L *starts at* v_0 and *ends at* v_k, or more simply that L goes *from* v_0 *to* v_k. Figure 1.1 shows the six lattice paths in $\mathbb{Z}^2$ from $(0,0)$ to $(2,2)$ with steps $(1,0)$ and $(0,1)$.

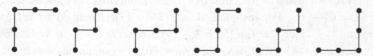

Figure 1.1 Six lattice paths.

1.2.1 Proposition. *Let $v = (a_1, \ldots, a_d) \in \mathbb{N}^d$, and let e_i denote the ith unit coordinate vector in $\mathbb{Z}^d$. The number of lattice paths in $\mathbb{Z}^d$ from the origin $(0, 0, \ldots, 0)$ to v with steps $e_1, \ldots, e_d$ is given by the multinomial coefficient $\binom{a_1 + \cdots + a_d}{a_1, \ldots, a_d}$.*

Proof. Let $v_0, v_1, \ldots, v_k$ be a lattice path being counted. Then the sequence $v_1 - v_0, v_2 - v_1, \ldots, v_k - v_{k-1}$ is simply a sequence consisting of $a_i \, e_i$'s in some order. The proof follows from equation (1.22). $\square$

Proposition 1.2.1 is the most basic result in the vast subject of *lattice path enumeration*. Further results in this area will appear throughout this book.

1.3 Cycles and Inversions

Permutations of sets and multisets are among the richest objects in enumerative combinatorics. A basic reason for this fact is the wide variety of ways to *represent* a permutation combinatorially. We have already seen that we can represent a set permutation either as a *word* or a *function*. In fact, for any set S, the function $w : [n] \to S$ given by $w(i) = w_i$ corresponds to the word $w_1 w_2 \cdots w_n$. Several additional representations will arise in Section 1.5. Many of the basic results derived here will play an important role in later analysis of more complicated objects related to permutations.

A second reason for the richness of the theory of permutations is the wide variety of interesting "statistics" of permutations. In the broadest sense, a statistic on some class $\mathcal{C}$ of combinatorial objects is just a function $f : \mathcal{C} \to S$, where S is any set (often taken to be $\mathbb{N}$). We want $f(x)$ to capture some combinatorially interesting feature of x. For instance, if x is a (finite) set, then $f(x)$ could be its number of elements. We can think of f as *refining* the enumeration of objects in $\mathcal{C}$. For instance, if $\mathcal{C}$ consists of all subsets of an n-set S and $f(x) = \#x$, then f refines the number 2^n of subsets of S into a sum $2^n = \sum_k \binom{n}{k}$, where $\binom{n}{k}$ is the number of subsets of S with k elements. In this section and the next two, we will discuss a number of different statistics on permutations.

Cycle Structure

If we regard a set permutation w as a bijection $w : S \to S$, then it is natural to consider for each $x \in S$ the sequence $x, w(x), w^2(x), \ldots$. Eventually (since w is a bijection and S is assumed finite) we must return to x. Thus for some unique $\ell \geq 1$, we have that $w^\ell(x) = x$ and that the elements $x, w(x), \ldots, w^{\ell-1}(x)$ are distinct. We call the sequence $(x, w(x), \ldots, w^{\ell-1}(x))$ a *cycle* of w of length ℓ. The cycles $(x, w(x), \ldots, w^{\ell-1}(x))$ and $(w^i(x), w^{i+1}(x), \ldots, w^{\ell-1}(x), x, \ldots, w^{i-1}(x))$ are considered the same. Every element of S then appears in a unique cycle of w, and we may regard w as a disjoint union or *product* of its distinct cycles $C_1, \ldots, C_k$, written $w = C_1 \cdots C_k$. For instance, if $w : [7] \to [7]$ is defined by $w(1) = 4$, $w(2) = 2$, $w(3) = 7$, $w(4) = 1$, $w(5) = 3$, $w(6) = 6$, $w(7) = 5$ (or $w = 4271365$ as a word), then $w = (14)(2)(375)(6)$. Of course this representation of w in disjoint cycle notation is not unique; we also have for instance $w = (753)(14)(6)(2)$.

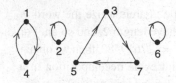

Figure 1.2 The digraph of the permutation (14)(2) (375)(6).

A geometric or graphical representation of a permutation w is often useful. A finite *directed graph* or *digraph* D is a triple (V, E, ϕ), where $V = \{x_1, \ldots, x_n\}$ is a set of *vertices*, E is a finite set of (directed) *edges* or *arcs*, and ϕ is a map from E to $V \times V$. If ϕ is injective, then we call D a *simple* digraph, and we can think of E as a subset of $V \times V$. If e is an edge with $\phi(e) = (x, y)$, then we represent e as an arrow directed from x to y. If w is permutation of the set S, then define the *digraph D_w of w* to be the directed graph with vertex set S and edge set $\{(x, y) : w(x) = y\}$. In other words, for every vertex x, there is an edge from x to $w(x)$. Digraphs of permutations are characterized by the property that every vertex has one edge pointing out and one pointing in. The disjoint cycle decomposition of a permutation of a finite set guarantees that D_w will be a disjoint union of directed cycles. For instance, Figure 1.2 shows the digraph of the permutation $w = (14)(2)(375)(6)$.

We noted earlier that the disjoint cycle notation of a permutation is not unique. We can define a *standard representation* by requiring that (a) each cycle is written with its largest element first, and (b) the cycles are written in increasing order of their largest element. Thus, the standard form of the permutation $w = (14)(2)(375)(6)$ is $(2)(41)(6)(753)$. Define $\widehat{w}$ to be the word (or permutation) obtained from w by writing it in standard form and erasing the parentheses. For example, with $w = (2)(41)(6)(753)$, we have $\widehat{w} = 2416753$. Now observe that we can uniquely recover w from $\widehat{w}$ by inserting a left parenthesis in $\widehat{w} = a_1 a_2 \cdots a_n$ preceding every *left-to-right maximum* or *record* (also called *outstanding element*); that is, an element a_i such that $a_i > a_j$ for every $j < i$. Then insert a right parenthesis where appropriate; that is, before every internal left parenthesis and at the end. Thus, the map $w \mapsto \widehat{w}$ is a *bijection* from $\mathfrak{S}_n$ to itself, known as the *fundamental bijection*. Let us sum up this information as a proposition.

1.3.1 Proposition. *a. The map $\mathfrak{S}_n \overset{\wedge}{\to} \mathfrak{S}_n$ defined above is a bijection.*
b. If $w \in \mathfrak{S}_n$ has k cycles, then $\widehat{w}$ has k left-to-right maxima.

If $w \in \mathfrak{S}_S$ where $\#S = n$, then let $c_i = c_i(w)$ be the number of cycles of w of length i. Note that $n = \sum i c_i$. Define the *type* of w, denoted type(w), to be the sequence $(c_1, \ldots, c_n)$. The total number of cycles of w is denoted $c(w)$, so $c(w) = c_1(w) + \cdots + c_n(w)$.

1.3.2 Proposition. *The number of permutations $w \in \mathfrak{S}_S$ of type $(c_1, \ldots, c_n)$ is equal to $n!/1^{c_1} c_1! 2^{c_2} c_2! \cdots n^{c_n} c_n!$.*

Proof. Let $w = w_1 w_2 \cdots w_n$ be any permutation of S. Parenthesize the word w so that the first c_i cycles have length 1, the next c_2 have length 2, and so on. For instance, if $(c_1, \ldots, c_9) = (1, 2, 0, 1, 0, 0, 0, 0, 0)$ and $w = 427619583$, then we obtain $(4)(27)(61)(9583)$. In general, we obtain the disjoint cycle decomposition of a permutation w' of type $(c_1, \ldots, c_n)$. Hence, we have defined a map $\Phi : \mathfrak{S}_S \to \mathfrak{S}_S^c$, where $\mathfrak{S}_S^c$ is the set of all $u \in \mathfrak{S}_S$ of type $c = (c_1, \ldots, c_n)$. Given $u \in \mathfrak{S}_S^c$, we claim that there are $1^{c_1} c_1! 2^{c_2} c_2! \cdots n^{c_n} c_n!$ ways to write it in disjoint cycle notation so that the cycle lengths are weakly increasing from left to right. Namely, order the cycles of length i in $c_i!$ ways, and choose the first elements of these cycles in i^{c_i} ways. These choices are all independent, so the claim is proved. Hence for each $u \in \mathfrak{S}_S^c$, we have $\#\Phi^{-1}(u) = 1^{c_1} c_1! 2^{c_2} c_2! \cdots n^{c_n} c_n!$, and the proof follows since $\#\mathfrak{S}_S = n!$. $\qquad\square$

NOTE. The proof of Proposition 1.3.2 can easily be converted into a bijective proof of the identity

$$n! = 1^{c_1} c_1! 2^{c_2} c_2! \cdots n^{c_n} c_n! \left(\#\mathfrak{S}_S^c \right),$$

analogous to our bijective proof of equation (1.16).

Proposition 1.3.2 has an elegant and useful formulation in terms of generating functions. Suppose that $w \in \mathfrak{S}_n$ has type $(c_1, \ldots, c_n)$. Write

$$t^{\text{type}(w)} = t_1^{c_1} t_2^{c_2} \cdots t_n^{c_n},$$

and define the *cycle indicator* or *cycle index* of $\mathfrak{S}_n$ to be the polynomial

$$Z_n = Z_n(t_1, \ldots, t_n) = \frac{1}{n!} \sum_{w \in \mathfrak{S}_n} t^{\text{type}(w)}. \tag{1.25}$$

(Set $Z_0 = 1$.) For instance,

$$Z_1 = t_1,$$

$$Z_2 = \frac{1}{2}(t_1^2 + t_2),$$

$$Z_3 = \frac{1}{6}(t_1^3 + 3t_1 t_2 + 2t_3),$$

$$Z_4 = \frac{1}{24}(t_1^4 + 6t_1^2 t_2 + 8t_1 t_3 + 3t_2^2 + 6t_4).$$

1.3.3 Theorem. *We have*

$$\sum_{n \geq 0} Z_n x^n = \exp\left(t_1 x + t_2 \frac{x^2}{2} + t_3 \frac{x^3}{3} + \cdots \right). \tag{1.26}$$

Proof. We give a naive computational proof. For a more conceptual proof, see Example 5.2.10. Let us expand the right-hand side of equation (1.26):

$$\exp\left(\sum_{i\geq 1} t_i \frac{x^i}{i}\right) = \prod_{i\geq 1} \exp\left(t_i \frac{x^i}{i}\right)$$

$$= \prod_{i\geq 1} \sum_{j\geq 0} t_i^j \frac{x^{ij}}{i^j\, j!}. \qquad (1.27)$$

Hence, the coefficient of $t_1^{c_1} \cdots t_n^{c_n} x^n$ is equal to 0 unless $\sum i c_i = n$, in which case it is equal to

$$\frac{1}{1^{c_1} c_1! \, 2^{c_2} c_2! \cdots} = \frac{1}{n!} \frac{n!}{1^{c_1} c_1! \, 2^{c_2} c_2! \cdots}.$$

Comparing with Proposition 1.3.2 completes the proof. □

Let us give two simple examples of the use of Theorem 1.3.3. For some additional examples, see Exercises 5.10 and 5.11. A more general theory of cycle indicators based on symmetric functions is given in Section 7.24. Write $F(t;x) = F(t_1, t_2, \ldots; x)$ for the right-hand side of equation (1.26).

1.3.4 Example. Let $e_6(n)$ be the number of permutations $w \in \mathfrak{S}_n$ satisfying $w^6 = 1$. A permutation w satisfies $w^6 = 1$ if and only if all its cycles have length 1,2,3 or 6. Hence,

$$e_6(n) = n!\, Z_n(t_i = 1 \text{ if } i\,|\,6,\ t_i = 0 \text{ otherwise}).$$

There follows

$$\sum_{n\geq 0} e_6(n) \frac{x^n}{n!} = F(t_i = 1 \text{ if } i\,|\,6,\ t_i = 0 \text{ otherwise})$$

$$= \exp\left(x + \frac{x^2}{2} + \frac{x^3}{3} + \frac{x^6}{6}\right).$$

For the obvious generalization to permutations w satisfying $w^r = 1$, see equation (5.31).

1.3.5 Example. Let $E_k(n)$ denote the expected number of k-cycles in a permutation $w \in \mathfrak{S}_n$. It is understood that the expectation is taken with respect to the uniform distribution on $\mathfrak{S}_n$, so

$$E_k(n) = \frac{1}{n!} \sum_{w\in\mathfrak{S}_n} c_k(w),$$

where $c_k(w)$ denotes the number of k-cycles in w. Now note that from the definition (1.25) of Z_n we have

$$E_k(n) = \frac{\partial}{\partial t_k} Z_n(t_1, \ldots, t_n)|_{t_i=1}.$$

Hence,

$$\sum_{n \geq 0} E_k(n) x^n = \frac{\partial}{\partial t_k} \exp\left(t_1 x + t_2 \frac{x^2}{2} + t_3 \frac{x^3}{3} + \cdots\right)\Bigg|_{t_i = 1}$$

$$= \frac{x^k}{k} \exp\left(x + \frac{x^2}{2} + \frac{x^3}{3} + \cdots\right)$$

$$= \frac{x^k}{k} \exp\log(1-x)^{-1}$$

$$= \frac{x^k}{k} \frac{1}{1-x}$$

$$= \frac{x^k}{k} \sum_{n \geq 0} x^n.$$

It follows that $E_k(n) = 1/k$ for $n \geq k$. Can the reader think of a simple explanation (Exercise 1.120)?

Now define $c(n,k)$ to be the number of permutations $w \in \mathfrak{S}_n$ with exactly k cycles. The number $s(n,k) := (-1)^{n-k} c(n,k)$ is known as a *Stirling number of the first kind*, and $c(n,k)$ is called a *signless Stirling number of the first kind*.

1.3.6 Lemma. *The numbers $c(n,k)$ satisfy the recurrence*

$$c(n,k) = (n-1)c(n-1,k) + c(n-1,k-1), \quad n,k \geq 1,$$

with the initial conditions $c(n,k) = 0$ if $n < k$ or $k = 0$, except $c(0,0) = 1$.

Proof. Choose a permutation $w \in \mathfrak{S}_{n-1}$ with k cycles. We can insert the symbol n after any of the numbers $1, 2, \ldots, n-1$ in the disjoint cycle decomposition of w in $n-1$ ways, yielding the disjoint cycle decomposition of a permutation $w' \in \mathfrak{S}_n$ with k cycles for which n appears in a cycle of length at least 2. Hence, there are $(n-1)c(n-1,k)$ permutations $w' \in \mathfrak{S}_n$ with k cycles for which $w'(n) \neq n$.

On the other hand, if we choose a permutation $w \in \mathfrak{S}_{n-1}$ with $k-1$ cycles, we can extend it to a permutation $w' \in \mathfrak{S}_n$ with k cycles satisfying $w'(n) = n$ by defining

$$w'(i) = \begin{cases} w(i), & \text{if } i \in [n-1] \\ n, & \text{if } i = n. \end{cases}$$

Thus there are $c(n-1,k-1)$ permutations $w' \in \mathfrak{S}_n$ with k cycles for which $w'(n) = n$, and the proof follows. $\qquad\square$

Most of the elementary properties of the numbers $c(n,k)$ can be established using Lemma 1.3.6 together with mathematical induction. However, combinatorial proofs are to be preferred whenever possible. An illuminating illustration of the various techniques available to prove elementary combinatorial identities is provided by the next result.

1.3.7 Proposition. *Let t be an indeterminate and fix $n \geq 0$. Then*

$$\sum_{k=0}^{n} c(n,k)t^k = t(t+1)(t+2)\cdots(t+n-1). \qquad (1.28)$$

First Proof. This proof may be regarded as "semi-combinatorial" since it is based directly on Lemma 1.3.6, which had a combinatorial proof. Let

$$F_n(t) = t(t+1)\cdots(t+n-1) = \sum_{k=0}^{n} b(n,k)t^k.$$

Clearly $b(n,k) = 0$ if $n = 0$ or $k = 0$, except $b(0,0) = 1$ (an empty product is equal to 1). Moreover, since

$$F_n(t) = (t+n-1)F_{n-1}(t)$$

$$= \sum_{k=1}^{n} b(n-1,k-1)t^k + (n-1)\sum_{k=0}^{n-1} b(n-1,k)t^k,$$

there follows $b(n,k) = (n-1)b(n-1,k) + b(n-1,k-1)$. Hence $b(n,k)$ satisfies the same recurrence and initial conditions as $c(n,k)$, so they agree. □

Second Proof. Our next proof is a straightforward argument using generating functions. In terms of the cycle indicator Z_n, we have

$$\sum_{k=0}^{n} c(n,k)t^k = n!Z_n(t,t,t,\dots).$$

Hence substituting $t_i = t$ in equation (1.26) gives

$$\sum_{n\geq 0}\sum_{k=0}^{n} c(n,k)t^k\frac{x^n}{n!} = \exp t(x + \frac{x^2}{2} + \frac{x^3}{3} + \cdots)$$

$$= \exp t(\log(1-x)^{-1})$$

$$= (1-x)^{-t}$$

$$= \sum_{n\geq 0}(-1)^n\binom{-t}{n}x^n$$

$$= \sum_{n\geq 0} t(t+1)\dots(t+n-1)\frac{x^n}{n!},$$

and the proof follows from taking coefficient of $x^n/n!$. □

Third Proof. The coefficient of t^k in $F_n(t)$ is

$$\sum_{1\leq a_1 < a_2 < \cdots < a_{n-k} \leq n-1} a_1 a_2 \cdots a_{n-k}, \qquad (1.29)$$

where the sum is over all $\binom{n-1}{n-k}$ $(n-k)$-subsets $\{a_1,\ldots,a_{n-k}\}$ of $[n-1]$. (Though irrelevant here, it is interesting to note that this sum is just the $(n-k)$th elementary symmetric function of $1,2,\ldots,n-1$.) Clearly (1.29) counts the number of pairs (S,f), where $S \in \binom{[n-1]}{n-k}$ and $f : S \to [n-1]$ satisfies $f(i) \le i$. Thus, we seek a bijection $\phi : \Omega \to \mathfrak{S}_{n,k}$ between the set Ω of all such pairs (S,f), and the set $\mathfrak{S}_{n,k}$ of $w \in \mathfrak{S}_n$ with k cycles.

Given $(S,f) \in \Omega$ where $S = \{a_1,\ldots,a_{n-k}\}_< \subseteq [n-1]$, define $T = \{j \in [n] : n - j \notin S\}$. Let the elements of $[n] - T$ be $b_1 > b_2 > \cdots > b_{n-k}$. Define $w = \phi(S,f)$ to be that permutation that when written in standard form satisfies: (i) the first (= greatest) elements of the cycles of w are the elements of T, and (ii) for $i \in [n-k]$ the number of elements of w preceding b_i and larger than b_i is $f(a_i)$. We leave it to the reader to verify that this construction yields the desired bijection. $\qquad\square$

1.3.8 Example. Suppose that in the preceding proof $n = 9, k = 4, S = \{1,3,4,6,8\}$, $f(1) = 1, f(3) = 2, f(4) = 1, f(6) = 3, f(8) = 6$. Then $T = \{2,4,7,9\}$, $[9] - T = \{1,3,5,6,8\}$, and $w = (2)(4)(753)(9168)$.

Fourth Proof of Proposition 1.3.7. There are two basic ways of giving a combinatorial proof that two polynomials are equal: (i) showing that their coefficients are equal and (ii) showing that they agree for sufficiently many values of their variable(s). We have already established Proposition 1.3.7 by the first technique; here we apply the second. If two polynomials in a single variable t (over the complex numbers, say) agree for all $t \in \mathbb{P}$, then they agree as polynomials. Thus, it suffices to establish (1.28) for all $t \in \mathbb{P}$.

Let $t \in \mathbb{P}$, and let $C(w)$ denote the set of cycles of $w \in \mathfrak{S}_n$. The left-hand side of (1.28) counts all pairs (w,f), where $w \in \mathfrak{S}_n$ and $f : C(w) \to [t]$. The right-hand side counts integer sequences $(a_1,a_2,\ldots,a_n)$ where $0 \le a_i \le t+n-i-1$. (There are historical reasons for this restriction of a_i, rather than, say, $1 \le a_i \le t+i-1$.) Given such a sequence $(a_1,a_2,\ldots,a_n)$, the following simple algorithm may be used to define (w,f). First, write down the number n and regard it as starting a cycle C_1 of w. Let $f(C_1) = a_n + 1$. Assuming $n, n-1, \ldots, n-i+1$ have been inserted into the disjoint cycle notation for w, we now have two possibilities:

i. $0 \le a_{n-i} \le t - 1$. Then start a new cycle C_j with the element $n - i$ to the left of the previously inserted elements, and set $f(C_j) = a_{n-i} + 1$.

ii. $a_{n-i} = t + k$ where $0 \le k \le i - 1$. Then insert $n - i$ into an old cycle so that it is not the leftmost element of any cycle, and so that it appears to the right of $k + 1$ of the numbers previously inserted.

This procedure establishes the desired bijection. $\qquad\square$

1.3.9 Example. Suppose $n = 9$, $t = 4$, and $(a_1, \ldots, a_9) = (4, 8, 5, 0, 7, 5, 2, 4, 1)$. Then w is built up as follows:

$$(9)$$
$$(98)$$
$$(7)(98)$$
$$(7)(968)$$
$$(7)(9685)$$
$$(4)(7)(9685)$$
$$(4)(73)(9685)$$
$$(4)(73)(96285)$$
$$(41)(73)(96285).$$

Moreover, $f(96285) = 2$, $f(73) = 3$, $f(41) = 1$.

Note that if we set $t = 1$ in the preceding proof, we obtain a combinatorial proof of the following result.

1.3.10 Proposition. *Let $n, k \in \mathbb{P}$. The number of integer sequences $(a_1, \ldots, a_n)$ such that $0 \le a_i \le n - i$ and exactly k values of a_i equal 0 is $c(n, k)$*

Note that because of Proposition 1.3.1, we obtain "for free" the enumeration of permutations by left-to-right maxima.

1.3.11 Corollary. *The number of $w \in \mathfrak{S}_n$ with k left-to-right maxima is $c(n, k)$.*

Corollary 1.3.11 illustrates one benefit of having different ways of representing the same object (here a permutation) – different enumerative problems involving the object turn out to be equivalent.

Inversions

The fourth proof of Proposition 1.3.7 (in the case $t = 1$) associated a permutation $w \in \mathfrak{S}_n$ with an integer sequence $(a_1, \ldots, a_n)$, $0 \le a_i \le n - i$. There is a different method for accomplishing this which is perhaps more natural. Given such a vector $(a_1, \ldots, a_n)$, assume that $n, n - 1, \ldots, n - i + 1$ have been inserted into w, expressed this time as a *word* (rather than a product of cycles). Then insert $n - i$ so that it has a_{n-i} elements to its left. For example, if $(a_1, \ldots, a_9) = (1, 5, 2, 0, 4, 2, 0, 1, 0)$, then w is built up as follows:

$$9$$
$$98$$
$$798$$
$$7968$$
$$79685$$
$$479685$$
$$4739685$$
$$47396285$$
$$417396285.$$

Clearly a_i is the number of entries j of w to the left of i satisfying $j > i$. A pair (w_i, w_j) is called an *inversion* of the permutation $w = w_1 w_2 \cdots w_n$ if $i < j$ and $w_i > w_j$. The sequence $I(w) = (a_1, \dots, a_n)$ is called the *inversion table* of w. The preceding algorithm for constructing w from its inversion table $I(w)$ establishes the following result.

1.3.12 Proposition. *Let*

$$\mathcal{T}_n = \{(a_1, \dots, a_n) : 0 \le a_i \le n - i\} = [0, n-1] \times [0, n-2] \times \cdots \times [0, 0].$$

The map $I : \mathfrak{S}_n \to \mathcal{T}_n$ *that sends each permutation to its inversion table is a bijection.*

Therefore, the inversion table $I(w)$ is yet another way to represent a permutation w. Let us also mention that the *code* of a permutation w is defined by $\text{code}(w) = I(w^{-1})$. Equivalently, if $w = w_1 \cdots w_n$ and $\text{code}(w) = (c_1, \dots, c_n)$, then c_i is equal to the number of elements w_j to the right of w_i (i.e., $i < j$) such that $w_i > w_j$. The question of whether to use $I(w)$ or $\text{code}(w)$ depends on the problem at hand and is clearly only a matter of convenience. Often it makes no difference which is used, such as in obtaining the next corollary.

1.3.13 Corollary. *Let* $\text{inv}(w)$ *denote the number of inversions of the permutation* $w \in \mathfrak{S}_n$. *Then*

$$\sum_{w \in \mathfrak{S}_n} q^{\text{inv}(w)} = (1+q)(1+q+q^2)\cdots(1+q+q^2+\cdots+q^{n-1}). \qquad (1.30)$$

Proof. If $I(w) = (a_1, \dots, a_n)$ then $\text{inv}(w) = a_1 + \cdots + a_n$. Hence,

$$\sum_{w \in \mathfrak{S}_n} q^{\text{inv}(w)} = \sum_{a_1=0}^{n-1} \sum_{a_2=0}^{n-2} \cdots \sum_{a_n=0}^{0} q^{a_1 + a_2 + \cdots + a_n}$$

$$= \left(\sum_{a_1=0}^{n-1} q^{a_1} \right) \left(\sum_{a_2=0}^{n-2} q^{a_2} \right) \cdots \left(\sum_{a_n=0}^{0} q^{a_n} \right),$$

as desired. $\qquad\qquad\qquad\qquad\qquad\qquad\qquad\qquad\qquad\qquad\qquad\qquad\qquad\square$

The polynomial $(1+q)(1+q+q^2)\cdots(1+q+\cdots+q^{n-1})$ is called "the q-analogue of $n!$" and is denoted $(\boldsymbol{n})!$. Moreover, we denote the polynomial $1 + q + \cdots + q^{n-1} = (1-q^n)/(1-q)$ by $(\boldsymbol{n})$ and call it "the q-analogue of n," so that

$$(\boldsymbol{n})! = (\boldsymbol{1})(\boldsymbol{2}) \cdots (\boldsymbol{n}).$$

In general, a *q-analogue* of a mathematical object is an object depending on the variable q that "reduces to" (an admittedly vague term) the original object when we set $q = 1$. To be a "satisfactory" q-analogue more is required, but there is no precise definition of what is meant by "satisfactory." Certainly one desirable property is that the original object concerns finite sets, while the q-analogue can

be interpreted in terms of subspaces of finite-dimensional vector spaces over the
finite field $\mathbb{F}_q$. For instance, $n!$ is the number of sequences $\emptyset = S_0 \subset S_1 \subset \cdots \subset S_n = [n]$ of subsets of $[n]$. (The symbol $\subset$ denotes strict inclusion, so $\#S_i = i$.)
Similarly if q is a prime power then $(n)!$ is the number of sequences $0 = V_0 \subset V_1 \subset \cdots \subset V_n = \mathbb{F}_q^n$ of subspaces of the n-dimensional vector space $\mathbb{F}_q^n$ over $\mathbb{F}_q$
(so dim $V_i = i$). For this reason $(n)!$ is regarded as a satisfactory q-analogue of
$n!$. We can also regard an i-dimensional vector space over $\mathbb{F}_q$ as the q-analogue
of an i-element set. Many more instances of q-analogues will appear throughout
this book, especially in Section 1.10. The theory of binomial posets developed
in Section 3.18 gives a partial explanation for the existence of certain classes of
q-analogues including $(n)!$.

We conclude this section with a simple but important property of the statistic
inv.

1.3.14 Proposition. *For any $w = w_1 w_2 \cdots w_n \in \mathfrak{S}_n$, we have* $\mathrm{inv}(w) = \mathrm{inv}(w^{-1})$.

Proof. The pair (i, j) is an inversion of w if and only if (w_j, w_i) is an inversion
of w^{-1}. □

1.4 Descents

In addition to cycle type and inversion table, there is one other fundamental statistic
associated with a permutation $w \in \mathfrak{S}_n$. If $w = w_1 w_2 \cdots w_n$ and $1 \le i \le n - 1$, then
i is a *descent* of w if $w_i > w_{i+1}$, while i is an *ascent* if $w_i < w_{i+1}$. (Sometimes
it is desirable to also define n to be a descent, but we will adhere to the previous
definition.) Define the *descent set* $D(w)$ of w by

$$D(w) = \{i : w_i > w_{i+1}\} \subseteq [n-1].$$

If $S \subseteq [n-1]$, then denote by $\alpha(S)$ (or $\alpha_n(S)$ if necessary) the number of permutations $w \in \mathfrak{S}_n$ whose descent set is contained in S, and by $\beta(S)$ (or $\beta_n(S)$) the
number whose descent set is equal to S. In symbols,

$$\alpha(S) = \#\{w \in \mathfrak{S}_n : D(w) \subseteq S\}, \tag{1.31}$$

$$\beta(S) = \#\{w \in \mathfrak{S}_n : D(w) = S\}. \tag{1.32}$$

Clearly,

$$\alpha(S) = \sum_{T \subseteq S} \beta(T). \tag{1.33}$$

As explained in Example 2.2.4, we can invert this relationship to obtain

$$\beta(S) = \sum_{T \subseteq S} (-1)^{\#(S-T)} \alpha(T). \tag{1.34}$$

1.4.1 Proposition. *Let $S = \{s_1, \ldots, s_k\}_< \subseteq [n-1]$. Then*

$$\alpha(S) = \binom{n}{s_1, s_2 - s_1, s_3 - s_2, \ldots, n - s_k}. \tag{1.35}$$

Proof. To obtain a permutation $w = w_1 w_2 \cdots w_n \in \mathfrak{S}_n$ satisfying $D(w) \subseteq S$, first choose $w_1 < w_2 < \cdots < w_{s_1}$ in $\binom{n}{s_1}$ ways. Then choose $w_{s_1+1} < w_{s_1+2} < \cdots < w_{s_2}$ in $\binom{n-s_1}{s_2-s_1}$ ways, and so on. We therefore obtain

$$\alpha(S) = \binom{n}{s_1}\binom{n-s_1}{s_2-s_1}\binom{n-s_2}{s_3-s_2}\cdots\binom{n-s_k}{n-s_k}$$

$$= \binom{n}{s_1, s_2-s_1, s_3-s_2, \ldots, n-s_k},$$

as desired.   □

1.4.2 Example. Let $n \geq 9$. Then

$$\beta_n(3,8) = \alpha_n(3,8) - \alpha_n(3) - \alpha_n(8) + \alpha_n(\emptyset)$$

$$= \binom{n}{3,5,n-8} - \binom{n}{3} - \binom{n}{8} + 1.$$

Two closely related descent sets are of special combinatorial interest. We say that a permutation $w = w_1 w_2 \cdots w_n \in \mathfrak{S}_n$ (or more generally any sequence of distinct numbers) is *alternating* (or *zigzag* or *down-up*) if $w_1 > w_2 < w_3 > w_4 < \cdots$. Equivalently, $D(w) = \{1,3,5,\ldots\} \cap [n-1]$. The alternating permutations in $\mathfrak{S}_4$ are 2143, 3142, 3241, 4132, 4231. Similarly, w is *reverse alternating* (or *up-down*) if $w_1 < w_2 > w_3 < w_4 > \cdots$. Equivalently, $D(w) = \{2,4,6,\ldots\} \cap [n-1]$. The reverse alternating permutations in $\mathfrak{S}_4$ are 1324, 1423, 2314, 2413, 3412. The number of alternating permutations $w \in \mathfrak{S}_n$ is denoted E_n (with $E_0 = 1$) and is called an *Euler number*. (Originally, $(-1)^n E_{2n}$ was called an Euler number.) Since w is alternating if and only if $n+1-w_1, n+1-w_2, \ldots, n+1-w_n$ is reverse alternating, it follows that E_n is also the number of reverse alternating permutations in $\mathfrak{S}_n$. We will develop some properties of alternating permutations and Euler numbers in various subsequent sections, especially Section 1.6.

NOTE. Some mathematicians define alternating permutations to be our reverse alternating permutations, while others define them to be permutations which are either alternating or reverse alternating according to our definition.

For the remainder of this section, we discuss some additional permutation statistics based on the descent set. The first of these is the *number of descents* of w, denoted $\mathrm{des}(w)$. Thus, $\mathrm{des}(w) = \#D(w)$. Let

$$A_d(x) = \sum_{w \in \mathfrak{S}_d} x^{1+\mathrm{des}(w)} \tag{1.36}$$

$$= \sum_{k=1}^{d} A(d,k)x^k.$$

Hence $A(d,k)$ is the number of permutations $w \in \mathfrak{S}_d$ with exactly $k - 1$ descents. The polynomial $A_d(x)$ is called an *Eulerian polynomial*, while $A(d,k)$ is an *Eulerian number*. We set $A(0,k) = \delta_{0k}$. The first few Eulerian polynomials are

$$A_0(x) = 1$$

$$A_1(x) = x$$

$$A_2(x) = x + x^2$$

$$A_3(x) = x + 4x^2 + x^3$$

$$A_4(x) = x + 11x^2 + 11x^3 + x^4$$

$$A_5(x) = x + 26x^2 + 66x^3 + 26x^4 + x^5$$

$$A_6(x) = x + 57x^2 + 302x^3 + 302x^4 + 57x^5 + x^6$$

$$A_7(x) = x + 120x^2 + 1191x^3 + 2416x^4 + 1191x^5 + 120x^6 + x^7$$

$$A_8(x) = x + 247x^2 + 4293x^3 + 15619x^4 + 15619x^5 + 4293x^6$$
$$+ 247x^7 + x^8.$$

The bijection $w \mapsto \hat{w}$ of Proposition 1.3.1 yields an interesting alternative description of the Eulerian numbers. Suppose that

$$w = (a_1, a_2, \ldots, a_{i_1})(a_{i_1+1}, a_{i_1+2}, \ldots, a_{i_2}) \cdots (a_{i_{k-1}+1}, a_{i_{k-1}+2}, \ldots, a_d)$$

is a permutation in $\mathfrak{S}_d$ written in standard form. Thus, $a_1, a_{i_1+1}, \ldots, a_{i_{k-1}+1}$ are the largest elements of their cycles, and $a_1 < a_{i_1+1} < \cdots < a_{i_{k-1}+1}$. It follows that if $w(a_i) \neq a_{i+1}$, then $a_i < a_{i+1}$. Hence, $a_i < a_{i+1}$ or $i = d$ if and only if $w(a_i) \geq a_i$, so that

$$d - \mathrm{des}(\hat{w}) = \#\{i \in [d] : w(i) \geq i\}.$$

A number i for which $w(i) \geq i$ is called a *weak excedance* of w, while a number i for which $w(i) > i$ is called an *excedance* of w. One easily sees that a permutation $w = w_1 w_2 \cdots w_d$ has k weak excedances if and only if the permutation $u_1 u_2 \cdots u_d$ defined by $u_i = d + 1 - w_{d-i+1}$ has $d - k$ excedances. Moreover, w has $d - 1 - j$ descents if and only if $w_d w_{d-1} \cdots w_1$ has j descents. We therefore obtain the following result.

1.4.3 Proposition. *The number of permutations $w \in \mathfrak{S}_d$ with k excedances, as well as the number with $k + 1$ weak excedances, is equal to the Eulerian number $A(d, k + 1)$.*

The next result gives a fundamental property of Eulerian polynomials related to generating functions.

1.4.4 Proposition. *For every $d \geq 0$, we have*

$$\sum_{m \geq 0} m^d x^m = \frac{A_d(x)}{(1-x)^{d+1}}. \tag{1.37}$$

Proof. The proof is by induction on d. Since $\sum_{m\geq 0} x^m = 1/(1-x)$, the assertion is true for $d = 0$. Now assume that equation (1.37) holds for some $d \geq 0$. Differentiate with respect to x and multiply by x to obtain

$$\sum_{m\geq 0} m^{d+1} x^m = \frac{x(1-x)A_d'(x) + (d+1)xA_d(x)}{(1-x)^{d+2}}. \tag{1.38}$$

Hence, it suffices to show that

$$A_{d+1}(x) = x(1-x)A_d'(x) + (d+1)xA_d(x).$$

Taking coefficients of x^k on both sides and simplifying yields

$$A(d+1,k) = kA(d,k) + (d-k+2)A(d,k-1). \tag{1.39}$$

The left-hand side of equation (1.39) counts permutations in $\mathfrak{S}_{d+1}$ with $k-1$ descents. We can obtain such a permutation uniquely in one of two ways. For the first way, choose a permutation $w = w_1 \cdots w_d \in \mathfrak{S}_d$ with $k-1$ descents, and insert $d+1$ after w_i if $i \in D(w)$, or insert $d+1$ at the end. There are k ways to insert $d+1$, so we obtain by this method $kA(d,k)$ permutations in $\mathfrak{S}_{d+1}$ with $k-1$ descents. For the second way, choose $w = w_1 \cdots w_d \in \mathfrak{S}_d$ with $k-2$ descents, and insert $d+1$ after w_i if $i \notin D(w)$, or insert $d+1$ at the beginning. There are $d-k+2$ ways to insert $d+1$, so we obtain a further $(d-k+2)A(d,k-1)$ permutations in $\mathfrak{S}_{d+1}$ with $k-1$ descents. We have verified that the recurrence (1.39) holds, so the proof follows by induction. $\square$

The appearance of the expression m^d in equation (1.37) suggests that there might be a more conceptual proof involving functions $f : [d] \to [m]$. We give such a proof at the end of this section.

We can also give a formula for the exponential generating function of the Eulerian polynomials themselves. For this purpose, define $A_0(x) = 1$.

1.4.5 Proposition. *We have*

$$\sum_{d\geq 0} A_d(x) \frac{t^d}{d!} = \frac{1-x}{1 - xe^{(1-x)t}}. \tag{1.40}$$

Proof. Perhaps the simplest proof at this point is to multiply equation (1.37) by $t^d/d!$ and sum on $d \geq 0$. We get (using the convention $0^0 = 1$, which is often "correct" in enumerative combinatorics)

$$\sum_{d\geq 0} \frac{A_d(x)}{(1-x)^{d+1}} \frac{t^d}{d!} = \sum_{d\geq 0}\sum_{m\geq 0} m^d x^m \frac{t^d}{d!}$$

$$= \sum_{m\geq 0} x^m e^{mt}$$

$$= \frac{1}{1 - xe^t}.$$

Now multiply both sides by $1 - x$ and substitute $(1 - x)t$ for t to complete the proof. (A more conceptual proof will be given in Section 3.19.) □

A further interesting statistic associated with the descent set $D(w)$ is the *major index* (originally called the *greater index*), denoted maj(w) (originally $\iota(w)$) and defined to be the sum of the elements of $D(w)$:

$$\text{maj}(w) = \sum_{i \in D(w)} i.$$

We next give a bijective proof of the remarkable result that inv and maj are *equidistributed*, that is, for any k,

$$\#\{w \in \mathfrak{S}_n : \text{inv}(w) = k\} = \#\{w \in \mathfrak{S}_n : \text{maj}(w) = k\}. \tag{1.41}$$

Note that in terms of generating functions, equation (1.41) takes the form

$$\sum_{w \in \mathfrak{S}_n} q^{\text{inv}(w)} = \sum_{w \in \mathfrak{S}_n} q^{\text{maj}(w)}.$$

1.4.6 Proposition. *We have*

$$\sum_{w \in \mathfrak{S}_n} q^{\text{maj}(w)} = (n)!. \tag{1.42}$$

Proof. We will recursively define a bijection $\varphi : \mathfrak{S}_n \to \mathfrak{S}_n$ as follows. Let $w = w_1 \cdots w_n \in \mathfrak{S}_n$. We will define words (or sequences) $\gamma_1, \ldots, \gamma_n$, where γ_k is a permutation of $\{w_1, \ldots, w_k\}$.

First, let $\gamma_1 = w_1$. Assume that γ_k has been defined for some $1 \le k < n$. If the last letter of γ_k (which turns out to be w_k) is greater (respectively, smaller) than w_{k+1}, then split γ_k after each letter greater (respectively, smaller) than w_{k+1}. These splits divide γ_k into compartments. Cyclically shift each compartment of γ_k one unit to the right, and place w_{k+1} at the end. Let γ_{k+1} be the word thus obtained. Set $\varphi(w) = \gamma_n$.

1.4.7 Example. Before analyzing the map φ, let us first give an example. Let $w = 683941725 \in \mathfrak{S}_9$. Then $\gamma_1 = 6$. It is irrelevant at this point whether $6 < w_2$ or $6 > w_2$ since there can be only one compartment, and $\gamma_2 = 68$. Now $8 > w_3 = 3$, so we split 68 after numbers greater than 3, getting $6|8$. Cyclically shifting the two compartments of length one leaves them unchanged, so $\gamma_3 = 683$. Now $3 < w_4 = 9$, so we split 683 after numbers less than 9. We get $6|8|3$ and $\gamma_4 = 6839$. Now $9 > w_5 = 4$, so we split 6839 after numbers greater than 4, giving $6|8|39$. The cyclic shift of 39 is 93, so $\gamma_5 = 68934$. Continuing in this manner gives the following sequence of γ_i's and compartments:

$$
\begin{array}{llllllllll}
6 \\
6| & 8 \\
6| & 8| & 3 \\
6| & 8| & 3 & 9 \\
6| & 8| & 9| & 3| & 4 \\
6| & 8 & 9 & 3| & 4| & 1 \\
6| & 3| & 8| & 9| & 4| & 1 & 7 \\
6 & 3| & 8 & 9 & 4| & 7 & 1| & 2 \\
3 & 6 & 4 & 8 & 9 & 1 & 7 & 2 & 5
\end{array}
$$

Hence, $\varphi(w) = 364891725$. Note that $\mathrm{maj}(w) = \mathrm{inv}(\varphi(w)) = 18$.

Returning to the proof of Proposition 1.4.6, we claim that φ is a bijection transforming maj to inv, that is,

$$\mathrm{maj}(w) = \mathrm{inv}(\varphi(w)). \qquad (1.43)$$

We have defined inv and maj for permutations $w \in \mathfrak{S}_n$, but precisely the same definition can be made for *any* sequence $w = w_1 \cdots w_n$ of integers. Namely,

$$\mathrm{inv}(w) = \#\{(i,j) : i < j,\ w_i > w_j\},$$

$$\mathrm{maj}(w) = \sum_{i\, :\, w_i > w_{i+1}} i.$$

Let $\eta_k = w_1 w_2 \cdots w_k$. We then prove by induction on k that $\mathrm{inv}(\gamma_k) = \mathrm{maj}(\eta_k)$, from which the proof follows by letting $k = n$.

Clearly $\mathrm{inv}(\gamma_1) = \mathrm{maj}(\eta_1) = 0$. Assume that $\mathrm{inv}(\gamma_k) = \mathrm{maj}(\eta_k)$ for some $k < n$. First, suppose that the last letter w_k of γ_k is greater than w_{k+1}. Thus, $k \in D(w)$, so we need to show that $\mathrm{inv}(\gamma_{k+1}) = k + \mathrm{inv}(\gamma_k)$. The last letter of any compartment C of γ_k is the largest letter of the compartment. Hence, when we cyclically shift this compartment, we create $\#C - 1$ new inversions. Each compartment contains exactly one letter larger than w_{k+1}, so when we append w_{k+1} to the end of γ_k, the number of new inversions $(i, k+1)$ is equal to the number m of compartments. Thus, altogether we have created

$$\sum_C (\#C - 1) + m = k$$

new inversions, as desired. The proof for the case $w_k < w_{k+1}$ is similar and will be omitted.

It remains to show that φ is a bijection. To do so we define φ^{-1}. Let $v = v_1 v_2 \cdots v_n \in \mathfrak{S}_n$. We want to find a (unique) $w = w_1 w_2 \cdots w_n \in \mathfrak{S}_n$ so that $\varphi(w) = v$. Let $\delta_{n-1} = v_1 v_2 \cdots v_{n-1}$ and $w_n = v_n$. Now suppose that δ_k and $w_{k+1}, w_{k+2}, \ldots, w_n$ have been defined for some $1 \le k < n$. If the *first* letter of δ_k

is greater (respectively, smaller) than w_{k+1}, then split δ_k before each letter greater (respectively, smaller) than w_{k+1}. Then in each compartment of δ_k thus formed, cyclically shift the letters one unit to the *left*. Let the last letter of the word thus formed be w_k, and remove this last letter to obtain δ_{k-1}. It is easily verified that this procedure simply reverses the procedure used to obtain $v = \varphi(w)$ from w, completing the proof. $\qquad\qquad\square$

Proposition 1.4.6 establishes the equidistribution of inv and maj on $\mathfrak{S}_n$. Whenever we have two equidistributed statistics $f, g : S \to \mathbb{N}$ on a set S, we can ask whether a stronger result holds, namely, whether f and g have a *symmetric joint distribution*. This means that for all j, k we have

$$\#\{x \in S : f(x) = j, \, g(x) = k\} = \#\{x \in S : f(x) = k, \, g(x) = j\}. \qquad (1.44)$$

This condition can be restated in terms of generating functions as

$$\sum_{x \in S} q^{f(x)} t^{g(x)} = \sum_{x \in S} q^{g(x)} t^{f(x)}.$$

The best way to prove (1.44) is to find a bijection $\psi : S \to S$ such that for all $x \in S$, we have $f(x) = g(\psi(x))$ and $g(x) = f(\psi(x))$. In other words, ψ interchanges the two statistics f and g.

Our next goal is to show that inv and maj have a symmetric joint distribution on $\mathfrak{S}_n$. We will not give an explicit bijection $\psi : \mathfrak{S}_n \to \mathfrak{S}_n$ interchanging inv and maj, but rather we will deduce it from a surprising property of the bijection φ defined in the proof of Proposition 1.4.6. To explain this property, define the *inverse descent set* $\mathrm{ID}(w)$ of $w \in \mathfrak{S}_n$ by $\mathrm{ID}(w) = D(w^{-1})$. Alternatively, we may think of $\mathrm{ID}(w)$ as the "reading set" of w as follows. We read the numbers $1, 2, \ldots, n$ in w from left-to-right in their standard order, going back to the beginning of w when necessary. For instance, if $w = 683941725$, then we first read 12, then 345, then 67, and finally 89. The cumulative number of elements in these reading sequences, excluding the last, form the reading set of w. It is easy to see that this reading set is just $\mathrm{ID}(w)$. For instance, $\mathrm{ID}(683941725) = \{2, 5, 7\}$.

We can easily extend the definition of $\mathrm{ID}(w)$ to arbitrary sequences $w_1 w_2 \cdots w_n$ of distinct integers. (We can even drop the condition that the w_i's are distinct, but we have no need here for such generality.) Simply regard $w = w_1 w_2 \cdots w_n$ as a permutation of its elements written in increasing order, that is, if $S = \{w_1, \ldots, w_n\} = \{u_1, \ldots, u_n\}_<$, then identify w with the permutation of S defined by $w(u_i) = w_i$. We can then write w^{-1} as a word in the same way as w and hence can define $\mathrm{ID}(w)$ as the descent set of w^{-1} written as a word. For instance, if $w = 74285$, then $w^{-1} = 54827$ and $\mathrm{ID}(w) = \{1, 3\}$. We can obtain the same result by reading w in the increasing order of its elements as before, obtaining reading sequences $u_1 u_2 \cdots u_{i_1}, \ u_{i_1+1} \cdots u_{i_2}, \ldots, u_{i_j+1} \cdots u_{i_n}$, and then obtaining $\mathrm{ID}(w) = \{i_1, i_2, \ldots, i_j\}$ (the cumulative numbers of elements in the reading sequences). For instance, with $w = 74285$ the reading sequences are 2, 45, 78, giving $\mathrm{ID}(w) = \{1, 3\}$ as before.

1.4.8 Theorem. *Let φ be the bijection defined in the proof of Proposition 1.4.6. Then for all $w \in \mathfrak{S}_n$, $\mathrm{ID}(w) = \mathrm{ID}(\varphi(w))$. In other words, φ preserves the inverse descent set.*

Proof. Preserve the notation of the proof of Proposition 1.4.6. We prove by induction on k that $\mathrm{ID}(\gamma_k) = \mathrm{ID}(\eta_k)$, from which the proof follows by setting $k = n$. Clearly, $\mathrm{ID}(\gamma_1) = \mathrm{ID}(\eta_1) = \emptyset$. Assume that $\mathrm{ID}(\gamma_k) = \mathrm{ID}(\eta_k)$ for some $k < n$. First, suppose that the last letter w_k of γ_k is greater than w_{k+1}, so that the last letter of any compartment C of γ_k is the unique letter in the compartment larger than w_{k+1}. Consider the reading of η_{k+1}. It will proceed just as for η_k until we encounter the largest letter of η_k less than w_{k+1}, in which case we next read w_{k+1} and then return to the beginning. Exactly the same is true for reading γ_{k+1}, so by the induction hypothesis, the reading sets of η_{k+1} and γ_{k+1} are the same up to this point. Let L be the set of remaining letters to be read. The letters in L are those greater than w_{k+1}. The reading words of these letters are the same for η_k and γ_k by the induction hypothesis. But the letters of L appear in the same order in η_k and η_{k+1} by definition of η_j. Moreover, they also appear in the same order in γ_k and γ_{k+1} since each such letter appears in exactly one compartment, so cyclic shifts (or indeed any permutations) within each compartment of γ_k does not change their order in γ_{k+1}. Hence, the reading words of the letters in L are the same for η_{k+1} and γ_{k+1}, so the proof follows for the case $w_k > w_{k+1}$. The case $w_k < w_{k+1}$ is similar and will be omitted. $\qquad\square$

Let $\mathrm{imaj}(w) = \mathrm{maj}(w^{-1}) = \sum_{i \in \mathrm{ID}(w)} i$. As an immediate corollary to Theorem 1.4.8, we get the symmetric joint distribution of three pairs of permutations statistics including $(\mathrm{inv}, \mathrm{maj})$, thereby improving Proposition 1.4.6. For further information about the bidistribution of $(\mathrm{maj}, \mathrm{imaj})$, see Exercise 4.47 and Corollary 7.23.9.

1.4.9 Corollary. *The three pairs of statistics $(\mathrm{inv}, \mathrm{maj})$, $(\mathrm{inv}, \mathrm{imaj})$, and $(\mathrm{maj}, \mathrm{imaj})$ have symmetric joint distributions.*

Proof. Let f be any statistic on $\mathfrak{S}_n$, and define g by $g(w) = f(w^{-1})$. Clearly (f, g) have a symmetric joint distribution, of which $(\mathrm{maj}, \mathrm{imaj})$ is a special case. By Theorem 1.4.8, φ transforms maj to inv while preserving imaj, so $(\mathrm{inv}, \mathrm{imaj})$ have a symmetric joint distribution. It then follows from Proposition 1.3.14 that $(\mathrm{inv}, \mathrm{maj})$ have a symmetric joint distribution. $\qquad\square$

We conclude this section by discussing a connection between permutations $w \in \mathfrak{S}_n$ and functions $f : [n] \to \mathbb{N}$ (the set $\mathbb{N}$ could be replaced by any totally ordered set) in which the descent set plays a leading role.

1.4.10 Definition. Let $w = w_1 w_2 \cdots w_n \in \mathfrak{S}_n$. We say that the function $f : [n] \to \mathbb{N}$ is *w-compatible* if the following two conditions hold.

(a) $f(w_1) \geq f(w_2) \geq \cdots \geq f(w_n)$

(b) $f(w_i) > f(w_{i+1})$ if $w_i > w_{i+1}$ (i.e., if $i \in D(w)$)

1.4.11 Lemma. *Given $f : [n] \to \mathbb{N}$, there is a unique permutation $w \in \mathfrak{S}_n$ for which f is w-compatible.*

Proof. An *ordered partition* or *set composition* of a (finite) set S is a vector $(B_1, B_2, \ldots, B_k)$ of subsets $B_i \subseteq S$ such that $B_i \neq \emptyset$, $B_i \cap B_j = \emptyset$ for $i \neq j$, and $B_1 \cup \cdots \cup B_k = S$. Clearly there is a unique ordered partition $(B_1, \ldots, B_k)$ of $[n]$ such that f is constant on each B_i and $f(B_1) > f(B_2) > \cdots > f(B_k)$ (where $f(B_i)$ means $f(m)$ for any $m \in B_i$). Then w is obtained by arranging the elements of B_1 in increasing order, then the elements of B_2 in increasing order, and so on. □

The enumeration of certain natural classes of w-compatible functions is closely related to the statistics des and maj, as shown by the next lemma. Further enumerative results concerning w-compatible functions appear in Subsection 3.15.1. For $w \in \mathfrak{S}_n$, let $\mathcal{A}(w)$ denote the set of all w-compatible functions $f : [n] \to \mathbb{N}$; and for $w \in \mathfrak{S}_d$, let $\mathcal{A}_m(w)$ denote the set of w-compatible functions $f : [d] \to [m]$, i.e., $\mathcal{A}_m(w) = \mathcal{A}(w) \cap [m]^{[d]}$, where in general if X and Y are sets then Y^X denotes the set of all functions $f : X \to Y$. Note that $\mathcal{A}_0(w) = \emptyset$.

1.4.12 Lemma. (a) *For $w \in \mathfrak{S}_d$ and $m \geq 0$, we have*

$$\#\mathcal{A}_m(w) = \binom{m+d-1-\mathrm{des}(w)}{d} = \left(\binom{m-\mathrm{des}(w)}{d} \right) \tag{1.45}$$

and

$$\sum_{m \geq 1} \#\mathcal{A}_m(w) \cdot x^m = \frac{x^{1+\mathrm{des}(w)}}{(1-x)^{d+1}}. \tag{1.46}$$

(If $0 \leq m < \mathrm{des}(w)$, then we set $\left(\binom{m-\mathrm{des}(w)}{d} \right) = 0$.)

(b) *For $f : [n] \to \mathbb{N}$, write $|f| = \sum_{i=1}^n f(i)$. Then for $w \in \mathfrak{S}_n$, we have*

$$\sum_{f \in \mathcal{A}(w)} q^{|f|} = \frac{q^{\mathrm{maj}(w)}}{(1-q)(1-q^2)\cdots(1-q^n)}. \tag{1.47}$$

Proof. The basic idea of both proofs is to convert "partially strictly decreasing" sequences to weakly decreasing sequences similar to our first direct proof in Section 1.2 of the formula $\left(\binom{n}{k} \right) = \binom{n+k-1}{k}$. We will give "proofs by example" that should make the general case clear.

(a) Let $w = 4632715$. Then $f \in \mathcal{A}_m(w)$ if and only if

$$m \geq f(4) \geq f(6) > f(3) > f(2) \geq f(7) > f(1) \geq f(5) \geq 1. \tag{1.48}$$

Let $g(5) = f(5), g(1) = f(1), g(7) = f(7) - 1, g(2) = f(2) - 1, g(3) = f(3) - 2,$ $g(6) = f(6) - 3, g(4) = f(4) - 3$. In general, $g(j) = f(j) - h_j$, where h_j is the number of descents of w to the right of j. Equation (1.48) becomes

$$m - 3 \geq g(4) \geq g(6) \geq g(3) \geq g(2) \geq g(7) \geq g(1) \geq g(5) \geq 1.$$

Clearly the number of such g is $\left(\binom{m-3}{7}\right) = \left(\binom{m-\text{des}(w)}{d}\right)$, and (1.45) follows. There are numerous ways to obtain equation (1.46) from equation (1.45), for example, by observing that

$$\left(\binom{m - \text{des}(w)}{d}\right) = (-1)^{m-\text{des}(w)-1}\binom{-(d+1)}{m - \text{des}(w) - 1}$$

and using (1.20).

(b) Let $w = 4632715$ as in (a). Then $f \in \mathcal{A}(w)$ if and only

$$f(4) \geq f(6) > f(3) > f(2) \geq f(7) > f(1) \geq f(5) \geq 0. \qquad (1.49)$$

Defining g as in (a), equation (1.49) becomes

$$g(4) \geq g(6) \geq g(3) \geq g(2) \geq g(7) \geq g(1) \geq g(5) \geq 0.$$

Moreover, $\sum f(i) = \sum g(i) + 10 = \sum g(i) + \text{maj}(w)$. Hence,

$$\sum_{f \in \mathcal{A}(w)} q^{|f|} = q^{\text{maj}(w)} \sum_{g(4) \geq g(6) \geq g(3) \geq g(2) \geq g(7) \geq g(1) \geq g(5) \geq 0} q^{g(4) + \cdots + g(5)}.$$

The latter sum is easy to evaluate in a number of ways, for example, as an iterated geometric progression (i.e., first sum on $g(4) \geq g(6)$, then on $g(6) \geq g(3)$, etc.). It also is equivalent to equation (1.76). The proof follows. $\qquad \square$

Let $\mathbb{N}^{[n]}$ denote the set of all functions $f : [n] \to \mathbb{N}$, and let $\mathcal{A}(w)$ denote those $f \in \mathbb{N}^{[n]}$ that are compatible with $w \in \mathfrak{S}_n$. Lemma 1.4.11 then says that we have a disjoint union

$$\mathbb{N}^{[n]} = \bigcup_{w \in \mathfrak{S}_n} \mathcal{A}(w). \qquad (1.50)$$

It also follows that

$$[m]^{[d]} = \bigcup_{w \in \mathfrak{S}_d} \mathcal{A}_m(w). \qquad (1.51)$$

We now are in a position to give more conceptual proofs of Propositions 1.4.4 and 1.4.6. Take the cardinality of both sides of (1.51), multiply by x^m, and sum on $m \geq 0$. We get

$$\sum_{m \geq 0} m^d x^m = \sum_{w \in \mathfrak{S}_d} \#\mathcal{A}_m(w) \cdot x^m.$$

The proof of Proposition 1.4.4 now follows from equation (1.46). Similarly, by (1.50) we have

$$\sum_{f \in \mathbb{N}^{[n]}} q^{|f|} = \sum_{w \in \mathfrak{S}_n} \sum_{f \in \mathcal{A}(w)} q^{|f|}.$$

The left-hand side is clearly $1/(1-q)^n$, whereas by equation (1.47) the right-hand side is

$$\sum_{w \in \mathfrak{S}_n} \frac{q^{\text{maj}(w)}}{(1-q)(1-q^2)\cdots(1-q^n)}.$$

Hence

$$\frac{1}{(1-q)^n} = \frac{\sum_{w \in \mathfrak{S}_n} q^{\text{maj}(w)}}{(1-q)(1-q^2)\cdots(1-q^n)}.$$

Multiplying by $(1-q)(1-q^2)\cdots(1-q^n)$ and simplifying gives Proposition 1.4.6.

1.5 Geometric Representations of Permutations

We have seen that a permutation can be regarded as either a function, a word, or a sequence (the inversion table). In this section, we will consider four additional ways of representing permutations and will illustrate the usefulness of each such representation.

The first representation is the most obvious, namely, a permutation matrix. Specifically, if $w \in \mathfrak{S}_n$, then define the $n \times n$ matrix P_w, with rows and columns indexed by $[n]$, as follows:

$$(P_w)_{ij} = \begin{cases} 1, & \text{if } w(i) = j \\ 0, & \text{otherwise.} \end{cases}$$

The matrix P_w is called the *permutation matrix* corresponding to w. Clearly, a square $(0,1)$-matrix is a permutation matrix if and only if it has exactly one 1 in every row and column. Sometimes it is more convenient to replace the 0's and 1's with some other symbols. For instance, the matrix P_w could be replaced by a $n \times n$ grid, where each square indexed by $(i, w(i))$ is filled in. Figure 1.3 shows the matrix P_w corresponding to $w = 795418362$, together with the equivalent representation as a grid with certain squares filled in.

To illustrate the use of permutation matrices as geometric objects per se, define a *decreasing subsequence* of *length k* of a permutation $w = w_1 \cdots w_n \in \mathfrak{S}_n$ to be a subsequence $w_{i_1} > w_{i_2} > \cdots > w_{i_k}$ (so $i_1 < i_2 < \cdots < i_k$ by definition of subsequence). (*Increasing subsequence* is similarly defined, though we have no need for this concept in the present example.) Let $f(n)$ be the number of permutations $w \in \mathfrak{S}_n$ with no decreasing subsequence of length three. For instance,

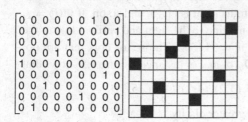

Figure 1.3 The permutation matrix of the permutation $w = 795418362$.

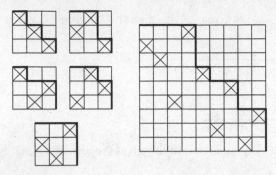

Figure 1.4 Lattice paths corresponding to 321-avoiding permutations.

$f(3) = 5$ since 321 is the only excluded permutation. Let w be a permutation with no decreasing subsequence of length three, and let P_w be its permutation matrix, where for better visualization we replace the 1's in P_w by X's. Draw a lattice path L_w from the upper-left corner of P_w to the lower-right corner, where each step is one unit to the right (east) or down (south), and where the "outside corners" (consisting of a right step followed by a down step) of L_w occur at the top and right of each square on or above the main diagonal containing an X. We trust that Figure 1.4 will make this definition clear; it shows the five paths for $w \in \mathfrak{S}_3$ as well as the path for $w = 412573968$. It is not hard to see that the lattice paths so obtained are exactly those that do not pass below the main diagonal. Conversely, it is also not hard to see that given a lattice path L not passing below the main diagonal, there is a unique permutation $w \in \mathfrak{S}_n$ with no decreasing subsequence of length three for which $L = L_w$.

We have converted our permutation enumeration problem to a much more tractable lattice path counting problem. It is shown in Corollary 6.2.3 that the number of such paths is the *Catalan number* $C_n = \frac{1}{n+1}\binom{2n}{n}$, so we have shown that

$$f(n) = C_n. \tag{1.52}$$

The growth diagrams discussed in Section 7.13 show a more sophisticated use of permutation matrices.

NOTE. The Catalan numbers form one of the most interesting and ubiquitous sequences in enumerative combinatorics; see Chapter 6, especially Corollary 6.2.3 and Exercise 6.19, for further information.

An object closely related to the permutation matrix P_w is the diagram of $w \in \mathfrak{S}_n$. Represent the set $[n] \times [n]$ as an $n \times n$ array of dots, using matrix coordinates, so the upper-left dot represents $(1,1)$, the dot to its right is $(1,2)$, and so on. If $w(i) = j$, then from the point (i, j) draw a horizontal line to the right and vertical line to the bottom. Figure 1.5 illustrates the case $w = 314652$. The set of dots that are not covered by lines is called the *diagram D_w* of w. For instance, Figure 1.5 shows that

$$D_{314652} = \{(1,1), (1,2), (3,2), (4,2), (4,5), (5,2)\}.$$

The dots of the diagram are circled for greater clarity.

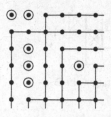

Figure 1.5 The diagram of the permutation $w = 314652$.

It is easy to see that if a_j denotes the number of elements of D_w in column j, then the inversion table of w is given by $I(w) = (a_1, a_2, \ldots, a_n)$. Similarly, if c_i is the number of elements in the ith row of D_w then $\text{code}(w) = (c_1, c_2, \ldots, c_n)$. If D_w^t denotes the transpose (relection about the main diagonal) of D_w, then $D_w^t = D_{w^{-1}}$.

As an illustration of the use of the diagram D_w, define a permutation $w = w_1 \cdots w_n \in \mathfrak{S}_n$ to be *132-avoiding* if there does not exist $i < j < k$ with $w_i < w_k < w_j$. In other words, no subsequence of w of length three has its terms in the same relative order as 132. Clearly, this definition can be generalized to define *u-avoiding* permutations, where $u \in \mathfrak{S}_k$. For instance, the previously considered permutations with no decreasing subsequence of length three are just 321-avoiding permutations.

It is not hard to see that w is 132-avoiding if and only if there exists integers $\lambda_1 \geq \lambda_2 \geq \cdots \geq 0$ such that for all $i \geq 0$ the ith row of D_w consists of the first λ_i dots in that row. In symbols,

$$D_w = \{(i,j) : 1 \leq j \leq \lambda_i\}.$$

Equivalently, if $(i,j) \in D_w$ and $i' \leq i$, $j' \leq j$, then $(i', j') \in D_w$. In the terminology of Section 1.7, the sequence $\lambda = (\lambda_1, \lambda_2, \ldots)$ is a *partition* of $\sum \lambda_i = \text{inv}(w)$, and D_w is the *Ferrers diagram* of λ. In this sense, diagrams of permutations are generalizations of diagrams of partitions. Note that in any $n \times n$ diagram D_w, where $w \in \mathfrak{S}_n$, there are at least i dots in the ith row that do not belong to D_w. Hence if w is 132-avoiding then the corresponding partition $\lambda = (\lambda_1, \ldots \lambda_n)$ satisfies $\lambda_i \leq n - i$. Conversely, it is easy to see that if λ satisfies $\lambda_i \leq n - i$, then the Ferrers diagram of λ is the diagram of a (necessarily 132-avoiding) permutation $w \in \mathfrak{S}_n$. Hence, the number of 132-avoiding permutations in $\mathfrak{S}_n$ is equal to the number of integer sequences $\lambda_1 \geq \cdots \geq \lambda_n \geq 0$ such that $\lambda_i \leq n - i$. It follows from Exercise 6.19(s) that the number of such sequences is the Catalan number $C_n = \frac{1}{n+1}\binom{2n}{n}$. (There is also a simple bijection with the lattice paths that we put in one-to-one correspondence with 321-avoiding permutations. In fact, the lattice path construction we applied to 321-avoiding permutations works equally well for 132-avoiding permutations if our paths go from the upper right to lower left; see Figure 1.6.) Hence by equation (1.52) the number of 132-avoiding permutations in $\mathfrak{S}_n$ is the same as the number of 321-avoiding permutations in $\mathfrak{S}_n$ (i.e., permutations in $\mathfrak{S}_n$ with no decreasing subsequence of length three). Simple symmetry

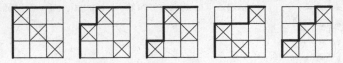

Figure 1.6 Lattice paths corresponding to 132-avoiding permutations in $\mathfrak{S}_3$.

Figure 1.7 The definition of $T(w)$.

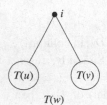

arguments (e.g., replacing $w_1 w_2 \cdots w_n$ with $w_n \cdots w_2 w_1$) then show that, for *any* $u \in \mathfrak{S}_3$, the number of u-avoiding permutations $w \in \mathfrak{S}_n$ is C_n.

Since $\#D_w = \mathrm{inv}(w)$, the preceding characterization of diagrams of 132-avoiding permutations $w \in \mathfrak{S}_n$ yields the following refinement of the enumeration of such w.

1.5.1 Proposition. *Let $\mathcal{S}_{132}(n)$ denote the set of 132-avoiding $w \in \mathfrak{S}_n$. Then*

$$\sum_{w \in \mathcal{S}_{132}(n)} q^{\mathrm{inv}(w)} = \sum_{\lambda} q^{|\lambda|},$$

where λ ranges over all integer sequences $\lambda_1 \geq \cdots \geq \lambda_n \geq 0$ satisfying $\lambda_i \leq n - i$, and where $|\lambda| = \sum \lambda_i$.

For further information on the sums appearing in Proposition 1.5.1, see Exercise 6.34(a).

We now consider two ways to represent a permutation w as a tree T and discuss how the structure of T interacts with the combinatorial properties of w. Let $w = w_1 w_2 \cdots w_n$ be any word on the alphabet $\mathbb{P}$ with no repeated letters. Define a binary tree $T(w)$ as follows. If $w = \emptyset$, then $T(w) = \emptyset$. If $w \neq \emptyset$, then let i be the least element (letter) of w. Thus, w can be factored uniquely in the form $w = uiv$. Now let i be the root of $T(w)$, and let $T(u)$ and $T(v)$ be the left and right subtrees of i; see Figure 1.7. This procedure yields an inductive definition of $T(w)$. The left successor of a vertex j is the least element k to the left of j in w such that all elements of w between k and j (inclusive) are $\geq j$, and similarly for the right successor.

1.5.2 Example. Let $w = 57316284$. Then $T(w)$ is given by Figure 1.8.

The correspondence $w \mapsto T(w)$ is a bijection between $\mathfrak{S}_n$ and *increasing binary trees* on n vertices; that is, binary trees with n vertices labeled $1, 2, \ldots, n$ such that the labels along any path from the root are increasing. To obtain w from $T(w)$,

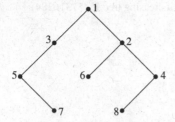

Figure 1.8 The increasing binary tree $T(57316284)$.

read the labels of w in *symmetric order*, that is, first the labels of the left subtree (in symmetric order, recursively), then the label of the root, and then the labels of the right subtree.

Let $w = w_1 w_2 \cdots w_n \in \mathfrak{S}_n$. Define the element w_i of w to be

$$
\begin{aligned}
&\text{a } \textit{double rise} \text{ or } \textit{double ascent}, &&\text{if } w_{i-1} < w_i < w_{i+1} \\
&\text{a } \textit{double fall} \text{ or } \textit{double descent}, &&\text{if } w_{i-1} > w_i > w_{i+1} \\
&\text{a } \textit{peak}, &&\text{if } w_{i-1} < w_i > w_{i+1} \\
&\text{a } \textit{valley}, &&\text{if } w_{i-1} > w_i < w_{i+1},
\end{aligned}
$$

where we set $w_0 = w_{n+1} = 0$. It is easily seen that the property listed below of an element i of w corresponds to the given property of the vertex i of $T(w)$.

Element i of w	Vertex i of $T(w)$ has precisely the following successors
double rise	right
double fall	left
valley	left and right
peak	none

From this discussion of the bijection $w \mapsto T(w)$, a large number of otherwise mysterious properties of increasing binary trees can be trivially deduced. The following proposition gives a sample of such results. Exercise 1.61 provides a further application of $T(w)$.

1.5.3 Proposition. *(a) The number of increasing binary trees with n vertices is n!.*

(b) The number of such trees for which exactly k vertices have left successors is the Eulerian number $A(n, k+1)$.

(c) The number of complete (i.e., every vertex is either an endpoint or has two successors) increasing binary trees with $2n + 1$ vertices is equal to the number E_{2n+1} of alternating permutations in $\mathfrak{S}_{2n+1}$.

Let us now consider a second way to represent a permutation by a tree. Given $w = w_1 w_2 \cdots w_n \in \mathfrak{S}_n$, construct an (unordered) tree $T'(w)$ with vertices $0, 1, \ldots, n$

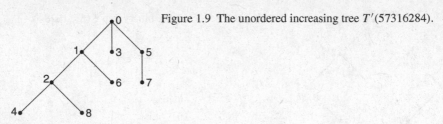

Figure 1.9 The unordered increasing tree $T'(57316284)$.

by defining vertex i to be the successor of the rightmost element j of w which precedes i and which is less than i. If there is no such element j, then let i be the successor of the root 0.

1.5.4 Example. Let $w = 57316284$. Then $T'(w)$ is given by Figure 1.9.

The correspondence $w \mapsto T'(w)$ is a bijection between $\mathfrak{S}_n$ and increasing trees on $n+1$ vertices. It is easily seen that the successors of 0 are just the *left-to-right minima* (or *retreating elements*) of w (i.e., elements w_i such that $w_i < w_j$ for all $j < i$, where $w = w_1 w_2 \cdots w_n$). Moreover, the endpoints of $T'(w)$ are just the elements w_i for which $i \in D(w)$ or $i = n$. Thus, in analogy to Proposition 1.5.3 (using Proposition 1.3.1 and the obvious symmetry between left-to-right maxima and left-to-right minima), we obtain the following result.

1.5.5 Proposition. *(a) The number of unordered increasing trees on $n+1$ vertices is $n!$.*

(b) The number of such trees for which the root has k successors is the signless Stirling number $c(n,k)$.]

(c) The number of such trees with k endpoints is the Eulerian number $A(n,k)$.

1.6 Alternating Permutations, Euler Numbers, and the cd-Index of $\mathfrak{S}_n$

In this section we consider enumerative properties of alternating permutations, as defined in Section 1.4. Recall that a permutation $w \in \mathfrak{S}_n$ is alternating if $D(w) = \{1,3,5,\ldots\} \cap [n-1]$, and reverse alternating if $D(w) = \{2,4,6,\ldots\} \cap [n-1]$.

1.6.1 Basic Properties

Recall that E_n denotes the number of alternating permutations (or reverse alternating permutations) $w \in \mathfrak{S}_n$ (with $E_0 = 1$) and is called an Euler number. The exponential generating function for Euler numbers is very elegant and surprising.

1.6.1 Proposition. *We have*

$$\sum_{n \geq 0} E_n \frac{x^n}{n!} = \sec x + \tan x$$

$$= 1 + x + \frac{x^2}{2!} + 2\frac{x^3}{3!} + 5\frac{x^4}{4!} + 16\frac{x^5}{5!} + 61\frac{x^6}{6!} + 272\frac{x^7}{7!} + 1385\frac{x^8}{8!} + \cdots.$$

Note that $\sec x$ is an even function (i.e., $\sec(-x) = \sec x$), while $\tan x$ is odd ($\tan(-x) = -\tan x$). It follows from Proposition 1.6.1 that

$$\sum_{n \geq 0} E_{2n} \frac{x^{2n}}{(2n)!} = \sec x, \tag{1.53}$$

$$\sum_{n \geq 0} E_{2n+1} \frac{x^{2n+1}}{(2n+1)!} = \tan x. \tag{1.54}$$

For this reason E_{2n} is sometimes called a *secant number* and E_{2n+1} a *tangent number.*

Proof of Proposition 1.6.1. Let $0 \leq k \leq n$. Choose a k-subset S of $[n]$ in $\binom{n}{k}$ ways, and set $\overline{S} = [n] - S$. Choose a reverse alternating permutation u of S in E_k ways, and choose a reverse alternating permutation v of $\overline{S}$ in E_{n-k} ways. Let w be the concatenation $u^r, n+1, v$, where u^r denotes the reverse of u (i.e., if $u = u_1 \cdots u_k$, then $u^r = u_k \cdots u_1$). When $n \geq 2$, we obtain in this way every alternating and every reverse alternating permutation w exactly once. Since there is a bijection between alternating and reverse alternating permutations of any finite (ordered) set, the number of w obtained is $2E_{n+1}$. Hence,

$$2E_{n+1} = \sum_{k=0}^{n} \binom{n}{k} E_k E_{n-k}, \ n \geq 1. \tag{1.55}$$

Set $y = \sum_{n \geq 0} E_n x^n / n!$. Taking into account the initial conditions $E_0 = E_1 = 1$, equation (1.55) becomes the differential equation

$$2y' = y^2 + 1, \ y(0) = 1.$$

The unique solution is $y = \sec x + \tan x$. □

NOTE. The clever counting of both alternating and reverse alternating permutations in the proof of Proposition 1.6.1 can be avoided at the cost of a little elegance. Namely, by considering the position of 1 in an alternating permutation w, we obtain the recurrence

$$E_{n+1} = \sum_{\substack{1 \leq j \leq n \\ j \text{ odd}}} \binom{n}{j} E_j E_{n-j}, \ n \geq 1.$$

This recurrence leads to a system of differential equations for the power series $\sum_{n \geq 0} E_{2n} x^{2n} / (2n)!$ and $\sum_{n \geq 0} E_{2n+1} x^{2n+1} / (2n+1)!$.

Note that equations (1.53) and (1.54) could in fact be used to *define* $\sec x$ and $\tan x$ in terms of alternating permutations. We can then try to develop as much trigonometry as possible (e.g., the identity $1 + \tan^2 x = \sec^2 x$) using this definition, thereby defining the subject of *combinatorial trigonometry*. For the first steps in this direction, see Exercise 5.7.

It is natural to ask whether Proposition 1.6.1 has a more conceptual proof. The proof preceding does not explain why we ended up with such a simple generating function. To be even more clear about this point, rewrite equation (1.53) as

$$\sum_{n\geq 0} E_{2n}\frac{x^{2n}}{(2n)!} = \frac{1}{\sum_{n\geq 0}(-1)^n\dfrac{x^{2n}}{(2n)!}}. \tag{1.56}$$

Compare this equation with the exponential generating function for the number of permutations in $\mathfrak{S}_n$ with descent set $[n-1]$:

$$\sum_{n\geq 0}\frac{x^n}{n!} = \frac{1}{\sum_{n\geq 0}(-1)^n\dfrac{x^n}{n!}}. \tag{1.57}$$

Could there be a reason why having descents in every second position corresponds to taking every second term in the denominator of (1.57) and keeping the signs alternating? Possibly the similarity between (1.56) and (1.57) is just a coincidence. All doubts are dispelled, however, by the following generalization of equation (1.56). Let $f_k(n)$ denote the number of permutations $w \in \mathfrak{S}_n$ satisfying

$$D(w) = \{k, 2k, 3k, \dots\} \cap [n-1]. \tag{1.58}$$

Then

$$\sum_{n\geq 0} f_k(kn)\frac{x^{kn}}{(kn)!} = \frac{1}{\sum_{n\geq 0}(-1)^n\dfrac{x^{kn}}{(kn)!}}. \tag{1.59}$$

Such a formula cries out for a more conceptual proof. One such proof is given in Section 3.19. Exercise 2.22 gives a further proof for $k = 2$ (easily extended to any k) based on Inclusion–Exclusion. Another enlightening proof, less elegant but more straightforward than the one in Section 3.19, is the following.

Proof of equation (1.59). We have

$$\frac{1}{\sum_{n\geq 0}(-1)^n\dfrac{x^{kn}}{(kn)!}} = \frac{1}{1 - \sum_{n\geq 1}(-1)^{n-1}\dfrac{x^{kn}}{(kn)!}}$$

$$= \sum_{j\geq 0}\left(\sum_{n\geq 1}(-1)^{n-1}\frac{x^{kn}}{(kn)!}\right)^j$$

$$= \sum_{j\geq 0}\sum_{N\geq j}\sum_{\substack{a_1+\cdots+a_j=N\\a_i\geq 1}}\binom{kN}{ka_1,\dots,ka_j}(-1)^{N-j}\frac{x^{kN}}{(kN)!}.$$

Comparing (carefully) with equations (1.34) and (1.35) completes the proof. $\square$

A similar proof can be given of equation (1.54) and its extension to permutations in $\mathfrak{S}_{kn+i}$ with descent set $\{k, 2k, 3k, \dots\} \cap [kn + i - 1]$ for $1 \le i \le k - 1$. Details are left as an exercise (Exercise 1.146).

1.6.2 Flip Equivalence of Increasing Binary Trees

Alternating permutations appear as the number of equivalence classes of certain naturally defined equivalence relations. (For an example unrelated to this section, see Exercise 3.127(b).) We will give an archetypal example in this subsection. In the next subsection, we will give a similar result, which has an application to the numbers $\beta_n(S)$ of permutations $w \in \mathfrak{S}_n$ with descent set S.

Recall that in Section 1.5 we associated an increasing binary tree $T(w)$ with a permutation $w \in \mathfrak{S}_n$. The *flip* of a binary tree at a vertex v is the binary tree obtained by interchanging the left and right subtrees of v. Define two increasing binary trees T and T' on the vertex set $[n]$ to be *equivalent* if T' can be obtained from T by a sequence of flips. Clearly, this definition of equivalence is an equivalence relation, and the number of increasing binary trees equivalent to T is $2^{n-\epsilon(T)}$, where $\epsilon(T)$ is the number of endpoints of T. The equivalence classes are in an obvious bijection with increasing *(1-2)-trees* on the vertex set $[n]$, that is, increasing (rooted) trees so that every non-endpoint vertex has one or two children. (These are not plane trees, i.e., the order in which we write the children of a vertex is irrelevant.) Figure 1.10 shows the five increasing (1,2)-trees on four vertices, so $f(4) = 5$. Let $f(n)$ denote the number of equivalence classes (i.e., the number of increasing (1,2)-trees on the vertex set $[n]$).

1.6.2 Proposition. *We have* $f(n) = E_n$ *(an Euler number).*

Proof. Perhaps the most straightforward proof is by generating functions. Let

$$y = \sum_{n \ge 1} f(n) \frac{x^n}{n!} = x + \frac{x^2}{2} + 2\frac{x^3}{6} + \cdots.$$

Then $y' = \sum_{n \ge 0} f(n+1) x^n / n!$. Every increasing (1,2)-tree with $n+1$ vertices either (a) is a single vertex ($n = 0$), (b) has one subtree of the root, which is an increasing (1,2)-tree with n vertices, or (c) has two subtrees of the root, each of which is an increasing (1,2)-tree, with n vertices in all. The order of the two

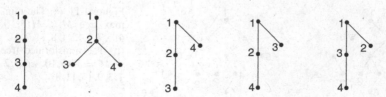

Figure 1.10 The five increasing (1,2)-trees with four vertices.

subtrees is irrelevant. From this observation, we obtain the differential equation $y' = 1 + y + \frac{1}{2}y^2$, $y(0) = 0$. The unique solution is $y = \sec x + \tan x - 1$, and the proof follows from Proposition 1.6.1. $\square$

ALGEBRAIC NOTE. Let $\mathcal{T}_n$ be the set of all increasing binary tree with vertex set $[n]$. For $T \in \mathcal{T}_n$ and $1 \le i \le n$, let $\omega_i T$ be the flip of T at vertex i. Then clearly the ω_i's generate a group isomorphic to $(\mathbb{Z}/2\mathbb{Z})^n$ acting on $\mathcal{T}_n$, and the orbits of this action are the flip equivalence classes.

1.6.3 Min-Max Trees and the cd-Index

We now consider a variant of the bijection $w \mapsto T(w)$ between permutations and increasing binary trees defined in Section 1.5 that has an interesting application to descent sets of permutations. We will just sketch the basic facts and omit most details of proofs (all of which are fairly straightforward). We define the *min-max tree* $M(w)$ associated with a sequence $w = a_1 a_2 \cdots a_n$ of distinct integers as follows. First, $M(w)$ is a binary tree with vertices labeled $a_1, a_2, \ldots, a_n$. Let j be the least integer for which *either* $a_j = \min\{a_1, \ldots, a_n\}$ or $a_j = \max\{a_1, \ldots, a_n\}$. Define a_j to be the root of $M(w)$. Then define (recursively) $M(a_1, \ldots, a_{j-1})$ to be the left subtree of a_j, and $M(a_{j+1}, \ldots, a_n)$ to be the right subtree. Figure 1.11(a) shows $M(5, 10, 4, 6, 7, 2, 12, 1, 8, 11, 9, 3)$. Note that no vertex of a min-max tree $M(w)$ has only a left successor. Note also that every vertex v is either the minimum or maximum element of the subtree with root v.

Given the min-max tree $M(w)$ where $w = a_1 \cdots a_n$, we will define operators ψ_i, $1 \le i \le n$, that permute the labels of $M(w)$, creating a new min-max tree $\psi_i M(w)$. The operator ψ_i only permutes the label of the vertex of $M(w)$ labeled a_i and the labels of the right subtree of this vertex. (Note that the vertex labeled a_i depends only on i and the tree $M(w)$, not on the permutation w.) All other vertices are fixed by ψ_i. In particular, if a_i is an endpoint, then $\psi_i M(w) = M(w)$. We denote by M_{a_i} the subtree of $M(w)$ consisting of a_i and the right subtree of a_i. Thus, a_i is either the minimum or maximum element of M_{a_i}. Suppose that a_i is the minimum element of M_{a_i}. Then replace a_i with the *largest* element of M_{a_i}, and permute the remaining elements of M_{a_i} so that they keep their same relative order. This defines $\psi_i M(w)$. Similarly, suppose that a_i is the maximum element of the subtree M_{a_i} with root a_i. Then replace a_i with the *smallest* element of M_{a_i}, and permute the

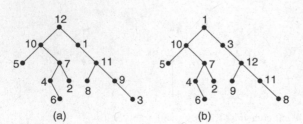

(a)   (b)

Figure 1.11 (a) The min-max tree $M = M(5, 10, 4, 6, 7, 2, 12, 1, 8, 11, 9, 3)$; (b) The transformed tree $\psi_7 M = M(5, 10, 4, 6, 7, 2, 1, 3, 9, 12, 11, 8)$.

remaining elements of M_{a_i} so that they keep their same relative order. Again this defines $\psi_i M(w)$. Figure 1.11(b) shows that $\psi_7 M(5, 10, 4, 6, 7, 2, 12, 1, 8, 11, 9, 3) = M(5, 10, 4, 6, 7, 2, 1, 3, 9, 12, 11, 8)$. We have $a_7 = 12$, so ψ_7 permutes vertex 12 and the vertices on the right subtree of 12. Vertex 12 is replaced by 1, the smallest vertex of the right subtree. The remaining elements $1, 3, 8, 9, 11$ get replaced with $3, 8, 9, 11, 12$ in that order.

Fact #1. *The operators* ψ_i *are commuting involutions and hence generate an (abelian) group* $\mathfrak{G}_w$ *isomorphic to* $(\mathbb{Z}/2\mathbb{Z})^{\iota(w)}$, *where* $\iota(w)$ *is the number of internal vertices of* $M(w)$. *Those* ψ_i *for which* a_i *is not an endpoint are a minimal set* G_w *of generators for* $\mathfrak{G}_w$. *Hence, there are precisely* $2^{\iota(w)}$ *different trees* $\psi M(w)$ *for* $\psi \in \mathfrak{G}_w$, *given by* $\psi_{i_1} \cdots \psi_{i_j} M(w)$, *where* $\{\psi_{i_1}, \ldots, \psi_{i_j}\}$ *ranges over all subsets of* G_w.

Given a permutation $w \in \mathfrak{S}_n$ and an operator $\psi \in \mathfrak{G}_w$, we define the permutation ψw by $\psi M(w) = M(\psi w)$. Define two permutations $v, w \in \mathfrak{S}_n$ to be *M-equivalent*, denoted $v \overset{M}{\sim} w$, if $v = \psi w$ for some $\psi \in \mathfrak{G}_w$. Clearly $\overset{M}{\sim}$ is an equivalence relation, and by Fact #1 the size of the equivalence class $[w]$ containing w is $2^{\iota(w)}$.

There is a simple connection between the descent sets of w and $\psi_i w$.

Fact #2. *Let* a_i *be an internal vertex of* $M(w)$ *with only a right child. Then*

$$D(\psi_i w) = \begin{cases} D(w) \cup \{i\}, & \text{if } i \notin D(w), \\ D(w) - \{i\}, & \text{if } i \in D(w). \end{cases}$$

Let a_i *be an internal vertex of* $M(w)$ *with both a left and right child. Then exactly one of* $i - 1, i$ *belongs to* $D(w)$, *and we have*

$$D(\psi_i w) = \begin{cases} (D(w) \cup \{i\}) - \{i - 1\}, & \text{if } i \notin D(w), \\ (D(w) \cup \{i - 1\}) - \{i\}, & \text{if } i \in D(w). \end{cases}$$

Note that if a_i is a vertex with two children, then a_{i-1} will always be an endpoint on the left subtree of a_i. It follows that the changes in the descent sets described by Fact #2 take place independently of each other. (In fact, this independence is equivalent to the commutativity of the ψ_i's.) The different descent sets $D(w)$, where w ranges over an M-equivalence class, can be conveniently encoded by a noncommutative polynomial in the letters a and b. Given a set $S \subseteq [n-1]$, define its *characteristic monomial* (or *variation*) to be the noncommutative monomial

$$u_S = e_1 e_2 \cdots e_{n-1}, \tag{1.60}$$

where

$$e_i = \begin{cases} a, & \text{if } i \notin S \\ b, & \text{if } i \in S. \end{cases}$$

For instance, $u_{D(37485216)} = ababbba$.

Now let $w = a_1 a_2 \cdots a_n \in \mathfrak{S}_n$, and let c, d, e be noncommutative indeterminates. For $1 \leq i \leq n$, define

$$
f_i = f_i(w) = \begin{cases} c, & \text{if } a_i \text{ has only a right child in } M(w), \\ d, & \text{if } a_i \text{ has a left and right child,} \\ e, & \text{if } a_i \text{ is an endpoint.} \end{cases}
$$

Let $\Phi'_w = \Phi'_w(c, d, e) = f_1 f_2 \cdots f_n$, and let $\Phi_w = \Phi_w(c, d) = \Phi'(c, d, 1)$, where 1 denotes the empty word. In other words, Φ_w is obtained from Φ'_w by deleting the e's. For instance, consider the permutation $w = 5, 10, 4, 6, 7, 2, 12, 1, 8, 11, 9, 3$ of Figure 1.11. The degrees (number of children) of the vertices $a_1, a_2, \ldots, a_{12}$ are $0, 2, 1, 0, 2, 0, 2, 1, 0, 2, 1, 0$, respectively. Hence,

$$
\Phi'_w = edcededcedce,
$$

$$
\Phi_w = dcddcdc. \tag{1.61}
$$

It is clear that if $v \overset{M}{\sim} w$, then $\Phi'_v = \Phi'_w$ and $\Phi_v = \Phi_w$, since Φ'_w depends only on $M(w)$ regarded as an *unlabeled* tree.

From Fact #2, we obtain the following result.

Fact #3. *Let $w \in \mathfrak{S}_n$, and let $[w]$ be the M-equivalence class containing w. Then*

$$
\Phi_w(a + b, ab + ba) = \sum_{v \in [w]} u_{D(v)}. \quad \square \tag{1.62}
$$

For instance, from equation (1.61), we have

$$
\sum_{v \in [w]} u_{D(v)} = (ab + ba)(a + b)(ab + ba)(ab + ba)(a + b)(ab + ba)(a + b).
$$

As a further example, Figure 1.12 shows the eight trees $M(v)$ in the M-equivalence class [315426], together with corresponding characteristic monomial $u_{D(v)}$. We see that

$$
\sum_{v \in [315426]} u_{D(v)} = babba + abbba + baaba + babab + ababa + abbab
$$

$$
+ baaab + abaab = (ab + ba)(a + b)(ab + ba),
$$

whence $\Phi_w = dcd$.

Fact #4. *Each equivalence class $[w]$ contains exactly one alternating permutation (as well as one reverse alternating permutation). Hence, the number of M-equivalence classes of permutations $w \in \mathfrak{S}_n$ is the Euler number E_n.*

It is not difficult to prove Fact #4 directly from the definition of the tree $M(w)$ and the group $\mathfrak{S}_w$, but it is also immediate from Fact #3. For in the expansion of $\Phi_w(a + b, ab + ba)$, there will be exactly one alternating term $bababa \cdots$ and one term $ababab \cdots$.

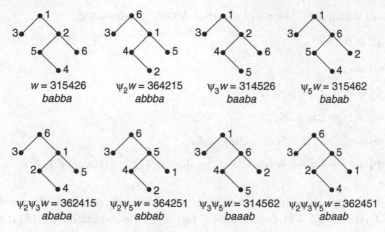

Figure 1.12 The M-equivalence class [315426].

Now consider the generating function

$$\Psi_n = \Psi_n(a,b) = \sum_{w \in \mathfrak{S}_n} u_{D(w)}$$

$$= \sum_{S \subseteq [n-1]} \beta(S) u_S. \qquad (1.63)$$

Thus, Ψ_n is a noncommutative generating function for the numbers $\beta(S)$. For instance, $\Psi_3 = aa + 2ab + 2ba + bb$. The polynomial Ψ_n is called the *ab-index* of the symmetric group $\mathfrak{S}_n$. (In the more general context of Section 3.17, Ψ_n is called the *ab*-index of the boolean algebra B_n.) We can group the terms of Ψ_n according to the M-equivalence classes $[w]$, that is,

$$\Psi_n = \sum_{[w]} \sum_{v \in [w]} u_{D(v)}, \qquad (1.64)$$

where the outer sum ranges over all distinct M-equivalence classes $[w]$ of permutations in $\mathfrak{S}_n$. Now by equation (1.62) the inner sum is just $\Phi_w(a+b, ab+ba)$. Hence, we have established the following result.

1.6.3 Theorem. *The ab-index* Ψ_n *can be written as a polynomial* Φ_n *in the variables* $c = a + b$ *and* $d = ab + ba$. *This polynomial is a sum of* E_n *monomials.*

The polynomial Φ_n is called the *cd-index* of the symmetric group $\mathfrak{S}_n$ (or boolean algebra B_n). It is a surprisingly compact way of codifying the numbers $\beta_n(S)$. The number of distinct terms in Φ_n is the Fibonacci number F_n (the number of *cd*-monomials of degree n, where $\deg c = 1$ and $\deg d = 2$; see Exercise 1.35(c)), compared with the 2^{n-1} terms of the ab-index Ψ_n. For instance,

$$\Phi_1 = 1,$$

$$\Phi_2 = c,$$

$$\Phi_3 = c^2 + d,$$

$$\Phi_4 = c^3 + 2cd + 2dc,$$

$$\Phi_5 = c^4 + 3c^2d + 5cdc + 3dc^2 + 4d^2,$$

$$\Phi_6 = c^5 + 4c^3d + 9c^2dc + 9cdc^2 + 4dc^3 + 12cd^2 + 10dcd + 12d^2c.$$

One nice application of the cd-index concerns inequalities among the number $\beta_n(S)$. Given $S \subseteq [n-1]$, define $\omega(S) \subseteq [n-2]$ by the condition $i \in \omega(S)$ if and only if exactly one of i and $i+1$ belongs to S, for $1 \le i \le n-2$. For instance, if $n = 9$ and $S = \{2,4,5,8\}$, then $\omega(S) = \{1,2,3,5,7\}$. Note that

$$\omega(S) = [n-2] \iff S = \{1,3,5,\dots\} \cap [n-1] \text{ or } S = \{2,4,6,\dots\} \cap [n-1]. \tag{1.65}$$

1.6.4 Proposition. *Let $S, T \subseteq [n-1]$. If $\omega(S) \subset \omega(T)$, then $\beta_n(S) < \beta_n(T)$.*

Proof. Let $w \in \mathfrak{S}_n$ and $\Phi'_w = f_1 f_2 \cdots f_n$, so each $f_i = c, d$, or e. Define

$$S_w = \{i - 1 : f_i = d\}.$$

It is easy to see that

$$\Phi_w = \sum_{\omega(X) \supseteq S_w} u_X.$$

Since Φ_n has nonnegative coefficients, it follows that if $\omega(S) \subseteq \omega(T)$, then $\beta_n(S) \le \beta_n(T)$. Now assume that S and T are any subsets of $[n-1]$ for which $\omega(S) \subset \omega(T)$ (strict containment). We can easily find a cd-word Φ_w for which $\omega(T) \supseteq S_w$ but $\omega(S) \not\supseteq S_w$. For instance, if $i \in \omega(T) - \omega(S)$, then let $\Phi_w = c^{i-1}dc^{n-2-i}$, so $S_w = \{i\}$. It follows that $\beta_n(S) < \beta_n(T)$. $\qquad\square$

1.6.5 Corollary. *Let $S \subseteq [n-1]$. Then $\beta_n(S) \le E_n$, with equality if and only if $S = \{1,3,5,\dots\} \cap [n-1]$ or $S = \{2,4,6,\dots\} \cap [n-1]$.*

Proof. Immediate from Proposition 1.6.4 and equation (1.65). $\qquad\square$

1.7 Permutations of Multisets

Much of what we have done concerning permutations of sets can be generalized to multisets. For instance, there are *two* beautiful theories of cycle decomposition for permutations of multisets (see Exercise 1.62 for one of them, and its solution for a reference to the other). In this section, however, we will only discuss some topics that will be of use later.

First, it is clear that we can define the descent set $D(w)$ of a permutation w of a (finite) multiset M on a totally ordered set (such as $\mathbb{P}$) exactly as we did for sets. Namely, if $w = w_1 w_2 \cdots w_n$, then

$$D(w) = \{i \, : \, w_i > w_{i+1}\}.$$

Thus we also have the concept of $\alpha(S) = \alpha_M(S)$ and $\beta(S) = \beta_M(S)$ for a multiset M, as well as the number $\mathrm{des}(w)$ of descents, the major index $\mathrm{maj}(w)$ and the multiset Eulerian polynomial

$$A_M(x) = \sum_{w \in \mathfrak{S}_M} x^{1+\mathrm{des}(w)},$$

and so on. In Section 4.4.5 we will consider a vast generalization of these concepts. Note for now that there is no obvious analogue of Proposition 1.4.1—that is, an explicit formula for the number $\alpha_M(S)$ of permutations $w \in \mathfrak{S}_M$ with descent set contained in S.

We can also define an *inversion* of $w = w_1 w_2 \cdots w_n \in \mathfrak{S}_M$ as a 4-tuple (i, j, w_i, w_j) for which $i < j$ and $w_i > w_j$, and as before we define $\mathrm{inv}(w)$ to be the number of inversions of w. Note that unlike the case for permutations we shouldn't define an inversion to be just the pair (w_i, w_j) since we can have $(w_i, w_j) = (w_k, w_l)$ but $(i, j) \neq (k, l)$. We wish to generalize Corollary 1.3.13 to multisets. To do so we need a fundamental definition. If $(a_1, \ldots, a_m)$ is a sequence of nonnegative integers summing to n, then define the *q-multinomial coefficient*

$$\binom{n}{a_1, \ldots, a_m} = \frac{(n)!}{(a_1)! \cdots (a_m)!}.$$

It is immediate from the definition that $\binom{n}{a_1, \ldots, a_m}$ is a rational function of q which, when evaluated at $q = 1$, becomes the ordinary multinomial coefficient $\binom{n}{a_1, \ldots, a_m}$. In fact, it is not difficult to see that $\binom{n}{a_1, \ldots, a_m}$ is a polynomial in q whose coefficients are nonnegative integers. One way to do this is as follows. Write $\binom{n}{k}$ as short for $\binom{n}{k, n-k}$ (exactly in analogy with the notation $\binom{n}{k}$ for binomial coefficients). The expression $\binom{n}{k}$ is a called a *q-binomial coefficient* (or *Gaussian polynomial*). It is straightforward to verify that

$$\binom{n}{a_1, \ldots, a_m} = \binom{n}{a_1}\binom{n-a_1}{a_2}\binom{n-a_1-a_2}{a_3} \cdots \binom{a_m}{a_m} \tag{1.66}$$

and

$$\binom{n}{k} = \binom{n-1}{k} + q^{n-k}\binom{n-1}{k-1}. \tag{1.67}$$

From these equations and the "initial conditions" $\binom{n}{0} = 1$, it follows by induction that $\binom{n}{a_1, \ldots, a_m}$ is a polynomial in q with nonnegative integer coefficients.

1.7.1 Proposition. *Let $M = \{1^{a_1}, \ldots, m^{a_m}\}$ be a multiset of cardinality $n = a_1 + \cdots + a_m$. Then*

$$\sum_{w \in \mathfrak{S}_M} q^{\mathrm{inv}(w)} = \binom{n}{a_1, \ldots, a_m}. \tag{1.68}$$

First Proof. Denote the left-hand side of (1.68) by $P(a_1, \ldots, a_m)$ and write $Q(n, k) = P(k, n-k)$. Clearly $Q(n, 0) = 1$. Hence in view of (1.66) and (1.67) it suffices to show that

$$P(a_1, \ldots, a_m) = Q(n, a_1) P(a_2, a_3, \ldots, a_m), \tag{1.69}$$

$$Q(n, k) = Q(n-1, k) + q^{n-k} Q(n-1, k-1). \tag{1.70}$$

If $w \in \mathfrak{S}_M$, then let w' be the permutation of $M' = \{2^{a_2}, \ldots, m^{a_m}\}$ obtained by removing the 1's from w, and let w'' be the permutation of $M'' = \{1^{a_1}, 2^{n-a_1}\}$ obtained from w by changing every element greater than 2 to 2. Clearly w is uniquely determined by w' and w'', and $\mathrm{inv}(w) = \mathrm{inv}(w') + \mathrm{inv}(w'')$. Hence,

$$P(a_1, \ldots, a_m) = \sum_{w' \in \mathfrak{S}_{M'}} \sum_{w'' \in \mathfrak{S}_{M''}} q^{\mathrm{inv}(w') + \mathrm{inv}(w'')}$$

$$= Q(n, a_1) P(a_2, a_3, \ldots, a_m),$$

which is (1.69).

Now let $M = \{1^k, 2^{n-k}\}$. Let $\mathfrak{S}_{M,i}$, $1 \leq i \leq 2$, consist of those $w \in \mathfrak{S}_M$ whose last element is i, and let $M_1 = \{1^{k-1}, 2^{n-k}\}$, $M_2 = \{1^k, 2^{n-k-1}\}$. If $w \in \mathfrak{S}_{M,1}$ and $w = u1$, then $u \in \mathfrak{S}_{M_1}$ and $\mathrm{inv}(w) = n - k + \mathrm{inv}(u)$. If $w \in \mathfrak{S}_{M,2}$ and $w = v2$, then $v \in \mathfrak{S}_{M_2}$ and $\mathrm{inv}(w) = \mathrm{inv}(v)$. Hence,

$$Q(n, k) = \sum_{u \in \mathfrak{S}_{M_1}} q^{\mathrm{inv}(u) + n - k} + \sum_{v \in \mathfrak{S}_{M_2}} q^{\mathrm{inv}(v)}$$

$$= q^{n-k} Q(n-1, k-1) + Q(n-1, k),$$

which is (1.70). $\square$

Second Proof. Define a map

$$\phi : \mathfrak{S}_M \times \mathfrak{S}_{a_1} \times \cdots \times \mathfrak{S}_{a_m} \to \mathfrak{S}_n$$

$$(w_0, w_1, \ldots, w_m) \mapsto w$$

by converting the a_i i's in w_0 to the numbers $a_1 + \cdots + a_{i-1} + 1, a_1 + \cdots + a_{i-1} + 2, a_1 + \cdots + a_{i-1} + a_i$ in the order specified by w_i. For instance $(21331223, 21, 231, 312) \mapsto 42861537$. We have converted 11 to 21 (preserving the relative order of the terms of $w_1 = 21$), 222 to 453 (preserving the order 231), and 333 to 867 (preserving 312). It is easily verified that ϕ is a bijection, and that

$$\mathrm{inv}(w) = \mathrm{inv}(w_0) + \mathrm{inv}(w_1) + \cdots + \mathrm{inv}(w_m). \tag{1.71}$$

By Corollary 1.3.13, we conclude

$$\left(\sum_{w \in \mathfrak{S}_M} q^{\mathrm{inv}(w)} \right) (a_1)! \cdots (a_m)! = (n)!,$$

and the proof follows. □

NOTE. If $w_1, \ldots, w_m$ are all identity permutations, then we obtain a map $\psi : \mathfrak{S}_M \to \mathfrak{S}_n$ known as *standardization*. For instance, $\psi(14214331) = 17428563$. Standardization is a very useful technique for reducing problems about multisets to sets. For a significant example, see Lemma 7.11.6.

The first proof of Proposition 1.7.1 can be classified as "semicombinatorial." We did not give a direct proof of (1.68) itself, but rather of the two recurrences (1.69) and (1.70). At this stage it would be difficult to give a direct combinatorial proof of (1.68) since there is no "obvious" combinatorial interpretation of the coefficients of $\binom{n}{a_1, \ldots, a_m}$ nor of the value of this polynomial at $q \in \mathbb{N}$. Thus, we will now discuss the problem of giving a combinatorial interpretation of $\binom{n}{k}$ for certain $q \in \mathbb{N}$, which will lead to a combinatorial proof of (1.68) when $m = 2$. Combined with our proof of (1.69), this yields a combinatorial proof of (1.68) in general. The reader unfamiliar with finite fields may skip the rest of this section, except for the brief discussion of partitions.

Let q be a prime power, and denote by $\mathbb{F}_q$ a finite field with q elements (all such fields are of course isomorphic) and by $\mathbb{F}_q^n$ the n-dimensional vector space of all n-tuples $(\alpha_1, \ldots, \alpha_n)$, where $\alpha_i \in \mathbb{F}_q$.

1.7.2 Proposition. *The number of k-dimensional subspaces of $\mathbb{F}_q^n$ is $\binom{n}{k}$.*

Proof. Denote the number in question by $G(n, k)$, and let $N = N(n, k)$ equal the number of ordered k-tuples $(v_1, \ldots, v_k)$ of linearly independent vectors in $\mathbb{F}_q^n$. We may choose v_1 in $q^n - 1$ ways, then v_2 in $q^n - q$ ways, and so on, yielding

$$N = (q^n - 1)(q^n - q) \cdots (q^n - q^{k-1}). \tag{1.72}$$

On the other hand, we may choose $(v_1, \ldots, v_k)$ by first choosing a k-dimensional subspace W of $\mathbb{F}_q^n$ in $G(n, k)$ ways, and then choosing $v_1 \in W$ in $q^k - 1$ ways, $v_2 \in W$ in $q^k - q$ ways, and so on. Hence,

$$N = G(n, k)(q^k - 1)(q^k - q) \cdots (q^k - q^{k-1}). \tag{1.73}$$

Comparing (1.72) and (1.73) yields

$$G(n, k) = \frac{(q^n - 1)(q^n - q) \cdots (q^n - q^{k-1})}{(q^k - 1)(q^k - q) \cdots (q^k - q^{k-1})}$$

$$= \frac{(n)!}{(k)!(n-k)!} = \binom{n}{k}. \qquad \square$$

Note that the above proof is completely analogous to the proof we gave in Section 1.2 that $\binom{n}{k} = \frac{n!}{k!(n-k)!}$. We may consider our proof of Proposition 1.7.2 to be the "q-analogue" of the proof that $\binom{n}{k} = \frac{n!}{k!(n-k)!}$.

Now define a *partition* of $n \in \mathbb{N}$ to be a sequence $\lambda = (\lambda_1, \lambda_2, \dots)$ of integers λ_i satisfying $\sum \lambda_i = n$ and $\lambda_1 \geq \lambda_2 \geq \cdots \geq 0$. We also write $\lambda = (\lambda_1, \dots, \lambda_k)$ if $\lambda_{k+1} = \lambda_{k+2} = \cdots = 0$. Thus, for example,

$$(3,3,2,1,0,0,0,\dots) = (3,3,2,1,0,0) = (3,3,2,1),$$

as partitions of 9. We may also informally regard a partition $\lambda = (\lambda_1, \dots, \lambda_k)$ of n (say with $\lambda_k > 0$) as a way of writing n as a sum $\lambda_1 + \cdots + \lambda_k$ of positive integers, *disregarding the order of the summands* (since there is a unique way of writing the summands in weakly decreasing order, where we don't distinguish between equal summands). Compare with the definition of a composition of n, in which the order of the parts is essential. If λ is a partition of n, then we write either $\lambda \vdash n$ or $|\lambda| = n$. The nonzero terms λ_i are called the *parts* of λ, and we say that λ has k parts where $k = \#\{i : \lambda_i > 0\}$. The number of parts of λ is also called the *length* of λ and denoted $\ell(\lambda)$. If the partition λ has m_i parts equal to i, then we write $\lambda = \langle 1^{m_1}, 2^{m_2}, \dots \rangle$, where terms with $m_i = 0$ and the superscript $m_i = 1$ may be omitted. For instance,

$$(4,4,2,2,2,1) = \langle 1^1, 2^3, 3^0, 4^2 \rangle = \langle 1, 2^3, 4^2 \rangle \vdash 15. \tag{1.74}$$

We also write $p(n)$ for the total number of partitions of n, $p_k(n)$ for the number of partitions of n with exactly k parts, and $p(j,k,n)$ for the number of partitions of n into at most k parts, with largest part at most j. For instance, there are seven partitions of 5, given by (omitting parentheses and commas from the notation) $5, 41, 32, 311, 221, 2111, 11111$, so $p(5) = 7$, $p_1(5) = 1$, $p_2(5) = 2$, $p_3(5) = 2$, $p_4(5) = 1$, $p_5(5) = 1$, $p(3,3,5) = 3$, and so on. By convention we agree that $p_0(0) = p(0) = 1$. Note that $p_n(n) = 1$, $p_{n-1}(n) = 1$ if $n > 1$, $p_1(n) = 1$, $p_2(n) = \lfloor n/2 \rfloor$. It is easy to verify the recurrence

$$p_k(n) = p_{k-1}(n-1) + p_k(n-k),$$

which provides a convenient method for making a table of the numbers $p_k(n)$ for n, k small.

Let $(\lambda_1, \lambda_2, \dots) \vdash n$. The *Ferrers diagram* or *Ferrers graph* of λ is obtained by drawing a left-justified array of n dots with λ_i dots in the ith row. For instance, the Ferrers diagram of the partition 6655321 is given by Figure 1.13(a). If we replace the dots by juxtaposed squares, then we call the resulting diagram the *Young diagram* of λ. For instance, the Young diagram of 6655321 is given by Figure 1.13(b). We will have more to say about partitions in various places throughout this book and especially in the next two sections. However, we will not attempt a systematic investigation of this enormous and fascinating subject.

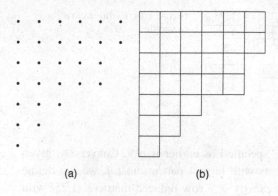

Figure 1.13 The Ferrers diagram and Young diagram of the partition 6655321.

(a)　　　　　　(b)

The next result shows the relevance of partitions to the q-binomial coefficients.

1.7.3 Proposition. *Fix* $j,k \in \mathbb{N}$. *Then*

$$\sum_{n \geq 0} p(j,k,n)q^n = \binom{j+k}{j}.$$

Proof. While it is not difficult to give a proof by induction using (1.67), we prefer a direct combinatorial proof based on Proposition 1.7.2. To this end, let $m = j + k$ and recall from linear algebra that any k-dimensional subspace of $\mathbb{F}_q^m$ (or of the m-dimensional vector space F^m over *any* field F) has a unique ordered basis $(v_1, \ldots, v_k)$ for which the matrix

$$M = \begin{bmatrix} v_1 \\ \vdots \\ v_k \end{bmatrix} \tag{1.75}$$

is in *row-reduced echelon form*. This means: (a) the first nonzero entry of each v_i is a 1; (b) the first nonzero entry of v_{i+1} appears in a column to the right of the first nonzero entry of v_i; and (c) in the column containing the first nonzero entry of v_i, all other entries are 0.

Now suppose that we are given an integer sequence $1 \leq a_1 < a_2 < \cdots < a_k \leq m$, and consider all row-reduced echelon matrices (1.75) over $\mathbb{F}_q$ for which the first nonzero entry of v_i occurs in the a_ith column. For instance, if $m = 7$, $k = 4$, and $(a_1, \ldots, a_4) = (1,3,4,6)$, then M has the form

$$\begin{bmatrix} 1 & * & 0 & 0 & * & 0 & * \\ 0 & 0 & 1 & 0 & * & 0 & * \\ 0 & 0 & 0 & 1 & * & 0 & * \\ 0 & 0 & 0 & 0 & 0 & 1 & * \end{bmatrix}$$

where the symbol $*$ denotes an arbitrary entry of $\mathbb{F}_q$. The number λ_i of $*$'s in row i is $j - a_i + i$, and the sequence $(\lambda_1, \lambda_2, \ldots, \lambda_k)$ defines a partition of some integer $n = \sum \lambda_i$ into at most k parts, with largest part at most j. The total number

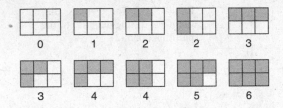

Figure 1.14 Partitions in a 2×3 rectangle.

of matrices (1.75) with $a_1, \ldots, a_k$ specified as earlier is $q^{|\lambda|}$. Conversely, given any partition λ into at most k parts with largest part at most j, we can define $a_i = j - \lambda_i + i$, and there exists exactly $q^{|\lambda|}$ row-reduced matrices (1.75) with $a_1, \ldots, a_k$ having their meaning as earlier.

Since the number of row-reduced echelon matrices (1.75) is equal to the number $\binom{j+k}{k}$ of k-dimensional subspaces of $\mathbb{F}_q^m$, we get

$$\binom{j+k}{k} = \sum_{\substack{\lambda \\ \leq k \text{ parts} \\ \text{largest part } \leq j}} q^{|\lambda|} = \sum_{n \geq 0} p(j,k,n) q^n.$$

$\square$

For readers familiar with this area, let us remark that the proof of Proposition 1.7.3 essentially constructs the well-known cellular decomposition of the Grassmann variety G_{km}.

The partitions λ enumerated by $p(j,k,n)$ may be described as those partitions of n whose Young diagram fits in a $k \times j$ rectangle. For instance, if $k = 2$ and $j = 3$, then Figure 1.14 shows the $\binom{5}{2} = 10$ partitions that fit in a 2×3 rectangle. The value of $|\lambda|$ is written beneath the diagram. It follows that

$$\binom{5}{2} = 1 + q + 2q^2 + 2q^3 + 2q^4 + q^5 + q^6.$$

It remains to relate Propositions 1.7.1 and 1.7.3 by showing that $p(j,k,n)$ is the number of permutations w of the multiset $M = \{1^j, 2^k\}$ with n inversions. Given a partition λ of n with at most k parts and largest part at most j, we will describe a permutation $w = w(\lambda) \in \mathfrak{S}_M$ with n inversions, leaving to the reader the easy proof that this correspondence is a bijection. Regard the Young diagram Y of λ as being contained in a $k \times j$ rectangle, and consider the lattice path L from the upper-right-hand corner to the lower-left-hand corner of the rectangle that travels along the boundary of Y. Walk along L, and write down a 1 whenever one takes a horizontal step and a 2 whenever one takes a vertical step. This process yields the desired permutation w. For instance, if $k = 3$, $j = 5$, $\lambda = 431$, then Figure 1.15 shows that path L and its labeling by 1's and 2's. We can also describe w by the condition that the 2's appear in positions $j - \lambda_i + i$, where $\lambda = (\lambda_1, \ldots, \lambda_k)$.

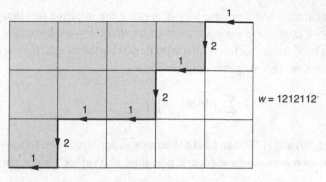

Figure 1.15 The lattice path associated with the partition 431.

1.8 Partition Identities

In the previous section, we defined a partition λ of $n \in \mathbb{N}$ and described its Ferrers diagram and Young diagram. In this section, we develop further the theory of partitions, in particular, the fascinating interaction between generating function identities and bijective proofs.

Let us begin by describing a fundamental involution on the set of partitions of n. Namely, if $\lambda \vdash n$, then define the *conjugate* partition λ' to be the partition whose Ferrers (or Young) diagram is obtained from that of λ by interchanging rows and columns. Equivalently, the diagram (Ferrers or Young) of λ' is the reflection of that of λ about the main diagonal. If $\lambda = (\lambda_1, \lambda_2, \dots)$, then the number of parts of λ' that equal i is $\lambda_i - \lambda_{i+1}$. This description of λ' provides a convenient method of computing λ' from λ without drawing a diagram. For instance, if $\lambda = (4, 3, 1, 1, 1)$, then $\lambda' = (5, 2, 2, 1)$.

Recall that $p_k(n)$ denotes the number of partitions of n into k parts. Similarly, let $p_{\leq k}(n)$ denote the number of partitions of n into at most k parts, that is, $p_{\leq k}(n) = p_0(n) + p_1(n) + \cdots + p_k(n)$. Now λ has at most k parts if and only if λ' has largest part at most k. This observation enables us to compute the generating function $\sum_{n \geq 0} p_{\leq k}(n) q^n$. A partition of n with largest part at most k may be regarded as a solution in nonnegative integers to $m_1 + 2m_2 + \cdots + km_k = n$. Here m_i is the number of times that the part i appears in the partition λ, that is, $\lambda = \langle 1^{m_1} 2^{m_2} \cdots k^{m_k} \rangle$. Hence,

$$
\begin{aligned}
\sum_{n \geq 0} p_{\leq k}(n) q^n &= \sum_{n \geq 0} \sum_{m_1 + \cdots + km_k = n} q^n \\
&= \sum_{m_1 \geq 0} \sum_{m_2 \geq 0} \cdots \sum_{m_k \geq 0} q^{m_1 + 2m_2 + \cdots + km_k} \\
&= \left(\sum_{m_1 \geq 0} q^{m_1} \right) \left(\sum_{m_2 \geq 0} q^{2m_2} \right) \cdots \left(\sum_{m_k \geq 0} q^{km_k} \right) \\
&= \frac{1}{(1-q)(1-q^2) \cdots (1-q^k)}.
\end{aligned}
\tag{1.76}
$$

This computation is just a precise way of writing the intuitive fact that the most natural way of computing the coefficient of q^n in $1/(1-q)(1-q^2)\cdots(1-q^k)$ entails computing all the partitions of n with largest part at most k. If we let $k \to \infty$, then we obtain the famous generating function

$$\sum_{n \geq 0} p(n)q^n = \prod_{i \geq 1} \frac{1}{1-q^i}. \tag{1.77}$$

Equations (1.76) and (1.77) can be considerably generalized. The following result, although by no means the most general possible, will suffice for our purposes.

1.8.1 Proposition. *For each $i \in \mathbb{P}$, fix a set $S_i \subseteq \mathbb{N}$. Let $\mathcal{S} = (S_1, S_2, \ldots)$, and define $P(\mathcal{S})$ to be the set of all partitions λ such that if the part i occurs $m_i = m_i(\lambda)$ times, then $m_i \in S_i$. Define the generating function in the variables $\boldsymbol{q} = (q_1, q_2, \ldots)$,*

$$F(\mathcal{S}, \boldsymbol{q}) = \sum_{\lambda \in P(\mathcal{S})} q_1^{m_1(\lambda)} q_2^{m_2(\lambda)} \cdots .$$

Then

$$F(\mathcal{S}, \boldsymbol{q}) = \prod_{i \geq 1} \left(\sum_{j \in S_i} q_i^j \right). \tag{1.78}$$

Proof. The reader should be able to see the validity of this result by "inspection." The coefficient of $q_1^{m_1} q_2^{m_2} \cdots$ in the right-hand side of (1.78) is 1 if each $m_i \in \mathcal{S}_j$, and 0 otherwise, which yields the desired result. $\qquad \square$

1.8.2 Corollary. *Preserve the notation of the previous proposition, and let $p(\mathcal{S}, n)$ denote the number of partitions of n belonging to $P(\mathcal{S})$, that is,*

$$p(\mathcal{S}, n) = \#\{\lambda \vdash n : \lambda \in P(\mathcal{S})\}.$$

Then

$$\sum_{n \geq 0} p(\mathcal{S}, n)q^n = \prod_{i \geq 1} \left(\sum_{j \in S_i} q^{ij} \right).$$

Proof. Put each $q_i = q^i$ in Proposition 1.8.1. $\qquad \square$

Let us now give a sample of some of the techniques and results from the theory of partitions. First, we give an idea of the usefulness of Young diagrams and Ferrers diagrams.

1.8.3 Proposition. *For any partition $\lambda = (\lambda_1, \lambda_2, \ldots)$ we have*

$$\sum_{i \geq 1} (i-1)\lambda_i = \sum_{i \geq 1} \binom{\lambda_i'}{2}. \tag{1.79}$$

Proof. Place an $i-1$ in each square of row i of the Young diagram of λ. For instance, if $\lambda = 5322$ we get

0	0	0	0	0
1	1	1		
2	2			
3	3			

If we add up all the numbers in the diagram by rows, then we obtain the left-hand side of (1.79). If we add up by columns, then we obtain the right-hand side. $\square$

1.8.4 Proposition. *Let $c(n)$ denote the number of self-conjugate partitions λ of n, that is, $\lambda = \lambda'$. Then*

$$\sum_{n \geq 0} c(n) q^n = (1+q)(1+q^3)(1+q^5) \cdots. \tag{1.80}$$

Proof. Let λ be a self-conjugate partition. Consider the "diagonal hooks" of the Ferrers diagram of $\lambda \vdash n$, as illustrated in Figure 1.16 for the partition $\lambda = 54431$. The number of dots in each hook form a partition μ of n into distinct odd parts. For Figure 1.16 we have $\mu = 953$. The map $\lambda \mapsto \mu$ is easily seen to be a bijection from self-conjugate partitions of n to partitions of n into distinct odd parts. The proof now follows from the special case $S_i = \{0,1\}$ if i is odd, and $S_i = \{0\}$ if i is even, of Corollary 1.8.2 (though it should be obvious by inspection that the right-hand side of (1.80) is the generating function for the number of partitions of n into distinct odd parts). $\square$

There are many results in the theory of partitions that assert the equicardinality of two classes of partitions. The quintessential example is given by the following result.

1.8.5 Proposition. *Let $q(n)$ denote the number of partitions of n into distinct parts and $p_{\text{odd}}(n)$ the number of partitions of n into odd parts. Then $q(n) = p_{\text{odd}}(n)$ for all $n \geq 0$.*

Figure 1.16 The diagonal hooks of the self-conjugate partition 54431.

First Proof (generating functions). Setting each $S_i = \{0, 1\}$ in Corollary 1.8.2 (or by direct inspection), we have

$$\sum_{n \geq 0} q(n)q^n = (1+q)(1+q^2)(1+q^3)\cdots$$

$$= \frac{1-q^2}{1-q} \cdot \frac{1-q^4}{1-q^2} \cdot \frac{1-q^6}{1-q^3}\cdots$$

$$= \frac{\prod_{n \geq 1}(1-q^{2n})}{\prod_{n \geq 1}(1-q^n)}$$

$$= \frac{1}{(1-q)(1-q^3)(1-q^5)\cdots}. \tag{1.81}$$

Again by Corollary 1.8.2 or by inspection, we have

$$\frac{1}{(1-q)(1-q^3)(1-q^5)\cdots} = \sum_{n \geq 0} p_{\text{odd}}(n)q^n,$$

and the proof follows. $\qquad\qquad\qquad\qquad\qquad\qquad\qquad\qquad\qquad\qquad\qquad\square$

Second Proof (bijective). Naturally a combinatorial proof of such a simple and elegant result is desired. Perhaps the simplest is the following. Let λ be a partition of n into odd parts, with the part $2j - 1$ occurring r_j times. Define a partition μ of n into distinct parts by requiring that the part $(2j-1)2^k$, $k \geq 0$, appears in μ if and only the binary expansion of r_j contains the term 2^k. We leave the reader to check the validity of this bijection, which rests on the fact that every positive integer can be expressed uniquely as a product of an odd positive integer and a power of 2. For instance, if $\lambda = \langle 9^5, 5^{12}, 3^2, 1^3 \rangle \vdash 114$, then

$$114 = 9(1+4) + 5(4+8) + 3(2) + 1(1+2)$$

$$= 9 + 36 + 20 + 40 + 6 + 1 + 2,$$

so $\mu = (40, 36, 20, 9, 6, 2, 1)$. $\qquad\qquad\qquad\qquad\qquad\qquad\qquad\qquad\qquad\qquad\square$

Third Proof (bijective). There is a completely different bijective proof which is a good example of "diagram cutting." Identify a partition λ into odd parts with its Ferrers diagram. Take each row of λ, convert it into a self-conjugate hook, and arrange these hooks diagonally in decreasing order. Now connect the upper-left-hand corner u with all dots in the "shifted hook" of u, consisting of all dots directly to the right of u and directly to the southeast of u. For the dot v directly below u (when $|\lambda| > 1$), connect it to all the dots in the conjugate shifted hook of u. Now take the northwest-most remaining dot above the main diagonal and connect it to its shifted hook, and similarly connect the northwest-most dot below the main diagonal with its conjugate shifted hook. Continue until all the entire diagram has been partitioned into shifted hooks and conjugate shifted hooks. The number of dots in these hooks form the parts of a partition μ of n into distinct parts.

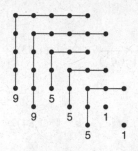

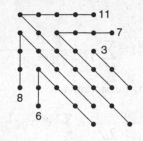

Figure 1.17 A second bijective proof that $q(n) = p_{\text{odd}}(n)$.

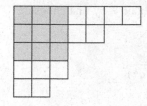

Figure 1.18 The Durfee square of the partition 75332.

Figure 1.17 shows the case $\lambda = 9955511$ and $\mu = (11,8,7,6,3)$. We trust that this figure will make the preceding rather vague description of the map $\lambda \mapsto \mu$ clear. It is easy to check that this map is indeed a bijection from partitions of n into odd parts to partitions of n into distinct parts. $\qquad\square$

There are many combinatorial identities asserting that a product is equal to a sum that can be interpreted in terms of partitions. We give three of the simplest in proposition 1.8.6, relegating some more interesting and subtle identities to the exercises. The second identity is related to the concept of the *rank* rank(λ) of a partition $\lambda = (\lambda_1, \lambda_2, \dots)$, defined to be the largest i for which $\lambda_i \geq i$. Equivalently, rank(λ) is the length of the main diagonal in the (Ferrers or Young) diagram of λ. It is also the side length of the largest square in the diagram of λ. We can place this square to include the first dot or box in the first row of the diagram, in which case it is called the *Durfee square* of λ. Figure 1.18 shows the Young diagram of the partition $\lambda = 75332$ of rank 3, with the Durfee square shaded.

1.8.6 Proposition. *(a) We have*

$$\frac{1}{\displaystyle\prod_{i \geq 1}(1 - xq^i)} = \sum_{k \geq 0} \frac{x^k q^k}{(1-q)(1-q^2)\cdots(1-q^k)}. \qquad (1.82)$$

(b) We have

$$\frac{1}{\displaystyle\prod_{i \geq 1}(1 - xq^i)} = \sum_{k \geq 0} \frac{x^k q^{k^2}}{(1-q)\cdots(1-q^k)(1-xq)\cdots(1-xq^k)}.$$

(c) We have

$$\prod_{i \geq 1}(1 + xq^i) = \sum_{k \geq 0} \frac{x^k q^{\binom{k+1}{2}}}{(1-q)(1-q^2)\cdots(1-q^k)}. \tag{1.83}$$

Proof. (a) It should be clear by inspection that

$$\frac{1}{\displaystyle\prod_{i \geq 1}(1 - xq^i)} = \sum_{\lambda} x^{\ell(\lambda)} q^{|\lambda|}, \tag{1.84}$$

where λ ranges over all partitions of all $n \geq 0$. We can obtain λ by first choosing $\ell(\lambda) = k$. It follows from equation (1.76) that

$$\sum_{\substack{\lambda \\ \ell(\lambda)=k}} q^{|\lambda|} = \frac{q^k}{(1-q)(1-q^2)\cdots(1-q^k)},$$

and the proof follows.

We should also indicate how (1.82) can be proved nonbijectively, since the technique is useful in other situations. Let

$$F(x,q) = \frac{1}{\displaystyle\prod_{i \geq 1}(1 - xq^i)}.$$

Clearly, $F(x,q) = F(xq,q)/(1-xq)$, and $F(x,q)$ is uniquely determined by this functional equation and the initial condition $F(0,q) = 1$. Now let

$$G(x,q) = \sum_{k \geq 0} \frac{x^k q^k}{(1-q)(1-q^2)\cdots(1-q^k)}.$$

Then

$$G(xq,q) = \sum_{k \geq 0} \frac{x^k q^{2k}}{(1-q)(1-q^2)\cdots(1-q^k)}$$

$$= \sum_{k \geq 0} \frac{x^k q^k}{(1-q)(1-q^2)\cdots(1-q^{k-1})} \left(\frac{1}{1-q^k} - 1\right)$$

$$= G(x,q) - xqG(x,q)$$

$$= (1-xq)G(x,q).$$

Since $G(x,0) = 1$, the proof follows.

(b) Again we use (1.84), but now the terms on the right-hand side will correspond to rank(λ) rather than $\ell(\lambda)$. If rank$(\lambda) = k$, then when we remove the Durfee square from the diagram of λ, we obtain disjoint diagrams of partitions μ and ν such that $\ell(\mu) \leq k$ and $\nu_1 = \ell(\nu') \leq k$. (For the partition $\lambda = 75332$ of Figure 1.18 we have

$\mu = 42$ and $\nu = 32$.) Every λ of rank k is obtained uniquely from such μ and ν. Moreover, $|\lambda| = k^2 + |\mu| + |\nu|$ and $\ell(\lambda) = k + \ell(\nu)$. It follows that

$$\sum_{\substack{\lambda \\ \text{rank}(\lambda) = k}} x^{\ell(\lambda)} q^{|\lambda|} = x^k q^{k^2} \frac{1}{(1-q)\cdots(1-q^k)} \cdot \frac{1}{(1-xq)\cdots(1-xq^k)}.$$

Summing over all $k \geq 0$ completes the proof.

(c) Now the coefficient of $x^k q^n$ in the left-hand side is the number of partitions of n into k *distinct* parts $\lambda_1 > \cdots > \lambda_k > 0$. Then $(\lambda_1 - k, \lambda_2 - k + 1, \ldots, \lambda_k - 1)$ is a partition of $n - \binom{k+1}{2}$ into at most k parts, from which the proof follows easily. $\qquad\square$

The generating function (obtained, e.g., from (1.82) by substituting x/q for x, or by a simple modification of either of our two proofs of Proposition 1.8.6(a))

$$\frac{1}{\prod_{i > 0}(1 - xq^i)} = \sum_{k \geq 0} \frac{x^k}{(1-q)(1-q^2)\cdots(1-q^k)}$$

is known as the *q-exponential function*, since $(1-q)(1-q^2)\cdots(1-q^n) = (1-q)^n (n)!$. We could even replace x with $(1-q)x$, getting

$$\frac{1}{\prod_{i \geq 0}(1 - x(1-q)q^i)} = \sum_{k \geq 0} \frac{x^k}{(k)!}. \tag{1.85}$$

The right-hand side reduces to e^x upon setting $q = 1$, though we cannot also substitute $q = 1$ on the left-hand side to obtain e^x. It is an instructive exercise (Exercise 1.101) to work out why this is the case. In other words, why does substituting $(1-q)x$ for x and then setting $q = 1$ in two expressions for the same power series not maintain the equality of the two series?

A generating function of the form

$$F(x) = \sum_{n \geq 0} a_n \frac{x^n}{(1-q)(1-q^2)\cdots(1-q^n)}$$

is called an *Eulerian* or *q-exponential* generating function. It is the natural *q*-analogue of an exponential generating function. We could just as well use

$$F(x(1-q)) = \sum_{n \geq 0} a_n \frac{x^n}{(n)!} \tag{1.86}$$

in place of $F(x)$. The use of $F(x)$ is traditional, though $F(x(1-q))$ is more natural combinatorially and has the virtue that setting $q = 1$ in the right-hand side of (1.86) gives an exponential generating function. We will see especially in the general theory of generating functions developed in Section 3.18 why the right-hand side of (1.86) is combinatorially "natural."

Propositions 1.8.6(a) and (c) have interesting "finite versions," where in addition to the number of parts we also restrict the largest part. Recall that $p(j,k,n)$ denotes the number of partitions $\lambda \vdash n$ for which $\lambda_1 \leq j$ and $\ell(\lambda) \leq k$. The proof of Proposition 1.8.6(a) then generalizes *mutatis mutandis* to give the following formula:

$$\frac{1}{\displaystyle\prod_{i=0}^{j}(1-xq^i)} = \sum_{k \geq 0} x^k \sum_{n \geq 0} p(j,k,n)q^n$$

$$= \sum_{k \geq 0} x^k \binom{j+k}{j}.$$

By exactly the same reasoning, using the proof of Proposition 1.8.6(c), we obtain

$$\prod_{i=0}^{j-1}(1+xq^i) = \sum_{k=0}^{j} x^k q^{\binom{k}{2}} \binom{j}{k}. \tag{1.87}$$

Equation (1.87) is known as the *q-binomial theorem*, since setting $q = 1$ gives the usual binomial theorem. It is a good illustration of the difficulty of writing down a q-analogue of an identity by inspection; it is difficult to predict without any prior insight why the factor $q^{\binom{k}{2}}$ appears in the terms on the right.

Of course, there are many other ways to prove the q-binomial theorem, including a straightforward induction on j. We can also give a finite field proof, where we regard q as a prime power. For each factor $1 + xq^i$ of the left-hand side of (1.87), choose either the term 1 or xq^i. If the latter, then choose a row vector γ_i of length j whose first nonzero coordinate is a 1, which occurs in the $(j - i)$th position. Thus, there are q^i choices for γ_i. After making this choice for all i, let V be the span in $\mathbb{F}_q^j$ of the chosen γ_i's. If we chose k of the γ_i's, then $\dim V = k$. Let M be the $k \times j$ matrix whose rows are the γ_i's in decreasing order of the index i. There is a unique $k \times k$ upper unitriangular matrix T (i.e., T is upper triangular with 1's on the main diagonal) for which TM is in row-reduced echelon form. Reversing these steps, for each k-dimensional subspace V of $\mathbb{F}_q^j$, let A be the unique $k \times j$ matrix in row-reduced echelon form whose row space is V. There are $q^{\binom{k}{2}}$ $k \times k$ upper unitriangular matrices T^{-1}, and for each of them the rows of $M = T^{-1}A$ define a choice of γ_i's. It follows that we obtain every k-dimensional subspace of $\mathbb{F}_q^j$ as a span of γ_i's exactly $q^{\binom{k}{2}}$ times, and the proof follows.

Variant. There is a slight variant of the previous finite field proof of (1.87), which has less algebraic significance but is more transparent combinatorially. Namely, once we have chosen the $k \times j$ matrix M, change every entry above the first 1 in any row to 0. We then obtain a matrix in row-reduced echelon form. There are $\binom{k}{2}$ entries of M that are changed to 0, so we get every row-reduced echelon matrix with k rows exactly $q^{\binom{k}{2}}$ times. The proof follows as before.

For yet another proof of equation (1.87) based on finite fields, see Exercise 3.119.

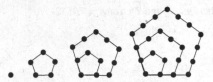

Figure 1.19 The pentagonal numbers 1, 5, 12, 22.

We next turn to a remarkable product expansion related to partitions. It is the archetype for a vast menagerie of similar results. We will give only a bijective proof; it is also an interesting challenge to find an algebraic proof. The result is called the *pentagonal number formula* or *pentagonal number theorem* because of the appearance of the numbers $k(3k-1)/2$, which are known as *pentagonal numbers*. See Figure 1.19 for an explanation of this terminology.

1.8.7 Proposition. *We have*

$$\prod_{k\geq 1}(1-x^k) = \sum_{n\in\mathbb{Z}}(-1)^n x^{n(3n-1)/2} \tag{1.88}$$

$$= 1 + \sum_{n\geq 1}(-1)^n \left(x^{n(3n-1)/2} + x^{n(3n+1)/2}\right) \tag{1.89}$$

$$= 1 - x - x^2 + x^5 + x^7 - x^{12} - x^{15} + x^{22} + x^{26} - \cdots.$$

Proof. Let $f(n) = q_e(n) - q_o(n)$, where $q_e(n)$ (respectively, $q_o(n)$) is the number of partitions of n into an even (respectively, odd) number of distinct parts. It should be clear that

$$\prod_{k\geq 1}(1-x^k) = \sum_{n\geq 0} f(n)x^n.$$

Hence we need to show that

$$f(n) = \begin{cases} (-1)^k, & \text{if } n = k(3k\pm 1)/2, \\ 0, & \text{otherwise.} \end{cases} \tag{1.90}$$

Let $Q(n)$ denote the set of all partitions of n into distinct parts. We will prove (1.90) when $n \neq k(3k\pm 1)/2$ by defining an involution $\varphi : Q(n) \to Q(n)$ such that $\ell(\lambda) \not\equiv \ell(\varphi(\lambda)) \pmod{2}$ for all $\lambda \in Q(n)$. When $n = k(3k\pm 1)/2$, we will define a partition $\mu \in Q(n)$ and an involution $\varphi : Q(n) - \{\mu\} \to Q(n) - \{\mu\}$ such that $\ell(\lambda) \not\equiv \ell(\varphi(\lambda)) \pmod{2}$ for all $\lambda \in Q(n) - \{\mu\}$, and moreover $\ell(\mu) = k$. Such a method of proof is called a *sign-reversing involution* argument. The involution φ changes the sign of $(-1)^{\ell(\lambda)}$ and hence cancels out all terms in the expansion

$$\sum_{\lambda\in Q(n)}(-1)^{\ell(\lambda)}$$

except those terms indexed by partitions λ not in the domain of φ. These partitions form a much smaller set that can be analyzed separately.

The definition of φ is quite simple. Let L_λ denote the last row of the Ferrers diagram of λ, and let D_λ denote the set of last elements of all rows i for which

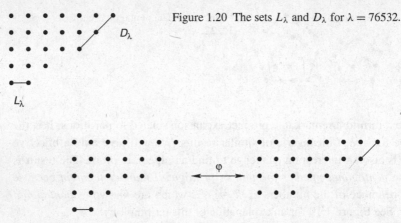

Figure 1.20 The sets L_λ and D_λ for $\lambda = 76532$.

Figure 1.21 The involution φ from the proof of the Pentagonal Number Formula.

$\lambda_i = \lambda_1 - i + 1$. Figure 1.20 shows L_λ and D_λ for $\lambda = 76532$. If $\#D_\lambda < \#L_\lambda$, define $\varphi(\lambda)$ to be the partition obtained from (the Ferrers diagram of) λ by removing D_λ and replacing it under L_λ to form a new row. Similarly, if $\#L_\lambda \leq \#D_\lambda$, define $\varphi(\lambda)$ to be the partition obtained from (the Ferrers diagram of) λ by removing L_λ and replacing it parallel and to the right of D_λ, beginning at the top row. Clearly, $\varphi(\lambda) = \mu$ if and only if $\varphi(\mu) = \lambda$. See Figure 1.21 for the case $\lambda = 76532$ and $\mu = 8753$. It is evident that φ is an involution where it is defined; the problem is that the diagram defined by $\varphi(\lambda)$ may not be a valid Ferrers diagram. A little thought shows that there are exactly two situations when this is the case. The first case occurs when λ has the form $(2k-1, 2k-2, \ldots, k)$. In this case $|\lambda| = k(3k-1)/2$ and $\ell(\lambda) = k$. The second bad case is $\lambda = (2k, 2k-1, \ldots, k+1)$. Now $|\lambda| = k(3k+1)/2$ and $\ell(\lambda) = k$. Hence, φ is a sign-reversing involution on all partitions λ, with the exception of a single partition of length k of numbers of the form $k(3k \pm 1)/2$, and the proof follows. $\qquad\qquad\qquad\qquad\qquad\qquad\qquad\qquad\qquad\qquad\qquad\qquad\square$

We can rewrite the Pentagonal Number Formula (1.88) in the form

$$\left(\sum_{n \geq 0} p(n)x^n \right) \left(\sum_{n \in \mathbb{Z}} (-1)^n x^{n(3n-1)/2} \right) = 1.$$

If we equate coefficients of x^n on both sides, then we obtain a recurrence for $p(n)$:

$$p(n) = p(n-1) + p(n-2) - p(n-5) - p(n-7) + p(n-12) + p(n-15) - \cdots.$$
$$(1.91)$$

It is understood that $p(n) = 0$ for $n < 0$, so the number of terms on the right-hand side is roughly $2\sqrt{2n/3}$. For instance,

$$p(20) = p(19) + p(18) - p(15) - p(13) + p(8) + p(5)$$

$$= 490 + 385 - 176 - 101 + 22 + 7$$

$$= 627.$$

Equation (1.91) affords the most efficient known method to compute all the numbers $p(1), p(2), \ldots, p(n)$ for given n. Much more sophisticated methods (discussed briefly later) are known for computing $p(n)$ that don't involve computing smaller values. It is known, for instance, that

$$p(10^4) = 36167251325636293988820471890953695495016030339315650422088186860588795256875406642059231055605290691643 5144.$$

In fact, $p(10^{15})$ can be computed exactly, a number with exactly 35,228,031 decimal digits.

It is natural to ask for the rate of growth of $p(n)$. To this end we mention without proof the famous asymptotic formula

$$p(n) \sim \frac{e^{\pi\sqrt{2n/3}}}{4\sqrt{3}n}. \tag{1.92}$$

For instance, when $n = 100$ the ratio of the right-hand side to the left is $1.0457\cdots$, whereas when $n = 1000$, it is $1.0141\cdots$. When $n = 10000$ the ratio is $1.00444\cdots$. There is in fact an asymptotic series for $p(n)$ that actually converges rapidly to $p(n)$. (Typically, an asymptotic series is divergent.) This asymptotic series is the best-known method for evaluating $p(n)$ for large n.

1.9 The Twelvefold Way

In this section we will be concerned with counting functions between two sets. Let N and X be finite sets with $\#N = n$ and $\#X = x$. We wish to count the number of functions $f : N \to X$ subject to certain restrictions. There will be three restrictions on the functions themselves and four restrictions on when we consider two functions to be the same. This gives a total of twelve counting problems, whose solution is called the *Twelvefold Way*.

The three restrictions on the functions $f : N \to X$ are the following.

(a) f is arbitrary (no restriction).
(b) f is injective (one-to-one).
(c) f is surjective (onto).

The four interpretations as to when two functions are the same (or *equivalent*) come about from regarding the elements of N and X as "distinguishable" or "indistinguishable." Think of N as a set of balls and X as a set of boxes. A function $f : N \to X$ consists of placing each ball into some box. If we can tell the balls apart, then the elements of N are called *distinguishable*, otherwise *indistinguishable*. Similarly if we can tell the boxes apart, then then elements of X are called *distinguishable*, otherwise *indistinguishable*. For example, suppose $N = \{1, 2, 3\}$,

$X = \{a,b,c,d\}$, and define functions $f,g,h,i : N \to X$ by

$$f(1) = f(2) = a, \quad f(3) = b,$$
$$g(1) = g(3) = a, \quad g(2) = b,$$
$$h(1) = h(2) = b, \quad h(3) = d,$$
$$i(2) = i(3) = b, \quad i(1) = c.$$

If the elements of both N and X are distinguishable, then the functions have the "pictures" shown by Figure 1.22. All four pictures are different, and the four functions are inequivalent. Now suppose that the elements of N (but not X) are indistinguishable. This assumption corresponds to erasing the labels on the balls. The pictures for f and g both become as shown in Figure 1.23, so f and g are equivalent. However, f, h, and i remain inequivalent. If the elements of X (but not N) are indistinguishable, then we erase the labels on the boxes. Thus, f and h both have the picture shown in Figure 1.24. (The order of the boxes is irrelevant if we can't tell them apart.) Hence, f and h are equivalent, but f, g, and i are inequivalent. Finally, if the elements of both N and X are indistinguishable, then all four functions have the picture shown in Figure 1.25, so all four are equivalent.

A rigorous definition of the above notions of equivalence is desirable. Two functions $f,g : N \to X$ are said to be *equivalent with N indistinguishable* if there is a bijection $u : N \to N$ such that $f(u(a)) = g(a)$ for all $a \in N$. Similarly, f and g are *equivalent with X indistinguishable* if there is a bijection $v : X \to X$ such that $v(f(a)) = g(a)$ for all $a \in N$. Finally, f and g are *equivalent with N and X indistinguishable* if there are bijections $u : N \to N$ and $v : X \to X$ such

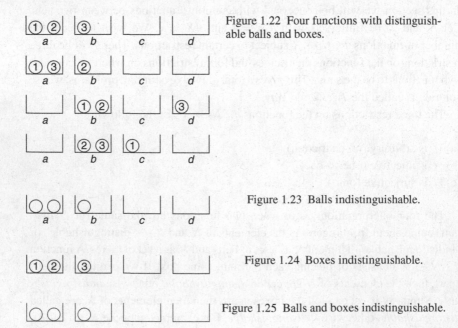

Figure 1.22 Four functions with distinguishable balls and boxes.

Figure 1.23 Balls indistinguishable.

Figure 1.24 Boxes indistinguishable.

Figure 1.25 Balls and boxes indistinguishable.

that $v(f(u(a))) = g(a)$ for all $a \in N$. These three notions of equivalence are all equivalence relations, and the number of "different" functions with respect to one of these equivalences simply means the number of equivalence classes. If f and g are equivalent (in any of the foregoing ways), then f is injective (respectively, surjective) if and only if g is injective (respectively, surjective). We, therefore, say that the notions of injectivity and surjectivity are *compatible* with the equivalence relation. By the "number of inequivalent injective functions $f : N \rightarrow X$," we mean the number of equivalence classes all of whose elements are injective.

We are now ready to present the Twelvefold Way. The twelve entries are numbered and will be discussed individually. The table gives the number of inequivalent functions $f : N \rightarrow X$ of the appropriate type, where $\#N = n$ and $\#X = x$.

The Twelvefold Way

Elements of N	Elements of X	Any f	Injective f	Surjective f
dist.	dist.	1. x^n	2. $(x)_n$	3. $x! S(n,x)$
indist.	dist.	4. $\left(\!\binom{x}{n}\!\right)$	5. $\binom{x}{n}$	6. $\left(\!\binom{x}{n-x}\!\right)$
dist.	indist.	7. $S(n,0) + S(n,1) + \cdots + S(n,x)$	8. $\begin{array}{ll} 1 & \text{if } n \leq x \\ 0 & \text{if } n > x \end{array}$	9. $S(n,x)$
indist.	indist.	10. $p_0(n) + p_1(n) + \cdots + p_x(n)$	11. $\begin{array}{ll} 1 & \text{if } n \leq x \\ 0 & \text{if } n > x \end{array}$	12. $p_x(n)$

Discussion of Twelvefold Way Entries

1. For each $a \in N$, $f(a)$ can be any of the x elements of X. Hence, there are x^n functions.
2. Say $N = \{a_1, \ldots, a_n\}$. Choose $f(a_1)$ in x ways, then $f(a_2)$ in $x-1$ ways, and so on, giving $x(x-1) \cdots (x-n+1) = (x)_n$ choices in all.
3. * A *partition* of a finite set N is a collection $\pi = \{B_1, B_2, \ldots, B_k\}$ of subsets of N such that
   a. $B_i \neq \emptyset$ for each i,
   b. $B_i \cap B_j = \emptyset$ if $i \neq j$,
   c. $B_1 \cup B_2 \cup \cdots \cup B_k = N$.

(Contrast this definition with that of an *ordered partition* in the proof of Lemma 1.4.11, for which the subsets $B_1, \ldots, B_k$ are linearly ordered.) We call B_i a *block* of π, and we say that π has k blocks, denoted $|\pi| = k$. Define $S(n,k)$ to be the number of partitions of an n-set into k-blocks. The number $S(n,k)$ is called a *Stirling number of the second kind*. (Stirling numbers of the first kind were defined preceding Lemma 1.3.6.) By convention, we put $S(0,0) = 1$. We use

---

* Discussion of entry 4 begins on page 79.

notation such as 135-26-4 to denote the partition of $[6]$ with blocks $\{1,3,5\}$, $\{2,6\}$, $\{4\}$. For instance, $S(4,2) = 7$, corresponding to the partitions 123-4, 124-3, 134-2, 234-1, 12-34, 13-24, 14-23. The reader should check that for $n \geq 1$, $S(n,k) = 0$ if $k > n$, $S(n,0) = 0$, $S(n,1) = 1$, $S(n,2) = 2^{n-1} - 1$, $S(n,n) = 1$, $S(n,n-1) = \binom{n}{2}$, and $S(n,n-2) = \binom{n}{3} + 3\binom{n}{4}$. (See Exercise 43.)

NOTE. There is a simple bijection between the equivalence relations $\sim$ on a set X (which may be infinite) and the partitions of X, namely, the equivalence classes of $\sim$ form a partition of X.

The Stirling numbers of the second kind satisfy the following basic recurrence:

$$S(n,k) = kS(n-1,k) + S(n-1,k-1). \tag{1.93}$$

Equation (1.93) is proved as follows. To obtain a partition of $[n]$ into k blocks, we can partition $[n-1]$ into k blocks and place n into any of these blocks in $kS(n-1,k)$ ways, or we can put n in a block by itself and partition $[n-1]$ into $k-1$ blocks in $S(n-1,k-1)$ ways. Hence, (1.93) follows. The recurrence (1.93) allows one to prove by induction many results about the numbers $S(n,k)$, though frequently there will be preferable combinatorial proofs. The *total* number of partitions of an n-set is called a *Bell number* and is denoted $B(n)$. Thus, $B(n) = \sum_{k=1}^{n} S(n,k)$, $n \geq 1$. The values of $B(n)$ for $1 \leq n \leq 10$ are given by the following table.

n	1	2	3	4	5	6	7	8	9	10
$B(n)$	1	2	5	15	52	203	877	4140	21147	115975

The following is a list of some basic formulas concerning $S(n,k)$ and $B(n)$:

$$S(n,k) = \frac{1}{k!} \sum_{i=0}^{k} (-1)^{k-i} \binom{k}{i} i^n, \tag{1.94a}$$

$$\sum_{n \geq k} S(n,k) \frac{x^n}{n!} = \frac{1}{k!} (e^x - 1)^k, \quad k \geq 0, \tag{1.94b}$$

$$\sum_{n \geq k} S(n,k) x^n = \frac{x^k}{(1-x)(1-2x)\cdots(1-kx)}, \tag{1.94c}$$

$$x^n = \sum_{k=0}^{n} S(n,k)(x)_k, \tag{1.94d}$$

$$B(n+1) = \sum_{i=0}^{n} \binom{n}{i} B(i), \quad n \geq 0, \tag{1.94e}$$

$$\sum_{n \geq 0} B(n) \frac{x^n}{n!} = \exp(e^x - 1). \tag{1.94f}$$

We now indicate the proofs of (1.94a)–(1.94f). For all except (1.94d), we describe noncombinatorial proofs, though with a bit more work combinatorial proofs can be given (see e.g. Example 5.2.4). Let $F_k(x) = \sum_{n \geq k} S(n,k) x^n/n!$. Clearly, $F_0(x) = 1$. From (1.93) we have

$$F_k(x) = k \sum_{n \geq k} S(n-1,k) \frac{x^n}{n!} + \sum_{n \geq k} S(n-1,k-1) \frac{x^n}{n!}.$$

Differentiate both sides to obtain

$$F_k'(x) = k F_k(x) + F_{k-1}(x). \tag{1.95}$$

Assume by induction that $F_{k-1}(x) = \frac{1}{(k-1)!}(e^x - 1)^{k-1}$. Then the unique solution to (1.95) whose coefficient of x^k is $1/k!$ is given by $F_k(x) = \frac{1}{k!}(e^x - 1)^k$. Hence (1.94b) is true by induction. To prove (1.94a), write

$$\frac{1}{k!}(e^x - 1)^k = \frac{1}{k!} \sum_{j=0}^{k} (-1)^{k-j} \binom{k}{j} e^{jx}$$

and extract the coefficient of x^n. To prove (1.94f), sum (1.94b) on k to obtain

$$\sum_{n \geq 0} B(n) \frac{x^n}{n!} = \sum_{k \geq 0} \frac{1}{k!}(e^x - 1)^k = \exp(e^x - 1).$$

Equation (1.94e) may be proved by differentiating (1.94f) and comparing coefficients, and it is also quite easy to give a direct combinatorial proof (Exercise 107). Equation (1.94c) is proved analogously to our proof of (1.94b) and can also be given a proof analogous to that of Proposition 1.3.7 (Exercise 45). It remains to prove (1.94d), and this will be done following the next paragraph.

We now verify entry 3 of the Twelvefold Way. We have to show that the number of surjective functions $f : N \to X$ is $x! S(n,x)$. Now $x! S(n,x)$ counts the number of ways of partitioning N into x blocks and then linearly ordering the blocks, say $(B_1, B_2, \ldots, B_x)$. Let $X = \{b_1, b_2, \ldots, b_x\}$. We associate the ordered partition $(B_1, B_2, \ldots, B_x)$ with the surjective function $f : N \to X$ defined by $f(i) = b_j$ if $i \in B_j$. (More succinctly, we can write $f(B_j) = b_j$.) This establishes the desired correspondence.

We can now give a simple combinatorial proof of (1.94d). The left-hand side is the total number of functions $f : N \to X$. Each such function is surjective onto a unique subset $Y = f(N)$ of X satisfying $\#Y \leq n$. If $\#Y = k$, then there are $k! S(n,k)$ such functions, and there are $\binom{x}{k}$ choices of subsets Y of X with $\#Y = k$. Hence,

$$x^n = \sum_{k=0}^{n} k! S(n,k) \binom{x}{k} = \sum_{k=0}^{n} S(n,k)(x)_k. \tag{1.96}$$

Equation (1.94d) has the following additional interpretation. The set $\mathcal{P} = K[x]$ of all polynomials in the indeterminate x with coefficients in the field K forms a vector space over K. The sets $B_1 = \{1, x, x^2, \ldots\}$ and $B_2 = \{1, (x)_1, (x)_2, \ldots\}$ are both bases for $\mathcal{P}$. Then (1.94d) asserts that the (infinite) matrix $S = [S(n,k)]_{k,n \in \mathbb{N}}$

is the transition matrix between the basis B_2 and the basis B_1. Now consider again equation (1.28) from Section 1.3. If we change x to $-x$ and multiply by $(-1)^n$, we obtain

$$\sum_{k=0}^{n} s(n,k)x^k = (x)_n.$$

Thus, the matrix $s = [s(n,k)]_{k,n\in\mathbb{N}}$ is the transition matrix from B_1 to B_2 and is, therefore, the *inverse* to the matrix S.

The assertion that the matrices S and s are inverses leads to the following result.

1.9.1 Proposition. *a. For all $m,n \in \mathbb{N}$, we have*

$$\sum_{k\geq 0} S(m,k)s(k,n) = \delta_{mn}.$$

b. Let $a_0,a_1,\dots$ and $b_0,b_1,\dots$ be two sequences of elements of a field K. The following two conditions are equivalent:
  i. For all $n \in \mathbb{N}$,

$$b_n = \sum_{k=0}^{n} S(n,k)a_k.$$

  ii. For all $n \in \mathbb{N}$,

$$a_n = \sum_{k=0}^{n} s(n,k)b_k.$$

Proof.

a. This is just the assertion that the product of the two matrices S and s is the identity matrix $[\delta_{mn}]$.
b. Let a and b denote the (infinite) column vectors $(a_0,a_1,\dots)^t$ and $(b_0,b_1,\dots)^t$, respectively (where ^{t} denotes transpose). Then (i) asserts that $Sa = b$. Multiply on the left by s to obtain $a = sb$, which is (ii). Similarly (ii) implies (i).

$\square$

The matrices S and s look as follows:

$$S = \begin{bmatrix} 1 & 0 & 0 & 0 & 0 & 0 & 0 & 0 \\ 0 & 1 & 0 & 0 & 0 & 0 & 0 & 0 \\ 0 & 1 & 1 & 0 & 0 & 0 & 0 & 0 \\ 0 & 1 & 3 & 1 & 0 & 0 & 0 & 0 \\ 0 & 1 & 7 & 6 & 1 & 0 & 0 & 0 \\ 0 & 1 & 15 & 25 & 10 & 1 & 0 & 0 \\ 0 & 1 & 31 & 90 & 65 & 15 & 1 & 0 \\ 0 & 1 & 63 & 301 & 350 & 140 & 21 & 1 \end{bmatrix} \cdots$$

$$s = \begin{bmatrix} 1 & 0 & 0 & 0 & 0 & 0 & 0 & 0 \\ 0 & 1 & 0 & 0 & 0 & 0 & 0 & 0 \\ 0 & -1 & 1 & 0 & 0 & 0 & 0 & 0 \\ 0 & 2 & -3 & 1 & 0 & 0 & 0 & 0 \\ 0 & -6 & 11 & -6 & 1 & 0 & 0 & 0 \\ 0 & 24 & -50 & 35 & -10 & 1 & 0 & 0 \\ 0 & -120 & 274 & -225 & 85 & -15 & 1 & 0 \\ 0 & 720 & -1764 & 1624 & -735 & 175 & -21 & 1 \\ & & & & \vdots & & & \end{bmatrix} \cdots$$

Equations (1.28) and (1.94d) also have close connections with the *calculus of finite differences*, about which we will say a very brief word here. Given a function $f : \mathbb{Z} \to K$ (or possibly $f : \mathbb{N} \to K$), where K is a field of characteristic 0, define a new function Δf, called the *first difference* of f, by

$$\Delta f(n) = f(n+1) - f(n).$$

We call Δ the first *difference operator*, and a succinct but greatly oversimplified definition of the calculus of finite differences would be that it is the study of the operator Δ. We may iterate Δ k times to obtain the *k-th difference operator*,

$$\Delta^k f = \Delta(\Delta^{k-1} f).$$

The field element $\Delta^k f(0)$ is called the *k-th difference of f at 0*. Define another operator E, called the *shift operator*, by $Ef(n) = f(n+1)$. Thus, $\Delta = E - 1$, where 1 denotes the identity operator. We now have

$$\Delta^k f(n) = (E - 1)^k f(n)$$

$$= \sum_{i=0}^{k} (-1)^{k-i} \binom{k}{i} E^i f(n)$$

$$= \sum_{i=0}^{k} (-1)^{k-i} \binom{k}{i} f(n+i). \tag{1.97}$$

In particular,

$$\Delta^k f(0) = \sum_{i=0}^{k} (-1)^{k-i} \binom{k}{i} f(i), \tag{1.98}$$

which gives an explicit formula for $\Delta^k f(0)$ in terms of the values $f(0), f(0), \ldots, f(k)$. We can easily invert (1.97) and express $f(n)$ in terms of the numbers $\Delta^i f(0)$. Namely,

$$f(n) = E^n f(0)$$

$$= (1 + \Delta)^n f(0)$$

$$= \sum_{k=0}^{n} \binom{n}{k} \Delta^k f(0). \tag{1.99}$$

NOTE. The operator Δ is a "discrete analogue" of the derivative operator $D = \frac{d}{dx}$. It is an instructive exercise to find finite difference analogues of concepts and results from calculus. For instance, the finite difference analogue of e^x is 2^n, since $De^x = e^x$ and $\Delta 2^n = 2^n$. Similarly, the finite difference analogue of x^n is $(x)_n$, since $Dx^n = nx^{n-1}$ and $\Delta(x)_n = n(x)_{n-1}$. The finite difference analogue of the Taylor series expansion

$$f(x) = \sum_{k \geq 0} \frac{1}{k!}(D^k f(0))x^k$$

is just equation (1.99), where we should write $\binom{n}{k} = \frac{1}{k!}(n)_k$ to make the analogy even more clear. A unified framework for working with operators such as D and Δ is provided by Exercise 5.37.

Now given the function $f : \mathbb{Z} \to K$, write on a line the values

$$\cdots f(-2) \; f(-1) \; f(0) \; f(1) \; f(2) \cdots.$$

If we write below the space between any two consecutive terms $f(i), f(i+1)$ their difference $f(i+1) - f(i) = \Delta f(i)$, then we obtain the sequence

$$\cdots \Delta f(-2) \; \Delta f(-1) \; \Delta f(0) \; \Delta f(1) \; \Delta f(2) \cdots.$$

Iterating this procedure yields the *difference table* of the function f. The kth row (regarding the top row as row 0) consists of the values $\Delta^k f(n)$. The diagonal beginning with $f(0)$ and extending down and to the right consists of the differences at 0 (i.e., $\Delta^k f(0)$). For instance, let $f(n) = n^4$ (where $K = \mathbb{Q}$, say). The difference table (beginning with $f(0)$) looks like

$$
\begin{array}{ccccccccccccc}
0 & & 1 & & 16 & & 81 & & 256 & & 625 & & \cdots \\
& 1 & & 15 & & 65 & & 175 & & 369 & & \\
& & 14 & & 50 & & 110 & & 194 & & \\
& & & 36 & & 60 & & 84 & & \\
& & & & 24 & & 24 & & \\
& & & & & 0 & & \\
& & & & & & \ddots
\end{array}
$$

Hence by (1.99),

$$n^4 = \binom{n}{1} + 14\binom{n}{2} + 36\binom{n}{3} + 24\binom{n}{4} + 0\binom{n}{5} + \cdots.$$

In this case, since n^4 is a polynomial of degree 4 and $\binom{n}{k}$, for fixed k, is a polynomial of degree k, the preceding expansion stops after the term $24\binom{n}{4}$, that is, $\Delta^k 0^4 = 0$ if $k > 4$ (or more generally, $\Delta^k n^4 = 0$ if $k > 4$). Note that by (1.94d) we have

$$n^4 = \sum_{k=0}^{4} k! S(4,k) \binom{n}{k},$$

so we conclude $1!S(4,1) = 1$, $2!S(4,2) = 14$, $3!S(4,3) = 36$, $4!S(4,1) = 24$.

There was of course nothing special about the function n^4 in the preceding discussion. The same reasoning establishes the following result.

1.9.2 Proposition. *Let K be a field of characteristic 0.*

(a) *A function $f : \mathbb{Z} \to K$ is a polynomial of degree at most d if and only if $\Delta^{d+1} f(n) = 0$ (or $\Delta^d f(n)$ is constant).*

(b) *If the polynomial $f(n)$ of degree at most d is expanded in terms of the basis $\binom{n}{k}$, $0 \le k \le d$, then the coefficients are $\Delta^k f(0)$; that is,*

$$f(n) = \sum_{i=0}^{d} \Delta^k f(0) \cdot \binom{n}{k}.$$

(c) *In the special case $f(n) = n^d$, we have*

$$\Delta^k 0^d = k! S(d,k).$$

1.9.3 Corollary. *Let $f : \mathbb{Z} \to K$ be a polynomial of degree d, where $char(K) = 0$. A necessary and sufficient condition that $f(n) \in \mathbb{Z}$ for all $n \in \mathbb{Z}$ is that $\Delta^k f(0) \in \mathbb{Z}$, $0 \le k \le d$. (In algebraic terms, the abelian group of all polynomials $f : \mathbb{Z} \to \mathbb{Z}$ of degree at most d is free with basis $\binom{n}{0}, \binom{n}{1}, \dots, \binom{n}{d}$.)*

Let us now proceed to the next entry of the Twelvefold Way.

4. The "balls" are indistinguishable, so we are only interested in *how many* balls go into each box $b_1, b_2, \dots, b_x$. If $v(b_i)$ balls go into box b_i, then v defines an n-element multiset on X. The number of such multisets is $\left(\!\binom{x}{n}\!\right)$.

5. This is similar to 4, except that each box contains at most one ball. Thus our multiset becomes a set, and there are $\binom{x}{n}$ n-element subsets of X.

6. Each box b_i must contain at least one ball. If we remove one ball from each box, then we obtain an $(n - x)$-element multiset on X. The number of such multisets is $\left(\!\binom{x}{n-x}\!\right)$. Alternatively, we can clearly regard a ball placement as a composition of n into x parts, whose number is $\binom{n-1}{x-1} = \left(\!\binom{x}{n-x}\!\right)$.

7. Since the boxes are indistinguishable, a function $f : N \to X$ is determined by the nonempty sets $f^{-1}(b)$, $b \in X$, where $f^{-1}(b) = \{a \in N : f(a) = b\}$. These sets form a partition π of N, called the *kernel* or *coimage* of f. The only restriction on π is that it can contain no more than x blocks. The number of partitions of N into at most x blocks is $S(n,0) + S(n,1) + \cdots + S(n,x)$.

8. Each block of the coimage π of f must have one element. There is one such π if $x \ge n$; otherwise, there is no such π.

9. If f is surjective, then none of the sets $f^{-1}(b)$ is empty. Hence, the coimage π contains exactly x blocks. The number of such π is $S(n,x)$.

10. Let $p_k(n)$ denote the number of partitions of n into k parts, as defined in Section 1.7. A function $f : N \to X$ with N and X both indistinguishable is

determined only by the *number of elements* in each block of its coimage π. The actual elements themselves are irrelevant. The only restriction on these numbers is that they be positive integers summing to n, and that there can be no more than x of them. In other words, the numbers form a partition of n into at most x parts. The number of such partitions is $p_0(n) + p_1(n) + \cdots + p_x(n)$. Note that this number is equal to $p_x(n + x)$ (Exercise 66).

11. Same argument as 8.

12. Analogous argument to 9. If $f : N \rightarrow X$ is surjective, then the coimage π of f has exactly x blocks, so their cardinalities form a partition of n into exactly x parts.

There are many possible generalizations of the Twelvefold Way and its individual entries. See the Notes for an extension of the Twelvefold Way to a "Thirtyfold Way." Another very natural generalization of some of the Twelvefold Way entries is the following. Let $\alpha = (\alpha_1, \ldots, \alpha_m) \in \mathbb{N}^m$ and $\beta = (\beta_1, \ldots, \beta_n) \in \mathbb{N}^n$. Suppose that we have α_i balls of color i, $1 \le i \le m$. Balls of the same color are indistinguishable. We also have n distinguishable boxes $B_1, \ldots, B_n$. In how many ways can we place the balls into the boxes so that box B_j has exactly β_j balls? Call this number $N_{\alpha\beta}$. Similarly define $M_{\alpha\beta}$ to be the number of such placements with the further condition that each box can contain at most one ball of each color. Clearly, $N_{\alpha\beta} = M_{\alpha\beta} = 0$ unless $\sum \alpha_i = \sum \beta_j$ (the total number of balls). Given a placement of the balls into the boxes, let A be the $m \times n$ matrix such that A_{ij} is the number of balls colored i that are placed in box B_j. It is easy to see that this placement is enumerated by $N_{\alpha\beta}$ if and only if the ith row sum of A is α_i and the jth column sum is β_j. In other words, A has *row sum vector* $\mathrm{row}(A) = \alpha$ and *column sum vector* $\mathrm{col}(A) = \beta$. Thus, $N_{\alpha\beta}$ is the number of $m \times n$ $\mathbb{N}$-matrices with $\mathrm{row}(A) = \alpha$ and $\mathrm{col}(A) = \beta$. Similarly, $M_{\alpha\beta}$ is the number of $m \times n$ $(0, 1)$-matrices with $\mathrm{row}(A) = \alpha$ and $\mathrm{col}(A) = \beta$. In general, there is no simple formula for $N_{\alpha\beta}$ or $M_{\alpha\beta}$, but there are many interesting special cases, generating functions, algebraic connections, and the like. See for instance Proposition 4.6.2, Proposition 5.5.8–Corollary 5.5.11, and the many appearances of $N_{\alpha\beta}$ and $M_{\alpha\beta}$ in Chapter 7.

1.10 Two q-Analogues of Permutations

We have seen that the vector space $\mathbb{F}_q^n$ is a good q-analogue of the n-element set $[n]$, and a k-dimensional subspace of $\mathbb{F}_q^n$ is a good q-analogue of a k-element subset of $[n]$. See in particular the finite field proofs of Proposition 1.7.3 and the q-binomial theorem (equation (1.87)). In this section, we pursue this line of thought further by considering the q-analogue of a permutation of the set $[n]$. It turns out that there are *two* good q-analogues that are closely related. This section involves some linear algebra over finite fields and is unrelated to the rest of the text; it may be omitted without significant loss of continuity.

1.10.1 A q-Analogue of Permutations as Bijections

A permutation w of the set $[n]$ may be regarded as an automorphism of $[n]$ (i.e., a bijection $w : [n] \to [n]$ preserving the "structure" of $[n]$). Since $[n]$ is being regarded simply as a set, any bijection $w : [n] \to [n]$ preserves the structure. Hence, one q-analogue of a permutation w is a bijection $A : \mathbb{F}_q^n \to \mathbb{F}_q^n$ preserving the structure of $\mathbb{F}_q^n$. The structure under consideration is that of a vector space, so A is simply an invertible linear transformation on $\mathbb{F}_q^n$. The set of all such linear transformations is denoted $\mathrm{GL}(n,q)$, the *general linear group* of degree n over $\mathbb{F}_q$. Thus, $\mathrm{GL}(n,q)$ is a q-analogue of the symmetric group $\mathfrak{S}_n$. We will sometimes identify a linear transformation $A \in \mathrm{GL}(n,q)$ with its matrix with respect to the standard basis $e_1, \ldots, e_n$ of $\mathbb{F}_q^n$, that is, e_i is the ith unit coordinate vector $(0, 0, \ldots, 0, 1, 0, \ldots, 0)$ (with 1 in the ith coordinate). Hence, $\mathrm{GL}(n,q)$ may be identified with the group of all $n \times n$ invertible matrices over $\mathbb{F}_q$.

For any of the myriad properties of permutations, we can try to find a corresponding property of linear transformations over $\mathbb{F}_q$. Here we will consider the following two properties: the total number of permutations in $\mathfrak{S}_n$, and the distribution of permutations by cycle type. The total number of elements (i.e., the order) of $\mathrm{GL}(n,q)$ is straightforward to compute.

1.10.1 Proposition. *We have*

$$\#\mathrm{GL}(n,q) = (q^n - 1)(q^n - q)(q^n - q^2) \cdots (q^n - q^{n-1}) \qquad (1.100)$$

$$= q^{\binom{n}{2}}(q-1)^n (\boldsymbol{n})!.$$

Proof. Regard $A \in \mathrm{GL}(n,q)$ as an $n \times n$ matrix. An arbitrary $n \times n$ matrix over $\mathbb{F}_q$ is invertible if and only if its rows are linearly independent. There are, therefore, $q^n - 1$ choices for the first row; it can be any nonzero element of $\mathbb{F}_q^n$. There are q vectors in $\mathbb{F}_q^n$ linearly dependent on the first row, so there are $q^n - q$ choices for the second row. Since the first two rows are linearly independent, they span a subspace V of $\mathbb{F}_q^n$ of dimension 2. The third row can be any vector in $\mathbb{F}_q^n$ not in V, so there are $q^n - q^2$ choices for the third row. Continuing this line of reasoning, there will be $q^n - q^{i-1}$ choices for the ith row, so we obtain (1.100). $\square$

The q-analogue of the cycle type of a permutation is more complicated. Two elements $u, v \in \mathfrak{S}_n$ have the same cycle type if and only if they are *conjugate* in $\mathfrak{S}_n$ (i.e., if and only if there exists a permutation $w \in \mathfrak{S}_n$ such that $v = wuw^{-1}$). Hence, a reasonable analogue of cycle type for $\mathrm{GL}(n,q)$ is the conjugacy class of an element of $\mathrm{GL}(n,q)$. In this context, it is better to work with *all* $n \times n$ matrices over $\mathbb{F}_q$ and then specialize to invertible matrices. Let $\mathrm{Mat}(n,q)$ denote the set (in fact, an $\mathbb{F}_q$-algebra of dimension n^2) of all $n \times n$ matrices over $\mathbb{F}_q$. We briefly review the theory of the adjoint action of $\mathrm{GL}(n,q)$ on $\mathrm{Mat}(n,q)$. The proper context for understanding this theory is commutative algebra, so we first review the relevant background. There is nothing special about finite fields in this theory, so we work over any field K, letting $\mathrm{GL}(n, K)$ (respectively, $\mathrm{Mat}(n, K)$) denote the set of invertible (respectively, arbitrary) $n \times n$ matrices over K.

Let R be a principal ideal domain (PID) that is not a field, and let M be a finitely generated R-module. Two irreducible (= prime, for PIDs) elements $x, y \in R$ are *equivalent* if $xR = yR$ (i.e., if $y = ex$ for some unit e). Let $\mathcal{P}$ be a maximal set of inequivalent irreducible elements of R. The structure theorem for finitely generated modules over PIDs asserts that M is isomorphic to a (finite) direct sum of copies of R and $R/x^i R$ for $x \in \mathcal{P}$ and $i \geq 1$. Moreover, the terms in this direct sum are unique up to the order of summands. Thus, there is a unique $k \geq 0$, and for each $x \in \mathcal{P}$ there is a unique partition $\lambda(x) = (\lambda_1(x), \lambda_2(x), \dots)$ (which may be the empty partition) such that

$$M \cong R^k \oplus \bigoplus_{x \in \mathcal{P}} \bigoplus_{i \geq 1} R/x^{\lambda_i} R.$$

If moreover M has finite length d (i.e., d is the largest integer j for which there is a proper chain $M_0 \subset M_1 \subset \cdots \subset M_j$ of submodules of M), then $k = 0$.

Now consider the case where $R = K[u]$, well-known to be a PID. Let $\mathcal{I} = \mathcal{I}(K)$ (abbreviated to $\mathcal{I}(q)$ when $K = \mathbb{F}_q$) denote the set of all nonconstant monic irreducible polynomials $f(u)$ over K, and let Par denote the set of all partitions of all nonnegative integers. Given $M \in \mathrm{Mat}(n, K)$, then M defines a $K[u]$-module structure on K^n, where the action of u is that of M. Let us denote this $K[u]$-module by $K[M]$. Since $K[M]$ has finite length as a $K[u]$-module (or even as a vector space over K), we have an isomorphism

$$K[M] \cong \bigoplus_{f \in \mathcal{I}(K)} \bigoplus_{i \geq 1} K[u]/\left(f(u)^{\lambda_i(f)}\right). \tag{1.101}$$

Moreover, the characteristic polynomial $\det(zI - M)$ of M is given by

$$\det(zI - M) = \prod_{f \in \mathcal{I}(K)} f(z)^{|\lambda(f)|}.$$

Now $\mathrm{GL}(n, K)$ acts on $\mathrm{Mat}(n, K)$ by conjugation, that is, if $A \in \mathrm{GL}(n, K)$ and $M \in \mathrm{Mat}(n, q)$, then $A \cdot M = AMA^{-1}$. (This action is called the *adjoint representation* or *adjoint action* of $\mathrm{GL}(n, K)$.) Moreover, two matrices M and N in $\mathrm{Mat}(n, K)$ are in the same orbit of this action if and only if $K[M]$ and $K[N]$ are isomorphic as $K[u]$-modules. Hence, by equation (1.101), we can index the orbit of M by a function

$$\Phi_M : \mathcal{I}(K) \to \mathrm{Par},$$

where

$$\sum_{f \in \mathcal{I}(K)} |\Phi_M(f)| \cdot \deg(f) = n, \tag{1.102}$$

namely, $\Phi_M(f) = \lambda(f)$. We call the function $\Phi = \Phi_M$ the *orbit type* of M. It is the analogue for $\mathrm{Mat}(n, K)$ of the cycle type of a permutation $w \in \mathfrak{S}_n$.

We now restrict ourselves to the case $K = \mathbb{F}_q$. As a first application of the description of the orbits of $\mathrm{GL}(n,q)$ acting adjointly on $\mathrm{Mat}(n,q)$, we can find the number of orbits. To do so, define $\beta(n,q) = \beta(n)$ to be the number of monic irreducible polynomials $f(z)$ of degree n over $\mathbb{F}_q$. It is well known (see Exercise 2.7) that

$$\beta(n,q) = \frac{1}{n}\sum_{d|n} \mu(d)q^{n/d}. \tag{1.103}$$

1.10.2 Proposition. *Let $\omega(n,q)$ denote the number of orbits of the adjoint action of $\mathrm{GL}(n,q)$ on $\mathrm{Mat}(n,q)$, or equivalently, the number of different functions $\Phi :$ $\mathcal{I}(q) \to$ Par satisfying (1.102). Then*

$$\omega(n,q) = \sum_j p_j(n)q^j,$$

where $p_j(n)$ denotes the number of partitions of n into j parts. Equivalently,

$$\sum_{n\geq 0} \omega(n,q)x^n = \prod_{j\geq 1}(1-qx^j)^{-1}.$$

Proof. We have

$$\sum_{n\geq 0} \omega(n,q)x^n = \sum_{\Phi:\mathcal{I}\to\mathrm{Par}} x^{\sum_{f\in\mathcal{I}}|\Phi(f)|\cdot\deg(f)}$$

$$= \prod_{f\in\mathcal{I}}\left(\sum_{\lambda\in\mathrm{Par}} x^{|\lambda|\cdot\deg(f)}\right)$$

$$= \prod_{f\in\mathcal{I}}\prod_{j\geq 1}\left(1-x^{j\cdot\deg(f)}\right)^{-1} \quad \text{(by (1.77))}$$

$$= \prod_{n\geq 1}\prod_{j\geq 1}(1-x^{jn})^{-\beta(n)}.$$

Now using the formula (1.103) for $\beta(n)$, we get

$$\log\sum_{n\geq 0}\omega(n,q)x^n = \sum_{n\geq 1}\sum_{j\geq 1}\beta(n)\log(1-x^{jn})^{-1}$$

$$= \sum_{n\geq 1}\sum_{j\geq 1}\frac{1}{n}\sum_{d|n}\mu(n/d)q^d\sum_{i\geq 1}\frac{x^{ijn}}{i}.$$

Extract the coefficient $c(d,N)$ of $q^d x^N$. Clearly, $c(d,N) = 0$, when $d \nmid N$, so assume $d|N$. We get

$$c(d,N) = \sum_{i|N}\frac{1}{i}\sum_{n:d|n|\frac{N}{i}}\frac{1}{n}\mu(n/d)$$

$$= \sum_{i|\frac{N}{d}}\frac{1}{i}\sum_{m|\frac{N}{id}}\frac{1}{dm}\mu(m).$$

An elementary and basic result of number theory asserts that

$$\sum_{k|r} \frac{\mu(k)}{k} = \frac{\phi(r)}{r},$$

where ϕ denotes the Euler phi-function. Hence,

$$c(d,N) = \frac{1}{d} \sum_{i|\frac{N}{d}} \frac{\phi(N/id)}{N/d}.$$

Another standard result of elementary number theory states that

$$\sum_{k|r} \phi(r/k) = \sum_{k|r} \phi(k) = r,$$

so we finally obtain

$$c(d,N) = \frac{1}{d} \frac{N/d}{N/d} = \frac{1}{d}.$$

On the other hand, we have

$$\log \prod_{n \geq 1} (1 - qx^n)^{-1} = \sum_{n \geq 1} \sum_{d \geq 1} \frac{q^d x^{nd}}{d}.$$

The coefficient $c'(d,N)$ of $q^d x^N$ is 0 unless $d|N$; otherwise, it is $1/d$. Hence, $c(d,n) = c'(d,n)$, and the proof follows. $\qquad\square$

NOTE. Proposition 1.10.2 shows that, insofar as the number of conjugacy classes is concerned, the "correct" q-analogue of $\mathfrak{S}_n$ is not the group $\mathrm{GL}(n,q)$ itself, but rather its adjoint action on $\mathrm{Mat}(n,q)$. The number of orbits $\omega(n,q)$ is a completely satisfactory q-analogue of $p(n)$, the number of conjugacy classes in $\mathfrak{S}_n$, since $\omega(n,q)$ is a polynomial in q with nonnegative integer coefficients satisfying $\omega(n,1) = p(n)$. Note that if $\omega^*(n,q)$ denotes the number of conjugacy classes in $\mathrm{GL}(n,q)$, then $\omega^*(n,q)$ is a polynomial in q satisfying $\omega^*(n,1) = 0$ (Exercise 1.190). For more conceptual proofs of Proposition 1.10.2, see Exercise 1.191.

We next define a "cycle indicator" of $M \in \mathrm{Mat}(n,q)$ that encodes the orbit of M. For every $f \in \mathcal{I}$ and every partition $\lambda \neq \emptyset$, let $t_{f,\lambda}$ be an indeterminate. If $\lambda = \emptyset$, then set $t_{f,\lambda} = 1$. Let $\Phi_M : \mathcal{I} \to \mathrm{Par}$ be the orbit type of M. Define

$$t^{\Phi_M} = \prod_{f \in \mathcal{I}} t_{f,\Phi_M(f)}.$$

Set

$$\gamma(n) = \gamma(n,q) = \#\mathrm{GL}(n,q).$$

We now define the *cycle indicator* (or *cycle index*) of $\mathrm{Mat}(n, q)$ to be the polynomial

$$Z_n(t; q) = Z_n(\{t_{f,\lambda}\}; q) = \frac{1}{\gamma(n)} \sum_{M \in \mathrm{Mat}(n,q)} t^{\Phi M}.$$

(Set $Z_0(t; q) = 1$.)

1.10.3 Example. (a) Let M be the diagonal matrix $\mathrm{diag}(1, 1, 3)$. Then $t^{\Phi M} = t_{z-1,(1,1)} t_{z-3,(1)}$ if $q \neq 2^m$; otherwise, $t^{\Phi M} = t_{z-1,(1,1,1)}$.

(b) Let $n = q = 2$. Then

$$Z_2(t; 2) = \frac{1}{6} \Big(t_{z,(1,1)} + 3t_{z,(2)} + 6t_{z,(1)} t_{z+1,(1)} + t_{z+1,(1,1)}$$

$$+ 3t_{z+1,(2)} + 2t_{z^2+z+1,(1)} \Big). \tag{1.104}$$

We now give a q-analogue of Theorem 1.3.3, in other words, a generating function for the polynomials $Z_n(t; q)$. To see the analogy more clearly, recall from equation (1.27) that

$$\sum_{n \geq 0} Z_n(t; q) x^n = \prod_{i \geq 1} \sum_{j \geq 0} t_i^j \frac{x^{ij}}{i^j j!}.$$

The denominator $i^j j!$ is the number of permutations $w \in \mathfrak{S}_{ij}$ that commute with a fixed permutation with j i-cycles.

1.10.4 Theorem. *We have*

$$\sum_{n \geq 0} Z_n(t; q) x^n = \prod_{f \in \mathcal{I}} \sum_{\lambda \in \mathrm{Par}} \frac{t_{f,\lambda} x^{|\lambda| \cdot \deg(f)}}{c_f(\lambda)}, \tag{1.105}$$

where $c_f(\lambda)$ is the number of matrices in $\mathrm{GL}(n, q)$ commuting with a fixed matrix M of size $|\lambda(f)| \cdot \deg(f)$ satisfying

$$\Phi_M(g) = \begin{cases} \lambda, & g = f, \\ \emptyset, & g \neq f. \end{cases}$$

Equivalently, $c_f(\lambda)$ is the number of $\mathbb{F}_q$-linear automorphisms of the ring

$$\mathbb{F}_q[M] \cong \bigoplus_{i \geq 1} \mathbb{F}_q[u] / \left(f(u)^{\lambda_i(f)} \right)$$

appearing in equation (1.101).

Proof (sketch). Let G be a finite group acting on a finite set X. For $a \in X$, let $Ga = \{g \cdot a : g \in G\}$, the *orbit* of G containing a. Also let $G_a = \{g \in G : g \cdot a = a\}$, the *stabilizer* of a. A basic and elementary result in group theory asserts that $\#Ga \cdot \#G_a = \#G$. Consider the present situation, where $G = \mathrm{GL}(n, q)$ is acting on

Mat(n, q). Let $M \in \text{Mat}(n, q)$. Then $A \in G_M$ if and only if $AMA^{-1} = M$ (i.e., if and only if A and M commute). Hence,

$$\#GM = \frac{\#G}{c_G(M)}, \tag{1.106}$$

where $c_G(M)$ is the number of elements of G commuting with M.

We have a unique direct sum decomposition

$$\mathbb{F}_q^n = \bigoplus_{f \in \mathcal{I}} V_f,$$

where

$$V_f = \{v \in \mathbb{F}_q^n : f(M)^r(v) = 0 \text{ for some } r \geq 1\}.$$

Thus, $M = \bigoplus_{f \in \mathcal{I}} M_f$, where $M_f V_f \subseteq V_f$ and $M_f V_g = \{0\}$ if $g \neq f$. A matrix A commuting with M respects this decomposition (i.e., $AV_f \subseteq V_f$ for all $f \in \mathcal{I}$). Thus, $A = \bigoplus_{f \in \mathcal{I}} A_f$, where $A_f V_f \subseteq V_f$ and $A_f V_g = \{0\}$ if $g \neq f$. Then A commutes with M if and only if A_f commutes with M_f for all f. In particular,

$$c_G(M) = \prod_{f \in \mathcal{I}} c_f(\Phi_M(f)).$$

It follows from equation (1.106) that the number of conjugates of M (i.e., the size of the orbit GM) is given by

$$\#GM = \frac{\gamma(n)}{\prod_f c_f(\Phi_M(f))}. \tag{1.107}$$

This number is precisely the coefficient of $t^{\Phi_M}/\gamma(n)$ in equation (1.105), and the proof follows. □

In order for Theorem 1.10.4 to be of any use, it is necessary to find a formula for the numbers $c_f(\lambda)$. There is one special case that is quite simple.

1.10.5 Lemma. *Let $f(z) = z - a$ for some $a \in \mathbb{F}_q$, and let $\langle 1^k \rangle$ denote the partition with k parts equal to 1. Then $c_f(\langle 1^k \rangle) = \gamma(k)$.*

Proof. We are counting matrices $A \in \text{GL}(k, q)$ that commute with a $k \times k$ diagonal matrix with a's on the diagonal, so A can be any matrix in $\text{GL}(k, q)$. □

1.10.6 Corollary. *Let d_n denote the number of diagonalizable (over $\mathbb{F}_q$) matrices $M \in \text{Mat}(n, q)$. Then*

$$\sum_{n \geq 0} d_n \frac{x^n}{\gamma(n)} = \left(\sum_{k \geq 0} \frac{x^k}{\gamma(k)} \right)^q.$$

Proof. A matrix M is diagonalizable over $\mathbb{F}_q$ if and only if its corresponding orbit type $\Phi_M : \mathcal{I} \to \text{Par}$ satisfies $\Phi_M(f) = \emptyset$ unless $f = z - a$ for $a \in \mathbb{F}_q$, and

$\Phi_M(z-a) = \langle 1^k \rangle$ in the notation of equation (1.74) (where we may have $k = 0$, i.e., a is not an eigenvalue of M). Hence,

$$d_n = \gamma(n) \, Z_n(t;q)|_{t_{z-a,\langle 1^k \rangle} = 1, \, t_{f,\lambda} = 0 \, \text{otherwise}} \cdot$$

Making the substitution $t_{z-a,\langle 1^k \rangle} = 1, t_{f,\lambda} = 0$ otherwise into Theorem 1.10.4 yields

$$\sum_{n \geq 0} d_n \frac{x^n}{\gamma(n)} = \prod_{a \in \mathbb{F}_q} \sum_{k \geq 0} \frac{x^k}{c_{z-a}(\langle 1^k \rangle)}.$$

The proof follows from Lemma 1.10.5. □

The evaluation of $c_f(\lambda)$ for arbitrary f and λ is more complicated. It may be regarded as the q-analogue of Proposition 1.3.2, since equation (1.107) shows that the number of conjugates of a matrix M is determined by the numbers $c_f(\Phi_M(f))$. This formula for $c_f(\lambda)$ is a fundamental enumerative result on enumerating classes of matrices in $\mathrm{Mat}(n,q)$, from which a host of other enumerative results can be derived. Let $\lambda' = (\lambda_1', \lambda_2', \dots)$ denote the conjugate partition to λ, and let $m_i = m_i(\lambda) = \lambda_i' - \lambda_{i+1}'$ be the number of parts of λ of size i. Set

$$h_i = \lambda_1' + \lambda_2' + \cdots + \lambda_i',$$

and let $d = \deg(f)$.

1.10.7 Theorem. *We have*

$$c_f(\lambda) = \prod_{i \geq 1} \prod_{j=1}^{m_i} \left(q^{h_i d} - q^{(h_i - j)d} \right). \tag{1.108}$$

1.10.8 Example. (a) Let $\lambda = (4,2,2,2,1)$, so $\lambda' = (5,4,1,1)$, $h_1 = 5$, $h_2 = 9$, $h_3 = 10$, $h_4 = 11$, $m_1 = 1$, $m_2 = 3$, $m_4 = 1$. Thus, for $\deg(f) = 1$, we have

$$c_f(4,2,2,2,1) = (q^5 - q^4)(q^9 - q^8)(q^9 - q^7)(q^9 - q^6)(q^{11} - q^{10}).$$

(b) Let $\lambda = (k)$, so $\lambda' = \langle 1^k \rangle$, $h_i = i$ for $1 \leq i \leq k$, and $m_k = 1$. For $\deg(f) = 1$, we get $c_f(k) = q^k - q^{k-1}$. Indeed, we are asking for the number of matrices $A \in \mathrm{GL}(k,q)$ commuting with a $k \times k$ Jordan block. Such matrices are easily seen to be upper triangular with constant diagonals (parallel to the main diagonal). There are $q - 1$ choices for the main diagonal and q choices for each of the $k - 1$ diagonals above the main diagonal, giving $(q-1)q^{k-1} = q^k - q^{k-1}$ choices in all.

Proof of Theorem 1.10.7. The proof is analogous to that of Proposition 1.3.2. We write down some data that determine a linear transformation $M \in \mathrm{Mat}(nd,q)$ for which $\Phi_M(f) = \lambda \vdash n$, and then we count in how many ways we obtain the same linear transformation M. Let $\ell = \ell(\lambda)$, the number of parts of λ, and similarly $k = \ell(\lambda') = \lambda_1$.

Now let
$$v = \{v_{ij} : 1 \le i \le \ell,\ 1 \le j \le d\lambda_i\}$$
be a basis B for $\mathbb{F}_q^{nd}$, together with the indexing v_{ij} of the basis elements. Thus, the number $N(n,d,q)$ of possible v is the number of *ordered* bases of $\mathbb{F}_q^{nd}$, namely,

$$N(n,d,q) = (q^{nd} - 1)(q^{nd} - q) \cdots (q^{nd} - q^{nd-1}) = \#GL(nd,q). \qquad (1.109)$$

Let $M = M_v$ be the unique linear transformation satisfying the following three properties:

- The characteristic polynomial $\det(zI - M)$ of M is $f(z)^n$.
- For all $1 \le i \le \ell$ and $1 \le j < \lambda_i d$, we have $M(v_{ij}) = v_{i,j+1}$.
- For all $1 \le i \le \ell$, we have that $M(v_{i,\lambda_i d})$ is a linear combination of the v_{ij}'s for $1 \le j \le \lambda_i d$.

It is not hard to see that M is indeed unique and that $\Phi_M(f) = \lambda$.

We now consider how many indexed bases $v = (v_{ij})$ determine the same linear transformation M. Given M, define

$$V_i = \{v \in \mathbb{F}_q^{nd} : f(M)^i(v) = 0\},\ \ 1 \le i \le k.$$

It is clear that

$$V_1 \subset V_2 \subset \cdots \subset V_k,$$
$$\dim V_i = (\lambda_1' + \lambda_2' + \cdots + \lambda_i')d = h_i d,$$
$$\dim(V_i/V_{i-1}) = \lambda_i' d.$$

If B is a subset of $\mathbb{F}_q^n$, then set

$$f(M)B = \{f(M)v : v \in B\}.$$

There are $q^{\dim(V_k)d} - q^{\dim(V_{k-1})d} = q^{h_k d} - q^{h_{k-1}d}$ choices for v_{11} (since v_{11} can be any vector in V_k not in V_{k-1}), after which all other v_{ij} are determined. There are then $q^{h_k d} - q^{(h_{k-1}+1)d}$ choices for v_{21} (since v_{21} can be any vector in V_k not in the span of V_{k-1} and $\{v_{11}, v_{12}, \ldots, v_{1d}\}$), and so on, down to $q^{h_k d} - q^{(h_{k-1}+m_k)d}$ choices for $v_{m_k,1}$.

Let
$$B_1 := \{v_{i1}, v_{i2}, \ldots, v_{id} : 1 \le i \le \lambda_k'\}.$$

Thus, B_1 is a subset of V_k whose image in V_k/V_{k-1} is a basis for V_k/V_{k-1}. Now $v_{m_k+1,1}$ $(= v_{\lambda_k'+1,1})$ can be any vector in V_{k-1} not in the span of $f(M)B_1 \cup V_{k-2}$, so there are

$$q^{\dim(V_{k-1})} - q^{\#B_1 + \dim(V_{k-2})} = q^{h_{k-1}d} - q^{m_k d + h_{k-2}d} = q^{h_{k-1}d} - q^{(h_{k-1}-m_{k-1})d}$$

choices for $v_{\lambda_k'+1,1}$. There are then $q^{h_{k-1}d} - q^{(h_{k-1}-m_{k-1}+1)d}$ choices for $v_{\lambda_k'+2,1}$, then $q^{h_{k-1}d} - q^{(h_{k-1}-m_{k-1}+2)d}$ choices for $v_{\lambda_k'+3,1}$, etc., down to $q^{h_{k-1}d} - q^{(h_{k-1}-1)d}$ choices for $v_{\lambda_{k-1}',1}$.

Let

$$B_2 = \{v_{i1}, v_{i2}, \ldots, v_{id} : \lambda'_k + 1 \le i \le \lambda'_{k-1}\},$$

so $B_2 = \emptyset$ if $\lambda'_k = \lambda'_{k-1}$. Then $f(M)(B_1 \cup B_2)$ is a subset of V_{k-1} whose image in V_{k-1}/V_{k-2} is a basis for V_{k-1}/V_{k-2}. Now $v_{\lambda'_{k-1}+1,1}$ can be any vector in V_{k-2} not in the span of $f(M)(B_1 \cup B_2) \cup V_{k-3}$, so there are

$$q^{\dim(V_{k-2})} - q^{\#B_1 + \#B_2 + \dim(V_{k-3})} = q^{h_{k-2}d} - q^{m_k d + m_{k-1}d + h_{k-3}d}$$

$$= q^{h_{k-2}d} - q^{(h_{k-2} - m_{k-2})d}$$

choices for $v_{\lambda'_{k-1}+1,1}$. There are then $q^{h_{k-2}d} - q^{(h_{k-2} - m_{k-2}+1)d}$ choices for $v_{\lambda'_{k-1}+2,1}$, then $q^{h_{k-2}d} - q^{(h_{k-2} - m_{k-2}+2)d}$ choices for $v_{\lambda'_{k-1}+3,1}$, and so on, down to $q^{h_{k-2}d} - q^{(h_{k-2}-1)d}$ choices for $v_{\lambda'_{k-2},1}$.

Continuing in this manner shows that the total number of choices for v is given by the right-hand side of equation (1.108).

We have shown that each indexed basis v of $\mathbb{F}_q^{nd}$ defines a matrix $M \in \mathrm{Mat}(nd, q)$ with $\Phi_M(f) = \lambda$. Moreover, every matrix satisfying $\Phi_M(f) = \lambda$ occurs the same number $L(n, d, q)$ of times, given by the right-hand side of (1.108). Since by (1.109) the number of indexed bases is $\#\mathrm{GL}(nd, q)$, we get that the number of matrices M satisfying $\Phi_M(f) = \lambda$ is equal to $\mathrm{GL}(nd, q)/L(nd, q)$. It follows from equation (1.106) that $L(nd, q) = c_f(\lambda)$, completing the proof.

As a slight variation, we can see directly that $L(nd, q) = c_f(\lambda)$ as follows. Let $v = (v_{ij})$ be a fixed indexed basis for $\mathbb{F}_q^{nd}$ with $M = M(v)$. Let $v' = (v'_{ij})$ be another indexed basis satisfying $M = M(v')$. Then the linear transformation $A \in \mathrm{GL}(nd, q)$ satisfying $A(v_{ij}) = v'_{ij}$ for all i, j commutes with M, and all matrices commuting with M arise in this way. Hence once again $L(nd, q) = c_f(\lambda)$. □

Even with the preceding formula for $c_f(\lambda)$, equation (1.105) is difficult to work with in its full generality. However, if we specialize each variable $t_{f,\lambda}$ to $t_f^{|\lambda|}$, then the following lemma allows a simplification of (1.105).

1.10.9 Lemma. *For any $f \in \mathcal{I}$ of degree d we have*

$$\sum_{\lambda \in \mathrm{Par}} \frac{x^{|\lambda|}}{c_f(\lambda)} = \prod_{r \ge 1} \left(1 - \frac{x}{q^{rd}}\right)^{-1}.$$

Proof. By Theorem 1.10.7, it suffices to assume $d = 1$. Our computations take place in the ring $\mathbb{C}(q)[[x]]$ (i.e., power series in x whose coefficients are rational functions in q with complex coefficients). It follows from Proposition 1.8.6(c) that

$$\prod_{r \ge 1} \left(1 - \frac{x}{q^r}\right)^{-1} = \sum_{n \ge 0} \frac{x^n q^{-n}}{(1 - q^{-1}) \cdots (1 - q^{-n})}$$

$$= \sum_{n \ge 0} \frac{(-1)^n x^n q^{\binom{n}{2}}}{(1 - q)(1 - q^2) \cdots (1 - q^n)}.$$

Hence, by Theorem 1.10.7, we need to prove that

$$\sum_{\lambda \vdash n} \prod_{i \geq 1} \prod_{j=1}^{m_i(\lambda)} \frac{1}{q^{h_i(\lambda)} - q^{h_i(\lambda)-j}} = \frac{(-1)^n q^{\binom{n}{2}}}{(1-q)(1-q^2)\cdots(1-q^n)}. \qquad (1.110)$$

Substitute $1/q$ for q in equation (1.110). We will simply write $h_i = h_i(\lambda)$ and $m_i = m_i(\lambda)$. Since

$$\frac{1}{q^{-h_i} - q^{-(h_i - j)}} = \frac{q^{h_i}}{1 - q^j},$$

the left-hand side of (1.110) becomes

$$\sum_{\lambda \vdash n} \prod_{i \geq 1} \frac{q^{m_i h_i}}{(1-q)\cdots(1-q^{m_i})}.$$

It is easy to see that

$$\sum_{i \geq 1} m_i h_i = \sum_{i \geq 1} (\lambda_i')^2,$$

which we denote by $\langle \lambda', \lambda' \rangle$.

Under the substitution $q \to 1/q$, the right-hand side of (1.110) becomes $q^n/(1 - q)\cdots(1 - q^n)$. Thus, we are reduced to proving that

$$\sum_{\lambda \vdash n} q^{\langle \lambda', \lambda' \rangle} \prod_{i \geq 1} \frac{1}{(1-q)\cdots(1-q^{m_i})} = \frac{q^n}{(1-q)\cdots(1-q^n)}. \qquad (1.111)$$

We can replace $\langle \lambda', \lambda' \rangle$ by $\langle \lambda, \lambda \rangle$ since this substitution merely permutes the terms in the sum. Set $m_i' = m_i(\lambda') = \lambda_i - \lambda_{i+1}$. Then

$$\sum_{\lambda \vdash n} q^{\langle \lambda, \lambda \rangle} \prod_{i \geq 1} \frac{1}{(1-q)\cdots(1-q^{m_i'})} = \sum_{\lambda \vdash n} \frac{q^{\langle \lambda, \lambda \rangle}}{(1-q)\cdots(1-q^{\lambda_1})} \binom{\lambda_1}{\lambda_2} \binom{\lambda_2}{\lambda_3} \cdots.$$

The coefficient of q^k in the right-hand side of (1.111) is equal to $p_n(k)$, the number of partitions of k with largest part n. Given such a partition $\mu = (\mu_1, \mu_2, \ldots)$, associate a partition $\lambda \vdash \mu_1$ by taking the rank (= length of the Durfee square) of μ, then the rank of the partition whose diagram is to the right of the Durfee square of μ, and so on. For instance, if $\mu = (7, 7, 5, 4, 3, 2)$, then $\lambda = (4, 2, 1)$ as indicated by Figure 1.26. Given λ, the generating function $\sum_\mu q^{|\mu|}$ for all corresponding μ is

$$\frac{q^{\langle \lambda, \lambda \rangle}}{(1-q)\cdots(1-q^{\lambda_1})} \binom{\lambda_1}{\lambda_2} \binom{\lambda_2}{\lambda_3} \cdots,$$

as indicated by Figure 1.27 (using Proposition 1.7.3), and the proof follows. □

Now let

$$\widehat{Z}_n(t;q) = Z_n(t;q)\big|_{t_{f,\lambda} = t_f^{|\lambda|}}.$$

For instance, from equation (1.104) we have

$$\widehat{Z}_2(t;2) = \frac{1}{6}\left(4t_z^2 + 4t_{z+1}^2 + 6t_z t_{z+1} + 2t_{z^2+z+1}\right).$$

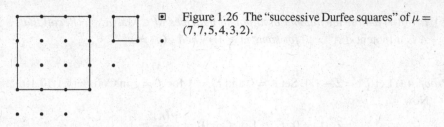

Figure 1.26 The "successive Durfee squares" of $\mu = (7,7,5,4,3,2)$.

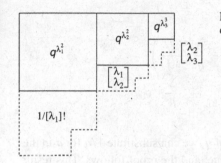

Figure 1.27 The "successive Durfee square decomposition" of λ.

Let $f = \prod_{f_i \in \mathcal{I}} f_i^{a_i}$, with $\deg f = n$. Then the coefficient of $t_{f_1}^{a_1} t_{f_2}^{a_2} \cdots$ in $\gamma(n,q)\widehat{Z}(t;q)$ is just the number of matrices $M \in \mathrm{Mat}(n,q)$ with characteristic polynomial f. Note that in general if we define $\deg(t_f) = \deg(f)$, then $\widehat{Z}_n(t;q)$ is homogeneous of degree n.

The following corollary is an immediate consequence of Theorem 1.10.4 and Lemma 1.10.9.

1.10.10 Corollary. *We have*

$$\sum_{n \geq 0} \widehat{Z}_n x^n = \prod_{f \in \mathcal{I}} \prod_{r \geq 1} \left(1 - \frac{t_f x^{\deg(f)}}{q^{r \deg(f)}}\right)^{-1}.$$

Many interesting enumerative results can be obtained from Theorem 1.10.4 and Corollary 1.10.10. We give a couple here and some more in the Exercises (193–195). Let $\beta^*(n,q)$ denote the number of monic irreducible polynomials $f(z) \neq z$ of degree n over $\mathbb{F}_q$. It follows from (1.103) that

$$\beta^*(n,q) = \begin{cases} q-1, & n=1, \\ \frac{1}{n}\sum_{d|n} \mu(d) q^{n/d}, & n>1. \end{cases} \tag{1.112}$$

1.10.11 Corollary. *(a) We have*

$$\frac{1}{1-x} = \prod_{n \geq 1} \prod_{r \geq 1} \left(1 - q^{rn} x^n\right)^{-\beta^*(n,q)}$$

(b) Let $g(n)$ denote the number of nilpotent matrices $M \in \text{Mat}(n,q)$. (Recall that A is nilpotent *if $A^m = 0$ for some $m \geq 1$.) Then $g(n) = q^{n(n-1)}$.*

Proof. (a) Let $\mathcal{I}^* = \mathcal{I} - \{z\}$. Set $t_z = 0$ and $t_f = 1$ for $f \neq z$ in Corollary 1.10.10. Now

$$\widehat{Z}_n(t_z = 0, \ t_f = 1 \text{ if } f \neq 0) = \frac{\gamma(n)}{\gamma(n)} = 1,$$

so we get

$$\frac{1}{1-x} = \prod_{f \in \mathcal{I}^*} \prod_{r \geq 1} \left(1 - \frac{x^{\deg(f)}}{q^{r \deg(f)}} \right)$$

$$= \prod_{n \geq 1} \prod_{r \geq 1} \left(1 - q^{-rn} x^n \right)^{-\beta^*(n,q)}.$$

Since the left-hand side is independent of q, we can substitute $1/q$ for q in the right-hand side without changing its value, and the proof follows. This result can also be proved by taking the logarithm of both sides and using the explicit formula for $\beta^*(n,q)$ given by equation (1.112).

(b) A matrix is nilpotent if and only if all its eigenvalues are 0. Hence,

$$g(n) = \gamma(n) \left. \widehat{Z}(t;q) \right|_{t_z=1, \ t_f=0 \text{ if } f \neq z}.$$

By Corollary 1.10.10 and Proposition 1.8.6(a) there follows

$$\sum_{n \geq 0} g(n) \frac{x^n}{\gamma(n)} = \prod_{r \geq 1} \left(1 - \frac{x}{q^r} \right)^{-1}$$

$$= \sum_{k \geq 0} \frac{q^{-k} x^k}{(1-q^{-1}) \cdots (1-q^{-k})}$$

$$= \sum_{k \geq 0} q^{k(k-1)} \frac{x^k}{\gamma(k)},$$

and the proof follows. (For a more direct proof, see Exercise 1.188.) $\qquad \square$

1.10.2 A q-Analogue of Permutations as Words

We now discuss a second q-analogue of permutations (already discussed briefly after Corollary 1.3.13) and then connect it with one discussed earlier (matrices in $\text{GL}(n,q)$). Rather than regarding permutations of $1,2,\ldots,n$ as bijections

$w : [n] \to [n]$, we may regard them as words $a_1 a_2 \cdots a_n$. Equivalently, we can identify w with the maximal chain (or *(complete) flag*)

$$\emptyset = S_0 \subset S_1 \subset \cdots \subset S_n = [n] \tag{1.113}$$

of subsets of $[n]$, by the rule $\{a_i\} = S_i - S_{i-1}$. For instance, the flag $\emptyset \subset \{2\} \subset \{2,4\} \subset \{1,2,4\} \subset \{1,2,3,4\}$ corresponds to the permutation $w = 2413$. The natural q-analogue of a flag (1.113) is a maximal chain or *(complete) flag* of subspaces

$$\{0\} = V_0 \subset V_1 \subset \cdots \subset V_n = \mathbb{F}_q^n \tag{1.114}$$

of subspaces of $\mathbb{F}_q^n$, so dim $V_i = i$. It is easy to count the number of such flags (as mentioned after Corollary 1.3.13).

1.10.12 Proposition. *The number $f(n,q)$ of complete flags* (1.114) *is given by*

$$f(n,q) = (n)! = (1+q)(1+q+q^2)\cdots(1+q+\cdots+q^{n-1}).$$

Proof. There are $\binom{n}{1} = (n)$ choices for V_1, then $\binom{n-1}{1}$ choices for V_2 (since the quotient space $\mathbb{F}_q^n/V_1$ is an $(n-1)$-dimensional vector space), etc. $\qquad\square$

Comparing Corollary 1.3.13 with Proposition 1.10.12, we see that

$$f(n,q) = \sum_{w \in \mathfrak{S}_n} q^{\mathrm{inv}(w)}.$$

We can ask whether there is a bijective proof of this fact analogous to our proof of Proposition 1.7.3. In other words, letting $\mathcal{F}(n,q)$ denote the set of all flags (1.114), we want to find a map $\varphi : \mathcal{F}(n,q) \to \mathfrak{S}_n$ such that $\#\varphi^{-1}(w) = q^{\mathrm{inv}(w)}$ for all $w \in \mathfrak{S}_n$. Such a map can be defined as follows. Let $F \in \mathcal{F}(n,q)$ be the flag (1.114). It is not hard to see that there is a unique ordered basis $\boldsymbol{v} = \boldsymbol{v}(F) = (v_1, v_2, \ldots, v_n)$ for $\mathbb{F}_q^n$ (where we regard each v_i as a column vector) satisfying the two conditions:

- $V_i = \mathrm{span}\{v_1, \ldots, v_i\}$, $1 \le i \le n$
- There is a unique permutation $\varphi(F) = w \in \mathfrak{S}_n$ for which the matrix $A = [v_1, \ldots, v_n]^t$ satisfies (a) $A_{i,w(i)} = 1$ for $1 \le i \le n$, (b) $A_{i,j} = 0$ if $j > w(i)$, and (c) $A_{j,w(i)} = 0$ if $j > i$. In other words, A can be obtained from the permutation matrix P_w (as defined in Section 1.5) by replacing the entries A_{ij} for $(i,j) \in D_w$ (as defined in Section 1.5) by any elements of $\mathbb{F}_q$. We call A a w-*reduced* matrix.

For instance, suppose that $w = 314652$. Figure 1.5 shows that the possible matrices A have the form

$$A = \begin{bmatrix} * & * & 1 & 0 & 0 & 0 \\ 1 & 0 & 0 & 0 & 0 & 0 \\ 0 & * & 0 & 1 & 0 & 0 \\ 0 & * & 0 & 0 & * & 1 \\ 0 & * & 0 & 0 & 1 & 0 \\ 0 & 1 & 0 & 0 & 0 & 0 \end{bmatrix}.$$

Let Ω_w be the set of flags $F \in \mathcal{F}(n,q)$ for which $\varphi(F) = w$. Thus,

$$\mathcal{F}(n,q) = \bigcup\nolimits_{w \in \mathfrak{S}_n} \cdot \Omega_w. \tag{1.115}$$

Since $\#D_w = \mathrm{inv}(w)$, we have $\#\Omega_w = q^{\mathrm{inv}(w)}$, so we have found the desired combinatorial interpretation of Proposition 1.10.12. The sets Ω_w are known as *Schubert cells*, and equation (1.115) gives the cellular decomposition of the flag variety $\mathcal{F}(n,q)$, completely analogous to the cellular decomposition of the Grassmann variety G_{km} given in the proof of Proposition 1.7.3. The canonical ordered basis $v(F)$ is the "flag analogue" of row-reduced echelon form, which gives a canonical ordered basis for a *subspace* (rather than a flag) of $\mathbb{F}_q^n$.

1.10.3 The Connection Between the Two q-Analogues

The order $\gamma(n,q)$ of $\mathrm{GL}(n,q)$ and the number $f(n,q)$ of complete flags is related by

$$\gamma(n,q) = q^{\binom{n}{2}}(q-1)^n f(n,q).$$

Can we find a simple combinatorial explanation? We would like to find a map $\psi : \mathrm{GL}(n,q) \to \mathcal{F}(n,q)$ satisfying $\#\psi^{-1}(F) = q^{\binom{n}{2}}(q-1)^n$ for all $F \in \mathcal{F}(n,q)$. The definition of ψ is quite simple: if $A = [v_1, \ldots, v_n]^t$, then let $\psi(F)$ be the flag $\{0\} = V_0 \subset V_1 \subset \cdots \subset V_n = \mathbb{F}_q^n$ given by $V_i = \mathrm{span}\{v_1, \ldots, v_i\}$. Given F, there are $q-1$ choices for v_1, then $q^2 - q$ choices for v_2, then $q^3 - q^2$ choices for v_3, and so on, showing that $\#\psi^{-1}(F) = q^{\binom{n}{2}}(q-1)^n$ as desired.

We have constructed maps $\mathrm{GL}(n,q) \overset{\psi}{\twoheadrightarrow} \mathcal{F}(n,q) \overset{\varphi}{\twoheadrightarrow} \mathfrak{S}_n$. Given $w \in \mathfrak{S}_n$, let $\Gamma_w = \psi^{-1}\varphi^{-1}(w)$. Thus,

$$\mathrm{GL}(n,q) = \bigcup\nolimits_{w \in \mathfrak{S}_n} \cdot \Gamma_w, \tag{1.116}$$

the *Bruhat decomposition* of $\mathrm{GL}(n,q)$. (The Bruhat decomposition is usually defined more abstractly and in greater generality than we have done.) It is immediate from the formulas $\#\Omega_w = q^{\mathrm{inv}(w)}$ and $\#\psi^{-1}(F) = q^{\binom{n}{2}}(q-1)^n$ that $\#\Gamma_w = q^{\binom{n}{2}}(q-1)^n q^{\mathrm{inv}(w)}$ and

$$\gamma(n,q) = q^{\binom{n}{2}}(q-1)^n \sum_{w \in \mathfrak{S}_n} q^{\mathrm{inv}(w)}. \tag{1.117}$$

Together with Corollary 1.3.13, equation (1.117) gives a second combinatorial proof of Proposition 1.10.1.

It is not difficult to give a concrete description of the "Bruhat cells" Γ_w. Namely, every element A of Γ_w can be uniquely written in the form $A = LM$, where L is a lower-triangular matrix in $\mathrm{GL}(n,q)$ and M is a w-reduced matrix. We omit the straightforward proof.

1.10.13 Example. (a) Every matrix $A \in \mathrm{GL}(2,q)$ can be uniquely written in one of the two forms

$$
\begin{bmatrix} a & 0 \\ b & c \end{bmatrix} \begin{bmatrix} 1 & 0 \\ 0 & 1 \end{bmatrix} = \begin{bmatrix} a & 0 \\ b & c \end{bmatrix}
$$

$$
\begin{bmatrix} a & 0 \\ b & c \end{bmatrix} \begin{bmatrix} \alpha & 1 \\ 1 & 0 \end{bmatrix} = \begin{bmatrix} \alpha a & a \\ \alpha b + c & b \end{bmatrix},
$$

where $b, \alpha \in \mathbb{F}_q$, $a, c \in \mathbb{F}_q^* = \mathbb{F}_q - \{0\}$.

(b) The cell Γ_{3142} consists of all matrices of the form

$$
\begin{bmatrix} a & 0 & 0 & 0 \\ b & c & 0 & 0 \\ d & e & f & 0 \\ g & h & i & j \end{bmatrix} \begin{bmatrix} \alpha & \beta & 1 & 0 \\ 1 & 0 & 0 & 0 \\ 0 & \gamma & 0 & 1 \\ 0 & 1 & 0 & 0 \end{bmatrix} = \begin{bmatrix} \alpha a & \beta a & a & 0 \\ \alpha b + c & \beta b & b & 0 \\ \alpha d + e & \beta d + \gamma f & d & f \\ \alpha g + h & \beta g + \gamma i + j & g & i \end{bmatrix},
$$

where $b, d, e, g, h, i, \alpha, \beta, \gamma \in \mathbb{F}_q$ and $a, c, f, j \in \mathbb{F}_q^*$.

The Bruhat decomposition (1.116) can be a useful tool for counting certain subsets S of $\mathrm{GL}(n,q)$, by computing each $\#(S \cap \Gamma_w)$ and summing over all $w \in \mathfrak{S}_n$. Proposition 1.10.15 illustrates this technique. First we need a simple enumerative lemma.

1.10.14 Lemma. *Fix q, and for any integer $n \geq 0$, let*

$$
a_n = \#\{(\alpha_1, \ldots, \alpha_n) \in (\mathbb{F}_q^*)^n : \sum \alpha_i = 0\}.
$$

Then $a_0 = 1$, and $a_n = \frac{1}{q}((q-1)^n + (q-1)(-1)^n)$ for $n > 0$.

Proof. Define $b_n = \sum_{k=0}^n \binom{n}{k} a_k$. Since every sequence $(\alpha_1, \ldots, \alpha_n) \in \mathbb{F}_q^n$ satisfying $\sum \alpha_i = 0$ can be obtained by first specifying $n - k$ terms to be 0 in $\binom{n}{k}$ ways and then specifying the remaining k terms in a_k ways, we have

$$
b_n = \begin{cases} 1, & n = 0 \\ q^{n-1}, & n \geq 1. \end{cases}
$$

There are many ways to see (e.g., equations (2.9) and (2.10)) that we can invert this relationship between the a_n's and b_n's to obtain

$$
a_n = \sum_{k=0}^n (-1)^{n-k} \binom{n}{k} b_k
$$

$$
= \frac{1}{q} \left[\sum_{k=0}^n (-1)^{n-k} \binom{n}{k} q^k + (q-1)(-1)^n \right]
$$

$$
= \frac{1}{q}((q-1)^n + (q-1)(-1)^n). \qquad \square
$$

1.10.15 Proposition. *Let* $GL_0(n,q) = \{A \in GL(n,q) : tr(A) = 0\}$, *where* $tr(A)$
denotes the trace of A, and set $\gamma_0(n,q) = \#GL_0(n,q)$. *Then*

$$\gamma_0(n,q) = \frac{1}{q}\left(\gamma(n,q) + (-1)^n(q-1)q^{\binom{n}{2}}\right).$$

Proof. Let id denote the identity permutation $1,2,\ldots,n$, so $inv(id) = 0$. We will
show that

$$\#(GL_0(n,q) \cap \Gamma_w) = \frac{1}{q}\#\Gamma_w, \ w \neq id,$$

$$\#(GL_0(n,q) \cap \Gamma_{id}) = \frac{1}{q}\left(\#\Gamma_{id} + (-1)^n(q-1)q^{\binom{n}{2}}\right),$$

from which the proof follows since $\sum_w \#\Gamma_w = \gamma(n,q)$.

Suppose that $w \neq id$. Let r be the least integer for which there exists an element
$(r,s) \in D_w$, where D_w denotes the diagram of w. It is easy to see that then $(r,r) \in$
D_w. Consider a general element $A = LM$ of Γ_w, so the entries L_{ij} satisfy $L_{ii} \in \mathbb{F}_q^*$,
$L_{ij} \in \mathbb{F}_q$ if $i > j$, and $L_{ij} = 0$ if $i < j$. Similarly, $M_{i,w(i)} = 1$, $M_{ij} \in \mathbb{F}_q$ if $(i,j) \in D_w$,
and $M_{ij} = 0$ otherwise. Thus, A_{rr} will be a polynomial in the L_{ij}'s and M_{ij}'s with
a term $L_{rr}M_{rr}$. (In fact, it is not hard to see that $A_{rr} = L_{rr}M_{rr}$, though we don't
need this stronger fact here.) There is no other occurrence of M_{rr} in a main diagonal
term of A. If we choose all the free entries of L and M except M_{rr} (subject to the
preceding conditions), then we can solve uniquely for M_{rr} (since its coefficient is
$L_{rr} \neq 0$) so that $tr(A) = 0$. Thus, rather than q choices for M_{rr} for *any* $A \in \Gamma_w$,
there is only one choice, so $\#(GL_0(n,q) \cap \Gamma_w) = \frac{1}{q}\#\Gamma_w$ as claimed.

Example. Consider the cell Γ_{3142} of Example 1.10.13(b). We have that
$\#(GL_0(4,q) \cap \Gamma_{3142})$ is the number of 13-tuples $(a,\ldots,j,\alpha,\beta,\gamma)$ such that
$b,d,e,g,h,i,\alpha,\beta,\gamma \in \mathbb{F}_q$ and $a,c,f,j \in \mathbb{F}_q^*$, satisfying

$$\alpha a + \beta b + d + i = 0. \tag{1.118}$$

We have $r = 1$, so we can specify all 13 variables except α in $q^8(q-1)^4$ ways,
and then solve equation (1.118) uniquely for α. Hence $\#(GL_0(4,q) \cap \Gamma_{3142}) =$
$q^8(q-1)^4 = \frac{1}{q}\#\Gamma_{3142}$.

Now let $w = id$, so $A = L$. Hence, we can choose the elements of A below the
diagonal in $q^{\binom{n}{2}}$ ways, while the number of choices for the diagonal elements is
the number a_n of Lemma 1.10.14. Hence from Lemma 1.10.14 we get

$$\#\Gamma_{id} = q^{\binom{n}{2}}a_n$$

$$= q^{\binom{n}{2}}\frac{1}{q}((q-1)^n + (q-1)(-1)^n),$$

and the proof follows. $\square$

Notes

It is not our intention here to trace the development of the basic ideas and results of enumerative combinatorics. It is interesting to note, however, that according to Heath [1.40, p. 319], a result of Xenocrates of Chalcedon (396–314 BCE) possibly "represents the first attempt on record to solve a difficult problem in permutations and combinations." (See also Biggs [1.8, p. 113].) Moveover, Exercise 1.203 shows that Hipparchus (c. 190–after 127 BCE) certainly was successful in solving such a problem. We should also point out that the identity of Example 1.1.17 is perhaps the oldest of all binomial coefficient identities. It is called by such names as the *Chu-Vandermonde identity* or *Vandermonde's theorem*, after Chu Shih-Chieh (Zhū Shìjié in Pinyin and 朱世杰 in simplified Chinese characters) (c. 1260–c. 1320) and Alexandre-Théophile Vandermonde (1735–1796).

Two valuable sources for the history of enumeration are Biggs [1.8] and Stein [1.71]. Knuth [1.49, §7.2.1.7] has written a fascinating history of the generation of combinatorial objects (such as all permutations of a finite set). Later we will give mostly references and comments not readily available in [1.8] and [1.71].

For further information on formal power series from a combinatorial viewpoint, see, for example Niven [1.60] and Tutte [1.73]. A rigorous algebraic approach appears in Bourbaki [1.12, Ch. IV, §5], and a further paper of interest is Bender [1.5]. Wilf [1.76] is a nice introduction to generating functions at the undergraduate level.

To illustrate the misconceptions (or at least infelicitous language) that can arise in dealing with formal power series, we offer the following quotations (anonymously) from the literature.

Since the sum of an infinite series is really not used, our viewpoint can be either rigorous or formal.

(1.3) demonstrates the futility of seeking a generating function, even an exponential one, for $IU(n)$; for it is so big that

$$F(z) = \sum_n IU(n)z^n/n!$$

fails to converge if $z \neq 0$. Any closed equation for F therefore has no solutions, and when manipulated by Taylor expansion, binomial theorem, etc., is bound to produce a heap of eggs (single -0- or double -∞-yolked). Try finding a generating function for 2^{2^n}.

Sometimes we have difficulties with convergence for some functions whose coefficients a_n grow too rapidly; then instead of the regular generating function we study the *exponential* generating function.

An analyst might at least raise the point that the only general techniques available for estimating the rate of growth of the coefficients of a power series require convergence (so that e.g. the apparatus of complex variable theory is available). There are, however, methods for estimating the coefficients of a divergent power series;

see Bender [1.6, §5] and Odlyzko [1.61, §7]. For further information on estimating coefficients of power series, see for instance Flajolet and Sedgewick [1.22], Odlyzko [1.61] and Pemantle and Wilson [1.63]. In particular, the asymptotic formula (1.12), due to Moser and Wyman [1.57], appears in [1.61, (8.49)].

The technique of representing combinatorial objects such as permutations by "models" such as words and trees has been extensively developed. A pioneering work in this area in the monograph [1.26] of Foata and Schützenberger. In particular, the "transformation fondamentale" on pp. 13–15 of this reference is essentially our map $w \mapsto \hat{w}$ of Proposition 1.3.1. Note, however, that this bijection was earlier used by Alfréd Rényi [1.67, §4] to prove Proposition 1.3.1. The history of the generating function for the cycle indicator of $\mathfrak{S}_n$ (Theorem 1.3.3) is discussed in the first paragraph of the Notes to Chapter 5. The generating function for permutations by number of inversions (Corollary 1.3.13) appears in Rodrigues [1.68] and Netto [1.58, p. 73]. The generalization to multisets (Proposition 1.7.1) is due to MacMahon [1.54, §1]. It was rediscovered by Carlitz [1.16]. The second proof given here was suggested by A. Björner and M. L. Wachs [1.9, §3]. The cellular decomposition of the Grassmann variety (the basis for our second proof of Proposition 1.7.3) is discussed by S. L. Kleiman and D. Laksov [1.46]. For some further historical information on the results of Rodriques and MacMahon, see the book review by Johnson [1.44]. The major index of a permutation was first considered by MacMahon [1.53], who used the term "greater index." The terminology "major index" was introduced by Foata [1.25] in honor of MacMahon, who was a major in the British army. MacMahon's main result on the major index is the equidistribution of $\text{inv}(w)$ and $\text{maj}(w)$ for $w \in \mathfrak{S}_n$. He gives the generating function (1.42) for $\text{maj}(w)$ in [1.53, §6] (where in fact w is a permutation of a multiset), and in [1.54] he shows the equidistribution with $\text{inv}(w)$. The bijective proof we have given here (proof of Proposition 1.4.6) appears in seminal papers [1.23][1.24] of Foata, which helped lay the groundwork for the modern theory of bijective proofs. The strengthening of Foata's result given by Corollary 1.4.9 is due to Foata and Schützenberger [1.28].

The investigation of the descent set and number of descents of a permutation (of a set or multiset) was begun by MacMahon [1.52]. MacMahon apparently did not realize that the number of permutations of $[n]$ with k descents is an Eulerian number. The first written statement connecting Eulerian numbers with descents seems to have been by Carlitz and Riordan [1.17] in 1953. The fundamental Lemma 1.4.11 is due to MacMahon [1.53, p. 287]. Eulerian numbers occur in some unexpected contexts, such as cube slicing (Exercise 1.51), juggling sequences [1.15], and the statistics of carrying in the standard algorithm for adding integers (Exercise 1.52). MacMahon [1.55, vol. 1, p. 186] was also the first person to consider the excedance of a permutation (though he did not give it a name) and showed the equidistribution of the number of descents with the number of excedances (Proposition 1.4.3).

We will not attempt to survey the vast subject of representing permutations by other combinatorial objects, but let us mention that an important generalization of the representation of permutations by plane trees is the paper of Cori [1.19]. The first result on pattern avoidance seems to be the proof of MacMahon [1.55, §97] that the number of 321-avoiding permutations $w \in \mathfrak{S}_n$ is the Catalan number C_n. MacMahon states his result not in terms of pattern avoidance, but rather in terms of permutations that are a union of two decreasing sequences. MacMahon's result was rediscovered by J. M. Hammersley [1.38], who stated it without proof. Proofs were later given by D. E. Knuth [1.48, §5.1.4] and D. Rotem [1.69]. For further information on 321-avoiding and 132-avoiding permutations, see Exercise 6.19(ee,ff) and the survey of Claesson and Kitaev [1.18]. For further information on pattern avoidance in general, see Exercises 57–59, as well as books by M. Bóna [1.11, Chs. 4–5] and by S. Heubach and T. Mansour [1.42].

Alternating permutations were first considered by D. André [1.1], who obtained the basic and elegant Proposition 1.6.1. (Note however that Ginsburg [1.34] asserts without giving a reference that Binet was aware before André that the coefficients of sec x count alternating permutations.) A combinatorial proof of Proposition 1.6.2 on flip equivalence is due to R. Donaghey [1.20]. Further information on the connection between alternating permutations and increasing trees appears in a paper of Kuznetsov, Pak, and Postnikov [1.51].

The cd-index $\Phi_n(c, d)$ was first considered by Foata and Schützenberger [1.27], who defined it in terms of certain permutations they called *André permutations*. Their term for the cd-index was "non-commutative André polynomial." Foata and Strehl [1.29][1.30] further developed the theory of André polynomials, André permutations, and their connection with permutation statistics. Meanwhile Jonathan Fine [1.21] defined a noncommutative polynomial $\Phi_P(c, d)$ associated with certain partially ordered sets (posets) P. This polynomial was first systematically investigated by Bayer and Klapper [1.4] and later by Stanley [1.70], who extended the class of posets P which possessed a cd-index $\Phi_P(c, d)$ to Eulerian posets. The basic theory of the cd-index of an Eulerian poset is covered in Section 3.17. M. Purtill [1.65, Thm. 6.1] showed that the cd-index Φ_n that we have defined is just the cd-index Φ_{B_n} (in the sense of Fine and Bayer-Klapper) of the boolean algebra B_n (the poset of all subsets of $[n]$, ordered by inclusion). The approach to the cd-index Φ_n given here, based on min-max trees, is due to G. Hetyei and E. Reiner [1.41]. For some additional properties of min-max trees, see Bóna [1.10]. Corollary 1.6.5 was first proved by Niven [1.59] by a complicated induction. De Bruijn [1.13] gave a simpler proof and extended it to Proposition 1.6.4. A further proof is due to Viennot [1.75]. The proof we have given based on the cd-index appears in Stanley [1.70, pp. 495–496]. For a generalization see Exericse 3.55.

The theory of partitions of an integer originated in the work of Euler, if we ignore some unpublished work of Leibniz that was either trival or wrong (see Knobloch [1.47]). An excellent introduction to this subject is the text by Andrews [1.2]. For a masterful survey of bijective proofs of partition identities, see Pak [1.62]. The latter two references provide historical information on the results appearing in

Section 1.8. The asymptotic formula (1.92) is due to Hardy and Ramanujan [1.39], and the asymptotic series mentioned after equation (1.92) is due to Rademacher [1.66]. More recently J. H. Bruinier and K. Ono,

⟨http://www.aimath.org/news/partition/brunier-ono⟩,

have given an explicit finite formula for $p(n)$. For an exposition of partition asymptotics, see Andrews [1.2, Ch. 5].

The idea of the Twelvefold Way (Section 1.9) is due to G.-C. Rota (in a series of lectures), while the terminology "Twelvefold Way" was suggested by Joel Spencer. An extension of the Twelvefold Way to a "Thirtyfold Way" (and suggestion of even more entries) is due to R. Proctor [1.64]. An interesting popular account of Bell numbers appears in an article by M. Gardner [1.33]. In particular, pictorial representations of the 52 partitions of a 5-element set are used as "chapter headings" for all but the first and last chapters of certain editions of *The Tale of Genji* by Lady Murasaki (c. 978–c. 1031 CE). A standard reference for the calculus of finite difference is the text by C. Jordan [1.45].

The cycle indicator $Z_n(t;q)$ of $GL(n,q)$ was first explicitly defined by Kung [1.50]. The underlying algebra was known much earlier; for instance, according to Green [1.35, p. 407] the basic Theorem 1.10.7 is due to P. Hall [1.36] and is a simple consequence of earlier work of Frobenius (see Jacobson [1.43, Thm. 19, p. 111]). Green himself sketches a proof on page 409, op. cit. Further work on the cycle indicator of $GL(n,q)$ was done by Stong [1.72] and Fulman [1.31]. A nice survey of enumeration of matrices over $\mathbb{F}_q$ was given by Morrison [1.56], whom we have followed for Exercises 1.193–1.195. Our proof of Lemma 1.10.9 is equivalent to one given by P. Hall [1.37].

The cellular decomposition (1.115) of the flag variety $\mathcal{F}(n,q)$ and the Bruhat decomposition (1.116) of $GL(n,K)$ (for any field K) are standard topics in Lie theory. See for instance Fulton and Harris [1.32, §23.4]. A complicated recursive description of the number of matrices in $GL(n,q)$ with trace 0 and a given rank r was given by Buckheister [1.14]. Bender [1.7] used this recurrence to give a closed-form formula. The proof we have given of the case $k = 0$ (Proposition 1.10.15) based on Bruhat decomposition is new. For a generalization see Exercise 1.196.

Bibliography

[1] D. André, Développement de sec$x$ and tg$x$, *C. R. Math. Acad. Sci. Paris* **88** (1879), 965–979.

[2] G. E. Andrews, *The Theory of Partitions*, Addison-Wesley, Reading, Mass., 1976.

[3] G. E. Andrews, ed., *Percy Alexander MacMahon, Collected Papers*, vol. 1, M.I.T. Press, Cambridge, Mass. 1978.

[4] M. M. Bayer and A. Klapper, A new index for polytopes, *Discrete Comput. Geom.* **6** (1991), 33–47.

[5] E. A. Bender, A lifting theorem for formal power series, *Proc. Amer. Math. Soc.* **42** (1974), 16–22.

[6] E. A. Bender, Asymptotic methods in enumeration, *SIAM Review* **16** (1974), 485–515. Errata, *SIAM Review* **18** (1976), 292.

[7] E. A. Bender, On Buckheister's enumeration of $n \times n$ matrices, *J. Combinatorial Theory, Ser. A* **17** (1974), 273–274.

[8] N. L. Biggs, The roots of combinatorics, *Historia Math.* **6** (1979), 109–136.

[9] A. Björner and M. L. Wachs, q-Hook length formulas for forests, *J. Combinatorial Theory Ser. A* **52** (1989), 165–187.

[10] M. Bóna, A combinatorial proof of a result of Hetyei and Reiner on Foata-Strehl type permutation trees, *Ann. Combinatorics* **1** (1997), 119–122.

[11] M. Bóna, *Combinatorics of Permutations*, Chapman and Hall/CRC, Boca Raton, 2004.

[12] N. Bourbaki, *Éléments de Mathématique, Livre II, Algèbre*, Ch. 4–5, 2e ed., Hermann, Paris, 1959.

[13] N. G. de Bruijn, Permutations with given ups and downs, *Nieuw Arch. Wisk.* **18** (1970), 61–65.

[14] P. G. Buckheister, The number of $n \times n$ matrices of rank r and trace α over a finite field, *Duke Math. J.* **39** (1972), 695–699.

[15] J. Buhler, D. Eisenbud, R. Graham, and C. Wright, Juggling drops and descents, *Amer. Math. Monthly* **101** (1994), 507–519.

[16] L. Carlitz, Sequences and inversions, *Duke Math. J.* **37** (1970), 193–198.

[17] L. Carlitz and J. Riordan, Congruences for Eulerian numbers, *Duke Math. J.* **20** (1953), 339–344.

[18] A. Claesson and S. Kitaev, Classification of bijections between 321- and 132-avoiding permutations, *Sém. Lothar. Combinatoire* **60** (2008), B60d.

[19] R. Cori, Une code pour les graphes planaires et ses applications, *Astérisque*, no. 27, Société Mathématique de France, Paris, 1975.

[20] R. Donaghey, Alternating permutations and binary increasing trees, *J. Combinatorial Theory Ser. A* **18** (1975), 141–148.

[21] J. Fine, Morse theory for convex polytopes, unpublished manuscript, 1985.

[22] P. Flajolet and R. Sedgewick, *Analytic Combinatorics*, Cambridge University Press, Cambridge, 2009.

[23] D. Foata, Sur un énoncé de MacMahon, *C. R. Acad. Sci. Paris* **258** (1964), 1672–1675.

[24] D. Foata, On the Netto inversion number of a sequence, *Proc. Amer. Math. Soc.* **19** (1968), 236–240.

[25] D. Foata, Distributions eulériennes et mahoniennes sur le groupe des permutations, in *Higher Combinatorics (Proc. NATO Advanced Study Inst., Berlin, 1976)*, Reidel, Dordrecht-Boston, Mass., 1977, pp. 27–49.

[26] D. Foata and M.-P. Schützenberger, Théorie géométrique des polynômes Eulériens, *Lecture Notes in Math.*, no. 138, Springer, Berlin, 1970.

[27] D. Foata and M.-P. Schützenberger, Nombres d'Euler et permutations alternantes, in *A Survey of Combinatorial Theory* (J. N. Srivistava et al., eds.), North-Holland, Amsterdam, 1973, pp. 173–187.

[28] D. Foata and M.-P. Schützenberger, Major index and inversion number of permutations, *Math. Machr.* **83** (1978), 143–159.

[29] D. Foata and V. Strehl, Euler numbers and variations of permutations, in *Colloquio Internazionale sulle Teorie Combinatorie (Roma, 1973)*, Tomo I, Atti dei Convegni Lincei, No. 17, Accad. Naz. Lincei, Rome, 1976, pp. 119–131.

[30] D. Foata and V. Strehl, Rearrangements of the symmetric group and enumerative properties of the tangent and secant numbers, *Math. Z.* **137** (1974), 257–264.

[31] J. Fulman, Random matrix theory over finite fields, *Bull. Amer. Math. Soc. (N.S.)* **39** (2002), 51–85.

[32] W. Fulton and J. Harris, *Representation Theory*, Springer-Verlag, New York, 1991.

[33] M. Gardner, Mathematical games, *Scientific American* **238** (May 1978), 24–30; reprinted (with an addendum) in *Fractal Music, Hypercards, and More*, Freeman, New York, 1992, pp. 24–38.

[34] J. Ginsburg, Stirling numbers, *Encyclopedia Britannica*, 1965.

[35] J. A. Green, The characters of the finite general linear groups, *Trans. Amer. Math. Soc.* **80** (1955), 402–447.

[36] P. Hall, Abelian p-groups and related modules, unpublished manuscript.

[37] P. Hall, A partition formula connected with Abelian groups, *Comm. Math. Helv.* **11** (1938/39), 126–129.

[38] J. M. Hammersley, A few seedlings of research, in *Proc. Sixth Berkeley Symposium on Mathematical Statistics and Probability (Berkeley, 1970/1971), Vol. 1: Theory of statistics*, Univ. California Press, Berkeley, Calif., 1972, pp. 345–394.

[39] G. H. Hardy and S. Ramanujan, Asymptotic formulae in combinatory analysis, *Proc. London Math. Soc. (2)* **17** (1918), 75–117. Also in *Collected Papers of S. Ramanunan*, Cambridge University Press, London and New York, 1927; reprinted by Chelsea, New York, 1962.

[40] T. L. Heath, *A History of Greek Mathematics*, vol. 1, Dover, New York, 1981.

[41] G. Hetyei and E. Reiner, Permutations trees and variation statistics, *European J. Combinatorics* **19** (1998), 847–866.

[42] S. Heubach and T. Mansour, *Combinatorics and Compositions on Words*, Chapman and Hall/CRC, Boca Raton, 2010.

[43] N. Jacobson, *Lectures in Abstract Algebra*, vol. II—Linear Algebra, D. van Nostrand Co., Princeton, N.J., 1953.

[44] W. P. Johnson, Review of *Mathematics and Social Utopias in France: Olinde Rodrigues and His Times*, *Amer. Math. Monthly* **114** (2007), 752–758.

[45] C. Jordan, *Calculus of Finite Differences*, Chelsea, New York, 1965.

[46] S. L. Kleiman and D. Laksov, Schubert calculus, *Amer. Math. Monthly* **79** (1972), 1061–1082.

[47] E. Knobloch, Leibniz on combinatorics, *Historia Math.* **1** (1974), 409–430.

[48] D. E. Knuth, *The Art of Computer Programming*, vol. 3, *Sorting and Searching*. Addison-Wesley, Reading, Mass., 1973; 2nd ed., 1998.

[49] D. E. Knuth, *The Art of Computer Programming*, vol. 4, *Fascicle 4*, Addison-Wesley, Upper Saddle River, N.J., 2006.

[50] J. P. S. Kung, The cycle structure of a linear transformation over a finite field, *Linear Algebra Appl.* **36** (1981), 141–155.

[51] A. G. Kuznetsov, I. M. Pak, and A. E. Postnikov, Increasing trees and alternating permutations, *Russian Math. Surveys* **49**:6 (1994), 79–114; translated from *Uspekhi Mat. Nauk* **49**:6 (1994), 79–110.

[52] P. A. MacMahon, Second memoir on the compositions of integers, *Phil. Trans.* **207** (1908), 65–134; reprinted in [1.3], pp. 687–756.

[53] P. A. MacMahon, The indices of permutations and the derivation therefrom of functions of a single variable associated with the permutations of any assemblage of objects, *Amer. J. Math.* **35** (1913), 281–322; reprinted in [1.3], pp. 508–549.

[54] P. A. MacMahon, Two applications of general theorems in combinatory analysis: (1) to the theory of inversions of permutations; (2) to the ascertainment of the numbers of terms in the development of a determinant which has amongst its elements an arbitrary number of zeros, *Proc. London Math. Soc. (2)* **15** (1916), 314–321; reprinted in [1.3], pp. 556–563.

[55] P. A. MacMahon, *Combinatory Analysis*, vols.1 and 2, Cambridge University Press, 1916; reprinted by Chelsea, New York, 1960, and by Dover, New York, 2004.

[56] K. E. Morrison, Integer sequences and matrices over finite fields, *J. Integer Sequences* (electronic) **9** (2006), Article 06.2.1.

[57] L. Moser and M. Wyman, On the solutions of $x^d = 1$ in symmetric groups, *Canad. J. Math.* **7** (1955), 159–168.

[58] E. Netto, *Lehrbuch der Combinatorik*, Teubner, Leipzig, 1900.

[59] I. Niven, A combinatorial problem of finite sequences, *Nieuw Arch. Wisk.* **16** (1968), 116–123.

[60] I. Niven, Formal power series, *Amer. Math. Monthly* **76** (1969), 871–889.

[61] A. Odlyzko, Asymptotic enumeration methods, in *Handbook of Combinatorics*, vol. 2 (R. L. Graham, M. Groetschel, and L. Lovász, eds.), Elsevier, Amsterdam, 1995, pp. 1063–1229.

[62] I. Pak, Partition bijections. A survey, *Ramanujan J.* **12** (2006), 5–76.

[63] R. Pemantle and M. Wilson, Twenty combinatorial examples of asymptotics derived from multivariate generating functions, *SIAM Review* **50** (2008), 199–272.

[64] R. Proctor, Let's expand Rota's Twelvefold Way for counting partitions!, preprint; arXiv:math/0606404.

[65] M. Purtill, André permutations, lexicographic shellability and the *cd*-index of a convex polytope, *Trans. Amer. Math. Soc.* **338** (1993), 77–104.

[66] H. Rademacher, On the partition function $p(n)$, *Proc. London Math. Soc. (2)* **43** (1937), 241–254.

[67] A. Rényi, Théorie des éléments saillants d'une suite d'observations, in *Colloq. Combinatorial Methods in Probability Theory, August 1–10, 1962*, Matematisk Institut, Aarhus Universitet, 1962, pp. 104–115.

[68] O. Rodrigues, Note sur les inversions, ou dérangements produits dans les permutations, *J. Math. Pures Appl.* **4** (1839), 236–240.

[69] D. Rotem, On a correspondence between binary trees and a certain type of permutation, *Inf. Proc. Letters* **4** (1975/76), 58–61.

[70] R. P. Stanley, Flag f-vectors and the *cd*-index, *Math. Z.* **216** (1994), 483–499.

[71] P. R. Stein, A brief history of enumeration, in *Science and Computers, a volume dedicated to Nicolas Metropolis* (G.-C. Rota, ed.), Academic Press, New York, 1986, pp. 169–206.

[72] R. A. Stong, Some asymptotic results on finite vector spaces, *Advances in Applied Math.* **9** (1988), 167–199.

[73] W. T. Tutte, On elementary calculus and the Good formula, *J. Combinatorial Theory* **18** (1975), 97–137.

[74] L. G. Valiant, The complexity of enumeration and reliability problems, *SIAM J. Comput.* **8** (1979), 410–421.

[75] G. Viennot, Permutations ayant une forme donnée, *Discrete Math.* **26** (1979), 279–284.

[76] H. S. Wilf, *Generatingfunctionology*, 3rd ed., A K Peters, Wellesley, Mass., 2006.

A Note About the Exercises

Each exercise is given a difficulty rating, as follows.

1. Routine, straightforward
2. Somewhat difficult or tricky
3. Difficult
4. Horrendously difficult
5. Unsolved

Further gradations are indicated by + and −. Thus, [1−] denotes an utterly trivial problem, and [5−] denotes an unsolved problem that has received little attention and may not be too difficult. A rating of [2+] denotes about the hardest problem that could be reasonably assigned to a class of graduate students. A few students may be capable of solving a [3−] problem, whereas almost none could solve a [3] in a reasonable period of time. Of course the ratings are subjective, and there is always the possibility of an overlooked simple proof that would lower the rating. Some problems (seemingly) require results or techniques from other branches of mathematics that are not usually associated with combinatorics. Here the rating is less meaningful – it is based on an assessment of how likely the reader is to discover for herself or himself the relevance of these outside techniques and results. An asterisk after the difficulty rating indicates that no solution is provided.

Exercises for Chapter 1

1. [1−] Let S and T be disjoint one-element sets. Find the number of elements of their union $S \cup T$.

2. [1+] We continue with a dozen simple numerical problems. Find as simple a solution as possible.

a. How many subsets of the set $[10] = \{1,2,\ldots,10\}$ contain at least one odd integer?

b. In how many ways can seven people be seated in a circle if two arrangements are considered the same whenever each person has the same neighbors (not necessarily on the same side)?

c. How many permutations $w : [6] \to [6]$ satisfy $w(1) \neq 2$?

d. How many permutations of $[6]$ have exactly two cycles (i.e., find $c(6,2)$)?

e. How many partitions of $[6]$ have exactly three blocks (i.e., find $S(6,3)$)?

f. There are four men and six women. Each man marries one of the women. In how many ways can this be done?

g. Ten people split up into five groups of two each. In how many ways can this be done?

h. How many compositions of 19 use only the parts 2 and 3?

i. In how many different ways can the letters of the word MISSISSIPPI be arranged if the four S's cannot appear consecutively?

j. How many sequences $(a_1,a_2,\ldots,a_{12})$ are there consisting of four 0's and eight 1's, if no two consecutive terms are both 0's?

k. A box is filled with three blue socks, three red socks, and four chartreuse socks. Eight socks are pulled out, one at a time. In how many ways can this be done? (Socks of the same color are indistinguishable.)

l. How many functions $f : [5] \to [5]$ are at most two-to-one, that is, $\#f^{-1}(n) \le 2$ for all $n \in [5]$?

3. Give *combinatorial* proofs of the following identities, where x,y,n,a,b are nonnegative integers.

a. [2–] $\displaystyle\sum_{k=0}^{n} \binom{x+k}{k} = \binom{x+n+1}{n}$

b. [1+] $\displaystyle\sum_{k=0}^{n} k\binom{n}{k} = n2^{n-1}$

c. [3] $\displaystyle\sum_{k=0}^{n} \binom{2k}{k}\binom{2(n-k)}{n-k} = 4^n$

d. [3–] $\displaystyle\sum_{k=0}^{m} \binom{x+y+k}{k}\binom{y}{a-k}\binom{x}{b-k} = \binom{x+a}{b}\binom{y+b}{a}$, where $m = \min(a,b)$

e. [1] $\displaystyle 2\binom{2n-1}{n} = \binom{2n}{n}$

f. [2–] $\displaystyle\sum_{k=0}^{n}(-1)^k\binom{n}{k} = 0, n \ge 1$

g. [2+] $\displaystyle\sum_{k=0}^{n} \binom{n}{k}^2 x^k = \sum_{j=0}^{n}\binom{n}{j}\binom{2n-j}{n}(x-1)^j$

h. [3–] $\displaystyle\sum_{i+j+k=n} \binom{i+j}{i}\binom{j+k}{j}\binom{k+i}{k} = \sum_{r=0}^{n}\binom{2r}{r}$, where $i,j,k \in \mathbb{N}$

4. [2]* Fix $j,k \in \mathbb{Z}$. Show that

$$\sum_{n\ge 0} \frac{(2n-j-k)!x^n}{(n-j)!(n-k)!(n-j-k)!n!} = \left[\sum_{n\ge 0}\frac{x^n}{n!(n-j)!}\right]\left[\sum_{n\ge 0}\frac{x^n}{n!(n-k)!}\right].$$

Any term with $(-r)!$ in the denominator, where $r > 0$, is set equal to 0.

5. [2]* Show that

$$\sum_{n_1,\dots,n_k \geq 0} \min(n_1,\dots,n_k) x_1^{n_1} \cdots x_k^{n_k} = \frac{x_1 \cdots x_k}{(1-x_1)\cdots(1-x_k)(1-x_1 x_2 \cdots x_k)}.$$

6. [3–]* For $n \in \mathbb{Z}$ let

$$J_n(2x) = \sum_{k \in \mathbb{Z}} \frac{(-1)^k x^{n+2k}}{k!(n+k)!},$$

where we set $1/j! = 0$ for $j < 0$. Show that

$$e^x = \sum_{n \geq 0} L_n J_n(2x),$$

where $L_0 = 1, L_1 = 1, L_2 = 3, L_{n+1} = L_n + L_{n-1}$ for $n \geq 2$. (The numbers L_n for $n \geq 1$ are *Lucas numbers*.)

7. [2]* Let

$$e^{x+\frac{x^2}{2}} = \sum_{n \geq 0} f(n) \frac{x^n}{n!}.$$

Find a simple expression for $\sum_{i=0}^{n} (-1)^{n-i} \binom{n}{i} f(i)$. (See equation (1.13).)

8. **a.** [2–] Show that

$$\frac{1}{\sqrt{1-4x}} = \sum_{n \geq 0} \binom{2n}{n} x^n.$$

b. [2–] Find $\sum_{n \geq 0} \binom{2n-1}{n} x^n$.

9. Let $f(m,n)$ be the number of paths from $(0,0)$ to $(m,n) \in \mathbb{N} \times \mathbb{N}$, where each step is of the form $(1,0)$, $(0,1)$, or $(1,1)$.
   a. [1+]* Show that $\sum_{m \geq 0} \sum_{n \geq 0} f(m,n) x^m y^n = (1-x-y-xy)^{-1}$.
   b. [3–] Find a simple explicit expression for $\sum_{n \geq 0} f(n,n) x^n$.

10. [2+] Let $f(n,r,s)$ denote the number of subsets S of $[2n]$ consisting of r odd and s even integers, with no two elements of S differing by 1. Give a bijective proof that $f(n,r,s) = \binom{n-r}{s}\binom{n-s}{r}$.

11. **a.** [2+] Let $m,n \in \mathbb{N}$. Interpret the integral

$$B(m+1,n+1) = \int_0^1 u^m (1-u)^n \, du,$$

as a probability and evaluate it by combinatorial reasoning.
   b. [3+] Let $n \in \mathbb{P}$ and $r,s,t \in \mathbb{N}$. Let x, y_k, z_k and a_{ij} be indeterminates, with $1 \leq k \leq n$ and $1 \leq i < j \leq n$. Let M be the multiset with n occurrences of x, r occurrences of each y_k, s occurrences of each z_k, and $2t$ occurrences of each a_{ij}. Let $f(n,r,s,t)$ be the number of permutations w of M such that (i) all y_k's appear before the kth x (reading the x's from left-to-right in w), (ii) all z_k's appear after the kth x, and (iii) all a_{ij}'s appear between the ith x and jth x. Show that

$$f(n,r,s,t) = \frac{[(r+s+1)n + tn(n-1)]!}{n! \, r!^n \, s!^n \, t!^n \, (2t)!^{\binom{n}{2}}}$$

$$\cdot \prod_{j=1}^{n} \frac{(r+(j-1)t)!(s+(j-1)t)!(jt)!}{(r+s+1+(n+j-2)t)!}. \tag{1.119}$$

   c. [3–] Consider the following chess position.

R. Stanley
Suomen Tehtäväniekat, 2005

Black is to make 14 consecutive moves, after which White checkmates Black in one move. Black may not move into check, and may not check White (except possibly on his last move). Black and White are *cooperating* to achieve the aim of checkmate. (In chess problem parlance, this problem is called a *serieshelpmate in 14.*) How many different solutions are there?

12. [2+]* Choose n points on the circumference of a circle in "general position." Draw all $\binom{n}{2}$ chords connecting two of the points. ("General position" means that no three of these chords intersect in a point.) Into how many regions will the interior of the circle be divided? Try to give an elegant proof avoiding induction, finite differences, generating functions, summations, and the like.

13. [2] Let p be prime and $a \in \mathbb{P}$. Show *combinatorially* that $a^p - a$ is divisible by p. (A combinatorial proof would consist of exhibiting a set S with $a^p - a$ elements and a partition of S into pairwise disjoint subsets, each with p elements.)

14. **a.** [2+] Let p be a prime, and let $n = \sum a_i p^i$ and $m = \sum b_i p^i$ be the p-ary expansions of the positive integers m and n. Show that

$$\binom{n}{m} \equiv \binom{a_0}{b_0}\binom{a_1}{b_1} \cdots \pmod{p}.$$

b. [3–] Use (a) to determine when $\binom{n}{m}$ is odd. For what n is $\binom{n}{m}$ odd for all $0 \le m \le n$? In general, how many coefficients of the polynomial $(1 + x)^n$ are odd?

c. [2+] It follows from (a), and is easy to show directly, that $\binom{pa}{pb} \equiv \binom{a}{b} \pmod{p}$. Give a *combinatorial proof* that in fact $\binom{pa}{pb} \equiv \binom{a}{b} \pmod{p^2}$.

d. [3–] If $p \ge 5$, then show in fact

$$\binom{pa}{pb} \equiv \binom{a}{b} \pmod{p^3}.$$

Is there a combinatorial proof?

e. [3–] Give a simple description of the largest power of p dividing $\binom{n}{m}$.

15. **a.** [2] How many coefficients of the polynomial $(1+x+x^2)^n$ are not divisible by 3?
    b. [3–] How many coefficients of the polynomial $(1+x+x^2)^n$ are odd?
    c. [2+] How many coefficients of the polynomial $\prod_{1\le i<j\le n}(x_i+x_j)$ are odd?

16. [3–]*

    a. Let p be a prime, and let A be the matrix $A = \left[\binom{j+k}{k}\right]_{j,k=0}^{p-1}$, taken over the field $\mathbb{F}_p$. Show that $A^3 = I$, the identity matrix. (Note that A vanishes below the main antidiagonal, i.e., $A_{jk} = 0$ if $j+k \ge p$.)

    b. How many eigenvalues of A are equal to 1?

17. **a.** [1+]* Let $m,n \in \mathbb{N}$. Prove the identity $\left(\!\binom{n}{m}\!\right) = \left(\!\binom{m+1}{n-1}\!\right)$.

    b. [2–] Give a combinatorial proof.

18. [2+]* Find a *simple* description of all $n \in \mathbb{P}$ with the following property: There exists $k \in [n]$ such that $\binom{n}{k-1}$, $\binom{n}{k}$, $\binom{n}{k+1}$ are in arithmetic progression.

19. **a.** [2+] Let $a_1,\dots,a_n \in \mathbb{N}$. Show that when we expand the product

$$\prod_{\substack{i,j=1 \\ i\neq j}}^{n}\left(1-\frac{x_i}{x_j}\right)^{a_i}$$

as a Laurent polynomial in $x_1,\dots,x_n$ (i.e., negative exponents allowed), then the constant term is the multinomial coefficient $\binom{a_1+\cdots+a_n}{a_1,\dots,a_n}$.

Hint: First prove the identity

$$1 = \sum_{i=1}^{n}\prod_{j\neq i}\left(1-\frac{x_i}{x_j}\right)^{-1}. \tag{1.120}$$

b. [2–] Put $n = 3$ to deduce the identity

$$\sum_{k=-a}^{a}(-1)^k\binom{a+b}{a+k}\binom{b+c}{b+k}\binom{c+a}{c+k} = \binom{a+b+c}{a,b,c}.$$

(Set $\binom{m}{i} = 0$ if $i < 0$.) Note that if we specialize $a = b = c$, then we obtain

$$\sum_{k=0}^{2a}(-1)^k\binom{2a}{k}^3 = \binom{3a}{a,a,a}.$$

c. [3+] Let q be an additional indeterminate. Show that when we expand the product

$$\prod_{1\le i<j\le k}\left(1-q\frac{x_i}{x_j}\right)\left(1-q^2\frac{x_i}{x_j}\right)\cdots\left(1-q^{a_i}\frac{x_i}{x_j}\right)$$

$$\cdot\left(1-\frac{x_j}{x_i}\right)\left(1-q\frac{x_j}{x_i}\right)\cdots\left(1-q^{a_j-1}\frac{x_j}{x_i}\right) \tag{1.121}$$

as a Laurent polynomial in $x_1,\dots,x_n$ (whose coefficients are now polynomials in q), then the constant term is the q-multinomial coefficient $\binom{a_1+\cdots+a_n}{a_1,\dots,a_n}$.

d. [3+] Let $k \in \mathbb{P}$. When the product

$$\prod_{1\le i<j\le n}\left[\left(1-\frac{x_i}{x_j}\right)\left(1-\frac{x_j}{x_i}\right)(1-x_ix_j)\left(1-\frac{1}{x_ix_j}\right)\right]^k$$

is expanded as earlier, show that the constant term is

$$\binom{k}{k}\binom{3k}{k}\binom{5k}{k}\cdots\binom{(2n-3)k}{k}\cdot\binom{(n-1)k}{k}.$$

e. [3–] Let $f(a_1,a_2,\ldots,a_n)$ denote the constant term of the Laurent polynomial

$$\prod_{i=1}^{n}(q^{-a_i}+q^{-a_i+1}+\cdots+q^{a_i}),$$

where each $a_i \in \mathbb{N}$. Show that

$$\sum_{a_1,\ldots,a_n\geq0} f(a_1,\ldots,a_n)x_1^{a_1}\cdots x_n^{a_n}$$

$$=(1+x_1)\cdots(1+x_n)\sum_{i=1}^{n}\frac{x_i^{n-1}}{(1-x_i^2)\prod_{j\neq i}(x_i-x_j)(1-x_ix_j)}.$$

20. [2]* How many $m \times n$ matrices of 0's and 1's are there, such that every row and column contains an even number of 1's? An odd number of 1's?

21. [2]* Fix $n \in \mathbb{P}$. In how many ways (as a function of n) can one choose a composition α of n, and then choose a composition of each part of α? (Give an elegant combinatorial proof.)

22. **a.** [2] Find the number of compositions of $n > 1$ with an even number of even parts. Naturally a combinatorial proof is preferred.
   b. [2+] Let $e(n), o(n)$, and $k(n)$ denote, respectively, the number of partitions of n with an even number of even parts, with an odd number of even parts, and that are self-conjugate. Show that $e(n) - o(n) = k(n)$. Is there a simple combinatorial proof?

23. [2] Give a simple "balls into boxes" proof that the total number of parts of all compositions of n is equal to $(n+1)2^{n-2}$. (The simplest argument expresses the answer as a sum of two terms.)

24. [2+] Let $1 \leq k < n$. Give a combinatorial proof that among all 2^{n-1} compositions of n, the part k occurs a total of $(n-k+3)2^{n-k-2}$ times. For instance, if $n = 4$ and $k = 2$, then the part 2 appears once in $2+1+1$, $1+2+1$, $1+1+2$, and twice in $2+2$, for a total of five times.

25. [2+] Let $n - r = 2k$. Show that the number $f(n,r,s)$ of compositions of n with r odd parts and s even parts is given by $\binom{r+s}{r}\binom{r+k-1}{r+s-1}$. Give a generating function proof and a bijective proof.

26. [2]* Let $\bar{c}(m,n)$ denote the number of compositions of n with largest part at most m. Show that

$$\sum_{n\geq0}\bar{c}(m,n)x^n = \frac{1-x}{1-2x+x^{m+1}}.$$

27. [2+] Find a simple explicit formula for the number of compositions of $2n$ with largest part exactly n.

28. [2]* Let $\kappa(n,j,k)$ be the number of weak compositions of n into k parts, each part less than j. Give a generating function proof that

$$\kappa(n,j,k) = \sum_{r+sj=n}(-1)^s\binom{k+r-1}{r}\binom{k}{s},$$

where the sum is over all pairs $(r,s) \in \mathbb{N}^2$ satisfying $r+sj=n$.

29. [2]* Fix $k, n \in \mathbb{P}$. Show that

$$\sum a_1 \cdots a_k = \binom{n+k-1}{2k-1},$$

where the sum ranges over all compositions $(a_1, \ldots, a_k)$ of n into k parts.

30. [2] Fix $1 \le k \le n$. How many integer sequences $1 \le a_1 < a_2 < \cdots < a_k \le n$ satisfy $a_i \equiv i \pmod 2$ for all i?

31. [2+]
    a. Let $\#N = n$, $\#X = x$. Find a simple explicit expression for the number of ways of choosing a function $f : N \to X$ and then linearly ordering each block of the coimage of f. (The elements of N and X are assumed to be distinguishable.)
    b. How many ways as in (a) are there if f must be surjective? (Give a simple explicit answer.)
    c. How many ways as in (a) are there if the elements of X are indistinguishable? (Express your answer as a finite sum.)

32. [2] Fix positive integers n and k. Let $\#S = n$. Find the number of k-tuples $(T_1, T_2, \ldots, T_k)$ of subsets T_i of S subject to each of the following conditions *separately*, that is, the three parts are independent problems (all with the same general method of solution).
    a. $T_1 \subseteq T_2 \subseteq \cdots \subseteq T_k$.
    b. The T_i's are pairwise disjoint.
    c. $T_1 \cup T_2 \cup \cdots \cup T_k = S$.

33. a. [2-]* Let $k, n \ge 1$. Find the number of sequences $\emptyset = S_0, S_1, \ldots, S_k$ of subsets of $[n]$ if for all $1 \le i \le k$ we have either (i) $S_{i-1} \subset S_i$ and $|S_i - S_{i-1}| = 1$, or (ii) $S_i \subset S_{i-1}$ and $|S_{i-1} - S_i| = 1$.
    b. [2+]* Suppose that we add the additional condition that $S_k = \emptyset$. Show that now the number $f_k(n)$ of sequences is given by

$$f_k(n) = \frac{1}{2^n} \sum_{i=0}^{n} \binom{n}{i} (n - 2i)^k.$$

    Note that $f_k(n) = 0$ if k is odd.

34. [2] Fix $n, j, k \in \mathbb{P}$. How many integer sequences are there of the form $1 \le a_1 < a_2 < \cdots < a_k \le n$, where $a_{i+1} - a_i \ge j$ for all $1 \le i \le k - 1$?

35. The *Fibonacci numbers* are defined by $F_1 = 1$, $F_2 = 1$, $F_n = F_{n-1} + F_{n-2}$ if $n \ge 3$. Express the following numbers in terms of the Fibonacci numbers.
    a. [2-] The number of subsets S of the set $[n] = \{1, 2, \ldots, n\}$ such that S contains no two consecutive integers.
    b. [2] The number of compositions of n into parts greater than 1.
    c. [2-] The number of compositions of n into parts equal to 1 or 2.
    d. [2] The number of compositions of n into odd parts.
    e. [2] The number of sequences $(\varepsilon_1, \varepsilon_2, \ldots, \varepsilon_n)$ of 0's and 1's such that $\varepsilon_1 \le \varepsilon_2 \ge \varepsilon_3 \le \varepsilon_4 \ge \varepsilon_5 \le \cdots$.
    f. [2+] $\sum a_1 a_2 \cdots a_k$, where the sum is over all 2^{n-1} compositions $a_1 + a_2 + \cdots + a_k = n$.
    g. [2+] $\sum (2^{a_1 - 1} - 1) \cdots (2^{a_k - 1} - 1)$, summed over the same set as in (f).
    h. [2+] $\sum 2^{\#\{i : a_i = 1\}}$, summed over the same set as (f).
    i. [2+] $\sum (-1)^{k-1} (5^{a_1 - 1} + 1) \cdots (5^{a_k - 1} + 1)$, summed over the same set as (f).
    j. [2+]* The number of sequences $(\delta_1, \delta_2, \ldots, \delta_n)$ of 0's, 1's, and 2's such that 0 is never immediately followed by 1.

k. [2+] The number of distinct terms of the polynomial

$$P_n = \prod_{j=1}^{n}(1 + x_j + x_{j+1}).$$

For instance, setting $x_1 = a$, $x_2 = b$, $x_3 = c$, we have $P_2 = 1 + a + 2b + c + ab + b^2 + ac + bc$, which has eight distinct terms.

36. [2] Fix $k, n \in \mathbb{P}$. Find a simple expression involving Fibonacci numbers for the number of sequences $(T_1, T_2, \ldots, T_k)$ of subsets T_i of $[n]$ such that

$$T_1 \subseteq T_2 \supseteq T_3 \subseteq T_4 \supseteq \cdots.$$

37. [2] Show that

$$F_{n+1} = \sum_{k=0}^{n}\binom{n-k}{k}. \tag{1.122}$$

38. [2]* Show that the number of permutations $w \in \mathfrak{S}_n$ fixed by the fundamental transformation $\mathfrak{S}_n \xrightarrow{\wedge} \mathfrak{S}_n$ of Proposition 1.3.1 (i.e., $w = \hat{w}$) is the Fibonacci number F_{n+1}.

39. [2+] Show that the number of ordered pairs (S, T) of subsets of $[n]$ satisfying $s > \#T$ for all $s \in S$ and $t > \#S$ for all $t \in T$ is equal to the Fibonacci number F_{2n+2}.

40. [2]* Suppose that n points are arranged on a circle. Show that the number of subsets of these points containing no two that are consecutive is the Lucas number L_n. This result shows that the Lucas number L_n may be regarded as a "circular analogue" of the Fibonacci number F_{n+2} (via Exercise 1.35(a)). For further explication, see Example 4.7.16.

41. **a.** [2] Let $f(n)$ be the number of ways to choose a subset $S \subseteq [n]$ and a permutation $w \in \mathfrak{S}_n$ such that $w(i) \notin S$ whenever $i \in S$. Show that $f(n) = F_{n+1}n!$.
b. [2+] Suppose that in (a) we require w to be an n-cycle. Show that the number of ways is now $g(n) = L_n(n-1)!$, where L_n is a Lucas number.

42. [3] Let

$$F(x) = \prod_{n \geq 2}(1 - x^{F_n}) = (1-x)(1-x^2)(1-x^3)(1-x^5)(1-x^8)\cdots$$

$$= 1 - x - x^2 + x^4 + x^7 - x^8 + x^{11} - x^{12} - x^{13} + x^{14} + x^{18} + \cdots.$$

Show that every coefficient of $F(x)$ is equal to $-1, 0$, or 1.

43. [2−] Using only the combinatorial definitions of the Stirling numbers $S(n,k)$ and $c(n,k)$, give formulas for $S(n,1)$, $S(n,2)$, $S(n,n)$, $S(n,n-1)$, $S(n,n-2)$ and $c(n,1)$, $c(n,2)$, $c(n,n)$, $c(n,n-1)$, $c(n,n-2)$. For the case $c(n,2)$, express your answer in terms of the harmonic number $H_m = 1 + \frac{1}{2} + \frac{1}{3} + \cdots + \frac{1}{m}$ for suitable m.

44. **a.** [2]* Show that the total number of cycles of all even permutations of $[n]$ and the total number of cycles of all odd permutations of $[n]$ differ by $(-1)^n(n-2)!$. Use generating functions.
b. [3−]* Give a bijective proof.

45. [2+] Let $S(n,k)$ denote a Stirling number of the second kind. The generating function $\sum_n S(n,k)x^n = x^k/(1-x)(1-2x)\cdots(1-kx)$ implies the identity

$$S(n,k) = \sum 1^{a_1-1}2^{a_2-1}\cdots k^{a_k-1}, \tag{1.123}$$

the sum being over all compositions $a_1 + \cdots + a_k = n$. Give a *combinatorial* proof of (1.123) analogous to the second proof of Proposition 1.3.7. That is, we want to associate with each partition π of $[n]$ into k blocks a composition $a_1 + \cdots + a_k = n$ such that exactly $1^{a_1-1}2^{a_2-1} \cdots k^{a_k-1}$ partitions π are associated with this composition.

46. **a.** [2] Let $n, k \in \mathbb{P}$, and let $j = \lfloor k/2 \rfloor$. Let $S(n,k)$ denote a Stirling number of the second kind. Give a generating function proof that

$$S(n,k) \equiv \binom{n-j-1}{n-k} \pmod{2}.$$

b. [3–] Give a combinatorial proof.
c. [2] State and prove an analogous result for Stirling numbers of the first kind.

47. Let D be the operator $\frac{d}{dx}$.
a. [2]* Show that $(xD)^n = \sum_{k=0}^{n} S(n,k)x^k D^k$.
b. [2]* Show that

$$x^n D^n = xD(xD-1)(xD-2) \cdots (xD-n+1) = \sum_{k=0}^{n} s(n,k)(xD)^k.$$

c. [2+]* Find the coefficients $a_{n,i,j}$ in the expansion

$$(x + D)^n = \sum_{i,j} a_{n,i,j} x^i D^j.$$

48. **a.** [3] Let $P(x) = a_0 + a_1 x + \cdots + a_n x^n$, $a_i \geq 0$, be a polynomial all of whose zeros are negative real numbers. Regard $a_k/P(1)$ as the probability of choosing k, so we have a probability distribution on $[0,n]$. Let $\mu = \frac{1}{P(1)} \sum_k k a_k = P'(1)/P(1)$, the *mean* of the distribution; and let m be the *mode* (i.e., $a_m = \max_k a_k$). Show that

$$|\mu - m| < 1.$$

More precisely, show that

$$m = k, \quad \text{if } k \leq \mu < k + \frac{1}{k+2},$$
$$m = k, \text{ or } k+1, \text{ or both}, \quad \text{if } k + \frac{1}{k+2} \leq \mu \leq k+1 - \frac{1}{n-k+1},$$
$$m = k+1, \quad \text{if } k+1 - \frac{1}{n-k+1} < \mu \leq k+1.$$

b. [2] Fix n. Show that the signless Stirling number $c(n,k)$ is maximized at $k = \lfloor 1 + \frac{1}{2} + \frac{1}{3} + \cdots + \frac{1}{n} \rfloor$ or $k = \lceil 1 + \frac{1}{2} + \frac{1}{3} + \cdots + \frac{1}{n} \rceil$. In particular, $k \sim \log(n)$.
c. [3] Let $S(n,k)$ denote a Stirling number of the second kind, and define K_n by $S(n, K_n) \geq S(n,k)$ for all k. Let t be the solution of the equation $te^t = n$. Show that for sufficiently large n (and probably all n), either $K_n + 1 = \lfloor e^t \rfloor$ or $K_n + 1 = \lceil e^t \rceil$.

49. **a.** [2+] Deduce from equation (1.38) that all the (complex) zeros of $A_d(x)$ are real and simple. (Use Rolle's theorem.)
b. [2–]* Deduce from Exercise 1.133(b) that the polynomial $\sum_{k=1}^{n} k! S(n,k)x^k$ has only real zeros.

50. A sequence $\alpha = (a_0, a_1, \ldots, a_n)$ of real numbers is *unimodal* if for some $0 \leq j \leq n$ we have $a_0 \leq a_1 \leq \cdots \leq a_j \geq a_{j+1} \geq a_{j+2} \geq \cdots \geq a_n$, and is *log-concave* if $a_i^2 \geq a_{i-1}a_{i+1}$ for $1 \leq i \leq n - 1$. We also say that α has *no internal zeros* if there does not exist $i < j < k$ with $a_i \neq 0$, $a_j = 0$, $a_k \neq 0$, and that α is *symmetric* if $a_i = a_{n-i}$ for all i.

Define a polynomial $P(x) = \sum a_i x^i$ to be unimodal, log-concave, and so on, if the sequence $(a_0, a_1, \ldots, a_n)$ of coefficients has that property.

a. [2–]* Show that a log-concave sequence of nonnegative real numbers with no internal zeros is unimodal.

b. [2+] Let $P(x) = \sum_{i=0}^n a_i x^i = \sum_{i=0}^n \binom{n}{i} b_i x^i \in \mathbb{R}[x]$. Show that if all the zeros of $P(x)$ are real, then the sequence $(b_0, b_1, \ldots, b_n)$ is log-concave. (When all $a_i \geq 0$, this statement is stronger than the assertion that $(a_0, a_1, \ldots, a_n)$ is log-concave.)

c. [2+] Let $P(x) = \sum_{i=0}^m a_i x^i$ and $Q(x) = \sum_{i=0}^n b_i x^i$ be symmetric, unimodal, and have nonnegative coefficients. Show that the same is true for $P(x)Q(x)$.

d. [2+] Let $P(x)$ and $Q(x)$ be log-concave with no internal zeros and nonnegative coefficients. Show that the same is true for $P(x)Q(x)$.

e. [2] Show that the polynomials $\sum_{w \in \mathfrak{S}_n} x^{\mathrm{des}(w)}$ and $\sum_{w \in \mathfrak{S}_n} x^{\mathrm{inv}(w)}$ are symmetric and unimodal.

f. [4–] Let $1 \leq p \leq n-1$. Given $w = a_1 \cdots a_n \in \mathfrak{S}_n$, define

$$\mathrm{des}_p(w) = \#\{(i,j) : i < j \leq i+p,\ a_i > a_j\}.$$

Thus $\mathrm{des}_1 = \mathrm{des}$ and $\mathrm{des}_{n-1} = \mathrm{inv}$. Show that the polynomial $\sum_{w \in \mathfrak{S}_n} x^{\mathrm{des}_p(w)}$ is symmetric and unimodal.

g. [2+] Let S be a subset of $\{(i,j) : 1 \leq i < j \leq n\}$. An S-*inversion* of $w = a_1 \cdots a_n \in \mathfrak{S}_n$ is a pair $(i,j) \in S$ for which $a_i > a_j$. Let $\mathrm{inv}_S(w)$ denote the number of S-inversions of w. Find a set S (for a suitable value of n) for which the polynomial $P_S(x) := \sum_{w \in \mathfrak{S}_n} x^{\mathrm{inv}_S(w)}$ is not unimodal.

51. [3–] Let $k, n \in \mathbb{P}$ with $k \leq n$. Let $V(n,k)$ denote the volume of the region $\mathcal{R}_{nk}$ in $\mathbb{R}^n$ defined by

$$0 \leq x_i \leq 1, \text{ for } 1 \leq i \leq n$$

$$k - 1 \leq x_1 + x_2 + \cdots + x_n \leq k.$$

Show that $V(n,k) = A(n,k)/n!$, where $A(n,k)$ is an Eulerian number.

52. [3–] Fix $b \geq 2$. Choose n random N-digit integers in base b (allowing intial digits equal to 0). Add these integers using the usual addition algorithm. For $0 \leq j \leq n-1$, let $f(j)$ be the number of times that we carry j in the addition process. For instance, if we add 71801, 80914, and 62688 in base 10, then $f(0) = 1$ and $f(1) = f(2) = 2$. Show that as $N \to \infty$, the expected value of $f(j)/N$ (i.e., the expected proportion of the time we carry a j) approaches $A(n, j+1)/n!$, where $A(n,k)$ is an Eulerian number.

53. a. [2]* The *Eulerian Catalan number* is defined by $EC_n = A(2n+1, n+1)/(n+1)$. The first few Eulerian Catalan numbers, beginning with $EC_0 = 1$, are 1, 2, 22, 604, 31238. Show that $EC_n = 2A(2n, n+1)$, whence $EC_n \in \mathbb{Z}$.

b. [3–]* Show that EC_n is the number of permutations $w = a_1 a_2 \cdots a_{2n+1}$ with n descents, such that every left factor $a_1 a_2 \cdots a_i$ has at least as many ascents as descents. For $n = 1$ we are counting the two permutations 132 and 231.

54. [2]* How many n-element multisets on $[2m]$ are there satisfying: (i) $1, 2, \ldots, m$ appear at most once each, and (ii) $m+1, m+2, \ldots, 2m$ appear an even number of times each?

55. [2–]* If $w = a_1 a_2 \cdots a_n \in \mathfrak{S}_n$ then let $w^r = a_n \cdots a_2 a_1$, the reverse of w. Express $\mathrm{inv}(w^r)$, $\mathrm{maj}(w^r)$, and $\mathrm{des}(w^r)$ in terms of $\mathrm{inv}(w)$, $\mathrm{maj}(w)$, and $\mathrm{des}(w)$, respectively.

56. [2+] Let M be a finite multiset on $\mathbb{P}$. Generalize equation (1.41) by showing that

$$\sum_{w \in \mathfrak{S}_M} q^{\mathrm{inv}(w)} = \sum_{w \in \mathfrak{S}_M} q^{\mathrm{maj}(w)},$$

where inv(w) and maj(w) are defined in Section 1.7. Try to give a proof based on results in Section 1.4 rather than generalizing the proof of (1.41).

57. [2+] Let $w = w_1 w_2 \cdots w_n \in \mathfrak{S}_n$. Show that the following conditions are equivalent.
   (i) Let $C(i)$ be the set of indices j of the columns C_j that intersect the ith row of the diagram $D(w)$ of w. For instance, if $w = 314652$ as in Figure 1.5, then $C(1) = \{1,2\}$, $C(3) = \{2\}$, $C(4) = \{2,5\}$, $C(5) = \{2\}$, and all other $C(i) = \emptyset$. Then for every i, j, either $C(i) \subseteq C(j)$ or $C(j) \subseteq C(i)$.
   (ii) Let $\lambda(w)$ be the entries of the inversion table $I(w)$ of w written in decreasing order. For instance, $I(52413) = (3,1,2,1,0)$ and $\lambda(52413) = (3,2,1,1,0)$. Regard λ as a partition of inv(w). Then $\lambda(w^{-1}) = \lambda(w)'$, the conjugate partition to $\lambda(w)$.
   (iii) The permutation w is 2143-*avoiding* (i.e., there do not exist $a < b < c < d$ for which $w_b < w_a < w_d < w_c$).

58. For $u \in \mathfrak{S}_k$, let $s_u(n) = \#S_u(n)$, the number of permutations $w \in \mathfrak{S}_n$ avoiding u. If also $v \in \mathfrak{S}_k$, then write $u \sim v$ if $s_u(n) = s_v(n)$ for all $n \geq 0$ (an obvious equivalence relation). Thus by the discussion preceding Proposition 1.5.1, $u \sim v$ for all $u, v \in \mathfrak{S}_3$.
   a. [2]* Let $u, v \in \mathfrak{S}_k$. Suppose that the permutation matrix P_v can be obtained from P_u by one of the eight dihedral symmetries of the square. For instance, $P_{u^{-1}}$ and be obtained from P_u by reflection in the main diagonal. Show that $u \sim v$. We then say that u and v are *equivalent by symmetry*, denoted $u \approx v$. Thus $\approx$ is a finer equivalence relation than $\sim$. What are the $\approx$ equivalence classes for $\mathfrak{S}_3$?
   b. [3] Show that there are exactly three $\sim$ equivalence classes for $\mathfrak{S}_4$. The equivalence classes are given by $\{1234, 1243, 2143, \ldots\}$, $\{3142, 1342, \ldots\}$, and $\{1342, \ldots\}$, where the omitted permutations are obtained by $\approx$ equivalence.

59. [3] Let $s_u(n)$ have the meaning of the previous exercise. Show that $c_u := \lim_{n \to \infty} s_u(n)^{1/n}$ exists and satisfies $1 < c_u < \infty$.

60. [2+] Define two permutations in $\mathfrak{S}_n$ to be *equivalent* if one can be obtained from the other by interchanging adjacent letters that differ by at least two, an obvious equivalence relation. For instance, when $n = 3$ we have the four equivalence classes $\{123\}, \{132, 312\}, \{213, 231\}, \{321\}$. Describe the equivalence classes in terms of more familiar objects. How many equivalence classes are there?

61. a. [3−] Let $w = w_1 \cdots w_n$. Let

$$F(x; a, b, c, d) = \sum_{n \geq 1} \sum_{w \in \mathfrak{S}_n} a^{v(w)} b^{p(w)-1} c^{r(w)} d^{f(w)} \frac{x^n}{n!},$$

where $v(w)$ denotes the number of valleys w_i of w for $1 \leq i \leq n$ (where $w_0 = w_{n+1} = 0$ as preceding Proposition 1.5.3), $p(w)$ the number of peaks, $r(w)$ the number of double rises, and $f(w)$ the number of double falls. For instance, if $w = 32451$, then 3 is a peak, 2 is a valley, 4 is a double rise, 5 is a peak, and 1 is a double fall. Thus,

$$F(x; a, b, c, d) = x + (c+d)\frac{x^2}{2!} + (c^2 + d^2 + 2ab + 2cd)\frac{x^3}{3!}$$

$$+ (c^3 + d^3 + 3cd^2 + 3c^2 d + 8abc + 8abd)\frac{x^4}{4!} + \cdots.$$

Show that

$$F(x; a, b, c, d) = \frac{e^{vx} - e^{ux}}{ve^{ux} - ue^{vx}}, \tag{1.124}$$

where $uv = ab$ and $u + v = c + d$. In other words, u and v are zeros of the polynomial $z^2 - (c+d)z + ab$; it makes no difference which zero we call u and which v.

b. [2–] Let $r(n,k)$ be the number of permutations $w \in \mathfrak{S}_n$ with k peaks. Show that

$$\sum_{n\geq 0}\sum_{k\geq 0} r(n,k)t^k \frac{x^n}{n!} = \frac{1+u\tan(xu)}{1-\frac{\tan(xu)}{u}}, \qquad (1.125)$$

where $u = \sqrt{t-1}$.

c. [2+] A *proper double fall* or *proper double descent* of a permutation $w = a_1 a_2 \cdots a_n$ is an index $1 < i < n$ for which $a_{i-1} > a_i > a_{i+1}$. (Compare with the definition of a double fall or double descent, where we also allow $i = 1$ and $i = n$ with the convention $a_0 = a_{n+1} = 0$.) Let $f(n)$ be the number of permutations $w \in \mathfrak{S}_n$ with no proper double descents. Show that

$$\sum_{n\geq 0} f(n)\frac{x^n}{n!} = \frac{1}{\displaystyle\sum_{j\geq 0}\left(\frac{x^{3j}}{(3j)!} - \frac{x^{3j+1}}{(3j+1)!}\right)} \qquad (1.126)$$

$$= 1 + x + 2\frac{x^2}{2!} + 5\frac{x^3}{3!} + 17\frac{x^4}{4!} + 70\frac{x^5}{5!} + 349\frac{x^6}{6!}$$

$$+ 2017\frac{x^7}{7!} + 13358\frac{x^8}{8!} + \cdots.$$

62. In this exercise we consider one method for generalizing the disjoint cycle decomposition of permutations of sets to multisets. A *multiset cycle* of $\mathbb{P}$ is a sequence $C = (i_1, i_2, \ldots, i_k)$ of positive integers with repetitions allowed, where we regard $(i_1, i_2, \ldots, i_k)$ as equivalent to $(i_j, i_{j+1}, \ldots, i_k, i_1, \ldots, i_{j-1})$ for $1 \leq j \leq k$. Introduce indeterminates $x_1, x_2, \ldots$, and define the *weight* of C by $w(C) = x_{i_1} \cdots x_{i_k}$. A *multiset permutation* or *multipermutation* of a multiset M is a multiset of multiset cycles, such that M is the multiset of all elements of the cycles. For instance, the multiset $\{1, 1, 2\}$ has the following four multipermutations: $(1)(1)(2)$, $(11)(2)$, $(12)(1)$, (112). The *weight* $w(\pi)$ of a multipermutation $\pi = C_1 C_2 \cdots C_j$ is given by $w(\pi) = w(C_1) \cdots w(C_j)$.

a. [2–]* Show that

$$\prod_C (1 - w(C))^{-1} = \sum_\pi w(\pi),$$

where C ranges over all multiset cycles on $\mathbb{P}$ and π over all (finite) multiset permutations on $\mathbb{P}$.

b. [2+] Let $p_k = x_1^k + x_2^k + \cdots$. Show that

$$\prod_C (1 - w(C))^{-1} = \prod_{k\geq 1} (1 - p_k)^{-1}.$$

c. [1+] Let $f_k(n)$ denote the number of multiset permutations on $[k]$ of total size n. For instance, $f_2(3) = 14$, given by $(1)(1)(1)$, $(1)(1)(2)$, $(1)(2)(2)$, $(2)(2)(2)$, $(11)(1)$, $(11)(2)$, $(12)(1)$, $(12)(2)$, $(22)(1)$, $(22)(2)$, (111), (112), (122), (222). Deduce from (b) that

$$\sum_{n\geq 0} f_k(n)x^n = \prod_{i\geq 1} (1 - kx^i)^{-1}.$$

d. [3–] Find a direct combinatorial proof of (b) or (c).

63. **a.** [2–] We are given n square envelopes of different sizes. In how many different ways can they be arranged by inclusion? For instance, if $n = 3$ there are six ways;

namely, label the envelopes A, B, C with A the largest and C the smallest, and let $I \in J$ mean that envelope I is contained in envelope J. Then the six ways are: (1) $\emptyset$, (2) $B \in A$, (3) $C \in A$, (4) $C \in B$, (5) $B \in A, C \in A$, (6) $C \in B \in A$.

b. [2] How many arrangements have exactly k envelopes that are not contained in another envelope? That don't contain another envelope?

64. a. [2] Let $f(n)$ be the number of sequences $a_1, \ldots, a_n$ of positive integers such that for each $k > 1$, k only occurs if $k - 1$ occurs before the last occurrence of k. Show that $f(n) = n!$. (For $n = 3$ the sequences are $111, 112, 121, 122, 212, 123$.)

b. [2] Show that $A(n, k)$ of these sequences satisfy $\max\{a_1, \ldots, a_n\} = k$.

65. [3] Let $y = \prod_{n \geq 1}(1 - x^n)^{-1}$. Show that

$$4y^3 y'' + 5xy^3 y''' + x^2 y^3 y^{(iv)} - 16y^2 y'^2 - 15xy^2 y' y'' + 20x^2 y^2 y' y'''$$

$$-19x^2 y^2 y''^2 + 10xyy'^3 + 12x^2 yy'^2 y'' + 6x^2 y'^4 = 0. \tag{1.127}$$

66. [2–]* Let $p_k(n)$ denote the number of partitions of n into k parts. Give a bijective proof that

$$p_0(n) + p_1(n) + \cdots + p_k(n) = p_k(n + k).$$

67. [2–]* Express the number of partitions of n with no part equal to 1 in terms of values $p(k)$ of the partition function.

68. [2]* Let $n \geq 1$, and let $f(n)$ be the number of partitions of n such that for all k, the part k occurs at most k times. Let $g(n)$ be the number of partitions of n such that no part has the form $i(i + 1)$ (i.e., no parts equal to $2, 6, 12, 20, \ldots$). Show that $f(n) = g(n)$.

69. [2]* Let $f(n)$ denote the number of self-conjugate partitions of n all of whose parts are even. Express the generating function $\sum_{n \geq 0} f(n)x^n$ as a simple product.

70. a. [2] Find a bijection between partitions $\lambda \vdash n$ of rank r and integer arrays

$$A_\lambda = \begin{pmatrix} a_1 & a_2 & \cdots & a_r \\ b_1 & b_2 & \cdots & b_r \end{pmatrix}$$

such that $a_1 > a_2 > \cdots > a_r \geq 0$, $b_1 > b_2 > \cdots > b_r \geq 0$, and $r + \sum(a_i + b_i) = n$.

b. [2+] A *concatenated spiral self-avoiding walk* (CSSAW) on the square lattice is a lattice path in the plane starting at $(0,0)$, with steps $(\pm 1, 0)$ and $(0, \pm 1)$ and first step $(1, 0)$, with the following three properties: (i) the path is *self-avoiding* (i.e, it never returns to a previously visited lattice point), (ii) every step after the first must continue in the direction of the previous step or turn right, and (iii) at the end of the walk it must be possible to turn right and walk infinitely many steps in the direction faced without intersecting an earlier part of the path. For instance, writing $N = (0, 1)$, etc., the five CSSAW's of length four are $NNNN$, $NNNE$, $NNEE$, $NEEE$, and $NESS$. Note for instance that $NEES$ is not a CSSAW since continuing with steps $WWW \cdots$ will intersect $(0, 0)$. Show that the number of CSSAW's of length n is equal to $p(n)$, the number of partitions of n.

71. [2+] How many pairs (λ, μ) of partitions of integers are there such that $\lambda \vdash n$, and the Young diagram of μ is obtained from the Young diagram of λ by adding a single square? Express your answer in terms of the partition function values $p(k)$ for $k \leq n$. Give a simple combinatorial proof.

72. **a.** [3–] Let $\lambda = (\lambda_1, \lambda_2, \ldots)$ and $\mu = (\mu_1, \mu_2, \ldots)$ be partitions. Define $\mu \leq \lambda$ if $\mu_i \leq \lambda_i$ for all i. Show that

$$\sum_{\mu \leq \lambda} q^{|\mu| + |\lambda|} = \frac{1}{(1-q)(1-q^2)^2(1-q^3)^2(1-q^4)^2 \cdots}. \qquad (1.128)$$

b. [3–] Show that the number of pairs (λ, μ) such that λ and μ have *distinct* parts, $\mu \leq \lambda$ as in (a), and $|\lambda| + |\mu| = n$, is equal to $p(n)$, the number of partitions of n. For instance, when $n = 5$ we have the seven pairs $(\emptyset, 5)$, $(\emptyset, 41)$, $(\emptyset, 32)$, $(1,4)$, $(2,3)$, $(1,31)$, and $(2,21)$.

73. [2] Let λ be a partition. Show that

$$\sum_i \left\lceil \frac{\lambda_{2i-1}}{2} \right\rceil = \sum_i \left\lceil \frac{\lambda'_{2i-1}}{2} \right\rceil,$$

$$\sum_i \left\lfloor \frac{\lambda_{2i-1}}{2} \right\rfloor = \sum_i \left\lceil \frac{\lambda'_{2i}}{2} \right\rceil,$$

$$\sum_i \left\lfloor \frac{\lambda_{2i}}{2} \right\rfloor = \sum_i \left\lfloor \frac{\lambda'_{2i}}{2} \right\rfloor.$$

74. [2] Let $p_k(n)$ denote the number of partitions of n into k parts. Fix $t \geq 0$. Show that as $n \to \infty$, $p_{n-t}(n)$ becomes eventually constant. What is this constant $f(t)$? What is the least value of n for which $p_{n-t}(n) = f(t)$? Your arguments should be combinatorial.

75. [2–] Let $p_k(n)$ be as in Exercise 1.74, and let $q_k(n)$ be the number of partitions of n into k distinct parts. For example, $q_3(8) = 2$, corresponding to $(5,2,1)$ and $(4,3,1)$. Give a simple combinatorial proof that $q_k\left(n + \binom{k}{2}\right) = p_k(n)$.

76. [2] Prove the partition identity

$$\prod_{i \geq 1}(1 + qx^{2i-1}) = \sum_{k \geq 0} \frac{x^{k^2} q^k}{(1-x^2)(1-x^4)\cdots(1-x^{2k})}. \qquad (1.129)$$

77. [3–] Give a "subtraction-free" bijective proof of the pentagonal number formula by proving directly the identity

$$1 + \frac{\sum_{n \text{ odd}}\left(x^{n(3n-1)/2} + x^{n(3n+1)/2}\right)}{\prod_{j \geq 1}(1-x^j)} = \frac{1 + \sum_{n \text{ even}}\left(x^{n(3n-1)/2} + x^{n(3n+1)/2}\right)}{\prod_{j \geq 1}(1-x^j)}.$$

78. **a.** [2] The *logarithmic derivative* of a power series $F(x)$ is $\frac{d}{dx}\log F(x) = F'(x)/F(x)$. By logarithmically differentiating the power series $\sum_{n \geq 0} p(n)x^n = \prod_{i \geq 1}(1 - x^i)^{-1}$, derive the recurrence

$$n \cdot p(n) = \sum_{i=1}^{n} \sigma(i)p(n-i),$$

where $\sigma(i)$ is the sum of the divisors of i.

b. [2+] Give a combinatorial proof.

79. **a.** [2+] Given a set $S \subseteq \mathbb{P}$, let $p_S(n)$ (resp. $q_S(n)$) denote the number of partitions of n (resp. number of partitions of n into distinct parts) whose parts belong to S. (These are special cases of the function $p(\mathcal{S}, n)$ of Corollary 1.8.2.) Call a pair (S, T), where $S, T \subseteq \mathbb{P}$, an *Euler pair* if $p_S(n) = q_T(n)$ for all $n \in \mathbb{N}$. Show that (S, T) is an Euler pair if and only if $2T \subseteq T$ (where $2T = \{2i : i \in T\}$) and $S = T - 2T$.
b. [1+] What is the significance of the case $S = \{1\}$, $T = \{1, 2, 4, 8, \dots\}$?

80. [2+] If λ is a partition of an integer n, let $f_k(\lambda)$ be the number of times k appears as a part of λ, and let $g_k(\lambda)$ be the number of distinct parts of λ that occur at least k times. For example, $f_2(4, 2, 2, 2, 1, 1) = 3$ and $g_2(4, 2, 2, 2, 1, 1) = 2$. Show that $\sum f_k(\lambda) = \sum g_k(\lambda)$, where $k \in \mathbb{P}$ is fixed and both sums range over all partitions λ of a fixed integer $n \in \mathbb{P}$.

81. [2+] A *perfect partition* of $n \geq 1$ is a partition $\lambda \vdash n$ which "contains" precisely one partition of each positive integer $m \leq n$. In other words, regarding λ as the multiset of its parts, for each $m \leq n$ there is a unique submultiset of λ whose parts sum to m. Show that the number of perfect partitions of n is equal to the number of *ordered* factorizations (with any number of factors) of $n + 1$ into integers ≥ 2.
Example. The perfect partitions of 5 are $(1, 1, 1, 1, 1)$, $(3, 1, 1)$, and $(2, 2, 1)$. The ordered factorizations of 6 are $6 = 2 \cdot 3 = 3 \cdot 2$.

82. [3] Show that the number of partitions of $5n + 4$ is divisible by 5.

83. [3–] Let $\lambda = (\lambda_1, \lambda_2, \dots) \vdash n$. Define

$$\alpha(\lambda) = \sum_i \lceil \lambda_{2i-1}/2 \rceil,$$

$$\beta(\lambda) = \sum_i \lfloor \lambda_{2i-1}/2 \rfloor,$$

$$\gamma(\lambda) = \sum_i \lceil \lambda_{2i}/2 \rceil,$$

$$\delta(\lambda) = \sum_i \lfloor \lambda_{2i}/2 \rfloor.$$

Let a, b, c, d be (commuting) indeterminates, and define

$$w(\lambda) = a^{\alpha(\lambda)} b^{\beta(\lambda)} c^{\gamma(\lambda)} d^{\delta(\lambda)}.$$

For instance, if $\lambda = (5, 4, 4, 3, 2)$, then $w(\lambda)$ is the product of the entries of the diagram

$$
\begin{array}{ccccc}
a & b & a & b & a \\
c & d & c & d & \\
a & b & a & b & \\
c & d & c & & \\
a & b & & &
\end{array}
$$

Show that

$$\sum_{\lambda \in \text{Par}} w(\lambda) = \prod_{j \geq 1} \frac{(1 + a^j b^{j-1} c^{j-1} d^{j-1})(1 + a^j b^j c^j d^{j-1})}{(1 - a^j b^j c^j d^j)(1 - a^j b^j c^{j-1} d^{j-1})(1 - a^j b^{j-1} c^j d^{j-1})}, \quad (1.130)$$

where Par denotes the set of all partitions λ of all integers $n \geq 0$.

84. [2]* Show that the number of partitions of n in which each part appears exactly 2, 3, or 5 times is equal to the number of partitions of n into parts congruent to ± 2, ± 3, $6 \pmod{12}$.

85. [2+]* Prove that the number of partitions of n in which no part appears exactly once equals the number of partitions of n into parts not congruent to $\pm 1 \pmod 6$.

86. [3] Prove that the number of partitions of n into parts congruent to 1 or 5 (mod 6) equals the number of partitions of n in which the difference between all parts is at least 3 and between multiples of 3 is at least 6.

87. [3–]* Let $A_k(n)$ be the number of partitions of n into odd parts (repetition allowed) such that exactly k distinct parts occur. For instance, when $n = 35$ and $k = 3$, one of the partitions being enumerated is $(9,9,5,3,3,3,3)$. Let $B_k(n)$ be the number of partitions $\lambda = (\lambda_1,\ldots,\lambda_r)$ of n such that the sequence $\lambda_1,\ldots,\lambda_r$ is composed of exactly k noncontiguous sequences of one or more consecutive integers. For instance, when $n = 44$ and $k = 3$, one of the partitions being enumerated is $(10,9,8,7,5,3,2)$, which is composed of $10,9,8,7$ and 5 and $3,2$. Show that $A_k(n) = B_k(n)$ for all k and n. Note that summing over all k gives Proposition 1.8.5 (i.e., $p_{\text{odd}}(n) = q(n)$).

88. **a.** [3] Prove the identities

$$\sum_{n\geq 0}\frac{x^{n^2}}{(1-x)(1-x^2)\cdots(1-x^n)} = \frac{1}{\prod_{k\geq 0}(1-x^{5k+1})(1-x^{5k+4})},$$

$$\sum_{n\geq 0}\frac{x^{n(n+1)}}{(1-x)(1-x^2)\cdots(1-x^n)} = \frac{1}{\prod_{k\geq 0}(1-x^{5k+2})(1-x^{5k+3})}.$$

b. [2] Show that the identities in (a) are equivalent to the following combinatorial statements:
- The number of partitions of n into parts $\equiv \pm 1 \pmod 5$ is equal to the number of partitions of n whose parts differ by at least 2.
- The number of partitions of n into parts $\equiv \pm 2 \pmod 5$ is equal to the number of partitions of n whose parts differ by at least 2 and for which 1 is not a part.

c. [2]* Let $f(n)$ be the number of partitions $\lambda \vdash n$ satisfying $\ell(\lambda) = \text{rank}(\lambda)$. Show that $f(n)$ is equal to the number of partitions of n whose parts differ by at least 2.

89. [3] A *lecture hall partition* of length k is a partition $\lambda = (\lambda_1,\ldots,\lambda_k)$ (some of whose parts may be 0) satisfying

$$0 \leq \frac{\lambda_k}{1} \leq \frac{\lambda_{k-1}}{2} \leq \cdots \leq \frac{\lambda_1}{k}.$$

Show that the number of lecture hall partitions of n of length k is equal to the number of partitions of n whose parts come from the set $1,3,5,\ldots,2k-1$ (with repetitions allowed).

90. [3] Let $f(n)$ be the number of partitions of n all of whose parts are Lucas numbers L_{2n+1} of odd index. For instance, $f(12) = 5$, corresponding to

$$1+1+1+1+1+1+1+1+1+1+1+1$$
$$4+1+1+1+1+1+1+1+1$$
$$4+4+1+1+1+1$$
$$4+4+4$$
$$11+1$$

Let $g(n)$ be the number of partitions $\lambda = (\lambda_1, \lambda_2, \ldots)$ such that $\lambda_i / \lambda_{i+1} > \frac{1}{2}(3 + \sqrt{5})$ whenever $\lambda_{i+1} > 0$. For instance, $g(12) = 5$, corresponding to

$$12, \quad 11 + 1, \quad 10 + 2, \quad 9 + 3, \quad 8 + 3 + 1.$$

Show that $f(n) = g(n)$ for all $n \geq 1$.

91. **a.** [3–] Show that

$$\sum_{n \in \mathbb{Z}} x^n q^{n^2} = \prod_{k \geq 1} (1 - q^{2k})(1 + x q^{2k-1})(1 + x^{-1} q^{2k-1}).$$

b. [2] Deduce from (a) the Pentagonal Number Formula (Proposition 1.8.7).

c. [2] Deduce from (a) the two identities

$$\prod_{k \geq 1} \frac{1 - q^k}{1 + q^k} = \sum_{n \in \mathbb{Z}} (-1)^n q^{n^2}, \tag{1.131}$$

$$\prod_{k \geq 1} \frac{1 - q^{2k}}{1 - q^{2k-1}} = \sum_{n \geq 0} q^{\binom{n+1}{2}}. \tag{1.132}$$

d. [2+] Deduce from (a) the identity

$$\prod_{k \geq 1} (1 - q^k)^3 = \sum_{n \geq 0} (-1)^n (2n + 1) q^{n(n+1)/2}.$$

Hint. First substitute $-x q^{-1/2}$ for x and $q^{1/2}$ for q.

92. [3] Let $\mathcal{S} \subseteq \mathbb{P}$ and let $p(\mathcal{S}, n)$ denote the number of partitions of n whose parts belong to $\mathcal{S}$. Let

$$\mathcal{S} = \{n : n \text{ odd or } n \equiv \pm 4, \pm 6, \pm 8, \pm 10 \,(\mathrm{mod}\,32)\},$$

$$\mathcal{T} = \{n : n \text{ odd or } n \equiv \pm 2, \pm 8, \pm 12, \pm 14 \,(\mathrm{mod}\,32)\}.$$

Show that $p(\mathcal{S}, n) = p(\mathcal{T}, n-1)$ for all $n \geq 1$. Equivalently, we have the remarkable identity

$$\prod_{n \in \mathcal{S}} \frac{1}{1 - x^n} = 1 + x \prod_{n \in \mathcal{T}} \frac{1}{1 - x^n}. \tag{1.133}$$

93. [3] Let

$$\mathcal{S} = \pm \{1, 4, 5, 6, 7, 9, 11, 13, 16, 21, 23, 28 \,(\mathrm{mod}\,66)\},$$

$$\mathcal{T} = \pm \{1, 4, 5, 6, 7, 9, 11, 14, 16, 17, 27, 29 \,(\mathrm{mod}\,66)\},$$

where

$$\pm \{a, b, \ldots \,(\mathrm{mod}\,m)\} := \{n \in \mathbb{P} : n \equiv \pm a, \pm b, \ldots \,(\mathrm{mod}\,m)\}.$$

Show that $p(\mathcal{S}, n) = p(\mathcal{T}, n)$ for all $n \geq 1$ *except* $n = 13$. Equivalently, we have another remarkable identity similar to equation (1.133):

$$\prod_{n \in \mathcal{S}} \frac{1}{1 - x^n} = x^{13} + \prod_{n \in \mathcal{T}} \frac{1}{1 - x^n}.$$

94. **a.** [3–] Let $n \geq 0$. Show that the following numbers are equal.
- The number of solutions to $n = \sum_{i \geq 0} a_i 2^i$, where $a_i = 0, 1$, or 2.
- Then number of odd integers k for which the Stirling number $S(n+1, k)$ is odd.
- The number of odd binomial coefficients of the form $\binom{n-k}{k}$, $0 \leq k \leq n$.
- The number of ways to write b_n as a sum of distinct Fibonacci numbers F_n, where

$$\prod_{i \geq 0}(1 + x^{F_{2i}}) = \sum_{n \geq 0} x^{b_n}, \quad b_0 < b_1 < \cdots .$$

b. [2–] Denote by a_{n+1} the number being counted by (a), so $(a_1, a_2, \ldots, a_{10}) = (1, 1, 2, 1, 3, 2, 3, 1, 4, 3)$. Deduce from (a) that

$$\sum_{n \geq 0} a_{n+1} x^n = \prod_{i \geq 0}(1 + x^{2^i} + x^{2^{i+1}}).$$

c. [2] Deduce from (a) that $a_{2n} = a_n$ and $a_{2n+1} = a_n + a_{n+1}$.

d. [3–] Show that every positive rational number can be written in exactly one way as a fraction a_n/a_{n+1}.

95. [3] At time $n = 1$ place a line segment (toothpick) of length one on the xy-plane, centered at $(0,0)$ and parallel to the y-axis. At time $n > 1$, place additional line segments that are centered at the end and perpendicular to an exposed toothpick end, where an *exposed end* is the end of a toothpick that is neither the end nor the midpoint of another toothpick. Figure 1.28 shows the configurations obtained for times $n \leq 6$. Let $f(n)$ be the total number of toothpicks that have been placed up to time n, and let

$$F(x) = \sum_{n \geq 1} f(n) x^n.$$

Figure 1.28 shows that

$$F(x) = x + 3x^2 + 7x^3 + 11x^4 + 15x^5 + 23x^6 + \cdots .$$

Show that

$$F(x) = \frac{x}{(1-x)(1-2x)} \left(1 + 2x \prod_{k \geq 0} \left(1 + x^{2^k - 1} + 2x^{2^k}\right) \right).$$

96. Define

$$x \prod_{n \geq 1}(1 - x^n)^{24} = \sum_{n \geq 1} \tau(n) x^n$$

$$= x - 24x^2 + 252x^3 - 1472x^4 + 4830x^5 - 6048x^6 - 16744x^7 + \cdots .$$

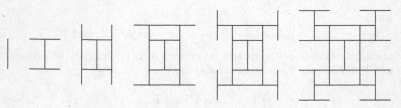

Figure 1.28 The growth of toothpicks.

a. [3+] Show that $\tau(mn) = \tau(m)\tau(n)$ if m and n are relatively prime.

b. [3+] Show that if p is prime and $n \geq 1$ then

$$\tau(p^{n+1}) = \tau(p)\tau(p^n) - p^{11}\tau(p^{n-1}).$$

c. [4] Show that if p is prime, then $|\tau(p)| < 2p^{11/2}$. Equivalently, write

$$\sum_{n \geq 0} \tau(p^n)x^n = \frac{P_p(x)}{1 - \tau(p)x + p^{11}x^2},$$

so by (b) and Theorem 4.4.1.1 the numerator $P_p(x)$ is a polynomial. Then the zeros of the denominator are not real.

d. [5] Show that $\tau(n) \neq 0$ for all $n \geq 1$.

97. [3−] Let $f(n)$ be the number of partitions of $2n$ whose Ferrers diagram can be covered by n edges, each connecting two adjacent dots. For instance, $(4,3,3,3,1)$ can be covered as follows:

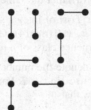

Show that $\sum_{n \geq 0} f(n)x^n = \prod_{i \geq 1}(1 - x^i)^{-2}$.

98. [2+] Let $n, a, k \in \mathbb{N}$ and $\zeta = e^{2\pi i/n}$. Show that

$$\binom{na}{k}_{q=\zeta} = \begin{cases} \binom{a}{b}, & k = nb \\ 0, & \text{otherwise.} \end{cases}$$

99. [2] Let $0 \leq k \leq n$ and $f(q) = \binom{n}{k}$. Compute $f'(1)$. Try to avoid a lot of computation.

100. [2+] State and prove a q-analogue of the Chu–Vandermonde identity

$$\sum_{i=0}^{n} \binom{a}{i}\binom{b}{n-i} = \binom{a+b}{n}$$

(Example 1.1.17).

101. [2]* Explain why we cannot set $q = 1$ on both sides of equation (1.85) to obtain the identity

$$1 = \sum_{k \geq 0} \frac{x^k}{k!}.$$

102. a. [2]* Let x and y be variables satisfying the commutation relation $yx = qxy$, where q commutes with x and y. Show that

$$(x+y)^n = \sum_{k=0}^{n} \binom{n}{k} x^k y^{n-k}.$$

b. [2]* Generalize to $(x_1 + x_2 + \cdots + x_m)^n$, where $x_i x_j = q x_j x_i$ for $i > j$.

c. [2+]* Generalize further to $(x_1 + x_2 + \cdots + x_m)^n$, where $x_i x_j = q_j x_j x_i$ for $i > j$, and where the q_j's are variables commuting with all the x_i's and with each other.

103. a. [3+] Given a partition λ (identified with its Young diagram) and $u \in \lambda$, let $a(u)$ (called the *arm length* of u) denote the number of squares directly to the right of u, counting u itself exactly once. Similarly, let $l(u)$ (called the *leg length* of u) denote the number of squares directly below u, counting u itself once. Thus, if $u = (i, j)$, then $a(u) = \lambda_i - j + 1$ and $l(u) = \lambda'_j - i + 1$. Define

$$\gamma(\lambda) = \#\{u \in \lambda : a(u) - l(u) = 0 \text{ or } 1\}.$$

Show that

$$\sum_{\lambda \vdash n} q^{\gamma(\lambda)} = \sum_{\lambda \vdash n} q^{\ell(\lambda)}, \tag{1.134}$$

where $\ell(\lambda)$ denotes the length (number of parts) of λ.

b. [2]* Clearly the coefficient of x^n in the right-hand side of equation (1.134) is 1. Show directly (without using (a)) that the same is true for the left-hand side.

104. [2+] Let $n \geq 1$. Find the number $f(n)$ of integer sequences $(a_1, a_2, \ldots, a_n)$ such that $0 \leq a_i \leq 9$ and $a_1 + a_2 + \cdots + a_n \equiv 0 \pmod{4}$. Give a simple explicit formula (no sums) that depends on the congruence class of n modulo 4.

105. a. [3–] Let $n \in \mathbb{P}$, and let $f(n)$ denote the number of subsets of $\mathbb{Z}/n\mathbb{Z}$ (the integers modulo n) whose elements sum to 0 in $\mathbb{Z}/n\mathbb{Z}$. For instance, $f(4) = 4$, corresponding to $\emptyset$, $\{0\}$, $\{1, 3\}$, $\{0, 1, 3\}$. Show that

$$f(n) = \frac{1}{n} \sum_{\substack{d \mid n \\ d \text{ odd}}} \phi(d) 2^{n/d},$$

where ϕ denotes Euler's totient function.

b. [5–] When n is odd, it can be shown using (a) (see Exercise 7.112) that $f(n)$ is equal to the number of necklaces (up to cyclic rotation) with n beads, each bead colored black or white. Give a combinatorial proof. (This is easy if n is prime.)

c. [5–] Generalize. For instance, investigate the number of subsets S of $\mathbb{Z}/n\mathbb{Z}$ satisfying $\sum_{i \in S} p(i) \equiv \alpha \pmod{n}$, where p is a fixed polynomial and $\alpha \in \mathbb{Z}/n\mathbb{Z}$ is fixed.

106. [2] Let $f(n, k)$ be the number of sequences $a_1 a_2 \cdots a_n$ of positive integers such that the largest number occurring is k and such that the first occurrence of i appears before the first occurrence of $i + 1$ ($1 \leq i \leq k - 1$). Express $f(n, k)$ in terms of familiar numbers. Give a combinatorial proof. (It is assumed that every number $1, 2, \ldots, k$ occurs at least once.)

107. [1+]* Give a direct combinatorial proof of equation (1.94e), namely,

$$B(n + 1) = \sum_{i=0}^{n} \binom{n}{i} B(i), \quad n \geq 0.$$

108. a. [2+] Give a combinatorial proof that the number of partitions of $[n]$ such that no two consecutive integers appear in the same block is the Bell number $B(n - 1)$.

b. [2+]* Give a combinatorial proof that the number of partitions of $[n]$ such that no two *cyclically consecutive* integers (i.e., two integers i, j for which $j \equiv i + 1 \pmod{n}$) is equal to the number of partitions of $[n]$ with no singleton blocks.

109. [2+]

   a. Show that the number of permutations $a_1 \cdots a_n \in \mathfrak{S}_n$ for which there is no $1 \le i < j \le n - 1$ satisfying $a_i < a_j < a_{j+1}$ is equal to the Bell number $B(n)$.

   b. Show that the same conclusion holds if the condition $a_i < a_j < a_{j+1}$ is replaced with $a_i < a_{j+1} < a_j$.

   c. Show that the number of permutations $w \in \mathfrak{S}_n$ satisfying the conditions of *both* (a) and (b) is equal to the number of involutions in $\mathfrak{S}_n$.

110. [3–] Let $f(n)$ be the number of partitions π of $[n]$ such that the union of no proper subset of the blocks of π is an interval $[a, b]$. For instance, $f(4) = 2$, corresponding to the partitions 13-24 and 1234, while $f(5) = 6$. Set $f(0) = 1$. Let

$$F(x) = \sum_{n \ge 0} f(n) x^n = 1 + x + x^2 + x^3 + 2x^4 + 6x^5 + \cdots .$$

Find the coefficients of $(x / F(x))^{\langle -1 \rangle}$.

111. [3–] Let $f(n)$ be the number of partitions π of $[n]$ such that no block of π is an interval $[a, b]$ (allowing $a = b$). Thus, $f(1) = f(2) = f(3) = 0$ and $f(4) = 1$, corresponding to the partition 13-24. Let

$$F(x) = \sum_{n \ge 0} f(n) x^n = 1 + x^4 + 5x^5 + 21x^6 + \cdots .$$

Express $F(x)$ in terms of the *ordinary* generating function $G(x) = \sum_{n \ge 0} B(n) x^n = 1 + x + 2x^2 + 5x^3 + 15x^4 + \cdots$.

112. [2]* How many permutations $w \in \mathfrak{S}_n$ have the same number of cycles as weak excedances?

113. [2–]* Fix $k, n \in \mathbb{P}$. How many sequences $(T_1, \ldots, T_k)$ of subsets T_i of $[n]$ are there such that the *nonempty* T_i form a partition of $[n]$?

114. a. [2–]* How many permutations $w = a_1 a_2 \cdots a_n \in \mathfrak{S}_n$ have the property that for all $1 \le i < n$, the numbers appearing in w between i and $i + 1$ (whether i is to the left or right of $i + 1$) are all less than i? An example of such a permutation is 976412358.

   b. [2–]* How many permutations $a_1 a_2 \cdots a_n \in \mathfrak{S}_n$ satisfy the following property: If $2 \le j \le n$, then $|a_i - a_j| = 1$ for some $1 \le i < j$? Equivalently, for all $1 \le i \le n$, the set $\{a_1, a_2, \ldots, a_i\}$ consists of consecutive integers (in some order). For example, for $n = 3$, there are the four permutations $123, 213, 231, 321$. More generally, find the number of such permutations with descent set $S \subseteq [n - 1]$.

115. [3–] Let $n = 2^{17} + 2$ and define $Q_n(t) = \sum_{S \subseteq [n-1]} t^{\beta_n(S)}$. Show that $e^{2\pi i / n}$ is (at least) a double root of $Q_n(t)$.

116. a. [2]* Show that the expected number of cycles of a random permutation $w \in \mathfrak{S}_n$ (chosen from the uniform distribution) is given by the harmonic number $H_n = 1 + \frac{1}{2} + \frac{1}{3} + \cdots + \frac{1}{n} \sim \log n$.

   b. [3] Let $f(n)$ be the expected length of the longest cycle of a random permutation $w \in \mathfrak{S}_n$ (again from the uniform distribuiton). Show that

$$\lim_{n \to \infty} \frac{f(n)}{n} = \int_0^\infty \exp\left(-x - \int_x^\infty \frac{e^{-y}}{y} dy \right) dx = 0.62432965 \cdots .$$

117. [2+] Let w be a random permutation of $1, 2, \ldots, n$ (chosen from the uniform distribution). Fix a positive integer $1 \le k \le n$. What is the probability p_{nk} that in the disjoint cycle decomposition of w, the length of the cycle containing 1 is k? In other words, what is the probability that k is the least positive integer for which $w^k(1) = 1$? Give a simple proof avoiding generating functions, induction, and so on.

118. **a.** [2]* Let w be a random permutation of $1, 2, \ldots, n$ (chosen from the uniform distribution), $n \geq 2$. Show that the probability that 1 and 2 are in the same cycle of w is $1/2$.

   b. [2+] Generalize (a) as follows. Let $2 \leq k \leq n$, and let $\lambda = (\lambda_1, \lambda_2, \ldots, \lambda_\ell) \vdash k$, where $\lambda_\ell > 0$. Choose a random permutation $w \in \mathfrak{S}_n$. Let P_λ be the probability that $1, 2, \ldots, \lambda_1$ are in the same cycle C_1 of w, and $\lambda_1 + 1, \ldots, \lambda_1 + \lambda_2$ are in the same cycle C_2 of w different from C_1, and so on. Show that

$$P_\lambda = \frac{(\lambda_1 - 1)! \cdots (\lambda_\ell - 1)!}{k!}.$$

   c. [3–] Same as (b), except now we take w uniformly from the alternating group $\mathfrak{A}_n$. Let the resulting probability be Q_λ. Show that

$$Q_\lambda = \frac{(\lambda_1 - 1)! \cdots (\lambda_\ell - 1)!}{(k-2)!} \left(\frac{1}{k(k-1)} + (-1)^{n-\ell} \frac{1}{n(n-1)} \right).$$

119. [2+] Let P_n denote the probability that a random permutation (chosen from the uniform distribution) in $\mathfrak{S}_{2n}$ has all cycle lengths at most n. Show that $\lim_{n \to \infty} P_n = 1 - \log 2 = 0.306852819 \cdots$.

120. [2+] Let $E_k(n)$ denote the expected number of k-cycles of a permutation $w \in \mathfrak{S}_n$, as discussed in Example 1.3.5. Give a simple combinatorial explanation of the formula $E_k(n) = 1/k$, $n \geq k$.

121. **a.** [2]* Let $f(n)$ denote the number of fixed-point free involutions $w \in \mathfrak{S}_{2n}$ (i.e., $w^2 = 1$, and $w(i) \neq i$ for all $i \in [2n]$). Find a simple expression for $\sum_{n \geq 0} f(n) x^n / n!$. (Set $f(0) = 1$.)

   b. [2–]* If $X \subseteq \mathbb{P}$, then write $-X = \{-i : i \in X\}$. Let $g(n)$ be the number of ways to choose a subset X of $[n]$, and then choose fixed point free involutions w on $X \cup (-X)$ and $\bar{w}$ on $\bar{X} \cup (-\bar{X})$, where $\bar{X} = \{i \in [n] : i \notin X\}$. Use (a) to find a simple expression for $g(n)$.

   c. [2+]* Find a combinatorial proof for the formula obtained for $g(n)$ in (b).

122. [2–]* Find $\sum_w x^{\mathrm{exc}(w)}$, where w ranges over all fixed-point free involutions in $\mathfrak{S}_{2n}$ and $\mathrm{exc}(w)$ denotes the number of excedances of w.

123. [2]* Let $\mathfrak{A}_n$ denote the alternating group on $[n]$ (i.e., the group of all permutations with an even number of cycles of even length). Define the *augmented cycle indicator* $\tilde{Z}_{\mathfrak{A}_n}$ of $\mathfrak{A}_n$ by

$$\tilde{Z}_{\mathfrak{A}_n} = \sum_{w \in \mathfrak{A}_n} t^{\mathrm{type}(w)},$$

as in equation (1.25). Show that

$$\sum_{n \geq 0} \tilde{Z}_{\mathfrak{A}_n} \frac{x^n}{n!} = \exp\left(t_1 x + t_3 \frac{x^3}{3} + t_5 \frac{x^5}{5} + \cdots \right) \cdot \cosh\left(t_2 \frac{x^2}{2} + t_4 \frac{x^4}{4} + t_6 \frac{x^6}{6} + \cdots \right).$$

124. **a.** [2] Let $f_k(n)$ denote the number of permutations $w \in \mathfrak{S}_n$ with k inversions. Show combinatorially that for $n \geq k$,

$$f_k(n+1) = f_k(n) + f_{k-1}(n+1).$$

b. [1+] Deduce from (a) that for $n \geq k$, $f_k(n)$ is a polynomial in n of degree k and leading coefficient $1/k!$. For instance, $f_2(n) = \frac{1}{2}(n+1)(n-2)$ for $n \geq 2$.

c. [2+] Let $g_k(n)$ be the polynomial that agrees with $f_k(n)$ for $n \geq k$. Find $\Delta^j g_k(-n)$; that is, find the coefficients a_j in the expansion

$$g_k(-n) = \sum_{j=0}^{k} a_j \binom{n}{j}.$$

125. [2+]* Find the number $f(n)$ of binary sequences $w = a_1 a_2 \cdots a_k$ (where k is arbitrary) such that $a_1 = 1$, $a_k = 0$, and $\mathrm{inv}(w) = n$. For instance, $f(4) = 5$, corresponding to the sequences 10000, 11110, 10110, 10010, 1100. How many of these sequences have exactly j 1's?

126. [2+]* Show that

$$\sum_w q^{\mathrm{inv}(w)} = q^n \prod_{j=0}^{n-1} (1 + q^2 + q^4 + \cdots + q^{4j}),$$

where w ranges over all fixed-point free involutions in $\mathfrak{S}_{2n}$, and where $\mathrm{inv}(w)$ denotes the number of inversions of w. Give a simple combinatorial proof analogous to the proof of Corollary 1.3.13.

127. [2]

a. Let $w \in \mathfrak{S}_n$, and let $R(w)$ be the set of positions of the records (or left-to-right maxima) of w. For instance, $R(3265174) = \{1, 3, 6\}$. For any finite set S of positive integers, set $x^S = \prod_{i \in S} x_i$. Show that

$$\sum_{w \in \mathfrak{S}_n} q^{\mathrm{inv}(w)} x^{R(w)} = x_1 (x_2 + q)(x_3 + q + 1) \cdots (x_n + q + q^2 + \cdots + q^{n-1}). \quad (1.135)$$

b. Let $V(w)$ be the set of the records themselves (e.g., $V(3265174) = \{3, 6, 7\}$). Show that

$$\sum_{w \in \mathfrak{S}_n} q^{\mathrm{inv}(w)} x^{V(w)} = (x_1 + q + q^2 + \cdots + q^{n-1})(x_2 + q + q^2 + \cdots + q^{n-2}) \cdots$$

$$\times (x_{n-1} + q) x_n. \quad (1.136)$$

128. a. [2] A permutation $a_1 \cdots a_n$ of $[n]$ is called *indecomposable* or *connected* if n is the least positive integer j for which $\{a_1, a_2, \ldots, a_j\} = \{1, 2, \ldots, j\}$. Let $f(n)$ be the number of indecomposable permutations of $[n]$, and set $F(x) = \sum_{n \geq 0} n! x^n$. Show that

$$\sum_{n \geq 1} f(n) x^n = 1 - \frac{1}{F(x)}. \quad (1.137)$$

b. [2+] If $a_1 \cdots a_n$ is a permutation of $[n]$, then a_i is called a *strong fixed point* if (1) $j < i \Rightarrow a_j < a_i$, and (2) $j > i \Rightarrow a_j > a_i$ (so in particular $a_i = i$). Let $g(n)$ be the number of permutations of $[n]$ with no strong fixed points. Show that

$$\sum_{n \geq 0} g(n) x^n = \frac{F(x)}{1 + x F(x)}.$$

c. [2+] A permutation $w \in \mathfrak{S}_n$ is *stabilized-interval-free* (SIF) if there does not exist $1 \le i < j \le n$ for which $w \cdot [i,j] = [i,j]$ (as sets). For instance, 615342 fails to be SIF since $w \cdot [3,5] = [3,5]$. Let $h(n)$ be the number of SIF permutations $w \in \mathfrak{S}_n$, and set

$$H(x) = \sum_{n \ge 0} h(n)x^n = 1 + x + x^2 + 2x^3 + 7x^4 + 34x^5 + 206x^6 + \cdots.$$

Show that

$$H(x) = \frac{x}{\left(\sum_{n \ge 0} n!x^{n+1}\right)^{\langle -1 \rangle}},$$

where $\langle -1 \rangle$ denotes compositional inverse (§5.4 of Vol. II). Equivalently, by the Lagrange inversion formula (Theorem 5.4.2 of Vol. II), $H(x)$ is uniquely defined by the condition

$$[x^{n-1}]H(x)^n = n!, \ n \ge 1.$$

d. [2+] A permutation $w \in \mathfrak{S}_n$ is called *simple* if it maps no interval $[i,j]$ of size $1 < j - i + 1 < n$ into another such interval. For instance, 3157462 is not simple, since it maps $[3,6]$ into $[4,7]$ (as sets). Let $k(n)$ be the number of simple permutations $w \in \mathfrak{S}_n$, and set

$$K(x) = \sum_{n \ge 1} k(n)x^n = x + 2x^2 + 2x^4 + 6x^5 + 46x^7 + 338x^8 + \cdots.$$

Show that

$$K(x) = \frac{2}{1+x} - \left(\sum_{n \ge 1} n!x^n\right)^{\langle -1 \rangle}.$$

129. a. [2]* Let $f_k(n)$ be the number of indecomposable permutations $w \in \mathfrak{S}_n$ with k inversions. Generalizing equation (1.137), show that

$$\sum_{n \ge 1} f_k(n)q^k x^n = 1 - \frac{1}{F(q,x)},$$

where $F(q,x) = \sum_{n \ge 0} (n)!x^n$. As usual, $(n)! = (1+q)(1+q+q^2)\cdots(1+q+\cdots+q^{n-1})$.

b. [2] Write $1/F(q,x) = \sum_{n \ge 0} g_n(q)x^n$, where $g_n(q) \in \mathbb{Z}[q]$. Show that $\sum_{n \ge 0} g_n(q)$ is a well-defined formal power series, even though it makes no sense to substitute directly $x = 1$ in $1/F(q,x)$.

c. [3] Write $1/F(q,x)$ in a form where it does make sense to substitute $x = 1$.

130. [2+] Let $u(n)$ be the number of permutations $w = a_1 \cdots a_n \in \mathfrak{S}_n$ such that $a_{i+1} \ne a_i \pm 1$ for $1 \le i \le n-1$. Equivalently, $f(n)$ is the number of ways to place n nonattacking kings on an $n \times n$ chessboard, no two on the same file or rank. Set

$$U(x) = \sum_{n \ge 0} u(n)x^n = 1 + x + 2x^4 + 14x^5 + 90x^6 + 646x^7 + 5242x^8 + \cdots.$$

Show that

$$U(x) = F\left(\frac{x(1-x)}{1+x}\right), \qquad (1.138)$$

where $F(x) = \sum_{n \ge 0} n!x^n$ as in Exercise 1.128.

131. [2+]* An n-dimensional cube K_n has $2n$ facets (or $(n-1)$-dimensional faces), which come in n antipodal pairs. A *shelling* of K_n is equivalent to a linear ordering $F_1, F_2, \ldots, F_{2n}$ of its facets such that for all $1 \leq i \leq n-1$, the set $\{F_1, \ldots, F_{2i}\}$ does not consist of i antipodal pairs. Let $f(n)$ be the number of shellings of K_n. Show that

$$\sum_{n \geq 1} f(n) \frac{x^n}{n!} = 1 - \left(\sum_{n \geq 0} (2n)! \frac{x^n}{n!} \right)^{-1}.$$

132. [1+]* Let $w \in \mathfrak{S}_n$. Which of the following items doesn't belong?
- $\mathrm{inv}(w) = 0$
- $\mathrm{maj}(w) = 0$
- $\mathrm{des}(w) = 0$
- $\mathrm{maj}(w) = \mathrm{des}(w) = \mathrm{inv}(w)$
- $D(w) = \emptyset$
- $c(w) = n$ (where $c(w)$ denotes the number of cycles of w)
- $w^5 = w^{12} = 1$

133. a. [2+] Let $A_n(x)$ be the Eulerian polynomial. Give a combinatorial proof that $\frac{1}{2} A_n(2)$ is equal to the number of *ordered* set partitions (i.e., partitions whose blocks are linearly ordered) of an n-element set.

  b. [2+]* More generally, show that

$$\frac{A_n(x)}{x} = \sum_{k=0}^{n-1} (n-k)! S(n, n-k)(x-1)^k.$$

  Note that $(n-k)! S(n, n-k)$ is the number of ordered partitions of an n-set into $n-k$ blocks.

134. [3−] Show that

$$A_n(x) = \sum_w x^{1+\mathrm{des}(w)} (1+x)^{n-1-2\mathrm{des}(w)},$$

where w ranges over all permutations in $\mathfrak{S}_n$ with no proper double descents (as defined in Exercise 1.61) and with no descent at the end. For instance, when $n = 4$, the permutations are 1234, 1324, 1423, 2134, 2314, 2413, 3124, 3412, 4123.

135. a. [2] Let $A_n(x)$ be the Eulerian polynomial. Show that

$$A_n(-1) = \begin{cases} (-1)^{(n+1)/2} E_n, & n \text{ odd}, \\ 0, & n \text{ even}. \end{cases}$$

  b. [3−] Give a combinatorial proof of (a) when n is odd.

136. [2+] What sequence $c = (c_1, \ldots, c_n) \in \mathbb{N}^n$ with $\sum i c_i = n$ maximizes the number of $w \in \mathfrak{S}_n$ of type c? For instance, when $n = 4$ the maximizing sequence is $(1, 0, 1, 0)$.

137. [3−] Let ℓ be a prime number and write $n = a_0 + a_1 \ell + a_2 \ell^2 + \cdots$, with $0 \leq a_i < \ell$ for all $i \geq 0$. Let $\kappa_\ell(n)$ denote the number of sequences $c = (c_1, c_2, \ldots, c_n) \in \mathbb{N}^n$ with $\sum i c_i = n$, such that the number of permutations $w \in \mathfrak{S}_n$ of type c is relatively prime to ℓ. Show that

$$\kappa_\ell(n) = p(a_0) \prod_{i \geq 1} (a_i + 1),$$

where $p(a_0)$ is the number of partitions of a_0. In particular, the number of c such that an odd number of $w \in \mathfrak{S}_n$ have type c is 2^b, where $\lfloor n/2 \rfloor$ has b 1's in its binary expansion.

138. [2+]* Find a simple formula for the number of alternating permutations $a_1 a_2 \cdots a_{2n} \in \mathfrak{S}_{2n}$ satisfying $a_2 < a_4 < a_6 < \cdots < a_{2n}$.

139. [2+] An *even tree* is a (rooted) tree such that every vertex has an even number of children. (Such a tree must have an odd number of vertices.) Note that these are *not* plane trees (i.e., we don't linearly order the subtrees of a vertex). Express the number $g(n)$ of increasing even trees with $2n + 1$ vertices in terms of Euler numbers. Use generating functions.

140. [3−] Define a *simsun permutation* to be a permutation $w \in \mathfrak{S}_n$ such that w has no proper double descents (as defined in Exercise 1.61(c)) and such that for all $0 \le k \le n - 1$, if we remove $n, n - 1, \cdots, n - k$ from w (written as a word) then the resulting permutation also has no proper double descents. For instance, $w = 3241$ is not simsun since if we remove 4 from w we obtain 321, which has a proper double descent. Show that the number of simsun permutations in $\mathfrak{S}_n$ is equal to the Euler number E_{n+1}.

141. a. [2+] Let $E_{n,k}$ denote the number of alternating permutations of $[n + 1]$ with first term $k + 1$. For instance, $E_{n,n} = E_n$. Show that

$$E_{0,0} = 1, \quad E_{n,0} = 0 \ (n \ge 1), \quad E_{n+1,k+1} = E_{n+1,k} + E_{n,n-k} \ (n \ge k \ge 0). \quad (1.139)$$

Note that if we place the $E_{n,k}$'s in the triangular array

$$
\begin{array}{ccccccccc}
 & & & & E_{00} & & & & \\
 & & & E_{10} & \to & E_{11} & & & \\
 & & E_{22} & \leftarrow & E_{21} & \leftarrow & E_{20} & & \\
 & E_{30} & \to & E_{31} & \to & E_{32} & \to & E_{33} & \\
E_{44} & \leftarrow & E_{43} & \leftarrow & E_{42} & \leftarrow & E_{41} & \leftarrow & E_{40} \\
 & & & & \cdots & & & &
\end{array}
\qquad (1.140)
$$

and read the entries in the direction of the arrows from top-to-bottom (the so-called *boustrophedon* or *ox-plowing* order), then the first number read in each row is 0, and each subsequent entry is the sum of the previous entry and the entry above in the previous row. The first seven rows of the array are as follows:

$$
\begin{array}{ccccccccccc}
 & & & & & 1 & & & & & \\
 & & & & 0 & \to & 1 & & & & \\
 & & & 1 & \leftarrow & 1 & \leftarrow & 0 & & & \\
 & & 0 & \to & 1 & \to & 2 & \to & 2 & & \\
 & 5 & \leftarrow & 5 & \leftarrow & 4 & \leftarrow & 2 & \leftarrow & 0 & \\
0 & \to & 5 & \to & 10 & \to & 14 & \to & 16 & \to & 16 \\
61 & \leftarrow & 61 & \leftarrow & 56 & \leftarrow & 46 & \leftarrow & 32 & \leftarrow & 16 & \leftarrow & 0. \\
 & & & & & \cdots & & & & &
\end{array}
$$

b. [3−] Define

$$[m,n] = \begin{cases} m, & m + n \text{ odd}, \\ n, & m + n \text{ even}. \end{cases}$$

Show that

$$\sum_{m \ge 0} \sum_{n \ge 0} E_{m+n,[m,n]} \frac{x^m}{m!} \frac{y^n}{n!} = \frac{\cos x + \sin x}{\cos(x + y)}. \qquad (1.141)$$

142. [3–] Define polynomials $f_n(a)$ for $n \geq 0$ by $f_0(a) = 1$, $f_n(0) = 0$ if $n \geq 1$, and $f'_n(a) = f_{n-1}(1-a)$. Thus,

$$f_1(a) = a,$$

$$f_2(a) = \frac{1}{2}(-a^2 + 2a),$$

$$f_3(a) = \frac{1}{3!}(-a^3 + 3a),$$

$$f_4(a) = \frac{1}{4!}(a^4 - 4a^3 + 8a),$$

$$f_5(a) = \frac{1}{5!}(a^5 - 10a^3 + 25a),$$

$$f_6(a) = \frac{1}{6!}(-a^6 + 6a^5 - 40a^3 + 96a).$$

Show that $\sum_{n \geq 0} f_n(1)x^n = \sec x + \tan x$.

143. a. [2–] Let $\mathrm{fix}(w)$ denote the number of fixed points (cycles of length 1) of the permutation $w \in \mathfrak{S}_n$. Show that

$$\sum_{w \in \mathfrak{S}_n} \mathrm{fix}(w) = n!.$$

Try to give a combinatorial proof, a generating function proof, and an algebraic proof.

b. [3+] Let Alt_n (respectively, Ralt_n) denote the set of alternating (respectively, reverse alternating) permutations $w \in \mathfrak{S}_n$. Define

$$f(n) = \sum_{w \in \mathrm{Alt}_n} \mathrm{fix}(w)$$

$$g(n) = \sum_{w \in \mathrm{Ralt}_n} \mathrm{fix}(w).$$

Show that

$$f(n) = \begin{cases} E_n - E_{n-2} + E_{n-4} - \cdots + (-1)^{(n-1)/2}E_1, & n \text{ odd}, \\ E_n - 2E_{n-2} + 2E_{n-4} - \cdots + (-1)^{(n-2)/2}2E_2 + (-1)^{n/2}, & n \text{ even}. \end{cases}$$

$$g(n) = \begin{cases} E_n - E_{n-2} + E_{n-4} - \cdots + (-1)^{(n-1)/2}E_1, & n \text{ odd}, \\ E_n - (-1)^{n/2}, & n \text{ even}. \end{cases}$$

144. a. [2] Let

$$F(x) = 2 \sum_{n \geq 0} q^n \frac{\prod_{j=1}^{n}(1 - q^{2j-1})}{\prod_{j=1}^{2n+1}(1 + q^j)},$$

where $q = \left(\frac{1-x}{1+x}\right)^{2/3}$. Show that $F(x)$ is well-defined as a formal power series. Note that $q(0) = 1 \neq 0$, so some special argument is needed.

b. [3+] Let $F(x)$ be defined by (a), and write

$$F(x) = \sum_{n \geq 0} f(n)x^n = 1 + x + x^2 + 2x^3 + 5x^4 + 17x^5 + 72x^6$$

$$+ 367x^7 + 2179x^8 + \cdots.$$

Show that $f(n)$ is equal to the number of alternating fixed-point free involutions in $\mathfrak{S}_{2n}$ (i.e., the number of permutations $w \in \mathfrak{S}_{2n}$ that are alternating permutations and have n cycles of length two). For instance, when $n = 3$ we have the two permutations 214365 and 645321, and when $n = 4$ we have the five permutations 21436587, 21867453, 64523187, 64827153, and 84627351.

145. [3–] Solve the following chess problem, where the condition "serieshelpmate" is defined in Exercise 1.11(c).

A. Karttunen, 2006

Serieshelpmate in 9: how many solutions?

146. [2+] Let $f_k(n)$ denote the number of permutations $w \in \mathfrak{S}_n$ such that

$$D(w) = \{k, 2k, 3k, \dots\} \cap [n-1],$$

as in equation (1.58). Let $1 \le i \le k$. Show that

$$\sum_{m \ge 0} f_k(mk+i) \frac{x^{mk+i}}{(mk+i)!} = \frac{\sum_{m \ge 0}(-1)^m \frac{x^{mk+i}}{(mk+i)!}}{\sum_{m \ge 0}(-1)^m \frac{x^{mk}}{(mk)!}}.$$

Note that when $i = k$ we can add 1 to both sides and obtain equation (1.59).

147. [2+] Call two permutations $u, v \in \mathfrak{S}_n$ *equivalent* if their min-max trees $M(u)$ and $M(v)$ are isomorphic as *unlabeled* binary trees. This notion of equivalence is clearly an equivalence relation. Show that the number of equivalence classes is the Motzkin number M_{n-1} defined in Exercise 6.37 and further explicated in Exercise 6.38.

148. [2+] Let $\Phi_n = \Phi_n(c,d)$ denote the cd-index of $\mathfrak{S}_n$, as defined in Theorem 1.6.3. Thus, $c = a+b$ and $d = ab+ba$. Let $S \subseteq [n-1]$, and let u_S be the variation of S as defined by equation (1.60). Show that

$$\Phi_n(a+2b, ab+ba+2b^2) = \sum_{S \subseteq [n-1]} \alpha(S) u_S,$$

where $\alpha(S)$ is given by equation (1.31).

149. [3–] If $F(x)$ is any power series with noncommutative coefficients such that $F(0) = 0$, then define $(1 - F(x))^{-1}$ to be the unique series $G(x)$ satisfying

$$(1 - F(x))G(x) = G(x)(1 - F(x)) = 1.$$

Equivalently, $G(x) = 1 + F(x) + F(x)^2 + \cdots$. Show that

$$\sum_{n \geq 1} \Phi_n(c,d) \frac{x^n}{n!} = \frac{\sinh(a-b)x}{a-b} \left[1 - \frac{1}{2} \left(\frac{c \cdot \sinh(a-b)x}{a-b} - \cosh(a-b)x + 1 \right) \right]^{-1}.$$

(1.142)

Note that the series on the right involves only *even* powers of $a - b$. Since $(a-b)^2 = c^2 - 2d$, it follows that the coefficients of this series are indeed polynomials in c and d.

150. **a.** [3–]* Let $f(n)$ (respectively, $g(n)$) be the total number of c's (respectively, d's) that appear when we write the cd-index $\Phi_n(c,d)$ as a sum of monomials. For instance, $\Phi_4(c,d) = c^3 + 2cd + 2dc$, so $f(4) = 7$ and $g(4) = 4$. Show using generating functions that $f(n) = 2E_{n+1} - (n+1)E_n$ and $g(n) = nE_n - E_{n+1}$.
b. [5–] Is there a combinatorial proof?

151. [3–] Let μ be a monomial of degree $n - 1$ in the noncommuting variables c,d, where $\deg(c) = 1$ and $\deg(d) = 2$. Show that $[\mu]\Phi_n(c,d)$ is the number of sequences $\mu = v_0, v_1, \ldots, v_{n-1} = 1$, where v_i is obtained from v_{i-1} by removing a c or changing a d to c. For instance, if $\mu = dcc$, there are three sequences: $(dcc, ccc, cc, c, 1)$, $(dcc, dc, cc, c, 1)$, $(dcc, dc, d, c, 1)$.

152. [3–] Continue the notation from the previous exercise. Replace each c in μ with 0, each d with 10, and remove the final 0. We get the characteristic vector of a set $S_\mu \subseteq [n-2]$. For instance, if $\mu = cd^2c^2d$ then we get the characteristic vector 01010001 of the set $S_\mu = \{2,4,8\}$. Show that $[\mu]\Phi_n(c,d)$ is equal to the number of simsun permutations (defined in Exercise 1.140) in $\mathfrak{S}_{n-1}$ with descent set S_μ.

153. **a.** [2] Let $f(n)$ denote the coefficient of d^n in the cd-index Φ_{2n+1}. Show that $f(n) = 2^{-n}E_{2n+1}$.
b. [3] Show that $f(n)$ is the number of permutations w of the multiset $\{1^2, 2^2, \ldots, (n+1)^2\}$ beginning with 1 such that between the two occurrences of i $(1 \leq i \leq n)$ there is exactly one occurrence of $i + 1$. For instance, $f(2) = 4$, corresponding to 123123, 121323, 132312, 132132.

154. **a.** [1+] Let $F(x) = \sum_{n \geq 0} f(n)x^n/n!$. Show that

$$e^{-x}F(x) = \sum_{n \geq 0} [\Delta^n f(0)]x^n/n!.$$

b. [2] Find the unique function $f : \mathbb{P} \to \mathbb{C}$ satisfying $f(1) = 1$ and $\Delta^n f(1) = f(n)$ for all $n \in \mathbb{P}$.
c. [2] Generalize (a) by showing that

$$e^{-x}F(x+t) = \sum_{n \geq 0} \sum_{k \geq 0} \Delta^n f(k) \frac{x^n}{n!} \frac{t^k}{k!}.$$

155. **a.** [1+] Let $F(x) = \sum_{n \geq 0} f(n)x^n$. Show that

$$\frac{1}{1+x} F\left(\frac{x}{1+x}\right) = \sum_{n \geq 0} [\Delta^n f(0)]x^n.$$

b. [2+] Find the unique functions $f, g : \mathbb{N} \to \mathbb{C}$ satisfying $\Delta^n f(0) = g(n)$, $\Delta^{2n}g(0) = f(n)$, $\Delta^{2n+1}g(0) = 0$, $f(0) = 1$.
c. [2+] Find the unique functions $f, g : \mathbb{N} \to \mathbb{C}$ satisfying $\Delta^n f(1) = g(n)$, $\Delta^{2n}g(0) = f(n)$, $\Delta^{2n+1}g(0) = 0$, $f(0) = 1$.

156. [2+] Let A be the abelian group of all polynomials $p : \mathbb{Z} \to \mathbb{C}$ such that $D^k p : \mathbb{Z} \to \mathbb{Z}$ for all $k \in \mathbb{N}$. (D^k denotes the kth derivative, and $D^0 p = p$.) Then A has a basis of the form $p_n(x) = c_n \binom{x}{n}$, $n \in \mathbb{N}$, where c_n is a constant depending only on n. Find c_n explicitly.

157. [2] Let λ be a complex number (or indeterminate), and let

$$y = 1 + \sum_{n \geq 1} f(n)x^n, \quad y^\lambda = \sum_{n \geq 0} g(n)x^n.$$

Show that

$$g(n) = \frac{1}{n} \sum_{k=1}^{n} [k(\lambda + 1) - n] f(k) g(n - k), \quad n \geq 1.$$

This formula affords a method of computing the coefficients of y^λ much more efficiently than using (1.5) directly.

158. [2+] Let $f_1, f_2, \ldots$ be a sequence of complex numbers. Show that there exist unique complex numbers $a_1, a_2, \ldots$ such that

$$F(x) := 1 + \sum_{n \geq 1} f_n x^n = \prod_{i \geq 1} (1 - x^i)^{-a_i}.$$

Set $\log F(x) = \sum_{n \geq 1} g_n x^n$. Find a formula for a_i in terms of the g_n's. What are the a_i's when $F(x) = 1 + x$ and $F(x) = e^{x/(1-x)}$?

159. [2] Let $F(x) = 1 + a_1 x + \cdots \in K[[x]]$, where K is a field satisfying $\mathrm{char}(K) \neq 2$. Show that there exist unique series $A(x), B(x)$ satisfying $A(0) = B(0) = 1$, $A(x) = A(-x)$, $B(x)B(-x) = 1$, and $F(x) = A(x)B(x)$. Find simple formulas for $A(x)$ and $B(x)$ in terms of $F(x)$.

160. a. [2] Let $0 \leq j < k$. The (k, j)-*multisection* of the power series $F(x) = \sum_{n \geq 0} a_n x^n$ is defined by

$$\Psi_{k,j} F(x) = \sum_{m \geq 0} a_{km+j} x^{km+j}.$$

Let $\zeta = e^{2\pi i/k}$ (where $i^2 = -1$). Show that

$$\Psi_{k,j} F(x) = \frac{1}{k} \sum_{r=0}^{k-1} \zeta^{-jr} F(\zeta^r x).$$

b. [2] As a simple application of (a), let $0 \leq j < k$, and let $f(n, k, j)$ be the number of permutations $w \in \mathfrak{S}_n$ satisfying $\mathrm{maj}(w) \equiv j \pmod{k}$. Show that $f(n, k, j) = n!/k$ if $n \geq k$.

c. [2+] Show that

$$f(k - 1, k, 0) = \frac{(k - 1)!}{k} + \sum_{\zeta} \frac{1}{(1 - \xi)^{k-1}},$$

where ξ ranges over all primitive kth roots of unity. Can this expression be simplified?

161. a. [2]* Let $F(x) = a_0 + a_1 x + \cdots \in K[[x]]$, with $a_0 = 1$. For $k \geq 2$ define $F_k(x) = \Phi_{k,0}(x) = \sum_{m \geq 0} a_{km} x^{km}$. Show that for $n \geq 1$,

$$[x^{km}] \frac{F(x)}{\Phi_{k,0} F(x)} = 0.$$

b. [2+] Let char $K \neq 2$. Given $G(x) = 1 + H(x)$ where $H(-x) = -H(x)$ (i.e., $H(x)$ has only odd exponents), find the general solution $F(x) = 1 + a_1 x + \cdots$ to $F(x)/F_2(x) = G(x)$. Express your answer in the form $F(x) = \Phi(G(x))E(x)$, where $\Phi(x)$ is a function independent from $G(x)$, and where $E(x)$ ranges over some class $\mathcal{E}$ of power series, also independent from $G(x)$.

162. [3–] Let $g(x) \in \mathbb{C}[[x]]$, $g(0) = 0$, $g(x) = g(-x)$. Find all power series $f(x)$ such that $f(0) = 0$ and

$$\frac{f(x) + f(-x)}{1 - f(x)f(-x)} = g(x).$$

Express your answer as an explicit algebraic function of $g(x)$ and a power series $h(x)$ (independent from $g(x)$) taken from some class of power series.

163. Let $f(x) \in \mathbb{C}[[x]]$, $f(x) = x +$ higher order terms. We say that $F(x,y) \in \mathbb{C}[[x,y]]$ is a *formal group law* or *addition law* for $f(x)$ if $f(x + y) = F(f(x), f(y))$.

a. [2–] Show that for every $f(x) \in \mathbb{C}[[x]]$ with $f(x) = x + \cdots$, there is a unique $F(x,y) \in \mathbb{C}[[x,y]]$ which is a formal group law for $f(x)$.

b. [3] Show that $F(x,y)$ is a formal group law if and only if $F(x,y) = x + y +$ higher order terms, and

$$F(F(x,y),z) = F(x,F(y,z)).$$

c. [2] Find $f(x)$ so that $F(x,y)$ is a formal group law for $f(x)$ in the following cases:
- $F(x,y) = x + y$.
- $F(x,y) = x + y + xy$.
- $F(x,y) = (x + y)/(1 - xy)$.
- $F(x,y) = x\sqrt{1 - y^2} + y\sqrt{1 - x^2}$.

d. [2+] Using equation (5.128), show that the formal group law for $f(x) = xe^{-x}$ is given by

$$F(x,y) = x + y - \sum_{n \geq 1} (n-1)^{n-1} \frac{x^n y + xy^n}{n!},$$

where we interpret $0^0 = 1$ in the summand indexed by $n = 1$.

e. [3] Find the formal group law for the function

$$f(x) = \int_0^x \frac{dt}{\sqrt{1 - t^4}}.$$

164. [3–] Solve the following equation for the power series $F(x,y) \in \mathbb{C}[[x,y]]$:

$$(xy^2 + x - y)F(x,y) = xF(x,0) - y.$$

The point is to make sure that your solution has a power series expansion at $(0,0)$.

165. [2+] Find a simple description of the coefficients of the power series $F(x) = x + \cdots \in \mathbb{C}[[x]]$ satisfying the functional equation

$$F(x) = (1 + x)F(x^2) + \frac{x}{1 - x^2}.$$

166. [2] Let $n \in \mathbb{P}$. Find a power series $F(x) \in \mathbb{C}[[x]]$ satisfying $F(F(x))^n = 1 + F(x)^n$, $F(0) = 1$.

167. [2] Let $F(x) \in \mathbb{C}[[x]]$. Find a simple expression for the exponential generating function of the derivatives of $F(x)$, that is,

$$\sum_{n \geq 0} D^n F(x) \frac{t^n}{n!}, \tag{1.143}$$

where $D = d/dx$.

168. Let K be a field satisfying $\text{char}(K) \neq 2$. If $A(x) = x + \sum_{n \geq 2} a_n x^n \in K[[x]]$, then let $A^{\langle -1 \rangle}(x)$ denote the compositional inverse of A; that is, $A^{\langle -1 \rangle}(A(x)) = A(A^{\langle -1 \rangle}(x)) = x$.

   a. [3–] Show that we can specify $a_2, a_4, \ldots$ arbitrarily, and they then determine uniquely $a_3, a_5, \ldots$ so that $A(-A(-x)) = x$. For instance,

$$a_3 = a_2^2,$$

$$a_5 = 3a_4 a_2 - 2a_2^4,$$

$$a_7 = 13a_2^6 - 18a_4 a_2^3 + 2a_4^2 + 4a_2 a_6.$$

   NOTE. Let $E(x) = A(-x)$. Then the conditions $A(x) = x + \cdots$ and $A(-A(-x)) = x$ are equivalent to $E(x) = -x + \cdots$ and $E(E(x)) = x$.

   b. [5–] What are the coefficients when a_{2n+1} is written as a polynomial in $a_2, a_4, \ldots$ as in (a)?

   c. [2+]* Show that $A(-A(-x)) = x$ if and only if there is a $B(x) = x + \sum_{n \geq 2} b_n x^n$ such that $A(x) = B^{\langle -1 \rangle}(-B(-x))$.

   d. [2+] Show that if $A(-A(-x)) = x$, then there is a unique $B(x)$ as in (c) of the form $B(x) = x + \sum_{n \geq 1} b_{2n} x^{2n}$. For instance,

$$b_2 = -\frac{1}{2} a_2,$$

$$b_4 = \frac{1}{8} \left(5a_2^3 - 4a_4 \right),$$

$$b_6 = -\frac{1}{16} \left(49a_2^5 - 56a_2^2 a_4 + 8a_6 \right).$$

   e. [5–] What are the coefficients when b_{2n} is written as a polynomial in $a_2, a_4, \ldots$ as in (d)?

   f. [2+] For any $C(x) = x + c_2 x^2 + c_3 x^3 + \cdots$, show that there are unique power series

$$A(x) = x + a_2 x^2 + a_3 x^3 + \cdots,$$
$$D(x) = x + d_3 x^3 + d_5 x^5 + \cdots,$$

such that $A(-A(-x)) = x$ and $C(x) = D(A(x))$. For instance,

$$a_2 = c_2,$$
$$d_3 = c_3 - c_2^2,$$
$$a_4 = c_4 - 3c_3 c_2 + 3c_2^3,$$
$$d_5 = c_5 + 3c_2^2 c_3 - 3c_2 c_4 - c_2^4.$$

   g. [2+] Find $A(x)$ and $D(x)$ as in (f) when $C(x) = -\log(1-x)$.

   h. [5–] What are the coefficients when a_{2n} and d_{2n+1} are written as a polynomial in $c_2, c_3, \ldots$ as in (f)?

i. [2+] Note that if $A(x) = x/(1+2x)$, then $A(-A(-x)) = x$. Show that

$$B^{\langle -1 \rangle}(-B(-x)) = x/(1+2x)$$

if and only if $e^{-x} \sum_{n \geq 0} b_{n+1} x^n/n!$ is an even function of x (i.e., has only even exponents).

j. [2+] Identify the coefficients b_{2n} of the unique $B(x) = x + \sum_{n \geq 1} b_{2n} x^{2n}$ satisfying $B^{\langle -1 \rangle}(-B(-x)) = x/(1+2x)$.

169. [2] Find a closed-form expression for the following generating functions:

a. $\displaystyle\sum_{n \geq 0} (n+2)^2 x^n$.

b. $\displaystyle\sum_{n \geq 0} (n+2)^2 \frac{x^n}{n!}$.

c. $\displaystyle\sum_{n \geq 0} (n+2)^2 \binom{2n}{n} x^n$.

170. a. [2−] Given $a_0 = \alpha$, $a_1 = \beta$, $a_{n+1} = a_n + a_{n-1}$ for $n \geq 1$, compute $y = \sum_{n \geq 0} a_n x^n$.

b. [2+] Given $a_0 = 1$ and $a_{n+1} = (n+1)a_n - \binom{n}{2}a_{n-2}$ for $n \geq 0$, compute $y = \sum_{n \geq 0} a_n x^n/n!$.

c. [2] Given $a_0 = 1$ and $2a_{n+1} = \sum_{i=0}^{n} \binom{n}{i} a_i a_{n-i}$ for $n \geq 0$, compute $\sum_{n \geq 0} a_n x^n/n!$ and find a_n explicitly. Compare equation (1.55), where (in the notation of the present exercise), $a_1 = 1$ and the recurrence holds for $n \geq 1$.

d. [3] Let $a_k(0) = \delta_{0k}$, and for $1 \leq k \leq n+1$ let

$$a_k(n+1) = \sum_{j=0}^{n} \binom{n}{j} \sum_{\substack{2r+s=k-1 \\ r,s \geq 0}} (a_{2r}(j) + a_{2r+1}(j)) a_s(n-j).$$

Compute $A(x,t) := \sum_{k,n \geq 0} a_k(n) t^k x^n/n!$.

171. Given a sequence $a_0, a_1, \ldots$ of complex numbers, let $b_n = a_0 + a_1 + \cdots + a_n$.

a. [1+]* Let $A(x) = \sum_{n \geq 0} a_n x^n$ and $B(x) = \sum_{n \geq 0} b_n x^n$. Show that

$$B(x) = \frac{A(x)}{1-x}.$$

b. [2+] Let $A(x) = \sum_{n \geq 0} a_n \frac{x^n}{n!}$ and $B(x) = \sum_{n \geq 0} b_n \frac{x^n}{n!}$. Show that

$$B(x) = \left(I(e^{-x} A'(x)) + a_0 \right) e^x, \qquad (1.144)$$

where I denotes the *formal integral*, that is,

$$I\left(\sum_{n \geq 0} c_n x^n \right) = \sum_{n \geq 0} c_n \frac{x^{n+1}}{n+1} = \sum_{n \geq 1} c_{n-1} \frac{x^n}{n}.$$

172. [3−] The *Legendre polynomial* $P_n(x)$ is defined by

$$\frac{1}{\sqrt{1 - 2xt + t^2}} = \sum_{n \geq 0} P_n(x) t^n.$$

Show that $(1-x)^n P_n((1+x)/(1-x)) = \sum_{k=0}^{n} \binom{n}{k}^2 x^k$.

173. [2+] Find simple closed expressions for the coefficients of the power series (expanded about $x = 0$):

a. $\sqrt{\dfrac{1+x}{1-x}}$.

b. $2\left(\sin^{-1}\dfrac{x}{2}\right)^2$.

c. $\sin(t\,\sin^{-1}x)$.

d. $\cos(t\,\sin^{-1}x)$.

e. $\sin(x)\sinh(x)$.

f. $\sin(x)\sin(\omega x)\sin(\omega^2 x)$, where $\omega = e^{2\pi i/3}$.

g. $\cos(\log(1+x))$ (express the answer as the real part of a complex number)

174. [1–] Find the order (number of elements) of the finite field $\mathbb{F}_2$.

175. [2+]* For $i, j \geq 0$ and $n \geq 1$, let $f_n(i,j)$ denote the number of pairs (V, W) of subspaces of $\mathbb{F}_q^n$ such that $\dim V = i$, $\dim W = j$, and $V \cap W = \{0\}$. Find a formula for $f_n(i,j)$ which is a power of q times a q-multinomial coefficient.

176. [2+] A sequence of vectors $v_1, v_2, \ldots$ is chosen uniformly and independently from $\mathbb{F}_q^n$. Let $E(n)$ be the expected value of k for which $v_1, \ldots, v_k$ span $\mathbb{F}_q^n$ but $v_1, \ldots, v_{k-1}$ don't span $\mathbb{F}_q^n$. For instance

$$E(1) = \frac{q}{q-1},$$

$$E(2) = \frac{q(2q+1)}{(q-1)(q+1)},$$

$$E(3) = \frac{q(3q^3 + 4q^2 + 3q + 1)}{(q-1)(q+1)(q^2+q+1)}.$$

Show that

$$E(n) = \sum_{i=1}^{n} \frac{q^i}{q^i - 1}.$$

177. a. [2+]* Let $f(n,q)$ denote the number of matrices $A \in \mathrm{Mat}(n,q)$ satisfying $A^2 = 0$. Show that

$$f(n,q) = \sum_{2i+j=n} \frac{\gamma_n(q)}{q^{i(i+2j)}\gamma_i(q)\gamma_j(q)},$$

where $\gamma_m(q) = \#\mathrm{GL}(m,q)$. (The sum ranges over all pairs $(i,j) \in \mathbb{N} \times \mathbb{N}$ satisfying $2i + j = n$.)

b. [2]* Write $f(n,q) = g(n,q)(q-1)^k$ so that $g(n,1) \neq 0, \infty$. Thus, $f(n,q)$ may be regarded as a q-analogue of $g(n,1)$. Show that

$$\sum_{n\geq 0} g(n,1)\frac{x^n}{n!} = e^{x^2+x}.$$

c. [5–] Is there an intuitive explanation of why $f(n,q)$ is a "good" q-analogue of $g(n,1)$?

178. [2+]* Let $f(n)$ be the number of pairs (A, B) of matrices in $\mathrm{Mat}(n,q)$ satisfying $AB = 0$. Show that

$$f(n) = \sum_{k=0}^{n} q^{n(n-k)} \binom{n}{k} (q^n - 1)(q^n - q) \cdots (q^n - q^{n-1}).$$

179. [2–]* How many pairs (A, B) of matrices in $\mathrm{Mat}(n, q)$ satisfy $A + B = AB$?

180. [5–] How many matrices $A \in \mathrm{Mat}(n, q)$ have square roots, i.e., $A = B^2$ for some $B \in \mathrm{Mat}(n, q)$? The $q = 1$ situation is Exercise 5.11(a).

181. [2]* Find a simple formula for the number $f(n)$ of matrices $A = (A_{ij}) \in \mathrm{GL}(n, q)$ such that $A_{11} = A_{1n} = A_{n1} = A_{nn} = 0$.

182. [2+] Let $f(n, q)$ denote the number of matrices $A = (A_{ij}) \in \mathrm{GL}(n, q)$ such that $A_{ij} \neq 0$ for all i, j. Let $g(n, q)$ denote the number of matrices $B = (B_{ij}) \in \mathrm{GL}(n - 1, q)$ such that $B_{ij} \neq 1$ for all i, j. Show that

$$f(n, q) = (q - 1)^{2n-1} g(n, q).$$

183. [2] Prove the identity

$$\frac{1}{1 - qx} = \prod_{d \geq 1} \left(1 - x^d\right)^{-\beta(d)}, \tag{1.145}$$

where $\beta(d)$ is given by equation (1.103).

184. a. [2]* Let $f_q(n)$ denote the number of monic polynomials $f(x)$ of degree n over $\mathbb{F}_q$ that do not have a zero in $\mathbb{F}_q$ (i.e., for all $\alpha \in \mathbb{F}_q$ we have $f(\alpha) \neq 0$). Find a simple formula for $F(x) = \sum_{n \geq 0} f_q(n) x^n$. Your answer should not involve any infinite sums or products.

NOTE. The constant polynomials $f(x) = \beta$ for $0 \neq \beta \in \mathbb{F}_q$ are included in the enumeration, but not the polynomial $f(x) = 0$.

  b. [2]* Use (a) to find a simple explicit formula for $f(n, q)$ when n is sufficiently large (depending on q).

185. a. [1]* Show that the number of monic polynomials of degree n over $\mathbb{F}_q$ is q^n.

  b. [2+] Recall that the *discriminant* of a polynomial $f(x) = (x - \theta_1) \cdots (x - \theta_n)$ is defined by

$$\mathrm{disc}(f) = \prod_{1 \leq i < j \leq n} (\theta_i - \theta_j)^2.$$

Show that the number $D(n, 0)$ of monic polynomials $f(x)$ over $\mathbb{F}_q$ with discriminant 0 (equivalently, $f(x)$ has an irreducible factor of multiplicity greater than 1) is q^{n-1}, $n \geq 2$.

  c. [2+] Generalize (a) and (b) as follows. Fix $k \geq 1$, and let X be any subset of $\mathbb{N}^k$ containing $(0, 0, \ldots, 0)$. If $f_1, \ldots, f_k$ is a sequence of monic polynomials over $\mathbb{F}_q$, then set $f = (f_1, \ldots, f_k)$ and $\deg(f) = (\deg(f_1), \ldots, \deg(f_k))$. Given an irreducible polynomial $p \in \mathbb{F}_q[x]$, let $\mathrm{mult}(p, f) = (\mu_1, \ldots, \mu_k)$, where μ_i is the multiplicity of p in f_i. Given $\beta \in \mathbb{N}^k$, let $N(\beta)$ be the number of k-tuples $f = (f_1, \ldots, f_k)$ of monic polynomials over $\mathbb{F}_q$ such that $\deg(f) = \beta$ and such that for any irreducible polynomial p over $\mathbb{F}_q$ we have $\mathrm{mult}(p, f) \in X$. By a straightforward generalization of Exercise 1.158 to the multivariate case, there are unique $a_\alpha \in \mathbb{Z}$ such that

$$F_X(x) := \sum_{\alpha \in X} x^\alpha = \prod_{\substack{\alpha \in \mathbb{N}^k \\ \alpha \neq (0,0,\ldots,0)}} (1 - x^\alpha)^{a_\alpha}, \tag{1.146}$$

where if $\alpha = (\alpha_1, \ldots, \alpha_k)$, then $x^\alpha = x_1^{\alpha_1} \cdots x_k^{\alpha_k}$. Show that

$$\sum_{\beta \in \mathbb{N}^k} N(\beta) x^\beta = \prod_{\substack{\alpha \in \mathbb{N}^k \\ \alpha \neq (0,0,\ldots,0)}} (1 - q x^\alpha)^{a_\alpha}.$$

Note that if $k = 1$ and $X = \mathbb{N}$, then $N(\beta)$ is the total number of monic polynomials of degree β. We have $f_{\mathbb{N}}(x) = 1/(1-x)$ and $\sum_{\beta \in \mathbb{N}} N(\beta)x^\beta = 1/(1-qx) = \sum_{n \geq 0} q^n x^n$, agreeing with (a).

186. Deduce from Exercise 1.185(c) the following results.

    a. [2] The number $N_r(n)$ of monic polynomials $f \in \mathbb{F}_q[x]$ of degree n with no factor of multiplicity at least r is given by

$$N_r(n) = q^n - q^{n-r+1}, \quad n \geq r. \qquad (1.147)$$

    Note that the case $r = 2$ is equivalent to (b).

    b. [2] Let $N(m,n)$ be the number of pairs (f,g) of monic relatively prime polynomials over $\mathbb{F}_q$ of degrees m and n. In other words, f and g have nonzero resultant. Then

$$N(m,n) = q^{m+n-1}, \quad m,n \geq 1. \qquad (1.148)$$

    c. [2+] A polynomial f over a field K is *powerful* if every irreducible factor of f occurs with multiplicity at least two. Let $P(n)$ be the number of powerful monic polynomials of degree n over $\mathbb{F}_q$. Show that

$$P(n) = q^{\lfloor n/2 \rfloor} + q^{\lfloor n/2 \rfloor - 1} - q^{\lfloor (n-1)/3 \rfloor}, \quad n \geq 2. \qquad (1.149)$$

187. a. [3–] Let q be an odd prime power. Show that as f ranges over all monic polynomials of degree $n > 1$ over $\mathbb{F}_q$, $\mathrm{disc}(f)$ is just as often a nonzero square in $\mathbb{F}_q$ as a nonsquare.

    b. [2+] For $n > 1$ and $a \in \mathbb{F}_q$, let $D(n,a)$ denote the number of monic polynomials of degree n over $\mathbb{F}_q$ with discriminant a. Thus by Exercise 1.185(b), we have $D(n,0) = q^{n-1}$. Show that if $(n(n-1), q-1) = 1$ (so $q = 2^m$) or $(n(n-1), q-1) = 2$ (so q is odd) then $D(n,a) = q^{n-1}$ for all $a \in \mathbb{F}_q$. (Here (r,s) denotes the greatest common divisor of r and s.)

    c. [5–] Investigate further the function $D(n,a)$ for general n and a.

188. [3] Give a direct proof of Corollary 1.10.11 (i.e., the number of nilpotent matrices in $\mathrm{Mat}(n,q)$ is $q^{n(n-1)}$).

189. [3–] Let V be an $(m+n)$-dimensional vector space over $\mathbb{F}_q$, and let $V = V_1 \oplus V_2$, where $\dim V_1 = m$ and $\dim V_2 = n$. Let $f(m,n)$ be the number of nilpotent linear transformations $A \colon V \to V$ satisfying $A(V_1) \subseteq V_2$ and $A(V_2) \subseteq V_1$. Show that

$$f(m,n) = q^{m(n-1)+n(m-1)}(q^m + q^n - 1),$$

190. a. [2] Let $\omega^*(n,q)$ denote the number of conjugacy classes in the group $\mathrm{GL}(n,q)$. Show that $\omega^*(n,q)$ is a polynomial in q satisfying $\omega^*(n,1) = 0$. For instance,

$$\omega^*(1,q) = q - 1,$$
$$\omega^*(2,q) = q^2 - 1,$$
$$\omega^*(3,q) = q^3 - q,$$
$$\omega^*(4,q) = q^4 - q,$$
$$\omega^*(5,q) = q^5 - q^2 - q + 1,$$
$$\omega^*(6,q) = q^6 - q^2,$$
$$\omega^*(7,q) = q^7 - q^3 - q^2 + 1,$$
$$\omega^*(8,q) = q^8 - q^3 - q^2 + q.$$

b. [2+] Show that

$$\omega^*(n,q) = q^n - q^{\lfloor (n-1)/2 \rfloor} + O(q^{\lfloor (n-1)/2 \rfloor - 1}).$$

c. [3–] Evaluate the polynomial values $\omega^*(n,0)$ and $\omega^*(n,-1)$. When is $\omega^*(n,q)$ divisible by q^2?

191. [3–] Give a more conceptual proof of Proposition 1.10.2, that is, the number $\omega(n,q)$ of orbits of $\mathrm{GL}(n,q)$ acting adjointly on $\mathrm{Mat}(n,q)$ is given by

$$\omega(n,q) = \sum_j p_j(n) q^j.$$

192. a. [2]* Find a simple formula for the number of surjective linear transformations $A : \mathbb{F}_q^n \to \mathbb{F}_q^k$.
b. [2]* Show that the number of $m \times n$ matrices of rank k over $\mathbb{F}_q$ is given by

$$\binom{m}{k}(q^n - 1)(q^n - q) \cdots (q^n - q^{k-1}).$$

193. [2] Let p_n denote the number of projections $P \in \mathrm{Mat}(n,q)$ (i.e., $P^2 = P$). Show that

$$\sum_{n \geq 0} p_n \frac{x^n}{\gamma_n} = \left(\sum_{k \geq 0} \frac{x^k}{\gamma(k)} \right)^2,$$

where as usual $\gamma(k) = \gamma(k,q) = \#\mathrm{GL}(k,q)$.

194. [2+] Let r_n denote the number of regular (or cyclic) $M \in \mathrm{Mat}(n,q)$ (i.e., the characteristic and minimal polynomials of A are the same). Equivalently, there is a column vector $v \in \mathbb{F}_q^n$ such that the set $\{A^i v : i \geq 0\}$ spans $\mathbb{F}_q^n$ (where we set $A^0 = I$). Show that

$$\sum_{n \geq 0} r_n \frac{x^n}{\gamma(n)} = \prod_{d \geq 1} \left(1 + \frac{x^d}{(q^d - 1)(1 - (x/q)^d)} \right)^{\beta(d)}$$

$$= \frac{1}{1-x} \prod_{d \geq 1} \left(1 + \frac{x^d}{q^d(q^d - 1)} \right)^{\beta(d)}.$$

195. [2] A matrix A is *semisimple* if it can be diagonalized over the algebraic closure of the base field. Let s_n denote the number of semisimple matrices $A \in \mathrm{Mat}(n,q)$. Show that

$$\sum_{n \geq 0} s_n \frac{x^n}{\gamma(n,q)} = \prod_{d \geq 1} \left(\sum_{j \geq 0} \frac{x^{jd}}{\gamma(j,q^d)} \right)^{\beta(d)}.$$

196. a. [2+] Generalize Proposition 1.10.15 as follows. Let $0 \leq k \leq n$, and let $f_k(n)$ be the number of matrices $A = (a_{ij}) \in \mathrm{GL}(n,q)$ satisfying $a_{11} + a_{22} + \cdots + a_{kk} = 0$. Then

$$f_k(n) = \frac{1}{q} \left(\gamma(n,q) + (-1)^k (q-1) q^{\frac{1}{2}k(2n-k-1)} \gamma(n-k,q) \right). \tag{1.150}$$

b. [2+] Let H be any linear hyperplane in the vector space $\mathrm{Mat}(n,q)$. Find (in terms of certain data about H) a formula for $\#(\mathrm{GL}(n,q) \cap H)$.

197. [3] Let $f(n)$ be the number of matrices $A \in \mathrm{GL}(n,q)$ with zero diagonal (i.e., all diagonal entries are equal to 0). Show that

$$f(n) = q^{\binom{n-1}{2}-1}(q-1)^n \sum_{i=0}^{n} (-1)^i \binom{n}{i}(n-i)!.$$

For instance,

$$f(1) = 0,$$
$$f(2) = (q-1)^2,$$
$$f(3) = q(q-1)(q^4 - 4q^2 + 4q - 1),$$
$$f(4) = q^3(q-1)(q^8 - q^6 - 5q^5 + 3q^4 + 11q^3 - 14q^2 + 6q - 1).$$

198. a. [2+] Let $h(n,r)$ denote the number of $n \times n$ symmetric matrices of rank r over $\mathbb{F}_q$. Show that

$$h(n+1,r) = q^r h(n,r) + (q-1)q^{r-1}h(n,r-1) + (q^{n+1} - q^{r-1})h(n,r-2),$$
$$(1.151)$$

with the initial conditions $h(n,0) = 1$ and $h(n,r) = 0$ for $r > n$.

b. [2] Deduce that

$$h(n,r) = \begin{cases} \displaystyle\prod_{i=1}^{s} \frac{q^{2i}}{q^{2i}-1} \cdot \prod_{i=0}^{2s-1}(q^{n-i}-1), & 0 \le r = 2s \le n, \\[4mm] \displaystyle\prod_{i=1}^{s} \frac{q^{2i}}{q^{2i}-1} \cdot \prod_{i=0}^{2s}(q^{n-i}-1), & 0 \le r = 2s+1 \le n. \end{cases}$$

In particular, the number $h(n,n)$ of $n \times n$ invertible symmetric matrices over $\mathbb{F}_q$ is given by

$$h(n,n) = \begin{cases} q^{m(m-1)}(q-1)(q^3-1)\cdots(q^{2m-1}), & n = 2m-1, \\ q^{m(m+1)}(q-1)(q^3-1)\cdots(q^{2m-1}), & n = 2m. \end{cases}$$

199. a. [3] Show that the following three numbers are equal:
- The number of symmetric matrices in $\mathrm{GL}(2n,q)$ with zero diagonal,
- The number of symmetric matrices in $\mathrm{GL}(2n-1,q)$,
- The number of skew-symmetric matrices ($A = -A^t$) in $\mathrm{GL}(2n,q)$.

b. [5] Give a combinatorial proof of (a). (No combinatorial proof is known that two of these items are equal.)

200. [3] Let $C_n(q)$ denote the number of $n \times n$ upper-triangular matrices X over $\mathbb{F}_q$ satisfying $X^2 = 0$. Show that

$$C_{2n}(q) = \sum_j \left[\binom{2n}{n-3j} - \binom{2n}{n-3j-1} \right] \cdot q^{n^2-3j^2-j}$$

$$C_{2n+1}(q) = \sum_j \left[\binom{2n+1}{n-3j} - \binom{2n+1}{n-3j-1} \right] \cdot q^{n^2+n-3j^2-2j}.$$

201. This exercise and the next show that simply stated counting problems over $\mathbb{F}_q$ can have complicated solutions beyond the realm of combinatorics. (See also Exercise 4.39(a).)

a. [3] Let

$$f(q) = \#\{(x,y,z) \in \mathbb{F}_q^3 : x+y+z = 0,\ xyz = 1\}.$$

Show that $f(q) = q + a - 2$, where
- if $q \equiv 2 \pmod 3$, then $a = 0$,
- if $q \equiv 1 \pmod 3$, then a is the unique integer such that $a \equiv 1 \pmod 3$ and $a^2 + 27b^2 = 4q$ for some integer b.

b. [2+] Let

$$g(q) = \#\{A \in \mathrm{GL}(3,q) : \mathrm{tr}(A) = 0,\ \det(A) = 1.\}$$

Express $g(q)$ in terms of the function $f(q)$ of part (a).

202. [4–] Let p be a prime, and let N_p denote the number of solutions modulo p to the equation $y^2 + y = x^3 - x$. Let $a_p = p - N_p$. For instance, $a_2 = -2$, $a_3 = 1$, $a_5 = 1$, $a_7 = -2$, and so on. Show that if $p \neq 11$, then

$$a_p = [x^p] x \prod_{n \geq 1} (1 - x^n)^2 (1 - x^{11n})^2$$

$$= [x^p](x - 2x^2 - x^3 + 2x^4 + x^5 + 2x^6 - 2x^7 - 2x^9 + \cdots).$$

203. [3] The following quotation is from Plutarch's *Table-Talk* VIII. 9, 732: "Chrysippus says that the number of compound propositions that can be made from only ten simple propositions exceeds a million. (Hipparchus, to be sure, refuted this by showing that on the affirmative side there are 103,049 compound statements, and on the negative side 310,952.)"

According to T. L. Heath, *A History of Greek Mathematics*, vol. 2, p. 245, "it seems impossible to make anything of these figures."

Can in fact any sense be made of Plutarch's statement?

Solutions to Exercises

1. *Answer:* 2. There is strong evidence that human babies, chimpanzees, and even rats have an understanding of this problem. See S. Dehaene, *The Number Sense: How the Mind Creates Mathematics*, Oxford, New York, 1997, pp. 23–27, 52–56.

2. Here is one possible way to arrive at the answers. There may be other equally simple (or even simpler) ways to solve these problems.

a. $2^{10} - 2^5 = 992$

b. $\frac{1}{2}(7-1)! = 360$

c. $5 \cdot 5!$ (or $6! - 5!$) $= 600$

d. $\binom{6}{1}4! + \binom{6}{2}3! + \frac{1}{2}\binom{6}{3}2!^2 = 274$

e. $\binom{6}{4} + \binom{6}{1}\binom{5}{2} + \frac{1}{3!}\binom{6}{2}\binom{4}{2} = 90$

f. $(6)_4 = 360$

g. $1 \cdot 3 \cdot 5 \cdot 7 \cdot 9 = 945$

h. $\binom{7}{2} + \binom{8}{3} + \binom{9}{1} = 86$

i. $\binom{11}{1,2,4,4} - \binom{8}{1,1,2,4} = 33810$

j. $\binom{8+1}{4} = 126$

k. $2\binom{8}{1,3,4} + 3\binom{8}{2,3,3} + \binom{8}{2,2,4} = 2660$

l. $5! + \binom{5}{2}(5)_4 + \frac{1}{2}\binom{5}{1}\binom{4}{2}(5)_3 = 2220$

3.   **a.** Given any n-subset S of $[x+n+1]$, there is a largest k for which $\#(S \cap [x+k]) = k$. Given k, we can choose S to consist of any k-element subset in $\binom{x+k}{k}$ ways, together with $\{x+k+2, x+k+3, \ldots, x+n+1\}$.

        b. *First proof.* Choose a subset of $[n]$ and circle one of its elements in $\sum k\binom{n}{k}$ ways. Alternatively, circle an element of $[n]$ in n ways, and choose a subset of what remains in 2^{n-1} ways.

        Second proof (not quite so combinatorial, but nonetheless instructive). Divide the identity by 2^n. It then asserts that the average size of a subset of $[n]$ is $n/2$. This follows since each subset can be paired with its complement.

        c. To give a noncombinatorial proof, simply square both sides of the identity (Exercise 1.8(a))

$$\sum_{n \geq 0} \binom{2n}{n} x^n = \frac{1}{\sqrt{1-4x}}$$

and equate coefficients. The problem of giving a combinatorial proof was raised by P. Veress and solved by G. Hajos in the 1930s. For some published proofs, see D. J. Kleitman, *Studies in Applied Math.* **54** (1975), 289–292; M. Sved, *Math. Intelligencer* **6**(4) (1984), 44–45; and V. De Angelis, *Amer. Math. Monthly* **113** (2006), 642–644.

        d. G. E. Andrews, *Discrete Math.* **11** (1975), 97–106.

        e. Given an n-element subset S of $[2n-1]$, associate with it the two n-element subsets S and $[2n] - S$ of $[2n]$.

        f. What does it mean to give a combinatorial proof of an identity with minus signs? The simplest (but not the only) possibility is to rearrange the terms so that all signs are positive. Thus, we want to prove that

$$\sum_{k \text{ even}} \binom{n}{k} = \sum_{k \text{ odd}} \binom{n}{k}, \quad n \geq 1. \tag{1.152}$$

Let $\mathcal{E}_n$ (respectively $\mathcal{O}_n$) denote the sets of all subsets of $[n]$ of even (respectively, odd) cardinality. The left-hand side of equation (1.152) is equal to $\#\mathcal{E}_n$, while the right-hand side is $\#\mathcal{O}_n$. Hence, we want to give a bijection $\varphi : \mathcal{E}_n \to \mathcal{O}_n$. The definition of φ is very simple:

$$\varphi(S) = \begin{cases} S \cup \{n\}, & n \notin S, \\ S - \{n\}, & n \in S. \end{cases}$$

Another way to look at this proof is to consider φ as an involution on all of $2^{[n]}$. Every orbit of φ has two elements, and their contributions to the sum $\sum_{S \subseteq [n]} (-1)^{\#S}$ cancel out, that is, $(-1)^{\#S} + (-1)^{\#\varphi(S)} = 0$. Hence φ is a *sign-reversing involution* as in the proof of Proposition 1.8.7.

g. The left-hand side counts the number of triples (S, T, f), where $S \subseteq [n]$, $T \subseteq [n+1, 2n]$, $\#S = \#T$, and $f : S \to [x]$. The right-hand side counts the number of triples (A, B, g), where $A \subseteq [n]$, $B \in \binom{[2n]-A}{n}$, and $g : A \to [x-1]$. Given (S, T, f), define (A, B, g) as follows: $A = f^{-1}([x-1])$, $B = ([n]-S) \cup T$, and $g(i) = f(i)$ for $i \in [x-1]$.

h. We have that $\binom{i+j}{i}\binom{j+k}{j}\binom{k+i}{i}$ is the number of triples (α, β, γ), where (i) α is a sequence of $i+j+2$ letters a and b beginning with a and ending with b, with $i+1$ a's (and hence $j+1$ b's), (ii) $\beta = (\beta_1, \ldots, \beta_{j+1})$ is a sequence of $j+1$ positive integers with sum $j+k+1$, and (iii) $\gamma = (\gamma_1, \ldots, \gamma_{i+1})$ is a sequence of $i+1$ positive integers with sum $k+i+1$. Replace the rth a in α by the word $c^{\gamma_r} d$, and replace the rth b in α by the word $d^{\beta_r} c$. In this way, we obtain a word δ in c, d of length $2n+4$ with $n+2$ c's and $n+2$ d's. This word begins with c and ends with $d(dc)^m$ for some $m \geq 1$. Remove the prefix c and suffix $d(dc)^m$ from δ to obtain a word ϵ of length $2(n-m+1)$ with $n-m+1$ c's and $n-m+1$ d's. The map $(\alpha, \beta, \gamma) \mapsto \epsilon$ is easily seen to yield a bijective proof of (h). This argument is due to Roman Travkin (private communication, October 2007).

Example. Let $n = 8$, $i = 2$, $j = k = 3$, $\alpha = abbaabb$, $\beta = (2, 3, 1, 1)$, $\gamma = (2, 3, 1)$. Then

$$\delta = (c^2 d)(d^2 c)(d^3 c)(c^3 d)(cd)(dc)(dc),$$

so $\epsilon = cd^3 cd^3 c^4 dc$.

NOTE. Almost any binomial coefficient identity can be proved nowadays automatically by computer. For an introduction to this subject, see M. Petkovšek, H. S. Wilf, and D. Zeilberger, $A = B$, A K Peters, Wellesley, Mass., 1996. Of course, it is still of interest to find elegant bijective proofs of such identities.

8. a. We have $1/\sqrt{1-4x} = \sum_{n \geq 0} \binom{-1/2}{n} (-4)^n x^n$. Now

$$\binom{-1/2}{n}(-4)^n = \frac{\left(-\frac{1}{2}\right)\left(-\frac{3}{2}\right) \cdots \left(-\frac{2n-1}{2}\right)(-4)^n}{n!}$$

$$= \frac{2^n \cdot 1 \cdot 3 \cdots (2n-1)}{n!} = \frac{(2n)!}{n!^2}.$$

b. Note that $\binom{2n-1}{n} = \frac{1}{2}\binom{2n}{n}$, $n > 0$ (see Exercise 1.3(e)).

9. b. Powerful methods exist for solving this type of problem (see Example 6.3.8); however, we give here a "naive" solution. Suppose the path has k steps of the form $(0, 1)$, and therefore k $(1, 0)$'s and $n-k$ $(1, 1)$'s. These $n+k$ steps may be chosen

in any order, so

$$f(n,n) = \sum_k \binom{n+k}{n-k,k,k} = \sum_k \binom{n+k}{2k}\binom{2k}{k}.$$

$$\Rightarrow \sum_{n \geq 0} f(n,n)x^n = \sum_k \binom{2k}{k} \sum_{n \geq 0} \binom{n+k}{2k} x^n$$

$$= \sum_k \binom{2k}{k} \frac{x^k}{(1-x)^{2k+1}}$$

$$= \frac{1}{1-x} \left(1 - \frac{4x}{(1-x)^2}\right)^{-1/2}, \text{ by Exercise 1.8(a)}$$

$$= \frac{1}{\sqrt{1-6x+x^2}}.$$

10. Let the elements of S be $a_1 < a_2 < \cdots < a_{r+s}$. Then the multiset $\{a_1, a_2 - 2, a_3 - 4, \ldots, a_{r+s} - 2(r+s-1)\}$ consists of r odd numbers and s even numbers in $[2(n-r-s+1)]$. Conversely, we can recover S from any r odd numbers and s even numbers (allowing repetition) in $[2(n-r-s+1)]$. Hence,

$$f(n,r,s) = \left(\binom{n-r-s+1}{r}\right)\left(\binom{n-r-s+1}{s}\right) = \binom{n-r}{s}\binom{n-s}{r}.$$

This result is due to Jim Propp, private communication dated 29 July 2006. Propp has generalized the result to any modulus $m \geq 2$ and has also given a q-analogue.

11. **a.** Choose $m+n+1$ points uniformly and independently from the interval $[0,1]$. The integral is then the probability that the last chosen point u is greater than the first m of the other points and less than the next n points. There are $(m+n+1)!$ orderings of the points, of which exactly $m!n!$ of them have the first m chosen points preceding u and the next n following u. Hence,

$$B(m+1,n+1) = \frac{m!n!}{(m+n+1)!}.$$

The function $B(x,y)$ for $\mathrm{Re}(x), \mathrm{Re}(y) > 0$ is the *beta function*.

There are many more interesting examples of the combinatorial evaluation of integrals. Two of the more sophisticated ones are P. Valtr, *Discrete Comput. Geom.* **13** (1995), 637–643; and *Combinatorica* **16** (1996), 567–573.

b. Choose $(1+r+s)n + 2t\binom{n}{2}$ points uniformly and independently from $[0,1]$. Label the first n chosen points x, the next r chosen points y_1, and so on, so that the points are labeled by the elements of M. Let P be the probability that the order of the points in $[0,1]$ is a permutation of M that we are counting. Then

$$P = \frac{n!\, r!^n\, s!^n\, (2t)!^{\binom{n}{2}}}{((r+s+1)n + tn(n-1))!} f(n,r,s,t)$$

$$= \int_0^1 \cdots \int_0^1 (x_1 \cdots x_n)^r ((1-x_1) \cdots (1-x_n))^s \prod_{1 \leq i < j \leq n} (x_i - x_j)^{2t} dx_1 \cdots dx_n.$$

This integral is the famous *Selberg integral*; see for example G. E. Andrews, R. Askey, and R. Roy, *Special Functions*, Cambridge University Press, Cambridge/New York, 1999 (Chapter 8), and P. J. Forrester and S. O. Warnaar, *Bull.*

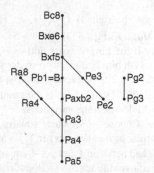

Figure 1.29 The solution poset for Exercise 1.11(c).

Amer. Math. Soc. **45** (2008), 489–534. The evaluation of this integral immediately gives equation (1.119). No combinatorial proof of (1.119) is known. Such a proof would be quite interesting since it would give a combinatorial evaluation of Selberg's integral.

c. One solution is 1.Pa5 2.Pa4 3.Pa3 4.Ra4 5.Ra8 6.Paxb2 7.Pb1=B 8.Pe2 9.Pe3 10.Bxf5 11.Bxe6 12.Bc8 13.Pg3 14.Pg2, after which White plays Bh2 mate. We attach indeterminates to each of the Black moves as follows: $1.a_{12}$ $2.a_{12}$ $3.x$ $4.a_{24}$ $5.a_{24}$ $6.a_{23}$ $7.a_{23}$ $8.a_{13}$ $9.a_{13}$ $10.x$ $11.a_{34}$ $12.a_{34}$ $13.a_{14}$ $14.a_{14}$. We also place an indeterminate x before Black's first move and after Black's last move. All solutions are then obtained by permutations of Black's 14 moves, together with x at the beginning and end, with the property that moves labeled by the same indeterminate must be played in the same order, and moves labeled a_{ij} must occur between the ith x and jth x. In the terminology of Chapter 3, the solutions correspond to the linear extensions of the poset shown in Figure 1.29. Hence the number of solutions is

$$f(4,0,0,1) = 54054.$$

For similar serieshelpmates (called *queue problems*) whose number of solutions has some mathematical significance, see Exercises 1.145, 6.23, and 7.18. Some references are given in the solution to Exercise 6.23. The present problem comes from the article R. Stanley, *Suomen Tehtäväniekat* **59**, no. 4 (2005), 193–203.

13. Let S consist of all p-tuples $(n_1, n_2, \ldots, n_p)$ of integers $n_i \in [a]$ such that not all the n_i's are equal. Hence, $\#S = a^p - a$. Define two sequences in S to be equivalent if one is a cyclic shift of the other (clearly an equivalence relation). Since p is prime each equivalence class contains exactly p elements, and the proof follows. For additional results of this nature, see I. M. Gessel, in *Enumeration and Design (Waterloo, Ont., 1982)*, Academic Press, Toronto, 1984, pp. 157–197, and G.-C. Rota and B. E. Sagan, *European J. Combin.* **1** (1980), 67–76.

14. a. We use the well-known and easily proved fact that $(x+1)^p \equiv x^p + 1 \pmod{p}$, meaning that each coefficient of the polynomial $(x+1)^p - (x^p+1)$ is divisible by p. Thus,

$$(x+1)^n = (x+1)^{\sum a_i p^i}$$

$$\equiv \prod_i \left(x^{p^i} + 1 \right)^{a_i} \pmod{p}$$

$$\equiv \prod_i \sum_{j=0}^{a_i} \binom{a_i}{j} x^{j p^i} \pmod{p}.$$

The coefficient of x^m on the left is $\binom{n}{m}$ and on the right is $\binom{a_0}{b_0}\binom{a_1}{b_1}\cdots$. This congruence is due to F. E. A. Lucas, *Bull. Soc. Math. France* **6** (1878), 49–54.

b. The binomial coefficient $\binom{n}{m}$ is odd if and only if the binary expansion of m is "contained" in that of n; that is, if m has a 1 in its ith binary digit, then so does n. Hence, $\binom{n}{m}$ is odd for all $0 \le m \le n$ if and only if $n = 2^k - 1$. More generally, the number of odd coefficients of $(1+x)^n$ is equal to $2^{b(n)}$, where $b(n)$ is the number of 1's in the binary expansion of n. See Exercise 1.15 for some variations.

c. Consider an $a \times p$ rectangular grid of squares. Choose pb of these squares in $\binom{pa}{pb}$ ways. We can choose the pb squares to consist of b entire rows in $\binom{a}{b}$ ways. Otherwise, in at least two rows, we will have picked between 1 and $p-1$ squares. For any choice of pb squares, cyclically shift the squares in each row independently. This partitions our choices into equivalence classes. Exactly $\binom{a}{b}$ of these classes contain one element; the rest contain a number of elements divisible by p^2.

d. Continue the reasoning of (c). If a choice of pb squares contains fewer than $b-2$ entire rows, then its equivalence class has cardinality divisible by p^3. From this we reduce the problem to the case $a = 2$, $b = 1$. Now

$$\binom{2p}{p} = \sum_{k=0}^{p}\binom{p}{k}^2$$

$$= 2 + p^2 \sum_{k=1}^{p-1} \frac{(p-1)^2(p-2)^2\cdots(p-k+1)^2}{k!^2}$$

$$\equiv 2 + p^2 \sum_{k=1}^{p-1} k^{-2} \pmod{p^3}.$$

But as k ranges from 1 to $p-1$, so does k^{-1} modulo p. Hence,

$$\sum_{k=1}^{p-1} k^{-2} \equiv \sum_{k=1}^{p-1} k^2 \pmod{p}.$$

Now use, for example, the identity

$$\sum_{k=1}^{n} k^2 = \frac{n(n+1)(2n+1)}{6}$$

to get

$$\sum_{k=1}^{p-1} k^2 \equiv 0 \pmod{p}, \quad p \ge 5.$$

e. The exponent of the largest power of p dividing $\binom{n}{m}$ is the number of carries needed to add m and $n-m$ in base p. See E. Kummer, *J. Math.* **44** (1852), 115–116, and L. E. Dickson, *Quart. J. Math.* **33** (1902), 378–384.

15. **a.** We have

$$1 + x + x^2 = \frac{1 - x^3}{1 - x} \equiv (1 - x)^2 \pmod{3}.$$

Hence, $(1 + x + x^2)^n \equiv (1 - x)^{2n} \pmod{3}$. It follows easily from Exercise 1.14(a) that if $2n$ has the ternary expansion $2n = \sum a_i 3^i$, then the number of coefficients of $(1 + x + x^2)^n$ not divisible by 3 is equal to $\prod(1 + a_i)$. This result was obtained in collaboration with T. Amdeberhan.

b. Let $f(n)$ be the desired number. First consider the case $n = 2^j(2^k - 1)$. Since $(1 + x + x^2)^{2^j} \equiv 1 + x^{2^j} + x^{2^{j+1}} \pmod{2}$, we have $f(n) = f(2^k - 1)$. Now

$$(1 + x + x^2)^{2^k - 1} \equiv \frac{1 + x^{2^k} + x^{2^{k+1}}}{1 + x + x^2} \pmod{2}.$$

It is easy to check that modulo 2 we have for k odd that

$$\frac{1 + x^{2^k} + x^{2^{k+1}}}{1 + x + x^2} = 1 + x + x^3 + x^4 + x^6 + x^7 + \cdots + x^{2^k - 2} + x^{2^k - 1} + x^{2^k}$$

$$+ x^{2^k + 2} + x^{2^k + 3} + x^{2^k + 5} + x^{2^k + 6} + \cdots + x^{2^{k+1} - 3} + x^{2^{k+1} - 2}.$$

It follows that $f(2^k - 1) = (2^{k+2} + 1)/3$. Similarly, when k is even we have

$$\frac{1 + x^{2^k} + x^{2^{k+1}}}{1 + x + x^2} = 1 + x + x^3 + x^4 + x^6 + x^7 + \cdots + x^{2^k - 4} + x^{2^k - 3} + x^{2^k - 1}$$

$$+ x^{2^k + 1} + x^{2^k + 2} + x^{2^k + 4} + x^{2^k + 5} + \cdots + x^{2^{k+1} - 3} + x^{2^{k+1} - 2}.$$

Hence, in this case, $f(2^k - 1) = (2^{k+2} - 1)/3$. For a generalization, see Exercise 4.25.

Now any positive integer n can be written uniquely as $n = \sum_{i=1}^r 2^{j_i}(2^{k_i} - 1)$, where $k_i \geq 1$, $j_1 \geq 0$, and $j_{i+1} > j_i + k_i$. We are simply breaking up the binary expansion of n into the maximal strings of consecutive 1's. The lengths of these strings are $k_1, \ldots, k_r$. Thus,

$$(1 + x + x^2)^n \equiv \prod_{i=1}^r (1 + x^{2^{j_i}} + x^{2^{j_i+1}})^{2^{k_i} - 1} \pmod{2}.$$

There is no cancellation among the coefficients when we expand this product since $j_{i+1} > j_i + 1$. Hence,

$$f(n) = \prod_{i=1}^r f(2^{k_i} - 1),$$

where $f(2^{k_i} - 1)$ is given earlier.

Example. The binary expansion of 6039 is 1011110010111. The maximal strings of consecutive 1's have lengths 1, 4, 1, and 3. Hence,

$$f(6039) = f(1) f(15) f(1) f(7) = 3 \cdot 21 \cdot 3 \cdot 11 = 2079.$$

c. We have

$$\prod_{1 \leq i < j \leq n} (x_i + x_j) \equiv \prod_{1 \leq i < j \leq n} (x_i - x_j) \pmod{2},$$

where the notation means that the corresponding coefficients of each side are congruent modulo 2. The latter product is just the value of the Vandermonde determinant $\det[x_i^{j-1}]_{i,j=1}^n$, so the number of odd coefficients is $n!$. This result can also be proved by a cancellation argument; see Exercise 2.34. A more subtle result, equivalent to Exercise 4.64(a), is that the number of *nonzero* coefficients of the polynomial $\prod_{1\le i<j\le n}(x_i+x_j)$ is equal to the number of forests on an n-element vertex set.

Some generalizations of the results of this exercise appear in T. Amdeberhan and R. Stanley, Polynomial coefficient enumeration, preprint dated 3 February 2008;

$$\langle\texttt{http://math.mit.edu/}\sim\texttt{rstan/papers/coef.pdf}\rangle.$$

See also Exercise 4.24.

16. a. This result was first given by N. Strauss as Problem 6527, *Amer. Math. Monthly* **93** (1986), 659, and later as the paper *Linear Algebra Appl.* **90** (1987), 65–72. An elegant solution to Strauss's problem was given by I. M. Gessel, *Amer. Math. Monthly* **95** (1988), 564–565, and by W. C. Waterhouse, *Linear Algebra Appl.* **105** (1988), 195–198. Namely, let V be the vector space of all functions $\mathbb{F}_p \to \mathbb{F}_p$. A basis for V consists of the functions $f_j(a) = a^j$, $0 \le j \le p-1$. Let $\Phi: V \to V$ be the linear transformation defined by $(\Phi f)(x) = (1-x)^{p-1} f(1/(1-x))$. Then it can be checked that A is just the matrix of Φ with respect to the basis f_j. It is now routine to verify that $A^3 = I$.

   b. *Answer*: $(p+2\epsilon)/3$, where $\epsilon = 1$ if $p \equiv 1 \pmod 3$ and $\epsilon = -1$ if $p \equiv -1 \pmod 3$. Both Strauss, op. cit., and Waterhouse, op. cit., in fact compute the Jordan normal form of A. Waterhouse uses the linear transformation Φ to give a proof similar to that given in (a).

17. b. Think of a choice of m objects from n with repetition allowed as a placement of $n-1$ vertical bars in the slots between m dots (including slots at the beginning and end). For example,

$$|..||...|..$$

corresponds to the multiset $\{1^0, 2^2, 3^0, 4^3, 5^2\}$. Now change the bars to dots and vice versa:

$$\cdot||..|||\cdot||$$

yielding $\{1^1, 2^0, 3^2, 4^0, 5^0, 6^1, 7^0, 8^0\}$. This procedure gives the desired bijection. (Of course a more formal description is possible but only seems to obscure the elegance and simplicity of the above bijection.)

19. a. One way to prove (1.120) is to recall the Lagrange interpolation formula. Namely, if $P(x)$ is a polynomial of degree less than n and $x_1, \ldots, x_n$ are distinct numbers (or indeterminates), then

$$P(x) = \sum_{i=1}^{n} P(x_i) \prod_{j\ne i} \frac{x-x_j}{x_i-x_j}.$$

Now set $P(x) = 1$ and $x = 0$.

Applying the hint, we see that the constant term $C(a_1, \ldots, a_n)$ satisfies the recurrence

$$C(a_1, \ldots, a_n) = \sum_{i=1}^{k} C(a_1, \ldots, a_i-1, \ldots, a_n),$$

if $a_i > 0$. If, on the other hand, $a_i = 0$, we have

$$C(a_1, \ldots, a_{i-1}, 0, a_{i+1}, \ldots, a_n) = C(a_1, \ldots, a_{i-1}, a_{i+1}, \ldots, a_n).$$

This is also the recurrence satisfied by $\binom{a_1+\cdots+a_n}{a_1,\ldots,a_n}$, and the initial conditions $C(0,\ldots,0) = 1$ and $\binom{0}{0,\ldots,0} = 1$ agree.

This result was conjectured by F. J. Dyson in 1962 and proved that same year by J. Gunson and K. Wilson. The elegant proof given here is due to I. J. Good in 1970. For further information and references, see [1.3, pp. 377–387].

b. This identity is due to A. C. Dixon, *Proc. London Math. Soc.* **35**(1)(1903), 285–289.

c. This is the "q-Dyson conjecture," due to G. E. Andrews, in *Theory and Application of Special Functions* (R. Askey, ed.), Academic Press, New York, 1975, pp. 191–224 (see §5). It was first proved by D. M. Bressoud and D. Zeilberger, *Discrete Math.* **54** (1985), 201–224. A more recent paper with many additional references is I. M. Gessel, L. Lv, G. Xin, and Y. Zhou, *J. Combinatorial Theory, Ser. A* **115** (2008), 1417–1435.

d. I. G. Macdonald conjectured a generalization of (a) corresponding to any root system R. The present problem corresponds to $R = D_n$, while (a) is the case $R = A_{n-1}$ (when all the a_i's are equal). After many partial results, the conjecture was proved for all root systems by E. Opdam, *Invent. Math.* **98** (1989), 1–18. Macdonald also gave a q-analogue of his conjecture, which was finally proved by I. Cherednik in 1993 and published in *Ann. Math.* **141** (1995), 191–216. For the original papers of Macdonald, see *Sem. d'Alg. Paul Dubriel et Marie-Paule Malliavin*, Lecture Notes in Math., no. 867, Springer, Berlin, pp. 90–97, and *SIAM J. Math. Anal.* **13** (1982), 988–1007.

e. Write

$$F(x) = F(x_1,\ldots,x_n) = \sum_{a_1,\ldots,a_n \geq 0} \left[\prod_{i=1}^n (q^{-a_i} + \cdots + q^{a_i}) \right] x_1^{a_1} \cdots x_n^{a_n}$$

$$= \prod_{i=1}^n \sum_{j \geq 0} (q^{-j} + \cdots + q^j) x_i^j$$

$$= \prod_{i=1}^n \sum_{j \geq 0} \left(\frac{q^{-j} - q^{j+1}}{1-q} \right) x_i^j$$

$$= \frac{1}{(1-q)^n} \prod_{i=1}^n \left[\frac{1}{1 - q^{-1}x_i} - \frac{q}{1 - qx_i} \right]$$

$$= \prod_{i=1}^n \frac{1 + x_i}{(1 - q^{-1}x_i)(1 - qx_i)}.$$

We seek the term $F_0(x)$ independent from q. By the Cauchy integral formula (letting each x_i be small),

$$F_0(x) = \frac{1}{2\pi i} \oint \frac{dq}{q} \prod_{i=1}^n \frac{1 + x_i}{(1 - q^{-1}x_i)(1 - qx_i)}$$

$$= \frac{(1+x_1)\cdots(1+x_n)}{2\pi i} \oint dq \prod_{i=1}^n \frac{q^{n-1}}{(q - x_i)(1 - qx_i)},$$

where the integral is around the circle $|q| = 1$. The integrand has a simple pole at $q = x_i$ with residue $x_i^{n-1}/(1 - x_i^2) \prod_{j \neq i} (x_i - x_j)(1 - x_i x_j)$, and the proof follows from the Residue Theorem.

NOTE. The complex analysis in the above proof can be replaced with purely formal computations using the techniques of Section 6.3.

22. **a.** Let $a_1 + \cdots + a_k$ be any composition of $n > 1$. If $a_1 = 1$, then associate the composition $(a_1 + a_2) + a_3 + \cdots + a_k$. If $a_1 > 1$, then associate $1 + (a_1 - 1) + a_2 + \cdots + a_k$. This defines an involution on the set of compositions of n that changes the parity of the number of even parts. Hence, the number in question is 2^{n-2}, $n \geq 2$. (Note the analogy with permutations: There are $\frac{1}{2}n!$ permutations with an even number of even cycles—namely, the elements of the alternating group.)

  b. It is easily seen that

  $$\sum_{n \geq 0} (e(n) - o(n)) x^n = \prod_{i \geq 1} (1 + (-1)^i x^i)^{-1}.$$

  In the first proof of Proposition 1.8.5, it was shown that

  $$\prod_{i \geq 1} (1 + x^i) = \prod_{i \geq 1} (1 - x^{2i-1})^{-1}.$$

  Hence (putting $-x$ for x and taking reciprocals),

  $$\prod_{i \geq 1} (1 + (-1)^i x^i)^{-1} = \prod_{i \geq 1} (1 + x^{2i+1})$$

  $$= \sum_{n \geq 0} k(n) x^n,$$

  by Proposition 1.8.4. A simple combinatorial proof of this exercise was given by the Cambridge Combinatorics and Coffee Club in December 1999.

23. Form all 2^{n-1} compositions of n as in (1.19). Each bar occurs in half the compositions, so there are $(n-1)2^{n-2}$ bars in all. The total number of parts is equal to the total number of bars plus the total number of compositions, so $(n-1)2^{n-2} + 2^{n-1} = (n+1)2^{n-2}$ parts in all. This argument is due to D. E. Knuth (private communication, 21 August 2007).

  Variant argument. Draw n dots in a row. Place a double bar before the first dot or in one of the $n-1$ spaces between the dots. Choose some subset of the remaining spaces between dots, and place a bar in each of these spaces. The double bar and the bars partition the dots into compartments that define a composition α of n as in equation (1.19). The compartment to the right of the double bar specifies one of the parts of α. Hence, the total number $f(n)$ of parts of all compositions of n is equal to the number of ways of choosing the double bar and bars as described previously. As an example, the figure

  $$..|.||..|...$$

  corresponds to the composition $(2, 1, 2, 3)$ of 8 with the third part selected.
  If we place the double bar before the first dot, then there are 2^{n-1} choices for the remaining bars. Otherwise there are $n-1$ choices for the double bar and then 2^{n-2} choices for the remaining bars. Hence, $f(n) = 2^{n-1} + (n-1)2^{n-2} = (n+1)2^{n-2}$.

24. Draw a line of n dots and circle k consecutive dots. Put a vertical bar to the left and right of the circled dots. For example, $n = 9$, $k = 3$; see Figure 1.30.
  Case 1. The circled dots don't include an endpoint. The procedure can then be done in $n - k - 1$ ways. Then there remain $n - k - 2$ spaces between uncircled dots. Insert at

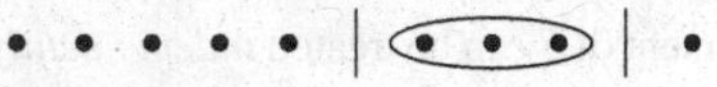

• • • • • | $\langle$ • • $\rangle$ | •

Figure 1.30 First step of the solution to Exercise 1.24.

• • • | • | • | $\langle$ • • $\rangle$ | •

Figure 1.31 Continuation of the solution to Exercise 1.24.

most one vertical bar in each space in 2^{n-k-2} ways. This defines a composition with one part equal to k circled. For example, if we insert bars as in Figure 1.31, then we obtain $3 + 1 + 1 + ③ + 1$.

Case 2. The circled dots include an endpoint. This happens in two ways, and now there are $n - k - 1$ spaces into which bars can be inserted in 2^{n-k-1} ways.

Hence, we get the answer

$$(n - k - 1)2^{n-k-2} + 2 \cdot 2^{n-k-1} = (n - k + 3)2^{n-k-2}.$$

25. It is clear that

$$\sum_{n,r,s} f(n,r,s)q^r t^s x^n = \sum_{j \geq 0} \left(\frac{qx}{1-x^2} + \frac{tx^2}{1-x^2} \right)^j.$$

The coefficient of $q^r t^s$ is given by

$$\binom{r+s}{r} \frac{x^{r+2s}}{(1-x^2)^{r+s}} = \binom{r+s}{r} \sum_{m \geq 0} \binom{m+r+s-1}{r+s-1} x^{2m+r+2s},$$

and the proof follows.

For a bijective proof, choose a composition of $r + k$ into $r + s$ parts in $\binom{r+k-1}{r+s-1}$ ways. Multiply r of these parts by 2 in $\binom{r+s}{r}$ ways. Multiply each of the other parts by 2 and subtract 1. We obtain each composition of n with r odd parts and s even parts exactly once, and the proof follows.

27. *Answer:* $(n+3)2^{n-2} - 1$.

30. Let $b_i = a_i - i + 1$. Then $1 \leq b_1 \leq b_2 \leq \cdots \leq b_k \leq n - k + 1$ and each b_i is odd. Conversely, given the b_i's we can uniquely recover the a_i's. Hence, setting $m = \lfloor (n-k+2)/2 \rfloor$, the number of odd integers in the set $[n - k + 1]$, we obtain the answer $\left(\binom{m}{k} \right) = \binom{m+k-1}{k} = \binom{q}{k}$, where $q = \lfloor (n+k)/2 \rfloor$.

This exercise is called *Terquem's problem.* For some generalizations, see M. Abramson and W. O. J. Moser, *J. Combinatorial Theory* **7** (1969), 171–180; S. M. Tanny, *Canad. Math. Bull.* **18** (1975), 769–770; J. de Biasi, *C. R. Acad. Sci. Paris Sér. A-B* **285** (1977), A89–A92; and I. P. Goulden and D. M. Jackson, *Discrete Math.* **22** (1978), 99–104. A further generalization is given by Exercise 1.10.

31. a. $x(x+1)(x+2) \cdots (x+n-1) = n! \left(\binom{n+1}{x-1} \right) = n! \left(\binom{x}{n} \right)$

b. $(n)_x (n-1)_{n-x} = n! \binom{n-1}{x-1}$

c. $\sum_{k=1}^{x} \frac{n!}{k!} \binom{n-1}{k-1}$

32. The key feature of this problem is that each element of S can be treated *independently*, as in Example 1.1.16.

   a. For each $x \in S$, we may specify the least i (if any) for which $x \in T_i$. There are $k+1$ choices for each x, so $(k+1)^n$ ways in all.

   b. Now each x can be in at most one T_i, so again there are $k+1$ choices for x and $(k+1)^n$ choices in all. (In fact, there is a very simple bijection between the sequences enumerated by (a) and (b).)

   c. Now each x can be in any subset of the T_i's except the subset $\emptyset$. Hence there are $2^k - 1$ choices for each x and $(2^k - 1)^n$ ways in all.

34. Let $b_i = a_i - (i-1)j$ to get $1 \le b_1 \le \cdots \le b_k \le n - (k-1)j$, so the number of sequences is $\left(\binom{n-(k-1)j}{k} \right)$.

35. **a.** Obtain a recurrence by considering those subsets S which do or do not contain n. *Answer:* F_{n+2}.

   b. Consider whether the first part is 2 or at least 3. *Answer:* F_{n-1}.

   c. Consider whether the first part is 1 or 2. *Answer:* F_{n+1}.

   d. Consider whether the first part is 1 or at least 3. *Answer:* F_n.

   e. Consider whether $\varepsilon = 0$ or 1. *Answer:* F_{n+2}.

   f. The following proof, as well as the proofs of (g) and (h), are due to Ira Gessel. Gessel (private communication, 2 May 2007) has developed a systematic approach to "Fibonacci composition formulas" based on factorization in free monoids as discussed in Section 4.7. The sum $\sum a_1 a_2 \cdots a_k$ counts the number of ways of inserting at most one vertical bar in each of the $n-1$ spaces separating a line of n dots, and then circling one dot in each compartment. An example is shown in Figure 1.32. Replace each bar by a 1, each uncircled dot by a 2, and each circled dot by a 1. For example, Figure 1.32 becomes

$$2\,1\,2\,2\,1\,1\,2\,1\,1\,1\,1\,2\,1\,1\,1\,2\,2\,1\,1\,1\,2\,1\,2.$$

We get a composition of $2n - 1$ into 1's and 2's, and this correspondence is invertible. Hence, by (c) the answer is F_{2n}.

A simple generating function proof can also be given using the identity

$$\sum_{k \ge 1} (x + 2x^2 + 3x^3 + \cdots)^k = \frac{x/(1-x)^2}{1 - x/(1-x)^2}$$

$$= \frac{x}{1 - 3x + x^2}$$

$$= \sum_{n \ge 1} F_{2n} x^n.$$

   g. Given a composition $(a_1,\ldots,a_k)$ of n, replace each part a_i with a composition α_i of $2a_i$ into parts 1 and 2, such that α_i begins with a 1, ends in a 2, and for all j the $2j$th 1 in α is followed by a 1, unless this $2j$th 1 is the last 1 in α. For instance, the part $a_i = 4$ can be replaced by any of the seven compositions 1111112, 111122, 111212, 11222, 121112, 12122, 12212. It can be checked that (i) every composition of $2n$ into parts 1 and 2, beginning with 1 and ending with 2, occurs exactly once by applying this procedure to all compositions of n, and (ii)

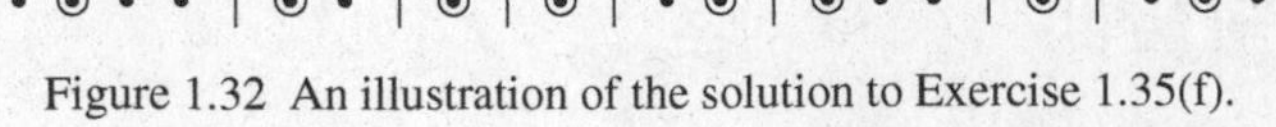

Figure 1.32 An illustration of the solution to Exercise 1.35(f).

the number of compositions that can replace a_i is $2^{a_i-1} - 1$. It follows from part (c) that the answer is F_{2n-2}. A generating function proof takes the form

$$\sum_{k\geq 1}(x^2 + 3x^3 + 7x^4 + \cdots)^k = \frac{x^2/(1-x)(1-2x)}{1 - x^2/(1-x)(1-2x)}$$

$$= \frac{x^2}{1 - 3x + x^2}$$

$$= \sum_{n\geq 2} F_{2n-2}x^n.$$

h. Given a composition $(a_1, \ldots, a_k)$ of n, replace each 1 with either 2 or $1,1$, and replace each $j > 1$ with $1, 2, \ldots, 2, 1$, where there are $j - 1$ 2's. Every composition of $2n$ with parts 1 and 2 is obtained in this way, so from part (c) we obtain the answer F_{2n+1}. A generating function proof takes the form

$$\frac{1}{1 - 2x - x^2 - x^3 - x^4 - \cdots} = \frac{1}{1 - x - \frac{x}{1-x}}$$

$$= \frac{1-x}{1 - 3x + x^2}$$

$$= \sum_{n\geq 0} F_{2n+1}x^n.$$

i. *Answer:* $2F_{3n-4}$ (with F_n defined for *all* $n \in \mathbb{Z}$ using the recurrence $F_n = F_{n-1} + F_{n-2}$), a consequence of the expansion

$$\frac{1}{1 + \frac{x}{1-5x} + \frac{x}{1-x}} = 1 - 2\sum_{n\geq 1} F_{3n-4}x^n.$$

A bijective proof is not known. This result is due to D. E. Knuth (private communication, August 21, 2007).

k. *Answer:* F_{2n+2}. Let $f(n)$ be the number in question. Now

$$P_n = P_{n-1} + P_{n-1}x_n + P_n x_{n+1}. \tag{1.153}$$

Each term of the above sum has $f(n - 1)$ terms when expanded as a polynomial in the x_i's. Since

$$P_{n-1} + P_{n-1}x_n = P_{n-2}(1 + x_{n-1} + x_n) + P_{n-2}(1 + x_{n-1} + x_n)x_n,$$

the only overlap between the three terms in equation (1.153) comes from $P_{n-2}x_n$, which has $f(n - 2)$ terms. Hence $f(n) = 3f(n - 1) - f(n - 2)$, from which the proof follows easily. This problem was derived from a conjecture of T. Amdeberhan (November 2007). For a variant, see Exercise 4.20.

36. Let $f_n(k)$ denote the answer. For each $i \in [n]$ we can decide which T_j contains i independently of the other $i' \in [n]$. Hence, $f_n(k) = f_k(1)^n$. But computing $f_k(1)$ is equivalent to Exercise 1.35(e). Hence, $f_n(k) = F_{k+2}^n$.

37. While it is not difficult to show that the right-hand side of equation (1.122) satisfies the Fibonacci recurrence and initial conditions, we prefer a more combinatorial proof. For instance, Exercise 1.34 in the case $j = 2$ shows that $\binom{n-k}{k}$ is the number of k-subsets of $[n - 1]$ containing no two consecutive integers. Now use Exercise 1.35(a).

39. *First Solution* (sketch). Let $a_{m,n}$ be the number of ordered pairs (S, T) with $S \subseteq [m]$ and $T \subseteq [n]$ satisfying $s > \#T$ for all $s \in S$ and $t > \#S$ for all $t \in T$. An easy bijection gives

$$a_{m,n} = a_{m-1,n} + a_{m-1,n-1}.$$

Using $a_{ij} = a_{ji}$, we get

$$a_{n,n} = a_{n,n-1} + a_{n-1,n-1}$$

$$a_{n,n-1} = a_{n-1,n-1} + a_{n-1,n-2},$$

from which it follows (using the initial conditions $a_{0,0} = 1$ and $a_{1,0} = 2$) that $a_{n,n} = F_{2n+2}$ and $a_{n,n-1} = F_{2n+1}$.

Second Solution (sketch). It is easy to see that

$$a_{m,n} = \sum_{\substack{i,j \geq 0 \\ i+j \leq \min\{m,n\}}} \binom{m-j}{i}\binom{n-i}{j}.$$

It can then be proved bijectively that $\sum_{\substack{i,j \geq 0 \\ i+j \leq n}} \binom{n-j}{i}\binom{n-i}{j}$ is the number of compositions of $2n+1$ with parts 1 and 2. The proof follows from Exercise 1.35(c).

This problem (for the case $n = 10$) appeared as Problem A-6 on the Fifty-First William Lowell Putnam Mathematical Competition (1990). These two solutions appear in K. S. Kedlaya, B. Poonen, and R. Vakil, *The William Lowell Putnam Mathematical Competition*, Mathematical Association of America, Washington, D.C., 2002, pp. 123–124.

41. **a.** Perhaps the most straightforward solution is to let $\#S = k$, giving

$$f(n) = \sum_{k=0}^{n} (n-k)_k (n-k)! \binom{n}{k}$$

$$= n! \sum_{k=0}^{n} \binom{n-k}{k}.$$

Now use Exercise 1.37. It is considerably trickier to give a direct bijective proof.

b. We now have

$$g(n) = \sum_{k=0}^{n-1} (n-k)_k (n-k-1)! \binom{n}{k}$$

$$= (n-1)! \sum_{k=0}^{n-1} \frac{n}{n-k} \binom{n-k}{k}.$$

There are a number ways to show that $L_n = \sum_{k=0}^{n-1} \frac{n}{n-k} \binom{n-k}{k}$, and the proof follows. This result was suggested by D. E. Knuth (private communication, 21 August 2007) upon seeing (a). A simple bijective proof was suggested by R. X. Du (private communication, 27 March 2011); namely, choose an n-cycle C in $(n-1)!$ ways, and regard the elements of C as n points on a circle. We can choose S to be any subset of the points, no two consecutive. By Exercise 1.40 this can be done in L_n ways, so the proof follows.

42. Let $\prod_{n\geq 2}(1-x^{F_n})=\sum_{k\geq 0}a_k x^k$. Split the interval $[F_n, F_{n+1}-1]$ into the three subintervals $[F_n, F_n+F_{n-3}-2], [F_n+F_{n-3}-1, F_n+F_{n-2}-1]$, and $[F_n+F_{n-2}, F_{n+1}-1]$. The following results can be shown by induction:

- The numbers $a_{F_n}, a_{F_n+1}, \ldots, a_{F_n+F_{n-3}-2}$ are equal to the numbers $(-1)^{n-1}a_{F_{n-3}-2}$, $(-1)^{n-1}a_{F_{n-3}-3}, \ldots, (-1)^{n-1}a_0$ in that order.
- The numbers $a_{F_n+F_{n-3}-1}, a_{F_n+F_{n-3}}, \ldots, a_{F_n+F_{n-2}-1}$ are equal to 0.
- The numbers $a_{F_n+F_{n-2}}, a_{F_n+F_{n-2}+1}, \ldots, a_{F_{n+1}-1}$ are equal to the numbers $a_0, a_1, \ldots, a_{F_{n-3}-1}$ in that order.

From these results the proof follows by induction.

N. Robbins, *Fibonacci Quart.* **34** (1996), 306–313, was the first to prove that the coefficients are $0, \pm 1$. The above explicit recursive description of the coefficents is due to F. Ardila, *Fibonacci Quart.* **42** (2004), 202–204. Another elegant proof was later given by Y. Zhao, The coefficients of a truncated Fibonacci series, *Fibonacci Quart.*, to appear, and a significant generalization by H. Diao, `arXiv:0802.1293`.

43. *Answer:*

$$S(n,1)=1 \quad c(n,1)=(n-1)!$$
$$S(n,2)=2^{n-1}-1 \quad c(n,2)=(n-1)!H_{n-1}$$
$$S(n,n)=1 \quad c(n,n)=1$$
$$S(n,n-1)=\binom{n}{2} \quad c(n,n-1)=\binom{n}{2}$$
$$S(n,n-2)=\binom{n}{3}+3\binom{n}{4} \quad c(n,n-2)=2\binom{n}{3}+3\binom{n}{4}.$$

An elegant method for computing $c(n,2)$ is the following. Choose a permutation $a_1 a_2 \cdots a_n \in \mathfrak{S}_n$ with $a_1=1$ in $(n-1)!$ ways. Choose $1 \leq j \leq n-1$ and let w be the permutation whose disjoint cycle form is $(a_1, a_2, \ldots, a_j)(a_{j+1}, a_{j+2}, \ldots, a_n)$. We obtain exactly j times every permutation with two cycles such that the cycle not containing 1 has length $n-j$. Hence, $c(n,2)=(n-1)!H_{n-1}$.

As a further example, let us compute $S(n,n-2)$. The block sizes of a partition of $[n]$ with $n-2$ blocks are either 3 (once) and 1 ($n-3$ times), or 2 (twice) and 1 ($n-4$ times). In the first case, there are $\binom{n}{3}$ ways of choosing the 3-element block. In the second case, there are $\binom{n}{4}$ ways of choosing the union of the two 2-element blocks, and then three ways to choose the blocks themselves. Hence, $S(n,n-2)=\binom{n}{3}+3\binom{n}{4}$ as claimed.

45. Define $a_{i+1}+a_{i+2}+\cdots+a_k$ to be the least r such that when $1,2,\ldots,r$ are removed from π, the resulting partition has i blocks.

46. **a.** We have by equation (1.94c) that

$$\sum_{n\geq 0}S(n,k)x^n = \frac{x^k}{(1-x)(1-2x)\cdots(1-kx)}$$
$$= \frac{x^k}{(1-x)^{\lceil k/2\rceil} (\mathrm{mod}\, 2)}.$$

b. The first of several persons to find a combinatorial proof were K. L. Collins and M. Hovey, *Combinatorica* **31** (1991), 31–32. For further congruence properties of $S(n,k)$, see L. Carlitz, *Acta Arith.* **10** (1965), 409–422.

c. Taking equation (1.28) modulo 2 gives

$$\sum_{k=0}^{n} c(n,k)t^k = t^{\lceil n/2 \rceil}(t+1)^{\lfloor n/2 \rfloor} \;(\mathrm{mod}\,2).$$

Hence,

$$c(n,k) \equiv \binom{\lfloor n/2 \rfloor}{k - \lceil n/2 \rceil} = \binom{\lfloor n/2 \rfloor}{n-k} \;(\mathrm{mod}\,2).$$

48. a. This remarkable result is due to J. N. Darroch, *Ann. Math. Stat.* **35** (1964), 1317–1321. For a nice exposition including much related work, see J. Pitman, *J. Combinatorial Theory Ser. A* **77** (1997), 279–303.

b. Let $P(x) = \sum_{k=0}^{n} c(n,k)x^k$. It is routine to compute from Proposition 1.3.7 that

$$\frac{P'(1)}{P(1)} = 1 + \frac{1}{2} + \frac{1}{3} + \cdots + \frac{1}{n},$$

and the proof follows from (a). For further information on the distribution of the number of cycles of a permutations $w \in \mathfrak{S}_n$, see Pitman, ibid., pp. 289–290.

c. This result is due to E. R. Canfield and C. Pomerance, *Integers* **2** (2002), A1 (electronic); *Corrigendum* **5**(1) (2005), A9, improving earlier expressions for K_n due to Canfield and Menon (independently). Previously it was shown by L. H. Harper, *Ann. Math. Stat.* **38** (1966), 410–414 (Lemma 1), that the polynomial $\sum_k S(n,k)x^k$ has real zeros. As Pitman points out in his paper cited in (a) (page 291), the result (a) of Darroch reduces the problem of estimating K_n to estimating the expected number of blocks of a random partition of $[n]$. For further discussion, see D. E. Knuth, *The Art of Computer Programming*, vol. 4, Fascicle 3, Addison-Wesley, Upper Saddle River, N.J., 2005 (Exercises 7.2.1.5–62 and 7.2.1.5–63(e)).

49. a. Let $F_d(x) = A_d(x)/(1-x)^{d+1}$. Differentiate equation (1.37) and multiply by x, yielding

$$F_{d+1}(x) = x\frac{d}{dx}F_d(x),$$

and so on.

b. The proof is by induction on d. Since $A_1(x) = x$, the assertion is true for $d = 1$. Assume the assertion for d. By Rolle's theorem, the function $f(x) = \frac{d}{dx}(1 - x)^{-d-1}A_d(x)$ has $d-1$ simple negative real zeros that interlace the zeros of $A_d(x)$. Since $\lim_{x \to -\infty} f(x) = 0$, there is an additional zero of $f(x)$ less than the smallest zero of $A_d(x)$. Using equation (1.38), we have accounted for d strictly negative simple zeros of $A_{d+1}(x)$, and $x = 0$ is an additional zero. The proof follows by induction. This result can be extended to permutations of a multiset; see R. Simion, *J. Combinatorial Theory, Ser. A* **36** (1984), 15–22.

50. b. Let $D = d/dx$. By Rolle's theorem, $Q(x) = D^{i-1}P(x)$ has real zeros, and thus also $R(x) = x^{n-i+1}Q(1/x)$. Again by Rolle's theorem, $D^{n-i-1}R(x)$ has real zeros. But one computes easily that

$$D^{n-i-1}R(x) = \frac{n!}{2}\left(b_{i-1}x^2 + 2b_i x + b_{i+1}\right).$$

In order for this quadratic polynomial to have real zeros, we must have $b_i^2 \geq b_{i-1}b_{i+1}$. This result goes back to I. Newton (see e.g. G. H. Hardy, J. E. Little-wood, and G. Pólya, *Inequalities*, 2nd ed., Cambridge University Press, Cambridge, England, 1952, p. 52).

c. Let us say that a polynomial $P(x) = \sum_{i=0}^{m} a_i x^i$ with coefficients satisfying $a_i = a_{m-i}$ has *center* $m/2$. (We don't assume that $\deg P(x) = m$, i.e., we may have $a_m = 0$.) Thus $P(x)$ has center $m/2$ if and only if $P(x) = x^m P(1/x)$. If also $Q(x) = x^n Q(1/x)$ (so $Q(x)$ has center $n/2$), then $P(x)Q(x) = x^{m+n} P(1/x)Q(1/x)$. Thus, $P(x)Q(x)$ has symmetric coefficients (with center $(m+n)/2$). It is also easy to show this simply by computing the coefficients of $P(x)Q(x)$ in terms of the coefficients of $P(x)$ and $Q(x)$.

Now assume that $P(x) = \sum_{i=0}^{m} a_i x^i$ has center $m/2$ and has unimodal coefficients, and similarly for $Q(x) = \sum_{i=0}^{n} b_i x^i$. Let $A_j(x) = x^j + x^{j+1} + \cdots + x^{m-j}$, a polynomial with center $m/2$, and similarly $B_j(x) = x^j + x^{j+1} + \cdots + x^{n-j}$. It is easy to see that

$$P(x) = \sum_{i=0}^{\lfloor m/2 \rfloor} (a_i - a_{i-1}) A_i(x),$$

$$Q(x) = \sum_{j=0}^{\lfloor n/2 \rfloor} (b_j - b_{j-1}) B_j(x).$$

Thus,

$$P(x)Q(x) = \sum_{i=0}^{\lfloor m/2 \rfloor} \sum_{j=0}^{\lfloor n/2 \rfloor} (a_i - a_{i-1})(b_j - b_{j-1}) A_i(x) B_j(x).$$

It is easy to check by explicit computation that $A_i(x)B_j(x)$ has unimodal coefficients and center $(m+n)/2$. Since $P(x)$ and $Q(x)$ have unimodal coefficients, we have

$$(a_i - a_{i-1})(b_j - b_{j-1}) \geq 0.$$

Hence, we have expressed $P(x)Q(x)$ as a nonnegative linear combination of unimodal polynomials, all with the same center $(m+n)/2$. It follows that $P(x)Q(x)$ is also unimodal (with center $(m+n)/2$).

d. Perhaps the most elegant proof (and one suggesting some nice generalizations) uses linear algebra. Write $P(x) = \sum_{i=0}^{m} a_i x^i$ and $Q(x) = \sum_{i=0}^{n} b_i x^i$. Set $a_i = 0$ if $i \notin [0,m]$, and similarly for b_i. If X and Y are $r \times r$ real matrices all of whose $k \times k$ minors are nonnegative, then the Cauchy-Binet theorem shows that the same is true for the matrix XY. Moreover, it is easily seen that if $c_0, c_1, \ldots, c_n$ is nonnegative and log-concave with no internal zeros, then $c_i c_j \geq c_{i-s} c_{j+s}$ whenever $i \leq j$ and $s \geq 0$. Now take $k = 2$, $X = [a_{j-i}]_{i,j=0}^{m+n}$, and $Y = [b_{j-i}]_{i,j=0}^{m+n}$, and the proof follows.

e. The symmetry of the two polynomials is easy to see in various ways. The polynomial $x \sum_{w \in \mathfrak{S}_n} x^{\text{des}(w)}$ is the Eulerian polynomial $A_n(x)$ by equation (1.36); now use (a), (b), and Exercise 1.49. The unimodality of the polynomial $\sum_{w \in \mathfrak{S}_n} x^{\text{inv}(w)}$ follows from (c) and the product formula (1.30). NOTE. A combinatorial proof of the unimodality of $\sum_{w \in \mathfrak{S}_n} x^{\text{inv}(w)}$ is implicit in the proof we have given, whereas a combinatorial proof of the log-concavity and unimodality of $A_n(x)$ is due to V. Gasharov, *J. Combinatorial Theory Ser. A* **82** (1998), 134–146 (§§4–5).

f. This result was proved by F. De Mari and M. Shayman, *Acta Appl. Math.* **12** (1988), 213–235, using the hard Lefschetz theorem from algebraic geometry. It would be interesting to give a more elementary proof. A related result was proved by M. Bóna, Generalized descents and normality, arXiv:0709.4483.

g. Let $n = 4$ and $S = \{(1,2),(2,3),(3,4),(1,4)\}$. Then

$$P_S(x) = x^4 + 8x^3 + 6x^2 + 8x + 1.$$

Note that part (f) asserts that $P_S(x)$ is unimodal for $S = \{(i,j) : 1 \leq i < j \leq n, \, j \leq i + p\}$. It seems likely (though this has not been checked) that the proof of De Mari and Shayman can be extended to the case $S = \{(i,j) : 1 \leq i < j \leq n, \, j \leq i + p_i\}$, where $p_1, \ldots, p_{n-1}$ are any nonnegative integers. Can anything further be said about those S for which $P_S(x)$ is unimodal?

For further information on the fascinating topic of unimodal and log-concave sequences, see R. Stanley, in *Graph Theory and Its Applications: East and West*, Ann. New York Acad. Sci., vol. 576, 1989, pp. 500–535, and the sequel by F. Brenti, in *Contemp. Math.* **178**, Amer. Math. Soc., Providence, R.I., 1994, pp. 71–89. For the unimodality of the q-binomial coefficient $\binom{n}{k}$ and related results, see Exercise 7.75.

51. This result goes back to P. S. de Laplace. The following proof is due to R. Stanley, in *Higher Combinatorics (Proc. NATO Advanced Study Inst., Berlin, 1976; M. Aigner, ed.)*, Reidel, Dordrecht/Boston, 1977, p. 49. Given $w \in \mathfrak{S}_n$, let $\mathcal{S}_w$ denote the region (a simplex) in $\mathbb{R}^n$ defined by

$$0 \leq x_{w(1)} \leq x_{w(2)} \leq \cdots \leq x_{w(n)} \leq 1.$$

Define $\mathcal{S}_{nk} = \bigcup_w \mathcal{S}_w$, where w ranges over all permutations in $\mathfrak{S}_n$ with exactly $k-1$ descents. It is easy to see that $\mathrm{vol}(\mathcal{S}_w) = 1/n!$, so $\mathrm{vol}(\mathcal{S}_{nk}) = A(n,k)/n!$. Define a map $\varphi : \mathcal{S}_{nk} \to \mathcal{R}_{nk}$ by $\varphi(x_1, \ldots, x_n) = (y_1, \ldots, y_n)$, where

$$y_i = \begin{cases} x_{i+1} - x_i, & \text{if } x_i < x_{i+1}, \\ 1 + x_{i+1} - x_i, & \text{if } x_i > x_{i+1}. \end{cases}$$

Here we set $x_{n+1} = 1$, and we leave φ undefined on the set of measure zero consisting of points where some $x_{i-1} = x_i$. One can check that φ is measure-preserving and a bijection up to a set of measure zero. Hence, $\mathrm{vol}(\mathcal{R}_{nk}) = \mathrm{vol}(\mathcal{S}_{nk}) = A(n,k)/n!$. For some additional proofs, see W. Meyer and R. von Randow, *Math. Annalen* **193** (1971), 315–321, S. M. Tanny, *Duke Math. J.* **40** (1973), 717–722; and J. W. Pitman, *J. Combinatorial Theory Ser. A* **77** (1997), 279–303 (pp. 295–296). For a refinement and further references, see R. Ehrenborg, M. A. Readdy, and E. Steingrímsson, *J. Combinatorial Theory Ser. A* **81** (1998), 121–126. For some related results, see Exercise 4.62.

52. This amusing result is due to J. Holte, *Amer. Math. Monthly* **104** (1997), 138–149. Holte derived this result in the setting of Markov chains and obtained many additional results about the combinatorics of carrying. Further work on this subject is due to P. Diaconis and J. Fulman, *Amer. Math. Monthly* **116** (2009), 788–803, and *Advances in Applied Math.* **43** (2009), 176–196, and A. Borodin, P. Diaconis, and J. Fulman, *Bull Amer. Math. Soc.* **47** (2009), 639–670. There is a simple intuitive reason, which is not difficult to make rigorous, why we get the Eulerian numbers. The probability that we carry j in a certain column is roughly the probability that if $i_1, \ldots, i_n$ are random integers in the interval $[0, b-1]$, then $bj \leq i_1 + \cdots + i_n < b(j+1)$. Now divide by b and use Exercise 1.51.

56. Let $\phi(w)$ denote the standardization (as defined in the second proof of Proposition 1.7.1) of $w \in \mathfrak{S}_M$. If $M = \{1^{m_1}, 2^{m_2}, \ldots\}$ and $\#M = n$, then $\{\phi(w) : w \in \mathfrak{S}_M\}$ consists of all permutations $v \in \mathfrak{S}_n$ such that $D(v^{-1}) = \{m_1, m_1 + m_2, \cdots\} \cap [n-1]$. It is easy to see that $\mathrm{inv}(w) = \mathrm{inv}(v)$ (a special case of (1.71)) and $\mathrm{maj}(w) = \mathrm{maj}(v)$. The proof now follows from equation (1.43) and Theorem 1.4.8. This result is due to P. A. MacMahon, stated explicitly on page 317 of his paper [1.54]. Some other classes of permutations that are equidistributed with respect to inv and maj are given by A. Björner and M. L. Wachs, *J. Combinatorial Theory Ser. A* **52** (1989), 165–187, and D. Foata and D. Zeilberger, *J. Comput. Applied Math.* **68** (1996), 79–101. See also the solution to Exercise 5.49(e).

57. Condition (i) does not hold if and only if there are indices $i < i'$ and $j < j'$ such that $(i, j) \in D(w)$, $(i', j') \in D(w)$, $(i, j') \notin D(w)$, $(i', j) \notin D(w)$. Let $w(i'') = j$ and $w(i''') = j'$. It is easy to check by drawing a diagram that $i < i'' < i' < i'''$ and $w(i'') < w(i) < w(i''') < w(i')$, so w is not 2143-avoiding. The steps are reversible, so (i) and (iii) are equivalent. The equivalence of (i) and (ii) follows from the fact that the jth term of $I(w)$ (respectively, $I(w^{-1})$) is the number of elements of $D(w)$ in column (respectively, row) j.

   The permutations of this exercise are called *vexillary*. For further information on their history and properties, see Exercise 7.22(d,e) of Vol. II.

58. **b.** The final step in obtaining this result was achieved by Z. Staṇkova, *Europ. J. Combin.* **17** (1996), 501–517. For further information, see H. S. Wilf, *Discrete Math.* **257** (2002), 575–583, and M. Bóna [1.11, §4.4].

59. This result is known as the *Stanley-Wilf conjecture*. It was shown by R. Arratia, *Electron. J. Combin.* **6**(1) (1999), N1, that the conjecture follows from the statement that there is a real number $c > 1$ (depending on u) for which $s_u(n) < c^n$ for all $n \geq 1$. This statement was given a surprisingly simple and elegant proof by A. Marcus and G. Tardos, *J. Combinatorial Theory Ser. A* **107** (2004), 153–160. A nice exposition of this proof due to D. Zeilberger is available at

   ⟨www.math.rutgers.edu/~zeilberg/mamarim/mamarimhtml/

   paramath.html⟩.

   Another nice exposition is given by M. Bóna [1.11, §4.5].

60. *Answer.* The equivalence classes consist of permutations whose inverses have a fixed descent set. The number of equivalence classes is therefore 2^{n-1}, the number of subsets of $[n-1]$.

   It is not difficult to prove this result directly but, it also can be understood in a nice way using the "Cartier-Foata theory" of Exercise 3.123.

61. **a.** By the properties of the bijection $w \mapsto T(w)$ discussed in Section 1.5, we have that

$$F(x; a, b, c, d) = \sum_{n \geq 1} \sum_{T} a^{\mathrm{lr}(T)} b^{e(T)-1} c^{r(T)} d^{l(T)} \frac{x^n}{n!},$$

   where T ranges over all increasing binary trees on the vertex set $[n]$, with $\mathrm{lr}(T)$ vertices with two children, $e(T)$ vertices that are endpoints, $l(T)$ vertices with just a left child, and $r(T)$ vertices with just a right child. By removing the root from T, we obtain the equation

$$\frac{\partial}{\partial x}(F - bx) = abF^2 + (c+d)F. \tag{1.154}$$

   Solving this equation (a Ricatti equation, with a well-known method of solution) with the initial condition $F(0; a, b, c, d) = 0$ yields equation (1.124).

   This result is due to L. Carlitz and R. Scoville, *J. reine angew. Math.* **265** (1974), 110–137 (§7). Our presentation follows Exercise 3.3.46 of I. P. Goulden and D. M. Jackson, *Combinatorial Enumeration*, John Wiley & Sons, New York, 1983; reprinted by Dover, Mineola, N.Y., 2004. This latter reference contains more details on solving the differential equation (1.154).

b. The generating function is given by $1 + tF(x; 1, t, 1, 1)$, which can be simplified to the right-hand side of equation (1.125).

The enumeration of permutations by number of peaks was first considered by F. N. David and D. E. Barton, *Combinatorial Chance*, Hafner, New York, 1962 (pp. 162–164). They obtain a generating function for $r(n, k)$ written in a different form from equation (1.125).

c. We have that $f(n)$ is the number of increasing binary trees on $[n]$ such that no vertex has only a left child except possibly the last vertex obtained by beginning with the root and taking right children. Let $g(n)$ be the number of increasing binary trees on $[n]$ such that no vertex has only a left child. Then

$$f(n+1) = \sum_{k=0}^{n} \binom{n}{k} f(k)g(n-k),$$

$$g(n+1) + \sum_{k=0}^{n-1} \binom{n}{k} g(k)g(n-k),$$

with $f(0) = g(0) = 1$. Setting $F(x) = \sum f(n)x^n/n!$ and $G(x) = \sum g(n)x^n/n!$, we obtain $F' = FG$ and $G + G' = G^2 + 1$. We can solve these differential equations to obtain equation (1.126). Goulden and Jackson, op. cit. (Exercise 5.2.17, attribution on page 306) attribute this result to P. Flajolet (private communication, 1982). The proof in Goulden and Jackson is based essentially on the Principle of Inclusion-Exclusion and is given here in Exercise 2.23.

62. **a.** First note that

$$p_k^n = \sum_{d \mid n} \sum_A d(w(A))^{nk/d}, \qquad (1.155)$$

where A ranges over all aperiodic cycles of length d (i.e., cycles of length d that are unequal to a proper cyclic shift of themselves). Now substitute (1.155) into the expansion of $\log \prod (1 - p_k)^{-1}$ and simplify.

This result is implicit in the work of R. C. Lyndon (see Lothaire [4.31, Thm. 5.1.5]). See also N. G. de Bruijn and D. A. Klarner, *SIAM J. Alg. Disc. Meth.* **3** (1982), 359–368. The result was stated explicitly by I. M. Gessel (unpublished). A different theory of cycles of multiset permutations, due to D. Foata, has a nice exposition in §5.1.2 of D. E. Knuth [1.48]. In Foata's theory, a multiset permutation has the meaning of Section 1.7.

b. Let $x_1 = \cdots = x_k = x$, and $x_j = 0$ if $j > k$.

c. Let $\sigma = (a_1, a_2, \ldots, a_{jk})$ be a multiset cycle of length jk, where k is the largest integer for which the word $u = a_1 a_2 \cdots a_{jk}$ has the form v^k for some word v of length j (where v^k denotes the concatenation of k copies of v). Let $\Gamma(\sigma) = p_k^j$. Given a multiset permutation $\pi = \sigma_1 \sigma_2 \cdots \sigma_m$ where each σ_i is a multiset cycle, define $\Gamma(\pi) = \Gamma(\sigma_1) \cdots \Gamma(\sigma_m)$. It can then be verified combinatorially that the number of multiset permutations π with fixed $w(\pi)$ and $\Gamma(\pi)$ is equal to the coefficient of $w(\pi)$ in $\Gamma(\pi)$, leading to the desired bijection.

63. Label the envelopes $1, 2, \ldots, n$ in decreasing order of size. Partially order an arrangement of envelopes by inclusion, and adjoin a root labeled 0 at the top. We obtain an (unordered) increasing tree on $n + 1$ vertices, and this correspondence is clearly invertible. Hence by Proposition 1.5.5, there are $n!$ arrangements in all, of which $c(n, k)$ have k envelopes not contained in another and $A(n, k)$ have k envelopes not containing another.

64. **a.** Let u be a sequence being counted, with m_i occurrences of i. Replace the 1's in u from right-to-left by $1, 2, \ldots, m_1$. Then replace the 2's from right-to-left by

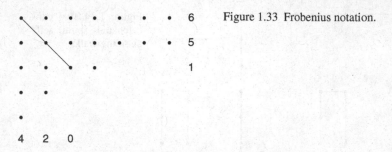

Figure 1.33 Frobenius notation.

$m_1 + 1, m_1 + 2, \ldots, m_1 + m_2$, and so on. This procedure gives a bijection with $\mathfrak{S}_n$. For instance, 13213312 corresponds to 38527614. Note that this bijection could also be described as $u \mapsto \rho\psi\rho(u)$, where $\rho(v)$ is the reversal of v, and ψ denotes standardization (defined after the second proof of Proposition 1.7.1).

b. The bijection in (a) has the property that $\max\{a_1, \ldots, a_n\} = \mathrm{des}(\rho(w)^{-1}) + 1$, and so on. This result was pointed out by D. E. Knuth (private communication, 21 August 2007) upon seeing (a).

65. It follows from a general theorem of Ramanujan (see D. Zagier, *The 1-2-3 of Modular Forms*, in (J. H. Bruinier, G. van der Geer, G. Harder and D. Zagier, eds.), Springer-Verlag, Berlin, 2008, Prop. 16, p. 49) that y satisfies a third order algebraic differential equation, but it is considerably more complicated than the fourth-degree equation (1.127). This equation was first computed by M. Rubey in 2010. See W. Hebisch and M. Rubey, *J. Symbolic Computation*, to appear.

70. a. Draw a line L along the main diagonal of the Ferrers diagram of λ. Then a_i is the number of dots in the ith row to the right of L, whereas b_i is the number of dots in the ith column below i. Figure 1.33 shows that $A_{77421} = \left(\begin{smallmatrix} 6 & 5 & 1 \\ 4 & 2 & 0 \end{smallmatrix}\right)$. This bijection is due to F. G. Frobenius, *Sitz. Preuss. Akad. Berlin* (1900), 516–534, and *Gesammelte Abh.* **3**, Springer, Berlin, 1969, pp. 148–166, and the array A_λ is called the *Frobenius notation* for λ.

b. Suppose that the path P consists of c_1 steps N, followed by c_2 steps E, then c_3 steps S, and so on, ending in c_ℓ steps. If $\ell = 2r$, then associate with P the partition λ whose Frobenius notation is

$$A_\lambda = \left(\begin{array}{ccccc} c_{\ell-1} & c_{\ell-3} & c_{\ell-5} & \cdots & c_1 \\ c_\ell - 1 & c_{\ell-2} - 1 & c_{\ell-4} - 1 & \cdots & c_2 - 1 \end{array} \right).$$

If $\ell = 2r - 1$ then associate with P the partition λ whose Frobenius notation is

$$A_\lambda = \left(\begin{array}{ccccc} c_{\ell-1} & c_{\ell-3} & \cdots & c_2 & 0 \\ c_\ell - 1 & c_{\ell-2} - 1 & \cdots & c_3 - 1 & c_1 - 1 \end{array} \right).$$

This sets up the desired bijection. For instance, the CSSAW of Figure 1.34(a) corresponds to the partition $\lambda = (8, 6, 5, 2, 1)$ with $A_\lambda = \left(\begin{smallmatrix} 7 & 4 & 2 \\ 4 & 2 & 0 \end{smallmatrix}\right)$, while Figure 1.34(b) corresponds to $\lambda = (4, 3, 3, 3, 2, 1, 1)$ with $A_\lambda = \left(\begin{smallmatrix} 3 & 1 & 0 \\ 6 & 3 & 1 \end{smallmatrix}\right)$. This result is due to A. J. Guttman and M. D. Hirschhorn, *J. Phys. A Math. Gen.* **17** (1984), 3613–3614. They give a combinatorial proof equivalent to the foregoing, though not stated in terms of Frobenius notation. The connection with Frobenius notation was given by G. E. Andrews, *Electronic J. Combinatorics* **18(2)** (2011), P6.

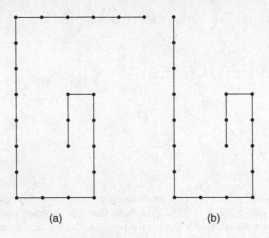

Figure 1.34 Two concatenated spiral self-avoiding walks.

(a)   (b)

71. *Answer.* $p(0) + p(1) + \cdots + p(n)$. Given $\nu \vdash k \leq n$, define λ to be ν with the part $n - k$ adjoined (in the correct position, so the parts remain weakly decreasing), and define μ to be ν with $n - k + 1$ adjoined. This yields the desired bijection. For some generalizations, see Theorem 3.21.11 and Exercise 3.150.

72. This exercise gives a glimpse of the fascinating subject of *plane partitions*, treated extensively in Sections 7.20–7.22 of Vol. II.
    a. Although equation (1.128) can be proved by ad hoc arguments, the "best" proof is a bijection using the RSK algorithm, the special case $q = 1, r = 2$, and $c \to \infty$ of Theorem 7.20.1 of Vol. II. A different generalization, but with a nonbijective proof, is given by Theorem 7.21.7 of Vol. II.
    b. This result is due to B. Gordon, *Proc. Amer. Math. Soc.* **13** (1962), 869–873. A bijective proof was given by C. Sudler, Jr., *Proc. Amer. Math. Soc.* **16** (1965), 161–168. This result can be generalized to a chain $\lambda^1 \subseteq \lambda^2 \subseteq \cdots \subseteq \lambda^k$ of any fixed number k of strict partitions, and with a fixed bound on the largest part of λ^k. See [7.146, Prop. 16.1] of Vol. II and G. E. Andrews, *Pacific J. Math.* **72** (1977), 283–291.

73. Consider for instance $\lambda = (5, 4, 4, 2, 1, 1)$, and put dots in the squares of the diagram of λ as follows:

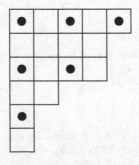

Count the total number of dots by rows and by columns to obtain the first identity. The other formulas are analogous. There are many further variations.

74. Subtract one from each part of a partition of n into $n - t$ parts to deduce that $p_{n-t}(n) = p(t)$ if and only if $n \geq 2t$.

75. The partition $\lambda_1 \geq \lambda_2 \geq \cdots \geq \lambda_k$ corresponds to $\lambda_1 + k - 1 > \lambda_2 + k - 2 > \cdots > \lambda_k$.

76. By the bijection illustrated in Figure 1.16, the coefficient of $q^k x^n$ in the left-hand side of equation (1.129) is equal to the number of self-conjugate partitions λ of n whose rank is k. If we remove the Durfee square from the diagram of λ, then we obtain two partitions μ and μ' (the conjugate of μ) with largest part at most k. Hence we obtain the right-hand side of (1.129).

One can also prove this identity by making the substitution $x \to x^2$ and $q \to qx^{-1}$ into equation (1.83).

77. Given $r \in \mathbb{Z}$, let λ be a partition satisfying $\lambda_1' + r \geq \lambda_1 - 1$. Define $\psi_r(\lambda)$ to be the partition obtained by removing the first column of (the diagram of) λ and adding a new row at the top of length $\lambda_1' + r$. We need to give a bijection

$$\gamma_n : \bigcup_{m \in 2\mathbb{Z}} \mathrm{Par}(n - m(3m - 1)/2) \to \bigcup_{m \in 1 + 2\mathbb{Z}} \mathrm{Par}(n - m(3m - 1)/2).$$

One can check that we can define γ_n as follows: for $\lambda \in \bigcup_{m \in 2\mathbb{Z}} \mathrm{Par}(n - m(3m - 1)/2)$, let

$$\gamma_n(\lambda) = \begin{cases} \psi_{-3m-1}(\lambda), & \text{if } \lambda_1 - \lambda_1' + 3m \leq 0, \\ \psi_{-3m+2}^{-1}(\lambda), & \text{if } \lambda_1 - \lambda_1' + 3m \geq 0. \end{cases}$$

This proof appears in D. M. Bressoud and D. Zeilberger, *Amer. Math. Monthly* **92** (1985), 54–55. Our presentation follows Pak [1.62, §5.4.1].

78. **a.** Some related results are due to Euler and recounted in [1.55, §303].

b. This problem was suggested by Dale Worley. For each $1 \leq i \leq n$, each partition λ of $n - i$, and each divisor d of i, we wish to associate a d-element multiset M of partitions of n so that every partition of n occurs exactly n times. Given i, λ, and d, simply associate d copies of the partition obtained by adjoining i/d d's to λ.

79. **a.** See [1.2, Cor. 8.6].

b. Clearly $p_S(n) = 1$ for all n, so the statement $q_S(n) = 1$ is just the uniqueness of the binary expansion of n.

80. For each partition λ of n and each part j of λ occurring at least k times, we need to associate a partition μ of n such that the total number of times a given μ occurs is the same as the number $f_k(\mu)$ of parts of μ that are equal to k. To do this, simply change k of the j's in λ to j k's. For example, $n = 6, k = 2$:

λ	j	μ
1 1 1 1 1 1	1	2 1 1 1 1
2 1 1 1 1	1	2 2 1 1
3 1 1 1	1	3 2 1
4 1 1	1	4 2
2 2 1 1	2	2 2 1 1
2 2 1 1	1	2 2 2
2 2 2	2	2 2 2
3 3	3	2 2 2.

This result was discovered by R. Stanley in 1972 and submitted to the Problems and Solutions section of *Amer. Math. Monthly*. It was rejected with the comment "A bit on the easy side, and using only a standard argument." Daniel I. A. Cohen learned of this result and included the case $k = 1$ as Problem 75 of Chapter 3 in his book *Basic Techniques of Combinatorial Theory*, Wiley, New York, 1978. For this reason, the case $k = 1$ is sometimes called "Stanley's theorem." The generalization from $k = 1$ to arbitrary k was independently found by Paul Elder in 1984, as reported by R. Honsberger, *Mathematical Gems III*, Mathematical Association of America, Washington, DC, 1985, p. 8. For this reason the general case is sometimes called "Elder's theorem." An independent proof of the general case was given by M. S. Kirdar and T. H. R. Skyrme, *Canad. J. Math.* **34** (1982), 194–195, based on generating functions. The bijection given here also appears in A. H. M. Hoare, *Amer. Math. Monthly* **93** (1986), 475–476. Another proof appears in L. Solomon, *Istituto Nazionale di Alta Matematica, Symposia Matematica*, vol. 13, Academic Press, London, 1974, pp. 453–466 (lemma on p. 461).

81. Given an ordered factorization $n + 1 = a_1 a_2 \cdots a_k$, set $a_0 = 1$ and let λ be the partition for which the part $a_0 a_1 \cdots a_{j-1}$ occurs with multiplicity $a_j - 1$, $1 \le j \le k$. For instance, if $24 = 3 \cdot 2 \cdot 4$, then we obtain the partition 666311 of 23. This procedure sets up a bijection with perfect partitions of n, due to P. A. MacMahon, *Messenger Math.* **20** (1891), 103–119; reprinted in [1.3, pp. 771–787]. Note that if we have a perfect partition λ of n with largest part m, then there are exactly two ways to add a part p to λ to obtain another perfect partition, namely, $p = m$ or $p = n + 1$.

82. This result is due to S. Ramanujan in 1919, who obtained the remarkable identity

$$\sum_{n \ge 0} p(5n + 4)x^n = 5 \frac{\prod_{k \ge 1}(1 - x^{5k})^5}{\prod_{k \ge 1}(1 - x^k)^6}.$$

F. J. Dyson conjectured in 1944 that for each $0 \le i \le 4$, exactly $p(5n+4)/5$ partitions λ of $5n + 4$ satisfy $\lambda_1 - \lambda_1' \equiv i \pmod 5$. This conjecture was proved by A. O. L. Atkin and H. P. F. Swinnerton-Dyer in 1953. Many generalizations of these results are known. For an introduction to the subject of partition congruences, see Andrews [1.2, Ch. 10]. For more recent work in this area, see K. Mahlburg, *Proc. Nat. Acad. Sci.* **102** (2005), 15373–15376.

83. *Some Hints.* Let A be the set of all partitions λ such that $\lambda_{2i-1} - \lambda_{2i} \le 1$ for all i, and let B be the set of all partitions λ such that λ' has only odd parts, each of which is repeated an even number of times. Verify the following statements.
- There is bijection $A \times B \to \mathrm{Par}$ satisfying $w(\mu)w(\nu) = w(\lambda)$ if $(\mu, \nu) \mapsto \lambda$.
- We have

$$\sum_{\lambda \in B} w(\lambda) = \prod_{j \ge 1} \frac{1}{1 - a^j b^j c^{j-1} d^{j-1}}.$$

- Let $\lambda \in A$. Then the pairs $(\lambda_{2i-1}, \lambda_{2i})$ fall into two classes: (a, a) (which can occur any number of times), and $(a + 1, a)$ (which can occur at most once). Deduce that

$$\sum_{\lambda \in A} w(\lambda) = \prod_{j \ge 1} \frac{(1 + a^j b^{j-1} c^{j-1} d^{j-1})(1 + a^j b^j c^j d^{j-1})}{(1 - a^j b^j c^j d^j)(1 - a^j b^j c^{j-1} d^{j-1})}.$$

This elegant bijective proof is due to C. Boulet, *Ramanujan J.* **3** (2006), 315–320, simplifying and generalizing previous work of G. E. Andrews, A. V. Sills, R. P. Stanley, and A. J. Yee.

86. This is *Schur's partition theorem*. See G. E. Andrews, in *q-Series: Their Development and Application in Analysis, Number Theory, Combinatorics, Physics, and Computer Algebra*, American Mathematical Society, Providence, R.I., 1986, pp. 53–58. For a bijective proof, see D. M. Bressoud, *Proc. Amer. Math. Soc.* **79** (1980), 338–340. It is surprising that Schur's partition theorem is easier to prove bijectively than the Rogers–Ramanujan identities (Exercise 1.88).

88. **a.** These are the famous *Rogers–Ramanujan identities*, first proved by L. J. Rogers, *Proc. London Math. Soc.* **25** (1894), 318–343, and later rediscovered by I. Schur,

Sitzungsber. Preuss. Akad. Wiss. Phys.-Math. Klasse (1917), 302–321, S. Ramanujan (sometime before 1913, without proof), and others. For a noncombinatorial proof, see, for example, [1.2, §7.3]. For an exposition and discussion of bijective proofs, see Pak [1.62, §7 and pp. 62–63]. For an interesting recent bijective proof, see C. Boulet and I. Pak, *J. Combinatorial Theory Ser. A* **113** (2006), 1019–1030. None of the known bijective proofs of the Rogers–Ramanujan identities can be considered "simple," comparable to the proof we have given of the pentagonal number formula (Proposition 1.8.7). An interesting reason for the impossibility of a nice proof was given by I. Pak, The nature of partition bijections II. Asymptotic stability, preprint.

b. These combinatorial interpretations of the Rogers–Ramanujan identities are due to P. A. MacMahon, [1.55, §§276–280]. They can be proved similarly to the proof of Proposition 1.8.6, based on the observation that $(\lambda_1, \lambda_2, \ldots, \lambda_k)$ is a partition of n with at most k parts if and only if $(\lambda_1 + 2k - 1, \lambda_2 + 2k - 3, \ldots, \lambda_k + 1)$ is a partition of $n + k^2$ whose parts differ by at least two and with exactly k parts, and similarly for $(\lambda_1 + 2k, \lambda_2 + 2k - 2, \ldots, \lambda_k + 2)$.

89. Let $\mu = (\mu_1, \ldots, \mu_k)$ be a partition of n into k odd parts less than $2k$. We begin with the lecture hall partition $\lambda^0 = (0, \ldots, 0)$ of length k and successively insert the parts $\mu_1, \mu_2, \ldots, \mu_k$ to build up a sequence of lecture hall partitions $\lambda^1, \lambda^2, \ldots, \lambda^k = \lambda$. The rule for inserting $\mu_i := 2\nu_i - 1$ into λ^{i-1} is the following. Add 1 to the parts of λ^{i-1} (allowing 0 as a part), beginning with the largest, until either (i) we have added 1 to μ_i parts of λ_{i-1}, or (ii) we encounter a value λ_{2c-1}^{i-1} for which

$$\frac{\lambda_{2c-1}^{i-1}}{n - 2c + 2} = \frac{\lambda_{2c}^{i-1}}{n - 2c + 1}.$$

In this case we add $\nu_i - c + 1$ to λ_{2c-1}^{i-1} and $\nu_i - c$ to λ_{2c}^{i-1}. It can then be checked that the map $\mu \mapsto \lambda$ gives the desired bijection.

Example. Let $k = 5$ and $\mu = (7, 5, 5, 3, 1)$. We have $\frac{\lambda_1^0}{5} = \frac{\lambda_2^0}{4} = 0$. Hence, $\lambda^1 = (4, 3, 0, 0, 0)$. We now have $\frac{\lambda_1^1}{5} \neq \frac{\lambda_2^1}{4}$, but $\frac{\lambda_3^1}{3} = \frac{\lambda_4^1}{2} = 0$. Hence, $\lambda^2 = (5, 4, 2, 1, 0)$. Continuing in this way, we get $\lambda^3 = (8, 6, 2, 1, 0)$, $\lambda^4 = (9, 7, 3, 1, 0)$, and $\lambda = \lambda^5 = (10, 7, 3, 1, 0)$.

Lecture hall partitions were introduced by M. Bousquet-Mélou and K. Eriksson, *Ramanujan J.* **1** (1997), 101–111, 165–185. They proved the result of this exercise as well as many generalizations and refinements. In our earlier sketch, we followed A. J. Yee, *Ramanujan J.* **5** (2001), 247–262. Her bijection is a simplified description of the bijection of Bousquet-Mélou and Eriksson. Much further work has been done in this area; see, for example, S. Corteel and C. D. Savage, *J. Combinatorial Theory Ser. A* **108** (2004), 217–245, for further information and references.

90. This curious result is connected with the theory of lecture hall partitions (Exercise 1.89). It was originally proved by M. Bousquet-Mélou and K. Eriksson, *Ramanujan J.* **1** (1997), 165–185 (end of Section 4). For a nice bijective proof of this result and related results, see C. D. Savage and A. J. Yee, *J. Combinatorial Theory Ser. A* **115** (2008), 967–996.

91. **a.** This famous result is the *Jacobi triple product identity*. It was first stated by C. F. Gauss (unpublished). The first published proof is due to C. G. J. Jacobi, *Fundamenta nova theoriae functionum ellipticarum*, Regiomonti, fratrum Bornträger, Paris, London, Amsterdam, St. Petersburg, 1829; reprinted in *Gesammelte Werke*, vol. 1, Reimer, Berlin, 1881, pp. 49–239. For a summary of its bijective proofs, see Pak [1.62, §6 and pp. 60–62].

b. Substitute $q^{3/2}$ for q and $-q^{1/2}$ for x, and simplify.

c. For the first, set $x = -1$ and use equation (1.81). For the second, substitute $q^{1/2}$ for both x and q. The right-hand side then has a factor equal to 2. Divide both sides by 2, and again use equation (1.81). These identities are due to Gauss, Zur Theorie der neuen Transscendenten II, *Werke*, Band III, Göttingen, 1866, pp. 436–445 (§4). For a cancellation proof, see Exercise 2.31. For another proof of equation (1.132) based on counting partitions of n with empty 2-core, see Exercise 7.59(g) of Vol. II.

d. After making the suggested substitution, we obtain

$$\sum_{n\in\mathbb{Z}}(-1)^n x^n q^{\binom{n}{2}} = \prod_{k\geq 1}(1-q^k)(1-xq^{k-1})(1-x^{-1}q^k).$$

Rewrite the left-hand side as

$$1+\sum_{n\geq 1}(-1)^n (x^{-n}+x^n)q^{\binom{n}{2}}.$$

Now divide both sides by $1-x$ and let $x \to 1$. The left-hand side becomes $\sum_{n\geq 0}(-1)^n(2n+1)q^{\binom{n}{2}}$. The right-hand side has a factor equal to $1-x$, so deleting this factor and then setting $x = 1$ gives

$$(1-q)^2\prod_{k\geq 2}(1-q^{k-1})(1-q^k)^2 = \prod_{k\geq 1}(1-q^k)^3,$$

and the proof follows. This identity is due to C. G. J. Jacobi, *Fundamenta Nova Theoriae Functionum Ellipticarum*, Regiomonti, Sumtibus fratrum Borntraeger, Königsberg, Germany, 1829, p. 90.

92. This identity is due to G. E. Andrews, *Amer. Math. Monthly* **94** (1987), 437–439. A simple proof based on the Jacobi triple product identity (Exercise 1.91) is due to F. G. Garvan, in *Number Theory for the Millenium, II* (Urbana, IL, 2000), A K Peters, Natick, MA, 2002, pp. 75–92 (§1). This paper contains many further similar identities. For a continuation, see F. G. Garvan and H. Yesilyurt, *Int. J. Number Theory* **3** (2007), 1–42. No bijective proofs are known of any of these identities.

93. This identity is due to F. G. Garvan, op. cit. This paper and the continuation by Garvan and Yesilyurt, op. cit., contain many similar identities. No bijective proofs are known of any of them.

94. The sequence $a_1, a_2, \ldots$ (sometimes prepended with $a_0 = 0$) is called *Stern's diatomic sequence*, after the paper by M. A. Stern, *J. Reine angew. Math.* **55** (1858), 193–220. For a survey of its remarkable properties, see S. Northshield, *Amer. Math. Monthly* **117** (2010), 581–598.

95. This remarkable result is due to D. Applegate, O. E. Pol, and N. J. A. Sloane, *Congressus Numerantium* **206** (2010), 157–191.

96. **a.** The function $\tau(n)$ is *Ramanujan's tau function*. The function

$$\Delta(t) = (2\pi)^{12}\sum_{n\geq 1}\tau(n)e^{2\pi i t}$$

plays an important role in the theory of modular forms; see, for example, T. Apostol, *Modular Forms and Dirichlet Series in Number Theory* 2nd ed., Springer-Verlag,

New York, 1997, p. 20, or J.-P. Serre, *A Course in Arithmetic*, Springer-Verlag, New York, 1973, §VII.4. The multiplicativity property of this exercise was conjectured by S. Ramanujan, *Trans. Cambridge Phil. Soc.* **22** (1916), 159–184, and proved by L. J. Mordell, *Proc. Cambridge Phil. Soc.* **19** (1917), 117–124.

b. This result was also conjectured by Ramanujan, op. cit., and proved by Mordell, op. cit.

c. This inequality was conjectured by Ramanujan, op. cit., and proved by P. R. Deligne, *Inst. Hautes Études Sci. Publ. Math.* **43** (1974), 273–307; **52** (1980), 137–252. Deligne deduced Ramanujan's conjecture (in a nontrivial way) from his proof of the Riemann hypothesis for varieties over finite fields (the most difficult part of the "Weil conjectures"). Deligne in fact proved a conjecture of Petersson generalizing Ramanujan's conjecture.

d. This inequality was conjectured by D. H. Lehmer, *Duke Math. J.* **14** (1947), 429–492. It is known to be true for (at least) $n < 2.2 \times 10^{16}$.

97. This result follows from the case $p = 2$ and $\mu = \emptyset$ of Exercise 7.59(e). Greta Panova (October 2007) observed that it can also be deduced from Exercise 1.83. Namely, first prove by induction that the Ferrers diagram of λ can be covered by edges if and only if the Young diagram of λ has the same number of white squares as black squares in the usual chessboard coloring. Thus, $f(n)$ is the coefficient of q^n in the right-hand side of equation (1.130) after substituting $a = d = q/y$ and $b = c = y$. Apply the Jacobi triple product identity (Exercise 1.91) to the numerator and then set $y = 0$ to get $\sum_{n \geq 0} f(n)q^n = 1/\prod_{j \geq 1}(1 - q^j)^2$.

98. Substitute na for j, $-x$ for x, and ζ for q in the q-binomial theorem (equation (1.87)). The proof follows straightforwardly from the identity

$$\prod_{m=0}^{na-1}(1 - \zeta^m x) = (1 - x^n)^a.$$

For a host of generalizations, see V. Reiner, D. Stanton, and D. White, *J. Combinatorial Theory Ser. A* **108** (2004), 17–50.

99. It is an immediate consequence of the identity $f(q) = q^{k(n-k)} f(1/q)$ that

$$f'(1) = \frac{1}{2}k(n - k)f(1) = \frac{1}{2}k(n - k)\binom{n}{k}.$$

100. The Chu–Vandermonde identity follows from $(1 + x)^{a+b} = (1 + x)^a(1 + x)^b$. Write $f_n(x) = (1 + x)(1 + qx) \cdots (1 + q^{n-1}x)$. The q-analogue of $(1 + x)^{a+b} = (1 + x)^a(1 + x)^b$ is $f_{a+b}(x) = f_b(x)f_a(q^b x)$. By the q-binomial theorem (equation (1.87)) we get

$$\sum_{n=0}^{a+b} q^{\binom{n}{2}}\binom{a+b}{n}x^n = \left(\sum_{k=0}^{b} q^{\binom{k}{2}}\binom{b}{k}x^k\right)\left(\sum_{k=0}^{a} q^{bk+\binom{k}{2}}\binom{a}{k}x^k\right).$$

Equating coefficients of x^n yields

$$q^{\binom{n}{2}}\binom{a+b}{n} = \sum_{k=0}^{n} q^{\binom{n-k}{2}+bk+\binom{k}{2}}\binom{b}{n-k}\binom{a}{k}$$

$$\Rightarrow \binom{a+b}{n} = \sum_{k=0}^{n} q^{k(k+b-n)}\binom{a}{k}\binom{b}{n-k}.$$

103. See Lemma 3.1 of K. Liu, C. H. F. Yan, and J. Zhou, *Sci. China, Ser. A* **45** (2002), 420–431, for a proof based on the Hilbert scheme of n points in the plane. A combinatorial proof of a continuous family of results including this exercise appears in N. Loehr and G. S. Warrington, *J. Combinatorial Theory Ser. A* **116** (2009), 379–403.

104. Let $f(x) = 1 + x + \cdots + x^9$ and $i^2 = 1$. It is not hard to see that

$$f(n) = \frac{1}{4}\left(f(1)^n + f(i)^n + f(-1)^n + f(-i)^n\right)$$

$$= \frac{1}{4}(10^n + (1+i)^n + (1-i)^n)$$

$$= \begin{cases} \frac{1}{4}(10^n + (-1)^k 2^{2k-1}), & n = 4k, \\[1ex] \frac{1}{4}(10^n + (-1)^k 2^{2k-1}), & n = 4k+1, \\[1ex] \frac{1}{4}10^n, & n = 4k+2, \\[1ex] \frac{1}{4}(10^n + (-1)^{k+1} 2^{2k}), & n = 4k+3. \end{cases}$$

105. a. Let $P(x) = (1+x)(1+x^2)\cdots(1+x^n) = \sum_{k\geq 0} a_k x^k$. Let $\zeta = e^{2\pi i/n}$ (or any primitive nth root of unity). Since for any integer k,

$$\sum_{j=1}^{n} \zeta^{kj} = \begin{cases} n, & \text{if } n|k, \\ 0, & \text{otherwise,} \end{cases}$$

we have

$$\frac{1}{n} \sum_{j=1}^{n} P(\zeta^j) = \sum_{j} a_{jn} = f(n).$$

Now if ζ^j is a primitive dth root of unity (so $d = n/(j,n)$), then

$$x^d - 1 = (x - \zeta^j)(x - \zeta^{2j})\cdots(x - \zeta^{dj}),$$

so putting $x = -1$ yields

$$(1+\zeta^j)(1+\zeta^{2j})\cdots(1+\zeta^{dj}) = \begin{cases} 2, & d \text{ odd}, \\ 0, & d \text{ even}. \end{cases}$$

Hence,

$$P(\zeta^j) = \begin{cases} 2^{n/d}, & d \text{ odd}, \\ 0, & d \text{ even}. \end{cases}$$

Since there are $\phi(d)$ values of $j \in [n]$ for which ζ^j is a primitive dth root of unity, we obtain

$$f(n) = \frac{1}{n} \sum_{j=1}^{n} P(\zeta^j) = \frac{1}{n} \sum_{\substack{d|n \\ d \text{ odd}}} \phi(d) 2^{n/d}.$$

This result appears in R. Stanley and M. F. Yoder, *JPL Technical Report 32-1526*, Deep Space Network **14** (1972), 117–123.

b. Suppose that n is an odd prime. Identify the beads of a necklace with $\mathbb{Z}/n\mathbb{Z}$ in an obvious way. Let $S \subseteq \mathbb{Z}/n\mathbb{Z}$ be the set of black beads. If $S \neq \emptyset$ and $S \neq \mathbb{Z}/n\mathbb{Z}$, then there is a unique $a \in \mathbb{Z}/n\mathbb{Z}$ for which

$$\sum_{x \in S} (x + a) = 0.$$

The set $\{x+a : x \in S\}$ represents the same necklace (up to cyclic symmetry), so we have associated with each nonmonochromatic necklace a subset of $\mathbb{Z}/n\mathbb{Z}$ whose elements sum to 0. Associate with the necklaces of all black beads and all white beads the subsets $S = \emptyset$ and and $S = \mathbb{Z}/n\mathbb{Z}$, and we have the desired bijection.

A proof for any odd n avoiding roots of unity and generating functions was given by Anders Kaseorg (private communication) in 2004, though the proof is not a direct bijection.

c. See A. M. Odlyzko and R. Stanley, *J. Number Theory* **10** (1978), 263–272.

106. We claim that $f(n,k)$ is just the Stirling number $S(n,k)$ of the second kind. We need to associate with a sequence $a_1 \cdots a_n$ being counted a partition of $[n]$ into k blocks. Simply put i and j in the same block when $a_i = a_j$. This yields the desired bijection. The sequences $a_1 \cdots a_n$ are called *restricted growth functions* or *restricted growth strings* (sometimes with 1 subtracted from each term). For further information, see S. Milne, *Advances in Math.* **26** (1977), 290–305.

108. a. Given a partition π of $[n-1]$, let $i, i+1, \ldots, j$ for $j > i$, be a maximal sequence of two or more consecutive integers contained in a block of π. Remove $j-1$, $j-3$, $j-5, \ldots$ from this sequence and put them in a block with n. Doing this for every such sequence $i, i+1, \ldots, j$ yields the desired bijection. See H. Prodinger, *Fibonacci Quart.* **19** (1981), 463–465, W. Y. C. Chen, E. Y. P. Deng, and R. R. X. Du, *Europ. J. Combin.* **26** (2005), 237–243, and W. Yang, *Discrete Math.* **156** (1996), 247–252.

Example. If $\pi = 1456\text{-}2378$, then the bijection gives 146-38-2579.

The preceding proof easily extends (as done in papers cited here) to show the following result: let $0 \le k \le n$, and let $B_k(n)$ be the number of partitions of $[n]$ so that if i and j are in a block then $|i - j| > k$. Then $B_k(n) = B(n-k)$.

109. a. Given a partition $\pi \in \Pi_n$, list the blocks in decreasing order of their smallest element. Then list the elements of each block with the least element first, followed by the remaining elements in decreasing order, obtaining a permutation $w \in \mathfrak{S}_n$. The map $\pi \mapsto w$ is bijection from Π_n to the permutations being enumerated. For instance, if $\pi = 13569 - 248 - 7$, then $w = 728419653$. To obtain π from w, break w before each left-to-right minimum. This result, as well as those in (b) and (c), is due to A. Claesson, *Europ. J. Combinatorics* **22** (2001), 961–971.

b. Now write the blocks in decreasing order of their smallest element, with the elements of each block written in increasing order.

c. Let w be the permutation corresponding to π as defined in (a). Then w also satisfies the condition of (b) if and only if each block of π has size one or two.

110. *Answer.* The coefficient of x^n is $B(n-1), n \ge 1$. See Proposition 2.6 of M. Klazar, *J. Combinatorial Theory Ser. A* **102** (2003), 63–87.

111. The number of ways to partition a k-element subset of $[n]$ into j intervals is $\binom{k-1}{j-1}\binom{n-k+j}{j}$, since we can choose the interval sizes from left-to-right in $\binom{k-1}{j-1}$ ways (the number of compositions of k into j parts), and then choose the intervals themselves in $\binom{n-k+j}{j}$ ways. Hence by the Principle of Inclusion-Exclusion (Theorem 2.1.1),

$$f(n) = B(n) + \sum_{k=1}^{n}\sum_{j=1}^{k} B(n-k)(-1)^j \binom{k-1}{j-1}\binom{n-k+j}{j}.$$

Now

$$\sum_{j=1}^{k}(-1)^j\binom{k-1}{j-1}\binom{n-k+j}{j}=(-1)^k\binom{n-k+1}{k}.$$

Hence

$$f(n)=\sum_{k=0}^{n}B(n-k)(-1)^k\binom{n-k+1}{k}$$

$$=\sum_{k=0}^{n}B(k)(-1)^{n-k}\binom{k+1}{n-k}.$$

(Is there some way to see this directly from Inclusion-Exclusion?) Now multiply by x^n and sum on $n \geq 0$. Since by the binomial theorem

$$\sum_{n\geq0}(-1)^{n-k}\binom{k+1}{n-k}x^n=x^k(1-x)^{k+1},$$

we get

$$F(x)=\sum_{k\geq0}B(k)x^k(1-x)^{k+1}$$

$$=(1-x)G(x(1-x)).$$

115. See D. Chebikin, R. Ehrenborg, P. Pylyavskyy, and M. A. Readdy, *J. Combinatorial Theory Ser. A* **116** (2009), 247–264. The polynomials $Q_n(t)$ are introduced in this paper and are shown to have many cyclotomic factors, but many additional such factors are not yet understood.

116. b. See L. A. Shepp and S. P. Lloyd, *Trans. Amer. Math. Soc.* **121** (1966), 340–357.

117. *Answer.* $p_{nk} = 1/n$ for $1 \leq k \leq n$. To see this, consider the permutations $v = b_1 \cdots b_{n+1}$ of $[n] \cup \{*\}$ beginning with 1. Put the elements to the left of $*$ in a cycle in the order they occur. Regard the elements to the right of $*$ as a word that defines a permutation of its elements (say with respect to the elements listed in increasing order). This defines a bijection between the permutations v and the permutations $w \in \mathfrak{S}_n$. The length of the cycle containing 1 is k if $b_{k+1} = *$. Since $*$ is equally likely to be any of $b_2, \ldots, b_{n+1}$, the proof follows.

Example. Let $v = 1652 * 4873$. Then w has the cycle $(1,6,5,2)$. The remaining elements are permuted as 4873 with respect to the increasing order 3478 (i.e., $w(3) = 4$, $w(4) = 8$, $w(7) = 7$, and $w(8) = 3$). In cycle form, we have $w = (1,6,5,2)(3,4,8)(7)$.

118. b. We compute equivalently the probability that $n, n-1, \ldots, n-\lambda_1+1$ are in the same cycle C_1, and $n-\lambda_1, \ldots, n-\lambda_1-\lambda_2+1$ are in the same cycle C_2 different from C_1, and so on. Apply the fundamental bijection of Proposition 1.3.1 to w, obtaining a permutation $v = b_1 \cdots b_n$. It is easy to check that w has the desired properties if and only if the restriction u of v to $n-k+1, n-k+2, \ldots, n$ has $n-k+\lambda_\ell$ appearing first, then the elements $n-k+1, n-k+2, \ldots, n-k+\lambda_\ell-1$ in some order, then $n-k+\lambda_{\ell-1}+\lambda_\ell$, then the elements $n-k+\lambda_\ell+$

$1, \ldots, n - k + \lambda_{\ell-1} + \lambda_{\ell} + 1$ in some order, then $n - k + \lambda_{\ell-2} + \lambda_{\ell-1} + \lambda_{\ell}$, and so on. Hence, of the $k!$ permutations of $n - k + 1, \ldots, n$ there are $(\lambda_1 - 1)! \cdots (\lambda_{\ell} - 1)!$ choices for u, and the proof follows. For a variant of this problem when the distribution isn't uniform, see O. Bernardi, R. X. Du, A. Morales, and R. Stanley, in preparation.

c. Let v be as in (b), and let $v' = b_2 b_1 b_3 b_4 \cdots b_n$. Exactly one of v and v' is even. Moreover, the condition in (b) on the restriction u is unaffected unless $b_1 = n - k + \lambda_{\ell}$ and $b_2 = n - k + i$ for some $1 \leq i \leq \lambda_{\ell} - 1$. In this case, v has exactly ℓ records, so w has exactly ℓ cycles. Hence, w is even if and only if $n - \ell$ is even. Moreover, the number of choices for u is

$$\frac{(n-2)!}{(k-2)!}(\lambda_1 - 1)! \cdots (\lambda_{\ell} - 1)!,$$

and the proof follows easily.

119. If a permutation $w \in \mathfrak{S}_{2n}$ has a cycle C of length $k > n$, then it has exactly one such cycle. There are $\binom{2n}{k}$ ways to choose the elements of C, then $(k-1)!$ ways to choose C, and finally $(2n-k)!$ ways to choose the remainder of w. Hence,

$$P_n = 1 - \frac{1}{(2n)!} \sum_{k=n+1}^{2n} \binom{2n}{k} (k-1)!(2n-k)!$$

$$= 1 - \sum_{k=n+1}^{2n} \frac{1}{k}$$

$$= 1 - \sum_{k=1}^{2n} \frac{1}{k} + \sum_{k=1}^{n} \frac{1}{k}$$

$$\sim 1 - \log(2n) + \log(n)$$

$$= 1 - \log 2,$$

and the proof follows. For an amusing application of this result, see P. M. Winkler, *Mathematical Mind-Benders*, A K Peters, Wellesley, MA, 2007 (pp. 12, 18–20).

120. *First Solution.* There are $\binom{n}{k}(k-1)!$ k-cycles, and each occurs in $(n-k)!$ permutations $w \in \mathfrak{S}_n$. Hence,

$$E_k(n) = \frac{1}{n!} \binom{n}{k}(k-1)!(n-k)! = \frac{1}{k}.$$

Second Solution. By Exercise 1.117 (for which we gave a simple bijective proof), the probability that some element $i \in [n]$ is in a k-cycle is $1/n$. Since there are n elements and each k-cycle contains k of them, the expected number of k-cycles is $(1/n)(n/k) = 1/k$.

124. a. Let $w = a_1 a_2 \cdots a_{n+1} \in \mathfrak{S}_{n+1}$ have k inversions, where $n \geq k$. There are $f_k(n)$ such w with $a_{n+1} = n + 1$. If $a_i = n + 1$ with $i < n + 1$, then we can interchange a_i and a_{i+1} to form a permutation $w' \in \mathfrak{S}_{n+1}$ with $k - 1$ inversions. Since $n \geq k$, every $w' = b_1 b_2 \cdots b_{n+1} \in \mathfrak{S}_{n+1}$ with $k - 1$ inversions satisfies $b_1 \neq n + 1$ and thus can be obtained from a $w \in \mathfrak{S}_{n+1}$ with k inversions as previously.

b. Use induction on k.

   c. By Corollary 1.3.13 we have

$$\sum_{k \geq 0} f_k(n)q^k = (1+q)(1+q+q^2)\cdots(1+q+\cdots+q^{n-1})$$

$$= \frac{(1-q)(1-q^2)\cdots(1-q^n)}{(1-q)^n}$$

$$= (1-q)(1-q^2)\cdots(1-q^n)\sum_{k \geq 0}\binom{-n}{k}(-1)^k q^k.$$

Hence, if $\prod_{i \geq 1}(1-q^i) = \sum_{j \geq 0} b_j q^j$, then

$$f_k(n) = \sum_{j=0}^{k}(-1)^j b_{k-j}, \quad n \geq k.$$

Moreover, it follows from the Pentagonal Number Formula (1.88) that

$$b_r = \begin{cases} (-1)^i, & \text{if } r = i(3i \pm 1)/2, \\ 0, & \text{otherwise.} \end{cases}$$

See pp. 15–16 of D. E. Knuth [1.48].

127. a. We can reason analogously to the proofs of Proposition 1.3.12 and Corollary 1.3.13. Given $w = w_1 w_2 \cdots w_n \in \mathfrak{S}_n$ and $1 \leq i \leq n$, define

$$r_i = \#\{j : j < i, \, w_j > w_i\}$$

and $\text{code}'(w) = (r_1, \ldots, r_n)$. For instance, $\text{code}'(3265174) = (0,1,0,1,4,0,3)$. Note that $\text{code}'(w)$ is just a variant of $\text{code}(w)$ and gives a bijection from $\mathfrak{S}_n$ to sequences $(r_1, \ldots, r_n)$ satisfying $0 \leq r_i \leq i-1$. Moreover, $\text{inv}(w) = \sum r_i$, and w_i is a left-to-right maximum if and only if $r_i = 0$. From these observations, equation (1.135) is immediate.

   b. Let $I(w) = (a_1, \ldots, a_n)$, the inversion table of w. Then $\text{inv}(w) = \sum a_i$ (as noted in the proof of Corollary 1.3.13), and i is the value of a record if and only if $a_i = 0$. From these observations, equation (1.136) is immediate.

128. a. First establish the recurrence

$$\sum_{j=1}^{n} f(j)(n-j)! = n!, \quad n \geq 1,$$

where we set $g(0) = 1$. Then multiply by x^n and sum on $n \geq 0$. This result appears in L. Comtet, *Comptes Rend. Acad. Sci. Paris* **A 275** (1972), 569–572, and is also considered by Comtet in his book *Advanced Combinatorics*, Reidel, Dordrecht/Boston, 1974 (Exercise VII.16). For an extension of this exercise and further references, see Exercise 2.13.

   b. (I. M. Gessel) Now we have

$$n! = g(n) + \sum_{j=1}^{n} g(j-1)(n-j)!, \quad n \geq 1,$$

where we set $g(0) = 1$.

   c. See D. Callan, *J. Integer Sequences* **7** (2004), article 04.1.8.

d. See M. H. Albert, M. D. Atkinson, and M. Klazar, *J. Integer Sequences* **6** (2003), article 02.4.4. For a survey of simple permutations, see R. Brignall, in *Permutation Patterns* (S. Linton, N. Ruškuc and V. Vatter, eds.), London Mathematical Society Lecture Note Series, vol. 376, Cambridge University Press, Cambridge, 2010, pp. 41–65. For some analogous results for set partitions, see M. Klazar, *J. Combinatorial Theory Ser. A* **102** (2003), 63–87.

129. a. It is easy to see that if w is an indecomposable permutation in $\mathfrak{S}_n$ with k inversions, then $n \le k+1$. (Moreover, there are exactly 2^{k-1} indecomposable permutations in $\mathfrak{S}_{k+1}$ with k inversions.) Hence, $g_n(q)$ has smallest term of degree $n-1$, and the proof follows.

b. *Answer.* We have the continued fraction

$$1 - \frac{1}{F(q,x)} = \cfrac{a_0}{1 - \cfrac{a_1}{1 - \cfrac{a_2}{1 - \cdots}}},$$

where
$$a_n = (q^{\lfloor (n+1)/2 \rfloor} + q^{\lfloor (n+1)/2 \rfloor + 1} + \cdots + q^n)x.$$

See A. de Medicis and X. G. Viennot, *Advances in Appl. Math.* **15** (1994), 262–304 (equations (1.24) and (1.25) and Theorem 5.3).

130. This result, stated in a less elegant form, is due to M. Abramson and W. O. J. Moser, *Ann. Math. Statist.* **38** (1967), 1245–1254. The solution in the form of equation (1.138) is due to L. W. Shapiro and A. B. Stephens, *SIAM J. Discrete Math.* **4** (1991), 275–280.

133. a. We have $\frac{1}{2}A_n(2) = \sum_{k=0}^{n-1} A(n,k+1)2^k$, where $A(n,k+1)$ permutations of $[n]$ have k descents. Thus we need to associate an ordered partition τ of $[n]$ with a pair (w,S), where $w \in \mathfrak{S}_n$ and $S \subseteq D(w)$. Given $w = a_1 a_2 \cdots a_n$, draw a vertical bar between a_i and a_{i+1} if $a_i < a_{i+1}$ or if $a_i > a_{i+1}$ and $i \in S$. The sets contained between bars (including the beginning and end) are read from left to right and define τ.

Example. Let $w = 724531968$ and $S = \{1,5\}$. Write $7|2|4|53|1|96|8$, so $\tau = (7,2,4,35,1,69,8)$.

134. See D. Foata and M.-P. Schützenberger, [1.26, Thm. 5.6]. For a vast generalization of this kind of formula, see E. Nevo and T. K. Petersen. *Discrete Computational Geometry* **45** (2011), 503–521.

135. a. Put $x = -1$ in equation (1.40) and compare with (1.54).

b. Let $n = 2m+1$. Since $\text{des}(w) = m$ if w is alternating, it suffices to show combinatorially that $\sum_w (-1)^{\text{des}(w)} = 0$, where w ranges over all nonalternating permutations in $\mathfrak{S}_n$. For a nonalternating permutation $w \in \mathfrak{S}_n$, let $T = T(w)$ be the increasing binary tree corresponding to w, as defined in Section 1.5. Since w is not alternating, it follows from the table preceding Proposition 1.5.3 that T has a vertex j with only one successor. For definiteness, choose the least such vertex j, and let T' be the flip of T at j, as defined in Section 1.6.2. Define $w' \in \mathfrak{S}_n$ by $T(w') = T'$. Clearly $w'' = w$, so we have defined an involution $w \mapsto w'$ on all nonalternating permutations in $\mathfrak{S}_n$. Since n is odd, it again follows from the table preceding Proposition 1.5.3 that $\text{des}(w)$ is the number of vertices of $T(w)$ with a left successor. Hence, $(-1)^{\text{des}(w)} + (-1)^{\text{des}(w')} = 0$, and the proof follows. For further aspects of this line of reasoning, see D. Foata and M.-P. Schützenberger, [1.26, Thm. 5.6].

136. *Answer.* $c_1 = c_{n-1} = 1$, all other $c_i = 0$.

137. The number of $w \in \mathfrak{S}_n$ of type c is $\tau(c) = n!/1^{c_1} c_1! \cdots n^{c_n} c_n!$. Let $n = a_0 + a_1 \ell$. It is not hard to see that $\tau(c)$ is prime to ℓ if and only if, setting $k = c_\ell$, we have $c_1 \geq (n_1 - k)\ell$ where $\binom{n_1}{k}$ is prime to ℓ It follows from Exercise 1.14 that the number of binomial coefficients $\binom{n_1}{k}$ prime to ℓ is $\prod_{i \geq 1}(a_i + 1)$. Since $(c_1 - (n_1 - k)\ell, c_2, \ldots, c_{\ell-1})$ can be the type of an arbitrary partition of a_0, the proof follows.

This result first appeared in I. G. Macdonald, *Symmetric Functions and Hall Polynomials*, Oxford University Press, Oxford, 1979; 2nd ed., 1995 (Ex. 10 of Ch. I.2). The proof given here appears on pp. 260–261 of R. Stanley, *Bull. Amer. Math. Soc.* **4** (1981), 254–265.

139. Let $z = \sum_{n \geq 1} g(n) x^n / n!$. Then $z' = 1 + \frac{1}{2} z^2 + \frac{1}{4!} z^4 + \cdots = \cosh(z)$. The solution to this differential equation satisfying $z(0) = 0$ is

$$z(x) = \log(\sec x + \tan x).$$

Since $z'(x) = \sec x$, it follows easily that $g(2n + 1) = E_{2n}$. For further information and a bijective proof, see Section 3 of A. G. Kuznetsov, I. M. Pak, and A. E. Postnikov, [1.51].

140. *Hint.* Let $f_k(n)$ be the number of simsun permutations in $\mathfrak{S}_n$ with k descents. By inserting $n + 1$ into a simsun permutation in $\mathfrak{S}_n$, establish the recurrence

$$f_k(n + 1) = (n - 2k + 2) f_{k-1} + (k + 1) f_k(n),$$

with the initial conditions $f_0(1) = 1$, $f_k(n) = 0$ for $k > \lfloor n/2 \rfloor$. Further details may be found in S. Sundaram, *Advances in Math.* **104** (1994), 225–296 (§3) in the context of symmetric functions. We can also give a bijective proof, as follows. Let $\mathcal{E}$ be a flip equivalence class of binary trees on the vertex set $[n + 1]$. There are E_{n+1} such flip equivalence classes (Proposition 1.6.2). There is a unique tree $T' \in \mathcal{E}$ such that (i) the path from the root 1 to $n + 1$ moves to the right, (ii) for every vertex not on this path with two children, the largest child is on the left, and (iii) any vertex with just one child has this child on the right. Let $w' \in \mathfrak{S}_{n+1}$ satisfy $T' = T(w')$ (as in Section 1.5). Then w' ends in $n + 1$; let $w \in \mathfrak{S}_n$ be w' with $n + 1$ removed. It is not hard to check that the map $\mathcal{E} \mapsto w$ gives a bijection between flip equivalence classes and simsun permutations. This proof is due to Maria Monks (October 2007).

Simsun permutations are named after Rodica Simion and Sheila Sundaram. They first appear in the paper S. Sundaram, ibid. (p. 267). They are variants of the *André permutations* of Foata and Schützenberger [1.27]. The terminology "simsun permutation" is due to S. Sundaram (after they were originally called "Sundaram permutations" by R. Stanley) in *J. Algebraic Combin.* **4** (1995), 69–92 (p. 75). For some further work on simsun permutations, see G. Hetyei, *Discrete Comput. Geom.* **16** (1996), 259–275.

141. a. *Hint.* Show that $E_{n+1,k}$ is the number of alternating permutations of $[n + 2]$ with first term $k + 1$ and second term unequal to k, and that $E_{n,n-k}$ is the number of alternating permutations of $[n + 2]$ with first term $k + 1$ and second term k.

The numbers $E_{n,k}$ are called *Entringer numbers*, after R. C. Entringer, *Nieuw. Arch. Wisk.* **14** (1966), 241–246. The triangular array (1.140) is due to L. Seidel, *Sitzungsber. Münch. Akad.* **4** (1877), 157–187 (who used the word "boustrophedon" to describe the triangle). It was rediscovered by A. Kempner, *Tôhoku Math. J.* **37** (1933), 347–362; R. C. Entringer, op. cit.; and V. I. Arnold, *Duke Math. J.* **63** (1991), 537–555. For further information and references, see J. Millar, N. J. A. Sloane, and N. E. Young, *J. Combinatorial Theory Ser. A* **76** (1996), 44–54. A more recent reference is R. Ehrenborg and S. Mahajan, *Ann. Comb.* **2** (1998),

111–129 (§2). The boustrephedon triangle was generalized to permutations with an arbitrary descent set by Viennot [1.75].

b. Rotate the triangle and change the sign of E_{mn} when $m+n \equiv 1,2 \,(\text{mod}\,4)$ to obtain the array

1		0		−1		0		5		0	⋯
	−1		−1		1		5		−5		
		0		2		4		−10			
			2		2		−14				
				0		−16					
					−16		⋯				

This array is just a difference table, as defined in Section 1.9. By (a) the exponential generating function for the first row is $\sec(ix) = \text{sech}(x)$. By Exercise 1.154(c) we get

$$\sum_{m\geq 0}\sum_{n\geq 0}(-1)^{\lfloor (2m+2n+3)/4\rfloor} E_{m+n,[m,n]}\frac{x^m}{m!}\frac{y^n}{n!} = e^{-x}\text{sech}(x+y).$$

If we convert all the negative coefficients to positive, it's not hard to see that the generating function becomes the right-hand side of equation (1.141), as claimed.

The transformation into a difference table that we have used here appears in Seidel, op. cit., and is treated systematically by D. Dumont, *Sém. Lotharingien de Combinatoire* **5** (1981), B05c (electronic). Equation (1.141) appears explicitly in R. L. Graham, D. E. Knuth, and O. Patashnik, *Concrete Mathematics*, 2nd ed., Addison-Wesley, Reading, Mass., 1994 (Exercise 6.75).

142. It is easy to verify that

$$\sum_{n\geq 0} f_n(a)x^n = (\sec x)(\cos(a-1)x + \sin ax),$$

and the proof follows. The motivation for this problem comes from the fact that for $0 \leq a \leq 1$, $f_n(a)$ is the volume of the convex polytope in $\mathbb{R}^n$ given by

$$x_i \geq 0 \ (1 \leq i \leq n), \ x_1 \leq a, \ x_i + x_{i+1} \leq 1 \ (1 \leq i \leq n-1).$$

For further information on the case $a = 1$, see Exercise 4.56(c).

143. a. *Combinatorial Proof.* Let $1 \leq i \leq n$. The number of permutations $w \in \mathfrak{S}_n$ fixing i is $(n-1)!$. Hence the total number of fixed points of all $w \in \mathfrak{S}_n$ is $n \cdot (n-1)! = n!$. *Generating Function Proof.* We have

$$f(n) := \sum_{w\in\mathfrak{S}_n}\text{fix}(w) = n!\frac{d}{dt_1}Z_n|_{t_i=1},$$

where Z_n is defined by (1.25). Hence by Theorem 1.3.3 we get

$$\sum_{n\geq 0}f(n)\frac{x^n}{n!} = \frac{d}{dt_1}\exp\left(t_1 x + t_2\frac{x^2}{2} + t_3\frac{x^3}{3} + \cdots\right)\bigg|_{t_i=1}$$

$$= x\exp\left(x + \frac{x^2}{2} + \frac{x^3}{3} + \cdots\right)$$

$$= \frac{x}{1-x}.$$

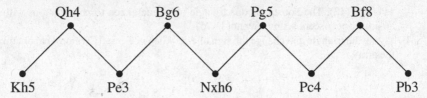

Figure 1.35 The solution poset for Exercise 1.145.

whence $f(n) = n!$.

Algebraic Proof. Let G be a finite group acting a set Y. By Burnside's lemma (Lemma 7.24.5, Vol. II), also called the Cauchy–Frobenius lemma, the average number of fixed points of $w \in G$ is the number of orbits of the action. Since the "defining representation" of $\mathfrak{S}_n$ on $[n]$ has one orbit, the proof follows.

b. This result is a straightforward consequence of Proposition 6.1 of R. Stanley, *J. Combinatorial Theory Ser. A* **114** (2007), 436–460. Is there a combinatorial proof?

144. a. It is in fact not hard to see that

$$2q^n \frac{\prod_{j=1}^n (1 - q^{2j-1})}{\prod_{j=1}^{2n+1}(1+q^j)} = \frac{2(2n-1)!!}{3^n} x^n + O(x^{n+1}),$$

where $(2n-1)!! = 1 \cdot 3 \cdot 5 \cdots (2n-1)$.

b. See page 450 of R. Stanley, *J. Combinatorial Theory Ser. A* **114** (2007), 436–460.

145. One solution is 1.Kh5 2.Pe3 3.Nxh6 4.Pc4 5.Pb3 6.Qh4 7.Bg6 8.Rg5 9.Bf8, followed by Nf6 mate. Label these nine Black moves as 1,3,5,7,9,2,4,6,8 in the order given. All solutions are a permutation of these nine moves. If $a_1, a_2, \dots, a_9$ is a permutation w of the *labels* of the moves, then they correspond to a solution if and only if w^{-1} is reverse alternating. (In other words, Qh4 must occur after both Kh5 and Pe3, Bg6 must occur after both Pe3 and Nxh6, etc.). In the terminology of Chapter 3, the solutions correspond to the linear extensions of the "zigzag poset" shown in Figure 1.35. Hence, the number of solutions is $E_9 = 7936$. For some properties of zigzag posets, see Exercise 3.66.

146. The proof is a straightforward generalization of the proof we indicated of equation (1.59). For a q-analogue, see Proposition 3.3.19.4 and the discussion following it.

147. A binary tree is an unlabeled min-max tree if and only if every nonendpoint vertex has a nonempty left subtree. Let f_n be the number of such trees on n vertices. Then

$$f_{n+1} = \sum_{k=1}^n f_k f_{n-k}, \quad n \geq 1.$$

Setting $y = \sum_{n \geq 0} f_n x^n$, we obtain

$$\frac{y - 1 - x}{x} = y^2 - y.$$

It follows that

$$y = \frac{1 + x - \sqrt{1 - 2x - 3x^2}}{2x}.$$

Comparing with the definition of M_n in Exercise 6.27 shows that $f_n = M_{n-1}, n \geq 1$.

148. It is easy to see from equations (1.33) and (1.63) that

$$\Psi_n(a+b,b) = \sum_{S \subseteq [n-1]} \alpha(S) u_S.$$

The proof follows from the formula $\Psi(a,b) = \Phi(a+b, ab+ba)$ (Theorem 1.6.3).

149. *Hint.* First establish the recurrence

$$2\Phi_n = \sum_{\substack{0 < i < n \\ n-i=2j-1}} \binom{n}{i} \Phi_i c(c^2 - 2d)^{j-1} - \sum_{\substack{0 < i < n \\ n-i=2j}} \binom{n}{i} \Phi_i (c^2 - 2d)^j$$

$$+ \begin{cases} 2(c^2 - 2d)^{k-1}, & n = 2k - 1, \\ 0, & n = 2k. \end{cases}$$

The generating function follows easily from multiplying this recurrence by $x^n/n!$ and summing on $n \geq 1$.

This result is due to R. Stanley, *Math. Z.* **216** (1994), 483–499 (Corollary 1.4).

151. This elegant result is due to R. Ehrenborg, private communication (2007), based on the Pyr operator of R. Ehrenborg and M. Readdy, *J. Algebraic Combin.* **8** (1998), 273–299. Using concepts from Chapter 3, the present exercise has the following interpretation. Let P be the poset whose elements are all cd-monomials. Define α to cover β in P if β is obtained from α by removing a c or changing a d to c. Then $[\mu]\Phi_n(c,d)$ is equal to the number of maximal chains of the interval $[1, \mu]$. The problem of counting such chains was considered by F. Bergeron, M. Bousquet-Mélou, and S. Dulucq, *Ann. Sci. Math. Québec* **19** (1995), 139–151. They showed that the total number of saturated chains from 1 to rank n is E_{n+1} (the sum of the coefficients of Φ_{n+1}), though they did not interpret the number of maximal chains in each interval. Further properties of the poset P (and some generalizations) were given by B. Sagan and V. Vatter, *J. Algebraic Combinatorics* **24** (2006), 117–136.

152. An analogous result where simsum permutations are replaced by "André permutations" was earlier proved by M. Purtill [1.65]. The result for simsun permutations was stated without proof by R. Stanley, *Math. Zeitschrift*, **216** (1994), 483–499 (p. 498), saying that it can be proved by "similar reasoning" to Purtill's. This assertion was further explicated by G. Hetyei, *Discrete Comput. Geom.* **16** (1996), 259–275 (Remark on p. 270).

153. a. *First Solution.* Put $c = 0$ and $d = 1$ in equation (1.142) (so $a - b = \sqrt{-2}$) and simplify. We obtain

$$\sum_{n \geq 0} f(n) \frac{x^{2n+1}}{(2n+1)!} = \sqrt{2} \tan(x/\sqrt{2}).$$

The proof follows from Proposition 1.6.1.

Second Solution. By equations (1.62) and (1.64) we have that $2^n f(n)$ is the number of complete (i.e., every internal vertex has two children) min-max trees with n internal vertices. A complete min-max tree with $n + 1$ internal vertices is obtained by placing either 1 or $2n + 3$ at the root, forming a left complete min-max subtree whose vertices are $2k + 1$ elements from $\{2, 3, \ldots, 2n+2\}$ ($0 \leq k \leq n$), and forming a right complete min-max subtree with the remaining elements. Hence setting $g(n) = 2^n f(n)$, we obtain the recurrence

$$g(n+1) = 2 \sum_{k=0}^{n} \binom{2n+1}{2k+1} g(k) g(n-k).$$

It is then straightforward to show that $g(n) = E_{2n+1}$. The result of this exercise was first proved by Foata and Schützenberger [1.27, Propriété 2.6] in the context of André polynomials.

R. Ehrenborg (private communication, 2007) points out that there is a similar formula for the coefficient of any monomial in Φ_n not containing two consecutive c's.

b. See R. L. Graham and N. Zang, Enumerating split-pair arrangements, preprint dated 10 January 2007. For some further combinatorial interpretations of F_n, see C. Poupard, *Europ. J. Combinatorics* **10** (1989), 369–374; A. G. Kuznetsov, I. M. Pak and A. E. Postnikov, *Uspekhi Mat. Nauk* **49** (1994), 79–110; and M. P. Develin and S. P. Sullivant, *Ann. Combinatorics* **7** (2003), 441–466 (Corollary 5.7).

154. a. Use equation (1.98).

b. By (a), $e^{-x} F'(x) = F(x)$, from which $F(x) = e^{e^x - 1}$, so $f(n)$ is the Bell number $B(n)$. The difference table in question looks like

$$
\begin{array}{ccccccc}
1 & 2 & 5 & 15 & 52 & 203 & \cdots \\
& 1 & 3 & 10 & 37 & 151 & \cdots \\
& & 2 & 7 & 27 & 114 & \cdots \\
& & & 5 & 20 & 87 & \cdots \\
& & & & 15 & 67 & \cdots \\
& & & & & 52 & \cdots \\
& & & & & & \ddots
\end{array}
$$

Note that the first row is identical to the leftmost diagonal below the first row. This "Bell number triangle" is due to C. S. Peirce, *Amer. J. Math.* **3** (1880), 15–57 (p. 48). It gained some popularity by appearing in the "Mathematical Games" column of M. Gardner [1.33, Fig. 13]. D. E. Knuth uses it to develop properties of Bell numbers in *The Art of Computer Programming*, vol. 4, Fascicle 3, Addison-Wesley, Upper Saddle River, N.J., 2005, Section 7.2.1.5, and gives some further properties in Exercises 7.2.1.5–26 to 7.2.1.5–31.

c. By Taylor's theorem and (a), we have

$$
\sum_{n \geq 0} \sum_{k \geq 0} \Delta^n f(k) \frac{x^n}{n!} \frac{t^k}{k!} = e^{-x} \left(F(x) + F'(x)t + F''(x) \frac{t^2}{2!} + \cdots \right)
$$

$$
= e^{-x} F(x+t).
$$

This result appears in D. Dumont and X. G. Viennot, *Ann. Discrete Math.* **6** (1980), 77–87, but is undoubtedly much older.

155. a. For further information related to this problem and Exercise 1.154(a), see D. Dumont, in *Séminaire Lotharingien de Combinatoire*, 5ème Session, Institut de Recherche Mathématique Avancée, Strasbourg, 1982, pp. 59–78.

b. One computes $f(0) = 1$, $f(1) = 2$, $f(2) = 6$, $f(3) = 20, \ldots$. Hence, guess $f(n) = \binom{2n}{n}$ and $F(x) := \sum f(n)x^n = (1 - 4x)^{-1/2}$. By (a) we then have $G(x) := \sum g(n)x^n = \frac{1}{1+x} F\left(\frac{x}{1+x}\right) = (1 - 2x - 3x^2)^{-1/2}$. To verify the guess, one must check that $\frac{1}{1+x} G\left(\frac{x}{1+x}\right) = F(x^2)$, which is routine.

c. (suggested by L. W. Shapiro) One computes $f(0) = 1$, $f(1) = 1$, $f(2) = 2$, $f(3) = 5$ $f(4) = 14, \ldots$. Hence, guess $f(n) = \frac{1}{n+1}\binom{2n}{n}$ (the Catalan number C_n) and

$F(x) := \sum f(n)x^n = \frac{1}{2x}(1 - (1 - 4x)^{1/2})$. Then

$$F_1(x) := \sum f(n+1)x^n = \frac{1}{x}(F(x) - 1) = \frac{1}{2x^2}(1 - 2x - (1 - 4x)^{1/2}),$$

so by (a),

$$G(x) := \sum g(n)x^n = \frac{1}{1+x}F_1\left(\frac{1}{1+x}\right) = \frac{1}{2x^2}(1 - x - (1 - 2x - 3x^2)^{-1/2}).$$

To verify this guess, one must check that $\frac{1}{1+x}G\left(\frac{1}{1+x}\right) = F(x^2)$, which is routine.

156. *Answer.* $c_n = \prod_p p^{\lfloor n/p \rfloor}$, where p ranges over all primes. Thus, $c_0 = 1$, $c_1 = 1$, $c_2 = 2$, $c_3 = 6$, $c_4 = 12$, $c_5 = 60$, $c_6 = 360$, and so on. See E. G. Strauss, *Proc. Amer. Math. Soc.* **2** (1951), 24–27. The sequence c_n can also be defined by the recurrence $c_0 = 1$ and $c_{n+1} = s_{n+1}c_n$, where s_{n+1} is the largest squarefree divisor of $n + 1$.

157. Let $z = y^\lambda$, and equate coefficients of x^{n-1} on both sides of $(\lambda + 1)y'z = (yz)'$. This result goes back to Euler and is discussed (with many similar methods for manipulating power series) in D. E. Knuth, *The Art of Computer Programming*, vol. 2, 3rd ed., Addison-Wesley, Upper Saddle River, N.J., 1997, Section 4.7. It was rediscovered by H. W. Gould, *Amer. Math. Monthly* **81** (1974), 3–14.

158. Let $\log F(x) = \sum_{n \geq 1} g_n x^n$. Then

$$\sum_{n \geq 1} g_n x^n = \sum_{i \geq 1}\sum_{j \geq 1} \frac{a_i x^{ij}}{j} = \sum_{n \geq 1}\frac{x^n}{n}\sum_{d|n} d a_d.$$

Hence,

$$n g_n = \sum_{d|n} d a_d,$$

so by the Möbius inversion formula of elementary number theory,

$$a_n = \frac{1}{n}\sum_{d|n} d g_d \mu(n/d). \tag{1.156}$$

We have $1 + x = (1 - x)^{-1}(1 - x^2)$ (no need to use (1.156)).

If $F(x) = e^{x/(1-x)}$ then $g_n = 1$ for all n, so by (1.156) we have $a_n = \phi(n)/n$, where $\phi(n)$ is Euler's totient function.

159. *Answer.* $A(x) = \sqrt{F(x)F(-x)}$, $B(x) = \sqrt{F(x)/F(-x)}$. This result is due to Marcelo Aguiar (private communication, 2006) as part of his theory of combinatorial Hopf algebras and noncommutative diagonalization.

160. **a.** This formula is a standard result of hoary provenance which follows readily from

$$\sum_{r=0}^{k-1} \zeta^{rj} = \begin{cases} 0, & 0 < j < k, \\ k, & j = 0. \end{cases}$$

b. Let $\zeta = e^{2\pi i/k}$. According to (a) and Proposition 1.4.6, we have

$$f(n,k,j) = \frac{1}{k}\sum_{r=0}^{k-1} \zeta^{-jr}(n)!\big|_{q=\zeta^r}. \tag{1.157}$$

If $n \geq k$ then at least one factor $1 + q + \cdots + q^m$ of $(n)!$ will vanish at $q = \zeta^r$ for $1 \leq r \leq k - 1$. Thus, the only surviving term of the sum is $(n)!\big|_{q=1} = n!$, and the proof follows.

c. When $n = k - 1$, we have $(n)!|_{q=\zeta^r} = 0$ unless $r = 0$ or ζ^r is a primitive kth root of unity. In the former case, we get the term $(k-1)!/k$. In the latter case, write $\xi = \zeta^r$. Then

$$(k-1)!|_{q=\xi} = \frac{(1-\xi)(1-\xi^2)\cdots(1-\xi^{k-1})}{(1-\xi)^{k-1}}. \qquad (1.158)$$

Now

$$\prod_{j=1}^{k-1}(q-\xi^j) = \frac{q^k - 1}{q - 1},$$

Letting $q \to 1$ gives $\prod_{j=1}^{k-1}(1-\xi^j) = k$. Hence from equation (1.158), we have

$$(k-1)!|_{q=\xi} = \frac{k}{(1-\xi)^{k-1}},$$

and the proof follows from setting $n = k - 1$ and $j = 0$ in equation (1.157).
NOTE. Let $\Phi_n(x)$ denote the (monic) nth cyclotomic polynomial, i.e., its zeros are the primitive nth roots of unity. It can be shown that if $n \geq 2$ then

$$f(n-1,n,0) = \frac{(n-1)!}{n} + (-1)^n(n-1)[x^{n-1}]\log\frac{\Phi_n(1+x)}{\Phi_n(1)}.$$

Let us also note that $\Phi_n(1) = p$ if n is the power of a prime p; otherwise $\Phi_n(1) = 1$.

161. b. We have

$$\frac{H(x)}{H(x)+H(-x)} = \frac{G(x)}{2}.$$

Hence,

$$\frac{H(-x)}{H(x)} = \frac{2}{G(x)} - 1$$

$$\Rightarrow \log H(-x) - \log H(x) = \log\left(\frac{2}{G(x)} - 1\right).$$

If we divide the left-hand side by -2, then we obtain the odd part of $\log H(x)$. Hence,

$$\log H(x) = -\frac{1}{2}\log\left(\frac{2}{G(x)} - 1\right) + E_1(x),$$

where $E_1(x)$ is any even power series in x with $E_1(0) = 0$. Thus, $E(x) := e^{E_1(x)}$ is an arbitrary even power series with $E(0) = 1$. Therefore, we get the general solution

$$H(x) = \left(\frac{2}{G(x)} - 1\right)^{-1/2} E(x).$$

162. Using the formulas

$$\tan(x+y) = \frac{\tan x + \tan y}{1 - (\tan x)(\tan y)},$$

$$\tan x/2 = \frac{\pm\sqrt{1+\tan^2 x} - 1}{\tan x},$$

we have $\tan(\tan^{-1} f(x) + \tan^{-1} f(-x)) = g(x)$

$$\Rightarrow \tan^{-1} f(x) = \frac{1}{2} \tan^{-1} g(x) + k(x), \quad k(x) = -k(-x)$$

$$\Rightarrow f(x) = \tan\left(\frac{1}{2} \tan^{-1} g(x) + k(x)\right)$$

$$= \frac{\tan \frac{1}{2} \tan^{-1} g(x) + \tan k(x)}{1 - (\tan \frac{1}{2} \tan^{-1} g(x)) \tan k(x)}$$

$$= \frac{\frac{\sqrt{1+g(x)^2}-1}{g(x)} + h(x)}{1 - \frac{\sqrt{1+g(x)^2}-1}{g(x)} h(x)}, \quad h(x) = -h(-x).$$

Choosing the correct sign gives

$$f(x) = \frac{-\sqrt{1+g(x)^2} - 1 + g(x)h(x)}{g(x) - (\sqrt{1+g(x)^2} - 1)h(x)},$$

where $h(x)$ is any even power series.

163. **a.** We have $F(x,y) = f(f^{\langle -1 \rangle}(x) + f^{\langle -1 \rangle}(y))$. The concept of a formal group law goes back to S. Bocher, *Ann. Math.* **47** (1946), 192–201.

 b. See for instance A. Fröhlich, *Lecture Notes in Math.*, no. 74, Springer-Verlag, Berlin/New York, 1968. For a combinatorial approach to formal groups via Hopf algebras, see C. Lenart, Ph.D. thesis, University of Manchester, 1996, and C. Lenart and N. Ray, Some applications of incidence Hopf algebras to formal group theory and algebraic topology, preprint, University of Manchester, 1995.

 c. $f(x) = x, e^x - 1, \tan x, \sin x$, respectively.

 d. Let $R(x) = (xe^{-x})^{\langle -1 \rangle}$. Thus,

$$F(x,y) = (R(x) + R(y))e^{-R(x)-R(y)}$$
$$= xe^{-R(y)} + ye^{-R(x)}.$$

 The proof follows from equation (5.128), which asserts that

$$e^{-R(x)} = 1 - \sum_{n \geq 1} (n-1)^{n-1} \frac{x^n}{n!}.$$

 e. Euler, Institutiones Calculi integralis, *Ac. Sc. Petropoli*, 1761, showed that

$$F(x,y) = \frac{x\sqrt{1-y^4} + y\sqrt{1-x^4}}{1 + x^2 y^2}.$$

164. Note that setting $x = 0$ is useless. Instead, write

$$F(x,y) = \frac{xF(x,0) - y}{xy^2 + x - y}.$$

The denominator factors as $x(y - \theta_1(x))(y - \theta_2(x))$, where

$$\theta_1(x), \theta_2(x) = \frac{1 \mp \sqrt{1-4x^2}}{2x}.$$

Now $y - \theta_1(x) \sim y - x$ as $x, y \to 0$, so the factor $1/(y - \theta_1(x))$ has no power series expansion about $(0,0)$. Since $F(x, y)$ has such an expansion, the factor $y - \theta_1(x)$ must appear in the numerator. Hence $x F(x, 0) = \theta_1(x)$, yielding

$$F(x, 0) = \frac{1 - \sqrt{1 - 4x^2}}{2x^2} = \sum_{n \geq 0} C_n x^{2n},$$

$$F(x, y) = \frac{2}{1 - 2xy + \sqrt{1 - 4x^2}}.$$

The solution to this exercise is a simple example of a technique known as the *kernel method*. This method originated in Exercise 2.2.1–.4 of Knuth's book *The Art of Computer Programming*, vol. 1, Addison-Wesley, Reading, Massachusetts, 1973, 3rd ed., 1997. The present exercise is the same as Knuth's (after omitting some preliminary steps). See Section 1 of H. Prodinger, *Sém. Lotharingien de Combinatoire* **50** (2004), article B50f, for further information and examples. An interesting variant of the kernel method applied to queuing theory appears in Chapter 14 of L. Flatto, *Poncelet's Theorem*, American Math. Society, Providence, R.I., 2009.

165. *Answer.* The coefficient $f(n)$ of $F(x)$ is the number of 1's in the binary expansion of n.

166. *Answer.* $F(x) = (1 + x^n)^{1/n} = \sum_{k \geq 0} \binom{1/n}{k} x^{kn}$.

167. Equation (1.143) is just the Taylor series expansion of $F(x + t)$ at $t = 0$.

168. a. It is not hard to check that for general $A(x) = x + a_2 x^2 + a_3 x^3 + \cdots$, we have

$$A(-A(-x)) = x + p_2 x^2 + p_3 x^3 + \cdots,$$

where p_{2n-1} and p_{2n} are polynomials in $a_2, a_3, \ldots, a_{2n-1}$. (It's easy to see that, in fact, $p_2 = 0$.) Moreover, the only term of p_{2n-1} involving a_{2n-1} is $2a_{2n-1}$. Hence, if $A(-A(-x)) = x$ then once $a_2, a_3, \ldots, a_{2n-2}$ are specified, we have that a_{2n-1} is uniquely determined. Thus, we need to show that if $a_2, a_4, \ldots, a_{2n-2}$ are specified, thereby determining $a_3, a_5, \ldots, a_{2n-1}$, then $p_{2n} = 0$. For instance, equating coefficients of x^3 in $A(-A(-x)) = x$ gives $a_3 = a_2^2$. Then

$$p_4 = a_2^3 - a_2 a_3 = a_2^3 - a_2(a_2^2) = 0.$$

We can reformulate the result we need to prove more algebraically. Given $A(x) = x + a_2 x^2 + \cdots$, let $B(x) = A(-A(-x)) = x + p_2 x^2 + \cdots$. Let $A^{\langle -1 \rangle}(x) = x + \alpha_2 x^2 + \alpha_3 x^3 + \cdots$. Then

$$A(-x) = B(-A^{\langle -1 \rangle}(x)) = A^{\langle -1 \rangle}(x) + p_2 A^{\langle -1 \rangle}(x)^2 - \cdots.$$

Taking the coefficient of x^{2n} gives

$$a_{2n} \equiv -\alpha_{2n} + p_{2n} \pmod{I}. \tag{1.159}$$

But also

$$-A(-x) = A^{\langle -1 \rangle}(B(x))$$
$$= B(x) + \alpha_2 B(x)^2 + \cdots.$$

Taking coefficients of x^{2n} yields

$$-a_{2n} \equiv p_{2n} + [x^{2n}] \sum_{i=2}^{2n} \alpha_i (x + p_{2n} x^{2n})^i \pmod{I}$$

$$\equiv p_{2n} + \alpha_{2n} \pmod{I}. \tag{1.160}$$

Equations (1.159) and (1.160) imply $p_{2n} \in I$, as desired. This proof was obtained in collaboration with Whan Ghang. NOTE. It was shown by Ghang that a_{2n+1} is a polynomial in $a_2, a_4, \ldots, a_{2n}$ with *integer* coefficients.

NOTE. An equivalent reformulation of the result of this item is the following. For any $A(x) = x + a_2 x^2 + \cdots \in K[[x]]$, either $A(-A(-x)) = x$ or $A(-A(-x)) - x$ has odd degree. This result can be considerably generalized. For instance, if $C(x) = -x + c_2 x^2 + \cdots$ and $C(C(x)) = x$, then (writing composition of functions as juxtaposition) either $ACAC(x) = x$ or $ACAC(x) - x$ has odd degree. More generally, if ζ is a primitive kth root of unity and $C(x) = \zeta x + c_2 x^2 + \cdots$, where $C^k = x$, then either $(AC)^k(x) = x$ or $(AC)^k(x) - x$ has degree $d \equiv 1 \pmod{k}$. The possibility of such a generalization was suggested by F. Bergeron (private communication, 2007).

b. Use induction on n.

c. Marcelo Aguiar (private communication, 2006) first obtained this result as part of his theory of combinatorial Hopf algebras and noncommutative diagonalization.

d. *Answer.* $A(x) = 2x/(2-x)$ and $D(x) = \log \frac{2+x}{2-x}$. This example is due to Aguiar.

e. First show the following.

- $$\sum_{n \geq 1} a_n \left(\frac{x}{1-x} \right)^n = \sum_{n \geq b_n} x^n \Longleftrightarrow e^x \sum_{j \geq 0} a_{j+1} \frac{x^j}{j!} = \sum_{j \geq 0} b_{j+1} \frac{x^j}{j!}.$$
  (See Exercises 154(a) and 155(a).)

- For any $F(x) = x + \sum_{n \geq 2} a_n x^n$ and $H(x) = x + \sum_{n \geq 2} b_n x^n$, we have

$$F^{\langle -1 \rangle}(-F(-x)) = H^{\langle -1 \rangle}(-H(-x)),$$

if and only if $F(x)/H(x)$ is odd.

f. *Answer.* $b_{2n} = (-1)^{n-1} E_{2n-1}$, where E_{2n-1} is an Euler number.

169. There are many possible methods. A uniform way to do all three parts is to note that for any power series $F(x) = \sum_{n \geq} a_n x^n$, we have

$$x D F(x) = \sum_{n \geq 0} n a_n x^n,$$

where $D = \frac{d}{dx}$. Hence,

$$(x D + 2)^2 F(x) = \sum_{n \geq 0} (n+2)^2 a_n x^n.$$

Letting $F(x) = 1/(1-x), e^x$, and $1/\sqrt{1-4x}$ yields, after some routine computation, the three answers:

$$\sum_{n \geq 0} (n+2)^2 x^n = \frac{4 - 3x + x^2}{(1-x)^3},$$

$$\sum_{n \geq 0} (n+2)^2 \frac{x^n}{n!} = (x^2 + 5x + 4)e^x,$$

$$\sum_{n \geq 0} (n+2)^2 \binom{2n}{n} x^n = \frac{4 - 22x + 36x^2}{(1-4x)^{3/2}}.$$

170. a. *Answer.* $y = (\alpha + (\beta - \alpha)x)/(1 - x - x^2)$. The general theory of linear recurrence relations with constant coefficients is developed in Sections 4.1–4.4.

b. The recurrence yields $y' = (xy)' - \frac{1}{2}xy^2$, $y(0) = 1$, from which we obtain

$$y = \frac{\exp\left(\frac{x}{2} + \frac{x^2}{4}\right)}{\sqrt{1-x}}.$$

For the significance of this generating function, see Example 5.2.9.

c. We obtain $2y' = y^2$, $y(0) = 1$, whence $y = 1/(1 - \frac{1}{2}x)$. Thus, $a_n = 2^{-n}n!$.

d. (sketch) Let $F_k(x) = \sum_{n\geq 0} a_k(n)x^n/n!$, so $A(x,t) = \sum_{k\geq 0} F_k(x)t^k$. The recurrence for $a_k(n)$ gives

$$F_k'(x) = \sum_{2r+s=k-1} (F_{2r}(x) + F_{2r+1}(x))F_s(x). \qquad (1.161)$$

Let $A_e(x) = \frac{1}{2}(A(x,t) + A(x,-t))$ and $A_o(x,t) = \frac{1}{2}(A(x,t) - A(x,-t))$. From equation (1.161) and some manipulations, we obtain the system of differential equations

$$\frac{\partial A_e}{\partial x} = tA_eA_o + A_o^2, \qquad (1.162)$$

$$\frac{\partial A_o}{\partial x} = tA_e^2 + A_eA_o.$$

To solve this system, note that

$$\frac{\partial A_e/\partial x}{\partial A_o/\partial x} = \frac{A_o}{A_e}.$$

Hence, $\frac{\partial}{\partial x}(A_e^2 - A_o^2) = 0$, so $A_e^2 - A_o^2$ is independent of x. Some experimentation suggests that $A_e^2 - A_o^2 = 1$, which together with (1.162) yields

$$\frac{\partial A_e}{\partial x} = tA_e\sqrt{A_e^2 - 1} + A_e^2 - 1.$$

This equation can be routinely solved by separation of variables (though some care must be taken to choose the correct branch of the resulting integral, including the correct sign of $\sqrt{A_e^2 - 1}$). A similar argument yields A_o, and we finally obtain the following expression for $A = A_e + A_o$:

$$A(x,t) = \sqrt{\frac{1-t}{1+t}}\left(\frac{2}{1 - \frac{1-\rho}{t}e^{\rho x}} - 1\right),$$

where $\rho = \sqrt{1-t^2}$. It can then be checked that this formula does indeed give the correct solution to the original differential equations, justifying the assumption that $A_e^2 - A_o^2 = 1$. For further details and motivation, see Section 2 of R. Stanley, *Michigan Math. J.*, **57** (2008), 675–687.

171. While this problem can be solved by the "brute force" method of computing the coefficients on the right-hand side of equation (1.144), it is better to note that $B'(x) - B(x) = A'(x)$ and then solve this differential equation for $B(x)$ with the initial condition $B(0) = a_0$. Alternatively, one could start with $A(x)$:

$$B(x) = (1 + I + I^2 + \cdots)A(x) = (1 - I)^{-1}A(x).$$

Multiplying by $1 - I$ and differentiating both sides results in the same differential equation $B'(x) - B(x) = A'(x)$. (It isn't difficult to justify these formal manipulations of the operator I.)

172. One method of proof is to first establish the three term recurrence

$$(n+1)P_{n+1}(x) = (2n+1)xP_n(x) - nP_{n-1}(x)$$

and then use induction.

173. a.

$$\sqrt{\frac{1+x}{1-x}} = (1+x)(1-x^2)^{-1/2}$$

$$= \sum_{n\geq 0} 4^{-n}\binom{2n}{n}(x^{2n}+x^{2n+1}).$$

b. $\displaystyle\sum_{n\geq 1}\frac{x^{2n}}{n^2\binom{2n}{n}}.$

c. $\displaystyle\sum_{n\geq 0}t(t^2-1^2)(t^2-3^2)\cdots(t^2-(2n-1)^2)\frac{x^{2n+1}}{(2n+1)!}.$

d. $\displaystyle\sum_{n\geq 0}t^2(t^2-2^2)(t^2-4^2)\cdots(t^2-(2n-2)^2)\frac{x^{2n}}{(2n)!}.$

e. $\displaystyle 2\sum_{n\geq 0}(-1)^n 2^{2n}\frac{x^{4n+2}}{(4n+2)!}.$ Similar results hold for $\cos(x)\cosh(x)$, $\cos(x)\sinh(x)$, and $\sin(x)\cosh(x)$.

f. $\displaystyle 6\sum_{n\geq 0}(-1)^n 2^{6n}\frac{x^{6n+3}}{(6n+3)!}.$ Similar results hold when any subset of the three sin's is replaced by cos. There seems to be no analogous result for *four* factors.

To do (c), for instance, first observe that the coefficient of $x^{2n+1}/(2n+1)!$ in $\sin(t\sin^{-1}x)$ is a polynomial $P_n(t)$ of degree $2n+1$ and leading coefficient $(-1)^n$. If $k\in\mathbb{Z}$, then $\sin(2k+1)\theta$ is an odd polynomial in $\sin\theta$ of degree $2k+1$. Hence, $P_n(\pm(2k+1))=0$ for $n>k$. Moreover, $\sin 0=0$ so $P_n(0)=0$. We now have sufficient information to determine $P_n(t)$ uniquely. To get (b), consider the coefficient of t^2 in (d). For (g), note that

$$\cos(\log(1+x)) = \Re(1+x)^i.$$

174. *Hint.* What is the number of elements of the set $\{0,1\}$?

176. Induction on n. We have $E(0)=0$. For $n\geq 1$, choose the first vector v_1 at random. If $v_1=0$, we expect $E(n)$ further steps, and this occurs with probability $1/q^n$. Otherwise, v_1 is not the zero vector. Consider the projection of our space to a subspace complementary to v_1. The uniform distribution over $\mathbb{F}_q^n$ projects to the uniform distribution over this copy of $\mathbb{F}_q^{n-1}$, and our sequence of vectors will span $\mathbb{F}_q^n$ precisely when the set of their projections spans $\mathbb{F}_q^{n-1}$. It follows that we expect $E(n-1)$ further steps, and so

$$E(n) = 1 + \frac{E(n)+(q^n-1)E(n-1)}{q^n}.$$

Solving this equation gives $E(n) = E(n-1)+q^n/(q^n-1)$, and so

$$E(n) = \sum_{i=1}^{n} q^i/(q^i-1).$$

This argument was suggested by J. Lewis (October 2009).

182. Suppose that $A \in \text{GL}(n, q)$ has no 0 entries. There are exactly $(q-1)^{2n-1}$ matrices of the form DAD', where D, D' are diagonal matrices in $\text{GL}(n, q)$. Exactly one of the matrices $C = DAD'$ has the first entry in every row and column equal to -1. Subtract the first column of C from every other column, obtaining a matrix D. Let B be obtained from D by removing the first row and column. Then B is a matrix in $\text{GL}(n-1, q)$ with no entry equal to 1, and every such matrix is obtained exactly once by this procedure.

183. *First Solution* (sketch). The identity asserts that each of the q^n monic polynomials of degree n can be written uniquely as a product of monic irreducible polynomials.
Second Solution (sketch). Take logarithms of both sides and simplify the right-hand side.

185. a. Note that $q^n - D(n, 0)$ is the number of monic polynomials of degree n over $\mathbb{F}_q$ with nonzero discriminant. In the same way that we obtained the first solution to Exercise 1.183, we get

$$\sum_{n \geq 0} (q^n - D(n, 0)) x^n = \prod_{d \geq 1} (1 + x^d)^{\beta(d)}.$$

Hence,

$$\sum_{n \geq 0} (q^n - D(n, 0)) x^n = \left(\frac{1 - x^2}{1 - x} \right)^{\beta(d)}$$

$$= \frac{1 - qx^2}{1 - qx},$$

the last step by Exercise 1.183. The proof follows easily. This result appears in D. E. Knuth, *The Art of Computer Programming*, vol. 2, 3rd ed., Addison-Wesley, Reading, Mass., 1997 (Exercise 4.6.2-2(b)) and is attributed to E. R. Berlekamp. Greta Panova (November 2007) showed that this problem can also be solved by establishing the recurrence

$$D(n, 0) = \sum_{k \geq 1} q^k (q^{2n-k} - D(n - 2k, 0)).$$

b. We have

$$\sum_{\beta \in \mathbb{N}^k} N(\beta) x^\beta = \prod_{d \geq 1} \left(\sum_{\alpha \in X} x^{\alpha d} \right)^{\beta(d)}$$

$$= \prod_{d \geq 1} \prod_{\substack{\alpha \in \mathbb{N}^k \\ \alpha \neq (0, 0, \dots, 0)}} (1 - x^{\alpha d})^{a_\alpha \beta(d)}.$$

The proof follows from Exercise 1.183.

186. a. Let $k = 1$ and $X = \{0, 1, \dots, r-1\}$ in Exercise 1.185(c). We get

$$\sum_{n \in X} x^n = 1 + x + \cdots + x^{r-1} = \frac{1 - x^r}{1 - x}.$$

Hence,

$$\sum_{n \geq 0} N_r(n) x^n = \frac{1 - qx^r}{1 - qx},$$

yielding equation (1.147). This result is stated in D. E. Knuth, *The Art of Computer Programming*, vol. 2, 3rd ed., Addison-Wesley, Reading, Mass., 1997 (solution to Exercise 4.6.2-2(b)).

b. Set $k = 2$ and $X = \{(m,0) : m \in \mathbb{N}\} \cup \{(0,n) : n \in \mathbb{P}\}$ to get

$$\sum_{m,n \geq 0} N(m,n)x^m y^n = \frac{1-qxy}{(1-qx)(1-qy)},$$

from which equation (1.148) follows.

c. Take $k = 1$ and $X = \mathbb{N} - \{1\}$. We get

$$\sum_{n \geq 0} P(n)x^n = \frac{1-qx^6}{(1-qx^2)(1-qx^3)},$$

via the identity

$$1 + \frac{t^2}{1-t} = \frac{1-t^6}{(1-t^2)(1-t^3)}.$$

Using the partial fraction expansion

$$\frac{1-qx^6}{(1-qx^2)(1-qx^3)} = -\frac{x}{q} + \frac{(1+q)(1+x)}{q(1-qx^2)} - \frac{1+qx+qx^2}{q(1-qx^3)},$$

it is routine to obtain equation (1.149). This result can also be obtained by noting that every monic powerful polynomial can be written uniquely in the form $f^2 g^3$, where f and g are monic and g is squarefree. Hence, $P(n) = \sum_{2i+3j=n} q^i(q^j - D(j,0))$, where $D(j,0)$ is defined in Exercise 1.185(b), and so on. This result is due to R. Stanley (proposer), Problem 11348, *Amer. Math. Monthly* **115** (2008), 262; R. Stong (solution), **117** (2010), 87–88.

NOTE. The term "powerful polynomial" is borrowed from the corresponding notion for integers. See for instance the Wikipedia entry "Powerful number" at

⟨http://en.wikipedia.org/wiki/Powerful_number⟩.

187. a. The *resultant* res(f,g) of two polynomials $f(x) = \prod(x-\theta_i)$ and $g(x) = \prod(x-\tau_j)$ over a field K is defined by

$$\text{res}(f,g) = \prod_{i,j}(\theta_i - \tau_j).$$

It is a standard fact (a consequence of the fact that res(f,g) is invariant under any permutation of the θ_i's and of the τ_j's) that res$(f,g) \in K$. Suppose that $f(x) = f_1(x) \cdots f_k(x)$ where each $f_i(x)$ is irreducible. Clearly,

$$\text{disc}(f) = \prod_{i=1}^{k} \text{disc}(f_i) \cdot \prod_{1 \leq i < j \leq k} \text{res}(f_i, f_j)^2. \tag{1.163}$$

A standard result from Galois theory states that the discriminant of an irreducible polynomial $g(x)$ of degree n over a field K is a square in K if and only if the Galois group of $g(x)$ (regarded as a group of permutations of the zeros of $g(x)$) is contained in the alternating group $\mathfrak{A}_n$. Now the Galois group of an irreducible polynomial of degree n over $\mathbb{F}_q$ is generated by an n-cycle and hence is contained in $\mathfrak{A}_n$ if and only if n is odd. It follows from equation (1.163) that if disc$(f) \neq 0$, then disc(f)

is a square in $\mathbb{F}_q$ if and only if $n - k$ is even. This result goes back to L. Stick-elberger, *Verh. Ersten Internationaler Mathematiker-Kongresses (Zürich, 1897)*, reprinted by Kraus Reprint Limited, Nendeln/Liechtenstein, 1967, pp. 182–193. A simplification of Stickelberger's argument was given by K. Dalen, *Math. Scand.* **3** (1955), 124–126. See also L. E. Dickson, *Bull. Amer. Math. Soc.* **13** (1906/07), 1–8, and R. G. Swan, *Pacific J. Math.* **12** (1962), 1099–1106 (Corollary 1). The foregoing proof is possibly new. NOTE: Swan, ibid. (§3), uses this result to give a simple proof of the law of quadratic reciprocity.

Now let $N_e(n)$ (respectively, $N_o(n)$) denote the number of monic polynomials of degree n which are a product of an even number (respectively, odd number) of distinct irreducible factors. It is easy to see (analogous to the solution to Exercise 1.183) that

$$\sum_{n \geq 0} (N_e(n) - N_o(n))x^n = \prod_{d \geq 1}(1 - x^d)^{\beta(d)}.$$

But

$$\prod_{d \geq 1}(1 - x^d)^{\beta(d)} = 1 - qx$$

by Exercise 1.183. Hence, $N_e(n) = N_o(n)$ for $n > 1$, and the proof follows.

b. Let $f(x) = \prod_{i=1}^{n}(x - \theta_i)$ be a monic polynomial of degree n over $\mathbb{F}_q$. For $a \in \mathbb{F}_q^* = \mathbb{F}_q - \{0\}$, write $f_a(x) = a^n f(x/a)$, so $f_a(x) = \prod_{i=1}^{n}(x - a\theta_i)$. It follows that

$$\mathrm{disc}(f_a(x)) = a^{n(n-1)}\mathrm{disc}(f(x)).$$

If $(n(n-1), q-1) = 1$ then the map $a \mapsto a^{n(n-1)}$ is a bijection on $\mathbb{F}_q^*$. Hence, if $\mathrm{disc}(f) \neq 0$, then we have $\{\mathrm{disc}(f_a) : a \in \mathbb{F}_q^*\} = \mathbb{F}_q^*$. It follows that $D(n,a) = D(n,b)$ for all $a, b \in \mathbb{F}_q^*$. Since $D(n,0) = q^{n-1}$, we have $D(n,a) = q^{n-1}$ for all $a \in \mathbb{F}_q$.

Now assume that $(n(n-1), q-1) = 2$. Thus, as a ranges over $\mathbb{F}_q^*$, $a^{n(n-1)}$ ranges over all squares in $\mathbb{F}_q^*$ twice each. Some care must be taken since we can have $f_a(x) = f_b(x)$ for $a \neq b$. (This issue did not arise in the case $(n(n-1), q-1) = 1$ since the $f_a(x)$'s had distinct discriminants.) Thus for each f, let P_f be the *multiset* of all f_a, $a \in \mathbb{F}_q^*$. The multiset union $\bigcup_f P_f$ contains each monic polynomial of degree n over $\mathbb{F}_q$ exactly $q - 1$ times. For each $a, b \in \mathbb{F}_q^*$ such that either both a, b or neither a, b are squares, the same number of polynomials (counting multiplicity) $g \in \bigcup_f P_f$ satisfy $\mathrm{disc}(g) = a$ as satisfy $\mathrm{disc}(g) = b$. Finally, by (a) it follows that the number of $g \in \bigcup_f P_f$ with square discriminants is the same as the number with nonsquare discriminants. Hence, $D(n,a) = D(n,b)$ for all $a, b \in \mathbb{F}_q^*$, and thus as previously for all $a, b \in \mathbb{F}_q$.

188. *First Solution.* Let V be an n-dimensional vector space over a field K, and fix an ordered basis $\mathbf{v} = (v_1, \ldots, v_n)$ of V. Let $\mathcal{N}_n$ denote the set of all nilpotent linear transformations $A : V \to V$. We will construct a bijection $\varphi : \mathcal{N}_n \to V^{n-1}$. Letting $V = \mathbb{F}_q$, it follows that $\#\mathcal{N}_n = \#(\mathbb{F}_q)^{n-1} = q^{n(n-1)}$.

The bijection is based on a standard construction in linear algebra known as *adapting* an ordered basis $\mathbf{w} = (w_1, \ldots, w_n)$ of a vector space V to an m-dimensional subspace U of V. It constructs from $\mathbf{w}$ in a canonical way a new ordered basis $w_{i_1}, \ldots, w_{i_{n-m}}, u_1, \ldots, u_m$ of V such that the first $n - m$ elements form a subsequence of $\mathbf{w}$ and the last m form an ordered basis of u. See, for example, M. C. Crabb, *Finite Fields and Their Applications* **12** (2006), 151–154 (p. 153) for further details.

Now let $A \in \mathcal{N}_n$, and write $V_i = A^i(V)$, $i \geq 0$. Let r be the least integer for which $V_r = 0$, so we have a strictly decreasing sequence

$$V = V_0 \supset V_1 \supset \cdots \supset V_r = 0.$$

Set $n_i = \dim V_i$ and $m_i = n_{i-1} - n_i$. Adapt the ordered basis v of V to V_1. Then adapt this new ordered basis to V_2, etc. After $r - 1$ steps we have constructed in a canonical way an ordered basis $y = (y_1, \ldots, y_n)$ such that $y_{n-n_i+1}, \ldots, y_n$ is a basis for V_i, $1 \leq i \leq r - 1$. We associate with A the $(n-1)$-tuple $\varphi(A) = (A(y_1), \ldots, A(y_{n-1})) \in V^{n-1}$. (Note that $A(y_n) = 0$.) It is straightforward to check that this construction gives a bijection $\varphi : \mathcal{N}_n \to V^{n-1}$ as desired.

This argument is due to M. C. Crabb, ibid., and we have closely followed his presentation (though with fewer details). As Crabb points out, this bijection can be regarded as a generalization of the Prüfer bijection (first proof of Proposition 5.3.2, specialized to rooted trees) for counting rooted trees on an n-element set. Further connections between the enumeration of trees and linear transformations were obtained by J.-B. Yin, Ph.D. thesis, M.I.T., 2009. For a further result of this nature, see Exercise 1.189.
Second Solution (sketch), due to Hansheng Diao, November 2007. Induction on n, the base case $n = 1$ being trivial. The statement is true for $k < n$. Let Q be the matrix in $\mathrm{Mat}(n,q)$ with 1's on the diagonal above the main diagonal and 0's elsewhere (i.e., a Jordan block of size n with eigenvalue 0). Let

$$\mathcal{A} = \{(M,N) \in \mathrm{Mat}(n,q) \times \mathrm{Mat}(n,q) : N \text{ is nilpotent, } QM = MN\}.$$

We compute $\#\mathcal{A}$ in two ways. Let $f(n)$ be the number of nilpotent matrices in $\mathrm{Mat}(n,q)$. We can choose N in $f(n)$ ways. Choose $v \in \mathbb{F}_q^n$ in q^n ways. Then there is a unique matrix $M \in \mathrm{Mat}(n,q)$ with first row v such that $QM = MN$. Hence,

$$\#\mathcal{A} = q^n f(n). \tag{1.164}$$

On the other hand, one can show that if M has rank r, then the number of choices for N so that $QM = MN$ is $f(n-r)q^{r(n-r)}$. Using Exercise 1.192(b) and induction, we get

$$\#\mathcal{A} = f(n) + \sum_{r=1}^{n} (q^n - 1) \cdots (q^n - q^{r-1}) q^{(n-r)(n-r-1)} \cdot q^{r(n-r)}$$

$$= f(n) + q^{n(n-1)}(q^n - 1).$$

Comparing with equation (1.164) completes the proof.
Third Solution (sketch), due to Greta Panova and Yi Sun (independently), November 2007. Count in two ways the number of $(n+1)$-tuples $(N, v_1, v_2, \ldots, v_n)$ with N nilpotent in $\mathrm{Mat}(n,q)$, and $v_i \in \mathbb{F}_q^n$ such that $N(v_i) = v_{i+1}$ ($1 \leq i \leq n - 1$) and $v_1 \neq 0$. On the one hand, there are $f(n)(q^n - 1)$ such $(n+1)$-tuples since they are determined by N and v_1. On the other hand, one can show that the number of such $(n+1)$-tuples such that $v_k \neq 0$ and $v_{k+1} = 0$ (with $v_{n+1} = 0$ always) is $f(n-k)q^{k(n-k)}(q^n - 1) \cdots (q^n - q^{k-1})$, yielding the recurrence

$$f(n)(q^n - 1) = \sum_{k=1}^{n} f(n-k)q^{k(n-k)}(q^n - 1) \cdots (q^n - q^{k-1}).$$

The proof follows straightforwardly by induction on n.
For some additional work on counting nilpotent matrices, see G. Lusztig, *Bull. London Math. Soc.* **8** (1976), 77–80.

189. See Proposition 4.27 of J. Yin, A q-Analogue of Spanning Trees: Nilpotent Transformations over Finite Fields, Ph.D. thesis, M.I.T., 2009. This result may be regarded as a q-analogue of the fact that the number of spanning trees of the complete bipartite graph K_{mn} is $m^{n-1}n^{m-1}$ (see Exercise 5.30, Vol. II).

190. a. We can imitate the proof of Proposition 1.10.2, using $\mathcal{I}^* := \mathcal{I} - \{x\}$ instead of $\mathcal{I}$ and β^* (defined by equation (1.112)) instead of β. We therefore get

$$\sum_{n \geq 0} \omega^*(n,q)x^n = \prod_{n \geq 1} \prod_{j \geq 1} (1 - x^{jn})^{-\beta^*(n)}$$

$$= \prod_{j \geq 1} \frac{1 - x^j}{1 - qx^j}, \tag{1.165}$$

from which the proof is immediate.

b. This result follows easily from the Pentagonal Number Formula (1.88) and Exercise 1.74. A more careful analysis shows that if $m = \lfloor (n-1)/2 \rfloor$, then

$$\omega^*(n,q) = q^n - q^m - q^{m-1} - q^{m-2} - \cdots - q^{2\lfloor (n+5)/6 \rfloor} + O(q^{\lfloor (n+5)/6 \rfloor - 1}).$$

c. It follows from the Pentagonal Number Formula and equation (1.165) that

$$\omega^*(n,0) = \begin{cases} (-1)^k, & \text{if } n = k(3k \pm 1)/2, \\ 0, & \text{otherwise,} \end{cases}$$

We also have

$$\omega^*(n,-1) = \begin{cases} 2(-1)^k, & \text{if } n = k^2, \\ 0, & \text{otherwise,} \end{cases}$$

a consequence of the identity (1.131) due to Gauss.

By differentiating (1.165) with respect to q and setting $q = 0$, it is not hard to see that $\omega^*(n,q)$ is divisible by q^2 if and only if

$$\frac{k(3k-1)}{2} < n < \frac{k(3k+1)}{2},$$

for some $k \geq 1$.

191. *First Solution* (in collaboration with G. Lusztig). Let F be an algebraic closure of $\mathbb{F}_q$. We claim that the set Ω of orbits of the adjoint representation of $\mathrm{GL}(n, F)$ has the structure

$$\Omega \cong \bigoplus_{\lambda \vdash n} F^{\ell(\lambda)}. \tag{1.166}$$

Let $\lambda = (\lambda_1, \lambda_2, \ldots, \lambda_k) \vdash n$, where $\lambda_k > 0$. Given $\alpha = (\alpha_1, \ldots, \alpha_k) \in F^k$, let $M = M(\lambda, \alpha) \in \mathrm{Mat}(n, F)$ be defined as follows: M is a direct sum of k Jordan blocks $J_1, \ldots, J_k$, with J_i containing λ_i main diagonal elements equal to α_i. We do yet have a set of orbit representatives, since if we have j blocks of the same size, then they can appear in any order. Hence, the different conjugacy classes formed by j blocks of size m has the structure $F^j/\mathfrak{S}_j$, where $\mathfrak{S}_j$ acts on F^j by permuting coordinates. But it is well known that $F^j/\mathfrak{S}_j \cong F^j$, namely, the elements of $F^j/\mathfrak{S}_j$ correspond to k-element multisets $\{\alpha_1, \ldots, \alpha_k\}$ of elements of F which we associate with $(\beta_1, \ldots, \beta_k) \in F^k$ by

$$\prod_{i=1}^{k} (x - \alpha_i) = x^k + \sum_{j=1}^{k} \beta_j x^{k-j}.$$

Hence, (1.166) follows. It is now a consequence of standard properties of the Frobenius map $\alpha \mapsto \alpha^q$ that the space Ω_q of orbits of the adjoint representation of $GL(n,q)$ has an analogous decomposition

$$\Omega_q \cong \bigoplus_{\lambda \vdash n} \mathbb{F}_q^{\ell(\lambda)},$$

and the proof follows.

Second Solution. Let $f(z) \in \mathbb{F}_q[z]$ be a monic polynomial of degree k. Let $f(z) = \prod f_i(z)^{r_i}$ be its factorization into irreducible factors (over $\mathbb{F}_q$). Let $M_f \in \mathrm{Mat}(n,q)$ be a matrix whose adjoint orbit is indexed by $\Phi : \mathcal{I}(q) \to \mathrm{Par}$ satisfying $\Phi(f_i) = (r_i)$ (the partition with one part equal to r_i). A specific example of such a matrix is the *companion matrix*

$$M_f = \begin{bmatrix} 0 & 0 & \cdots & 0 & -\beta_0 \\ 1 & 0 & \cdots & 0 & -\beta_1 \\ 0 & 1 & \cdots & 0 & -\beta_2 \\ \vdots & \vdots & \ddots & \vdots & \vdots \\ 0 & 0 & \cdots & 1 & -\beta_{k-1} \end{bmatrix},$$

where $f(z) = \beta_0 + \beta_1 z + \cdots + \beta_{k-1} z^{k-1} + z^k$. For fixed k, the space of all such M_f is just an affine space $\mathbb{F}_q^k$ (since it is isomorphic to the space of all monic polynomials of degree k). Now given a partition $\lambda \vdash n$ with conjugate $\lambda' = (\lambda_1', \lambda_2', \dots)$, choose polynomials $f_i(z) \in \mathbb{F}_q[z]$ such that $\deg f_i = \lambda_i'$ and $f_{i+1} | f_i$ for all $i \geq 1$. Let $M = M_{f_1} \oplus M_{f_2} \oplus \cdots \in \mathrm{Mat}(n,q)$. For fixed λ, the space of all such M has the structure $\mathbb{F}_q^{\lambda_1'} = \mathbb{F}_q^{\ell(\lambda)}$ (since once f_{i+1} is chosen, there are $q^{\lambda_{i+1}' - \lambda_i'}$ choices for f_i). It is easy to check that the M's form a cross-section of the orbits as λ ranges over all partitions of n, so the number of orbits is $\sum_{\lambda \vdash n} q^{\ell(\lambda)}$. This argument appears in J. Hua, *J. Combinatorial Theory Ser. A* **79** (1997), 105–117 (Theorem 11).

Third Solution, due to Gabriel Tavares Bujokas and Yufei Zhao (independently), November 2007. We want the number of functions $\Phi : \mathcal{I}(q) \to \mathrm{Par}$ satisfying $\sum_{f \in \mathcal{I}(q)} |\Phi_M(f)| \cdot \deg(f) = n$. For each $i \geq 1$, let

$$p_i = \prod_{f \in \mathcal{I}(q)} f^{m_i(\Phi(f))},$$

where $m_i(\Phi(f))$ denotes the number of parts of $\Phi(f)$ equal to i. Thus, the p_i's are arbitrary monic polynomials satisfying $\sum i \deg(p_i) = n$. First, choose $\lambda \vdash \langle 1^{d_1}, 2^{d_2}, \dots \rangle \vdash n$ and then each p_i so that $\deg p_i = d_i$. There are thus $q^{\sum d_i} = q^{\ell(\lambda)}$ choices for the p_i's, so

$$\omega(n,q) = \sum_{\lambda \vdash n} q^{\ell(\lambda)} = \sum_j p_j(n) q^j.$$

193. A matrix P is a projection if and only if $\Phi_P(z) = \langle 1^k \rangle$ for some k, $\Phi_P(z-1) = \langle 1^{n-k} \rangle$, and otherwise $\Phi_P(f) = \emptyset$. The proof now follows from Theorem 1.10.4 and Lemma 1.10.5 exactly as does Corollary 1.10.6.

194. A matrix M is regular if and only if for all $f \in \mathcal{I}(q)$ there is an integer $k \geq 0$ such that $\Phi_M(f) = (k)$. Write $c_f(k)$ for $c_f(\lambda)$ when $\lambda = (k)$. From Theorem 1.10.7, we have

$$c_f(k) = q^{kd} - q^{(k-1)d}, \ k \geq 1,$$

where $d = \deg(f)$. Substitute $t_{f,\lambda} = 1$ if $\lambda = (k)$ and $t_{f,\lambda} = 0$ otherwise in Theorem 1.10.4 to get

$$\sum_{n \geq 0} r_n \frac{x^n}{\gamma_n} = \prod_{f \in \mathcal{I}} \left(1 + \sum_{k \geq 1} \frac{x^{k \cdot \deg(f)}}{q^{k \cdot \deg(f)} - q^{(k-1) \cdot \deg(f)}} \right)$$

$$= \prod_{d \geq 1} \left(1 + \sum_{k \geq 1} \frac{x^{kd}}{q^{kd}(1 - q^{-d})} \right)^{\beta(d)}$$

$$= \prod_{d \geq 1} \left(1 + \frac{(x/q)^d}{(1 - q^{-d})(1 - (x/q)^d)} \right)^{\beta(d)}$$

$$= \prod_{d \geq 1} \left(1 + \frac{x^d}{(q^d - 1)(1 - (x/q)^d)} \right)^{\beta(d)}.$$

We can write this identity in the alternative form

$$\sum_{n \geq 0} r_n \frac{x^n}{\gamma_n} = \frac{1}{1-x} \prod_{d \geq 1} \left(1 + \frac{x^d}{q^d(q^d - 1)} \right)^{\beta(d)}$$

by using equation (1.145) with x/q substituted for x.

195. A matrix M is semisimple if and only if for all $f \in \mathcal{I}(q)$ there is an integer $k \geq 0$ such that $\Phi_M(f) = \langle 1^k \rangle$. The proof now follows from Theorem 1.10.4 and Lemma 1.10.5 exactly as does Corollary 1.10.6.

196. a. The proof parallels that of Proposition 1.10.15. We partition $\mathfrak{S}_n$ into two classes $\mathcal{A}$ and $\mathcal{B}$, where

$$\mathcal{A} = \{ w \in \mathfrak{S}_n : w \neq 12 \cdots ku \text{ for some } u \in \mathfrak{S}_{[k+1,n]} \},$$

$$\mathcal{B} = \{ w \in \mathfrak{S}_n : w = 12 \cdots ku \text{ for some } u \in \mathfrak{S}_{[k+1,n]} \}.$$

Let

$$G(n,k,q) = \{ A \in \mathrm{GL}(n,q) : A_{11} + \cdots + A_{kk} = 0 \}.$$

For $w \in \mathcal{A}$, we have

$$\#(\Gamma_w \cap G(n,k,q)) = \frac{1}{q} \#\Gamma_w.$$

For $w \in \mathcal{B}$, we have

$$\#(\Gamma_w \cap G(n,k,q)) = q^{\binom{n}{2} + \mathrm{inv}(w)} (q-1)^{n-k} a_k$$

$$= q^{\binom{n}{2} + \mathrm{inv}(w)} (q-1)^{n-k} ((q-1)^k + (-1)^k (q-1)).$$

Hence,

$$\sum_{w \in \mathcal{B}} \#(\Gamma_w \cap G(n,k,q)) = \frac{1}{q} \left(\sum_{w \in \mathcal{B}} q^{\binom{n}{2} + \mathrm{inv}(w)} (q-1)^n \right)$$

$$+ (-1)^k (q-1) q^{\binom{n}{2} - \binom{n-k}{2}} \sum_{u \in \mathfrak{S}_{[k+1,n]}} q^{\binom{n-k}{2} + \mathrm{inv}(u)} (q-1)^{n-k} \Bigg)$$

$$= \frac{1}{q} \Bigg(\sum_{w \in B} (\# \Gamma_w) + (-1)^k (q-1) q^{\frac{1}{2} k(2n-k-1)} \gamma_{n-k}(q) \Bigg),$$

and the proof follows.

b. The hyperplane H can be defined by $H = \{ M \in \mathrm{Mat}(n,q) : M \cdot N = 0 \}$, where N is a fixed nonzero matrix in $\mathrm{Mat}(n,q)$ and $M \cdot N = \mathrm{tr}(MN^t)$, the standard dot product in the vector space $\mathrm{Mat}(n,q)$. If $P, Q \in \mathrm{GL}(n,q)$, then $M \cdot N = 0$ if and only if $((P^t)^{-1} M (Q^t)^{-1}) \cdot (PNQ) = 0$. Since two matrices $N, N' \in \mathrm{Mat}(n,q)$ are related by $N' = PNQ$ for some $P, Q \in \mathrm{GL}(n,q)$ if and only if they have the same rank, it follows that $\#(\mathrm{GL}(n,q) \cap H)$ depends only on $\mathrm{rank}(N)$. If $\mathrm{rank}(N) = k$, then we may take

$$N_{ij} = \begin{cases} 1, & 1 \le i = j \le k, \\ 0, & \text{otherwise.} \end{cases}$$

Hence, $\#(\mathrm{GL}(n,q) \cap H)$ is given by the right-hand side of equation (1.150).

197. *Hint.* Let $f(n,k)$ be the number of $k \times n$ matrices over $\mathbb{F}_q$ of rank k with zero diagonal, where $1 \le k \le n$. Show that

$$f(n, k+1) = q^{k-1}(q-1)(f(n,k) \cdot (n-k) - f(n-1,k)),$$

with the initial condition $f(n,1) = q^{n-1} - 1$. The solution to this recurrence is

$$f(n,k) = q^{\binom{k-1}{2}-1}(q-1)^k \left(\sum_{i=0}^{k} (-1)^i \binom{k}{i} \frac{(n-i)!}{(n-k)!} \right).$$

Now set $k = n$.

This result is due to J. B. Lewis, R. I. Liu, A. H. Morales, G. Panova, S. V. Sam, and Y. Zhang, Matrices with restricted entries and q-analogues of permutations, $\mathrm{arXiv:1011.4539}$ (Proposition 2.2). This paper contains a host of other results about counting matrices over $\mathbb{F}_q$. A further result in this paper is given by Exercise 1.199.

198. a. An $(n+1) \times (n+1)$ symmetric matrix may be written as

$$N = \begin{bmatrix} \beta & y \\ y^t & M \end{bmatrix},$$

where M is an $n \times n$ symmetric matrix, $\beta \in \mathbb{F}_q$, and $y \in \mathbb{F}_q^n$. Elementary linear algebra arguments show that from a particular matrix M of rank r we obtain:

- $q^{n+1} - q^{r+1}$ matrices N of rank $r+2$,
- $(q-1)q^r$ matrices N of rank $r+1$,
- q^r matrices N of rank r,
- no matrices of other ranks.

The recurrence (1.151) follows. This recurrence (with more details of the proof) was given by J. MacWilliams, *Amer. Math. Monthly* **76** (1969), 152–164, and was used to prove (b). A simpler recurrence for $h(n,n)$ alone was given by G. Lusztig, *Transformation Groups* **10** (2005), 449–487 (end of §3.14).

b. We can simply verify that the stated formula for $h(n,r)$ satisfies the recurrence (1.151), together with the initial conditions. For some generalizations and further information, see R. Stanley, *Ann. Comb.* **2** (1998), 351–363; J. R. Stembridge, *Ann. Comb.* **2** (1998), 365–385; F. Chung and C. Yang, *Ann. Comb.* **4** (2000), 13–25; and P. Belkale and P. Brosnan, *Duke Math. J.* **116** (2003), 147–188.

NOTE. There is a less ad hoc way to compute the quantity $h(n,n)$. Namely, $GL(n,q)$ acts on $n \times n$ invertible symmetric matrices M over $\mathbb{F}_q$ by $A \cdot M = A^t M A$. This action has two orbits whose stabilizers are the two forms of the orthogonal group $O(n,q)$. The orbit sizes can be easily computed from standard facts about $O(n,q)$. For further details, see R. Stanley, op. cit. (§4).

199. a. The equality of the first two items when q is even is due to J. MacWilliams, *Amer. Math. Monthly* **76** (1969), 152–164 (Theorems 2, 3). The equality of the second two items appears in O. Jones, *Pacific J. Math.* **180** (1997), 89–100. For the remainder of the exercise, see Section 3 of the paper of Lewis–Liu–Morales–Panova–Sam–Zhang cited in the solution to Exercise 1.197.

200. This result was conjectured by A. A. Kirillov and A. Melnikov, in *Algèbre non commutative, groupes quantiques et invariants (Reims, 1995)*, Sémin. Congr. **2**, Soc. Math. France, Paris, 1997, pp. 35–42, and proved by S. B. Ekhad and D. Zeilberger, *Electron. J. Combin.* **3**(1) (1996), R2. No conceptual reason is known for such a simple formula.

201. a. The result follows from the theory of Gauss sums as developed, for example, in K. Ireland and M. Rosen, *A Classical Introduction to Modern Number Theory*, 2nd ed., Springer-Verlag, New York, 1990, and may have been known to Gauss or Eisenstein. This information was provided by N. Elkies (private communication, 1 August 2006).

b. The argument is analogous to the proof of Proposition 1.10.15. Let

$$G = \{A \in GL(3,q) : \text{tr}(A) = 0,\ \det(A) = 1\}.$$

If $123 \neq w \in \mathfrak{S}_3$, then $\#(\Gamma_2 \cap G) = \frac{1}{q(q-1)}\#\Gamma_w$. On the other hand, $\#(\Gamma_{123} \cap G) = q^3 f(q)$. Hence, we get

$$\#G = \frac{1}{q(q-1)}\left(\gamma(3,q) - \#\Gamma_{123}\right) + q^3 f(q).$$
$$= q^3(q-1)^2(q^2 + 2q + 2) + q^3 f(q).$$

202. This result is an instance of the Shimura–Taniyama–Weil conjecture, namely, every elliptic curve is modular. An important special case of the conjecture (sufficient to imply Fermat's Last Theorem) was proved by A. Wiles in 1993, with a gap fixed by Wiles and R. Taylor in 1994. The full conjecture was proved by Breuil, Conrad, Diamond, and Taylor in 1999. Our example follows H. Darmon, *Notices Amer. Math. Soc.* **46** (1999), 1397–1401, which has much additional information.

203. The statement about 103,049 was resolved in January 1994 when David Hough, then a graduate student at George Washington University, noticed that 103,049 is the total number of bracketings of a string of 10 letters. The problem of finding the number of bracketing of a string of n letters is known as *Schröder's second problem* and is discussed in Section 6.2, Vol. II. See also the Notes to Chapter 6, Vol. II, where also a possible interpretation of 310,952 is discussed. Hough's discovery was first published by R. Stanley, *Amer. Math. Monthly* **104** (1997), 344–350. A more scholarly account was given by F. Acerbi, *Archive History Exact Sci.* **57** (2003), 465–502.

2

Sieve Methods

2.1 Inclusion-Exclusion

Roughly speaking, a "sieve method" in enumerative combinatorics is a method for determining the cardinality of a set S that begins with a larger set and somehow subtracts off or cancels out unwanted elements. Sieve methods have two basic variations: (1) We can first approximate our answer with an overcount, and then subtract off an overcounted approximation of our original error, and so on, until after finitely many steps we have "converged" to the correct answer. This method is the combinatorial essence of the Principle of Inclusion-Exclusion, to which this section and the next four are devoted. (2) The elements of the larger set can be weighted in a natural combinatorial way so that the unwanted elements cancel out, leaving only the original set S. We discuss this technique in Sections 2.6 and 2.7.

The Principle of Inclusion-Exclusion is one of the fundamental tools of enumerative combinatorics. Abstractly, the Principle of Inclusion-Exclusion amounts to nothing more than computing the inverse of a certain matrix. As such, it is simply a minor result in linear algebra. The beauty of the principle lies not in the result itself, but rather in its wide applicability. We will give several example of problems that can be solved by Inclusion-Exclusion, some in a rather subtle way. First we state the principle in its purest form.

2.1.1 Theorem. *Let S be an n-set. Let V be the 2^n-dimensional vector space (over some field K) of all functions $f : 2^S \to K$. Let $\phi : V \to V$ be the linear transformation defined by*

$$\phi f(T) = \sum_{Y \supseteq T} f(Y), \text{ for all } T \subseteq S. \tag{2.1}$$

Then ϕ^{-1} exists and is given by

$$\phi^{-1} f(T) = \sum_{Y \supseteq T} (-1)^{\#(Y-T)} f(Y), \text{ for all } T \subseteq S. \tag{2.2}$$

Proof. Define $\psi : V \to V$ by $\psi f(T) = \sum_{Y \supseteq T} (-1)^{\#(Y-T)} f(Y)$. Then (composing functions right to left)

$$\phi \psi f(T) = \sum_{Y \supseteq T} (-1)^{\#(Y-T)} \phi f(Y)$$

$$= \sum_{Y \supseteq T} (-1)^{\#(Y-T)} \sum_{Z \supseteq Y} f(Z)$$

$$= \sum_{Z \supseteq T} \left(\sum_{Z \supseteq Y \supseteq T} (-1)^{\#(Y-T)} \right) f(Z).$$

Setting $m = \#(Z - T)$, we have

$$\sum_{\substack{Z \supseteq Y \supseteq T \\ (Z,T \text{ fixed})}} (-1)^{\#(Y-T)} = \sum_{i=0}^{m} (-1)^i \binom{m}{i} = \delta_{0m},$$

the latter equality by putting $x = -1$ in equation (1.18) or by Exercise 1.3(f), so $\phi \psi f(T) = f(T)$. Hence, $\phi \psi f = f$, so $\psi = \phi^{-1}$.   □

The following is the usual combinatorial situation involving Theorem 2.1.1. We think of S as being a set of properties that the elements of some given set A of objects may or may not have. For any subset T of S, let $f_=(T)$ be the number of objects in A that have *exactly* the properties in T (so they fail to have the properties in $\overline{T} = S - T$). More generally, if $w : A \to K$ is any weight function on A with values in a field (or abelian group) K, then one could set $f_=(T) = \sum_x w(x)$, where x ranges over all objects in A having exactly the properties in T. Let $f_\geq(T)$ be the number of objects in A that have *at least* the properties in T. Clearly then,

$$f_\geq(T) = \sum_{Y \supseteq T} f_=(Y). \tag{2.3}$$

Hence by Theorem 2.1.1,

$$f_=(T) = \sum_{Y \supseteq T} (-1)^{\#(Y-T)} f_\geq(Y). \tag{2.4}$$

In particular, the number of objects having *none* of the properties in S is given by

$$f_=(\emptyset) = \sum_{Y \supseteq T} (-1)^{\#Y} f_\geq(Y), \tag{2.5}$$

where Y ranges over all subsets S. In typical applications of the Principle of Inclusion-Exclusion, it will be relatively easy to compute $f_\geq(Y)$ for $Y \subseteq S$, so equation (2.4) will yield a formula for $f_=(T)$.

In equation (2.4) one thinks of $f_\geq(T)$ (the term indexed by $Y = T$) as being a first approximation to $f_=(T)$. We then subtract

$$\sum_{\substack{Y \supseteq T \\ \#(Y-T)=1}} f_\geq(Y),$$

to get a better approximation. Next we add back in

$$\sum_{\substack{Y \supseteq T \\ \#(Y-T)=2}} f_{\geq}(Y),$$

and so on, until finally reaching the explicit formula (2.4). This reasoning explains the terminology "Inclusion-Exclusion."

Perhaps the most standard formulation of the Principle of Inclusion-Exclusion is one that dispenses with the set S of properties per se, and just considers subsets of A. Thus, let $A_1, \ldots, A_n$ be subsets of a finite set A. For each subset T of $[n]$, let

$$A_T = \bigcap_{i \in T} A_i$$

(with $A_\emptyset = A$), and for $0 \leq k \leq n$ set

$$S_k = \sum_{\#T=k} \#A_T, \tag{2.6}$$

the sum of the cardinalities, or more generally the weighted cardinalities

$$w(A_T) = \sum_{x \in A_T} w(x),$$

of all k-tuple intersections of the A_i's. Think of A_i as defining a property P_i by the condition that $x \in A$ satisfies P_i if and only if $x \in A_i$. Then A_T is just the set of objects in A that have at least the properties in T, so by (2.5) the number $\#(\overline{A}_1 \cap \cdots \cap \overline{A}_n)$ of elements of A lying in *none* of the A_i's is given by

$$\#(\overline{A}_1 \cap \cdots \cap \overline{A}_n) = S_0 - S_1 + S_2 - \cdots + (-1)^n S_n, \tag{2.7}$$

where $S_0 = A_\emptyset = \#A$.

The Principle of Inclusion-Exclusion and its various reformulations can be dualized by interchanging $\cap$ and $\cup$, $\subseteq$ and $\supseteq$, and so on, throughout. The dual form of Theorem 2.1.1 states that if

$$\widetilde{\phi} f(T) = \sum_{Y \subseteq T} f(Y), \quad \text{for all } T \subseteq S,$$

then $\widetilde{\phi}^{-1}$ exists and is given by

$$\widetilde{\phi}^{-1} f(T) = \sum_{Y \subseteq T} (-1)^{\#(T-Y)} f(Y), \quad \text{for all } T \subseteq S.$$

Similarly, if we let $f_{\leq}(T)$ be the (weighted) number of objects of A having *at most* the properties in T, then

$$f_{\leq}(T) = \sum_{Y \subseteq T} f_{=}(Y),$$

$$f_{=}(T) = \sum_{Y \subseteq T} (-1)^{\#(T-Y)} f_{\leq}(Y). \tag{2.8}$$

A common special case of the Principle of Inclusion-Exclusion occurs when the function $f_=(T) = f_=(T')$ whenever $\#T = \#T'$. Thus also $f_\geq(T)$ depends only on $\#T$, and we set $a(n-i) = f_=(T)$ and $b(n-i) = f_\geq(T)$ whenever $\#T = i$. (CAVEAT. In many problems the set A of objects and S of properties will depend on a parameter p, and the functions $a(i)$ and $b(i)$ may depend on p. Thus, for example, $a(0)$ and $b(0)$ are the number of objects having *all* the properties, and this number may certainly depend on p. Proposition 2.2.2 is devoted to the situation when $a(i)$ and $b(i)$ are independent of p.) We thus obtain from equations (2.3) and (2.4) the equivalence of the formulas

$$b(m) = \sum_{i=0}^{m} \binom{m}{i} a(i), \quad 0 \leq m \leq n, \tag{2.9}$$

$$a(m) = \sum_{i=0}^{m} \binom{m}{i} (-1)^{m-i} b(i), \quad 0 \leq m \leq n. \tag{2.10}$$

In other words, the inverse of the $(n+1) \times (n+1)$ matrix whose (i,j)-entry $(0 \leq i, j \leq n)$ is $\binom{j}{i}$ has (i,j)-entry $(-1)^{j-i} \binom{j}{i}$. For instance,

$$\begin{bmatrix} 1 & 1 & 1 & 1 \\ 0 & 1 & 2 & 3 \\ 0 & 0 & 1 & 3 \\ 0 & 0 & 0 & 1 \end{bmatrix}^{-1} = \begin{bmatrix} 1 & -1 & 1 & -1 \\ 0 & 1 & -2 & 3 \\ 0 & 0 & 1 & -3 \\ 0 & 0 & 0 & 1 \end{bmatrix}.$$

Of course, we may let n approach ∞ so that (2.9) and (2.10) are equivalent for $n = \infty$.

Note that in language of the calculus of finite differences (see equation (1.98)), (2.10) can be rewritten as

$$a(m) = \Delta^m b(0), \quad 0 \leq m \leq n.$$

2.2 Examples and Special Cases

The canonical example of the use of the Principle of Inclusion-Exclusion is the following.

2.2.1 Example. (the "derangement problem" or "*problème des rencontres*"). How many permutations $w \in \mathfrak{S}_n$ have no fixed points, that is, $w(i) \neq i$ for all $i \in [n]$? Such a permutation is called a *derangement*. Call this number $D(n)$. Thus, $D(0) = 1, D(1) = 0, D(2) = 1, D(3) = 2$. Think of the condition $w(i) = i$ as the ith property of w. Now the number of permutations with *at least* the set $T \subseteq [n]$ of points fixed is $f_\geq(T) = b(n-i) = (n-i)!$, where $\#T = i$ (since we fix the elements of T and permute the remaining $n-i$ elements arbitrarily). Hence by (2.10), the number

$f_=(\emptyset) = a(n) = D(n)$ of permutations with *no* fixed points is

$$D(n) = \sum_{i=0}^{n} \binom{n}{i}(-1)^{n-i}i!. \tag{2.11}$$

This last expression may be rewritten

$$D(n) = n!\left(1 - \frac{1}{1!} + \frac{1}{2!} - \frac{1}{3!} + \cdots + (-1)^n\frac{1}{n!}\right). \tag{2.12}$$

Since $0.36787944\cdots = e^{-1} = \sum_{j\geq0}(-1)^j/j!$, it is clear from (2.12) that $n!/e$ is a good approximation to $D(n)$, and indeed it is not difficult to show that $D(n)$ is the nearest integer to $n!/e$. It also follows immediately from (2.12) that for $n \geq 1$,

$$D(n) = nD(n-1) + (-1)^n, \tag{2.13}$$

$$D(n) = (n-1)(D(n-1) + D(n-2)). \tag{2.14}$$

While it is easy to give a direct combinatorial proof of equation (2.14), considerably more work is necessary to prove (2.13) combinatorially. (See Exercise 2.8.) In terms of generating functions, we have that

$$\sum_{n\geq0} D(n)\frac{x^n}{n!} = \frac{e^{-x}}{1-x}.$$

The function $b(i) = i!$ has a very special property—it depends only on i, not on n. Equivalently, the number of permutations $w \in \mathfrak{S}_n$ that have at most the set $T \subseteq [n]$ of points *unfixed* depends only on $\#T$, not on n. This means that equation (2.11) can be rewritten in the language of the calculus of finite differences (see equation (1.98)) as

$$D(n) = \Delta^n x!\big|_{x=0},$$

which is abbreviated $\Delta^n0!$. Since the number $b(i)$ of permutations in $\mathfrak{S}_n$ that have at most some specified i-set of points unfixed depends only on i, the same is true of the number $a(i)$ of permutations in $\mathfrak{S}_n$ that have exactly some specified i set of points unfixed. It is clear combinatorially that $a(i) = D(i)$, and this fact is also evident from equations (2.10) and (2.11).

Let us state formally the general result that follows from the preceding considerations.

2.2.2 Proposition. *For each $n \in \mathbb{N}$, let B_n be a (finite) set, and let S_n be a set of n properties that elements of B_n may or may not have. Suppose that for every $T \subseteq S_n$, the number of $x \in B_n$ that lack at most the properties in T (i.e., that have at least the properties in $S - T$) depends only on $\#T$, not on n. Let $b(n) = \#B_n$, and let $a(n)$ be the number of objects $x \in B_n$ that have none of the properties in S_n. Then $a(n) = \Delta^n b(0)$.*

2.2.3 Example. Let us consider an example to which the previous proposition does *not* apply. Let $h(n)$ be the number of permutations of the multiset $M_n = \{1^2, 2^2, \ldots, n^2\}$ with no two consecutive terms equal. Thus, $h(0) = 1$, $h(1) = 0$, and $h(2) = 2$ (corresponding to the permutations 1212 and 2121). Let P_i, for $1 \leq i \leq n$, be the property that the permutation w of M_n has two consecutive i's. Hence we seek $f_=(\emptyset) = h(n)$. It is clear by symmetry that for fixed n, $f_\geq(T)$ depends only on $i = \#T$, so write $g(i) = f_\geq(T)$. Clearly $g(i)$ is equal to the number of permutations w of the multiset $\{1, 2, \ldots, i, (i+1)^2, \ldots, n^2\}$ (replace any $j \geq i$ appearing in w by two consecutive j's), so

$$g(i) = (2n - i)! 2^{-(n-i)},$$

a special case of equation (1.23). Note that $b(i) := g(n - i) = (n + i)! 2^{-i}$ is not a function of i alone, so that Proposition 2.2.2 is indeed inapplicable. However, we do get from (2.10) that

$$h(n) = \sum_{i=0}^{n} \binom{n}{i} (-1)^{n-i} (n+i)! 2^{-i} = \Delta^n (n+i)! 2^{-i} \big|_{i=0}.$$

Here the function $(n + i)! 2^{-i}$ to which Δ^n is applied depends on n.

We turn next to an example for which the final answer can be represented by a determinant.

2.2.4 Example. Recall that in Chapter 1 (Section 1.4) we defined the *descent set* $D(w)$ of a permutation $w = a_1 a_2 \cdots a_n$ of $[n]$ by $D(w) = \{i : a_i > a_{i+1}\}$. Our object here is to obtain an expression for the quantity $\beta_n(S)$, the number of permutations $w \in \mathfrak{S}_n$ with descent set S. Let $\alpha_n(S)$ be the number of permutations whose descent set is *contained* in S, as in equation (1.31). Thus (as pointed out in equation (1.31))

$$\alpha_n(S) = \sum_{T \subseteq S} \beta_n(T).$$

It was stated in equation (1.34) and follows from (2.8) that

$$\beta_n(S) = \sum_{T \subseteq S} (-1)^{\#(S-T)} \alpha_n(T).$$

Recall also that if the elements of S are given by $1 \leq s_1 < s_2 < \cdots < s_k \leq n - 1$, then by Proposition 1.4.1 we have

$$\alpha_n(S) = \binom{n}{s_1, s_2 - s_1, s_3 - s_2, \ldots, n - s_k}.$$

Therefore,

$$\beta_n(S) = \sum_{1 \leq i_1 < i_2 < \cdots < i_j \leq k} (-1)^{k-j} \binom{n}{s_{i_1}, s_{i_2} - s_{i_1}, \ldots, n - s_{i_j}}. \tag{2.15}$$

We can write (2.15) in an alternative form as follows. Let f be any function defined on $[0,k+1] \times [0,k+1]$ satisfying $f(i,i) = 1$ and $f(i,j) = 0$ if $i > j$. Then the terms in the sum

$$A_k = \sum_{1 \leq i_1 < i_2 < \cdots < i_j \leq k} (-1)^{k-j} f(0,i_1) f(i_1,i_2) \cdots f(i_j,k+1)$$

are just the nonzero terms in the expansion of the $(k+1) \times (k+1)$ determinant with (i,j)-entry $f(i,j+1)$, $(i,j) \in [0,k] \times [0,k]$. Hence, if we set $f(i,j) = 1/(s_j - s_i)!$ (with $s_0 = 0$, $s_{k+1} = n$), we obtain from (2.15) that

$$\beta_n(S) = n! \det[1/(s_{j+1} - s_i)!], \tag{2.16}$$

$(i,j) \in [0,k] \times [0,k]$. For instance, if $n = 8$ and $S = \{1,5\}$, then

$$\beta_n(S) = 8! \begin{vmatrix} \dfrac{1}{1!} & \dfrac{1}{5!} & \dfrac{1}{8!} \\[2ex] 1 & \dfrac{1}{4!} & \dfrac{1}{7!} \\[2ex] 0 & 1 & \dfrac{1}{3!} \end{vmatrix} = 217.$$

By an elementary manipulation (whose details are left to the reader), equation (2.16) can also be written in the form

$$\beta_n(S) = \det\left[\binom{n - s_i}{s_{j+1} - s_i}\right], \tag{2.17}$$

where $(i,j) \in [0,k] \times [0,k]$ as before.

2.2.5 Example. We can obtain a q-analogue of the previous example with very little extra work. We seek some statistic $s(w)$ of permutations $w \in \mathfrak{S}_n$ such that

$$\sum_{\substack{w \in \mathfrak{S}_n \\ D(w) \subseteq S}} q^{s(w)} = \binom{n}{s_1, s_2 - s_1, \ldots, n - s_k}, \tag{2.18}$$

where the elements of S are $1 \leq s_1 < s_2 < \cdots < s_k \leq n-1$ as above. We will then automatically obtain q-analogues of equations (2.15), (2.16), and (2.17). We claim that (2.18) holds when $s(w) = \text{inv}(w)$, the number of inversions of w. To see this, set $t_1 = s_1$, $t_2 = s_2 - s_1, \ldots, t_{k+1} = n - s_k$. Let $M = \{1^{t_1}, \ldots, (k+1)^{t_{k+1}}\}$. Recall from Proposition 1.7.1 that

$$\sum_{u \in \mathfrak{S}(M)} q^{\text{inv}(u)} = \binom{n}{t_1, t_2, \ldots, t_{k+1}}. \tag{2.19}$$

Now given $u \in \mathfrak{S}(M)$, let $v \in \mathfrak{S}_n$ be the standardization of u as defined after the second proof of Proposition 1.7.1, so $\text{inv}(u) = \text{inv}(v)$. We call v a *shuffle* of the sets $[1,s_1], [s_1+1,s_2], \ldots, [s_k+1,n]$. Now set $w = v^{-1}$. It is easy to see that v is a shuffle of $[1, s_1], [s_1 + 1, s_2], \ldots, [s_k + 1, n]$ if and only if $D(w) \subseteq \{s_1, s_2, \ldots, s_k\}$. Since $\mathrm{inv}(v) = \mathrm{inv}(w)$ by Proposition 1.3.14, we obtain

$$\sum_{\substack{w \in \mathfrak{S}_n \\ D(w) \subseteq S}} q^{\mathrm{inv}(w)} = \binom{n}{s_1, s_2 - s_1, \ldots, n - s_k}, \tag{2.20}$$

as desired.

Thus, set

$$\beta_n(S, q) = \sum_{\substack{w \in \mathfrak{S}_n \\ D(w) = S}} q^{\mathrm{inv}(w)}.$$

By simply mimicking the reasoning of Example 2.2.4, we obtain

$$\beta_n(S, q) = (n)! \det \left[1/(s_{j+1} - s_i)! \right]_0^k$$

$$= \det \left[\binom{n - s_i}{s_{j+1} - s_i} \right]_0^k. \tag{2.21}$$

For instance, if $n = 8$ and $S = \{1, 5\}$, then

$$\beta_n(S, q) = (8)! \begin{vmatrix} \dfrac{1}{(1)!} & \dfrac{1}{(5)!} & \dfrac{1}{(8)!} \\[2mm] 1 & \dfrac{1}{(4)!} & \dfrac{1}{(7)!} \\[2mm] 0 & 1 & \dfrac{1}{(3)!} \end{vmatrix}$$

$$= q^2 + 3q^3 + 6q^4 + 9q^5 + 13q^6 + 17q^7 + 21q^8 + 23q^9$$
$$+ 24q^{10} + 23q^{11} + 21q^{12} + 18q^{13} + 14q^{14} + 10q^{15}$$
$$+ 7q^{16} + 4q^{17} + 2q^{18} + q^{19}.$$

If we analyze the reason why we obtained a determinant in the previous two examples, then we get the following result.

2.2.6 Proposition. *Let* $S = \{P_1, \ldots, P_n\}$ *be a set of properties, and let* $T = \{P_{s_1}, \ldots, P_{s_k}\} \subseteq S$, *where* $1 \leq s_1 < \cdots < s_k \leq n$. *Suppose that* $f_{\leq}(T)$ *has the form*

$$f_{\leq}(T) = h(n) e(s_0, s_1) e(s_1, s_2) \cdots e(s_k, s_{k+1})$$

for certain functions h *on* $\mathbb{N}$ *and* e *on* $\mathbb{N} \times \mathbb{N}$, *where we set* $s_0 = 0$, $s_{k+1} = n$, $e(i, i) = 1$, *and* $e(i, j) = 0$ *if* $j < i$. *Then*

$$f_=(T) = h(n) \det \left[e(s_i, s_{j+1}) \right]_0^k.$$

2.3 Permutations with Restricted Position

The derangement problem asks for the number of permutations $w \in \mathfrak{S}_n$ where for each i, certain values of $w(i)$ are disallowed (namely, we disallow $w(i) = i$). We now consider a general theory of such permutations. It is traditionally described using terminology from the game of chess. Let $B \subseteq [n] \times [n]$, called a *board*. If $w \in \mathfrak{S}_n$, then define the graph $G(w)$ of w by

$$G(w) = \{(i, w(i)) : i \in [n]\}.$$

Now define

$$N_j = \#\{w \in \mathfrak{S}_n : j = \#(B \cap G(w))\},$$

$r_k = $ number of k-subsets of B such that no two

elements have a common coordinate

$= $ number of ways to place k nonattacking rooks on B.

Also define the *rook polynomial* $r_B(x)$ of the board B by

$$r_B(x) = \sum_k r_k x^k.$$

We may identify $w \in \mathfrak{S}_n$ with the placement of n nonattacking rooks on the squares $(i, w(i))$ of $[n] \times [n]$. Thus N_j is the number of ways to place n nonattacking rooks on $[n] \times [n]$ such that exactly j of these rooks lie in B. For instance, if $n = 4$ and $B = \{(1,1), (2,2), (3,3), (3,4), (4,4)\}$, then $N_0 = 6$, $N_1 = 9$, $N_2 = 7$, $N_3 = 1$, $N_4 = 1$, $r_0 = 1$, $r_1 = 5$, $r_2 = 8$, $r_3 = 5$, $r_4 = 1$. Our object is to describe the numbers N_j, and especially N_0, in terms of the numbers r_k. To this end, define the polynomial

$$N_n(x) = \sum_j N_j x^j.$$

2.3.1 Theorem. *We have*

$$N_n(x) = \sum_{k=0}^n r_k (n-k)!(x-1)^k. \tag{2.22}$$

In particular,

$$N_0 = N_n(0) = \sum_{k=0}^n (-1)^k r_k (n-k)!. \tag{2.23}$$

First proof. Let C_k be the number of pairs (w, C), where $w \in \mathfrak{S}_n$ and C is a k-element subset of $B \cap G(w)$. For each j, choose w in N_j ways so that $j = \#(B \cap G(w))$, and then choose C is $\binom{j}{k}$ ways. Hence, $C_k = \sum_j \binom{j}{k} N_j$. On the other hand, we first could choose C in r_k ways and then "extend" to w in $(n-k)!$ ways. Hence, $C_k = r_k(n-k)!$. Therefore,

$$\sum_j \binom{j}{k} N_j = r_k(n-k)!,$$

or equivalently (multiplying by y^k and summing on k),

$$\sum_j (y+1)^j N_j = \sum_k r_k(n-k)! y^k.$$

Putting $y = x - 1$ yields the desired formula.

Second proof. It suffices to assume $x \in \mathbb{P}$. The left-hand side of equation (2.22) counts the number of ways to place n nonattacking rooks on $[n] \times [n]$ and labeling each rook on B with an element of $[x]$. On the other hand, such a configuration can be obtained by placing k nonattacking rooks on B, labeling each of them with an element of $\{2, \ldots, x\}$, placing $n - k$ additional nonattacking rooks on $[n] \times [n]$ in $(n - k)!$ ways, and labeling the new rooks on B with 1. This argument establishes the desired bijection. $\square$

The two proofs of Theorem 2.3.1 provide another illustration of the principle enunciated in Chapter 1 (third proof of Proposition 1.3.7) about the two combinatorial methods for showing that two polynomials are identical. It is certainly also possible to prove (2.23) by a direct application of Inclusion-Exclusion, generalizing Example 2.2.1. Such a proof would not be considered combinatorial since we have not explicitly constructed a bijection between two sets (but see Section 2.6 for a method of making such a proof combinatorial). The two proofs we have given may be regarded as "semicombinatorial," since they yield by direct bijections formulas involving parameters y and x, respectively; and we then obtain (2.23) by setting $y = -1$ and $x = 0$, respectively. In general, a semicombinatorial proof of (2.5) can easily be given by first showing combinatorially that

$$\sum_X f_=(X) x^{\#X} = \sum_Y f_\geq(Y)(x-1)^{\#Y} \tag{2.24}$$

or

$$\sum_X f_=(X)(y+1)^{\#X} = \sum_Y f_\geq(Y) y^{\#Y}, \tag{2.25}$$

and then setting $x = 0$ and $y = -1$, respectively.

As a example of Theorem 2.3.1, take $B = \{(1,1),(2,2),(3,3),(3,4),(4,4)\}$ as earlier. Then

$$N_4(x) = 4! + 5 \cdot 3!(x-1) + 8 \cdot 2!(x-1)^2 + 5 \cdot 1!(x-1)^3 + (x-1)^4$$
$$= x^4 + x^3 + 7x^2 + 9x + 6.$$

2.3.2 Example (derangements revisted). Take $B = \{(1,1), (2,2), \ldots, (n,n)\}$. We want to compute $N_0 = D(n)$. Clearly, $r_k = \binom{n}{k}$, so

$$N_n(x) = \sum_{k=0}^{n} \binom{n}{k} (n-k)! (x-1)^k$$

$$= \sum_{k=0}^{n} \frac{n!}{k!} (x-1)^k$$

$$\Rightarrow N_0 = \sum_{k=0}^{n} (-1)^k \frac{n!}{k!}.$$

2.3.3 Example. (Problème des ménages). This famous problem is equivalent to asking for the number $M(n)$ of permutations $w \in \mathfrak{S}_n$ such that $w(i) \neq i, i + 1 \pmod{n}$ for all $i \in [n]$. In other words, we seek N_0 for the board

$$B = \{(1,1), (2,2), \ldots, (n,n), (1,2), (2,3), \ldots, (n-1,n), (n,1)\}.$$

By looking at a picture of B, we see that r_k is equal to the number of ways to choose k points, no two consecutive, from a collection of $2n$ points arranged in a circle.

2.3.4 Lemma. *The number of ways to choose k points, no two consecutive, from a collection of m points arranged in a circle is $\frac{m}{m-k} \binom{m-k}{k}$.*

First Proof. Let $f(m,k)$ be the desired number; and let $g(m,k)$ be the number of ways to choose k nonconsecutive points from m points arranged in a circle, next coloring the k points red, and then coloring one of the non-red points blue. Clearly $g(m,k) = (m-k)f(m,k)$. But we can also compute $g(m,k)$ as follows. First, color a point blue in m ways. We now need to color k points red, no two consecutive, from a *linear* array of $m - 1$ points. One way to proceed is as follows. (See also Exercise 1.34.) Place $m - 1 - k$ uncolored points on a line, and insert k red points into the $m - k$ spaces between the uncolored points (counting the beginning and end) in $\binom{m-k}{k}$ ways. Hence, $g(m,k) = m\binom{m-k}{k}$, so $f(m,k) = \frac{m}{m-k}\binom{m-k}{k}$. $\qquad\square$

This proof is based on a general principle of passing from "circular" to "linear" arrays. We will discuss this principle further in Chapter 4 (see Proposition 4.7.13).

Second Proof. Label the points $1, 2, \ldots, m$ in clockwise order. We wish to color k of them red, no two consecutive. First, we count the number of ways when 1 isn't colored red. Place $m - k$ uncolored points on a circle, label one of these 1, and insert k red points into the $m - k$ spaces between the uncolored points in $\binom{m-k}{k}$ ways. On the other hand, if 1 is to be colored red, then place $m - k - 1$ points on the circle, color one of these points red and label it 1, and then insert in $\binom{m-k-1}{k-1}$

ways $k - 1$ red points into the $m - k - 1$ allowed spaces. Hence,

$$f(m,k) = \binom{m-k}{k} + \binom{m-k-1}{k-1} = \frac{m}{m-k}\binom{m-k}{k}.$$

$\square$

2.3.5 Corollary. *The polynomial $N_n(x)$ for the board $B = \{(i,i),(i,i + 1)\,(\mathrm{mod}\,n) : 1 \le i \le n\}$ is given by*

$$N_n(x) = \sum_{k=0}^{n} \frac{2n}{2n-k}\binom{2n-k}{k}(n-k)!(x-1)^k.$$

In particular, the number N_0 of permutations $w \in \mathfrak{S}_n$ such that $w(i) \ne i, i + 1\,(\mathrm{mod}\,n)$ for $1 \le i \le n$ is given by

$$N_0 = \sum_{k=0}^{n} \frac{2n}{2n-k}\binom{2n-k}{k}(n-k)!(-1)^k.$$

Corollary 2.3.5 suggest the following question. Let $1 \le k \le n$, and let $B_{n,k}$ denote the board

$$B_{n,k} = \{(i,i),(i,i+1),\ldots,(i,i+k-1)\,(\mathrm{mod}\,n) : 1 \le i \le n\}.$$

Find the rook polynomial $R_{n,k}(x) = \sum_i r_i(n,k)x^i$ of $B_{n,k}$. Thus, by equation (2.23) the number $f(n,k)$ of permutations $w \in \mathfrak{S}_n$ satisfying $w(i) \ne i, i+1,\ldots,i+k-1\,(\mathrm{mod}\,n)$ is given by

$$f(n,k) = \sum_{i=0}^{n}(-1)^i r_i(n,k)(n-i)!.$$

Such permutations are called *k-discordant*. For instance, 1-discordant permutations are just derangements. When $k > 2$ there is no simple explicit expression for $r_i(n,k)$ as there was for $k = 1,2$. However, we shall see in Example 4.7.19 that there exists a polynomial $Q_k(x,y) \in \mathbb{Z}[x,y]$ such that

$$\sum_n R_{n,k}(x)y^n = \frac{-y\dfrac{\partial}{\partial y}Q_k(x,y)}{Q_k(x,y)},$$

provided that $R_{n,k}(x)$ is suitably interpreted when $n < k$. For instance,

$$Q_1(x,y) = 1 - (1+x)y,$$

$$Q_2(x,y) = (1 - (1+2x)y + x^2y^2)(1 - xy),$$

$$Q_3(x,y) = (1 - (1+2x)y - xy^2 + x^3y^3)(1 - xy).$$

2.4 Ferrers Boards

Given a particular board or class of boards B, we can ask whether the rook numbers r_i have any special properties of interest. Here we will discuss a class of boards called *Ferrers boards*. Given integers $0 \leq b_1 \leq \cdots \leq b_m$, the Ferrers board of shape $(b_1, \ldots, b_m)$ is defined by

$$B = \{(i, j) : 1 \leq i \leq m, \ 1 \leq j \leq b_i\},$$

where we are using ordinary cartesian coordinates so the $(1, 1)$ square is at the bottom left. The board B depends (up to translation) only on the *positive* b_i's. However, it will prove to be a technical convenience to allow $b_i = 0$. Note that B is just a reflection and rotation of the Young diagram of the partition $\lambda = (b_m, \ldots, b_1)$.

2.4.1 Theorem. *Let $\sum r_k x^k$ be the rook polynomial of the Ferrers board B of shape $(b_1, \ldots, b_m)$. Set $s_i = b_i - i + 1$. Then*

$$\sum_k r_k \cdot (x)_{m-k} = \prod_{i=1}^{m} (x + s_i).$$

Proof. Let $x \in \mathbb{N}$, and let B' be the Ferrers board of shape $(b_1 + x, \ldots, b_m + x)$. Regard $B' = B \cup C$, where C is an $x \times m$ rectangle placed below B. See Figure 2.1 for the case $(b_1, b_2, b_3) = (1, 2, 4)$. We count $r_m(B')$ in two ways.

1. Place k rooks on B in r_k ways, then $m - k$ rooks on C in $(x)_{m-k}$ ways, to get

$$r_m(B') = \sum_k r_k \cdot (x)_{m-k}.$$

2. Place a rook in the first column of B' in $x + b_1 = x - s_1$ ways, then a rook in the second column in $x + b_2 - 1 = x + s_2$ ways, and so on, to get

$$f_m(B') = \prod_{i=1}^{m} (x + s_i).$$

This completes the proof. $\qquad\square$

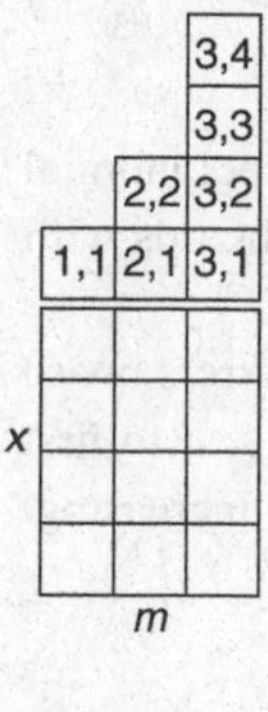

Figure 2.1 The Ferrers board of shape $(1, 2, 4)$ with a rectangle C underneath.

2.4.2 Corollary. *Let B be the triangular board (or* staircase*) of shape* $(0,1,2,\ldots,m-1)$. *Then* $r_k = S(m, m-k)$.

Proof. We have that each $s_i = 0$. Hence, by Theorem 2.4.1,

$$x^m = \sum r_k \cdot (x)_{m-k}.$$

It follows from equation (1.94d) that $r_k = S(m, m-k)$. $\square$

A combinatorial proof of Corollary 2.4.2 is clearly desirable. We wish to associate a partition of $[m]$ into $m-k$ blocks with a placement of k nonattacking rooks on $B = \{(i,j) : 1 \le i \le m,\ 1 \le j < i\}$. If a rook occupies (i,j), then define i and j to be in the same block of the partition. It is easy to check that this procedure yields the desired correspondence. $\square$

2.4.3 Corollary. *Two Ferrers boards, each with m columns (allowing empty columns), have the same rook polynomial if and only if their multisets of the numbers s_i are the same.*

Corollary 2.4.3 suggests asking for the number of Ferrers boards with a rook polynomial equal to that of a given board B.

2.4.4 Theorem. *Let $0 \le c_1 \le \cdots \le c_m$, and let $f(c_1,\ldots,c_m)$ be the number of Ferrers boards with no empty columns and having the same rook polynomial as the Ferrers board of shape $(c_1,\ldots,c_m)$. Add enough intial 0's to $c_1,\ldots,c_m$ to get a shape $(b_1,\ldots,b_t) = (0,0,\ldots,0,c_1,\ldots,c_m)$ such that if $s_i = b_i - i + 1$, then $s_1 = 0$ and $s_i < 0$ for $2 \le i \le t$. Suppose that a_i of the s_j's are equal to $-i$, so $\sum_{i \ge 1} a_i = t - 1$. Then*

$$f(c_1,\ldots,c_m) = \binom{a_1 + a_2 - 1}{a_2}\binom{a_2 + a_3 - 1}{a_3}\binom{a_3 + a_4 - 1}{a_4}\cdots.$$

Proof (sketch). By Corollary 2.4.3, we seek the number of permutations $d_1 d_2 \cdots d_{t-1}$ of the multiset $\{1^{a_1}, 2^{a_2}, \ldots\}$ such that $0 \ge d_1 - 1 \ge d_2 - 2 \ge \cdots \ge d_{t-1} - t + 1$. Equivalently, $d_1 = 1$ and d_i must be followed by a number $d_{i+1} \le d_i + 1$. Place the a_1 1's down in a line. The a_2 2's may be placed arbitrarily in the a_1 spaces following each 1 in $\left(\binom{a_1}{a_2}\right) = \binom{a_1 + a_2 - 1}{a_2}$ ways. Now the a_3's may be placed arbitrarily in the a_2 spaces following each 2 in $\left(\binom{a_2}{a_3}\right) = \binom{a_2 + a_3 - 1}{a_3}$ ways, and so on, completing the proof. $\square$

For instance, there are no other Ferrers boards with the same rook polynomial as the triangular board $(0,1,\ldots,n-1)$, while there are 3^{n-1} Ferrers boards with the same rook polynomial as the $n \times n$ chessboard $[n] \times [n]$.

If in the proof of Theorem 2.4.4 we want all the columns of our Ferrers board to have distinct lengths, then we must arrange the multiset $\{1^{a_1}, 2^{a_2}, \ldots\}$ to first strictly increase from 1 to its maximum in unit steps and then to be non-increasing. Hence, we obtain the following result.

2.4.5 Corollary. *Let B be a Ferrers board. Then there is a unique Ferrers board whose columns have distinct (nonzero) lengths and that has the same rook polynomial as B.*

For instance, the unique "increasing" Ferrers board with the same rook polynomial as $[n] \times [n]$ has shape $(1, 3, 5, \ldots, 2n - 1)$.

2.5 V-Partitions and Unimodal Sequences

We now give an example of a sieve process that cannot be derived (except in a very contrived way) using the Principle of Inclusion-Exclusion. By a *unimodal sequence of weight n* (also called an *n-stack*), we mean a $\mathbb{P}$-sequence $d_1 d_2 \cdots d_m$ such that

a. $\sum d_i = n$.
b. For some j, we have $d_1 \leq d_2 \leq \cdots \leq d_j \geq d_{j+1} \geq \cdots \geq d_m$.

Many interesting combinatorial sequences turn out to be unimodal. (See Exercise 1.50 for some examples.) In this section, we shall be concerned not with any particular sequence, but rather with counting the total number $u(n)$ of unimodal sequences of weight n. By convention we set $u(0) = 0$. For instance, $u(5) = 15$, since all 16 compositions of 5 are unimodal except 212. Now set

$$U(q) = \sum_{n \geq 0} u(n)q^n$$

$$= q + 2q^2 + 4q^3 + 8q^4 + 15q^5 + 27q^6 + 47q^7 + 79q^8 + \cdots.$$

Our object is to find a nice expression for $U(q)$. Write $[j]! = (1-q)(1-q^2) \cdots (1 - q^j)$. It is easy to see that the number of unimodal sequences of weight n with largest term k is the coefficient of q^n in $q^k / [k-1]![k]!$. Hence,

$$U(q) = \sum_{k \geq 1} \frac{q^k}{[k-1]![k]!} \tag{2.26}$$

This is analogous to the formula

$$\sum_{n \geq 0} p(n)q^n = \sum_{k \geq 0} \frac{q^k}{[k]!},$$

where $p(n)$ is the number of partitions of n. (Put $x = 1$ in equation (1.82).) What we want, however, is an analogue of equation (1.77), which states that

$$\sum_{n \geq 0} p(n)q^n = \prod_{i \geq 1} (1 - q^i)^{-1}.$$

It turns out to be easier to work with objects slightly different from unimodal sequences, and then relate them to unimodal sequences at the end. We define a *V-partition of n* to be an $\mathbb{N}$-array

$$\begin{bmatrix} & a_1 & a_2 & \cdots \\ c & & & \\ & b_1 & b_2 & \cdots \end{bmatrix} \tag{2.27}$$

such that $c + \sum a_i + \sum b_i = n$, $c \geq a_1 \geq a_2 \geq \cdots$ and $c \geq b_1 \geq b_2 \geq \cdots$. Hence, a V-partition may be regarded as a unimodal sequence "rooted" at one of its largest parts. Let $v(n)$ be the number of V-partitions of n, with $v(0) = 1$. Thus, for instance $v(4) = 12$, since there is one way of rooting 4, one way for 13, one for 31, two for 22, one for 211, one for 112, and four for 1111. Set

$$V(q) = \sum_{n \geq 0} v(n) q^n$$

$$= 1 + q + 3q^2 + 6q^3 + 12q^4 + 21q^5 + 38q^6 + 63q^7 + 106q^8 + \cdots.$$

Analogously to (2.26), we have

$$V(q) = \sum_{k \geq 0} \frac{q^k}{[k]!^2},$$

but as before we want a product formula for $V(q)$.

Let V_n be the set of all V-partitions of n, and let D_n be the set of all *double partitions* of n, that is, $\mathbb{N}$-arrays

$$\begin{bmatrix} a_1 & a_2 & \cdots \\ b_1 & b_2 & \cdots \end{bmatrix} \tag{2.28}$$

such that $\sum a_i + \sum b_i = n$, $a_1 \geq a_2 \geq \cdots$ and $b_1 \geq b_2 \geq \cdots$. If $d(n) \# D_n$, then clearly

$$\sum_{n \geq 0} d(n) q^n = \prod_{i \geq 1} (1 - q^i)^{-2}. \tag{2.29}$$

Now define $\Gamma_1 : D_n \to V_n$ by

$$\Gamma_1 \begin{bmatrix} a_1 & a_2 & \cdots \\ b_1 & b_2 & \cdots \end{bmatrix} = \begin{cases} \begin{bmatrix} & a_2 & a_3 & \cdots \\ a_1 & & & \\ & b_1 & b_2 & \cdots \end{bmatrix}, & \text{if } a_1 \geq b_1, \\ \begin{bmatrix} & a_1 & a_2 & \cdots \\ b_1 & & & \\ & b_2 & b_3 & \cdots \end{bmatrix}, & \text{if } b_1 > a_1. \end{cases}$$

Clearly, Γ_1 is surjective, but it is not injective. Every V-partition in the set

$$V_n^1 = \left\{ \begin{bmatrix} & a_1 & a_2 & \cdots \\ c & & & \\ & b_1 & b_2 & \cdots \end{bmatrix} \in V_n : c > a_1 \right\}$$

appears twice as a value of Γ_1, so

$$\#V_n = \#D_n - \#V_n^1.$$

Next define $\Gamma_2 \colon D_{n-1} \to V_n^1$ by

$$\Gamma_2 \begin{bmatrix} a_1 & a_2 & \cdots \\ b_1 & b_2 & \cdots \end{bmatrix} = \begin{cases} \begin{bmatrix} & a_2 & a_3 & \cdots \\ a_1+1 & & & \\ & b_1 & b_2 & \cdots \end{bmatrix}, & \text{if } a_1+1 \geq b_1, \\[2ex] \begin{bmatrix} & a_1+1 & a_2 & \cdots \\ b_1 & & & \\ & b_2 & b_3 & \cdots \end{bmatrix}, & \text{if } b_1 > a_1+1. \end{cases}$$

Again Γ_2 is surjective, but every V-partition in the set

$$V_n^2 = \left\{ \begin{bmatrix} & a_1 & a_2 & \cdots \\ c & & & \\ & b_1 & b_2 & \cdots \end{bmatrix} \in V_n : c > a_1 > a_2 \right\}$$

appears twice as a value of Γ_2. Hence, $\#V_n^1 = \#D_{n-1} - \#V_n^2$, so

$$\#V_n = \#D_n - \#D_{n-1} + \#V_n^2.$$

Next define $\Gamma_3 \colon D_{n-3} \to V_n^2$ by

$$\Gamma_3 \begin{bmatrix} a_1 & a_2 & \cdots \\ b_1 & b_2 & \cdots \end{bmatrix} = \begin{cases} \begin{bmatrix} & a_2+1 & a_3 & a_4 & \cdots \\ a_1+2 & & & & \\ & b_1 & b_2 & b_3 & \cdots \end{bmatrix}, & \text{if } a_1+2 \geq b_1, \\[2ex] \begin{bmatrix} & a_1+2 & a_2+1 & a_3 & \cdots \\ b_1 & & & & \\ & b_2 & b_3 & b_4 & \cdots \end{bmatrix}, & \text{if } b_1 > a_1+2. \end{cases}$$

We obtain

$$\#V_n = \#D_n - \#D_{n-1} + \#D_{n-3} - \#V_n^3,$$

where

$$V_n^3 = \left\{ \begin{bmatrix} & a_1 & a_2 & \cdots \\ c & & & \\ & b_1 & b_2 & \cdots \end{bmatrix} \in V_n : c > a_1 > a_2 > a_3 \right\}.$$

Continuing this process, we obtain maps $\Gamma_i \colon D_{n-\binom{i}{2}} \to V_n^{i-1}$. The process stops when $\binom{i}{2} > n$, so we obtain the sieve-theoretic formula

$$v(n) = d(n) - d(n-1) + d(n-3) - d(n-6) + \cdots,$$

where we set $d(m) = 0$ for $m < 0$. Thus, using equation (2.29), we obtain the following result.

2.5.1 Proposition. *We have*

$$V(q) = \left(\sum_{n \geq 0} (-1)^n q^{\binom{n+1}{2}} \right) \prod_{i \geq 1} (1-q^i)^{-2}.$$

We can now obtain an expression for $U(q)$ using the following result.

2.5.2 Proposition. *We have*

$$U(q) + V(q) = \prod_{i \geq 1} (1-q^i)^{-2}.$$

Proof. Let U_n be the set of all unimodal sequences of weight n. We need to find a bijection $D_n \to U_n \cup V_n$. Such a bijection is given by

$$
\begin{bmatrix} a_1 & a_2 & \cdots \\ b_1 & b_2 & \cdots \end{bmatrix} \mapsto
\begin{cases}
\begin{bmatrix} & a_2 & a_3 & \cdots \\ a_1 & & & \\ & b_1 & b_2 & \cdots \end{bmatrix}, & \text{if } a_1 \geq b_1 \\
\cdots a_2\, a_1\, b_1\, b_2 \cdots, & \text{if } b_1 > a_1
\end{cases}
$$

$\square$

2.5.3 Corollary. *We have*

$$
U(q) = \left(\sum_{n \geq 1} (-1)^{n-1} q^{\binom{n+1}{2}} \right) \prod_{i \geq 1} (1 - q^i)^{-2}.
$$

2.6 Involutions

Recall now the viewpoint of Section 1.1 that the best way to determine that two finite sets have the same cardinality is to exhibit a bijection between them. We will show how to apply this principle to the identity (2.5). (The seemingly more general (2.4) is done exactly the same way.) As it stands this identity does not assert that two sets have the same cardinality. Therefore, we rearrange terms so that all signs are positive. Thus, we wish to prove the identity

$$
f_=(\emptyset) + \sum_{\#Y \text{ odd}} f_{\geq}(Y) = \sum_{\#Y \text{ even}} f_{\geq}(Y), \tag{2.30}
$$

where $f_=(Y)$ (respectively, $f_{\geq}(Y)$) denotes the number of objects in a set A having exactly (respectively, at least) the properties in $T \subseteq S$. The left-hand side of (2.30) is the cardinality of the set $M \cup N$, where M is the set of objects x having none of the properties in S, and N is the set of ordered triples (x, Y, Z), where $x \in A$ has exactly the properties $Z \supseteq Y$ with $\#Y$ odd. The right-hand side of (2.30) is the cardinality of the set N' of ordered triples (x', Y', Z'), where $x' \in A$ has exactly the properties $Z' \supseteq Y'$ with $\#Y'$ even. Totally order the set S of properties, and define $\sigma: M \cup N \to N'$ as follows:

$$
\sigma(x) = (x, \emptyset, \emptyset), \text{ if } x \in M
$$

$$
\sigma(x, Y, Z) =
\begin{cases}
(x, Y - i, Z), & \text{if } (x, Y, Z) \in N \\
& \text{and } \min Y = \min Z = i, \\[2ex]
(x, Y \cup i, Z), & \text{if } (x, Y, Z) \in N \\
& \text{and } \min Z = i < \min Y.
\end{cases}
$$

It is easily seen that σ is a bijection with inverse

$$\sigma^{-1}(x,Y,Z) = \begin{cases} x \in M, & \text{if } Y = Z = \emptyset, \\[2mm] (x, Y - i, Z) \in N, & \text{if } Y \neq \emptyset \\ & \text{and } \min Y = \min Z = i, \\[2mm] (x, Y \cup i, Z) \in N, & \text{if } Z \neq \emptyset \text{ and} \\ & \min Z = i < \min Y \\ & (\text{where we set } \min Y = \infty \text{ if } Y = \emptyset). \end{cases}$$

This construction yields the desired bijective proof of (2.30).

Note that if in the definition of σ^{-1} we identify $x \in M$ with $(x, \emptyset, \emptyset) \in N'$ (so $\sigma^{-1}(x, \emptyset, \emptyset) = (x, \emptyset, \emptyset)$), then $\sigma \cup \sigma^{-1}$ is a function $\tau : N \cup N' \to N \cup N'$ satisfying: (a) τ is an *involution*; that is, $\tau^2 = \mathrm{id}$; (b) the fixed points of τ are the triples $(x, \emptyset, \emptyset)$, so are in one-to-one correspondence with M; and (c) if (x, Y, Z) is not a fixed point of τ and we set $\tau(x, Y, Z) = (x, Y', Z')$, then

$$(-1)^{\#Y} + (-1)^{\#Y'} = 0.$$

Thus, the involution τ selects terms from the right-hand side of (2.5) (or rather, terms from the right-hand side of (2.5) after each $f_{\geq}(Y)$ is written as a sum (2.3)) that add up to the left-hand side, and then τ cancels out the remaining terms.

We can put the preceding discussion in the following general context. Suppose that the finite set X is written as a disjoint union $X^+ \cup X^-$ of two subsets X^+ and X^-, called the "positive" and "negative" parts of X, respectively. Let τ be an involution on X that satisfies:

a. If $\tau(x) = y$ and $x \neq y$, then exactly one of x, y belongs to X^+ (so the other belongs to X^-).
b. If $\tau(x) = x$ then $x \in X^+$.

If we define a weight function w on X by

$$w(x) = \begin{cases} 1, & x \in X^+, \\ -1, & x \in X^-, \end{cases}$$

then clearly

$$\#\mathrm{Fix}(\tau) = \sum_{x \in X} w(x), \tag{2.31}$$

where $\mathrm{Fix}(\tau)$ denotes the fixed point set of τ. Just as in the previous paragraph, the involution τ has selected terms from the right-hand side of (2.31) which add up to the left-hand side, and has cancelled the remaining terms.

We now consider a more complicated situation. Suppose that we have another set $\widetilde{X}$ that is also expressed as a disjoint union $\widetilde{X} = \widetilde{X}^+ \cup \widetilde{X}^-$, and an involution $\widetilde{\tau}$ on $\widetilde{X}$ satisfying (a) and (b) Suppose that we also are given a sign-preserving bijection $f : X \to \widetilde{X}$, that is, $f(X^+) = \widetilde{X}^+$ and $f(X^-) = \widetilde{X}^-$. Clearly, then

#Fix(τ) = #Fix$(\tilde{\tau})$, since #Fix(τ) = #X^+ − #X^- and #Fix$(\tilde{\tau})$ = #$\tilde{X}^+$ − #$\tilde{X}^-$. We wish to construct in a canonical way a bijection g between Fix(τ) and Fix$(\tilde{\tau})$. This construction is known as the *involution principle* and is a powerful technique for converting noncombinatorial proofs into combinatorial ones.

The bijection $g\colon$ Fix$(\tau) \to$ Fix$(\tilde{\tau})$ is defined as follows. Let $x \in$ Fix(τ). It is easily seen, since X is finite, that there is a nonnegative integer n for which

$$f(\tau f^{-1}\tilde{\tau} f)^n(x) \in \text{Fix}(\tilde{\tau}). \qquad (2.32)$$

Define $g(x)$ to be $f(\tau f^{-1}\tilde{\tau} f)^n(x)$ where n is the *least* nonnegative integer for which (2.32) holds.

We leave it to the reader to verify rigorously that g is a bijection from Fix(τ) to Fix$(\tilde{\tau})$. There is, however, a nice geometric way to visualize the situation. Represent the elements of X and $\tilde{X}$ as vertices of a graph Γ. Draw an undirected edge between two distinct vertices x and y if (1) $x, y \in X$ and $\tau(x) = y$; or (2) $x, y \in \tilde{X}$ and $\tilde{\tau}(x) = y$; or (3) $x \in X$, $y \in \tilde{X}$, and $f(x) = y$. Every component of Γ will then be either a cycle disjoint from Fix(τ) and Fix$(\tilde{\tau})$, or a path with one endpoint z in Fix(τ) and the other endpoint $\tilde{z}$ in Fix$(\tilde{\tau})$. Then g is defined by $g(z) = \tilde{z}$. See Figure 2.2.

There is a variation of the involution principle that is concerned with "sieve-equivalence." We will mention only the simplest case here; see Exercise 2.36 for further development. Suppose that X and $\tilde{X}$ are (disjoint) finite sets. Let $Y \subseteq X$ and $\tilde{Y} \subseteq \tilde{X}$, and suppose that we are given bijections $f\colon X \to \tilde{X}$ and $g\colon Y \to \tilde{Y}$. Hence #$(X - Y)$ = #$(\tilde{X} - \tilde{Y})$, and we wish to construct an explicit bijection h between $X - Y$ and $\tilde{X} - \tilde{Y}$. Pick $x \in X - Y$. As in equation (2.32), there will be a nonnegative integer n for which

$$f(g^{-1}f)^n(x) \in \tilde{X} - \tilde{Y}. \qquad (2.33)$$

In this case, n is unique since if $x \in \tilde{X} - \tilde{Y}$, then $g^{-1}(y)$ is undefined. Define $h(x)$ to be $f(g^{-1}f)^n(x)$ where n satisfies (2.33). One easily checks that $h\colon X - Y \to \tilde{X} - \tilde{Y}$ is a bijection.

Let us consider a simple example of the bijection $h\colon X - Y \to \tilde{X} - \tilde{Y}$.

2.6.1 Example. Let Y be the set of all permutations $w \in \mathfrak{S}_n$ that fix 1, that is, $w(1) = 1$. Let $\tilde{Y}$ be the set of all permutations $w \in \mathfrak{S}_n$ with exactly one cycle.

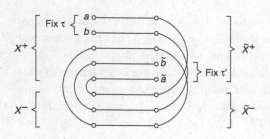

Figure 2.2 An illustration of the involution principle.

```
123 ─────────────── (1)(2)(3)       Figure 2.3 The bijection h: 𝔖_n − Y → 𝔖_n − Ỹ.
132 ─────────────── (1)(23)
213 ─────────────── (12)(3)
231 ─────────────── (123)
312 ─────────────── (132)
321 ─────────────── (13)(2)
```

Thus, $\#Y = \#\widetilde{Y} = (n-1)!$, so

$$\#(\mathfrak{S}_n - Y) = \#(\mathfrak{S}_n - \widetilde{Y}) = n! - (n-1)!.$$

It may not be readily apparent, however, how to construct a bijection h between $\mathfrak{S}_n - Y$ and $\mathfrak{S}_n - \widetilde{Y}$. On the other hand, it is easy to construct a bijection g between Y and $\widetilde{Y}$; namely, if $w = 1a_2 \cdots a_n \in Y$ (where w is written as a word, i.e., $w(i) = a_i$), then set $g(w) = (1, a_2, \ldots, a_n)$ (written as a cycle). Of course we choose the bijection $f: \mathfrak{S}_n \to \mathfrak{S}_n$ to be the identity. Then equation (2.33) defines the bijection $h: \mathfrak{S}_n - Y \to \mathfrak{S}_n - \widetilde{Y}$. For example, when $n = 3$ we depict f by solid lines and g by broken lines in Figure 2.3. Hence (writing permutations in the domain as words and in the range as products of cycles),

$$h(213) = (12)(3)$$

$$h(231) = (1)(2)(3)$$

$$h(312) = (1)(23)$$

$$h(321) = (13)(2).$$

It is natural to ask here (and in other uses of the involution and related principles) whether there is a more direct description of h. In this example there is little difficulty because Y and $\widetilde{Y}$ are *disjoint* subsets (when $n \geq 2$) of the same set $\mathfrak{S}_n$. This special situation yields

$$h(w) = \begin{cases} w, & \text{if } w \notin \widetilde{Y}, \\ g^{-1}(w), & \text{if } w \in \widetilde{Y}. \end{cases} \tag{2.34}$$

2.7 Determinants

In Proposition 2.2.6 we saw that a determinant $\det[a_{ij}]_0^n$, with $a_{ij} = 0$ if $j < i - 1$, can be interpreted combinatorially using the Principle of Inclusion-Exclusion. In this section we will consider the combinatorial significance of arbitrary determinants, by setting up a combinatorial problem in which the right-hand side of equation (2.31) is the expansion of a determinant.

We will consider lattice paths $L = (v_0, v_1, \ldots, v_k)$ in $\mathbb{N}^2$, as defined in Section 1.2, with steps $v_i - v_{i-1} = (1, 0)$ or $(0, -1)$. We picture L by drawing an edge between v_{i-1} and v_i, $1 \leq i \leq k$. For instance, the lattice path

y

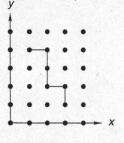

Figure 2.4 A lattice path in $\mathbb{N}^2$.

x

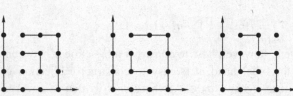

Figure 2.5 Three non-intersecting 2-paths.

$((1,4),(2,4),(2,3),(2,2),(3,2),(3,1))$ is drawn in Figure 2.4. An *n-path* is an n-tuple $L = (L_1,\ldots,L_n)$ of lattice paths. Let $\alpha,\beta,\gamma,\delta \in \mathbb{N}^n$. Then L is of *type* $(\alpha,\beta,\gamma,\delta)$ if L_i goes from (β_i,γ_i) to (α_i,δ_i). (Clearly then, $\alpha_i \geq \beta_i$ and $\gamma_i \geq \delta_i$.) The n-path L is *intersecting* if for some $i \neq j$ L_i and L_j have a point in common; otherwise, L is *nonintersecting*. Define the *weight* of a horizontal step from (i,j) to $(i+1,j)$ to be the indeterminate x_j, and the weight $\Lambda(L)$ of L to be the product of the weights of its horizontal steps. For instance, the path in Figure 2.4 has weight x_2x_4.

If $\alpha = (\alpha_1,\ldots,\alpha_n) \in \mathbb{N}^n$ and $w \in \mathfrak{S}_n$, then let $w(\alpha) = (\alpha_{w(1)},\ldots,\alpha_{w(n)})$. Let $\mathcal{A} = \mathcal{A}(\alpha,\beta,\gamma,\delta)$ be the set of all n-paths of type $(\alpha,\beta,\gamma,\delta)$, and let $A = A(\alpha,\beta,\gamma,\delta)$ be the sum of their weights. Consider a path from (β_i,γ_i) to (α_i,δ_i). Let $m = \alpha_i - \beta_i$. For each j satisfying $1 \leq j \leq m$, there is exactly one horizontal step of the form $(j-1+\beta_i,k_j) \to (j+\beta_i,k_j)$. The numbers $k_1,\ldots,k_m$ can be chosen arbitrarily provided

$$\gamma_i \geq k_1 \geq k_2 \geq \cdots \geq k_m \geq \delta_i. \tag{2.35}$$

Hence, if we define

$$h(m;\gamma_i,\delta_i) = \sum x_{k_1}x_{k_2}\cdots x_{k_m},$$

summed over all integer sequences (2.35), then

$$A(\alpha,\beta,\gamma,\delta) = \prod_{i=1}^{n} h(\alpha_i - \beta_i;\gamma_i,\delta_i). \tag{2.36}$$

(In the terminology of Section 7.4, $h(m;\gamma_i,\delta_i)$ is the complete homogeneous symmetric function $h_m(x_{\delta_i},x_{\delta_i+1},\ldots,x_{\gamma_i})$.)

Now let $\mathcal{B} = \mathcal{B}(\alpha,\beta,\gamma,\delta)$ be the set of all nonintersecting n-paths of type $(\alpha,\beta,\gamma,\delta)$, and let $B = B(\alpha,\beta,\gamma,\delta)$ be the sum of their weights. For instance, let $\alpha = (2,3)$, $\beta = (1,1)$, $\gamma = (2,3)$, $\delta = (1,0)$. Then $B(\alpha,\beta,\gamma,\delta) = x_2x_3^2 + x_1x_3^2 + x_1x_2x_3$, corresponding to the nonintersecting 2-paths shown in Figure 2.5.

2.7.1 Theorem. *Let* $\alpha, \beta, \gamma, \delta \in \mathbb{N}^n$ *such that for* $w \in \mathfrak{S}_n$, $\mathcal{B}(w(\alpha), \beta, \gamma, w(\delta))$ *is empty unless* w *is the identity permutation. (For example, this condition occurs if* $\alpha_i < \alpha_{i+1}$, $\beta_i < \beta_{i+1}$, $\gamma_i \le \gamma_{i+1}$, *and* $\delta_i \le \delta_{i+1}$ *for* $1 \le i \le n - 1$.) *Then*

$$B(\alpha, \beta, \gamma, \delta) = \det[h(\alpha_j - \beta_i; \gamma_i, \delta_j)]_1^n, \qquad (2.37)$$

where we set $h(\alpha_j - \beta_i; \gamma_i, \delta_j) = 0$ *whenever there are no sequences* (2.35).

Proof. When we expand the right-hand side of equation (2.37), we obtain

$$\sum_{w \in \mathfrak{S}_n} (\operatorname{sgn} w) A(w(\alpha), \beta, \gamma, w(\delta)). \qquad (2.38)$$

Let $\mathcal{A}_w = \mathcal{A}(w(\alpha), \beta, \gamma, w(\delta))$. We will construct a bijection $L \to L^*$ from $\left(\bigcup_{w \in \mathfrak{S}_n} \mathcal{A}_w \right) - \mathcal{B}$ to itself satisfying:

a. $L^{**} = L$; that is, $*$ is an involution.
b. $\Lambda(L^*) = \Lambda(L)$, that is, $*$ is weight-preserving.
c. If $L \in \mathcal{A}_u$ and $L^* \in \mathcal{A}_v$ then $\operatorname{sgn} u = -\operatorname{sgn} v$.

Then by grouping together terms of (2.38) corresponding to pairs (L, L^*) of intersecting n-paths, we see that all terms cancel except for those producing the desired result $B(\alpha, \beta, \gamma, \delta)$.

To construct the involution $*$, let L be an intersecting n-path. We need to single out some canonically defined pair (L_i, L_j) of paths from L that intersect, and then some canonically defined intersection point (x, y) of these paths. One of many ways to do this is the following. Let i be the least integer for which L_i and L_k intersect for some $k \neq i$, and let x be the least integer such that L_i intersects some L_k with $k > i$ at a point (x, y), and then of all such k let j be the minimum. Construct L_i^* by following L_i to its first intersection point $v = (x, y)$ with L_j, and then following L_j to the end. Construct L_j^* similarly by following L_j to v and then L_i to the end. For $k \neq i, j$ let $L_k^* = L_k$.

Property (a) follows since the triple (i, j, v) can be obtained from L^* by the same rule that L^* is obtained from L. Property (b) is immediate since the totality of single steps in L and L^* is identical. Finally, v is obtained from u by multiplication by the transposition (i, j), so (c) follows. □

Theorem 2.7.1 has important applications in the theory of symmetric functions (see the first proof of Theorem 7.16.1), but let us be content here with a simple example of its use.

2.7.2 Example. Let $r, s \in \mathbb{N}$ and let S be a subset of $[0, r] \times [0, s]$. How many lattice paths are there between $(0, r)$ and $(s, 0)$ that don't intersect S? Call this number $f(r, s, S)$. Let $S = \{(a_1, b_1), \dots, (a_k, b_k)\}$, and set

$$\alpha = (s, a_1, \dots, a_k), \quad \beta = (0, a_1, \dots, a_k),$$

$$\gamma = (r, b_1, \dots, b_k), \quad \delta = (0, b_1, \dots, b_k).$$

Then $f(r,s,S) = B(\alpha,\beta,\gamma,\delta)$, where we set each weight $x_m = 1$. Now

$$h(\alpha_j - \beta_i; \gamma_i, \delta_j)\big|_{x_m=1} = \binom{\alpha_j + \gamma_i - \beta_i - \delta_j}{\alpha_j - \beta_i}.$$

Hence by Theorem 2.7.1,

$$f(r,s,S) = \begin{vmatrix} \binom{r+s}{r} & \binom{r+a_1-b_1}{a_1} & \cdots & \binom{r+a_k-b_k}{a_k} \\[2mm] \binom{s-a_1+b_1}{s-a_1} & 1 & \cdots & \binom{a_k-b_k-a_1+b_1}{a_k-a_1} \\[2mm] & & \vdots & \\[2mm] \binom{s-a_k+b_k}{s-a_k} & \binom{a_1-b_1-a_k+b_k}{a_1-a_k} & \cdots & 1 \end{vmatrix},$$

where we set $\binom{i}{j} = 0$ if $j < 0$ or $i - j < 0$. When we expand this determinant we obtain a formula for $f(r,s,S)$ that can also be deduced directly from the Principle of Inclusion-Exclusion. Indeed, by a suitable permutation of rows and columns the above expression for $f(r,s,S)$ becomes a special case of Proposition 2.2.6. (In its full generality, however, Theorem 2.7.1 cannot be deduced from Proposition 2.2.6; indeed, the determinant (2.37) will in general have no zero entries.)

Notes

As P. Stein says in his valuable monograph [1.71], the Principle of Inclusion-Exclusion "is doubtless very old; its origin is probably untraceable." An extensive list of references is given by Takács [2.22], and exact citations for results listed here without reference may be found there. In probabilistic form, the Principle of Inclusion-Exclusion can be traced back to A. de Moivre and less clearly to J. Bernoulli, and is sometimes referred to as "Poincaré's theorem." The first statement in combinatorial terms may be due to C. P. da Silva and is sometimes attributed to Sylvester.

Example 2.2.1 (the derangement problem) was first solved by P. R. de Montmort (in probabilistic terms) and later independently investigated by Euler.

Example 2.2.4 (enumeration of permutations by descent set) was first obtained by MacMahon [1.55, vol. 1, p. 190] and has been rediscovered several times since. Example 2.2.5 first appears in Stanley [2.21]. The problème des ménages (or *menage problem*) (Example 2.3.3) was suggested by Tait to Cayley and Muir, but they did not reach a definitive answer. The problem was independently considered by Lucas and solved by him in a rather unsatisfactory form. The elegant formula given in Corollary 2.3.5 is due to Touchard. For references to more recent work see Comtet [2.3, p. 185] and Dutka [2.4]. The theory of rook polynomials in general is due to Kaplansky and Riordan [2.13]; see Riordan [2.17, Chs. 7–8]. Ferrers boards were first considered by D. Foata and M.-P. Schützenberger [2.5] and developed further by Goldman, Joichi, and White [2.8]–[2.11]. The proof given here of Theorem 2.4.4 was suggested by P. Leroux. There have been many further developments in the area of rook theory; see for instance Sjöstrand [2.18] and the references given there. The results of Section 2.5 first appeared in Stanley [2.19, Ch. IV.3] and were restated in [2.20, §23].

The involution principle was first stated by Garsia and Milne [2.6], where it was used to give a long-sought-for combinatorial proof of the Rogers-Ramanujan identities. (See Pak [1.62, §7] for more information.) For further discussion of the involution principle, sieve equivalence, and related results, see Cohen [2.2], Gordon [2.12], and Wilf [2.23]. The combinatorial proof of the Principle of Inclusion-Exclusion given in Section 2.6 appears implicitly in Remmel [2.16] and is made more explicit in Zeilberger [2.24]. Theorem 2.7.1 and its proof are anticipated by Chaundy [2.1], Karlin and McGregor [2.14], and Lindström [2.15], though the first explicit statement appears in a paper of Gessel and Viennot [2.7]. It was independently rediscovered several times since the paper of Gessel and Viennot. Our presentation closely follows that of Gessel and Viennot.

Bibliography

[1] T. W. Chaundy, Partition-generating functions, *Quart. J. Math. (Oxford)* **2** (1931), 234–240.

[2] D. I. A. Cohen, PIE-sums: A combinatorial tool for partition theory, *J. Combinatorial Theory, Ser. A* **31** (1981), 223–236.

[3] L. Comtet, *Advanced Combinatorics*, Reidel, Boston, 1974.

[4] J. Dutka, On the problème des ménages, *Math. Intelligencer* **8** (1986), no. 3, 18–25 and 33.

[5] D. Foata and M.-P. Schützenberger, On the rook polynomials of Ferrers relations, *Colloquia Mathematica Societatis Jano Bolyai, 4*, Combinatorial Theory and Its Applications, vol. 2 (P. Erdős, A. Renyi, and V. Sós, eds.), North-Holland, Amsterdam, 1970, pp. 413–436.

[6] A. M. Garsia and S. C. Milne, A Rogers-Ramanujan bijection, *J. Combinatorial Theory, Ser. A* **31** (1981), 289–339.

[7] I. M. Gessel and X. G. Viennot, Binomial determinants, paths, and hook-length formulas, *Advances in Math.* **58** (1985), 300–321.

[8] J. Goldman, J. Joichi, and D. White, Rook theory I: Rook equivalence of Ferrers boards, *Proc. Amer. Math. Soc.* **52** (1975), 485–492.

[9] J. Goldman, J. Joichi, and D. White, Rook polynomials, Möbius inversion, and the umbral calculus, *J. Combinatorial Theory, Ser. A* **21** (1976), 230–239.

[10] J. Goldman, J. Joichi, and D. White, Rook theory IV: Orthogonal sequences of rook polynomials, *Studies in Applied Math.* **56** (1977), 267–272.

[11] J. Goldman, J. Joichi, and D. White, Rook theory III: Rook polynomials and the chromatic structure of graphs, *J. Combinatorial Theory Ser. B* **25** (1978), 135–142.

[12] B. Gordon, Sieve-equivalence and explicit bijections, *J. Combinatorial Theory Ser. A* **34** (1983), 90–93.

[13] I. Kaplansky and J. Riordan, The problem of the rooks and its applications, *Duke Math. J.* **13** (1946), 259–268.

[14] S. Karlin and G. McGregor, Coincidence probabilities, *Pacific J. Math.* **9** (1959), 1141–1164.

[15] B. Lindström, On the vector representation of induced matroids, *Bull. London Math. Soc.* **5** (1973), 85–90.

[16] J. Remmel, Bijective proofs of some classical partition identities, *J. Combinatorial Theory Ser. A* **33** (1982), 273–286.

[17] J. Riordan, *An Introduction to Combinatorial Analysis*, Wiley, New York, 1958.

[18] J. Sjöstrand, Bruhat intervals as rooks on skew Ferrers boards, *J. Combinatorial Theory Ser. A* **114** (2007), 1182–1198.

[19] R. Stanley, *Ordered structures and partitions*, Ph.D. thesis, Harvard University, 1971.

[20] R. Stanley, Ordered structures and partitions, *Mem. Amer. Math. Soc.* **119** (1972), iii+104 pages.

[21] R. Stanley, Binomial posets, Möbius inversion, and permutation enumeration, *J. Combinatorial Theory* **20** (1976), 336–356.

[22] L. Takács, On the method of inclusion and exclusion, *J. Amer. Stat. Soc.* **62** (1967), 102–113.

[23] H. S. Wilf, Sieve-equivalence in generalized partition theory, *J. Combinatorial Theory, Ser. A* **34** (1983), 80–89.

[24] D. Zeilberger, Garsia and Milne's bijective proof of the inclusion-exclusion principle, *Discrete Math.* **51** (1984), 109–110.

Exercises for Chapter 2

1. [3] Explain why the Principle of Inclusion-Exclusion has the numerical value

$$8.539734222267356706546\cdots.$$

2. [2–]* Give a bijective proof of equation (2.24) or (2.25), that is,

$$\sum_X f_=(X)x^{\#X} = \sum_Y f_\ge(Y)(x-1)^{\#Y}$$

or

$$\sum_X f_=(X)(y+1)^{\#X} = \sum_Y f_\ge(Y)y^{\#Y}.$$

3. [2] Let $S = \{P_1,\dots,P_n\}$ be a set of properties, and let f_k (respectively, $f_{\ge k}$) denote the number of objects in a finite set A that have *exactly* k (respectively, *at least* k of the properties). Show that

$$f_k = \sum_{i=k}^n (-1)^{i-k} \binom{i}{k} g_i, \qquad (2.39)$$

and

$$f_{\ge k} = \sum_{i=k}^n (-1)^{i-k} \binom{i-1}{k-1} g_i, \qquad (2.40)$$

where

$$g_i = \sum_{\substack{T \subseteq S \\ \#T=i}} f_\ge(T).$$

4. **a.** [2] Let $A_1,\dots,A_n$ be subsets of a finite set A, and define S_k, $0 \le k \le n$, by (2.6). Show that

$$S_k - S_{k+1} + \cdots + (-1)^{n-k} S_n \ge 0, \quad 0 \le k \le n. \qquad (2.41)$$

b. [2+] Find necessary and sufficient conditions on a vector $(S_0, S_1,\dots,S_n) \in \mathbb{N}^{n+1}$ so that there exist subsets $A_1,\dots,A_n$ of a finite set A satisfying (2.6).

5. a. [2] Let

$$0 \to V_n \xrightarrow{\partial_n} V_{n-1} \xrightarrow{\partial_{n-1}} \cdots \xrightarrow{\partial_1} V_0 \xrightarrow{\partial_0} W \to 0 \qquad (2.42)$$

be an exact sequence of finite-dimensional vector spaces over some field; that is, the ∂_j's are linear transformations satisfying $\operatorname{im} \partial_{j+1} = \ker \partial_j$ (with ∂_n injective and ∂_0 surjective). Show that

$$\dim W = \sum_{i=0}^{n} (-1)^i \dim V_i. \qquad (2.43)$$

b. [2] Show that for $0 \le j \le n$,

$$\operatorname{rank} \partial_j = \sum_{i=j}^{n} (-1)^{i-j} \dim V_i, \qquad (2.44)$$

so in particular the quantity on the right-hand side is nonnegative.

c. [2] Suppose that we are given only that equation (2.42) is a *complex*; that is, $\partial_j \partial_{j+1} = 0$ for $0 \le j \le n-1$, or equivalently $\operatorname{im} \partial_{j+1} \subseteq \ker \partial_j$. Show that if equation (2.44) holds for $0 \le j \le n$, then (2.42) is exact.

d. [2+] Let $A_1, \ldots, A_n$ be subsets of a finite set A, and for $T \subseteq [n]$ set $A_T = \bigcap_{i \in T} A_i$. In particular, $A_\emptyset = A$. Let V_T be the vector space (over some field) with a basis consisting of all symbols $[a, T]$ where $a \in A_T$. Set $V_j = \bigoplus_{\#T=j} V_T$, and define for $1 \le i \le n$ linear transformations $\partial_j \colon V_j \to V_{j-1}$ by

$$\partial_j [a, T] = \sum_{i=1}^{j} (-1)^{i-1} [a, T - t_i], \qquad (2.45)$$

where the elements of T are $t_1 < \cdots < t_j$. Also, define W to be the vector space with basis $\{[a] : a \in \bar{A}_1 \cap \cdots \cap \bar{A}_n\}$, and define $\partial_0 \colon V_0 \to W$ by

$$\partial_0 [a, \emptyset] = \begin{cases} [a], & \text{if } a \in \bar{A}_1 \cap \cdots \cap \bar{A}_n, \\ 0, & \text{otherwise.} \end{cases}$$

(Here $\bar{A}_i = A - A_i$.) Show that (2.42) is an exact sequence.

e. [1+] Deduce equation (2.7) from (a) and (d).

f. [1+] Deduce Exercise 2.4(a) from (b) and (d).

6. In this exercise, we consider a *multiset* generalization of the Principle of Inclusion-Exclusion.

a. [2] Let N be a finite multiset, say $N = \{x_1^{a_1}, \ldots, x_k^{a_k}\}$. For each $1 \le r \le k$ and $1 \le i \le a_r$, let P_{ir} be some property that each of the elements of a set A may or may not have, with the condition that if $1 \le i \le j \le a_r$ then any object with property P_{jr} also has P_{ir}. (For instance, if A is a set of integers, then P_{ir} could be the property of being divisible by r^i.) For every submultiset $M \subseteq N$, let $f_=(M)$ be the number of objects in A with *exactly* the properties in M; in other words, if $M = \{x_1^{b_1}, \ldots, x_k^{b_k}\}$, then $f_=(M)$ counts those objects in A that have property $P_{b_r,r}$ but fail to have $P_{b_r+1,r}$ for $1 \le r \le k$. Similarly, define $f_\ge(M)$ so

$$f_\ge(M) = \sum_{Y \supseteq M} f_=(Y). \qquad (2.46)$$

Show that

$$f_=(M) = \sum_{\substack{Y \supseteq M \\ Y-M \text{ is a set}}} (-1)^{\#(Y-M)} f_\ge(Y). \qquad (2.47)$$

Dually, if

$$f_\le(M) = \sum_{Y \subseteq M} f_=(Y),$$ (2.48)

then

$$f_=(M) = \sum_{\substack{Y \subseteq M \\ M-Y \text{ is a set}}} (-1)^{\#(M-Y)} f_\le(Y).$$ (2.49)

b. [2] Suppose that we encode the multiset $N = \{x_1^{a_1}, \ldots, x_k^{a_k}\}$ by the integer $n = p_1^{a_1} \cdots p_k^{a_k}$, where $p_1, \ldots, p_k$ are distinct primes. Thus, submultisets M of N correspond to (positive) divisors d of n. What do equations (2.48) and (2.49) become in this setting?

7. [2] Fix a prime power q. Prove equation (1.103), namely, the number $\beta(n)$ of monic irreducible polynomials of degree n over the field $\mathbb{F}_q$ is given by

$$\beta(n) = \frac{1}{n} \sum_{d|n} \mu(d) q^{n/d}.$$

(Use Exercise 2.6(b).)

8. **a.** [3–] Give a direct combinatorial proof of equation (2.13); that is,

$$D(n) = nD(n-1) + (-1)^n.$$

b. [2] Let $\mathcal{E}(n)$ denote the set of permutations $w \in \mathfrak{S}_n$ whose first ascent is in an even position (where we always count n as an ascent). For instance, $\mathcal{E}(3) = \{213, 312\}$, and $\mathcal{E}(4) = \{2134, 2143, 3124, 3142, 3241, 4123, 4132, 4231, 4321\}$. Set $E(n) = \#\mathcal{E}(n)$. Show that $E(n) = nE(n-1) + (-1)^n$. Hence (since $E(1) = D(1) = 0$), we have $E(n) = D(n)$.

c. [2+] Give a bijection between the permutations being counted by $E(n)$ and the derangements of $[n]$.

9. [2–] Prove the formula $\Delta^k 0^d = k! S(d,k)$ of Proposition 1.9.2(c) (equivalent to equation (1.94a)) using the Principle of Inclusion-Exclusion.

10. **a.** [1+]* How many functions $f: [n] \to [n]$ have no fixed points?

b. [2] Let $E(n)$ be the number obtained in (a). Show that $\lim_{n\to\infty} E(n)/n! = 1/e$, the same as $\lim_{n\to\infty} D(n)/n!$ (Example 2.2.1). Which of $D(n)/n!$ and $E(n)/n!$ gives the better approximation to $1/e$?

11. [3–] Let $a_1, \ldots, a_k$ be positive integers with $\sum a_i = n$. Let $S = \{a_1, a_1 + a_2, \ldots, a_1 + a_2 + \cdots + a_{k-1}\}$. Show that the number of derangements in $\mathfrak{S}_n$ with descent set S is the coefficient of $x_1^{a_1} \cdots x_k^{a_k}$ in the expansion of

$$\frac{1}{(1+x_1)\cdots(1+x_k)(1-x_1-\cdots-x_k)}.$$

12. [2+] Let $\alpha = (\alpha_1, \ldots, \alpha_k) \in \mathbb{N}^k$, and let M_α be the multiset $\{1^{\alpha_1}, \ldots, k^{\alpha_k}\}$. A *derangement* of M_α is a permutation $a_1 a_2 \cdots a_n$ (where $n = \sum \alpha_i$) of M_α that disagrees in every position with the permutation we get by listing the elements of M in weakly increasing order. For instance, the multiset $\{1, 2^2, 3\}$ has the two derangements 2132 and 2312.

Let $D(\alpha)$ denote the number of derangments of M_α. Show that

$$\sum_{\alpha \in \mathbb{N}^k} D(\alpha)x^\alpha = \frac{1}{(1+x_1)\cdots(1+x_k)\left(1 - \frac{x_1}{1+x_1} - \cdots - \frac{x_k}{1+x_k}\right)}$$

$$= \frac{1}{1 - \sum_S(\#S - 1)\prod_{i\in S} x_i},$$

where S ranges over all nonempty subsets of $[n]$.

13. Let $w = a_1 a_2 \cdots a_n \in \mathfrak{S}_n$. The *connectivity set* $C(w)$ of w is defined by

$$C(w) = \{i : a_j < a_k \text{ for all } j \le i < k\} \subseteq [n-1].$$

In other words, $i \in C(w)$ if $\{a_1,\ldots,a_i\} = [i]$. For instance, $C(2314675) = \{3,4\}$. (Exercise 1.128(a) deals with the enumeration of permutations $w \in \mathfrak{S}_n$ satisfying $C(w) = \emptyset$.)

a. [2] If $S = \{i_1,\ldots,i_k\}_< \subset [n-1]$, then let

$$\eta(S) = i_1!(i_2 - i_1)!\cdots(i_k - i_{k-1})!(n - i_k)!.$$

Hence by Proposition 1.4.1 we have $\alpha(S) = n!/\eta(S)$, the number of permutations $w \in \mathfrak{S}_n$ with descent set $D(w) \subseteq S$. Show that

$$\#\{w \in \mathfrak{S}_n : S \subseteq C(w)\} = \eta(S).$$

b. [2+] Given $S, T \subseteq [n-1]$, let $\overline{S} = [n-1] - S$, and define

$$X_{ST} = \#\{w \in \mathfrak{S}_n : C(w) = \overline{S}, \ D(w) = T\},$$

$$Z_{ST} = \#\{w \in \mathfrak{S}_n : \overline{S} \subseteq C(w), \ T \subseteq D(w)\},$$

$$= \sum_{\substack{S' \supseteq S \\ T' \supseteq T}} X_{S'T'}.$$

For instance, for $n = 4$, we have the following table of X_{ST}.

$S\backslash T$	$\emptyset$	1	2	3	12	13	23	123
$\emptyset$	1							
1	0	1						
2	0	0	1					
3	0	0	0	1				
12	0	1	1	0	1			
13	0	0	0	0	0	1		
23	0	0	1	1	0	0	1	
123	0	1	2	1	2	4	2	1

Show that

$$Z_{ST} = \begin{cases} \eta(\overline{S})/\eta(\overline{T}), & \text{if } S \supseteq T, \\ 0, & \text{otherwise.} \end{cases} \tag{2.50}$$

c. [2−] Let $M = (M_{ST})$ be the matrix whose rows and columns are indexed by subsets $S, T \subseteq [n-1]$ (taken in some order), with

$$M_{ST} = \begin{cases} 1, & \text{if } S \supseteq T, \\ 0, & \text{otherwise.} \end{cases}$$

Let $D = (D_{ST})$ be the diagonal matrix with $D_{SS} = \eta(\overline{S})$. Let $Z = (Z_{ST})$, that is, the matrix whose (S,T)-entry is Z_{ST}. Show that equation (2.50) can be restated as follows:

$$Z = DMD^{-1}.$$

Similarly show that if $X = (X_{ST})$, then

$$MXM = Z.$$

d. [1+] For an invertible matrix $A = (A_{ST})$, write A_{ST}^{-1} for the (S,T)-entry of the inverse matrix A^{-1}. Show that the Principle of Inclusion-Exclusion (Theorem 2.1.1) is equivalent to

$$M_{ST}^{-1} = (-1)^{\#S + \#T} M_{ST}.$$

e. [2–] Define the matrix $Y = (Y_{ST})$ by

$$Y_{ST} = \#\{w \in \mathfrak{S}_n : \overline{S} \subseteq C(w), \ T = D(w)\}.$$

Show that $Y = MX = ZM^{-1}$.

f. [2+] Show that the matrices Z, Y, X have the following inverses:

$$Z_{ST}^{-1} = (-1)^{\#S + \#T} Z_{ST},$$

$$Y_{ST}^{-1} = (-1)^{\#S + \#T} \#\{w \in \mathfrak{S}_n : \overline{S} = C(w), \ T \subseteq D(w)\},$$

$$X_{ST}^{-1} = (-1)^{\#S + \#T} X_{ST}.$$

14. a. [2+]* Let $A_k(n)$ denote the number of k-element antichains in the boolean algebra B_n (i.e., the number of subsets S of $2^{[n]}$ such that no element of S is a subset of another). Show that

$$A_1(n) = 2^n,$$

$$A_2(n) = \frac{1}{2}\left(4^n - 2 \cdot 3^n + 2^n\right),$$

$$A_3(n) = \frac{1}{6}\left(8^n - 6 \cdot 6^n + 6 \cdot 5^n + 3 \cdot 4^n - 6 \cdot 3^n + 2 \cdot 2^n\right),$$

$$A_4(n) = \frac{1}{24}\left(16^n - 12 \cdot 12^n + 24 \cdot 10^n + 4 \cdot 9^n - 18 \cdot 8^n\right.$$

$$\left. +6 \cdot 7^n - 36 \cdot 6^n + 11 \cdot 4^n - 22 \cdot 3^n + 6 \cdot 2^n\right).$$

b. [2+]* Show that for fixed $k \in \mathbb{P}$ there exist integers $a_{k,2}, a_{k,3}, \ldots, a_{k,2^k}$ such that

$$A_k(n) = \frac{1}{k!} \sum_{i=2}^{2^k} a_{k,i} i^n.$$

Show in particular that $a_{k,2^k} = 1, a_{k,i} = 0$ if $3 \cdot 2^{k-2} < i < 2^k$, and $a_{k,3 \cdot 2^{k-2}} = k(k-1)$.

15. a. [2–] Given a permutation $w \in \mathfrak{S}_3$, let P_w denote the corresponding permutation matrix; that is, the (i,j)-entry of P_w is equal to $\delta_{w(i),j}$. Let α_w, where $w \in \mathfrak{S}_3$, be integers satisfying $\sum_w \alpha_w P_w = 0$. Show that

$$\alpha_{123} = \alpha_{231} = \alpha_{312} = -\alpha_{132} = -\alpha_{213} = -\alpha_{321}.$$

b. [2] Let $H_n(r)$ denote the number of $n \times n$ $\mathbb{N}$-matrices A for which every row and column sums to r. Assume the theorem that A is a sum of permutation matrices is known. Deduce from this result (for the case $n = 3$) and (a) that

$$H_3(r) = \binom{r+5}{5} - \binom{r+2}{5}. \tag{2.51}$$

c. [3–] Give a direct combinatorial proof that

$$H_3(r) = \binom{r+4}{4} + \binom{r+3}{4} + \binom{r+2}{4}.$$

16. [2] Fix $k \geq 1$. How many permutations of $[n]$ have no cycle of length k? If $f_k(n)$ denotes this number, then compute $\lim_{n \to \infty} f_k(n)/n!$.

17. a. [2] Let $f_2(n)$ be the number of permutations of the integers modulo n that consist of a single cycle $(a_1, a_2, \ldots, a_n)$ and for which $a_i + 1 \neq a_{i+1} \pmod{n}$ for all i (with $a_{n+1} = a_1$). For example, for $n = 4$, there is one such permutation; namely, $(1, 4, 3, 2)$. Set $f_2(0) = 1$ and $f_2(1) = 0$. Use the Principle of Inclusion-Exclusion to find a formula for $f_2(n)$.

b. [1+] Write the answer to (a) in the form $\Delta^n g(0)$ for some function g.

c. [2–] Find the generating function $\sum_{n \geq 0} f_2(n) x^n / n!$.

d. [2–] Express the derangment number $D(n)$ in terms of the numbers $f_2(k)$.

e. [2–] Show that

$$\lim_{n \to \infty} \frac{f_2(n)}{(n-1)!} = \frac{1}{e}.$$

f. [3–] Generalize (e) to show that $f_2(n)$ has the asymptotic expansion

$$\frac{f_2(n)}{(n-1)!} \sim \frac{1}{e} \left(1 - \frac{1}{n} + \frac{1}{n^3} + \frac{1}{n^4} - \frac{2}{n^5} - \frac{9}{n^6} + \cdots + \frac{a_i}{n^i} + \cdots \right), \tag{2.52}$$

where $\sum_{i \geq 0} a_i x^i / i! = \exp(1 - e^x)$. By definition, equation (2.52) means that for any $k \in \mathbb{N}$,

$$\lim_{n \to \infty} n^k \left[\frac{f_2(n)}{(n-1)!} - \frac{1}{e} \sum_{i=0}^{k} \frac{a_i}{n^i} \right] = 0.$$

18. [3] Let $k \geq 2$. Let $f_k(n)$ be the number of cycles as in Exercise 2.17 such that for no i do we have

$$w(i + j) \equiv w(i) + j \pmod{n}, \quad \text{for all } j = 1, 2, \ldots, k - 1,$$

where the argument $i + j$ is taken modulo n. Use the Principle of Inclusion-Exclusion to show that

$$\frac{f_3(n)}{(n-1)!} = 1 - \frac{1}{n} - \frac{3}{2} \frac{1}{n^2} - \frac{14}{3} \frac{1}{n^3} + O(n^{-4}),$$

$$\frac{f_4(n)}{(n-1)!} = 1 - \frac{1}{n^2} - \frac{5}{n^3} - \frac{29}{2} \frac{1}{n^4} + O(n^{-5}),$$

$$\frac{f_k(n)}{(n-1)!} = 1 - \frac{1}{n^{k-2}} - \frac{(k-2)(k+1)}{2} \frac{1}{n^{k-1}}$$

$$- \frac{k(k+1)(3k^2 - 5k - 10)}{24} \frac{1}{n^k} + O(n^{-k-1}),$$

for fixed $k \geq 5$.

In particular, for fixed $k \geq 3$ we have $\lim_{n\to\infty} f_k(n)/(n-1)! = 1$.

19. [2] Suppose that $2n$ persons are sitting in a circle. In how many ways can they form n pairs if no two adjacent persons can form a pair? Express your answer as a finite sum.

20. [2] Call two permutations of the $2n$-element set $S = \{a_1, a_2, \ldots, a_n, b_1, b_2, \ldots, b_n\}$ *equivalent* if one can be obtained from the other by interchanges of *consecutive* elements of the form $a_i b_i$ or $b_i a_i$. For example, $a_2 b_3 a_3 b_2 a_1 b_1$ is equivalent to itself and to $a_2 a_3 b_3 b_2 a_1 b_1$, $a_2 b_3 a_3 b_2 b_1 a_1$, and $a_2 a_3 b_3 b_2 b_1 a_1$. How many equivalence classes are there?

21. a. [2+]* Given numbers (or elements of a commutative ring with 1) a_i for $i \in \mathbb{Z}$, with $a_i = 0$ for $i < 0$ and $a_0 = 1$, let $f(k) = \det[a_{j-i+1}]_1^k$. In particular, $f(0) = 1$. Show that

$$\sum_{k \geq 0} f(k) x^k = \frac{1}{1 - a_1 x + a_2 x^2 - \cdots}.$$

b. [2] Suppose that in (a) we drop the condition $a_0 = 1$, say $a_0 = \alpha$. Deduce from (a) that

$$\sum_{k \geq 0} f(k) x^k = \frac{1}{1 + \sum_{i \geq 1} (-1)^i \alpha^{i-1} a_i x^i}.$$

c. [2+] Suppose that in (b) we let the first row of the matrix be arbitrary, that is, let $M_k = (m_{ij})_1^k$ be the $k \times k$ matrix defined by

$$m_{1j} = b_j,$$
$$m_{ij} = a_{j-i+1}, \ i \geq 2,$$

where $a_0 = \alpha$ and $a_i = 0$ for $i < 0$. Let $g(k) = \det M_k$. Show that

$$\sum_{k \geq 1} g(k) x^k = \frac{\sum_{j \geq 1} (-1)^{j-1} \alpha^{j-1} b_j x^j}{1 + \sum_{i \geq 1} (-1)^i \alpha^{i-1} a_i x^i}.$$

d. [2] Fix $0 < a \leq d$. Let $\beta(k) = \beta_{a+kd}(a, a+d, a+2d, \ldots, a+(k-1)d)$. Deduce from equation (2.16) that

$$\sum_{k \geq 0} \beta(k) \frac{x^k}{(a+kd)!} = \frac{\sum_{j \geq 0} (-1)^j \dfrac{x^j}{(a+jd)!}}{\sum_{i \geq 0} (-1)^i \dfrac{x^i}{(id)!}}.$$

Give a q-analogue based on Example 2.2.5.

e. [2]* Suppose that in Proposition 2.2.6 the function $e(i, j)$ has the form

$$e(i, j) = \alpha_{j-i}$$

for certain numbers α_k, with $\alpha_0 = 1$ and $\alpha_k = 0$ for $k < 0$. Show that $f_=(S)$ is equal to the coefficient of x^{n+1} in the power series

$$h(n)(1 - \alpha_1 x + \alpha_2 x^2 - \alpha_3 x^3 + \cdots)^{-1}.$$

22. a. [2+] Let E_{2n} denote the number of alternating permutations $w \in \mathfrak{S}_{2n}$. Thus by Proposition 1.6.1, we have

$$\left(\sum_{n \geq 0} E_{2n} \frac{x^{2n}}{(2n)!} \right) \left(1 - \frac{x^2}{2!} + \frac{x^4}{4!} - \cdots \right) = 1.$$

Equating coefficients of $x^{2n}/(2n)!$ on both sides gives

$$E_{2n} = \binom{2n}{2} E_{2n-2} - \binom{2n}{4} E_{2n-4} + \binom{2n}{6} E_{2n-6} - \cdots. \qquad (2.53)$$

Give a sieve-theoretic proof of equation (2.53).

b. [2+] State and prove a similar result for E_{2n+1}.

23. a. [2+] Give a sieve-theoretic proof of Exercise 1.61(c), that is, if $f(n)$ is the number of permutations $w \in \mathfrak{S}_n$ with no proper double descents, then

$$\sum_{n \geq 0} f(n) \frac{x^n}{n!} = \frac{1}{\displaystyle\sum_{j \geq 0} \left(\frac{x^{3j}}{(3j)!} - \frac{x^{3j+1}}{(3j+1)!} \right)}.$$

b. [2+]* Generalize (a) as follows. Let $f_r(n)$ be the number of permutations $w \in S_n$ with no r consecutive descents (where n is not considered a descent). Give a sieve-theoretic proof that

$$\sum_{n \geq 0} f_r(n) \frac{x^n}{n!} = \frac{1}{\displaystyle\sum_{j \geq 0} \left(\frac{x^{(r+1)j}}{((r+1)j)!} - \frac{x^{(r+1)j+1}}{((r+1)j+1)!} \right)}.$$

24. [2+]* Fix $j, k \geq 1$. For $n \geq 0$ let $f(n)$ be the number of integer sequences $a_1, a_2, \ldots, a_n$ such that $1 \leq a_i \leq k$ for $1 \leq i \leq n$, and $a_i \geq a_{i-1} - j$ for $2 \leq i \leq n$. Give a sieve-theoretic proof that

$$F(x) := \sum_{n \geq 0} f(n) x^n = \frac{1}{\displaystyle\sum_{i \geq 0} (-1)^i \binom{k - j(i-1)}{i} x^i}.$$

(Note that the denominator is actually a finite sum.)

25. a. [2]* Let $f_i(m, n)$ be the number of $m \times n$ matrices of 0's and 1's with at least one 1 in every row and column, and with a total of i 1's. Use the Principle of Inclusion-Exclusion to show that

$$\sum_i f_i(m, n) t^i = \sum_{k=0}^{n} (-1)^k \binom{n}{k} ((1+t)^{n-k} - 1)^m. \qquad (2.54)$$

b. [2]* Show that

$$\sum_{m,n \geq 0} \sum_{i \geq 0} f_i(m, n) y^i \frac{x^m y^n}{m! n!} = e^{-x-y} \sum_{i \geq 0} \sum_{j \geq 0} (1+t)^{ij} \frac{x^i y^j}{i! j!}.$$

Note that this formula, unlike equation (2.54), exhibits the symmetry between m and n.

26. [2+]* Let $\pi \in \Pi_n$, the set of partitions of $[n]$. Let $S(\pi, r)$ denote the number of $\sigma \in \Pi_n$ such that $|\sigma| = r$ and $\#(A \cap B) \leq 1$ for all $A \in \pi$ and $B \in \sigma$. (This last condition is equivalent to $\pi \wedge \sigma = \hat{0}$ in the lattice structure on Π_n defined in Example 3.10.4.) Show that

$$S(\pi, r) = \frac{1}{r!} \sum_{i=0}^{r} \binom{r}{i} (-1)^{r-i} \prod_{A \in \pi} (i)_{\#A}$$

$$= \frac{1}{r!} \Delta^r \prod_{A \in \pi} (n)_{\#A}|_{n=0}.$$

27. a. [3–] Let F be a forest, with $\ell = \ell(F)$ components, on the vertex set $[n]$. We say that F is *rooted* if we specify a root vertex for each connected component of F. Thus, if $c_1, \ldots, c_\ell$ are the number of vertices of the components of F (so $\sum c_i = n$), then the number $p(F)$ of ways to root F is $c_1 c_2 \cdots c_\ell$. Show that the number of k-component rooted forests on $[n]$ that contain F is equal to

$$p(F) \binom{\ell - 1}{\ell - k} n^{\ell - k}.$$

b. [2+] Given any graph G on $[n]$ with no multiple edges, define the polynomial

$$P(G, x) = \sum_F x^{\ell(F) - 1}, \tag{2.55}$$

summed over all rooted forests F on $[n]$ contained in G. Let $\overline{G}$ denote the complement of G; that is, $\{i, j\} \in \binom{[n]}{2}$ is an edge of G if and only if $\{i, j\}$ is not an edge of $\overline{G}$. Use (a) and the Principle of Inclusion-Exclusion to show that

$$P(\overline{G}, x) = (-1)^{n-1} P(G, -x - n). \tag{2.56}$$

In particular, the number $c(\overline{G})$ of spanning trees of $\overline{G}$ (i.e., subgraphs of $\overline{G}$ that are trees and that use all the vertices of $\overline{G}$) is given by

$$c(\overline{G}) = (-1)^{n-1} P(G, -n)/n. \tag{2.57}$$

c. [2] The *complete graph* K_n has vertex set $[n]$ and an edge between any two distinct vertices (so $\binom{n}{2}$ edges in all). The *complete bipartite graph* $K_{r,s}$ has vertex set $A \cup B$, where A and B are disjoint with $\#A = r$ and $\#B = s$, and with one edge between each vertex of A and each vertex of B (so rs edges in all). Use (b) to find the number of spanning trees of K_n and $K_{r,s}$.

28. [3] Let $r \geq 1$. An *r-stemmed V-partition* of n is an array

$$\begin{bmatrix} & & & & b_1 & b_2 & b_3 & \cdots \\ a_1 & a_2 & \cdots & a_r & & & & \\ & & & & c_1 & c_2 & c_3 & \cdots \end{bmatrix}$$

of nonnegative integers satisfying $a_1 \geq a_2 \geq \cdots \geq a_r \geq b_1 \geq b_2 \geq b_3 \geq \cdots$, $a_r \geq c_1 \geq c_2 \geq c_3 \geq \cdots$, and $\sum a_i + \sum b_i + \sum c_i = n$. Hence, a 1-stemmed V-partition is just a V-partition. Let $v_r(n)$ denote the number of r-stemmed V-partitions of n. Show that

$$\sum_{n \geq 0} v_r(n) x^n = \frac{p_r(x) T(x) - q_r(x)}{(1-x)(1-x^2) \cdots (1-x^{r-1}) \prod_{i \geq 1} (1-x^i)^2},$$

where

$$p_1(x) = 1, \ p_2(x) = 2, \ q_1(x) = 0, \ q_2(x) = 1,$$
$$p_r(x) = 2p_{r-1}(x) + (x^{r-2} - 1)p_{r-2}(x), \ r > 2,$$
$$q_r(x) = 2q_{r-1}(x) + (x^{r-2} - 1)q_{r-2}(x), \ r > 2,$$
$$T(x) = \sum_{i \geq 0} (-1)^i x^{\binom{i+1}{2}}.$$

29. a. [2]* A *concave composition* of n is a nonnegative integer sequence $a_1 > a_2 > \cdots > a_r = b_r < b_{r-1} < \cdots < b_1$ such that $\sum(a_i + b_i) = n$. For instance, the eight concave compositions of 6 are 33, 5001, 4002, 3003, 2112, 2004, 1005, and 210012. Let $f(n)$ denote the number of concave partitions of n. Give a combinatorial proof that $f(n)$ is even for $n \geq 1$.

b. [5–] Set

$$F(q) = \sum_{n \geq 0} f(n)q^n = 1 + 2q^2 + 2q^3 + 4q^4 + 4q^5 + 8q^6 + \cdots.$$

Give an Inclusion-Exclusion proof, analogous to the proof of Proposition 2.5.1, that

$$F(q) = \frac{1 - \sum_{n \geq 1} q^{n(3n-1)/2}(1 - q^n)}{(1 - q)(1 - q^2)(1 - q^3) \cdots}.$$

30. [3] Give a sieve-theoretic proof of the Pentagonal Number Formula (Proposition 1.8.7), namely,

$$\frac{1 + \sum_{n \geq 1}(-1)^n[x^{n(3n-1)/2} + x^{n(3n+1)/2}]}{\prod_{i \geq 1}(1 - x^i)} = 1.$$

Your sieve should start with all partitions of $n \geq 0$ and sieve out all but the empty partition of 0.

31. [3–] Give cancellation proofs, similar to our proof of the Pentagonal Number Formula (Proposition 1.8.7), of the two identities of Exercise 1.91(c), namely,

$$\prod_{k \geq 1} \frac{1 - q^k}{1 + q^k} = \sum_{n \in \mathbb{Z}} (-1)^n q^{n^2},$$

$$\prod_{k \geq 1} \frac{1 - q^{2k}}{1 - q^{2k-1}} = \sum_{n \geq 0} q^{\binom{n+1}{2}}.$$

32. [3–] Give a cancellation proof of the identity

$$\sum_{k=0}^{n} (-1)^k \binom{n}{k} = \begin{cases} (1-q)(1-q^3) \cdots (1-q^{n-1}), & n \text{ even}, \\ 0, & n \text{ odd}. \end{cases}$$

33. [2–] Deduce from equation (2.21) that

$$\det \left[\binom{n-i}{j-i+1} \right]_0^{n-1} = q^{\binom{n}{2}}. \tag{2.58}$$

34. A *tournament* T on the vertex set $[n]$ is a directed graph on $[n]$ with no loops such that each pair of distinct vertices is joined by exactly one directed edge. The *weight* $w(e)$ of a directed edge e from i to j (denoted $i \to j$) is defined to be x_j if $i < j$ and $-x_j$ if $i > j$. The weight of T is defined to be $w(T) = \prod_e w(e)$, where e ranges over all edges of T.

a. [2–] Show that

$$\sum_T w(T) = \prod_{1 \le i < j \le n} (x_j - x_i), \tag{2.59}$$

where the sum is over all $2^{\binom{n}{2}}$ tournaments on $[n]$.

b. [2–] The tournament T is *transitive* if there is a permutation $z \in \mathfrak{S}_n$ for which $z(i) < z(j)$ if and only if $i \to j$. Show that a nontransitive tournament contains a 3-cycle (i.e., a triple (t, u, v) of vertices for which $t \to u \to v \to t$).

c. [1+] If T and T' are tournaments on $[n]$ then write $T \leftrightarrow T'$ if T' can be obtained from T by reversing a 3-cycle; that is, replacing the edges $t \to u, u \to v, v \to t$ with $u \to t$, $v \to u, t \to v$, and leaving all other edges unchanged. Show that $w(T') = -w(T)$.

d. [2] Show that if $T \leftrightarrow T'$ then T and T' have the same number of 3-cycles.

e. [2+] Deduce from (a)–(d) that

$$\det\left[x_i^{j-1}\right]_1^n = \prod_{1 \le i < j \le n} (x_j - x_i),$$

by canceling out all terms in the left-hand side of (2.59) except those corresponding to transitive T.

35. a. [2] Let $f(x_1, \ldots, x_n)$ be a homogeneous polynomial of degree n over a field K. Show that

$$[x_1 x_2 \cdots x_n] f(x_1, \ldots, x_n) = \sum_{(\epsilon_1, \ldots, \epsilon_n) \in \{0,1\}^n} (-1)^{n - \sum \epsilon_i} f(\epsilon_1, \ldots, \epsilon_n). \tag{2.60}$$

(Regard each ϵ_i in the exponent of -1 as an integer and in the argument of f as an element of K.)

b. [2] Let $A = (a_{ij})$ be an $n \times n$ matrix. The *permanent* of A is defined by

$$\mathrm{per}(A) = \sum_{w \in \mathfrak{S}_n} a_{1,w(1)} a_{2,w(2)} \cdots a_{n,w(n)}.$$

In other words, the formula for $\mathrm{per}(A)$ is the same as the expansion of $\det(A)$ but with all signs positive. Show that

$$\mathrm{per}(A) = \sum_{S \subseteq [n]} (-1)^{n - \#S} \prod_{i=1}^n \sum_{j \in S} a_{ij}. \tag{2.61}$$

36. [3–] Let $A_1, \ldots, A_n$ be subsets of a finite set A, and $B_1, \ldots, B_n$ subsets of a finite set B. For each subset S of $[n]$, let $A_S = \bigcap_{i \in S} A_i$ and $B_S = \bigcap_{i \in S} B_i$. Given bijections $f_S \colon A_S \to B_S$ for each $S \subseteq [n]$, construct an explicit bijection $h \colon A - \bigcup_{i=1}^n A_i \to B - \bigcup_{i=1}^n B_i$. Your definition of h should depend only on the f_S's, and not on some ordering of the elements of A or on the labeling of the subsets $A_1, \ldots, A_n$ and $B_1, \ldots, B_n$.

37. [3–]* Given $a, b \in \mathbb{P}$ with $a < b$, let $C(b - a)$ denote the number of lattice paths in $\mathbb{Z}^2$ from $(2a, 0)$ to $(2b, 0)$ with steps $(1, 1)$ or $(1, -1)$ that never pass below the x-axis. (It follows from Corollary 6.2.3(iv) that $C(b - a)$ is the Catalan number $\frac{1}{b-a+1}\binom{2(b-a)}{b-a}$, but this fact is irrelevant here.) Now given $\{a_1, a_2, \ldots, a_{2n}\}_< \subset \mathbb{Z}$, let $C(a_1, a_2, \ldots, a_{2n})$ denote the number of ways to connect the points $(2a_1, 0)$, $(2a_2, 0), \ldots$, $(2a_{2n}, 0)$ with n pairwise disjoint lattice paths $L_1, \ldots, L_n$ of the type just described. (Thus, each L_i connects some $(2a_j, 0)$ to some $(2a_k, 0)$, $j \neq k$. If $i \neq j$ then L_i and L_j do not intersect, including endpoints, so each $(2a_i, 0)$ is an endpoint of exactly one L_i.)

Now given a triangular array $A = (a_{ij})$ with $1 \leq i < j \leq 2n$, define the *pfaffian* of A by

$$\mathrm{Pf}(A) = \sum \varepsilon(i_1, j_1, \ldots, i_n, j_n) a_{i_1 j_1} \cdots a_{i_n j_n},$$

where the summation is over all partitions $\{\{i_1, j_1\}_<, \ldots, \{i_n, j_n\}_<\}$ of $[2n]$ into 2-element blocks, and where $\varepsilon(i_1, j_1, \ldots, i_n, j_n)$ denotes the sign of the permutation (written in two-line form)

$$\begin{pmatrix} 1 & 2 & \cdots & 2n-1 & 2n \\ i_1 & j_1 & \cdots & i_n & j_n \end{pmatrix}.$$

(It is easy to see that $\varepsilon(i_1, j_1, \ldots, i_n, j_n)$ does not depend on the order of the n blocks.) Give a proof analogous to that of Theorem 2.7.1 of the formula

$$C(a_1, a_2, \ldots, a_{2n}) = \mathrm{Pf}(C(a_j - a_i)).$$

For instance,

$$C(0, 3, 5, 6) = \mathrm{Pf} \begin{vmatrix} C(3) & C(5) & C(6) \\ & C(2) & C(3) \\ & & C(1) \end{vmatrix}$$

$$= \mathrm{Pf} \begin{vmatrix} 5 & 42 & 132 \\ & 2 & 5 \\ & & 1 \end{vmatrix}$$

$$= 5 \cdot 1 + 132 \cdot 2 - 42 \cdot 5$$

$$= 59.$$

Solutions to Exercises

1. We have

$$\text{Principle of Inclusion-Exclusion} = \text{PIE}$$

$$= \pi e$$

$$= (3.141592653 \cdots)(2.718281828 \cdots)$$

$$= 8.53973422267356706546 \cdots.$$

3. We have

$$\sum_{i=k}^{n}(-1)^{i-k}\binom{i}{k}g_i = \sum_{i=k}^{n}(-1)^{i-k}\binom{i}{k}\sum_{T\subseteq S\#T=i}f_\geq(T)$$

$$= \sum_{i=k}^{n}(-1)^{i-k}\binom{i}{k}\sum_{\substack{T\subseteq R\subseteq S\\ \#T=i}}f_=(R)$$

$$= \sum_{R\subseteq S}f_=(R)\sum_{T\subseteq R}(-1)^{\#T-k}\binom{\#T}{k}.$$

If $\#R = r$ then the inner sum is equal to

$$\sum_{j=0}^{r}(-1)^{j-k}\binom{r}{j}\binom{j}{k} = \binom{r}{k}\sum_{j=0}^{r}(-1)^{j-k}\binom{r-k}{r-j} = \delta_{kr},$$

and the proof of equation (2.39) follows. The sum (2.40) is evaluated similarly. An extensive bibliography appears in Takács [2.22].

4. **a.** If we regard A_i as the set of elements having property P_i, then

$$A_T = f_\geq(T) = \sum_{Y\supseteq T}f_=(Y).$$

Hence,

$$S_k - S_{k-1} + \cdots + (-1)^{n-k}S_n = \sum_{\#T\geq k}(-1)^{\#T-k}f_\geq(T)$$

$$= \sum_{\#T\geq k}\sum_{Y\supseteq T}(-1)^{\#T-k}f_=(Y)$$

$$= \sum_{\#Y\geq k}f_=(Y)\sum_{\substack{T\subseteq Y\\ \#T\geq k}}(-1)^{\#T-k}$$

$$= \sum_{\#Y\geq k}f_=(Y)\sum_{i=k}^{\#Y}(-1)^{i-k}\binom{\#Y}{i}.$$

It is easy to see that $\sum_{i=k}^{m}(-1)^{i-k}\binom{m}{i} = \binom{m-1}{k-1} \geq 0$. Since $f_=(Y) \geq 0$, equation (2.41) follows.
Setting

$$S = f_=(\emptyset) = \#(\bar{A}_1 \cap \cdots \cap \bar{A}_n) = S_0 - S_1 + \cdots + (-1)^n S_n,$$

the inequality (2.41) can be rewritten

$$S \geq 0$$
$$S \leq S_0$$
$$S \geq S_0 - S_1$$
$$S \leq S_0 - S_1 + S_2$$
$$\vdots$$

In other words, the partial sums $S_0 - S_1 + \cdots + (-1)^k S_k$ successively overcount and undercount the value of S. In this form, equation (2.41) is due to Carlo Bonferroni (1892–1960), *Pubblic. Ist. Sup. Sc. Ec. Comm. Firenze* **8** (1936), 1–62. These inequalities sometimes make it possible to estimate S accurately when not all the S_i's can be computed explicitly.

b. *Answer.* $\sum_{i=1}^{k}(-1)^{i-k}\binom{i}{k}S_i \geq 0, \ 0 \leq k \leq n.$

5. **a.** The most straightforward proof is by induction on n, the case $n = 0$ being trivial (since when $n = 0$ exactness implies that $W \cong V_0$). The details are omitted.

   b. The sequence

$$0 \to V_n \xrightarrow{\partial_n} V_{n-1} \xrightarrow{\partial_{n-1}} \cdots \xrightarrow{\partial_{j+1}} V_j \xrightarrow{\partial_j} \operatorname{im}\partial_j \to 0$$

   is exact. But $\dim(\operatorname{im}\partial_j) = \operatorname{rank}\partial_j$, so the proof follows from (a).

   c. By equation (2.44), we have $\dim V_j = \operatorname{rank}\partial_j + \operatorname{rank}\partial_{j+1}$. On the other hand, $\operatorname{rank}\partial_{j+1} = \dim(\operatorname{im}\partial_{j+1})$ and $\operatorname{rank}\partial_j = \dim V_j - \dim(\ker\partial_j)$, so $\dim(\operatorname{im}\partial_{j+1}) = \dim(\ker\partial_j)$. Since $\operatorname{im}\partial_{j+1} \subseteq \ker\partial_j$, the proof follows.

   d. For fixed $a \in A$, let V_T^a be the span of the symbols $[a, T]$ if $a \in A_T$; otherwise, $V_T^a = 0$. Let $V_j^a = \bigoplus_{\#T=j} V_T^a$, and let W^a be the span of the single element $[a]$ if $a \in \bar{A}_1 \cap \cdots \cap \bar{A}_n$; otherwise, $W^a = 0$. Then $\partial_j : V_j^a \to V_{j-1}^a$, $j \geq 1$, and $\partial_0 : V_0^a \to W^a$. (Thus, the sequence (2.42) is the *direct sum* of such sequences for fixed a.) From this discussion, it follows that we may assume $A = \{a\}$.

   Clearly, ∂_0 is surjective, so exactness holds at W. It is straightforward to check that $\partial_j \partial_{j+1} = 0$, so (2.42) is a complex. Since $A = \{a\}$, we have $\dim V_j = \binom{n}{j}$ and $\sum_{i=j}^{n}(-1)^{i-j}\dim V_i = \binom{n-1}{j-1}$. There are several ways to show that $\operatorname{rank}\partial_j = \binom{n-1}{j-1}$, so the proof follows from (c).

   There are many other proofs, whose accessibility depends on background. For instance, the complex (2.42) in the case at hand (with $A = \{a\}$) is the tensor product of the complexes $C_i : 0 \to U_i \xrightarrow{\partial_0} W \to 0$, where U_i is spanned by $[a, \{t_i\}]$. Clearly, each C_i is exact; hence, so is (2.42). (The definition (2.45) was not plucked out of the air; it is a *Koszul relation*, and (2.42) (with $A = \{a\}$) is a *Koszul complex*. See almost any textbook on homological algebra for further information.).

   e,f. Follows from $\dim V_T = \#A_T$, whence $\dim V_j = S_j$.

6. **a.** Straightforward generalization of Theorem 2.1.1.

   b. We obtain the classical Möbius inversion formula (see Example 3.8.4). More specifically, let D_n denote the set of all divisors of n, and let $f, g : D_n \to K$. Equations (2.48) and (2.49) then assert that the following two formulas are equivalent:

$$g(m) = \sum_{d \mid m} f(d), \text{ for all } m \mid n,$$

$$f(m) = \sum_{d \mid m} \mu(m/d)g(d), \text{ for all } m \mid n. \qquad (2.62)$$

7. Each element $\alpha \in \mathbb{F}_{q^n}$ generates a subfield $\mathbb{F}_q(\alpha)$ of order q^d for some $d \mid n$. Thus, α is a zero of a unique monic irreducible polynomial $f_\alpha(x)$ of degree d over $\mathbb{F}_q$. Every

such polynomial has d distinct zeros, all belonging to $\mathbb{F}_{q^n}$. Hence,

$$q^n = \sum_{d|n} d\beta(d).$$

Möbius inversion (see equation (2.62)) gives

$$n\beta(n) = \sum_{d|n} \mu(n/d)q^d = \sum_{d|n} \mu(d)q^{n/d}.$$

8. **a.** See J. B. Remmel, *Europ. J. Combinatorics* **4** (1983), 371–374, and H. S. Wilf, *Math. Mag.* **57** (1984), 37–40.

    b. Note that the last entry a_n of a permutation $w = a_1 \cdots a_n \in \mathfrak{S}_n$ has no effect on the location of the first ascent unless $w = n, n-1, \ldots, 1$, in which case the contribution to $E(n)$ is $(-1)^n$. See J. Désarménien, *Sem. Lotharingien de Combinatoire* (electronic) **8** (1983), B08b; formerly *Publ. I.R.M.A. Strasbourg*, 229/S-08, 1984, pp. 11–16. For a generalization, see J. Désarménien and M. L. Wachs, *Sem. Lotharingien de Combinatoire* (electronic) **19** (1988), B19a; formerly *Publ. I.R.M.A. Strasbourg*, 361/S-19, 1988, pp. 13–21. See also Vol. II, Exercise 7.65 for a related result dealing with symmetric functions.

    NOTE. We can also see that $E(n) = D(n)$ by noting that the number of permutations $w \in \mathfrak{S}_n$, whose first ascent is in position k, is $n!\left(\frac{1}{k!} - \frac{1}{(k+1)!}\right)$ for $0 \leq k < n$ and is $1/n!$ for $k = n$, and then comparing with equation (2.12). Moreover, in an unpublished paper at

    ⟨http://people.brandeis.edu/~gessel/homepage/papers/

    color.pdf⟩,

    Gessel gives an elegant bijective proof in terms of "hook factorizations" of permutations that

$$\sum_{n \geq 0} E(n)\frac{x^n}{n!} = \frac{1}{1 - \sum_{k \geq 2}(k-1)\frac{x^k}{k!}} = \frac{e^{-x}}{1 - x}.$$

    c. Write a derangement w as a product of cycles. Arrange these cycles in decreasing order of their smallest element. Within each cycle, put the smallest element in the *second* position. Then erase the parentheses, obtaining another permutation w'. For instance, let $w = 974382651 = (85)(43)(627)(91)$; then $w' = 854362791$. It is not hard to check that the map $w \mapsto w'$ is a bijection from derangements in $\mathfrak{S}_n$ to $\mathcal{E}(n)$. This bijection is due to J. Désarménien, ibid.

    I am grateful to Ira Gessel for providing most of the information for this exercise.

9. We interpret $k!S(d,k)$ as the number of surjective functions $f: [d] \to [k]$. Let A be the set of all functions $f: [d] \to [k]$, and for $i \in [k]$ let P_i be the property that $i \notin \operatorname{im} f$. A function $f \in A$ lacks at most the properties $T \subseteq S \subseteq \{P_1, \ldots, P_k\}$ if and only if $\operatorname{im} f \subseteq \{i : P_i \in T\}$; hence, the number of such f is i^d, where $\#T = i$. The proof follows from Proposition 2.2.2.

10. b. We have

$$\frac{D(n)}{n!} - \frac{1}{e} = \left(1 - \frac{1}{1!} + \frac{1}{2!} - \cdots + (-1)^n \frac{1}{n!}\right)$$

$$- \left(1 - \frac{1}{1!} + \frac{1}{2!} - \cdots + (-1)^n \frac{1}{n!} + (-1)^{n+1} \frac{1}{(n+1)!} + \cdots\right)$$

$$= \frac{(-1)^n}{(n+1)!} + \cdots,$$

while

$$\left(1 - \frac{1}{n}\right)^n - \frac{1}{e} = e^{n \log(1 - \frac{1}{n})} - \frac{1}{e}$$

$$= e^{n(-\frac{1}{n} - \frac{1}{2n^2} + \cdots)} - \frac{1}{e}$$

$$= -\frac{1}{2ne} + \cdots.$$

Hence, $D(n)/n!$ is a *much* better approximation to $1/e$ than $E(n)/n!$.

11. This result was proved by G.-N. Han and G. Xin, *J. Combinatorial Theory Ser. A* **116** (2009), 449–459 (Theorems 1 and 9), using the theory of symmetric functions. A bijective proof was given by N. Eriksen, R. Freij, and J. Wästlund, *Electronic J. Combinatorics* **16**(1) (2009), #R32 (Theorem 2.1).

12. The Inclusion-Exclusion formula (2.11) for $D(n)$ generalizes straightforwardly to

$$D(\alpha) = \sum_{\beta_1=0}^{\alpha_1} \cdots \sum_{\beta_k=0}^{\alpha_k} \binom{\alpha_1}{\beta_1} \cdots \binom{\alpha_k}{\beta_k} (-1)^{\beta_1 + \cdots + \beta_k} \binom{\sum (\alpha_i - \beta_i)}{\alpha_1 - \beta_1, \ldots, \alpha_k - \beta_k}.$$

Let $\gamma_i = \alpha_i - \beta_i$. We get

$$\sum_\alpha D(\alpha) x^\alpha = \sum_{\beta, \gamma} \binom{\beta_1 + \gamma_1}{\beta_1} \cdots \binom{\beta_k + \gamma_k}{\beta_k} (-1)^{\sum \beta_i} \binom{\gamma_1 + \cdots + \gamma_k}{\gamma_1, \ldots, \gamma_k} x^{\sum (\beta_i + \gamma_i)}$$

$$= \sum_\gamma \binom{\gamma_1 + \cdots + \gamma_k}{\gamma_1, \ldots, \gamma_k} x^{\sum \gamma_i} \sum_\beta \binom{\beta_1 + \gamma_1}{\beta_1} \cdots \binom{\beta_k + \gamma_k}{\beta_k} (-1)^{\sum \beta_i} x^{\sum \beta_i}$$

$$= \sum_\gamma \binom{\gamma_1 + \cdots + \gamma_k}{\gamma_1, \ldots, \gamma_k} x^{\sum \gamma_i} (1 + x_1)^{-\gamma_1 - 1} \cdots (1 + x_k)^{-\gamma_k - 1}$$

$$= \frac{1}{(1 + x_1) \cdots (1 + x_k)} \sum_{n \geq 0} \left(\frac{x_1}{1 + x_1} + \cdots + \frac{x_k}{1 + x_k}\right)^n$$

$$= \frac{1}{(1 + x_1) \cdots (1 + x_k) \left(1 - \frac{x_1}{1 + x_1} - \cdots - \frac{x_k}{1 + x_k}\right)}.$$

This result appears as Exercise 4.5.5 in Goulden and Jackson [3.32], as a special case of the more general Exercise 4.5.4.

13. These results appear in R. Stanley, *J. Integer Sequences* **8** (2005), article 05.3.8. This paper also gives an extension to multisets and a q-analogue. For a generalization to arbitrary Coxeter groups, see N. Bergeron, C. Hohlweg, and M. Zabrocki, *J. Algebra* **303** (2006), 831–846, and M. Marietti, *Europ. J. Combinatorics* **29** (2008), 1555–1562.

15. a. The result follows easily after checking that any five of the matrices P_w are linearly independent.

b. Let A be a 3×3 $\mathbb{N}$-matrix for which every row and column sums to r. It is given that we can write

$$A = \sum_{w \in \mathfrak{S}_3} \alpha_w P_w, \qquad (2.63)$$

where $\alpha_w \in \mathbb{N}$ and $\sum \alpha_w = r$. By Section 1.2, the number of ways to choose $\alpha_w \in \mathbb{N}$ such that $\sum \alpha_w = r$ is $\binom{r+5}{5}$. By (a), the representation (2.63) is unique provided at least one of $\alpha_{213}, \alpha_{132}, \alpha_{321}$ is 0. The number of ways to choose $\alpha_{123}, \alpha_{231}, \alpha_{312} \in \mathbb{N}$ and $\alpha_{213}, \alpha_{132}, \alpha_{321} \in \mathbb{P}$ such that $\sum \alpha_w = r$ is equal to the number of weak compositions of $r - 3$ into six parts; that is, $\binom{r+2}{5}$. Hence, $H_3(r) = \binom{r+5}{5} - \binom{r+2}{5}$. Equation (2.51) appears in §407 of MacMahon [1.55], essentially with the previous proof. To evaluate $H_4(r)$ by a similar technique would be completely impractical, although it can be shown using the Hilbert syzygy theorem that such a computation could be done in principle. See R. Stanley, *Duke Math. J.* **40** (1973), 607–632. For a different approach toward evaluating $H_n(r)$ for any n, see Proposition 4.6.2. The theorem mentioned in the statement of (b) is called the Birkhoff–von Neumann theorem and is proved for general n in Lemma 4.6.1.

c. One can check that every matrix being counted can be represented in exactly one way in one of the forms

$$\begin{bmatrix} a+e & b+d & c \\ c+d & a & b+e \\ b & c+e & a+d \end{bmatrix}, \quad \begin{bmatrix} a & b+d & c+e+1 \\ c+d & a+e+1 & b \\ b+e+1 & c & a+d \end{bmatrix},$$

$$\begin{bmatrix} a+d+1 & b & c+e+1 \\ c & a+e+1 & b+d+1 \\ b+e+1 & c+d+1 & a \end{bmatrix},$$

where $a, b, c, d, e \in \mathbb{N}$, from which the proof is immediate. The idea behind this proof is to associate an indeterminate x_w to each $w \in \mathfrak{S}_3$, and then to use the identity

$$1 - x_{132}x_{213}x_{321} = (1 - x_{321}) + x_{321}(1 - x_{132}) + x_{321}x_{132}(1 - x_{213}).$$

Details are left to the reader. For yet another way to obtain $H_3(r)$, see M. Bóna, *Math. Mag.* **70** (1997), 201–203.

16. *Answer.*

$$f_k(n) = \sum_{i=0}^{\lfloor n/k \rfloor} (-1)^i \frac{n!}{i!\,k^i},$$

$$\lim_{n \to \infty} \frac{f_k(n)}{n!} = \sum_{i \geq 0} \frac{(-1)^i}{i!\,k^i} = e^{-1/k}.$$

17. a. $\displaystyle\sum_{i=0}^{n} (-1)^i \binom{n}{i}(n - i - 1)!$, provided we define $(-1)! = 1$.

b. $g(n) = (n-1)!$, with $g(0) = 1$.

c. $e^{-x}(1 - \log(1 - x))$.

d. $D(n) = f_2(n) + f_2(n+1)$.

This problem goes back to W. A. Whitworth, *Choice and Chance*, 5th ed. (and presumably earlier editions), Stechert, New York, 1934 (Prop. 34 and Ex. 217). For further information and references, including solutions to (e) and (f), see R. Stanley, *JPL Space Programs Summary 37–40*, vol. 4 (1966), 208–214, and S. M. Tanny, *J. Combinatorial Theory* 21 (1976), 196–202.

18. See R. Stanley, ibid.

19. *Answer.* $\sum_{k=0}^{n}(-1)^k \binom{2n-k}{k}(2n-2k-1)!!$, where $(2m-1)!! = 1 \cdot 3 \cdot 5 \cdots (2m-1)$.

20. Call a permutation *standard* if b_i is not immediately followed by a_i for $1 \le i \le n$. Clearly, each equivalence class contains exactly one standard permutation. A straightforward use of Inclusion-Exclusion shows that the number of standard permutations is equal to

$$\sum_{i=0}^{n} \binom{n}{i}(-1)^i (2n-i)! = \Delta^n (n+i)! \big|_{i=0}.$$

21. **b.** Let $A_k = [a_{j-i+1}]_1^k$. Let D be the diagonal matrix $\operatorname{diag}(\alpha, \alpha^2, \ldots, \alpha^k)$. Then $D^{-1}AD = [\alpha^{j-i} a_{j-i+1}]$. Since $\det A = \det D^{-1}AD$, the proof follows from (a).

  c. If we remove the first row and ith column from M, then we obtain a matrix

  $$M_i = \begin{bmatrix} B & C \\ 0 & A_{n-k} \end{bmatrix},$$ where B is an upper triangular $(i-1) \times (i-1)$ matrix with α's on the diagonal. Hence, when we expand $\det M$ along the first row, we get

  $$\det M = b_1(\det A_{k-1}) - \alpha b_2(\det A_{k-2}) + \cdots + (-1)^{k-1}\alpha^{k-1} b_k(\det A_0).$$

  The proof follows.

  d. For some alternative approaches and results related to this item, see Proposition 1.6.1, equation (1.59), Exercise 2.22, and equation (3.98).

22. **a.** Let S_k be the set of permutations $w = a_1 a_2 \cdots a_{2n} \in \mathfrak{S}_{2n}$ satisfying

  $$a_1 > a_2 < a_3 > a_4 < \cdots > a_{2n-2k}, \; a_{2n-2k+1} > a_{2n-2k+2} > \cdots > a_{2n},$$

  and let T_k be those permutations in S_k that also satisfy $a_{2n-2k} > a_{2n-2k+1}$. Hence, $S_1 - T_1$ consists of all alternating permutations in $\mathfrak{S}_n$. Moreover, $T_i = S_{i+1} - T_{i+1}$. Hence,

  $$E_n = \#(S_1 - T_1) = \#S_1 - \#(S_2 - T_2) = \cdots = \#S_1 - \#S_2 + \#S_3 - \cdots.$$

  A permutation in S_k is obtained by choosing $a_{2n-2k+1}, a_{2n-2k+2}, \ldots, a_{2n}$ in $\binom{2n}{2k}$ ways and then $a_1, a_2, \ldots, a_{2n-2k}$ in $E_{2(n-k)}$ ways. Hence, $\#S_k = \binom{2n}{2k} E_{2(n-k)}$, and the proof follows.

  b. The recurrence is

  $$E_{2n+1} = \binom{2n+1}{2} E_{2n-1} - \binom{2n+1}{4} E_{2n-3} + \binom{2n+1}{6} E_{2n-5} - \cdots + (-1)^n,$$

  proved similarly to (a) but with the additional complication of accounting for the term $(-1)^n$.

23. **a.** The argument is analogous to that of the previous exercise. Let S_k be the set of those permutations $a_1 a_2 \cdots a_n \in \mathfrak{S}_n$ such that $a_1 a_2 \cdots a_{n-k}$ has no proper double descents and $a_{n-k+1} > a_{n-k+1} > \cdots > a_n$. Let T_k consist of those permutations in S_k that also satisfy $a_{n-k-1} > a_{n-k} > a_{n-k+1}$. Let U_k consist of those permutations in S_k that also

satisfy $a_{n-k} > a_{n-k+1}$. Then $T_k = S_{k+2} - U_{k+2}$, $U_k = S_{k+1} - T_{k+1}$, and $S_0 = S_1 - T_1$. Hence,

$$
\begin{aligned}
f(n) = \#S_0 &= \#(S_1 - T_1) = \#S_1 - \#(S_3 - U_3) \\
&= \#S_1 - \#S_3 + \#(S_4 - T_4) = \#S_1 - \#S_3 + \#S_4 - \#(S_6 - U_6),
\end{aligned}
$$

and so on. Since $\#S_k = \binom{n}{k} f(n-k)$, the proof follows. This result (with a different proof) appears in F. N. David and D. E. Barton, *Combinatorial Chance*, Hafner, New York, 1962, pp. 156–157. See also I. M. Gessel, Ph.D. thesis, M.I.T. (Example 3, page 51), and I. P. Goulden and D. M. Jackson, *Combinatorial Enumeration*, John Wiley & Sons, New York, 1983; reprinted by Dover, Mineola, N.Y., 2004 (Exercise 5.2.17).

27. **a.** Follows easily from Proposition 5.3.2.
   b. Let $f_k(\overline{G})$ denote the coefficient of x^{k-1} in $P(\overline{G}, x)$; that is, $f_k(\overline{G})$ is equal to the number of k-component rooted forests F of $\overline{G}$. By the Principle of Inclusion-Exclusion,

$$
f_k(\overline{G}) = \sum_F (-1)^{n-\ell(F)} g_k(F),
$$

where F ranges over all spanning forests of G, and where $g_k(F)$ denotes the number of k-component rooted forests on $[n]$ that contain F. (Note that $n - \ell(F)$ is equal to the number of edges of F.) By (a), $g_k(F) = p(F)\binom{\ell-1}{\ell-k} n^{\ell-k}$, where $\ell = \ell(F)$. Hence,

$$
f_k(\overline{G}) = \sum_F (-1)^{n-\ell} p(F) \binom{\ell-1}{\ell-k} n^{\ell-k}. \tag{2.64}
$$

On the other hand, from equation (2.56) the coefficient of x^{k-1} in $(-1)^{n-1} P(G, -x-n)$ is equal to

$$
(-1)^{n-1} \sum_F (-1)^{\ell-1} p(F) \binom{\ell-1}{k-1} n^{\ell-1-(k-1)}, \tag{2.65}
$$

again summed over all spanning forests F of G, with $\ell = \ell(F)$. Since equations (2.64) and (2.65) agree, the result follows.

Equation (2.57) (essentially the case $x = 0$ of (2.56)) is implicit in H. N. V. Temperley, *Proc. Phys. Soc.* **83** (1984), 3–16. See also Theorem 6.2 of J. W. Moon, *Counting Labelled Trees*, Canadian Mathematical Monographs, no. 1, 1970. The general case (2.56) is due to S. D. Bedrosian, *J. Franklin Inst.* **227** (1964), 313–326. A subsequent proof of (2.56) using matrix techniques is due to A. K. Kelmans. See equation (2.19) in D. M. Cvetković, M. Doob, and H. Sachs, *Spectra of Graphs*, 2nd ed., Johann Ambrosius Barth Verlag, Heidelberg, 1995. A simple proof of (2.56) and additional references appear in J. W. Moon and S. D. Bedrosian, *J. Franklin Inst.* **316** (1983), 187–190.

Equation (2.56) may be regarded as a "reciprocity theorem" for rooted trees. It can be used, in conjunction with the obvious fact $P(G+H,x) = xP(G,x)P(H,x)$ (where $G + H$ denotes the disjoint union of G and H) to unify and simplify many known results involving the enumeration of spanning trees and forests. Part (c) illustrates this technique.

c. We have

$$P(K_1, x) = 1 \Rightarrow P(nK_1, x) = x^{n-1}$$
$$\Rightarrow P(K_n, x) = (x+n)^{n-1} \text{ (so } c(K_n) = n^{n-2})$$
$$\Rightarrow P(K_r + K_s, x) = x(x+r)^{r-1}(x+s)^{s-1}$$
$$\Rightarrow P(K_{r,s}, x) = (x+r+s)(x+s)^{r-1}(x+r)^{s-1}$$
$$\Rightarrow c(K_{r,s}) = s^{r-1} r^{s-1}.$$

28. This result appeared in R. Stanley [2.19, Ch. 5.3] and was stated without proof in [2.20, Prop. 23.8].

29. b. A generating function proof was given by G. E. Andrews, Concave compositions,

⟨http://www.math.psu.edu/andrews/pdf/277.pdf⟩.

30. See G. W. E. Andrews, in *The Theory of Arithmetic Functions* (A. A. Gioia and D. L. Goldsmith, eds.), Lecture Notes in Math., no. 251, Springer, Berlin, 1972, pp. 1–20. See also Chapter 9 in [1.2].

31. These identities are due to Gauss. See I. Pak [1.62, §5.5].

32. This identity is due to Gauss. A cancellation proof was given by W. Y. C. Chen, Q.-H. Hou and A. Lascoux, *J. Combinatorial Theory Ser. A* **102** (2003), 309–320, where several other proofs are also cited.

33. Let $S = \{1, 2, \ldots, n-1\}$ in (2.21). There is a unique $w \in \mathfrak{S}_n$ with $D(w) = S$, namely, $w = n, n-1, \ldots, 1$, and then $\text{inv}(w) = \binom{n}{2}$. Hence, $\beta_n(S, q) = q^{\binom{n}{2}}$. On the other hand, the right-hand side of (2.21) becomes the left-hand side of (2.58), and the proof follows.

34. This exercise is due to I. M. Gessel, *J. Graph Theory* **3** (1979), 305–307. Part (d) was first shown by M. G. Kendall and B. Babington Smith, *Biometrika* **33** (1940), 239–251. The crucial point in (e) is the following. Let G be the graph whose vertices are the tournaments T on $[n]$ and whose edges consist of pairs T, T' with $T \leftrightarrow T'$. Then from (c) and (d), we deduce that G is bipartite and that every connected component of G is regular, so the connected component containing the vertex T consists of a certain number of tournaments of weight $w(T)$ and an equal number of weight $-w(T)$.

Some far-reaching generalizations appear in D. Zeilberger and D. M. Bressoud, *Discrete Math.* **54** (1985), 201–224 (reprinted in *Discrete Math.* **306** (2006), 1039–1059); D. M. Bressoud, *Europ. J. Combinatorics* **8** (1987), 245–255; and R. M. Calderbank and P. J. Hanlon, *J. Combinatorial Theory Ser. A* **41** (1986), 228–245. The first of these references gives a solution to Exercise 1.19(c).

35. a. By linearity, it suffices to assume that f is a monomial of degree n. If the support of f (set of variables occurring in f) is S, then

$$f(\epsilon_1, \ldots, \epsilon_n) = \begin{cases} 1, & \epsilon_i = 1 \text{ for all } x_i \in S, \\ 0, & \text{otherwise.} \end{cases}$$

Hence,

$$\sum_{(\epsilon_1, \ldots, \epsilon_n) \in \{0,1\}^n} (-1)^{n - \sum \epsilon_i} f(\epsilon_1, \ldots, \epsilon_n) = \prod_{x_i \notin S} (1-1)$$

$$= \begin{cases} 1, & f = x_1 x_2 \cdots x_n, \\ 0, & \text{otherwise,} \end{cases}$$

and the proof follows.

b. Note that

$$\operatorname{per}(A) = [x_1 x_2 \cdots x_n] \prod_{i=1}^{n} (a_{i1}x_1 + a_{i2}x_2 + \cdots + a_{in}x_n),$$

and use (a). Equation (2.61) is due to H. J. Ryser, *Combinatorial Mathematics*, Math. Assoc. of America, 1963 (Chap. 2, Cor. 4.2). For further information on permanents, see H. Minc, *Permanents*, Encyclopedia of Mathematics and Its Applications, Vol. 6, Addison-Wesley, Reading, Mass., 1978; reprinted by Cambridge University Press, 1984.

36. See B. Gordon [2.12].

3

Partially Ordered Sets

3.1 Basic Concepts

The theory of partially ordered sets (or *posets*) plays an important unifying role in enumerative combinatorics. In particular, the theory of Möbius inversion on a partially ordered set is a far-reaching generalization of the Principle of Inclusion-Exclusion, and the theory of binomial posets provides a unified setting for various classes of generating functions. These two topics will be among the highlights of this chapter, though many other interesting uses of partially ordered sets will also be given.

To get a glimpse of the potential scope of the theory of partially ordered sets as it relates to the Principle of Inclusion-Exclusion, consider the following example. Suppose we have four finite sets A, B, C, D such that

$$D = A \cap B = A \cap C = B \cap C = A \cap B \cap C.$$

It follows from the Principle of Inclusion-Exclusion that

$$|A \cup B \cup C| = |A| + |B| + |C| - |A \cap B| - |A \cap C| - |B \cap C|$$
$$+ |A \cap B \cap C|$$
$$= |A| + |B| + |C| - 2|D|. \tag{3.1}$$

The relations $A \cap B = A \cap C = B \cap C = A \cap B \cap C$ collapsed the general seven-term expression for $|A \cup B \cup C|$ into a four-term expression, since the collection of intersections of A, B, C has only four distinct members. What is the significance of the coefficient -2 in equation (3.1)? Can we compute such coefficients efficiently for more complicated sets of equalities among intersections of sets $A_1, \ldots, A_n$? It is clear that the coefficient -2 depends only on the *partial order relation* among the distinct intersections A, B, C, D of the sets A, B, C – that is, on the fact that $D \subseteq A, D \subseteq B, D \subseteq C$ (where we continue to assume that $D = A \cap B = A \cap C = B \cap C = A \cap B \cap C$). In fact, we shall see that -2 is a certain value of the Möbius function of this partial order (with an additional element corresponding to the empty intersection adjoined). Hence, Möbius inversion results in a simplification of Inclusion-Exclusion under appropriate circumstances. However, we shall also see that the applications of Möbius inversion are much further-reaching than as a generalization of Inclusion-Exclusion.

Before plunging headlong into the theory of incidence algebras and Möbius functions, it is worthwhile to develop some feeling for the structure of finite partially ordered sets. Hence in the first five sections of this chapter, we collect together some of the basic definitions and results on the subject, though strictly speaking most of them are not needed in order to understand the theory of Möbius inversion.

A *partially ordered set* P (or *poset*, for short) is a set (which by abuse of notation we also call P), together with a binary relation denoted $\leq$ (or $\leq_P$ when there is a possibility of confusion), satisfying the following three axioms:

1. For all $t \in P$, $t \leq t$ (*reflexivity*).
2. If $s \leq t$ and $t \leq s$, then $s = t$ (*antisymmetry*).
3. If $s \leq t$ and $t \leq u$, then $s \leq u$ (*transitivity*).

We use the obvious notation $t \geq s$ to mean $s \leq t$, $s < t$ to mean $s \leq t$ and $s \neq t$, and $t > s$ to mean $s < t$. We say that two elements s and t of P are *comparable* if $s \leq t$ or $t \leq s$; otherwise, s and t are *incomparable**, denoted $s \parallel t$.

Before giving a rather lengthy list of definitions associated with posets, let us first look at some examples of posets of combinatorial interest that will later be considered in more detail.

3.1.1 Example. a. Let $n \in \mathbb{P}$. The set $[n]$ with its usual order forms an n-element poset with the special property that any two elements are comparable. This poset is denoted $\boldsymbol{n}$. Of course $\boldsymbol{n}$ and $[n]$ coincide as sets, but we use the notation $\boldsymbol{n}$ to emphasize the order structure.

b. Let $n \in \mathbb{N}$. We can make the set $2^{[n]}$ of all subsets of $[n]$ into a poset B_n by defining $S \leq T$ in B_n if $S \subseteq T$ as sets. One says that B_n consists of the subsets of $[n]$ "ordered by inclusion."

c. Let $n \in \mathbb{P}$. The set of all positive integer divisors of n can be made into a poset D_n in a "natural" way by defining $i \leq j$ in D_n if j is divisible by i (denoted $i \mid j$).

d. Let $n \in \mathbb{P}$. We can make the set Π_n of all partitions of $[n]$ into a poset (also denoted Π_n) by defining $\pi \leq \sigma$ in Π_n if every block of π is contained in a block of σ. For instance, if $n = 9$ and if π has blocks 137, 2, 46, 58, 9, and σ has blocks 13467, 2589, then $\pi \leq \sigma$. We then say that π is a *refinement* of σ and that Π_n consists of the partitions of $[n]$ "ordered by refinement."

e. In general, any collection of sets can be ordered by inclusion to form a poset. Some cases will be of special combinatorial interest. For instance, let $B_n(q)$ consist of all subspaces of the n-dimensional vector space $\mathbb{F}_q^n$, ordered by inclusion. We will see that $B_n(q)$ is a nicely behaved q-analogue of the poset B_n defined in (b).

---

* "Comparable" and "incomparable" are accented on the syllable "com."

We now list a number of basic definitions and results connected with partially ordered sets. Some readers may wish to skip directly to Section 3.6, and to consult the intervening material only when necessary.

Two posets P and Q are *isomorphic*, denoted $P \cong Q$, if there exists an *order-preserving bijection* $\phi \colon P \to Q$ whose inverse is order-preserving; that is,

$$s \leq t \text{ in } P \iff \phi(s) \leq \phi(t) \text{ in } Q.$$

For example, if B_S denotes the poset of all subsets of the set S ordered by inclusion, then $B_S \cong B_T$ whenever $\#S = \#T$.

Some care has to be taken in defining the notion of "subposet." By a *weak subposet* of P, we mean a subset Q of the elements of P and a partial ordering of Q such that if $s \leq t$ in Q, then $s \leq t$ in P. If Q is a weak subposet of P with $P = Q$ as sets, then we call P a *refinement* of Q. By an *induced subposet* of P, we mean a subset Q of P and a partial ordering of Q such that for $s, t, \in Q$ we have $s \leq t$ in Q if and only if $s \leq t$ in P. We then say the subset Q of P has the *induced order*. Thus, the finite poset P has exactly $2^{\#P}$ induced subposets. By a *subposet* of P, we will always mean an *induced* subposet. A special type of subposet of P is the (closed) *interval* $[s, t] = \{u \in P : s \leq u \leq t\}$, defined whenever $s \leq t$. (Thus, the empty set is *not* regarded as a closed interval.) The interval $[s, s]$ consists of the single point s. We similarly define the *open* interval $(s, t) = \{u \in P : s < u < t\}$, so $(s, s) = \emptyset$. If every interval of P is finite, then P is called a *locally finite* poset. We also define a subposet Q of P to be *convex* if $t \in Q$ whenever $s < t < u$ in P and $s, u \in Q$. Thus, an interval is convex.

If $s, t \in P$, then we say that t *covers* s or s is *covered by* t, denoted $s \lessdot t$ or $t \gtrdot s$, if $s < t$ and no element $u \in P$ satisfies $s < u < t$. Thus t covers s if and only if $s < t$ and $[s, t] = \{s, t\}$. A locally finite poset P is completely determined by its cover relations. The *Hasse diagram* of a finite poset P is the graph whose vertices are the elements of P, whose edges are the cover relations, and such that if $s \lessdot t$ then t is drawn "above" s (i.e., with a higher vertical coordinate). Figure 3.1 shows the Hasse diagrams of all posets (up to isomorphism) with at most four elements. Some care must be taken in "recognizing" posets from their Hasse diagrams. For instance,

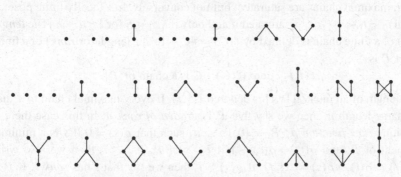

Figure 3.1 The posets with at most four elements.

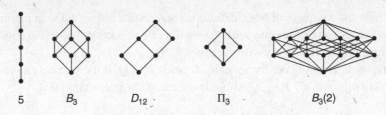

$$5 \qquad B_3 \qquad D_{12} \qquad \Pi_3 \qquad B_3(2)$$

Figure 3.2 Some examples of posets.

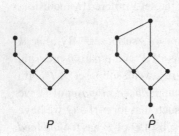

Figure 3.3 Adjoining a $\hat{0}$ and $\hat{1}$.

$$P \qquad\qquad \hat{P}$$

the graph ∧ is a perfectly valid Hasse diagram, yet appears to be missing from Figure 3.1. We trust the reader will resolve this anomaly. Similarly, why does the

graph ◇ not appear above? Figure 3.2 illustrates the Hasse diagrams of some of the posets considered in Example 3.1.1.

We say that P *has a* $\hat{0}$ if there exists an element $\hat{0} \in P$ such that $t \geq \hat{0}$ for all $t \in P$. Similarly, P *has a* $\hat{1}$ if there exists $\hat{1} \in P$ such that $t \leq \hat{1}$ for all $t \in P$. We denote by $\widehat{P}$ the poset obtained from P by adjoining a $\hat{0}$ and $\hat{1}$ (in spite of a $\hat{0}$ or $\hat{1}$ that P may already possess). See Figure 3.3 for an example.

A *chain* (or *totally ordered set* or *linearly ordered set*) is a poset in which any two elements are comparable. Thus, the poset n of Example 3.1.1(a) is a chain. A subset C of a poset P is called a *chain* if C is a chain when regarded as a subposet of P. The chain C of P is called *maximal* if it is not contained in a larger chain of P. The chain C of P is called *saturated* (or *unrefinable*) if there does not exist $u \in P - C$ such that $s < u < t$ for some $s, t \in C$ and such that $C \cup \{u\}$ is a chain. Thus, maximal chains are saturated, but not conversely. In a locally finite poset, a chain $t_0 < t_1 < \cdots < t_n$ is saturated if and only if $t_{i-1} \lessdot t_i$ for $1 \leq i \leq n$. The *length* $\ell(C)$ of a finite chain is defined by $\ell(C) = \#C - 1$. The length (or *rank*) of a finite poset P is

$$\ell(P) := \max\{\ell(C) : C \text{ is a chain of } P\}.$$

The length of an interval $[s, t]$ is denoted $\ell(s, t)$. If every maximal chain of P has the same length n, then we say that P is *graded of rank* n. In this case there is a unique *rank function* $\rho: P \to \{0, 1, \ldots, n\}$ such that $\rho(s) = 0$ if s is a minimal element of P, and $\rho(t) = \rho(s) + 1$ if $t \gtrdot s$ in P. If $s \leq t$, then we also write $\rho(s, t) = \rho(t) - \rho(s) = \ell(s, t)$. If $\rho(s) = i$, then we say that s has *rank* i. If P is graded of rank n and has p_i elements of rank i, then the polynomial

$$F(P,x) = \sum_{i=0}^{n} p_i x^i$$

is called the *rank-generating function* of P. For instance, all the posets $\boldsymbol{n}$, B_n, D_n, Π_n, and $B_n(q)$ are graded. The reader can check the entries of the following table (some of which will be discussed in more detail later).

Poset P	Rank of $t \in P$	Rank of P
$\boldsymbol{n}$	$t - 1$	$n - 1$
B_n	card t	n
D_n	number of prime divisors of t (counting multiplicity)	number of prime divisors of n (counting multiplicity)
Π_n	$n - |t|$	$n - 1$
$B_n(q)$	dim t	n

The rank-generating functions of these posets are as follows. For D_n, let $n = p_1^{a_1} \cdots p_k^{a_k}$ be the prime power factorization of n. We write, for example, $(\boldsymbol{n})_x$ for the q-analogue $(\boldsymbol{n})$ of n in the variable x, so

$$(\boldsymbol{n})_x = \frac{1 - x^n}{1 - x} = 1 + x + x^2 + \cdots + x^{n-1}.$$

$$F(\boldsymbol{n}, x) = (\boldsymbol{n})_x,$$

$$F(B_n, x) = (1 + x)^n,$$

$$F(D_n, x) = (a_1 + 1)_x \cdots (a_k + 1)_x,$$

$$F(\Pi_n, x) = \sum_{i=0}^{n-1} S(n, n-i) x^i,$$

$$F(B_n(q), x) = \sum_{i=0}^{n} \binom{n}{i} x^i.$$

We can extend the definition of a graded poset in an obvious way to certain infinite posets. Namely, we say that P is *graded* if it can be written $P = P_0 \cup P_1 \cup \cdots$ such that every maximal chain has the form $t_0 \lessdot t_1 \lessdot \cdots$, where $t_i \in P_i$. We then have a rank function $\rho : P \to \mathbb{N}$ just as in the finite case. If each P_i is finite then we also have a rank-generating function $F(P, q)$ as before, though now it may be a power series rather than a polynomial.

A *multichain* of the poset P is a chain with repeated elements; that is, a multiset whose underlying set is a chain of P. A *multichain of length n* may be regarded as a sequence $t_0 \le t_1 \le \cdots \le t_n$ of elements of P.

An *antichain* (or *Sperner family* or *clutter*) is a subset A of a poset P such that any two distinct elements of A are incomparable. An *order ideal* (or *semi-ideal* or *down-set* or *decreasing subset*) of P is a subset I of P such that if $t \in I$ and $s \leq t$, then $s \in I$. Similarly, a *dual order ideal* (or *up-set* or *increasing subset* or *filter*) is a subset I of P such that if $t \in I$ and $s \geq t$, then $s \in I$. When P is finite, there is a one-to-one correspondence between antichains A of P and order ideals I. Namely, A is the set of maximal elements of I, while

$$I = \{s \in P : s \leq t \text{ for some } t \in A\}. \tag{3.2}$$

The set of all order ideals of P, ordered by inclusion, forms a poset denoted $J(P)$. In Section 3.4 we shall investigate $J(P)$ in greater detail. If I and A are related as in equation (3.2), then we say that A *generates* I. If $A = \{t_1, \ldots, t_k\}$, then we write $I = \langle t_1, \ldots, t_k \rangle$ for the order ideal generated by A. The order ideal $\langle t \rangle$ is the *principal order ideal* generated by t, denoted Λ_t. Similarly V_t denotes the principal dual order ideal generated by t, that is, $V_t = \{s \in P : s \geq t\}$.

3.2 New Posets from Old

Various operations can be performed on one or more posets. If P and Q are posets on *disjoint* sets, then the *disjoint union* (or *direct sum*) of P and Q is the poset $P + Q$ on the union $P \cup Q$ such that $s \leq t$ in $P + Q$ if either (a) $s, t \in P$ and $s \leq t$ in P, or (b) $s, t \in Q$ and $s \leq t$ in Q. A poset that is not a disjoint union of two nonempty posets is said to be *connected*. The disjoint union of P with itself n times is denoted nP; hence, an n-element antichain is isomorphic to $n\mathbf{1}$. If P and Q are on disjoint sets as previously, then the *ordinal sum* of P and Q is the poset $P \oplus Q$ on the union $P \cup Q$ such that $s \leq t$ in $P \oplus Q$ if (a) $s, t \in P$ and $s \leq t$ in P, or (b) $s, t \in Q$ and $s \leq t$ in Q, or (c) $s \in P$ and $t \in Q$. Hence, an n-element chain is given by $n = \mathbf{1} \oplus \mathbf{1} \oplus \cdots \oplus \mathbf{1}$ (n times). Of the 16 four-element posets, exactly one of them cannot be built up from the poset $\mathbf{1}$ using the operations of disjoint union and ordinal sum. Posets that *can* be built up in this way are called *series-parallel posets*. (See Exercises 3.14, 3.15(c), and 5.39 for further information on such posets.)

If P and Q are posets, then the *direct* (or *cartesian*) *product* of P and Q is the poset $P \times Q$ on the set $\{(s, t) : s \in P \text{ and } t \in Q\}$ such that $(s, t) \leq (s', t')$ in $P \times Q$ if $s \leq s'$ in P and $t \leq t'$ in Q. The direct product of P with itself n times is denoted P^n. To draw the Hasse diagram of $P \times Q$ (when P and Q are finite), draw the Hasse diagram of P, replace each element t of P by a copy Q_t of Q, and connect corresponding elements of Q_s and Q_t (with respect to some isomorphism $Q_s \cong Q_t$) if s and t are connected in the Hasse diagram of P. For instance, the Hasse diagram of the direct product $\bigwedge \cdot \times \bigwedge$ is drawn as indicated in Figure 3.4.

It is clear from the definition of the direct product that $P \times Q$ and $Q \times P$ are isomorphic. However, the Hasse diagrams obtained by interchanging P and Q

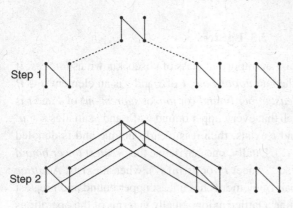

Figure 3.4 Drawing a direct product of posets.

in the above procedure in general look completely different, although they are of course isomorphic. If P and Q are graded with rank-generating functions $F(P,x)$ and $F(Q,x)$, then it is easily seen that $P \times Q$ is graded and

$$F(P \times Q,x) = F(P,x)F(Q,x). \tag{3.3}$$

A further operation on posets is the *ordinal product* $P \otimes Q$. This is the partial ordering on $\{(s,t) : s \in P \text{ and } t \in Q\}$ obtained by setting $(s,t) \leq (s',t')$ if (a) $s = s'$ and $t \leq t'$, or (b) $s < s'$. To draw the Hasse diagram of $P \otimes Q$ (when P and Q are finite), draw the Hasse diagram of P, replace each element t of P by a copy Q_t of Q, and then connect every maximal element of Q_s with every minimal element of Q_t whenever t covers s in P. If P and Q are graded and Q has rank r, then the analogue of equation (3.3) for ordinal products becomes

$$F(P \otimes Q,x) = F(P,x^{r+1})F(Q,x).$$

Note that in general $P \otimes Q$ and $Q \otimes P$ do not have the same rank-generating function, so in particular they are not isomorphic.

A further operation that we wish to consider is the *dual* of a poset P. This is the poset P^* on the same set as P, but such that $s \leq t$ in P^* if and only if $t \leq s$ in P. If P and P^* are isomorphic, then P is called *self-dual*. Of the 16 four-element posets, 8 are self-dual.

If P and Q are posets, then Q^P denotes the set of all order-preserving maps $f: P \to Q$; that is, $s \leq t$ in P implies $f(s) \leq f(t)$ in Q. We give Q^P the structure of a poset by defining $f \leq g$ if $f(t) \leq g(t)$ for all $t \in P$. It is an elementary exercise to check the validity of the following rules of *cardinal arithmetic* (for posets).

a. $+$ and $\times$ are associative and commutative.
b. $P \times (Q + R) \cong (P \times Q) + (P \times R)$.
c. $R^{P+Q} \cong R^P \times R^Q$.
d. $(R^P)^Q \cong R^{P \times Q}$.

3.3 Lattices

We now turn to a brief survey of an important class of posets known as *lattices*. If s and t belong to a poset P, then an *upper bound* of s and t is an element $u \in P$ satisfying $u \geq s$ and $u \geq t$. A *least upper bound* (or *join* or *supremum*) of s and t is an upper bound u of s and t such that every upper bound v of s and t satisfies $v \geq u$. If a least upper bound of s and t exists, then it is clearly unique and is denoted $s \vee t$ (read "s join t" or "s sup t"). Dually, one can define the *greatest lower bound* (or *meet* or *infimum*) $s \wedge t$ (read "s meet t" or "s inf t"), when it exists. A *lattice* is a poset L for which every pair of elements has a least upper bound and greatest lower bound. One can also define a lattice axiomatically in terms of the operations $\vee$ and $\wedge$, but for combinatorial purposes this is not necessary. The reader should check, however, that in a lattice L:

a. The operations $\vee$ and $\wedge$ are associative, commutative, and idempotent (i.e., $t \wedge t = t \vee t = t$);
b. $s \wedge (s \vee t) = s = s \vee (s \wedge t)$ (absorption laws);
c. $s \wedge t = s \Leftrightarrow s \vee t = t \Leftrightarrow s \leq t$.

Clearly all finite lattices have a $\hat{0}$ and $\hat{1}$. If L and M are lattices, then so are L^*, $L \times M$, and $L \oplus M$. However, $L + M$ will never be a lattice unless one of L or M is empty, but $\widehat{L + M}$ (i.e., $L + M$ with an $\hat{0}$ and $\hat{1}$ adjoined) is always a lattice. Figure 3.5 shows the Hasse diagrams of all lattices with at most six elements.

In checking whether a (finite) poset is a lattice, it is sometimes easy to see that meets, say, exist, but the existence of joins is not so clear. Thus, the criterion of the next proposition can be useful. If every pair of elements of a poset P has a

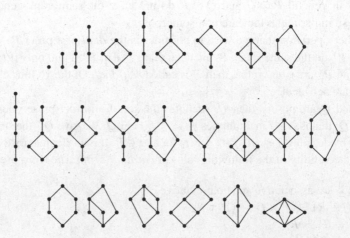

Figure 3.5 The lattices with at most six elements.

meet (respectively, join), then we say that P is a *meet-semilattice* (respectively, *join-semilattice*).

3.3.1 Proposition. *Let P be a finite meet-semilattice with $\hat{1}$. Then P is a lattice. (Of course, dually a finite join-semilattice with $\hat{0}$ is a lattice.)*

Proof. If $s, t \in P$, then the set $S = \{u \in P : u \geq s \text{ and } u \geq t\}$ is finite (since P is finite) and nonempty (since $\hat{1} \in P$). Clearly by induction, the meet of finitely many elements of a meet-semilattice exists. Hence, we have $s \vee t = \bigwedge_{u \in S} u$. □

Proposition 3.3.1 fails for infinite lattices because an *arbitrary* subset of L need not have a meet or join. (See Exercise 3.26.) If in fact every subset of L does have a meet and join, then L is called a *complete lattice*. Clearly, a complete lattice has a $\hat{0}$ and $\hat{1}$.

We now consider one of the types of lattices of most interest to combinatorics.

3.3.2 Proposition. *Let L be a finite lattice. The following two condtions are equivalent.*

  i. *L is graded, and the rank function ρ of L satisfies*

$$\rho(s) + \rho(t) \geq \rho(s \wedge t) + \rho(s \vee t)$$

  for all $s, t \in L$.
 ii. *If s and t both cover $s \wedge t$, then $s \vee t$ covers both s and t.*

Proof. (i)$\Rightarrow$(ii) Suppose that s and t cover $s \wedge t$. Then $\rho(s) = \rho(t) = \rho(s \wedge t) + 1$ and $\rho(s \vee t) > \rho(s) = \rho(t)$. Hence by (i), $\rho(s \vee t) = \rho(s) + 1 = \rho(t) + 1$, so $s \vee t$ covers both s and t.

(ii)$\Rightarrow$(i) Suppose that L is not graded, and let $[u, v]$ be an interval of L of minimal length that is not graded. Then there are elements s_1, s_2 of $[u, v]$ that cover u and such that all maximal chains of each interval $[s_i, v]$ have the same length ℓ_i, where $\ell_1 \neq \ell_2$. By (ii), there are saturated chains in $[s_i, v]$ of the form $s_i < s_1 \vee s_2 < t_1 < t_2 < \cdots < t_k = v$, contradicting $\ell_1 \neq \ell_2$. Hence, L is graded.

Now suppose that there is a pair $s, t \in L$ with

$$\rho(s) + \rho(t) < \rho(s \wedge t) + \rho(s \vee t), \tag{3.4}$$

and choose such a pair with $\ell(s \wedge t, s \vee t)$ minimal, and then with $\rho(s) + \rho(t)$ minimal. By (ii), we cannot have both s and t covering $s \wedge t$. Thus, assume that $s \wedge t < s' < s$, say. By the minimality of $\ell(s \wedge t, s \vee t)$ and $\rho(s) + \rho(t)$, we have

$$\rho(s') + \rho(t) \geq \rho(s' \wedge t) + \rho(s' \vee t). \tag{3.5}$$

Now $s' \wedge t = s \wedge t$, so equations (3.4) and (3.5) imply

$$\rho(s) + \rho(s' \vee t) < \rho(s') + \rho(s \vee t).$$

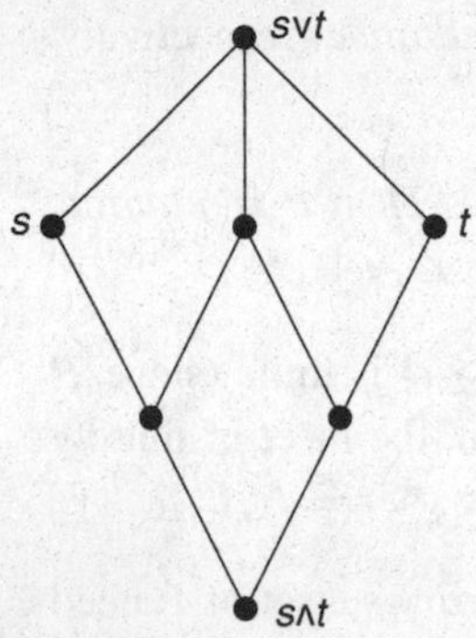

Figure 3.6 A semimodular but nonmodular lattice.

Clearly, $s \wedge (s' \vee t) \geq s'$ and $s \vee (s' \vee t) = s \vee t$. Hence, setting $S = s$ and $T = s' \vee t$, we have found a pair $S, T \in L$ satisfying $\rho(S) + \rho(T) < \rho(S \wedge T) + \rho(S \vee T)$ and $\ell(S \wedge T, S \vee T) < \ell(s \wedge t, s \vee t)$, a contradiction. This completes the proof.   □

A finite lattice satisfying either of the (equivalent) conditions of the previous proposition is called a *finite upper semimodular lattice*, or a just a *finite semimodular lattice*. The reader may check that of the 15 lattices with six elements, exactly 8 are semimodular.

A finite lattice L whose dual L^* is semimodular is called *lower semimodular*. A finite lattice that is both upper and lower semimodular is called a *modular lattice*. By Proposition 3.3.2, a finite lattice L is modular if and only if it is graded, and its rank function ρ satisfies

$$\rho(s) + \rho(t) = \rho(s \wedge t) + \rho(s \vee t) \ \text{ for all } s, t \in L. \tag{3.6}$$

For instance, the lattice $B_n(q)$ of subspaces (ordered by inclusion) of an n-dimensional vector space over the field $\mathbb{F}_q$ is modular, since the rank of a subspace is just its dimension, and equation (3.6) is then familiar from linear algebra. Every semimodular lattice with at most six elements is modular. There is a unique seven-element non-modular, semimodular lattice, which is shown in Figure 3.6. This lattice is not modular since $s \vee t$ covers s and t, but s and t don't cover $s \wedge t$. It can be shown that a finite lattice L is modular if and only if for all $s, t, u \in L$ such that $s \leq u$, we have

$$s \vee (t \wedge u) = (s \vee t) \wedge u. \tag{3.7}$$

This allows the concept of modularity to be extended to nonfinite lattices, though we will only be concerned with the finite case. Equation (3.7) also shows immediately that a sublattice of a modular lattice is modular. (A subset M of a lattice L is a *sublattice* if it is closed under the operations of $\wedge$ and $\vee$ in L.)

A lattice L with $\hat{0}$ and $\hat{1}$ is *complemented* if for all $s \in L$ there is a $t \in L$ such that $s \wedge t = \hat{0}$ and $s \vee t = \hat{1}$. If for all $s \in L$ the complement t is unique, then L is *uniquely complemented*. If every interval $[s, t]$ of L is itself complemented, then L is *relatively complemented*. An *atom* of a finite lattice L is an element covering

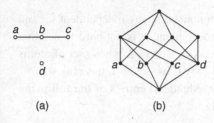

Figure 3.7 A subset S of the affine plane and its corresponding geometric lattice $L(S)$.

(a)   (b)

$\hat{0}$, and L is said to be *atomic* (or a *point lattice*) if every element of L is a join of atoms. (We always regard $\hat{0}$ as the join of an empty set of atoms.) Dually, a *coatom* is an element that $\hat{1}$ covers, and a *coatomic* lattice is defined in the obvious way. Another simple result of lattice theory, whose proof we omit, is the following.

3.3.3 Proposition. *Let L be a finite semimodular lattice. The following two conditions are equivalent.*

  i. L is relatively complemented.
  ii. L is atomic.

A finite semimodular lattice satisfying either of the two (equivalent) conditions (i) or (ii) is called a *finite geometric lattice*. A basic example is the following. Take any finite set S of points in some affine space (respectively, vector space) V over a field K (or even over a division ring). Then the subsets of S of the form $S \cap W$, where W is an affine subspace (respectively, linear subspace) of V, ordered by inclusion, form a geometric lattice $L(S)$. For instance, taking $S \subset \mathbb{R}^2$ (regarded as an affine space) to be as in Figure 3.7(a), then the elements of $L(S)$ consist of $\emptyset$, $\{a\}$, $\{b\}$, $\{c\}$, $\{d\}$, $\{a,d\}$, $\{b,d\}$, $\{c,d\}$, $\{a,b,c\}$, $\{a,b,c,d\}$. For this example, $L(S)$ is in fact modular and is shown in Figure 3.7(b).

    NOTE. A geometric lattice is intimately related to the subject of *matroid theory*. A (finite) *matroid* may be defined as a pair $(S, \mathcal{I})$, where S is a finite set and $\mathcal{I}$ is a collection of subsets of S satisfying the two conditions:

- If $F \in \mathcal{I}$ and $G \subseteq F$, then $G \in \mathcal{I}$. In other words, $\mathcal{I}$ is an order ideal of the boolean algebra B_S of all subsets of S (defined in Section 3.4).
- For any $T \subseteq S$, let $\mathcal{I}_T$ be the restriction of $\mathcal{I}$ to T, that is, $\mathcal{I}_T = \{F \in \mathcal{I} : F \subseteq T\}$. Then all maximal (under inclusion) elements of $\mathcal{I}_T$ have the same number of elements.

(There are several equivalent definitions of a matroid.) The elements of $\mathcal{I}$ are called *independent* sets. They are an abstraction of linear independent sets of a vector space or affinely independent subsets of an affine space. Indeed, if S is a finite subset of a vector space (respectively, affine subset of an affine space) and $\mathcal{I}$ is the collection of linearly independent (respectively, affinely independent) subsets of S, then $(S, \mathcal{I})$ is a matroid. A matroid is *simple* if every two-element subset of $\mathcal{I}$ is independent. Every matroid can be "simplified" (converted to a simple matroid) by removing all elements of S not contained in any independent set and by identifying any two points that are not independent. It is not hard to see that matroids on a set S are in bijection with geometric lattices L whose set of atoms is S, where a set $T \subseteq S$ is independent if and only if its join in L has rank $\#T$.

The reader may wish to verify the (partly redundant) entries of the following table concerning the posets of Example 3.1.1.

Poset P	Properties that P possesses	Properties that P lacks (n large)
n	modular lattice	complemented, atomic, coatomic, geometric
B_n	modular lattice, relatively complemented, uniquely complemented, atomic, coatomic, geometric	
D_n	modular lattice	complemented, atomic coatomic, geometric (unless n is squarefree, in which case $D_n \cong B_k$)
Π_n	geometric lattice	modular
$B_n(q)$	modular lattice, complemented, atomic, coatomic, geometric	uniquely complemented

3.4 Distributive Lattices

The most important class of lattices from the combinatorial point of view are the *distributive* lattices. These are defined by the distributive laws

$$s \vee (t \wedge u) = (s \vee t) \wedge (s \vee u),$$
$$s \wedge (t \vee u) = (s \wedge t) \vee (s \wedge u). \tag{3.8}$$

(One can prove that either of these laws implies the other.) If we assume $s \leq u$ in the first law, then we obtain equation (3.7), since $s \vee u = u$. Hence, every distributive lattice is modular. The lattices n, B_n, and D_n of Example 3.1.1 are distributive, while Π_n ($n \geq 3$) and $B_n(q)$ ($n \geq 2$) are not distributive. Further examples of distributive lattices are the lattices $J(P)$ of order ideals of the poset P. The lattice operations $\wedge$ and $\vee$ on order ideals are just ordinary intersection and union (as subsets of P). Since the union and intersection of order ideals is again an order ideal, it follows from the well-known distributivity of set union and intersection over one another that $J(P)$ is indeed a distributive lattice. The *fundamental theorem for finite distributive lattices* (FTFDL) states that the converse is true when P is finite.

3.4.1 Theorem (FTFDL). *Let L be a finite distributive lattice. Then there is a unique (up to isomorphism) poset P for which $L \cong J(P)$.*

Remark. For combinatorial purposes, it would in fact be best to *define* a finite distributive lattice as any poset of the form $J(P)$, P finite. However, to avoid conflict with established practices we have given the usual definition.

To prove Theorem 3.4.1, we first need to produce a candidate P and then show that indeed $L \cong J(P)$. Toward this end, define an element s of a lattice L to be *join-irreducible* if $s \neq \hat{0}$ and one cannot write $s = t \vee u$ where $t < s$ and $u < s$. (*Meet-irreducible* is defined dually.) In a finite lattice, an element is join-irreducible if and only if it covers exactly one element. An order ideal I of the finite poset P is join-irreducible in $J(P)$ if and only if it is a principal order ideal of P. Hence, there is a one-to-one correspondence between the join-irreducibles Λ_s of $J(P)$ and the elements s of P. Since $\Lambda_s \subseteq \Lambda_t$ if and only if $s \leq t$, we obtain the following result.

3.4.2 Proposition. *The set of join-irreducibles of $J(P)$, considered as an (induced) subposet of $J(P)$, is isomorphic to P. Hence, $J(P) \cong J(Q)$ if and only if $P \cong Q$.*

Proof of Theorem 3.4.1. Because of Proposition 3.4.2, it suffices to show that if P is the subposet of join-irreducibles of L, then $L \cong J(P)$. Given $t \in L$, let $I_t = \{s \in P : s \leq t\}$. Clearly $I_t \in J(P)$, so the mapping $t \mapsto I_t$ defines an order-preserving (in fact, meet-preserving) map $L \overset{\phi}{\to} J(P)$ whose inverse is order-preserving on $\phi(L)$. Moreover, ϕ is injective since $J(P)$ is a lattice. Hence, we need to show that ϕ is surjective. Let $I \in J(P)$ and $t = \bigvee \{s : s \in I\}$. We need to show that $I = I_t$. Clearly, $I \subseteq I_t$. Suppose that $u \in I_t$. Now

$$\bigvee \{s : s \in I\} = \bigvee \{s : s \in I_t\}. \tag{3.9}$$

Apply $\wedge u$ to equation (3.9). By distributivity, we get

$$\bigvee \{s \wedge u : s \in I\} = \bigvee \{s \wedge u : s \in I_t\}. \tag{3.10}$$

The right-hand side is just u, since one term is u and all others are $\leq u$. Since u is join-irreducible (being by definition an element of P), it follows from equation (3.10) that some $t \in I$ satisfies $t \wedge u = u$, that is, $u \leq t$. Since I is an order ideal we have $u \in I$, so $I_t \subseteq I$. Hence, $I = I_t$, and the proof is complete. $\square$

In certain combinatorial problems, infinite distributive lattices of a special type occur naturally. Thus we define a *finitary* distributive lattice to be a locally finite distributive lattice L with $\hat{0}$. It follows that L has a unique rank function $\rho : L \to \mathbb{N}$ given by letting $\rho(t)$ be the length of any saturated chain from $\hat{0}$ to t. If L has finitely many elements p_i of any given rank $i \in \mathbb{N}$, then we can define the *rank-generating function* $F(L, x)$ by

$$F(L, x) = \sum_{t \in L} x^{\rho(t)} = \sum_{i \geq 0} p_i x^i.$$

In this case, of course, $F(L, x)$ need not be a polynomial but in general is a formal power series. We leave to the reader to check that FTFDL carries over to finitary distributive lattices as follows.

3.4.3 Proposition. *Let P be a poset for which every principal order ideal is finite. Then the poset $J_f(P)$ of* finite *order ideals of P, ordered by inclusion, is a finitary distributive lattice. Conversely, if L is a finitary distributive lattice and P is its subposet of join-irreducibles, then every principal order ideal of P is finite and $L \cong J_f(P)$.*

3.4.4 Example. (a) If P is an infinite antichain, then $J_f(P)$ has infinitely many elements on each level, so $F(J_f(P), x)$ is undefined.

(b) Let $P = \mathbb{N} \times \mathbb{N}$. Then $J_f(P)$ is a very interesting distributive lattice known as *Young's lattice*, denoted Y. It is not hard to see that

$$F(Y, x) = \sum_{i \geq 0} p(i) x^i = \frac{1}{\prod_{n \geq 1}(1 - x^n)},$$

where $p(i)$ denotes the number of partitions of i (Sections 1.7 and 1.8). In fact, Y is isomorphic to the poset of all partitions $\lambda = (\lambda_1, \lambda_2, \dots)$ of all integers $n \geq 0$, ordered componentwise (or by containment of Young diagrams). For further information on Young's lattice, see Exercise 3.149, Section 3.21, and various places in Chapter 7.

We now turn to an investigation of the combinatorial properties of $J(P)$ (where P is finite) and of the relationship between P and $J(P)$. If I is an order ideal of P, then the elements of $J(P)$ that cover I are just the order ideals $I \cup \{t\}$, where t is a minimal element of $P - I$. From this observation we conclude the following result.

3.4.5 Proposition. *If P is an n-element poset, then $J(P)$ is graded of rank n. Moreover, the rank $\rho(I)$ of $I \in J(P)$ is just the cardinality $\#I$ of I, regarded as an order ideal of P.*

It follows from Propositions 3.4.2, 3.4.5, and FTFDL that there is a bijection between (nonisomorphic) posets P of cardinality n and (nonisomorphic) distributive lattices of rank n. This bijection sends P to $J(P)$, and the inverse sends $J(P)$ to its poset of join-irreducibles. In particular, the number of nonisomorphic posets of cardinality n equals the number of nonisomorphic distributive lattices of rank n.

If $P = \mathbf{n}$, an n-element chain, then $J(P) \cong \mathbf{n+1}$. At the other extreme, if $P = n\mathbf{1}$, an n-element antichain, then any subset of P is an order ideal, and $J(P)$ is just the set of subsets of P, ordered by inclusion. Hence $J(n\mathbf{1})$ is isomorphic to the poset B_n of Example 3.1.1(b), and we simply write $B_n = J(n\mathbf{1})$. We call B_n a *boolean algebra* of rank n. (The usual definition of a boolean algebra gives it more structure than merely that of a distributive lattice, but for our purposes we simply regard B_n as a certain distributive lattice.) It is clear from FTFDL (or otherwise) that the following conditions on a finite distributive lattice L are equivalent.

a. L is a boolean algebra.
b. L is complemented.

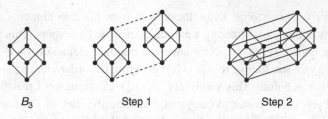

Figure 3.8 Drawing B_4 from B_3.

c. L is relatively complemented.

d. L is atomic.

e. $\hat{1}$ is a join of atoms of L.

f. L is a geometric lattice.

g. Every join-irreducible of L covers $\hat{0}$.

h. If L has n join-irreducibles (equivalently, rank$(L) = n$), then L has at least (equivalently, exactly) 2^n elements.

i. The rank-generating function of L is $(1+x)^n$ for some $n \in \mathbb{N}$.

Given an order ideal I of P, define a map $f_I : P \to \mathbf{2}$ by

$$f_I(t) = \begin{cases} 1, & t \in I \\ 2, & t \notin I. \end{cases}$$

Clearly, f is order-preserving, that is, $f \in \mathbf{2}^P$. Then $f_I \leq f_{I'}$ in $\mathbf{2}^P$ if and only if $I \supseteq I'$. Hence, $J(P)^* \cong \mathbf{2}^P$. Note also that $J(P^*) \cong J(P)^*$ and $J(P+Q) \cong J(P) \times J(Q)$. In particular, $B_n = J(n\mathbf{1}) \cong J(\mathbf{1})^n \cong \mathbf{2}^n$. This observation gives an efficient method for drawing B_n using the method of the previous section for drawing products. For instance, the Hasse diagram of B_3 is given by the first diagram in Figure 3.8. The other two diagrams show how to obtain the Hasse diagram of B_4.

If $I \leq I'$ in the distributive lattice $J(P)$, then the interval $[I, I']$ is isomorphic to $J(I' - I)$, where $I' - I$ is regarded as an (induced) subposet of P. In particular, $[I, I']$ is a distributive lattice. (More generally, any sublattice of a distributive lattice is distributive, an immediate consequence of the definition (3.8) of a distributive lattice.) It follows that there is a one-to-one correspondence between intervals $[I, I']$ of $J(P)$ isomorphic to B_k ($k \geq 1$) such that no interval $[K, I']$ with $K < I$ is a boolean algebra, and k-element antichains of P. Equivalently, k-element antichains in P correspond to elements of $J(P)$ that cover exactly k elements.

We can use these ideas to describe a method for drawing the Hasse diagram of $J(P)$, given P. Let I be the set of minimal elements of P, say of cardinality m. To begin with, draw $B_m \cong J(I)$. Now choose a minimal element of $P - I$, say t. Adjoin a join-irreducible to $J(I)$ covering the order ideal $\Lambda_t - \{t\}$. The set of joins of elements covering $\Lambda_t - \{t\}$ must form a boolean algebra, so draw in any new joins necessary to achieve this. Now there may be elements covering $\Lambda_t - \{t\}$ whose covers don't yet have joins. Draw these in to form boolean algebras. Continue until all sets of elements covering a particular element have joins. This yields the distributive lattice $J(I \cup \{t\})$. Now choose a minimal element u of $P - I - \{t\}$ and adjoin a join-irreducible to $J(I \cup \{t\})$ covering the order ideal $\lambda_u - \{u\}$. "Fill in" the covers as before. This yields $J(I \cup \{t, u\})$. Continue until reaching $J(P)$. The actual process is easier to carry out than describe. Let us illustrate with P given by Figure 3.9(a). We will denote subsets of P such as $\{a, b, d\}$ as abd. First, draw $B_3 = J(abc)$ as in Figure 3.9(b). Adjoin the order ideal $\Lambda_d = abd$ above ab (and label it d) (Figure 3.9(c)). Fill in the joins of the elements covering ab (Figure 3.9(d)). Adjoin bce above bc (Figure 3.9(e)). Fill in joins of elements covering bc (Figure 3.9(f)). Fill in joins of elements covering abc (Figure 3.9(g)). Adjoin cf above c (Figure 3.9(h)). Fill in joins of elements covering c. These joins (including the empty join c) form a rank three boolean algebra. The elements c, ac, bc, cf, and abc are already there, so we need the three additional elements acf, bcf, and $abcf$ (Figure 3.9(i)). Now fill in joins of elements covering bc (Figure 3.9(j)). Finally, fill in joins of elements covering abc (Figure 3.9(k)). With a little practice, this procedure yields a fairly efficient method for computing the rank-generating function $F(J(P), x)$ by hand. For the present example, we see that

$$F(J(P), x) = 1 + 3x + 4x^2 + 5x^3 + 4x^4 + 3x^5 + x^6.$$

For further information about "zigzag" posets (or fences) as in Figure 3.9, see Exercise 3.66.

3.5 Chains in Distributive Lattices

We have seen that many combinatorial properties of the finite poset P have simple interpretations in terms of $J(P)$. For instance, the number of k-element order ideals of P equals the number of elements of $J(P)$ of rank k, and the number of k-element antichains of P equals the number of elements of $J(P)$ that cover exactly k elements. We wish to discuss one further example of this nature.

3.5.1 Proposition. *Let P be a finite poset and $m \in \mathbb{N}$. The following quantities are equal:*

a. *The number of order-preserving maps $\sigma : P \to m$,*
b. *The number of multichains $\hat{0} = I_0 \le I_1 \le \cdots \le I_m = \hat{1}$ of length m in $J(P)$,*
c. *The cardinality of $J(P \times m - 1)$.*

Proof. Given $\sigma : P \to m$, define $I_j = \sigma^{-1}(j)$. Given $\hat{0} = I_0 \le I_1 \le \cdots \le I_m = \hat{1}$, define the order ideal I of $P \times m - 1$ by $I = \{(t, j) \in P \times m - 1 : t \in I_{m-j}\}$. Given the order ideal I of $P \times m - 1$, define $\sigma : P \to m$ by $\sigma(t) = \min\{m - j : (t, j) \in I\}$ if $(t, j) \in I$ for some j, and otherwise $\sigma(t) = m$. These constructions define the desired bijections. $\square$

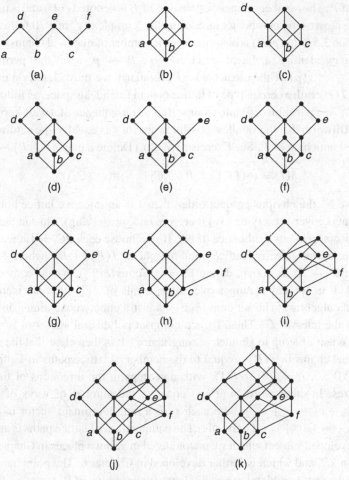

Figure 3.9 Drawing $J(P)$.

Note that the equivalence of (a) and (c) also follows from the computation

$$m^P \cong (2^{m-1})^P \cong 2^{m-1 \times P}.$$

As a modification of the preceding proposition, we have the following result.

3.5.2 Proposition. *Preserve the notation of Proposition 3.5.1. The following quantities are equal:*

a. The number of surjective *order-preserving maps* $\sigma : P \to m$,

b. The number of chains $\hat{0} = I_0 < I_1 < \cdots < I_m = \hat{1}$ *of length m in* $J(P)$.

Proof. Analogous to the proof of Proposition 3.5.1. ☐

One special case of Proposition 3.5.2 is of particular interest. If $\#P = p$, then an order-preserving bijection $\sigma : P \to p$ is called a *linear extension* or *topological sorting* of P. The number of linear extensions of P is denoted $e(P)$ and is probably the single most useful number for measuring the "complexity" of P. It follows from Proposition 3.5.2 that $e(P)$ is also equal to the number of maximal chains of $J(P)$.

We may identify a linear extension $\sigma: P \to p$ with the permutation $\sigma^{-1}(1), \ldots, \sigma^{-1}(p)$ of the elements of P. Similarly we may identify a maximal chain of $J(P)$ with a certain type of lattice path in Euclidean space, as follows. Let $C_1, \ldots, C_k$ be a partition of P into chains. (It is a consequence of a well-known theorem of Dilworth that the smallest possible value of k is equal to the cardinality of the largest antichain of P. See Exercise 3.77(d).) Define a map $\delta: J(P) \to \mathbb{N}^k$ by

$$\delta(I) = (\#(I \cap C_1), \#(I \cap C_2), \ldots, \#(I \cap C_k)).$$

If we give $\mathbb{N}^k$ the obvious product order, then δ is an injective lattice homomorphism that is cover-preserving (and therefore rank-preserving). Thus in particular, $J(P)$ is isomorphic to a sublattice of $\mathbb{N}^k$. If we choose each $\#C_i = 1$, then we get a rank-preserving injective lattice homomorphism $J(P) \to B_p$, where $\#P = p$. Given $\delta: P \to \mathbb{N}^k$ as above, define $\Gamma_\delta = \bigcup_T \mathrm{conv}(\delta(T))$, where conv denotes convex hull in $\mathbb{R}^k$ and T ranges over all intervals of $J(P)$ that are isomorphic to boolean algebras. (The set $\mathrm{conv}(\delta(T))$ is just a cube whose dimension is the length of the interval T.) Thus, Γ_δ is a compact polyhedral subset of $\mathbb{R}^k$, which is independent of δ (up to geometric congruence). It is then clear that the number of maximal chains in $J(P)$ is equal to the number of lattice paths in Γ_δ from the origin $(0, 0, \ldots, 0) = \delta(\hat{0})$ to $\delta(\hat{1})$, with unit steps in the directions of the coordinate axes. In other words, $e(P)$ is equal to the number of ways of writing $\delta(\hat{1}) = v_1 + v_2 + \cdots + v_p$, where each v_i is a unit coordinate vector in $\mathbb{R}^k$ and where $v_1 + v_2 + \cdots + v_i \in \Gamma_\delta$ for all i. The enumeration of lattice paths is an extensively developed subject which we encountered in various places in Chapter 1 and in Section 2.7, and which is further developed in Chapter 6. The point here is that certain lattice path problems are equivalent to determining $e(P)$ for some P. Thus, they are also equivalent to the problem of counting certain types of permutations.

3.5.3 Example. Let P be given by Figure 3.10(a). Take $C_1 = \{a, c\}$, $C_2 = \{b, d, e\}$. Then $J(P)$ has the embedding δ into $\mathbb{N}^2$ given by Figure 3.10(b). To get the polyhedral set Γ_δ, we simply "fill in" the squares in Figure 3.10(b), yielding the polyhedral set of Figure 3.10(c). There are nine lattice paths of the required type from $(0, 0)$ to $(2, 3)$ in Γ_δ, that is, $e(P) = 9$. The corresponding nine permutations of P are $abcde, bacde, abdce, badce, bdace, abdec, badec, bdaec, bdeac$.

3.5.4 Example. Let P be a disjoint union $C_1 + C_2$ of chains C_1 and C_2 of cardinalities m and n. Then Γ_δ is an $m \times n$ rectangle with vertices $(0, 0)$, $(m, 0)$, $(0, n)$, (m, n). As noted in Proposition 1.2.1, the number of lattice paths from $(0, 0)$ to (m, n) with steps $(1, 0)$ and $(0, 1)$ is just $\binom{m+n}{m} = e(C_1 + C_2)$. A linear extension $\sigma: P \to m + n$ is completely determined by the image $\sigma(C_1)$, which can be any m-element subset of $m + n$. Thus once again, we obtain $e(C_1 + C_2) = \binom{m+n}{m}$. More

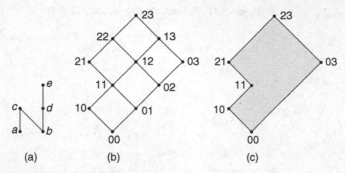

Figure 3.10 A polyhedral set associated with a finite distributive lattice.

generally, if $P = P_1 + P_2 + \cdots + P_k$ and $n_i = \#P_i$, then

$$e(P) = \binom{n_1 + \cdots + n_k}{n_1, \ldots, n_k} e(P_1)e(P_2) \cdots e(P_k).$$

3.5.5 Example. Let $P = 2 \times n$, and take $C_1 = \{(2, j) : j \in n\}$, $C_2 = \{(1, j) : j \in n\}$. Then $\delta(J(P)) = \{(i, j) \in \mathbb{N}^2 : 0 \leq i \leq j \leq n\}$. For example, the embedded poset $\delta(J(2 \times 3))$ is shown in Figure 3.11. Hence, $e(P)$ is equal to the number of lattice paths from $(0,0)$ to (n,n), with steps $(1,0)$ and $(0,1)$, that never fall below (or by symmetry, that never rise above) the main diagonal $x = y$ of the (x, y)-plane. These lattice paths arose in the enumeration of 321-avoiding permutations in Section 1.5, where it was mentioned that they are counted by the Catalan numbers $C_n = \frac{1}{n+1}\binom{2n}{n}$. It follows that $e(2 \times n) = C_n$. By the definition of $e(P)$, we see that this number is also equal to the number of $2 \times n$ matrices with entries the distinct integers $1, 2, \ldots, 2n$, such that every row and column is increasing. For instance, $e(2 \times 3) = 5$, corresponding to the matrices

$$\begin{matrix} 123 & 124 & 125 & 134 & 135 \\ 456 & 356 & 346 & 256 & 246. \end{matrix}$$

Such matrices are examples of *standard Young tableaux* (SYT), discussed extensively in Chapter 7.

We have now seen two ways of looking at the numbers $e(P)$: as counting certain order-preserving maps (or permutations) and as counting certain chains (or lattice paths). There is yet another way of viewing $e(P)$—as satisfying a certain *recurrence*. Regard e as a function on $J(P)$, that is, if $I \in J(P)$ then $e(I)$ is the number of linear extensions of I (regarded as a subposet of P). Thus, $e(I)$ is also the number of saturated chains from $\hat{0}$ to I in $J(P)$. From this observation it is clear that

Figure 3.11 The distributive lattice $J(\mathbf{2} \times \mathbf{3})$.

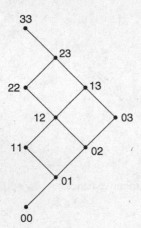

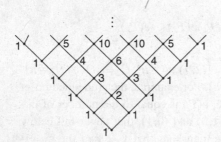

Figure 3.12 The distributive lattice $J_f(\mathbb{N} + \mathbb{N})$.

$$e(I) = \sum_{I' \lessdot I} e(I'), \tag{3.11}$$

where I' ranges over all elements of $J(P)$ that I covers. In other words, if we label the element $I \in J(P)$ by $e(I)$, then $e(I)$ is the sum of those $e(I')$ that lie "just below" I. This recurrence is analogous to the definition of Pascal's triangle, where each entry is the sum of the two "just above." Indeed, if we take P to be the infinite poset $\mathbb{N} + \mathbb{N}$ and let $J_f(P)$ be the lattice of finite order ideals of P, then $J_f(P) \cong \mathbb{N} \times \mathbb{N}$, and labeling the element $I \in J_f(P)$ by $e(I)$ yields precisely Pascal's triangle (though upside-down from the usual convention in writing it). Each finite order ideal I of $\mathbb{N} + \mathbb{N}$ has the form $\mathbf{m} + \mathbf{n}$ for some $m, n \in \mathbb{N}$, and from Example 3.5.4 we indeed have $e(\mathbf{m} + \mathbf{n}) = \binom{m+n}{m}$, the number of maximal chains in $\mathbf{m} \times \mathbf{n}$. See Figure 3.12.

Because of the previous example, we define a *generalized Pascal triangle* to be a finitary distributive lattice $L = J_f(P)$, together with the function $e: L \to \mathbb{P}$. The entries $e(I)$ of a generalized Pascal triangle thus have three properties in common with the usual Pascal triangle: (a) They count certain types of permutations, (b) they count certain types of lattice paths, and (c) they satisfy a simple recurrence.

3.6 Incidence Algebras

Let P be a locally finite poset, and let $\text{Int}(P)$ denote the set of (closed) intervals of P. (Recall that the empty set is not an interval.) Let K be a field. If $f : \text{Int}(P) \to K$, then we write $f(x, y)$ for $f([x, y])$.

3.6.1 Definition. The *incidence algebra* $I(P, K)$ (denoted $I(P)$ for short) of P over K is the K-algebra of all functions

$$f : \text{Int}(P) \to K$$

(with the usual structure of a vector space over K), where multiplication (or *convolution*) is defined by

$$fg(s, u) = \sum_{s \le t \le u} f(s, t) g(t, u).$$

This sum is finite (and hence fg is well-defined), since P is locally finite. It is easy to see that $I(P, K)$ is an associative algebra with (two-sided) identity, denoted δ or 1, defined by

$$\delta(s, t) = \begin{cases} 1, & \text{if } s = t, \\ 0, & \text{if } s \ne t. \end{cases}$$

One can think of $I(P, K)$ as consisting of all infinite linear combinations of symbols $[s, t]$, where $[s, t] \in \text{Int}(P)$. Convolution is defined uniquely by requiring that

$$[s, t] \cdot [u, v] = \begin{cases} [s, v], & \text{if } t = u, \\ 0, & \text{if } t \ne u, \end{cases}$$

and then extending to all of $I(P, K)$ by bilinearity (allowing infinite linear combinations of the $[s, t]$'s). The element $f \in I(P, K)$ is identified with the expression

$$f = \sum_{[s, t] \in \text{Int}(P)} f(s, t)[s, t].$$

If P if finite, then label the elements of P by $t_1, \ldots, t_p$ where $t_i < t_j \Rightarrow i < j$. (The number of such labelings is $e(P)$, the number of linear extensions of P.) Then $I(P, K)$ is isomorphic to the algebra of all upper triangular matrices $M = (m_{ij})$ over K, where $1 \le i, j \le p$, such that $m_{ij} = 0$ if $t_i \not\le t_j$. (*Proof.* Identify m_{ij} with $f(t_i, t_j)$.) For instance, if P is given by Figure 3.13, then $I(P)$ is isomorphic to the algebra of all matrices of the form

$$\begin{bmatrix} * & 0 & * & 0 & * \\ 0 & * & * & * & * \\ 0 & 0 & * & 0 & * \\ 0 & 0 & 0 & * & * \\ 0 & 0 & 0 & 0 & * \end{bmatrix}.$$

Partially Ordered Sets

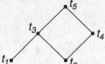

Figure 3.13 A five-element poset.

3.6.2 Proposition. *Let $f \in I(P)$. The following conditions are equivalent:*

a. *f has a left inverse.*
b. *f has a right inverse.*
c. *f has a two-sided inverse (which is necessarily the unique left and right inverse).*
d. *$f(t,t) \neq 0$ for all $t \in P$.*

Moreover, if f^{-1} exists, then $f^{-1}(s,u)$ depends only on the poset $[s,u]$.

Proof. The statement that $fg = \delta$ is equivalent to

$$f(s,s)g(s,s) = 1 \text{ for all } s \in P \tag{3.12}$$

and

$$g(s,u) = -f(s,s)^{-1} \sum_{s < t \leq u} f(s,t)g(t,u), \text{ for all } s < u \text{ in } P. \tag{3.13}$$

It follows that f has a right inverse g if and only if $f(s,s) \neq 0$ for all $s \in P$, and in that case $f^{-1}(s,u)$ depends only on $[s,u]$. Now the same reasoning applied to $hf = \delta$ shows that f has a left inverse h if and only if $f(s,s) \neq 0$ for all $s \in P$; that is, if and only if f has a right inverse. But from $fg = \delta$ and $hf = \delta$, we have that $g = h$, and the proof follows. $\square$

NOTE. The fact that a right-inverse of f is a two-sided inverse also follows from general algebraic reasoning. Namely, the restriction of f to Int($[s,u]$) satisfies a polynomial equation with nonzero constant term. An example of such an equation is the characteristic equation

$$\prod_{t \in [s,u]} (f - f(t,t)) = 0. \tag{3.14}$$

Hence, a right inverse of f is a polynomial in f and therefore commutes with f.

Let us now survey some useful functions in $I(P)$. The *zeta function* ζ is defined by

$$\zeta(t,u) = 1, \text{ for all } t \leq u \text{ in } P.$$

Thus,

$$\zeta^2(s,u) = \sum_{s \leq t \leq u} 1 = \#[s,u].$$

More generally, if $k \in \mathbb{P}$, then

$$\zeta^k(s,u) = \sum_{s=s_0 \leq s_1 \leq \cdots \leq s_k = u} 1,$$

the number of multichains of length k from s to u. Similarly,

$$(\zeta - 1)(s,u) = \begin{cases} 1, & \text{if } s < u, \\ 0, & \text{if } s = u. \end{cases}$$

Hence, if $k \in \mathbb{P}$, then $(\zeta - 1)^k(s,u)$ is the number of chains $s = s_0 < s_1 < \cdots < s_k = u$ of length k from s to u. By Propositions 3.5.1 and 3.5.2, we have additional interpretations of $\zeta^k(s,u)$ and $(\zeta - 1)^k(s,u)$ when P is a distributive lattice.

Now consider the element $2 - \zeta \in I(P)$. Thus,

$$(2 - \zeta)(s,t) = \begin{cases} 1, & \text{if } s = t, \\ -1, & \text{if } s < t. \end{cases}$$

By Proposition 3.6.2, $2 - \zeta$ is invertible. We claim that $(2 - \zeta)^{-1}(s,t)$ is equal to the *total* number of chains $s = s_0 < s_1 < \cdots < s_k = t$ from s to t. We sketch two justifications of this fact.

First Justification. Let ℓ be the length of the longest chain in the interval $[s,t]$. Then $(\zeta - 1)^{\ell+1}(u,v) = 0$ for all $s \leq u \leq v \leq t$. Thus, for $s \leq u \leq v \leq t$, we have

$$(2 - \zeta)[1 + (\zeta - 1) + (\zeta - 1)^2 + \cdots + (\zeta - 1)^\ell](u,v)$$

$$= [1 - (\zeta - 1)][1 + (\zeta - 1) + \cdots + (\zeta - 1)^\ell](u,v)$$

$$= [1 - (\zeta - 1)^{\ell+1}](u,v) = \delta(u,v).$$

Hence, $(2 - \zeta)^{-1} = 1 + (\zeta - 1) + \cdots + (\zeta - 1)^\ell$ when restricted to $\mathrm{Int}([s,t])$. But by the definition of ℓ, it is clear that $[1 + (\zeta - 1) + \cdots + (\zeta - 1)^\ell](s,t)$ is the total number of chains from s to t, as desired. $\square$

Second Justification. Our second justification is essentially equivalent to the first one, but it uses a little topology to avoid having to restrict our attention to an interval. The topological approach can be used to perform without effort many similar kinds of computations in $I(P)$. We define a topology on $I(P)$ (analogous to the topology on $\mathbb{C}[[x]]$ defined in Section 1.1) by saying that a sequence $f_1, f_2, \ldots$ of functions converges to f if for all $s \leq t$, there exists $n_0 = n_0(s,t) \in \mathbb{P}$ such that $f_n(s,t) = f(s,t)$ for all $n \geq n_0$. With this topology, the following computation is valid (because the infinite series converges):

$$(2 - \zeta)^{-1} = (1 - (\zeta - 1))^{-1} = \sum_{k \geq 0} (\zeta - 1)^k,$$

so

$$(2 - \zeta)^{-1}(s,t) = \sum_{k \geq 0} (\zeta - 1)^k (s,t)$$

$$= \sum_{k \geq 0} (\text{number of chains of length } k \text{ from } s \text{ to } t)$$

$$= \text{total number of chains from } s \text{ to } t.$$

□

Similarly to the above interpretation of $(2 - \zeta)^{-1}$, we leave to the reader to verify that $(1 - \eta)^{-1}(s,t)$ is equal to the total number of maximal chains in the interval $[s,t]$, where η is defined by

$$\eta(s,t) = \begin{cases} 1, & \text{if } t \text{ covers } s, \\ 0, & \text{otherwise.} \end{cases}$$

3.7 The Möbius Inversion Formula

It follows from Proposition 3.6.2 that the zeta function ζ of a locally finite poset is invertible; its inverse is called the *Möbius function* of P and is denoted μ (or μ_P if there is possible ambiguity). One can define μ recursively without reference to the incidence algebra. Namely, the relation $\mu \zeta = \delta$ is equivalent to

$$\mu(s,s) = 1, \text{ for all } s \in P,$$

$$\mu(s,u) = - \sum_{s \leq t < u} \mu(s,t), \text{ for all } s < u \text{ in } P. \qquad (3.15)$$

3.7.1 Proposition (Möbius inversion formula). *Let P be a poset for which every principal order ideal Λ_t is finite. Let $f, g \colon P \to K$, where K is a field. Then*

$$g(t) = \sum_{s \leq t} f(s), \text{ for all } t \in P, \qquad (3.16)$$

if and only if

$$f(t) = \sum_{s \leq t} g(s) \mu(s,t), \text{ for all } t \in P. \qquad (3.17)$$

Proof. The set K^P of all functions $P \to K$ forms a vector space on which $I(P,K)$ acts (on the right) as an algebra of linear transformations by

$$(f\xi)(t) = \sum_{s \leq t} f(s)\xi(s,t),$$

where $f \in K^P, \xi \in I(P,K)$. The Möbius inversion formula is then nothing but the statement

$$f\zeta = g \iff f = g\mu.$$

□

NOTE. It is also easy to give a naive computational proof of Proposition 3.7.1. Assuming (3.16), we have (for fixed $t \in P$)

$$\sum_{s \leq t} g(s)\mu(s,t) = \sum_{s \leq t} \mu(s,t) \sum_{u \leq s} f(u)$$

$$= \sum_{u \leq t} f(u) \sum_{u \leq s \leq t} \mu(s,t)$$

$$= \sum_{u \leq t} f(u)\delta(u,t)$$

$$= f(t),$$

which is (3.17). A completely analogous argument shows that (3.16) follows from (3.17).

A dual formulation of the Möbius inversion formula is sometimes convenient.

3.7.2 Proposition (Möbius inversion formula, dual form). *Let P be a poset for which every principal dual order ideal V_t is finite. Let $f, g: P \to K$. Then*

$$g(s) = \sum_{t \geq s} f(t), \quad \text{for all } s \in P,$$

if and only if

$$f(s) = \sum_{t \geq s} \mu(s,t)g(t), \quad \text{for all } s \in P.$$

Proof. Exactly as above, except now $I(P,K)$ acts on the *left* by

$$(\xi f)(s) = \sum_{t \geq s} \xi(s,t)f(t).$$

$\square$

As in the Principle of Inclusion-Exclusion, the purely abstract statement of the Möbius inversion formula as given here is just a trivial observation in linear algebra. What is important are the applications of the Möbius inversion formula. First, we show the Möbius inversion formula does indeed explain formulas such as equation (3.1).

Given n finite sets $S_1, \ldots, S_n$, let P be the poset of all their intersections ordered by inclusion, including the empty intersection $S_1 \cup \cdots \cup S_n = \hat{1}$. If $T \in P$, then let $f(T)$ be the number of elements of T that belong to no $T' < T$ in P, and let $g(T) = \#T$. We want an expression for $\#(S_1 \cup \cdots \cup S_n) = \sum_{T \leq \hat{1}} f(T) = g(\hat{1})$. Now $g(T) = \sum_{T' \leq T} f(T')$, so by Möbius inversion on P we have

$$0 = f(\hat{1}) = \sum_{T \in P} g(T)\mu(T,\hat{1}) \Rightarrow g(\hat{1}) = -\sum_{T < \hat{1}} \#T \cdot \mu(T,\hat{1}),$$

as desired. In the example given by equation (3.1), P is given by Figure 3.14. Indeed, $\mu(A,\hat{1}) = \mu(B,\hat{1}) = \mu(C,\hat{1}) = -1$ and $\mu(D,\hat{1}) = 2$, so (3.1) follows.

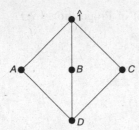

Figure 3.14 The poset related to equation (3.1).

3.8 Techniques for Computing Möbius Functions

In order for the Möbius inversion formula to be of any value, it is necessary to be able to compute the Möbius function of posets P of interest. We begin with a simple example that can be done by brute force.

3.8.1 Example. Let P be the chain $\mathbb{N}$. It follows directly from equation (3.15) that

$$\mu(i, j) = \begin{cases} 1, & \text{if } i = j, \\ -1, & \text{if } i + 1 = j, \\ 0, & \text{otherwise.} \end{cases}$$

The Möbius inversion formula takes the form

$$g(n) = \sum_{i=0}^{n} f(i) \text{ for all } n > 0$$

if and only if

$$f(0) = g(0), \text{ and } f(n) = g(n) - g(n-1) \text{ for all } n > 0.$$

In other words, the operations Σ and Δ (with Σ suitably initialized) are inverses of one another, the finite difference analogue of the "fundamental theorem of calculus."

Since only in rare cases can Möbius functions be computed by inspection as in Example 3.8.1, we need general techniques for their evaluation. We begin with the simplest result of this nature.

3.8.2 Proposition (the product theorem). *Let P and Q be locally finite posets, and let $P \times Q$ be their direct product. If $(s, t) \le (s', t')$ in $P \times Q$ then*

$$\mu_{P \times Q}((s, t), (s', t')) = \mu_P(s, s')\mu_Q(t, t').$$

Proof. Let $(s, t) \le (s', t')$. We have

$$\sum_{(s,t) \le (u,v) \le (s',t')} \mu_P(s, u)\mu_Q(t, v) = \left(\sum_{s \le u \le s'} \mu_P(s, u) \right) \left(\sum_{t \le v \le t'} \mu_Q(t, v) \right)$$

$$= \delta_{ss'}\delta_{tt'} = \delta_{(s,t),(s',t')}.$$

Comparing with equation (3.15), which determines μ uniquely, completes the proof. $\qquad\square$

For readers familiar with tensor products, we mention a more conceptual way of proving the previous proposition. Namely, one easily sees that

$$I(P \times Q, K) \cong I(P,K) \otimes_K I(Q,K)$$

and $\zeta_{P\times Q} = \zeta_P \otimes \zeta_Q$. Taking inverses gives $\mu_{P\times Q} = \mu_P \otimes \mu_Q$.

3.8.3 Example. Let $P = B_n$, the boolean algebra of rank n. Now $B_n \cong \mathbf{2}^n$, and the Möbius function of the chain $\mathbf{2} = \{1,2\}$ is given by $\mu(1,1) = \mu(2,2) = 1$, $\mu(1,2) = -1$. Hence if we identify B_n with the set of all subsets of an n-set X, then we conclude from the product theorem that

$$\mu(T,S) = (-1)^{\#(S-T)}.$$

Since $\#(S-T)$ is the length $\ell(T,S)$ of the interval $[T,S]$, in purely order-theoretic terms, we have

$$\mu(T,S) = (-1)^{\ell(T,S)}. \tag{3.18}$$

The Möbius inversion formula for B_n becomes the following statement. Let $f,g \colon B_n \to K$; then

$$g(S) = \sum_{T \subseteq S} f(T), \quad \text{for all } S \subseteq X,$$

if and only if

$$f(S) = \sum_{T \subseteq S} (-1)^{\#(S-T)} g(T), \quad \text{for all } S \subseteq X.$$

This is just equation (2.8). Hence, we can say that "Möbius inversion on a boolean algebra is equivalent to the Principle of Inclusion-Exclusion." Note that equation (2.8), together with the Möbius inversion formula (Proposition 3.7.1), actually *proves* (3.18), so now we have two proofs of this result.

3.8.4 Example. Let $n_1,\ldots,n_k$ be nonnegative integers, and let $P = (n_1 + 1) \times (n_2 + 1) \times \cdots \times (n_k + 1)$, a product of chains of lengths $n_1,\ldots,n_k$. Note that P is isomorphic to the distributive lattice $J(n_1 + n_2 + \cdots + n_k)$. Identify P with the set of all k-tuples $(a_1, a_2,\ldots,a_k) \in \mathbb{N}^k$ with $0 \le a_i \le n_i$, ordered componentwise. If $a_i \le b_i$ for all i, then the interval $[(a_1,\ldots,a_k),(b_1,\ldots,b_k)]$ in P is isomorphic to $(b_1 - a_1 + 1) \times \cdots \times (b_k - a_k + 1)$. Hence by Example 3.8.1 and Proposition 3.8.2, we have

$$\mu((a_1,\ldots,a_k),(b_1,\ldots,b_k)) = \begin{cases} (-1)^{\sum(b_i - a_i)}, & \text{if each } b_i - a_i = 0 \text{ or } 1, \\ 0, & \text{otherwise.} \end{cases}$$
$$\tag{3.19}$$

Equivalently,

$$\mu(s,t) = \begin{cases} (-1)^{\ell(s,t)}, & \text{if } [s,t] \text{ is a boolean algebra,} \\ 0, & \text{otherwise.} \end{cases}$$

(See Example 3.9.6 for a mild generalization.)

There are two further ways of interest to interpret the lattice $P = (n_1 + 1) \times \cdots \times (n_k + 1)$. First, P is isomorphic to the poset of submultisets of the multiset $\{x_1^{n_1}, \ldots, x_k^{n_k}\}$, ordered by inclusion. Second, if N is a positive integer of the form $p_1^{n_1} \cdots p_k^{n_k}$, where the p_i's are distinct primes, then P is isomorphic to the poset D_N defined in Example 3.1.1(c) of positive integral divisors of N, ordered by divisibility (i.e., $r \le s$ in D_N if $r|s$). In this latter context, equation (3.19) takes the form

$$\mu(r,s) = \begin{cases} (-1)^t, & \text{if } s/r \text{ is a product of } t \text{ distinct primes,} \\ 0, & \text{otherwise.} \end{cases}$$

In other words, $\mu(r,s)$ is just the classical number-theoretic Möbius function $\mu(s/r)$. The Möbius inversion formula becomes the classical one, namely,

$$g(n) = \sum_{d|n} f(d), \quad \text{for all } n|N,$$

if and only if

$$f(n) = \sum_{d|n} g(d)\mu(n/d), \quad \text{for all } n|N.$$

This example explains the termimology "Möbius function of a poset."

Rather than restricting ourselves to the divisors of a fixed integer N, it is natural to consider the poset P of *all* positive integers, ordered by divisibility. Since any interval $[r,s]$ of this poset appears as an interval in the lattice of divisors of s (or of any N for which $s|N$), the Möbius function remains $\mu(r,s) = \mu(s/r)$. Abstractly, the poset P is isomorphic to the finitary distributive lattice

$$J_f(\mathbb{P} + \mathbb{P} + \mathbb{P} + \cdots) = J_f\left(\sum_{n \ge 1} \mathbb{P}\right) \cong \prod_{n \ge 1} \mathbb{N}, \tag{3.20}$$

where the product $\prod_{n \ge 1} \mathbb{N}$ is the *restricted direct product*; that is, only finitely many components of an element of the product are nonzero. Alternatively, P can be identified with the lattice of all finite multisets of the set $\mathbb{P}$ (or any countably infinite set).

We now come to a very important way of computing Möbius functions.

3.8.5 Proposition (Philip Hall's theorem). *Let P be a finite poset, and let $\widehat{P}$ denote P with a $\hat{0}$ and $\hat{1}$ adjoined. Let c_i be the number of chains $\hat{0} = t_0 < t_1 < \cdots < t_i = \hat{1}$ of length i between $\hat{0}$ and $\hat{1}$. (Thus, $c_0 = 0$ and $c_1 = 1$.) Then*

$$\mu_{\widehat{P}}(\hat{0}, \hat{1}) = c_0 - c_1 + c_2 - c_3 + \cdots. \tag{3.21}$$

Proof. We have

$$
\begin{aligned}
\mu_{\widehat{P}}(\hat{0}, \hat{1}) &= (1 + (\zeta - 1))^{-1}(\hat{0}, \hat{1}) \\
&= (1 - (\zeta - 1) + (\zeta - 1)^2 - \cdots)(\hat{0}, \hat{1}) \\
&= \delta(\hat{0}, \hat{1}) - (\zeta - 1)(\hat{0}, \hat{1}) + (\zeta - 1)^2(\hat{0}, \hat{1}) - \cdots \\
&= c_0 - c_1 + c_2 - \cdots .
\end{aligned}
$$

$\square$

The significance of Proposition 3.8.5 is that it shows that $\mu(\hat{0}, \hat{1})$ (and thus $\mu(s, t)$ for any interval $[s, t]$) can be interpreted as an Euler characteristic, and therefore links the Möbius function of P with the powerful machinery of algebraic topology. To see the connection, recall that an (abstract) *simplicial complex* on a vertex set V is a collection Δ of subsets of V satisfying:

a. If $t \in V$ then $\{t\} \in \Delta$,
b. if $F \in \Delta$ and $G \subseteq F$, then $G \in \Delta$.

Thus, Δ is just an order ideal of the boolean algebra B_V that contains all one-element subsets of V. An element $F \in \Delta$ is called a *face* of Δ, and the *dimension* of F is defined to be $\#F - 1$. In particular, the empty set $\emptyset$ is always a face of Δ (provided $\Delta \neq \emptyset$), of dimension -1. Also define the *dimension* of Δ by

$$
\dim \Delta = \max_{F \in \Delta} (\dim F).
$$

If Δ is finite, then let f_i denote the number of i-dimensional faces of Δ. Define the *reduced Euler characteristic* $\widetilde{\chi}(\Delta)$ by

$$
\widetilde{\chi}(\Delta) = \sum_i (-1)^i f_i = -f_{-1} + f_0 - f_1 + \cdots . \tag{3.22}
$$

Note that $f_{-1} = 1$ unless $\Delta = \emptyset$. The simplicial complexes $\Delta_1 = \emptyset$ and $\Delta_2 = \{\emptyset\}$ are not the same; in particular, $\widetilde{\chi}(\Delta_1) = 0$ and $\widetilde{\chi}(\Delta_2) = -1$.

NOTE. The reduced Euler characteristic $\widetilde{\chi}(\Delta)$ is related to the ordinary Euler characteristic $\chi(\Delta)$ by $\widetilde{\chi}(\Delta) = \chi(\Delta) - 1$ (if $\Delta \neq \emptyset$). Thus, in computing $\chi(\Delta)$, the empty set is not considered as a face, whereas for $\widetilde{\chi}(\Delta)$ we do regard it as a face.

Now if P is any poset, then define a simplicial complex $\Delta(P)$ as follows: The vertices of $\Delta(P)$ are the elements of P, and the faces of $\Delta(P)$ are the chains of P. The simplicial complex $\Delta(P)$ is called the *order complex* of P. We then conclude from equations (3.21) and (3.22) the following result.

3.8.6 Proposition (Proposition 3.8.5, restated). *Let P be a finite poset. Then*

$$
\mu_{\widehat{P}}(\hat{0}, \hat{1}) = \widetilde{\chi}(\Delta(P)).
$$

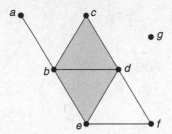

Figure 3.15 A geometric realization of a simplicial complex.

Proposition 3.8.5 gives an expression for $\mu(\hat{0},\hat{1})$ that is self-dual (i.e., remains unchanged if P is replaced by P^*). Thus, we see that in any locally finite poset P,

$$\mu_P(s,t) = \mu_{P^*}(t,s).$$

(One can also prove this fact using $\mu\zeta = \zeta\mu$.)

Let us recall that in topology one associates a topological space $|\Delta|$, called the *geometric realization* of Δ, with a simplicial complex Δ. (One also says that Δ is a *triangulation* of the space $|\Delta|$.) Informally, place the vertices of Δ in sufficiently general position (e.g., linearly independent) in some Euclidean space. Then

$$|\Delta| = \bigcup_{F \in \Delta} \operatorname{conv}(F),$$

where conv denotes convex hull. For instance, if the maximal faces of Δ are (abbreviating $\{a,b\}$ as ab, etc.) ab, bcd, bde, df, ef, g, then the geometric realization $|\Delta|$ is shown in Figure 3.15.

The reduced Euler characteristic $\tilde{\chi}(X)$ of the space $X = |\Delta|$ is defined by

$$\tilde{\chi}(X) = \sum_i (-1)^i \operatorname{rank} \tilde{H}_i(X;\mathbb{Z}),$$

where $\tilde{H}_i(X;\mathbb{Z})$ denotes the ith reduced homology group of X. One then has from elementary algebraic topology that

$$\tilde{\chi}(X) = \tilde{\chi}(\Delta), \tag{3.23}$$

so that $\mu_{\hat{P}}(\hat{0},\hat{1})$ depends only on the topological space $|\Delta(P)|$ of $\Delta(P)$. For instance, if $\Delta(P)$ is a triangulation of an n-dimensional sphere, then $\mu_{\hat{P}}(\hat{0},\hat{1}) = (-1)^n$.

3.8.7 Example (for readers familiar with some topology). A *finite regular cell complex* Γ is a finite set of nonempty pairwise-disjoint open cells $\sigma_i \subset \mathbb{R}^N$ such that

a. $(\bar{\sigma}_i, \bar{\sigma}_i - \sigma_i) \approx (\mathbb{B}^n, \mathbb{S}^{n-1})$, for some $n = n(i)$,
b. each $\bar{\sigma}_i - \sigma_i$ is a union of σ_j's.

Here $\bar{\sigma}_i$ denotes the closure of σ_i (in the usual topology on $\mathbb{R}^N$), $\approx$ denotes homeomorphism, $\mathbb{B}^n$ is the unit ball $\{(x_1,\ldots,x_n) \in \mathbb{R}^n : x_1^2 + \cdots + x_n^2 \leq 1\}$, and $\mathbb{S}^{n-1}$ is the unit sphere $\{(x_1,\ldots,x_n) \in \mathbb{R}^n : x_1^2 + \cdots + x_n^2 = 1\}$. Note that a cell σ_i may consist of a single point, corresponding to the case $n = 0$. Also, define the *underlying space* of Γ to be the topological space $|\Gamma| = \bigcup \sigma_i \subset \mathbb{R}^N$. Given a finite regular cell complex Γ, define its (first) *barycentric subdivision* sd(Γ) to be the abstract simplicial complex whose vertices consist of the closed cells $\bar{\sigma}_i$ of Γ, and whose faces consist of those sets $\{\bar{\sigma}_{i_1},\ldots,\bar{\sigma}_{i_k}\}$ of vertices forming a *flag* $\bar{\sigma}_{i_1} \subset \bar{\sigma}_{i_2} \subset \cdots \subset \bar{\sigma}_{i_k}$. The crucial property of a regular cell complex to concern us here is that the geometric realization $|\text{sd}(\Gamma)|$ of the simplicial complex sd(Γ) is homeomorphic to the underlying space $|\Gamma|$ of the cell complex Γ:

$$|\text{sd}(\Gamma)| \approx |\Gamma|. \tag{3.24}$$

Now given a finite regular cell complex Γ, let $P(\Gamma)$ be the poset of cells of Γ, ordered by defining $\sigma_i \leq \sigma_j$ if $\bar{\sigma}_i \subseteq \bar{\sigma}_j$. It follows from the definition of sd(Γ) that $\Delta(P(\Gamma)) = \text{sd}(\Gamma)$. From Proposition 3.8.6 and equations (3.23) and (3.24), we conclude the following.

3.8.8 Proposition. *Let Γ be a finite regular cell complex, and let $P = P(\Gamma)$. Then*

$$\mu_{\widehat{P}}(\hat{0},\hat{1}) = \widetilde{\chi}(|\Gamma|), \tag{3.25}$$

where $\widetilde{\chi}(|\Gamma|)$ is the reduced Euler characteristic of the topological space $|\Gamma|$.

Propositions 3.8.6 and 3.8.8 deal with the topological significance of the integer $\mu_{\widehat{P}}(\hat{0},\hat{1})$. We are also interested in other values $\mu_{\widehat{P}}(s,t)$, so we briefly discuss this point. Let Δ be any finite simplicial complex, and let $F \in \Delta$. The *link* of F is the subcomplex of Δ defined by

$$\text{lk}\, F = \{G \in \Delta : G \cap F = \emptyset \text{ and } G \cup F \in \Delta\}. \tag{3.26}$$

If P is a finite poset and $s < t$ in P, then choose saturated chains $s_1 \lessdot s_2 \lessdot \cdots \lessdot s_j = s$ and $t = t_1 \lessdot t_2 \lessdot \cdots \lessdot t_k$ in P such that s_1 is a minimal element and t_k is a maximal element of P. Let $F = \{s_1,\ldots,s_j,t_1,\ldots,t_k\} \in \Delta(P)$. Then lk F is just the order complex of the open interval $(s,t) = \{u \in P : s < u < t\}$, so by Proposition 3.8.6 we have

$$\mu(s,t) = \widetilde{\chi}(\text{lk}\, F). \tag{3.27}$$

Now suppose that Δ is an abstract simplicial complex that triangulates a manifold M, with or without boundary. (In other words, $|\Delta| \approx M$.) Let $\emptyset \neq F \in \Delta$. It is well known from algebraic topology that lk F has the same homology groups as a sphere or ball of dimension equal to $\dim(\text{lk}\, F) = \max_{G \in \text{lk}\, F}(\dim G)$. Moreover, lk F will have the homology groups of a ball precisely when F lies on the boundary $\partial \Delta$ of Δ. Equivalently, F is contained in some face F' such that $\dim F' = \dim \Delta - 1$ and

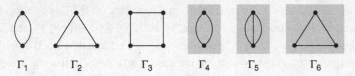

Figure 3.16 Some regular cell complexes.

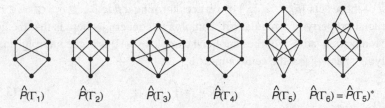

Figure 3.17 The face posets of the regular cell complexes of Figure 3.16.

F' is contained in a unique maximal face of Δ. (Somewhat surprisingly, lk F need not be simply connected and $|\text{lk } F|$ need not be a manifold!) Since $\tilde{\chi}(\mathbb{S}^n) = (-1)^n$ and $\tilde{\chi}(\mathbb{B}^n) = 0$, we deduce from equations (3.25) and (3.27) the following result.

3.8.9 Proposition. *Let Γ be a finite regular cell complex. Suppose that $|\Gamma|$ is a manifold, with or without boundary. Let $P = P(\Gamma)$. Then*

$$
\mu_{\widehat{P}}(s,t) = \begin{cases} 0, & \text{if } s \neq \hat{0},\, t = \hat{1}, \text{ and the cell } s \text{ lies on the} \\ & \text{boundary of } |\Gamma|, \\ \tilde{\chi}(|\Gamma|), & \text{if } (s,t) = (\hat{0}, \hat{1}), \\ (-1)^{\ell(s,t)}, & \text{otherwise.} \end{cases}
$$

Motivated by Proposition 3.8.9, we define a finite graded poset P with $\hat{0}$ and $\hat{1}$ to be *semi-Eulerian* if $\mu_P(s,t) = (-1)^{\ell(s,t)}$ whenever $(s,t) \neq (\hat{0}, \hat{1})$, and to be *Eulerian* if in addition $\mu_P(\hat{0}, \hat{1}) = (-1)^{\ell(\hat{0},\hat{1})}$. Thus, Proposition 3.8.9 implies that if $|\Gamma|$ is a manifold (without boundary), then $\widehat{P}(\Gamma)$ is semi-Eulerian. Moreover, if $|\Gamma|$ is a sphere, then $\widehat{P}(\Gamma)$ is Eulerian. By Example 3.8.3, boolean algebras B_n are Eulerian; indeed, $B_n = \widehat{P}(\Gamma)$, where Γ is the boundary complex of an $(n-1)$-simplex. Hence, $|\Delta(B_n)| \approx \mathbb{S}^{n-2}$, a vast topological strengthening of the mere computation of the Möbius function of B_n. Some interesting properties of Eulerian posets appear in Sections 3.16 and 3.17.

3.8.10 Example. a. The diagrams of Figure 3.16 represent finite regular cell complexes Γ such that $|\Gamma| \cong \mathbb{S}^1$ or $|\Gamma| \cong \mathbb{S}^2$. (Shaded regions represent 2-cells.) The corresponding Eulerian posets $\widehat{P}(\Gamma)$ are shown in Figure 3.17. Note that $\widehat{P}(\Gamma_2)$ and $\widehat{P}(\Gamma_3)$ are lattices. This is because in Γ_2 and Γ_3, any intersection $\bar{\sigma}_i \cap \bar{\sigma}_j$ is some $\bar{\sigma}_k$.

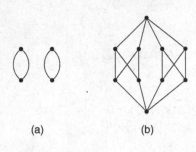

Figure 3.18 A nonspherical regular cell complex and its Eulerian face poset.

(a)   (b)

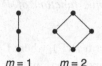

Figure 3.19 Face posets of some 0-dimensional manifolds.

$m = 1$ $m = 2$ $m = 3$

b. The diagram $\bigcirc$ represents a certain cell complex Γ that is *not* regular, since for the unique 1-cell σ we do not have $\bar{\sigma} - \sigma \approx \mathbb{S}^0$. (The sphere $\mathbb{S}^0$ consists of two points, while $\bar{\sigma} - \sigma$ is just a single point.) The corresponding poset $P = P(\Gamma)$ is the two-element chain, and $|\Delta(P)|$ is not homeomorphic to $|\Gamma|$. (We have $|\Gamma| \approx \mathbb{S}^1$ while $|\Delta(P)| \approx \mathbb{B}^1$.) Note that $\widehat{P}$ is not Eulerian even though $|\Gamma|$ is a sphere.

c. Let Γ be given by Figure 3.18(a). Then $|\Gamma|$ is a manifold without boundary with the same Euler characteristic as $\mathbb{S}^1$ (namely, 0), though $|\Gamma| \not\approx \mathbb{S}^1$. Hence, $\widehat{P}(\Gamma)$ is Eulerian even though Γ does not have the same homology groups as a sphere. See Figure 3.18(b).

d. If Γ is a disjoint union of m points then $|\Gamma|$ is a manifold with Euler characteristic m. Hence, $\widehat{P}(\Gamma)$ is semi-Eulerian, but not Eulerian if $m \neq 2$. See Figure 3.19.

For our final excursion into topology, let P be a finite graded poset with $\hat{0}$ and $\hat{1}$. We say that the Möbius function of P *alternates in sign* if

$$(-1)^{\ell(s,t)}\mu(s,t) \geq 0, \quad \text{for all } s \leq t \text{ in } P.$$

A finite poset P is said to be *Cohen-Macaulay* over an abelian group A if for every $s < t$ in $\widehat{P}$, the order complex $\Delta(s,t)$ of the open interval (s,t) satisfies

$$\widetilde{H}_i(\Delta(s,t); A) = 0, \quad \text{if } i < \dim \Delta(s,t). \tag{3.28}$$

Here $\widetilde{H}_i(\Delta(s,t); A)$ denotes reduced simplicial homology with coefficients in A. It follows from standard topological arguments that if P is Cohen-Macaulay over some group A, then P is Cohen-Macaulay over $\mathbb{Q}$. Hence, we may as well take $A = \mathbb{Q}$ to get the widest class of posets. One can easily show that a Cohen-Macaulay poset is graded. If equation (3.28) holds (say with $A = \mathbb{Q}$) and if $d = \dim \Delta(s,t)$, then equation (3.27) implies that

$$\mu_P(s,t) = \tilde{\chi}(\Delta(s,t)) = (-1)^d \dim_{\mathbb{Q}} \tilde{H}_d(\Delta(s,t);\mathbb{Q}) \geq 0.$$

Since $d = \ell(s,t) - 2$, we conclude that

$$(-1)^{\ell(s,t)} \mu_P(s,t) = \dim_{\mathbb{Q}} \tilde{H}_d(\Delta(s,t);\mathbb{Q}) \geq 0.$$

We have therefore proved the following result.

3.8.11 Proposition. *If P is Cohen-Macaulay, then the Möbius function of P alternates in sign.*

Examples of Cohen-Macaulay posets include those of the form $P(\Gamma)$, where Γ is a finite regular cell complex such that $|\Gamma|$ is a manifold of dimension d, with or without boundary, satisfying $\tilde{H}_i(\Gamma;\mathbb{Q}) = 0$ if $i < d$. It can be shown that for any finite regular cell complex Γ, the question of whether $P(\Gamma)$ is Cohen-Macaulay depends only on the space $|\Gamma|$. It can also be shown that if $\hat{P}$ is a finite semimodular lattice, then P is Cohen-Macaulay. Though we will not prove this fact here, we will later (Proposition 3.10.1) prove the weaker assertion that the Möbius function of a finite semimodular lattice alternates in sign.

3.9 Lattices and Their Möbius Functions

There are special methods for computing the Möbius function of a lattice that are inapplicable to general posets. We will develop these results in a unified way using the theory of Möbius algebras. While the applications to Möbius functions can also be proved without recourse to Möbius algebras, we prefer the convenience and elegance of the algebraic viewpoint.

3.9.1 Definition. Let L be a lattice and K a field. The *Möbius algebra* $A(L,K)$ is the semigroup algebra of L with the meet operation, over K. In other words, $A(L,K)$ is the vector space over K with basis L, with (bilinear) multiplication defined by $s \cdot t = s \wedge t$ for all $s,t \in L$.

The Möbius algebra $A(L,K)$ is commutative and has a vector space basis consisting of idempotents, namely, the elements of L. It follows from general ring-theoretic considerations (Wedderburn theory or otherwise) that when L is finite we have $A(L,K) \cong K^{\#L}$. We wish to make this isomorphism more explicit. To do so, define for $t \in L$ the element $\delta_t \in A(L,K)$ by

$$\delta_t = \sum_{s \leq t} \mu(s,t)s.$$

Hence by the Möbius inversion formula,

$$t = \sum_{s \leq t} \delta_s. \tag{3.29}$$

The number of δ_t's is equal to $\#L = \dim_K A(L, K)$, and equation (3.29) shows that they span $A(L, K)$. Hence, the δ_t's form a K-basis for $A(L, K)$.

3.9.2 Theorem. *Let L be a finite lattice and let $A'(L, K)$ be the abstract algebra $\bigoplus_{t \in L} K_t$, where each $K_t \cong K$. Denote by δ_t' the identity element of K_t, so $\delta_s' \delta_t' = \delta_{st} \delta_t'$ (where δ_{st} denotes the Kronecker delta). Define a linear transformation $\theta \colon A(L, K) \to A'(L, K)$ by setting $\theta(\delta_t) = \delta_t'$ and extending by linearity. Then θ is an isomorphism of algebras.*

Proof. If $t \in L$, then let $t' = \sum_{s \leq t} \delta_s' \in A'(L, K)$. Since θ is clearly a vector space isomorphism, we need only show that $s't' = (s \wedge t)'$. Now

$$s't' = \left(\sum_{u \leq s} \delta_u' \right) \left(\sum_{v \leq t} \delta_v' \right) = \sum_{\substack{u \leq s \\ v \leq t}} \delta_{uv} \delta_u'$$

$$= \sum_{w \leq s \wedge t} \delta_w' = (s \wedge t)'. \qquad \Box$$

3.9.3 Corollary (Weisner's theorem). *Let L be a finite lattice with at least two elements, and let $\hat{1} \neq a \in L$. Then*

$$\sum_{t \,:\, t \wedge a = \hat{0}} \mu(t, \hat{1}) = 0.$$

Proof. In the Möbius algebra $A(L, K)$ we have

$$a\delta_{\hat{1}} = \left(\sum_{b \leq a} \delta_b \right) \delta_{\hat{1}} = 0, \quad \text{if } a \neq \hat{1}. \tag{3.30}$$

On the other hand,

$$a\delta_{\hat{1}} = a \sum_{t \in L} \mu(t, \hat{1}) t = \sum_{t \in L} \mu(t, \hat{1})(a \wedge t). \tag{3.31}$$

Writing $a\delta_{\hat{1}} = \sum_{t \in L} c_t \cdot t$, we conclude from equation (3.30) that $c_{\hat{0}} = 0$ and from (3.31) that $c_{\hat{0}} = \sum_{t \,:\, t \wedge a = \hat{0}} \mu(t, \hat{1})$. $\qquad \Box$

Looking at the defining recurrence (3.15) for the Möbius function, we see that Corollary 3.9.3 gives a similar recurrence, but in general with many fewer terms. Some applications of Corollary 3.9.3 will be given soon. First, we give some other consequences of Theorem 3.9.2. The next result is known as the *Crosscut Theorem*.

3.9.4 Corollary (Crosscut Theorem). *Let L be a finite lattice, and let X be a subset of L such that (a) $\hat{1} \notin X$, and (b) if $s \in L$ and $s \neq \hat{1}$, then $s \leq t$ for some $t \in X$. Then*

$$\mu(\hat{0}, \hat{1}) = \sum_k (-1)^k N_k, \tag{3.32}$$

where N_k is the number of k-subsets of X whose meet is $\hat{0}$.

Proof. For any $t \in L$, we have in $A(L, K)$ that

$$\hat{1} - t = \sum_{s \le \hat{1}} \delta_s - \sum_{s \le t} \delta_s = \sum_{s \not\le t} \delta_s.$$

Hence by Theorem 3.9.2,

$$\prod_{t \in X} (\hat{1} - t) = \sum_s \delta_s,$$

where s ranges over all elements of L satisfying $s \not\le t$ for all $t \in X$. By hypothesis, the only such element is $\hat{1}$. Hence,

$$\prod_{t \in X} (\hat{1} - t) = \delta_{\hat{1}}.$$

If we now expand both sides as linear combinations of elements of L and equate coefficients of $\hat{0}$, the result follows. $\qquad\square$

NOTE (for topologists). Let Γ be the set of all subsets of X (as in the previous corollary) whose meet is *not* $\hat{0}$. Then Γ is a simplicial complex, and equation (3.32) asserts that $\tilde{\chi}(\Gamma) = \mu(\hat{0}, \hat{1})$. Let $P' = P - \{\hat{0}, \hat{1}\}$. Comparing with Proposition 3.8.6, which asserts that $\mu(\hat{0}, \hat{1}) = \tilde{\chi}(\Delta(P'))$, suggests that the two simplicial complexes Γ and $\Delta(P')$ might have deeper topological similarities than merely having the same Euler characteristic. Indeed, it can be shown that Γ and $\Delta(P')$ are homotopy equivalent, a good example of combinatorial reasoning leading to a stronger topological result.

It is clear that a subset X of L satisfies conditions (a) and (b) of Corollary 3.9.4 if and only if X contains the set A^* of all coatoms (= elements covered by $\hat{1}$) of L. To make the numbers N_k as small as possible, we should take $X = A^*$. Note that if $\hat{0}$ is not the meet of all the coatoms of L, then each $N_k = 0$. Hence, we obtain the following corollary.

3.9.5 Corollary. *If L is a finite lattice for which $\hat{0}$ is not a meet of coatoms, then $\mu(\hat{0}, \hat{1}) = 0$. Dually, if $\hat{1}$ is not a join of atoms, then again $\mu(\hat{0}, \hat{1}) = 0$.*

3.9.6 Example. Let $L = J(P)$ be a finite distributive lattice. The interval $[I, I']$ of L is a boolean algebra if and only if $I' - I$ is an antichain of P. More generally, the join of all atoms of the interval $[I, I']$ (regarded as a sublattice of L) is the order ideal $I \cup M$, where M is the set of minimal elements of the subposet $I' - I$ of P. Hence, I' is a join of atoms of $[I, I']$ if and only if $[I, I']$ is a boolean algebra. From Example 3.8.3 and Corollary 3.9.5, we obtain the Möbius function of L, namely,

$$\mu(I, I') = \begin{cases} (-1)^{\ell(I, I')} = (-1)^{\#(I' - I)}, & \text{if } [I, I'] \text{ is a boolean algebra (i.e., if} \\ & \qquad I' - I \text{ is an antichain of } P), \\ 0, & \text{otherwise.} \end{cases}$$

3.10 The Möbius Function of a Semimodular Lattice

We wish to apply the dualized form of Corollary 3.9.3 to a finite semimodular lattice L of rank n with rank function ρ. Pick a to be an atom of L. Suppose $a \vee t = \hat{1}$. If also $a \leq t$, then $t = \hat{1}$. Hence, either $t \wedge a = \hat{0}$ or $t = \hat{1}$. Now from the definition of semimodularity, we have $\rho(t) + \rho(a) \geq \rho(t \wedge a) + \rho(t \vee a)$, so either $t = \hat{1}$ or $\rho(t) + 1 \geq 0 + n$. Hence, either $t = \hat{1}$, or t is a coatom. From Corollary 3.9.3 (dualized) there follows

$$\mu(\hat{0}, \hat{1}) = - \sum_{\substack{\text{coatoms } t \\ \text{such that} \\ t \not\geq a}} \mu(\hat{0}, t). \tag{3.33}$$

Since every interval of a semimodular lattice is again semimodular (e.g., by Proposition 3.3.2), we conclude from equation (3.33) and induction on n the following result, mentioned at the end of Section 3.8.

3.10.1 Proposition. *The Möbius function of a finite semimodular lattice alternates in sign.*

Since $(-1)^{\ell(s,t)} \mu(s,t)$ is a nonnegative integer for any $s \leq t$ in a finite semimodular lattice L, we can ask whether this integer actually counts something associated with the structure of L. This question will be answered in Section 3.14.

We now turn to two of the most important examples of semimodular lattices.

3.10.2 Example. Let q be a prime power, and let $V_n = \mathbb{F}_q^n$, an n-dimensional vector space over the finite field $\mathbb{F}_q$. (Any n-dimensional vector space over $\mathbb{F}_q$ will do, but for definiteness we choose $\mathbb{F}_q^n$.) Let $B_n(q)$ denote the poset of all subspaces of V_n, ordered by inclusion, as defined in Example 3.1.1(e). We observed in Section 3.3 that $B_n(q)$ is a graded lattice of rank n, where the rank $\rho(W)$ of a subspace is just its dimension. We also mentioned that since any two subspaces W, W' of V satisfy the "modular equality"

$$\dim W + \dim W' = \dim(W \cap W') + \dim(W \cup W'),$$

it follows from equation (3.6) that $B_n(q)$ is in fact a *modular lattice*. Since every subspace of $B_n(q)$ is the span of its one-dimensional subspaces, $B_n(q)$ is also a *geometric lattice*. The interval $[W, W']$ of $B_n(q)$ is isomorphic to the lattice of subspaces of the quotient space W'/W, so $[W, W'] \cong B_m(q)$, where $m = \ell(W, W') = \dim W' - \dim W$. Hence, $\mu(W, W')$ depends only on the integer $\ell = \ell(W, W')$, so we write $\mu_\ell = \mu(W, W')$. It is now an easy task to compute μ_ℓ using equation (3.33). Let a be an element of $B_n(q)$ of rank 1 (i.e., an atom). Now $B_n(q)$ has a total of $\binom{n}{n-1} = q^{n-1} + q^{n-2} + \cdots + 1$ coatoms, of which $\binom{n-1}{n-2} = q^{n-2} + q^{n-3} + \cdots + 1$ lie above a. Hence, there are q^{n-1} coatoms t satisfying $t \not\geq a$, so from equation (3.33) we have

$$\mu_n = -q^{n-1} \mu_{n-1}.$$

Together with the initial condition $\mu_0 = 1$, there follows

$$\mu_n = (-1)^n q^{\binom{n}{2}}. \tag{3.34}$$

3.10.3 Example. We give one simple example of the use of equation (3.34). We wish to count the number of spanning subsets of V_n. For the purpose of this example, we say that the empty set $\varnothing$ spans no space, while the subset $\{0\}$ spans the zero-dimensional subspace $\{0\}$.* If $W \in B_n(q)$, then let $f(W)$ be the number of subsets of V_n whose span is W, and let $g(W)$ be the number whose span is contained in W. Hence, $g(W) = 2^{q^{\dim W}} - 1$, since $\varnothing$ has no span. Clearly,

$$g(W) = \sum_{T \leq W} f(T), \quad .$$

so by Möbius inversion in $B_n(q)$,

$$f(W) = \sum_{T \leq W} g(T) \mu(T, W).$$

Putting $W = V_n$, there follows

$$f(V_n) = \sum_{T \in B_n(q)} g(T) \mu(T, V_n)$$

$$= \sum_{k=0}^{n} \binom{n}{k} (-1)^{n-k} q^{\binom{n-k}{2}} \left(2^{q^k} - 1 \right).$$

3.10.4 Example. Let Π_S denote the set of all partitions of the finite set S, and write Π_n for $\Pi_{[n]}$. As in Example 3.1.1(d), we partially order Π_S by *refinement*; that is, define $\pi \leq \sigma$ if every block of π is contained in a block of σ. For instance, Π_1, Π_2, and Π_3 are shown in Figure 3.20. It is easy to check that Π_n is graded of rank $n - 1$. The rank $\rho(\pi)$ of $\pi \in \Pi_n$ is equal to $n - $ (number of blocks of π) $= n - \#\pi$. Hence, the rank-generating function of Π_n is given by

$$f(\Pi_n, x) = \sum_{k=0}^{n-1} S(n, n-k) x^k, \tag{3.35}$$

where $S(n, n - k)$ is a Stirling number of the second kind. If $\pi, \sigma \in \Pi_n$, then $\pi \wedge \sigma$ has as blocks the nonempty sets $B \cap C$, where $B \in \pi$ and $C \in \sigma$. Hence, Π_n is a meet-semilattice. Since the partition of $[n]$ with a single block $[n]$ is a $\hat{1}$ for Π_n, it follows from Proposition 3.3.1 that Π_n is a *lattice*.

Suppose that $\pi = \{B_1, \ldots, B_k\}$. Then the interval $[\pi, \hat{1}]$ is isomomorphic in an obvious way to Π_π, the lattice of partitions of the set $\{B_1, \ldots, B_k\}$. Hence, $[\pi, \hat{1}] \cong \Pi_k$. Now it is easy to see that in Π_k, the join of any two distinct atoms has rank two. Merever, any $\pi \in \Pi_n$ is the join of those atoms $\{B_1, \ldots, B_{n-1}\}$ such

---

* The standard convention is that the empty set spans $\{0\}$. If we wish to retain this convention, then we need to enlarge $B_n(q)$ by adding $\varnothing$ below $\{0\}$.

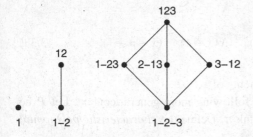

Figure 3.20 Small partition lattices.

that $\#B_1 = 2$ (so $\#B_i = 1$ for $2 \leq i \leq n-1$) and B_1 is a subset of some block of π. Hence Π_n is a *geometric lattice*.

The previous paragraph determined the structure of $[\pi, \hat{1}]$. Let us now consider the structure of any interval $[\sigma, \pi]$. Suppose that $\pi = \{B_1, \ldots, B_k\}$ and that B_i is partitioned into λ_i blocks in σ. We leave to the reader the easy argument that

$$[\sigma, \pi] \cong \Pi_{\lambda_1} \times \Pi_{\lambda_2} \times \cdots \times \Pi_{\lambda_k}.$$

In particular, $[\hat{0}, \pi] \cong \Pi_1^{a_1} \times \cdots \times \Pi_n^{a_n}$ if π has a_i blocks of size i.

NOTE. In analogy to the type of a permutation, we define the *type* of $\pi \in \Pi_n$ by $\text{type}(\pi) = (a_1, \ldots, a_n)$ if π has a_i blocks of size i, for $1 \leq i \leq n$. It is easy to prove in analogy with Proposition 1.3.2 that the number of partitions $\pi \in \Pi_n$ of type $(a_1, \ldots, a_n)$ is given by

$$\#\{\pi \in \Pi_n : \text{type}(\pi) = (a_1, \ldots, a_n)\} = \frac{n!}{1!^{a_1} a_1! \, 2!^{a_2} a_2! \cdots n!^{a_n} a_n!}. \tag{3.36}$$

As an example of the structure of $[\sigma, \pi]$, let $\sigma = 1\text{-}2\text{-}3\text{-}45\text{-}67\text{-}890$ and $\pi = 14567\text{-}2890\text{-}3$. Then

$$[\sigma, \pi] \cong \Pi_{\{1,45,67\}} \times \Pi_{\{2,890\}} \times \Pi_{\{3\}} \cong \Pi_3 \times \Pi_2 \times \Pi_1.$$

Now set $\mu_n = \mu(\hat{0}, \hat{1})$, where μ is the Möbius function of Π_n. If $[\sigma, \pi] = \Pi_{\lambda_1} \times \Pi_{\lambda_2} \times \cdots \times \Pi_{\lambda_k}$, then by Proposition 3.8.2 we have $\mu(\sigma, \pi) = \mu_{\lambda_1} \times \mu_{\lambda_2} \times \cdots \times \mu_{\lambda_k}$. Hence to determine μ completely, it suffices to compute μ_n. Although Π_n is geometric so that equation (3.33) applies, it is easier to appeal directly to Corollary 3.9.3. Pick a to be the partition with the two blocks $\{1, 2, \ldots, n-1\}$ and $\{n\}$. An element t of Π_n satisfies $t \wedge a = \hat{0}$ if and only if $t = \hat{0}$ or t is an atom whose unique two-element block has the form $\{i, n\}$ for some $i \in [n-1]$. The interval $[t, \hat{1}]$ is isomorphic to Π_{n-1}, so from Corollary 3.9.3 we have $\mu_n = -(n-1)\mu_{n-1}$. Since $\mu_0 = 1$, we conclude

$$\mu_n = (-1)^{n-1}(n-1)!. \tag{3.37}$$

There are many other ways to prove this important result, some of which we shall consider later. Let us simply point out here the more general result (which is proved in Example 3.11.11)

$$\sum_{\pi \in \Pi_n} \mu(\hat{0}, \pi) x^{\#\pi} = (x)_n = x(x-1)\cdots(x-n+1). \tag{3.38}$$

To get equation (3.37), equate coefficients of x.

Equation (3.38) can be put in the following more general context. Let P be a finite graded poset with $\hat{0}$, say of rank n. Define the *characteristic polynomial* $\chi_P(x)$ of P by

$$\chi_P(x) = \sum_{t \in P} \mu(\hat{0}, t) x^{n-\rho(t)}$$

$$= \sum_{k=0}^{n} w_k x^{n-k}, \text{ say.} \tag{3.39}$$

The coefficient w_k is called the kth *Whitney number of P of the first kind*:

$$w_k = \sum_{\substack{t \in P \\ \rho(t)=k}} \mu(\hat{0}, t).$$

In this context, the number of elements of P of rank k is denoted W_k and is called the kth *Whitney number of P of the second kind*. Thus the rank-generating function $F(P,x)$ of P is given by

$$F(P,x) = \sum_{t \in P} x^{\rho(t)}$$

$$= \sum_{k=0}^{n} W_k x^k.$$

It follows from equation (3.38) that

$$\chi_{\Pi_n}(x) = (x-1)(x-2)\cdots(x-n+1),$$

since Π_n has rank $n-1$ and $\#\pi = n - \rho(\pi)$. Hence from Proposition 1.3.7, we have $w_k = s(n, n-k)$, a Stirling number of the first kind. Moreover, equation (3.35) yields $W_k = S(n, n-k)$ for the lattice Π_n. For a poset-theoretic reason for the inverse relationship between $S(n,k)$ and $s(n,k)$ given by Proposition 1.9.1(a), see Exercise 3.130(a).

3.11 Hyperplane Arrangements

3.11.1 Basic Definitions

In this section, we give an interesting geometric application of Möbius functions which has a vast number of further applications and extensions. The basic geometric concept to concern us will be a (finite) *hyperplane arrangement* (or just *arrangement* for short), that is, a finite set $\mathcal{A}$ of affine hyperplanes in a finite-dimensional vector space $V \cong K^n$, where K is a field. To make sure that the definition of a hyperplane arrangement is clear, we define a *linear hyperplane* to be an $(n-1)$-dimensional subspace H of V, that is,

$$H = \{v \in V : \alpha \cdot v = 0\},$$

where α is a fixed nonzero vector in V and $\alpha \cdot v$ is the usual dot product (after identifying V with K^n):

$$(\alpha_1, \ldots, \alpha_n) \cdot (v_1, \ldots, v_n) = \sum \alpha_i v_i. \tag{3.40}$$

An *affine hyperplane* is a translate J of a linear hyperplane, that is,

$$J = \{v \in V : \alpha \cdot v = a\},$$

where α is a fixed nonzero vector in V and $a \in K$. The vector α is the *normal* to J, unique up to multiplication by a nonzero scalar.

Let $\mathcal{A}$ be an arrangement in the vector space V. The *dimension* $\dim(\mathcal{A})$ of $\mathcal{A}$ is defined to be $\dim(V)$ $(= n)$, while the *rank* $\mathrm{rank}(\mathcal{A})$ of $\mathcal{A}$ is the dimension of the space spanned by the normals to the hyperplanes in $\mathcal{A}$. We say that $\mathcal{A}$ is *essential* if $\mathrm{rank}(\mathcal{A}) = \dim(\mathcal{A})$. Suppose that $\mathrm{rank}(\mathcal{A}) = r$, and take $V = K^n$. Let Y be a complementary space in K^n to the subspace X spanned by the normals to hyperplanes in $\mathcal{A}$. Define

$$W = \{v \in V : v \cdot y = 0, \text{for all } y \in Y\}.$$

If $K = \mathbb{R}$, then we can simply take $W = X$. (More generally, if $\mathrm{char}(K) = 0$, then we can take $W = X$ provided that we modify the definition of the scalar product (3.40).) By elementary linear algebra, we have

$$\mathrm{codim}_W(H \cap W) = 1 \tag{3.41}$$

for all $H \in \mathcal{A}$. In other words, $H \cap W$ is a hyperplane of W, so the set $\mathcal{A}_W :=
\{H \cap W : H \in \mathcal{A}\}$ is an essential arrangement in W. Moreover, the arrangements $\mathcal{A}$ and $\mathcal{A}_W$ are "essentially the same," meaning in particular that they have the same intersection poset (as defined below in Subsection 3.11.2). Let us call $\mathcal{A}_W$ the *essentialization* of $\mathcal{A}$, denoted $\mathrm{ess}(\mathcal{A})$. When $K = \mathbb{R}$ and we take $W = X$, then the arrangement $\mathcal{A}$ is obtained from $\mathcal{A}_W$ by "stretching" the hyperplane $H \cap W \in \mathcal{A}_W$ orthogonally to W. Thus, if $W^\perp$ denotes the orthogonal complement to W in V, then $H' \in \mathcal{A}_W$ if and only if $H' \oplus W^\perp \in \mathcal{A}$. Note that in characteristic p this type of reasoning fails since the orthogonal complement of a subspace W can intersect W in a subspace of dimension greater than 0.

3.11.1 Example. Let $\mathcal{A}$ consist of the lines $x = a_1, \ldots, x = a_k$ in K^2 (with coordinates x and y). Then we can take W to be the x-axis, and $\mathrm{ess}(\mathcal{A})$ consists of the points $x = a_1, \ldots, x = a_k$ in K.

3.11.2 The Intersection Poset and Characteristic Polynomial

Let $\mathcal{A}$ be an arrangement in a vector space V, and let $L(\mathcal{A})$ be the set of all *nonempty* intersections of hyperplanes in $\mathcal{A}$, including V itself as the intersection over the empty set. Define $s \leq t$ in $L(\mathcal{A})$ if $s \supseteq t$ (as subsets of V). In other words, $L(\mathcal{A})$ is partially ordered by *reverse* inclusion. The vector space V is the $\hat{0}$ element of $L(\mathcal{A})$. We call $L(\mathcal{A})$ the *intersection poset* of $\mathcal{A}$. It is the fundamental combinatorial object associated with an arrangement.

An arrangement $\mathcal{A}$ is called *central* if $\bigcap_{H \in \mathcal{A}} H \neq \emptyset$. We can translate all the hyperplanes in a central arrangement by a fixed vector so that $\mathbf{0} \in \bigcap_{H \in \mathcal{A}} H$, where $\mathbf{0}$ denotes the origin of V. Thus, each hyperplane $H \in \mathcal{A}$ is a *linear* hyperplane and can therefore be identified with a point f_H in the dual space V^* (namely, if H is defined by $\alpha \cdot v = 0$, then H corresponds to the linear functional $f(v) = \alpha \cdot v$). Let L be the geometric lattice, as defined after Proposition 3.3.3, consisting of all intersections of $\{f_H : H \in \mathcal{A}\}$ with linear subspaces of V^*, ordered by inclusion. It is straightforward to see that $L \cong L(\mathcal{A})$. We have therefore proved the following result.

3.11.2 Proposition. *Let $\mathcal{A}$ be a (finite) hyperplane arrangement in a vector space V. If $\mathcal{A}$ is central, then $L(\mathcal{A})$ is a geometric lattice. For any $\mathcal{A}$, every interval of $L(\mathcal{A})$ is a geometric lattice.*

We now define the *characteristic polynomial* $\chi_{\mathcal{A}}(x)$ of the arrangement $\mathcal{A}$ by

$$\chi_{\mathcal{A}}(x) = \sum_{t \in L(\mathcal{A})} \mu(\hat{0}, t) x^{\dim(t)}.$$

Compare this definition with that of the characteristic polynomial $\chi_{L(\mathcal{A})}$ of the poset $L(\mathcal{A})$ itself (equation (3.39)). If $\mathcal{A}$ is essential, then $\chi_{\mathcal{A}}(x) = \chi_{L(\mathcal{A})}(x)$; in general,

$$\chi_{\mathcal{A}}(x) = x^{n-r} \chi_{L(\mathcal{A})}(x),$$

where $n = \dim(\mathcal{A})$ and $r = \operatorname{rank}(\mathcal{A})$. Since every interval of $L(\mathcal{A})$ is a geometric lattice, it follows from Proposition 3.10.1 that the coefficients of $\chi_{\mathcal{A}}(x)$ alternate in sign. More precisely, we have

$$\chi_{\mathcal{A}}(x) = x^n - a_1 x^{n-1} + \cdots + (-1)^{n-r} a_{n-r} x^{n-r},$$

where each $a_i > 0$. Note that $a_1 = \#\mathcal{A}$, the number of hyperplanes in $\mathcal{A}$.

We now use the Crosscut Theorem (Corollary 3.9.4) to give a formula (Proposition 3.11.3) for the characteristic polynomial $\chi_{\mathcal{A}}(x)$. Next we employ this formula for $\chi_{\mathcal{A}}(x)$ to give a recurrence (Proposition 3.11.5) for $\chi_{\mathcal{A}}(x)$. We then use this recurrence to give a formula (Theorem 3.11.7) for the number of regions and number of (relatively) bounded regions of a real arrangement.

Extending slightly the definition of a central arrangement, call any subset $\mathcal{B}$ of $\mathcal{A}$ central if $\bigcap_{H \in \mathcal{B}} H \neq \emptyset$.

3.11.3 Proposition. *Let $\mathcal{A}$ be an arrangement in an n-dimensional vector space. Then*

$$\chi_{\mathcal{A}}(x) = \sum_{\substack{\mathcal{B} \subseteq \mathcal{A} \\ \mathcal{B} \text{ central}}} (-1)^{\#\mathcal{B}} x^{n - \text{rank}(\mathcal{B})}. \qquad (3.42)$$

3.11.4 Example. Let $\mathcal{A}$ be the arrangement in $\mathbb{R}^2$ shown below.

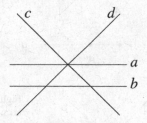

The following table shows all central subsets $\mathcal{B}$ of $\mathcal{A}$ and the values of $\#\mathcal{B}$ and rank$(\mathcal{B})$.

$\mathcal{B}$	$\#\mathcal{B}$	rank$(\mathcal{B})$
$\emptyset$	0	0
a	1	1
b	1	1
c	1	1
d	1	1
ac	2	2
ad	2	2
bc	2	2
bd	2	2
cd	2	2
acd	3	2

It follows that $\chi_{\mathcal{A}}(x) = x^2 - 4x + (5 - 1) = x^2 - 4x + 4$.

Proof of Proposition 3.11.3. Let $t \in L(\mathcal{A})$. Let

$$\Lambda_t = \{s \in L(\mathcal{A}) : s \le t\},$$

the principal order ideal generated by t. Define

$$\mathcal{A}_t = \{H \in \mathcal{A} : H \le t \text{ (i.e., } t \subseteq H)\}. \qquad (3.43)$$

By the Crosscut Theorem (Corollary 3.9.4), we have

$$\mu(\hat{0}, t) = \sum_k (-1)^k N_k(t),$$

where $N_k(t)$ is the number of k-subsets of $\mathcal{A}_t$ with join t. In other words,

$$\mu(\hat{0}, t) = \sum_{\substack{\mathcal{B} \subseteq \mathcal{A}_t \\ t = \bigcap_{H \in \mathcal{B}} H}} (-1)^{\#\mathcal{B}}.$$

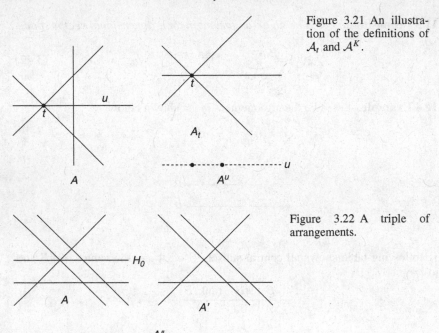

Figure 3.21 An illustration of the definitions of $\mathcal{A}_t$ and $\mathcal{A}^K$.

Figure 3.22 A triple of arrangements.

Note that $t = \bigcap_{H \in \mathcal{B}} H$ implies that $\mathrm{rank}(\mathcal{B}) = n - \dim t$. Now multiply both sides by $x^{\dim(t)}$ and sum over t to obtain equation (3.42).   □

The characteristic polynomial $\chi_{\mathcal{A}}(x)$ satisfies a fundamental recurrence, which we now describe. Let $\mathcal{A}$ be an arrangement in the vector space V. A *subarrangement* of $\mathcal{A}$ is a subset $\mathcal{B} \subseteq \mathcal{A}$. Thus, $\mathcal{B}$ is also an arrangement in V. If $t \in L(\mathcal{A})$, then let $\mathcal{A}_t$ be the subarrangement of equation (3.43). Also define an arrangement $\mathcal{A}^t$ in the affine subspace $t \in L(\mathcal{A})$ by

$$\mathcal{A}^t = \{t \cap H \neq \emptyset : H \in \mathcal{A} - \mathcal{A}_t\}. \tag{3.44}$$

Note that if $t \in L(\mathcal{A})$, then

$$L(\mathcal{A}_t) \cong \Lambda_t := \{s \in L(\mathcal{A}) : s \leq t\},$$
$$L(\mathcal{A}^t) \cong V_t := \{s \in L(\mathcal{A}) : s \geq t\}. \tag{3.45}$$

Figure 3.21 shows an arrangement $\mathcal{A}$, two elements $t, u \in L(\mathcal{A})$, and the arrangements $\mathcal{A}_t$ and $\mathcal{A}^u$.

Choose $H_0 \in \mathcal{A}$. Let $\mathcal{A}' = \mathcal{A} - \{H_0\}$ and $\mathcal{A}'' = \mathcal{A}^{H_0}$. We call $(\mathcal{A}, \mathcal{A}', \mathcal{A}'')$ a *triple* of arrangements with *distinguished hyperplane* H_0. An example is shown in Figure 3.22.

3.11.5 Proposition (Deletion-Restriction). *Let $(\mathcal{A}, \mathcal{A}', \mathcal{A}'')$ be a triple of real arrangements. Then*

$$\chi_{\mathcal{A}}(x) = \chi_{\mathcal{A}'}(x) - \chi_{\mathcal{A}''}(x).$$

Proof. Let $H_0 \in \mathcal{A}$ be the hyperplane defining the triple $(\mathcal{A}, \mathcal{A}', \mathcal{A}'')$. Split the sum on the right-hand side of (3.42) into two sums, depending on whether $H_0 \notin \mathcal{B}$ or $H_0 \in \mathcal{B}$. In the former case, we get

$$\sum_{\substack{H_0 \notin \mathcal{B} \subseteq \mathcal{A} \\ \mathcal{B} \text{ central}}} (-1)^{\#\mathcal{B}} x^{n - \text{rank}(\mathcal{B})} = \chi_{\mathcal{A}'}(x).$$

In the latter case, set $\mathcal{B}_1 = (\mathcal{B} - \{H_0\})^{H_0}$, a central arrangement in $H_0 \cong K^{n-1}$ and a subarrangement of $\mathcal{A}^{H_0} = \mathcal{A}''$. Suppose that S is a set of $r \geq 1$ hyperplanes in $\mathcal{A}$ that all have the same intersection with H_0. Then

$$\sum_{\emptyset \neq T \subseteq S} (-1)^{\#T} = -1,$$

the same result we would get if $r = 1$. Since $\#\mathcal{B}_1 = \#\mathcal{B} - 1$ and $\text{rank}(\mathcal{B}_1) = \text{rank}(\mathcal{B}) - 1$, we get

$$\sum_{\substack{H_0 \in \mathcal{B} \subseteq \mathcal{A} \\ \mathcal{B} \text{ central}}} (-1)^{\#\mathcal{B}} x^{n - \text{rank}(\mathcal{B})} = \sum_{\mathcal{B}_1 \in \mathcal{A}''} (-1)^{\#\mathcal{B}_1 + 1} x^{(n-1) - \text{rank}(\mathcal{B}_1)}$$

$$= -\chi_{\mathcal{A}''}(x),$$

and the proof follows. $\qquad\square$

3.11.3 Regions

Hyperplane arrangements have special combinatorial properties when $K = \mathbb{R}$, which we assume for the remainder of this subsection. A *region* of an arrangement $\mathcal{A}$ (defined over $\mathbb{R}$) is a connected component of the complement X of the hyperplanes:

$$X = \mathbb{R}^n - \bigcup_{H \in \mathcal{A}} H.$$

Let $\mathcal{R}(\mathcal{A})$ denote the set of regions of $\mathcal{A}$, and let

$$r(\mathcal{A}) = \#\mathcal{R}(\mathcal{A}),$$

the number of regions. For instance, the arrangement $\mathcal{A}$ of Figure 3.23 has $r(\mathcal{A}) = 14$.

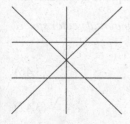

Figure 3.23 An arrangement with 14 regions and four bounded regions.

It is a simple exercise to show that every region $R \in \mathcal{R}(\mathcal{A})$ is open and convex (continuing to assume $K = \mathbb{R}$), and hence homeomorphic to the interior of an n-dimensional ball $\mathbb{B}^n$. Note that if W is the subspace of V spanned by the normals to the hyperplanes in $\mathcal{A}$, then the map $R \mapsto R \cap W$ is a bijection between $\mathcal{R}(\mathcal{A})$ and $\mathcal{R}(\mathcal{A}_W)$. We say that a region $R \in \mathcal{R}(\mathcal{A})$ is *relatively bounded* if $R \cap W$ is bounded. If $\mathcal{A}$ is essential, then relatively bounded is the same as bounded. We write $b(\mathcal{A})$ for the number of relatively bounded regions of $\mathcal{A}$. For instance, in Example 3.11.1, take $K = \mathbb{R}$ and $a_1 < a_2 < \cdots < a_k$. Then the relatively bounded regions are the regions $a_i < x < a_{i+1}$, $1 \le i \le k-1$. In ess$(\mathcal{A})$, they become the (bounded) open intervals (a_i, a_{i+1}). There are also two regions of $\mathcal{A}$ that are not relatively bounded, namely, $x < a_1$ and $x > a_k$. As another example, the arrangement of Figure 3.23 is essential and has four bounded regions.

3.11.6 Lemma. *Let $(\mathcal{A}, \mathcal{A}', \mathcal{A}'')$ be a triple of real arrangements with distinguished hyperplane H_0. Then*

$$r(\mathcal{A}) = r(\mathcal{A}') + r(\mathcal{A}''),$$

$$b(\mathcal{A}) = \begin{cases} b(\mathcal{A}') + b(\mathcal{A}''), & \text{if rank}(\mathcal{A}) = \text{rank}(\mathcal{A}'), \\ 0, & \text{if rank}(\mathcal{A}) = \text{rank}(\mathcal{A}') + 1. \end{cases}$$

NOTE. If rank$(\mathcal{A}) = $ rank$(\mathcal{A}')$, then also rank$(\mathcal{A}) = 1 + $ rank$(\mathcal{A}'')$. The following figure illustrates the situation when rank$(\mathcal{A}) = $ rank$(\mathcal{A}') + 1$.

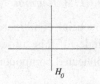

H_0

Proof. Note that $r(\mathcal{A})$ equals $r(\mathcal{A}')$ plus the number of regions of $\mathcal{A}'$ cut into two regions by H_0. Let R' be such a region of $\mathcal{A}'$. Then $R' \cap H_0 \in \mathcal{R}(\mathcal{A}'')$. Conversely, if $R'' \in \mathcal{R}(\mathcal{A}'')$, then points near R'' on either side of H_0 belong to the same region $R' \in \mathcal{R}(\mathcal{A}')$, since any $H \in \mathcal{R}(\mathcal{A}')$ separating them would intersect R''. Thus, R' is cut in two by H_0. We have established a bijection between regions of $\mathcal{A}'$ cut into two by H_0 and regions of $\mathcal{A}''$, establishing the first recurrence.

The second recurrence is proved analogously; the details are omitted. □

We come to one of the central theorems in the subject of hyperplane arrangements.

3.11.7 Theorem. *Let $\mathcal{A}$ be an arrangement in an n-dimensional real vector space. Then*

$$r(\mathcal{A}) = (-1)^n \chi_{\mathcal{A}}(-1), \tag{3.46}$$

$$b(\mathcal{A}) = (-1)^{\text{rank}(\mathcal{A})} \chi_{\mathcal{A}}(1). \tag{3.47}$$

Proof. Equation (3.46) holds for $\mathcal{A} = \emptyset$, since $r(\emptyset) = 1$ and $\chi_\emptyset(x) = x^n$. By Lemma 3.11.6 and Proposition 3.11.5, both $r(\mathcal{A})$ and $(-1)^n \chi_{\mathcal{A}}(-1)$ satisfy the same recurrence, so the proof of (3.46) follows.

Now consider equation (3.47). Again it holds for $\mathcal{A} = \emptyset$ since $b(\emptyset) = 1$. (Recall that $b(\mathcal{A})$ is the number of *relatively* bounded regions. When $\mathcal{A} = \emptyset$, the entire ambient space $\mathbb{R}^n$ is relatively bounded.) Now

$$\chi_{\mathcal{A}}(1) = \chi_{\mathcal{A}'}(1) - \chi_{\mathcal{A}''}(1).$$

Let $d(\mathcal{A}) = (-1)^{\text{rank}(\mathcal{A})} \chi_{\mathcal{A}}(1)$. If $\text{rank}(\mathcal{A}) = \text{rank}(\mathcal{A}') = \text{rank}(\mathcal{A}'') + 1$, then $d(\mathcal{A}) = d(\mathcal{A}') + d(\mathcal{A}'')$. If $\text{rank}(\mathcal{A}) = \text{rank}(\mathcal{A}') + 1$ then $b(\mathcal{A}) = 0$ [why?] and $L(\mathcal{A}') \cong L(\mathcal{A}'')$ [why?]. Thus, $d(\mathcal{A}) = 0$. Hence in all cases, $b(\mathcal{A})$ and $d(\mathcal{A})$ satisfy the same recurrence, so $b(\mathcal{A}) = d(\mathcal{A})$. $\square$

As an application of Theorem 3.11.7, we compute the number of regions of an arrangement whose hyperplanes are in *general position*, that is,

$$\{H_1, \ldots, H_p\} \subseteq \mathcal{A}, \ p \leq n \Rightarrow \dim(H_1 \cap \cdots \cap H_p) = n - p,$$
$$\{H_1, \ldots, H_p\} \subseteq \mathcal{A}, \ p > n \Rightarrow H_1 \cap \cdots \cap H_p = \emptyset.$$

For instance, if $n = 2$ then a set of lines is in general position if and only if no two are parallel and no three meet at a point.

3.11.8 Proposition (general position). *Let $\mathcal{A}$ be an n-dimensional arrangement of m hyperplanes in general position. Then*

$$\chi_{\mathcal{A}}(x) = x^n - mx^{n-1} + \binom{m}{2} x^{n-2} - \cdots + (-1)^n \binom{m}{n}.$$

In particular, if $\mathcal{A}$ is a real arrangement, then

$$r(\mathcal{A}) = 1 + m + \binom{m}{2} + \cdots + \binom{m}{n},$$

$$b(\mathcal{A}) = (-1)^n \left(1 - m + \binom{m}{2} - \cdots + (-1)^n \binom{m}{n} \right)$$
$$= \binom{m-1}{n}.$$

Proof. Every $\mathcal{B} \subseteq \mathcal{A}$ with $\#\mathcal{B} \leq n$ defines an element $x_{\mathcal{B}} = \bigcap_{H \in \mathcal{B}} H$ of $L(\mathcal{A})$. Hence, $L(\mathcal{A})$ is a *truncated boolean algebra*:

$$L(\mathcal{A}) \cong \{S \subseteq [m] : \#S \leq n\},$$

ordered by inclusion. If $t \in L(\mathcal{A})$ and rank$(t) = k$, then $[\hat{0}, t] \cong B_k$, a boolean algebra of rank k. By equation (3.18), there follows $\mu(\hat{0}, t) = (-1)^k$. Hence,

$$\chi_{\mathcal{A}}(x) = \sum_{\substack{S \subseteq [m] \\ \#S \leq n}} (-1)^{\#S} x^{n - \#S}$$

$$= x^n - m x^{n-1} + \cdots + (-1)^n \binom{m}{n}.$$

$\square$

3.11.4 The Finite Field Method

In this subsection we will describe a method based on finite fields for computing the characteristic polynomial of an arrangement defined over $\mathbb{Q}$. We will then give two examples; further examples may be found in Exercise 3.115.

Suppose that the arrangement $\mathcal{A}$ is defined over $\mathbb{Q}$. By multiplying each hyperplane equation by a suitable integer, we may assume $\mathcal{A}$ is defined over $\mathbb{Z}$. In that case, we can take coefficients modulo a prime p and get an arrangement $\mathcal{A}_q$ defined over the finite field $\mathbb{F}_q$, where $q = p^r$. We say that $\mathcal{A}$ has *good reduction* mod p (or over $\mathbb{F}_q$) if $L(\mathcal{A}) \cong L(\mathcal{A}_q)$.

For instance, let $\mathcal{A}$ be the affine arrangement in $\mathbb{Q}^1 = \mathbb{Q}$ consisting of the points 0 and 10. Then $L(\mathcal{A})$ contains three elements, namely, $\mathbb{Q}$, $\{0\}$, and $\{10\}$. If $p \neq 2, 5$ then 0 and 10 remain distinct, so $\mathcal{A}$ has good reduction. On the other hand, if $p = 2$ or $p = 5$ then $0 = 10$ in $\mathbb{F}_p$, so $L(\mathcal{A}_p)$ contains just two elements. Hence, $\mathcal{A}$ has bad reduction when $p = 2, 5$.

3.11.9 Proposition. *Let $\mathcal{A}$ be an arrangement defined over $\mathbb{Z}$. Then $\mathcal{A}$ has good reduction for all but finitely many primes p.*

Proof. Let $H_1, \ldots, H_j$ be affine hyperplanes, where H_i is given by the equation $\alpha_i \cdot x = a_i$ ($\alpha_i \in \mathbb{Z}^n, a_i \in \mathbb{Z}$). By linear algebra, we have $H_1 \cap \cdots \cap H_j \neq \emptyset$ if and only if

$$\text{rank} \begin{bmatrix} \alpha_1 & a_1 \\ \vdots & \vdots \\ \alpha_j & a_j \end{bmatrix} = \text{rank} \begin{bmatrix} \alpha_1 \\ \vdots \\ \alpha_j \end{bmatrix}. \tag{3.48}$$

Moreover, if (3.48) holds, then

$$\dim(H_1 \cap \cdots \cap H_j) = n - \text{rank} \begin{bmatrix} \alpha_1 \\ \vdots \\ \alpha_j \end{bmatrix}.$$

Now for any $r \times s$ matrix A, we have rank$(A) \geq t$ if and only if some $t \times t$ submatrix B satisfies $\det(B) \neq 0$. It follows that $L(\mathcal{A}) \not\cong L(\mathcal{A}_p)$ if and only if at least one member S of a certain finite collection $\mathcal{S}$ of subsets of integer matrices B satisfies the following condition:

$$(\forall B \in S) \ \det(B) \neq 0 \text{ but } \det(B) \equiv 0 \,(\text{mod } p).$$

This can only happen for finitely many p, namely, for certain B we must have $p \mid \det(B)$, so $L(\mathcal{A}) \cong L(\mathcal{A}_p)$ for p sufficiently large. $\qquad \square$

The main result of this subsection is the following. Like many fundamental results in combinatorics, the proof is easy but the applicability very broad.

3.11.10 Theorem. *Let $\mathcal{A}$ be an arrangement in $\mathbb{Q}^n$, and suppose that $L(\mathcal{A}) \cong L(\mathcal{A}_q)$ for some prime power q. Then*

$$\chi_{\mathcal{A}}(q) = \# \left(\mathbb{F}_q^n - \bigcup_{H \in \mathcal{A}_q} H \right)$$

$$= q^n - \# \bigcup_{H \in \mathcal{A}_q} H.$$

Proof. Let $t \in L(\mathcal{A}_q)$ so $\#t = q^{\dim(t)}$. Here $\dim(t)$ can be computed either over $\mathbb{Q}$ or $\mathbb{F}_q$. Define two functions $f, g : L(\mathcal{A}_q) \to \mathbb{Z}$ by

$$f(t) = \#t$$

$$g(t) = \# \left(t - \bigcup_{u > t} u \right).$$

In particular,

$$g(\hat{0}) = g(\mathbb{F}_q^n) = \# \left(\mathbb{F}_q^n - \bigcup_{H \in \mathcal{A}_q} H \right).$$

Clearly,

$$f(t) = \sum_{u \geq t} g(u).$$

Let μ denote the Möbius function of $L(\mathcal{A}) \cong L(\mathcal{A}_q)$. By the Möbius inversion formula (Proposition 3.7.1),

$$g(t) = \sum_{u \geq t} \mu(t, u) f(u)$$

$$= \sum_{u \geq t} \mu(t, u) q^{\dim(u)}.$$

Put $t = \hat{0}$ to get

$$g(\hat{0}) = \sum_{u} \mu(\hat{0}, u) q^{\dim(u)} = \chi_{\mathcal{A}}(q).$$

$\qquad \square$

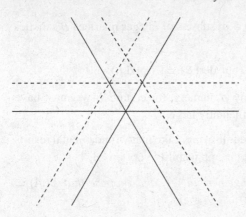

Figure 3.24 The Shi arrangement $\mathcal{S}_3$ in $\ker(x_1 + x_2 + x_3)$.

3.11.11 Example. The *braid arrangement* $\mathcal{B}_n$ of rank $n - 1$ is the arrangement in K^n with hyperplanes $x_i - x_j = 0$ for $1 \le i < j \le n$. The characteristic polynomial of $\mathcal{B}_n$ is particularly easy to compute by the finite field method. Namely, for a large prime p (actually, any prime) $\chi_{\mathcal{B}_n}(p)$ is equal to the number of vectors $(x_1, \ldots, x_n) \in \mathbb{F}_p^n$ such that $x_i \neq x_j$ for all $i < j$. There are p choices for x_1, then $p - 1$ choices for x_2, etc., giving $\chi_{\mathcal{B}_n}(p) = p(p-1)\cdots(p-n+1) = (p)_n$. Hence,

$$\chi_{\mathcal{B}_n}(x) = (x)_n. \tag{3.49}$$

In fact, it is not hard to see that $L_{\mathcal{B}_n} \cong \Pi_n$, the lattice of partitions of the set $[n]$. (See Exercise 3.108(b).) Thus in particular, we have proved equation (3.38).

3.11.12 Example. In this example, we consider a modification (or deformation) of the braid arrangement called the *Shi arrangement* and denoted $\mathcal{S}_n$. It consists of the hyperplanes

$$x_i - x_j = 0, 1, \quad 1 \le i < j \le n.$$

Thus, $\mathcal{S}_n$ has $n(n - 1)$ hyperplanes and $\operatorname{rank}(\mathcal{S}_n) = n - 1$. Figure 3.24 shows the Shi arrangement $\mathcal{S}_3$ in $\ker(x_1 + x_2 + x_3) \cong \mathbb{R}^2$ (i.e., the space $\{(x_1, x_2, x_3) \in \mathbb{R}^3 : x_1 + x_2 + x_3 = 0\}$).

3.11.13 Theorem. *The characteristic polynomial of $\mathcal{S}_n$ is given by*

$$\chi_{\mathcal{S}_n}(x) = x(x - n)^{n-1}.$$

Proof. Let p be a large prime. By Theorem 3.11.10, we have

$$\chi_{\mathcal{S}_n}(p) = \#\{(\alpha_1, \ldots, \alpha_n) \in \mathbb{F}_p^n : i < j \Rightarrow \alpha_i \neq \alpha_j \text{ and } \alpha_i \neq \alpha_j + 1\}.$$

Choose a weak ordered partition $\pi = (B_1, \ldots, B_{p-n})$ of $[n]$ into $p - n$ blocks (i.e., $\bigcup B_i = [n]$ and $B_i \cap B_j = \emptyset$ if $i \neq j$, such that $1 \in B_1$). ("Weak" means that we allow $B_i = \emptyset$.) For $2 \le i \le n$ there are $p - n$ choices for j such that $i \in B_j$, so $(p - n)^{n-1}$ choices in all. We will illustrate the following argument with the example $p = 11$, $n = 6$, and

$$\pi = (\{1, 4\}, \{5\}, \emptyset, \{2, 3, 6\}, \emptyset). \tag{3.50}$$

Arrange the elements of $\mathbb{F}_p$ clockwise on a circle. Place $1, 2, \ldots, n$ on some n of these points as follows. Place elements of B_1 consecutively (clockwise) in increasing order with 1 placed at some element $\alpha_1 \in \mathbb{F}_p$. Skip a space and place the elements of B_2 consecutively in increasing order. Skip another space and place the elements of B_3 consecutively in increasing order, and so on. For our example, (3.50), say $\alpha_1 = 6$. We then get the following placement of $1, 2, \ldots, 6$ on $\mathbb{F}_{11}$.

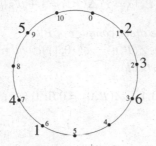

Let α_i be the position (element of $\mathbb{F}_p$) at which i was placed. For our example, we have

$$(\alpha_1, \alpha_2, \alpha_3, \alpha_4, \alpha_5, \alpha_6) = (6, 1, 2, 7, 9, 3).$$

It is easily verified that we have defined a bijection from the $(p-n)^{n-1}$ weak ordered partitions $\pi = (B_1, \ldots, B_{p-n})$ of $[n]$ into $p - n$ blocks such that $1 \in B_1$, together with the choice of $\alpha_1 \in \mathbb{F}_p$, to the set $\mathbb{F}_p^n - \cup_{H \in (\mathcal{S}_n)_p} H$. There are $(p-n)^{n-1}$ choices for π and p choices for α_1, so it follows from Theorem 3.11.10 that $\chi_{\mathcal{S}_n}(p) = p(p-n)^{n-1}$. Hence, $\chi_{\mathcal{S}_n}(x) = x(x-n)^{n-1}$. $\qquad\square$

We obtain the following corollary immediately from Theorem 3.11.7.

3.11.14 Corollary. *We have* $r(\mathcal{S}_n) = (n+1)^{n-1}$ *and* $b(\mathcal{S}_n) = (n-1)^{n-1}$.

NOTE. Since $r(\mathcal{S}_n)$ and $b(\mathcal{S}_n)$ have such simple formulas, it is natural to ask for a direct bijective proof of Corollary 3.11.14. A number of such proofs are known; a sketch that $r(\mathcal{S}_n) = (n+1)^{n-1}$ is given in Exercise 3.111.

3.12 Zeta Polynomials

Let P be a finite poset. If $n \geq 2$, then define $Z(P, n)$ to be the number of multi-chains $t_1 \leq t_2 \leq \cdots \leq t_{n-1}$ in P. We call $Z(P, n)$ (regarded as a function of n) the *zeta polynomial* of P. First we justify this nomenclature and collect together some elementary properties of $Z(P, n)$.

3.12.1 Proposition. *a. Let b_i be the number of chains $t_1 < t_2 < \cdots < t_{i-1}$ in P. Then $b_{i+2} = \Delta^i Z(P, 2)$, $i \geq 0$, where Δ is the finite difference operator. In other words,*

$$Z(P, n) = \sum_{i \geq 2} b_i \binom{n-2}{i-2}. \tag{3.51}$$

In particular, $Z(P,n)$ is a polynomial function of n whose degree d is equal to the length of the longest chain of P, and whose leading coefficient is $b_{d+2}/d!$. Moreover, $Z(P,2) = \#P$ (as is clear from the definition of $Z(P,n)$).

b. *Since $Z(P,n)$ is a polynomial for all integers $n \geq 2$, we can define it for all $n \in \mathbb{Z}$ (or even all $n \in \mathbb{C}$). Then*

$$Z(P,1) = \chi(\Delta(P)) = 1 + \mu_{\widehat{P}}(\hat{0},\hat{1}),$$

where $\Delta(P)$ denotes the order complex of P.

c. *If P has a $\hat{0}$ and $\hat{1}$, then $Z(P,n) = \zeta^n(\hat{0},\hat{1})$ for all $n \in \mathbb{Z}$ (explaining the term zeta polynomial). In particular,*

$$Z(P,-1) = \mu(\hat{0},\hat{1}), \quad Z(P,0) = 0 \text{ (if } \hat{0} \neq \hat{1}), \text{ and } Z(P,1) = 1.$$

Proof.

a. The number of $(n-1)$-element multichains with support $t_1 < t_2 < \cdots < t_{i-1}$ is $\left(\binom{i-1}{n-1-(i-1)}\right) = \binom{n-2}{i-2}$, from which equation (3.51) follows. The additional information about $Z(P,n)$ can be read off from (3.51).

b. Putting $n = 1$ in (3.51) yields

$$Z(P,1) = \sum_{i \geq 2} b_i \binom{-1}{i-2} = \sum_{i \geq 2} (-1)^i b_i.$$

Now use Proposition 3.8.5.

c. If P has a $\hat{0}$ and $\hat{1}$, then the number of multichains $t_1 \leq t_2 \leq \cdots \leq t_{n-1}$ is the same as the number of multichains $\hat{0} = t_0 \leq t_1 \leq t_2 \leq \cdots \leq t_{n-1} \leq t_n = \hat{1}$, which is $\zeta^n(\hat{0},\hat{1})$ for $n \geq 2$. There are several ways to see that $Z(P,n)$, as defined by (3.51) for all $n \geq 2$, is equal to $\zeta^n(\hat{0},\hat{1})$ for all $n \in \mathbb{Z}$. For instance, it follows from equation (3.14) that $\Delta^{d+1}\zeta^k|_{k=0} = 0$ (as linear transformations) [why?]. Multiplying by ζ^n gives $\Delta^{d+1}\zeta^n = 0$ for any $n \in \mathbb{Z}$. Hence by Proposition 1.9.2, $\zeta^n(\hat{0},\hat{1})$ is a polynomial function for all $n \in \mathbb{Z}$ and, thus, must agree with (3.51) for all $n \in \mathbb{Z}$. $\qquad \square$

If $m \in \mathbb{P}$, then let $\Omega_P(m)$ denote the number of order-preserving maps $\sigma : P \to \mathbf{m}$. It follows from Proposition 3.5.1 that $\Omega_P(m) = Z(J(P),m)$. Hence, $\Omega_P(m)$ is a polynomial function of m of degree $p = \#P$ and leading coefficient $e(P)/p!$. (This can easily be seen by a more direct argument.) We call $\Omega_P(m)$ the *order polynomial* of P. Thus, the order polynomial of P is the zeta polynomial of $J(P)$. For further information on order polynomials in a more general setting of labeled posets, see Section 3.15.3.

3.12.2 Example. Let $P = B_d$, the boolean algebra of rank d. Then $Z(B_d,n)$ for $n \geq 1$ is equal to the number of multichains $\emptyset = S_0 \subseteq S_1 \subseteq \cdots \subseteq S_n = S$ of subsets of a d-set S. For any $s \in S$, we can pick arbitrarily the least positive integer $i \in [n]$ for which $s \in S_i$. Hence, $Z(B_d,n) = n^d$. (We can also see this from $Z(B_d,n) = \Omega_{d\mathbf{1}}(n)$, since *any* map $\sigma : d\mathbf{1} \to \mathbf{n}$ is order-preserving.) Putting $n = -1$ yields $\mu_{B_d}(\hat{0}, \hat{1}) = (-1)^d$, a third proof of equation (3.18). This computation of $\mu(\hat{0}, \hat{1})$ is an interesting example of a "semicombinatorial" proof. We evaluate $Z(B_d, n)$ combinatorially for $n \geq 1$ and then substitute $n = -1$. Many other theorems involving Möbius functions of posets P can be proved in such a fashion, by proving combinatorially for $n \geq 1$ an appropriate result for $Z(P, n)$ and then letting $n = -1$.

3.13 Rank Selection

Let P be a finite graded poset of rank n, with rank function $\rho : P \to [0, n]$. If $S \subseteq [0, n]$ then define the subposet

$$P_S = \{t \in P : \rho(t) \in S\},$$

called the *S-rank-selected subposet* of P. For instance, $P_\emptyset = \emptyset$ and $P_{[0,n]} = P$. Now define $\alpha_P(S)$ (or simply $\alpha(S)$) to be the number of maximal chains of P_S. For instance, $\alpha(i)$ (short for $\alpha(\{i\})$) is just the number of elements of P of rank i. The function $\alpha_P : 2^{[0,n]} \to \mathbb{Z}$ is called the *flag f-vector* of P. Also define $\beta_P(S) = \beta(S)$ by

$$\beta(S) = \sum_{T \subseteq S} (-1)^{\#(S-T)} \alpha(T). \tag{3.52}$$

Equivalently, by the Principle of Inclusion-Exclusion,

$$\alpha(S) = \sum_{T \subseteq S} \beta(T). \tag{3.53}$$

The function β_P is called the *flag h-vector* of P.

NOTE. The reason for the terminology "flag f-vector" and "flag h-vector" is the following. Let Δ be a finite $(d - 1)$-dimensional simplicial complex with f_i i-dimensional faces. The vector $f(\Delta) = (f_0, f_1, \ldots, f_{d-1})$ is called the *f-vector* of Δ. Define integers $h_0, \ldots, h_d$ by the condition

$$\sum_{i=0}^{d} f_{i-1}(x - 1)^{d-i} = \sum_{i=0}^{d} h_i x^{d-i}.$$

(Recall that $f_{-1} = 1$ unless $\Delta = \emptyset$.) The vector $h(\Delta) = (h_0, h_1, \ldots, h_d)$ is called the *h-vector* of Δ and is often more convenient to work with than the f-vector. It is easy to check that for a finite graded poset P with order complex $\Delta = \Delta(P)$, we have

$$f_i(\Delta) = \sum_{\#S = i+1} \alpha_P(S),$$

$$h_i(\Delta) = \sum_{\#S = i} \beta_P(S).$$

Thus α_P and β_P extend in a natural way the counting of faces by dimension (or cardinality) to the counting of *flags* (or chains) of P (which are just faces of $\Delta(P)$) by the ranks of the elements of the flags.

Figure 3.25 A naturally labeled poset.

If μ_S denotes the Möbius function of the poset $\widehat{P}_S = P_S \cup \{\hat{0}, \hat{1}\}$, then it follows from Proposition 3.8.5 that

$$\beta_P(S) = (-1)^{\#S-1}\mu_S(\hat{0}, \hat{1}). \tag{3.54}$$

For this reason, the function β_P is also called the *rank-selected Möbius invariant* of P.

Suppose that P has a $\hat{0}$ and $\hat{1}$. It is then easily seen that

$$\alpha_P(S) = \alpha_P(S \cap [n-1]),$$

$$\beta_P(S) = 0, \text{ if } S \nsubseteq [n-1] \text{ (i.e., if } 0 \in S \text{ or } n \in S).$$

Hence, we lose nothing by restricting our attention to $S \subseteq [n-1]$. For this reason, if we know in advance that P has a $\hat{0}$ and $\hat{1}$ (e.g., if P is a lattice), then we will only consider $S \subseteq [n-1]$.

Equations (3.53) and (3.54) suggest a combinatorial method for interpreting the Möbius function of P. The numbers $\alpha(S)$ have a combinatorial definition. If we can define numbers $\gamma(S) \geq 0$ so that there is a combinatorial proof that $\alpha(S) = \sum_{T \subseteq S} \gamma(T)$, then it follows that $\gamma(S) = \beta(S)$ so $\mu_S(\hat{0}, \hat{1}) = (-1)^{\#S-1}\gamma(S)$. We cannot expect to define $\gamma(S)$ for any P since in general we need not have $\beta(S) \geq 0$. However, there are large classes of posets P for which $\gamma(S)$ can indeed be defined in a nice combinatorial manner. To introduce the reader to this subject, we will consider two special cases here, while the next section is concerned with a more general result of this nature.

Let $L = J(P)$ be a finite distributive lattice of rank n (so $\#P = n$). Let $\omega: P \to [n]$ be an order-preserving bijection (i.e., a linear extension of P). In the present context we call ω a *natural labeling* of P. We may identify a linear extension $\sigma: P \to [n]$ of P with a permutation $\omega(\sigma^{-1}(1)), \ldots, \omega(\sigma^{-1}(n))$ of the set $[n]$ of *labels* of P. (Compare Section 3.5, where we identified a linear extension with a permutation of the *elements* of P.) The set of all $e(P)$ permutations of $[n]$ obtained in this way is denoted $\mathcal{L}(P, \omega)$ and is called the *Jordan-Hölder set* of P. For instance, if (P, ω) is given by Figure 3.25, then $\mathcal{L}(P, \omega)$ consists of the five permutations 1234, 2134, 1243, 2143, 2413.

3.13.1 Theorem. *Let $L = J(P)$ as above, and let $S \subseteq [n-1]$. Then $\beta_L(S)$ is equal to the number of permutations $w \in \mathcal{L}(P, \omega)$ with descent set S.*

Proof. Let $S = \{a_1, a_2, \ldots, a_k\}_<$. It follows by definition that $\alpha_L(S)$ is equal to the number of chains $I_1 \subset I_2 \subset \cdots \subset I_k$ of order ideals of P such that $\#I_i = a_i$.

Given such a chain of order ideals, define a permutation $w \in \mathcal{L}(P, \omega)$ as follows: First, arrange the labels of the elements of I_1 in increasing order. To the right of these arrange the labels of the elements of $I_2 - I_1$ in increasing order. Continue until at the end we have the labels of the elements of $P - I_k$ is increasing order. This establishes a bijection between maximal chains of L_S and permutations $w \in \mathcal{L}(P, \omega)$ whose descent set is *contained in* S. Hence if $\gamma_L(S)$ denotes the number of $w \in \mathcal{L}(P, \omega)$ whose descent set *equals* S, then

$$\alpha_L(S) = \sum_{T \subseteq S} \gamma_L(T),$$

and the proof follows. $\qquad\qquad\square$

3.13.2 Corollary. *Let $L = B_n$, the boolean algebra of rank n, and let $S \subseteq [n-1]$. Then $\beta_L(S)$ is equal to the total number of permutations of $[n]$ with descent set S. Thus, $\beta_L(S) = \beta_n(S)$ as defined in Example 2.2.4.*

Just as Example 2.2.5 is a q-generalization of Example 2.2.4, so we can generalize the previous corollary.

3.13.3 Theorem. *Let $L = B_n(q)$, the lattice of subspaces of the vector space $\mathbb{F}_q^n$. Let $S \subseteq [n-1]$. Then*

$$\beta_L(S) = \sum_w q^{\mathrm{inv}(w)},$$

where the sum is over all permutations $w \in \mathfrak{S}_n$ with descent set S, and where $\mathrm{inv}(w)$ is the number of inversions of w.

Proof. Let $S = \{a_1, a_2, \ldots, a_k\}_<$. Then

$$\alpha_L(S) = \binom{n}{a_1}\binom{n-a_1}{a_2-a_1}\binom{n-a_2}{a_3-a_2}\cdots\binom{n-a_k}{n-a_k}$$

$$= \binom{n}{a_1, a_2 - a_1, \ldots, n - a_k}.$$

The proof now follows by comparing equation (2.20) from Chapter 2 with equation (3.53). $\qquad\qquad\square$

3.14 R-Labelings

In this section we give a wide class $\mathcal{A}$ of posets P for which the flag h-vector $\beta_P(S)$ has a direct combinatorial interpretation (and is therefore nonnegative). If $P \in \mathcal{A}$, then every interval of P will also belong to $\mathcal{A}$, so in particular the Möbius function of P alternates in sign.

Let $\mathcal{H}(P)$ denote the set of pairs (s, t) of elements of P for which t covers s. We may think of elements of $\mathcal{H}(P)$ as edges of the Hasse diagram of P.

3.14.1 Definition. Let P be a finite graded poset with $\hat{0}$ and $\hat{1}$. A function $\lambda \colon \mathcal{H}(P) \to \mathbb{Z}$ is called an *R-labeling* of P if, for every interval $[s,t]$ of P, there is a unique saturated chain $s = t_0 \lessdot t_1 \lessdot \cdots \lessdot t_\ell = t$ satisfying

$$\lambda(t_0,t_1) \leq \lambda(t_1,t_2) \leq \cdots \leq \lambda(t_{\ell-1},t_\ell). \tag{3.55}$$

A poset P possessing an R-labeling λ is called *R-labelable* or an *R-poset*, and the chain $s = t_0 \lessdot t_1 \lessdot \cdots \lessdot t_\ell = t$ satisfying equation (3.55) is called the *increasing chain* from s to t.

Note that if $Q = [s,t]$ is an interval of P, then the restriction of λ to $\mathcal{H}(Q)$ is an R-labeling of $\mathcal{H}(Q)$. Hence Q is also an R-poset, so any property satisfied by all R-posets P is also satisfied by any interval of P.

3.14.2 Theorem. *Let P be an R-poset of rank n. Let λ be an R-labeling of P, and let $S \subseteq [n-1]$. Then $\beta_P(S)$ is equal to the number of maximal chains* $\mathfrak{m} \colon \hat{0} = t_0 \lessdot t_1 \lessdot \cdots \lessdot t_n = \hat{1}$ *of P for which the sequence*

$$\lambda(\mathfrak{m}) := (\lambda(t_0,t_1), \lambda(t_1,t_2), \ldots, \lambda(t_{n-1},t_n))$$

has descent set S; that is, for which

$$D(\lambda(\mathfrak{m})) := \{i \,:\, \lambda(t_{i-1},t_i) > \lambda(t_i,t_{i+1})\} = S.$$

Proof. Let $\mathfrak{c} \colon \hat{0} < u_1 < \cdots < u_s < \hat{1}$ be a maximal chain in $\widehat{P_S}$. We claim there is a unique maximal chain $\mathfrak{m}$ of P containing $\mathfrak{c}$ and satisfying $D(\lambda(\mathfrak{m})) \subseteq S$. Let $\mathfrak{m} \colon \hat{0} = t_0 \lessdot t_1 \lessdot \cdots \lessdot t_n = \hat{1}$ be such a maximal chain (if one exists), and let $S = \{a_1, \ldots, a_s\}_<$. Thus, $t_{a_i} = u_i$. Since $\lambda(t_{a_i-1}, t_{a_i-1+1}) \leq \lambda(t_{a_i-1+1}, t_{a_i-1+2}) \leq \cdots \leq \lambda(t_{a_i-1}, t_{a_i})$ for $1 \leq i \leq s+1$ (where we set $a_0 = \hat{0}$, $a_{s+1} = \hat{1}$), we must take $t_{a_i-1}, t_{a_i-1+1}, \ldots, t_{a_i}$ to be the unique increasing chain of the interval $[u_{i-1}, u_i] = [t_{a_{i-1}}, t_{a_i}]$. Thus, M exists and is unique, as claimed.

It follows that the number $\alpha'_P(S)$ of maximal chains $\mathfrak{m}$ of P satisfying $D(\lambda(\mathfrak{m})) \subseteq S$ is just the number of maximal chains of P_S; that is, $\alpha'_P(S) = \alpha_P(S)$. If $\beta'_P(S)$ denotes the number of maximal chains $\mathfrak{m}$ of P satisfying $D(\lambda(\mathfrak{m})) = S$, then clearly

$$\alpha'_P(S) = \sum_{T \subseteq S} \beta'_P(T).$$

Hence from equation (3.53) we conclude $\beta'_P(S) = \beta_P(S)$. $\qquad\square$

3.14.3 Example. We now consider some examples of R-posets. Let P be a natural partial order on $[n]$, as in Theorem 3.13.1. Let $(I, I') \in \mathcal{H}(J(P))$, so I and I' are order ideals of P with $I \subset I'$ and $\#(I' - I) = 1$. Define $\lambda(I, I')$ to be the unique element of $I' - I$. For any interval $[K, K']$ of $J(P)$ there is a unique increasing chain $K = K_0 < K_1 < \cdots < K_\ell = K'$ defined by letting the sole element of $K_i - K_{i-1}$ be the least integer (in the usual linear order on $[n]$) contained in $K' - K_{i-1}$. Hence λ is an R-labeling, and indeed Theorems 3.13.1 and 3.14.2 coincide. We next mention without proof two generalizations of this example.

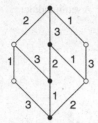

Figure 3.26 A supersolvable lattice.

3.14.4 Example. A finite lattice L is *supersolvable* if it possesses a maximal chain $\mathfrak{c}$, called an *M-chain*, such that the sublattice of L generated by $\mathfrak{c}$ and any other chain of L is distributive. Example of supersolvable lattices include modular lattices, the partition lattice Π_n, and the lattice of subgroups of a finite supersolvable group. For modular lattices, any maximal chain is an M-chain. For the lattice Π_n, a chain $\hat{0} = \pi_0 \lessdot \pi_1 \lessdot \cdots \lessdot \pi_{n-1} = \hat{1}$ is an M chain if and only if each partition π_i ($1 \leq i \leq n-1$) has exactly one block B_i with more than one element (so $B_1 \subset B_2 \subset \cdots \subset B_{n-1} = [n]$). The number of M-chains of Π_n is $n!/2$, $n \geq 2$. For the lattice L of subgroups of a supersolvable group G, an M-chain is given by a *normal series* $\{1\} = G_0 \lessdot G_1 \lessdot \cdots \lessdot G_n = G$; that is, each G_i is a normal subgroup of G, and each G_{i+1}/G_i is cyclic of prime order. (There may be other M-chains.)

If L is supersolvable with M-chain $\mathfrak{c}\colon \hat{0} = t_0 \lessdot t_1 \lessdot \cdots \lessdot t_n = \hat{1}$, then an R-labeling $\lambda\colon \mathcal{H}(P) \to \mathbb{Z}$ is given by

$$\lambda(s,t) = \min\{i \;:\; s \vee t_i = t \vee t_i\}. \tag{3.56}$$

If we restrict λ to the (distributive) lattice L' of L generated by $\mathfrak{c}$ and some other chain, then we obtain an R-labeling of L' that coincides with Example 3.14.3. Figure 3.26 shows a (nonsemimodular) supersolvable lattice L with an M-chain denoted by solid dots, and the corresponding R-labeling λ. There are five maximal chains, with labels 312, 132, 123, 213, 231 and corresponding descent sets $\{1\}$, $\{2\}$, $\emptyset$, $\{1\}$, $\{2\}$. Hence $\beta(\emptyset) = 1$, $\beta(1) = \beta(2) = 2$, $\beta(1,2) = 0$. Note that all maximal chain labels are permutations of [3]; for the significance of this fact see Exercise 3.125.

3.14.5 Example. Let L be a finite (upper) semimodular lattice. Let P be the subposet of join-irreducibles of L. Let $\omega\colon P \to [k]$ be an order-preserving bijection (so $\#P = k$), and write $t_i = \omega^{-1}(i)$. Define for $(s,t) \in \mathcal{H}(L)$,

$$\lambda(s,t) = \min\{i \;:\; s \vee t_i = t\}. \tag{3.57}$$

Then λ is an R-labeling, and hence semimodular lattices are R-posets. Figure 3.27 shows on the left a semimodular lattice L with the elements t_i denoted by i, and on the right the corresponding R-labeling λ. There are seven maximal chains, with labels 123, 132, 213, 231, 312, 321, 341, and corresponding descent sets $\emptyset$, $\{2\}$,$\{1\}$,$\{2\}$, $\{1\}$, $\{1,2\}$, $\{2\}$. Hence, $\beta(\emptyset) = 1$, $\beta(1) = 2$, $\beta(2) = 3$, $\beta(1,2) = 1$.

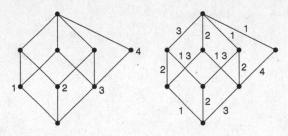

Figure 3.27 A semimodular lattice.

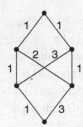

Figure 3.28 A nonlattice with an R-labeling.

Examples 3.14.4 and 3.14.5 both have the property that we can label certain elements of L as t_i (or just i) and then define λ by the similar formulas (3.56) and (3.57). Many additional R-lattices have this property, though not all of them do. Of course, equations (3.56) and (3.57) are meaningless for posets that are not lattices. Figure 3.28 illustrates a poset P that is not a lattice, together with an R-labeling λ.

3.15 (P,ω)-Partitions

3.15.1 The Main Generating Function

A (P,ω)-partition is a kind of interpolation between partitions and compositions. The poset P specifies inequalities among the parts, and the labeling ω specifies which of these inequalities are strict. There is a close connection with descent sets of permutations and the related statistics maj (the major index) and des (the number of descents).

Let P be a finite poset of cardinality p. Let $\omega: P \to [p]$ be a bijection, called a *labeling* of P.

3.15.1 Definition. A (P,ω)-*partition* is a map $\sigma: P \to \mathbb{N}$ satisfying the conditions:

- If $s \leq t$ in P, then $\sigma(s) \geq \sigma(t)$. In other words, σ is *order-reversing*.
- If $s < t$ and $\omega(s) > \omega(t)$, then $\sigma(s) > \sigma(t)$.

If $\sum_{t \in P} \sigma(t) = n$, then we say that σ is a (P,ω)-partition *of n*.

If ω is *natural* (i.e., $s < t \Rightarrow \omega(s) < \omega(t)$), then a (P,ω)-partition is just an order-reversing map $\sigma: P \to \mathbb{N}$. We then call σ simply a *P-partition*. Similarly, if ω is *dual natural* (i.e., $s < t \Rightarrow \omega(s) > \omega(t)$), then a (P,ω)-partition is a *strict order-reversing map* $\sigma : P \to \mathbb{N}$ (i.e., $s < t \Rightarrow \sigma(s) > \sigma(t)$). We then call σ a *strict P-partition*.

Let $P = \{t_1,\dots,t_p\}$. The fundamental generating function associated with (P,ω)-partitions is defined by

$$F_{P,\omega} = F_{P,\omega}(x_1,\dots,x_p) = \sum_{\sigma} x_1^{\sigma(t_1)} \cdots x_p^{\sigma(t_p)},$$

where σ ranges over all (P,ω)-partitions $\sigma : P \to \mathbb{N}$. If ω is natural, then we write simply F_P for $F_{P,\omega}$. The generating function $F_{P,\omega}$ essentially lists all (P,ω)-partitions and contains all possible information about them. Indeed, it is easy to recover the labeled poset (P,ω) if $F_{P,\omega}$ is known.

3.15.2 Example. (a) Suppose that P is a naturally labeled p-element chain $t_1 < \cdots < t_p$. Then

$$F_P = \sum_{a_1 \geq a_2 \geq \cdots \geq a_p \geq 0} x_1^{a_1} x_2^{a_2} \cdots x_p^{a_p}$$

$$= \frac{1}{(1-x_1)(1-x_1x_2)\cdots(1-x_1x_2\cdots x_p)}.$$

(b) Suppose that P is a dual naturally labeled p-element chain $t_1 < \cdots < t_p$. Then

$$F_{P,\omega} = \sum_{a_1 > a_2 > \cdots > a_p \geq 0} x_1^{a_1} x_2^{a_2} \cdots x_p^{a_p}$$

$$= \frac{x_1^{p-1} x_2^{p-2} \cdots x_{p-1}}{(1-x_1)(1-x_1x_2)\cdots(1-x_1x_2\cdots x_p)}.$$

(c) If P is a p-element antichain, then all labelings are natural. We get

$$F_P = \sum_{a_1,\dots,a_p \geq 0} x_1^{a_1} x_2^{a_2} \cdots x_p^{a_p}$$

$$= \frac{1}{(1-x_1)(1-x_2)\cdots(1-x_p)}.$$

(d) Suppose that P has a minimal element t_1 and two elements t_2, t_3 covering t_1, with the labeling $\omega(t_1) = 2$, $\omega(t_2) = 1$, $\omega(t_3) = 3$. Then

$$F_{P,\omega} = \sum_{b < a \geq c} x_1^a x_2^b x_3^c.$$

Let $\mathcal{L}(P,\omega)$ denote the set of linear extensions of P, regarded as permutations of the *labels* $\omega(t)$, as done in Section 3.13 for natural labelings. Thus, $\mathcal{L}(P,\omega) \subseteq \mathfrak{S}_p$. Again following Section 3.13, we call $\mathcal{L}(P,\omega)$ the *Jordan-Hölder set* of the labeled

Figure 3.29 A labeled poset.

poset (P,ω). For instance, if (P,ω) is given by Figure 3.29 (with the labels circled) then

$$\mathcal{L}(P,\omega) = \{3124, 3142, 1324, 1342, 1432\}.$$

Write $\mathcal{A}(P,\omega)$ for the set of all (P,ω)-partitions $\sigma: P \to \mathbb{N}$, and let $\sigma \in \mathcal{A}(P,\omega)$. Define $\sigma': [p] \to \mathbb{N}$ by

$$\sigma'(i) = \sigma(\omega^{-1}(i)).$$

In other words, σ and σ' are essentially the same function, but the argument of σ is an *element* $t \in P$, while the argument of σ' is the *label* $\omega(t)$ of that element. This distinction is particularly important in Subsection 3.15.3 (reciprocity), where we deal with two different labelings of the same poset. We know from Lemma 1.4.11 that there is a unique permutation $w \in \mathfrak{S}_p$ for which σ' is w-compatible. For any $w \in \mathfrak{S}_p$, we write S_w for the set of all functions $\sigma: P \to \mathbb{N}$ for which σ' is w-compatible. We come to the fundamental lemma on (P,ω)-partitions.

3.15.3 Lemma. *A function $\sigma: P \to \mathbb{N}$ is a (P,ω)-partition if and only if σ' is w-compatible with some (necessarily unique) $w \in \mathcal{L}(P,\omega)$. Equivalently, we have the disjoint union*

$$\mathcal{A}(P,\omega) = \bigcup_{w \in \mathcal{L}(P,\omega)} S_w.$$

Proof. We first show that if $\sigma \in \mathcal{A}(P,\omega) \cap S_w$, then $w \in \mathcal{L}(P,\omega)$. Let $w = w_1 w_2 \cdots w_p$. Suppose that $i < j$ and $\omega(s) = w_i$, $\omega(t) = w_j$. We need to show that we cannot have $s > t$. If $\sigma(s) = \sigma(t)$, then by definition of w-compatibility we have $w_i < w_{i+1} < \cdots < w_j$. Hence, by definition of (P,ω)-partition, we cannot have $s > t$. If instead $\sigma(s) > \sigma(t)$, then again by definition of (P,ω)-partition we cannot have $s > t$, so $w \in \mathcal{L}(P,\omega)$.

It remains to show that if $w \in \mathcal{L}(P,\omega)$ and σ' is w-compatible, then $\sigma \in \mathcal{A}(P,\omega)$. Clearly, σ is order-reversing, so we need to show that if $s < t$ and $\omega(s) = w_i > w_j = \omega(t)$, then $\sigma(s) > \sigma(t)$. Since $w_i > w_j$, somewhere in w between w_i and w_j is a descent $w_k > w_{k+1}$. Thus,

$$\sigma(s) = \sigma(w_i) \geq \sigma(w_{i+1}) \geq \cdots \geq \sigma(w_k) > \sigma(w_{k+1}) \geq \cdots \geq \sigma(w_j) = \sigma(t),$$

and the proof follows. $\square$

Given $w \in \mathfrak{S}_p$, let

$$F_w = \sum_{\sigma \in S_w} x_1^{\sigma(t_1)} \cdots x_p^{\sigma(t_p)}, \qquad (3.58)$$

be the generating function for all functions $\sigma : P \to \mathbb{N}$ for which σ' is w-compatible. The next result is a straightforward extension of Lemma 1.4.12. Write $w'_i = j$ if $w_i = \omega(t_j)$.

3.15.4 Lemma. *Let* $w = w_1 \cdots w_p \in \mathfrak{S}_p$. *Then*

$$F_w = \frac{\prod_{j \in D(w)} x_{w'_1} x_{w'_2} \cdots x_{w'_j}}{\prod_{i=1}^{p} \left(1 - x_{w'_1} x_{w'_2} \cdots x_{w'_i}\right)}. \tag{3.59}$$

Proof. Let $\sigma \in S_w$. Define numbers c_i, $1 \le i \le p$, by

$$c_i = \begin{cases} \sigma'(w_i) - \sigma'(w_{i+1}), & \text{if } i \notin D(w), \\ \sigma'(w_i) - \sigma'(w_{i+1}) - 1, & \text{if } i \in D(w), \end{cases} \tag{3.60}$$

where we set $\sigma'(w_{p+1}) = 0$. Note that $c_i \ge 0$ and that any choice of $c_1, c_2, \ldots, c_p \in \mathbb{N}$ defines a unique function $\sigma \in S_w$ satisfying equation (3.60). Then

$$x_1^{\sigma(t_1)} \cdots x_p^{\sigma(t_p)} = \prod_{i=1}^{p} \left(x_{w'_1} x_{w'_2} \cdots x_{w'_i}\right)^{c_i} \cdot \prod_{j \in D(w)} x_{w'_1} x_{w'_2} \cdots x_{w'_j}.$$

This sets up a one-to-one correspondence between the terms in the left- and right-hand sides of equation (3.59), so the proof follows. $\square$

Combining Lemmas 3.15.3 and 3.15.4, we obtain the main theorem on the generating function $F_{P,\omega}$.

3.15.5 Theorem. *Let* (P, ω) *be a labeled p-element poset. Then*

$$F_{P,\omega}(x_1, \ldots, x_p) = \sum_{w \in \mathcal{L}(P,\omega)} \frac{\prod_{j \in D(w)} x_{w'_1} x_{w'_2} \cdots x_{w'_j}}{\prod_{i=1}^{p} \left(1 - x_{w'_1} x_{w'_2} \cdots x_{w'_i}\right)}. \tag{3.61}$$

3.15.6 Example. Let (P, ω) be given by Figure 3.29. Then Lemma 3.15.3 says that every (P, ω)-partition $\sigma : P \to \mathbb{N}$ satisfies exactly one of the conditions

$$\begin{array}{ccccccc} \sigma'(3) & > & \sigma'(1) & \ge & \sigma'(2) & \ge & \sigma'(4), \\ \sigma'(1) & \ge & \sigma'(3) & > & \sigma'(2) & \ge & \sigma'(4), \\ \sigma'(3) & > & \sigma'(1) & \ge & \sigma'(4) & > & \sigma'(2), \\ \sigma'(1) & \ge & \sigma'(3) & \ge & \sigma'(4) & > & \sigma'(2), \\ \sigma'(1) & \ge & \sigma'(4) & > & \sigma'(3) & > & \sigma'(2). \end{array}$$

It follows that

$$F_{P,\omega}(x_1,x_2,x_3,x_4) = \frac{x_1}{(1-x_1)(1-x_1x_2)(1-x_1x_2x_3)(1-x_1x_2x_3x_4)}$$
$$+\frac{x_1x_2}{(1-x_2)(1-x_1x_2)(1-x_1x_2x_3)(1-x_1x_2x_3x_4)}$$
$$+\frac{x_1^2x_2x_4}{(1-x_1)(1-x_1x_2)(1-x_1x_2x_4)(1-x_1x_2x_3x_4)}$$
$$+\frac{x_1x_2x_4}{(1-x_2)(1-x_1x_2)(1-x_1x_2x_4)(1-x_1x_2x_3x_4)}$$
$$+\frac{x_1x_2^2x_4^2}{(1-x_2)(1-x_2x_4)(1-x_1x_2x_4)(1-x_1x_2x_3x_4)}.$$

This example illustrates the underlying combinatorial meaning behind the efficacy of the fundamental Lemma 3.15.3—it allows the set $\mathcal{A}(P,\omega)$ of all (P,ω)-partitions to be partitioned into finitely many (namely, $e(P)$) "simple" subsets, each of which can be handled separately.

3.15.2 Specializations

We now turn to two basic specializations of the generating function $F_{P,\omega}$. Let $a(n)$ denote the number of (P,ω)-partitions of n. Define the generating function

$$G_{P,\omega}(x) = \sum_{n\geq 0} a(n)x^n. \tag{3.62}$$

Clearly, $G_{P,\omega}(x) = F_{P,\omega}(x,x,\ldots,x)$. Moreover, $\prod_{j\in D(w)} x^j = x^{\mathrm{maj}(w)}$. Hence from Theorems 3.15.5, we obtain the following result.

3.15.7 Theorem. *The generating function $G_{P,\omega}(x)$ has the form*

$$G_{P,\omega}(x) = \frac{W_{P,\omega}(x)}{(1-x)(1-x^2)\cdots(1-x^p)}, \tag{3.63}$$

where $W_{P,\omega}(x)$ is a polynomial given by

$$W_{P,\omega}(x) = \sum_{w\in\mathcal{L}(P,\omega)} x^{\mathrm{maj}(w)}. \tag{3.64}$$

If we take P to be the antichain $p\mathbf{1}$ (with any labeling), then clearly $G_P(x) = (1-x)^{-p}$. Comparing with equation (3.64) yields

$$\sum_{w\in\mathfrak{S}_p} x^{\mathrm{maj}(w)} = (1+x)(1+x+x^2)\cdots(1+x+\cdots+x^{p-1}),$$

which is the same (up to a change in notation) as equation (1.42).

Recall from Section 3.12 that we defined the order polynomial $\Omega_P(m)$ to be the number of order-preserving maps $\sigma: P \to \mathbf{m}$. By replacing $\sigma(t)$ with $m+1-\sigma(t)$, we see that $\Omega_P(m)$ is also the number of order-*reversing* maps $P \to m$ (i.e., the number of P-partitions $P \to m$). We can therefore extend the definition to labeled posets by defining the (P,ω)-*order polynomial* $\Omega_{P,\omega}(m)$ for $m \in \mathbb{P}$ to be the number of (P,ω)-partitions $\sigma : P \to m$. Let $e_{P,\omega}(s)$ be the number of *surjective* (P,ω)-partitions $P \to s$. Note that for any ω we have $e_{P,\omega}(p) = e(P)$. By first choosing the image $\sigma(P)$ of the (P,ω)-partition $\sigma : P \to m$, it is clear that

$$\Omega_{P,\omega}(m) = \sum_{s=1}^{p} e_{P,\omega}(s)\binom{m}{s}.$$

It follows that $\Omega_{P,\omega}(m)$ is a polynomial in m of degree p and leading coefficient $e(P)/p!$.

Now define

$$H_{P,\omega}(x) = \sum_{m \geq 0} \Omega_{P,\omega}(m) x^m.$$

The fundamental property of order polynomials is the following.

3.15.8 Theorem. *We have*

$$H_{P,\omega}(x) = \frac{\sum_{w \in \mathcal{L}(P,\omega)} x^{1+\mathrm{des}(w)}}{(1-x)^{p+1}}. \tag{3.65}$$

Proof. Immediate from equation (1.46) and Lemma 3.15.3. $\qquad\square$

In analogy to equation (1.36), we write

$$A_{P,\omega}(x) = \sum_{w \in \mathcal{L}(P,\omega)} x^{1+\mathrm{des}(w)},$$

called the (P,ω)-*Eulerian polynomial*. As usual, when ω is natural, we just write $A_P(x)$ and call it the *P-Eulerian polynomial*. Note that Proposition 1.4.4 corresponds to the case $P = p\mathbf{1}$, when the order polynomial is just the Eulerian polynomial $A_p(x)$. Note also that if we take coefficients of x^m in equation (3.65) (or by equation (1.45)), then we obtain

$$\Omega_{P,\omega}(m) = \sum_{w \in \mathcal{L}(P,\omega)} \left(\left(\binom{m - \mathrm{des}(w)}{p} \right) \right).$$

3.15.3 Reciprocity

With a labeling ω of P, we can associate a certain "dual" labeling $\overline{\omega}$. The connection between ω and $\overline{\omega}$ will lead to a generalization of the reciprocity formula $\left(\binom{n}{k}\right) = (-1)^k \binom{-n}{k}$ of equation (1.21).

Let ω be a labeling of the p-element poset P. Define the *complementary labeling* $\overline{\omega}$ by

$$\overline{\omega}(t) = p + 1 - \omega(t).$$

For instance, if ω is natural so that a (P,ω)-partition is just a P-partition, then a $(P,\overline{\omega})$-partition is a strict P-partition. If $w = w_1 w_2 \cdots w_p \in \mathfrak{S}_p$, then let $\overline{w} = p+1-w_1, p+1-w_2, \ldots, p+1-w_p \in \mathfrak{S}_p$. Note that $D(\overline{w}) = [p-1] - D(w)$.

3.15.9 Lemma. *Let F_w be as in equation (3.58). Then as rational functions,*

$$x_1 x_2 \cdots x_p F_{\overline{w}}(x_1, \ldots, x_p) = (-1)^p F_w \left(\frac{1}{x_1}, \ldots, \frac{1}{x_p} \right).$$

Proof. Let $w = w_1 \cdots w_p$ and $w_i' = j$ if $w(i) = \omega(t_j)$ as before. We have by Lemma 3.15.4 that

$$F_w \left(\frac{1}{x_1}, \ldots, \frac{1}{x_p} \right) = \frac{\prod_{j \in D(w)} \left(x_{w_1'} x_{w_2'} \cdots x_{w_j'} \right)^{-1}}{\prod_{i=1}^{p} \left(1 - \left(x_{w_1'} x_{w_2'} \cdots x_{w_i'} \right)^{-1} \right)}$$

$$= (-1)^p \frac{x_{w_1'}^p x_{w_2'}^{p-1} \cdots x_{w_p'} \prod_{j \in D(w)} \left(x_{w_1'} x_{w_2'} \cdots x_{w_j'} \right)^{-1}}{\prod_{i=1}^{p} \left(1 - x_{w_1'} x_{w_2'} \cdots x_{w_i'} \right)}.$$

$$(3.66)$$

But

$$\left(\prod_{j \in D(w)} x_{w_1'} x_{w_2'} \cdots x_{w_j'} \right) \left(\prod_{k \in D(\overline{w})} x_{w_1'} x_{w_2'} \cdots x_{w_k'} \right) = \prod_{j=1}^{p-1} x_{w_1'} x_{w_2'} \cdots x_{w_j'}$$

$$= x_{w_1'}^{p-1} x_{w_2'}^{p-2} \cdots x_{w_{p-1}'}.$$

The proof now follows upon comparing equation (3.66) with Lemma 3.15.4 for $\overline{w}$. $\qquad \square$

3.15.10 Theorem (the reciprocity theorem for (P,ω)-partitions). *The rational functions $F_{P,\omega}(x_1, \ldots, x_p)$ and $F_{P,\overline{\omega}}(x_1, \ldots, x_p)$ are related by*

$$x_1 x_2 \cdots x_p F_{P,\overline{\omega}}(x_1, \ldots, x_p) = (-1)^p F_{P,\omega} \left(\frac{1}{x_1}, \ldots, \frac{1}{x_p} \right).$$

Proof. Immediate from Theorem 3.15.5 and Lemma 3.15.9. $\qquad \square$

The power series and polynomials $G_{P,\omega}(x)$, $W_{P,\omega}(x)$, $H_{P,\omega}(x)$, and $A_{P,\omega}(x)$ are well-behaved with respect to reciprocity. It is immediate from the preceding discussion that

$$x^p G_{P,\overline{\omega}}(x) = (-1)^p G_{P,\omega}(1/x),$$

$$W_{P,\overline{\omega}}(x) = x^{\binom{p}{2}} W_{P,\omega}(1/x),$$

$$H_{P,\overline{\omega}}(x) = (-1)^{p+1} H_{P,\omega}(1/x), \qquad (3.67)$$

$$A_{P,\overline{\omega}}(x) = x^{p+1} A_{P,\omega}(1/x).$$

There is also an elegant reciprocity result for the order polynomial $\Omega_{P,\omega}(m)$ itself (and not just its generating function $H_{P,\omega}(x)$). We first need the following lemma. It is a special case of Proposition 4.2.3 and Corollary 4.3.1, where more conceptual proofs are given than the naive argument below.

3.15.11 Lemma. *Let $f(m)$ be a polynomial over a field K of characteristic 0, with* $\deg f \le p$. *Let* $H(x) = \sum_{m \ge 0} f(m) x^m$. *Then there is a polynomial $P(x) \in K[x]$, with $\deg P \le p$, such that*

$$H(x) = \frac{P(x)}{(1-x)^{p+1}}. \tag{3.68}$$

Moreover, as rational functions we have

$$\sum_{m \ge 1} f(-m) x^m = -H(1/x). \tag{3.69}$$

Proof. By linearity it suffices to prove the result for some basis of the space of polynomials $f(m)$ of degree at most p. Choose the basis $\binom{m+i}{p}$, $0 \le i \le p$. Let

$$H_i(x) = \sum_{m \ge 0} \binom{m+i}{p} x^m$$

$$= \frac{x^{p-i}}{(1-x)^{p+1}},$$

establishing equation (3.68). Now

$$-H_i(1/x) = \frac{-x^{-p+i}}{(1-1/x)^{p+1}}$$

$$= \frac{(-1)^p x^{i+1}}{(1-x)^{p+1}}$$

$$= (-1)^p \sum_{m \ge 1} \binom{m+p-i-1}{p} x^m$$

$$= \sum_{m \ge 1} \binom{-m+i}{p} x^m,$$

and the proof of equation (3.69) follows. □

3.15.12 Corollary (the reciprocity theorem for order polynomials). *The polynomials $\Omega_{P,\bar{\omega}}(m)$ and $\Omega_{P,\omega}(m)$ are related by*

$$\Omega_{P,\bar{\omega}}(m) = (-1)^p \Omega_{P,\omega}(-m).$$

Proof. Immediate from equation (3.67) and Lemma 3.15.11. □

3.15.4 Natural Labelings

When ω is a natural labeling many properties of P dealing with the length of chains are closely connected with the generating functions we have been considering. Recall that we suppress the labeling ω from our notation when ω is natural, so for instance we write $\mathcal{L}(P)$ for $\mathcal{L}(P,\omega)$ when ω is natural. We also use an overline to denote that a labeling is dual natural; for instance, $\overline{G}_P(x)$ denotes $G_{P,\omega}(x)$ for ω dual natural. To begin, if $t \in P$, then define $\delta(t)$ to be the length ℓ of the longest chain $t = t_0 < t_1 < \cdots < t_\ell$ of P whose first element is t. Also define

$$\delta(P) = \sum_{t \in P} \delta(t).$$

3.15.13 Corollary. *Let $p = \#P$. Then the degree of the polynomial $W_P(x)$ is $\binom{p}{2} - \delta(P)$. Moreover, $W_P(x)$ is a monic polynomial. (See Corollary 4.2.4(ii) for the significance of these results.)*

Proof. By equation (3.64) we need to show that

$$\max_{w \in \mathcal{L}(P)} \mathrm{maj}(w) = \binom{p}{2} - \delta(P),$$

and that there is a unique w achieving this maximum. Let $w = a_1 a_2 \cdots a_p \in \mathcal{L}(P)$, and suppose that the longest chain of P has length ℓ. Given $0 \le i \le \ell$, let j_i be the largest integer for which $\delta(a_{j_i}) = i$. Clearly $j_1 > j_2 > \cdots > j_\ell$. Now for each $1 \le i \le \ell$, there is some element a_{k_i} of P satisfying $a_{j_i} < a_{k_i}$ in P (and thus also $a_{j_i} < a_{k_i}$ in $\mathbb{Z}$) and $\delta(a_{k_i}) = \delta(a_{j_i}) - 1$. It follows that $j_i < k_i \le j_{i-1}$. Hence, somewhere in w between positions j_i and j_{i-1} there is a pair $a_r < a_{r+1}$ in $\mathbb{Z}$, so

$$\mathrm{maj}(w) \le \binom{p}{2} - \sum_{i=1}^{\ell} j_i.$$

If δ_i denotes the number of elements t of P satisfying $\delta(t) = i$, then by definition $j_i \ge \delta_i + \delta_{i+1} + \cdots + \delta_\ell$. Hence,

$$\mathrm{maj}(w) \le \binom{p}{2} - \sum_{i=1}^{\ell} (\delta_i + \delta_{i+1} + \cdots + \delta_\ell)$$

$$= \binom{p}{2} - \sum_{i=1}^{\ell} i\delta_i$$

$$= \binom{p}{2} - \sum_{t \in P} \delta(t).$$

If equality holds, then the last δ_0 elements t of w satisfy $\delta(t) = 0$, the next δ_1 elements t from the right satisfy $\delta(t) = 1$, and so on. Moreover, the last δ_0 elements must be arranged in decreasing order as elements of $\mathbb{Z}$, the next δ_1 elements also in decreasing order, etc. Hence there is a unique w for which equality hold. $\square$

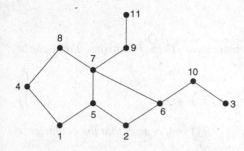

Figure 3.30 A naturally labeled poset
P with $\delta(P) = 19$.

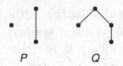

Figure 3.31 Nongraded posets satisfying the δ-chain condition.

3.15.14 Example. Let P be the naturally labeled poset shown in Figure 3.30. Then the unique $w \in \mathcal{L}(P)$ satisfying $\mathrm{maj}(w) = \binom{p}{2} - \delta(P)$ is given by

$$w = 2, 1, 6, 5, 7, 9, 4, 3, 11, 10, 8,$$

so $\mathrm{maj}(w) = 36$ and $\delta(P) = 19$.

For our next result concerning the polynomial $W_P(x)$, let $\mathcal{A}(P)$ (respectively, $\overline{\mathcal{A}}(P)$) denote the set of all P-partitions (respectively, strict P-partitions). Define a map (denoted $'$) $\mathcal{A}(P) \to \overline{\mathcal{A}}(P)$ by the formula

$$\sigma'(t) = \sigma(t) + \delta(t), \quad t \in P. \tag{3.70}$$

Clearly, this correspondence is injective.

We say that P satisfies the δ-*chain condition* if for all $t \in P$, all maximal chains of the principal dual order ideal $V_t = \{u \in P : u \geq t\}$ have the same length. If P has a $\hat{0}$, then this is equivalent to saying that P is graded. Note, however, that the posets P and Q of Figure 3.31 satisfy the δ-chain condition but are not graded.

3.15.15 Lemma. *The injection $\sigma \mapsto \sigma'$ is a bijection from $\mathcal{A}(P)$ to $\overline{\mathcal{A}}(P)$ if and only if P satisfies the δ-chain condition.*

Proof. The "if" part is easy to see. To prove the "only if" part, we need to show that if P fails to satisfy the δ-chain condition, then there is a $\tau \in \overline{\mathcal{A}}(P)$ such that $\tau - \delta \notin \mathcal{A}(P)$. Assume that P does not satisfy the δ-chain condition. Then there exist two elements t_0, t_1 of P such that t_1 covers t_0 and $\delta(t_0) > \delta(t_1) + 1$. Define τ by

$$\tau(t) = \begin{cases} \delta(t), & \text{if } t \geq t_0 \text{ and } t \neq t_1 \text{ (in } P), \\ \delta(t) + 1, & \text{if } t \not\geq t_0 \text{ or } t = t_1 \text{ (in } P). \end{cases}$$

It is easily seen that $\tau \in \overline{\mathcal{A}}(P)$, but

$$\tau(t_0) - \delta(t_0) = 0 < 1 = \tau(t_1) - \delta(t_1).$$

Since $t_0 < t_1$, $\tau - \delta \notin \mathcal{A}(P)$.   □

3.15.16 Theorem. *Let P be a p-element poset. Then P satisfies the δ-chain condition if and only if*

$$x^{\binom{p}{2}-\delta(P)} W_P(1/x) = W_P(x). \tag{3.71}$$

(Since $\deg W_P(x) = \binom{p}{2} - \delta(P)$, equation (3.71) simply says that the coefficients of $W_P(x)$ read the same backwards as forwards.)

Proof. Let $\sigma \in \mathcal{A}(P)$ with $|\sigma| = n$. Then the strict P-partition σ' defined by (3.70) satisfies $|\sigma'| = n + \delta(P)$. Hence from Lemma 3.15.15, it follows that P satisfies the δ-chain condition if an only if $a(n) = \overline{a}(n + \delta(P))$ for all $n \geq 0$. In terms of generating functions, this condition becomes $x^{\delta(P)} G_P(x) = \overline{G}_P(x)$. The proof now follows from Theorem 3.15.7.   □

Theorem 3.15.16 has an analogue for order polynomials. Recall that P is *graded* if all maximal chains of P have the same length. We say that P satisfies the λ-*chain condition* if every element of P is contained in a chain of maximum length. Clearly, a graded poset satisfies the λ-chain condition. The converse is false, as shown by Exercise 3.7(a).

Let $\mathcal{A}_m(P)$ (respectively, $\overline{\mathcal{A}}_m(P)$) denote the set of all order-reversing maps (respectively, strict order-reversing maps) $\sigma : P \to m$. The next result is the analogue of Lemma 3.15.15 for graded posets and for the λ-chain condition.

3.15.17 Lemma. *Let P be a finite poset with longest chain of length ℓ. For each $i \in \mathbb{P}$, define an injection $\theta_i : \mathcal{A}_i(P) \to \overline{\mathcal{A}}_{\ell+i}(P)$ by $\theta_i(\sigma) = \sigma + \delta$.*

a. *The map θ_1 is a bijection (i.e, $\#\overline{\mathcal{A}}_{\ell+1}(P) = 1$) if and only if P satisfies the λ-chain condition.*
b. *The maps θ_1 and θ_2 are both bijections if and only if P is graded. In this case θ_i is a bijection for all $i \in \mathbb{P}$.*

Proof. **a.** The "if" part is clear. To prove the converse, define $\delta^*(t)$ for $t \in P$ to be the length k of the longest chain $t_0 < t_1 < \cdots < t_k = t$ in P with top t. Thus, $\delta(t) + \delta^*(t)$ is the length of the longest chain of P containing t, and $\delta(t) + \delta^*(t) = \ell$ for all $t \in P$ if and only if P satisfies the λ-chain condition. Define $\sigma, \tau \in \overline{\mathcal{A}}_{\ell+1}(P)$ by $\sigma(t) = 1 + \delta(t)$ and $\tau(t) = \ell - \delta^*(t) + 1$. Then $\sigma \neq \tau$ if (and only if) P fails to satisfy the λ-chain condition, so in this case θ_1 is not a bijection.

b. Again the "if" part is clear. To prove the converse, assume that P is not graded. If P does not satisfy the λ-chain condition, then by (a) θ_1 is not a bijection. Hence, assume that P satisfies the λ-chain condition. Let $t_0 < t_1 < \cdots < t_m$ be a maximal chain of P with $m < \ell$. Let k be the greatest integer, $0 \leq k \leq m$, such that $\delta(t_k) > m - k$. Since P satisfies the λ-chain condition and t_0 is a minimal element of P, $\delta(t_0) = \ell > m$; so k always exists. Furthermore, $k \neq m$ since t_m is a maximal

Figure 3.32 A graded poset.

element of P. Define a map $\sigma : P \to [\ell + 2]$ as follows:

$$\sigma(t) = \begin{cases} 1 + \delta(t), & \text{if } t \not\leq t_{k+1}, \\ 1 + \max(\delta(t), \delta(t_{k+1}) + \lambda(t, t_{k+1}) + 1), & \text{if } t \leq t_{k+1}, \end{cases}$$

where $\lambda(t, t_{k+1})$ denotes the length of the longest chain in the interval $[t, t_{k+1}]$. It is not hard to see that $\sigma \in \overline{\mathcal{A}}_{\ell+2}(P)$. Moreover,

$$\sigma(t_k) - \delta(t_k) = 1, \quad \sigma(t_{k+1}) - \delta(t_{k+1}) = 2,$$

so $\sigma - \delta \notin \mathcal{A}(P)$. Hence, θ_2 is not a bijection, and the proof is complete. $\qquad\square$

3.15.18 Corollary. *Let P be a p-element poset with longest chain of length ℓ. Then*

$$\Omega_P(-1) = \Omega_P(-2) = \cdots = \Omega_P(-\ell) = 0.$$

Moreover:

a. *P satisfies the λ-chain condition if and only if $\Omega_P(-\ell - 1) = (-1)^p$.*
b. *The following three conditions are equivalent:*
  i. *P is graded.*
  ii. *$\Omega_P(-\ell - 1) = (-1)^p$ and $\Omega_P(-\ell - 2) = (-1)^p \Omega_P(2)$.*
  iii. *$\Omega_P(-\ell - m) = (-1)^p \Omega_P(m)$ for all $m \in \mathbb{Z}$.*

The following example illustrates the computational use of Corollary 3.15.18.

3.15.19 Example. Let P be given by Figure 3.32. Thus, $\Omega_P(m)$ is a polynomial of degree 6, and by the preceding corollary $\Omega_P(0) = \Omega_P(-1) = \Omega_P(-2) = 0$, $\Omega_P(1) = \Omega_P(-3) = 1$, $\Omega_P(2) = \Omega_P(-4)$. Thus as soon as we compute $\Omega_P(2)$, we know seven values of $\Omega_P(m)$, which suffice to determine $\Omega_P(m)$ completely. In fact, $\Omega_P(2) = 14$, from which we compute

$$\sum_{m \geq 0} \Omega_P(m) x^m = \frac{x + 7x^2 + 7x^3 + x^4}{(1 - x)^7}$$

and

$$\Omega_P(m) = \frac{1}{180}(4m^6 + 24m^5 + 55m^4 + 60m^3 + 31m^2 + 6m)$$

$$= \frac{1}{180} m(m + 1)^2 (m + 2)(2m + 1)(2m + 3).$$

3.16 Eulerian Posets

Let us recall the definition of an Eulerian poset following Proposition 3.8.9: A finite graded poset P with $\hat{0}$ and $\hat{1}$ is *Eulerian* if $\mu_P(s,t) = (-1)^{\ell(s,t)}$ for all $s \leq t$ in P. Eulerian posets enjoy many remarkable properties concerned with the enumeration of chains. In this section, we will consider several duality properties of Eulerian posets, while the next section deals with a generalization of the cd-index.

3.16.1 Proposition. *Let P be an Eulerian poset of rank n. Then $Z(P,-m) = (-1)^n Z(P,m)$.*

Proof. By Proposition 3.12.1(c) we have

$$Z(P,-m) = \mu^m(\hat{0},\hat{1})$$
$$= \sum \mu(t_0,t_1)\mu(t_1,t_2)\cdots\mu(t_{m-1},t_m),$$

summed over all multichains $\hat{0} = t_0 \leq t_1 \leq \cdots \leq t_m = \hat{1}$. Since P is Eulerian, $\mu(t_{i-1},t_i) = (-1)^{\ell(t_{i-1},t_i)}$. Hence, $\mu(t_0,t_1)\mu(t_1,t_2)\cdots\mu(t_{m-1},t_m) = (-1)^n$, so $Z(P,-m) = (-1)^n \zeta^m(\hat{0},\hat{1}) = (-1)^n Z(P,m)$. □

Define a finite poset P with $\hat{0}$ to be *simplicial* if each interval $[\hat{0},t]$ is isomorphic to a boolean algebra.

3.16.2 Proposition. *Let P be simplicial. Then $Z(P,m) = \sum_{i \geq 0} W_i(m-1)^i$, where*

$$W_i = \#\{t \in P : [\hat{0},t] \cong B_i\}.$$

In particular, if P is graded then $Z(P,q+1)$ is the rank-generating function of P.

Proof. Let $t \in P$, and let $Z_t(P,m)$ denote the number of multichains $t_1 \leq t_2 \leq \cdots \leq t_{m-1} = t$ in P. By Example 3.12.2, $Z_t(P,m) = (m-1)^i$ where $[\hat{0},t] \cong B_i$. But $Z(P,m) = \sum_{t \in P} Z_t(P,m)$, and the proof follows. □

Now suppose that P is Eulerian and $P' := P - \{\hat{1}\}$ is simplicial. By considering multichains in P that do not contain $\hat{1}$, we see that

$$Z(P',m+1) = Z(P,m+1) - Z(P,m) = \Delta Z(P,m).$$

Hence by Proposition 3.16.2,

$$\Delta Z(P,m) = \sum_{i=0}^{n-1} W_i m^i, \tag{3.72}$$

where P has W_i elements of rank i (and $n = \mathrm{rank}(P)$ as usual). On the other hand, by Proposition 3.16.1 we have $Z(P,-m) = (-1)^n Z(P,m)$, so $\Delta Z(P,-m) = (-1)^{n-1} \Delta Z(P,m-1)$. Combining with equation (3.72) yields

$$\sum_{i=0}^{n-1} W_i(m-1)^i = \sum_{i=0}^{n-1} (-1)^{n-1-i} W_i m^i. \tag{3.73}$$

Equation (3.73) imposes certain linear relations on the W_i's, known as the *Dehn-Sommerville equations*. In general, there will be $\lfloor n/2 \rfloor$ independent equations (in addition to $W_0 = 1$). We list below these equations for $2 \leq n \leq 6$, where we have set $W_0 = 1$.

$$n = 2: \qquad\qquad\qquad W_1 = 2,$$

$$n = 3: \qquad\qquad\qquad W_1 - W_2 = 0,$$

$$n = 4: \qquad\qquad\qquad W_1 - W_2 + W_3 = 2,$$

$$2W_2 - 3W_3 = 0,$$

$$n = 5: \qquad\qquad\qquad W_1 - W_2 + W_3 - W_4 = 0,$$

$$W_3 - 2W_4 = 0,$$

$$n = 6: \qquad\qquad W_1 - W_2 + W_3 - W_4 + W_5 = 0,$$

$$2W_2 - 3W_3 + 4W_4 - 5W_5 = 0,$$

$$2W_4 - 5W_5 = 0.$$

A more elegant way of stating these equations will be discussed in conjunction with Theorem 3.16.9.

A fundamental example of an Eulerian lattice L for which $L - \{\hat{1}\}$ is simplicial is the lattice of faces of a triangulation Δ of a sphere, with a $\hat{1}$ adjoined. In this case W_i is just the number of $(i - 1)$-dimensional faces of Δ.

Let us point out that although we have derived equation (3.73) as a special case of Proposition 3.16.1, one can also deduce Proposition 3.16.1 from (3.73). Namely, given an Eulerian poset P, apply (3.73) to the poset of chains of P with a $\hat{1}$ adjoined. The resulting formula is formally equivalent to Proposition 3.16.1.

Next we turn to a duality theorem for the numbers $\beta_P(S)$ when P is Eulerian.

3.16.3 Lemma. *Let P be a finite poset with $\hat{0}$ and $\hat{1}$, and let $t \in P - \{\hat{0}, \hat{1}\}$. Then*

$$\mu_{P-t}(\hat{0}, \hat{1}) = \mu_P(\hat{0}, \hat{1}) - \mu_P(\hat{0}, t)\mu_P(t, \hat{1}).$$

Proof. This result is a simple consequence of Proposition 3.8.5. □

3.16.4 Lemma. *Let P be as above, and let Q be any subposet of P containing $\hat{0}$ and $\hat{1}$. Then*

$$\mu_Q(\hat{0}, \hat{1}) = \sum (-1)^k \mu_P(\hat{0}, t_1)\mu_P(t_1, t_2) \cdots \mu_P(t_k, \hat{1}),$$

where the sum ranges over all chains $\hat{0} < t_1 < \cdots < t_k < \hat{1}$ in P such that $t_i \notin Q$ for all i. (The chain $\hat{0} < \hat{1}$ contributes $\mu(\hat{0}, \hat{1})$ to the sum.)

Proof. Iterate Lemma 3.16.3 by successively removing elements of $P - Q$ from P. □

3.16.5 Proposition. *Let P be Eulerian of rank n, and let Q be any subposet of P containing $\hat{0}$ and $\hat{1}$. Set $\overline{Q} = (P - Q) \cup \{\hat{0}, \hat{1}\}$. Then*

$$\mu_Q(\hat{0}, \hat{1}) = (-1)^{n-1} \mu_{\overline{Q}}(\hat{0}, \hat{1}).$$

Proof. Since P is Eulerian, we have

$$\mu_P(\hat{0}, t_1) \mu_P(t_1, t_2) \cdots \mu_P(t_k, \hat{1}) = (-1)^n$$

for all chains $\hat{0} < t_1 < \cdots < t_k < \hat{1}$ in P. Hence from Lemma 3.16.4, we have $\mu_Q(\hat{0}, \hat{1}) = \sum (-1)^{k+n}$, where the sum ranges over all chains $\hat{0} < t_1 < \cdots < t_k < \hat{1}$ in $\overline{Q}$. The proof follows from Proposition 3.8.5. $\square$

3.16.6 Corollary. *Let P be Eulerian of rank n, let $S \subseteq [n-1]$, and set $\overline{S} = [n-1] - S$. Then $\beta_P(S) = \beta_P(\overline{S})$.*

Proof. Apply Proposition 3.16.5 to the case $Q = P_S \cup \{\hat{0}, \hat{1}\}$ and use equation (3.54). $\square$

Topological digression

Proposition 3.16.5 provides an instructive example of the usefulness of interpreting the Möbius function as a (reduced) Euler characteristic and then considering the actual homology groups. In general, we expect that if we suitably strengthen the hypotheses to take into account the homology groups, then the conclusion will be similarly strengthened. Indeed, suppose that instead of merely requiring that $\mu_P(s, t) = (-1)^{\ell(s,t)}$, we assume that

$$\widetilde{H}_i(\Delta(s,t); K) = \begin{cases} 0, & i \neq \ell(s,t) - 2, \\ K, & i = \ell(s,t) - 2, \end{cases}$$

where K is a field (or any coefficient group), and $\Delta(s, t)$ denotes the order complex (as defined in Section 3.8) of the open interval (s, t). Equivalently, P is Eulerian and Cohen-Macaulay over K. (We then say that P is a *Gorenstein* poset* over K. The asterisk is part of the notation, not a footnote indicator.) Let $Q, \overline{Q}$ be as in Proposition 3.16.5, and set $Q' = Q - \{\hat{0}, \hat{1}\}$, $\overline{Q}' = \overline{Q} - \{\hat{0}, \hat{1}\}$. The *Alexander duality theorem* for simplicial complexes asserts in the present context that

$$\widetilde{H}_i(\Delta(Q'); K) \cong \widetilde{H}^{n-i-3}(\Delta(\overline{Q}'); K).$$

(When K is a field there is a (noncanonical) isomorphism $\widetilde{H}^j(\Delta; K) \cong \widetilde{H}_j(\Delta; K)$.) In particular, $\widetilde{\chi}(\Delta(Q')) = (-1)^{n-1} \widetilde{\chi}(\Delta(\overline{Q}'))$, which is equivalent to Proposition 3.16.5 (by Proposition 3.8.6). Hence, Proposition 3.16.5 may be regarded as the "Möbius-theoretic analogue" of the Alexander duality theorem.

Finally, we come to a remarkable "master duality theorem" for Eulerian posets P. We will associate with P two polynomials $f(P, x)$ and $g(P, x)$ defined next. Define $\widetilde{P}$ to be the set of all intervals $[\hat{0}, t]$ of P, ordered by inclusion. Clearly, the map $P \to \widetilde{P}$ defined by $t \mapsto [\hat{0}, t]$ is an isomorphism of posets. The polynomials f and g are defined inductively as follows.

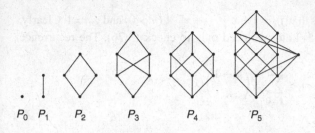

Figure 3.33 Some Eulerian posets.

P_0 P_1 P_2 P_3 P_4 P_5

1.

$$f(\mathbf{1},x) = g(\mathbf{1},x) = 1. \tag{3.74}$$

2. If $n+1 = \operatorname{rank} P > 0$, then $f(P,x)$ has degree n, say $f(P,x) = h_0 + h_1 x + \cdots + h_n x^n$. Then define

$$g(P,x) = h_0 + (h_1 - h_0)x + (h_2 - h_1)x^2 + \cdots + (h_m - h_{m-1})x^m, \tag{3.75}$$

where $m = \lfloor n/2 \rfloor$.

3. If $n+1 = \operatorname{rank} P > 0$, then define

$$f(P,x) = \sum_{\substack{Q \in \tilde{P} \\ Q \neq P}} g(Q,x)(x-1)^{n-\rho(Q)}. \tag{3.76}$$

We call $f(P,x)$ the *toric h-polynomial* of P, and we call $g(P,x)$ the *toric g-polynomial* of P. The sequence $(h_0, \ldots, h_n)$ of coefficients of $f(P,x)$ is called the *toric h-vector* of P. The *toric g-vector* is defined similarly.

3.16.7 Example. Consider the six Eulerian posets of Figure 3.33. Write f_i and g_i for $f(P_i,x)$ and $g(P_i,x)$, respectively. We compute recursively that

$$f_0 = g_0 = 1,$$
$$f_1 = g_0 = 1, \quad g_1 = 1,$$
$$f_2 = 2g_1 + g_0(x-1) = 1 + x, \quad g_2 = 1,$$
$$f_3 = 2g_2 + 2g_1(x-1) + (x-1)^2 = 1 + x^2, \quad g_3 = 1 - x,$$
$$f_4 = 3g_2 + 3g_1(x-1) + (x-1)^2 = 1 + x + x^2, \quad g_4 = 1,$$
$$f_5 = 2g_4 + g_3 + 4g_2(x-1) + 3g_1(x-1)^2 + (x-1)^3 = 1 + x^3,$$
$$g_5 = 1 - x.$$

3.16.8 Example. Write $f_n = f(B_n,x)$ and $g_n = g(B_n,x)$, where B_n is a boolean algebra of rank n. A simple computation yields

$$f_0 = 1, \quad g_0 = 1, \quad f_1 = 1, \quad g_1 = 1, \quad f_2 = 1 + x, \quad g_2 = 1,$$
$$f_3 = 1 + x + x^2, \quad g_3 = 1, \quad f_4 = 1 + x + x^2 + x^3, \quad g_4 = 1.$$

This computation suggests that $f_n = 1 + x + \cdots + x^{n-1}$ $(n > 0)$ and $g_n = 1$. Clearly, equations (3.74) and (3.75) hold; we need only to check (3.76). The recurrence (3.76) reduces to

$$f_{n+1} = \sum_{k=0}^{n} g_k \binom{n+1}{k} (x-1)^{n-k}.$$

Substituting $g_k = 1$ yields

$$f_{n+1} = \sum_{k=0}^{n} \binom{n+1}{k} (x-1)^{n-k}$$

$$= (x-1)^{-1} \left[((x-1)+1)^{n+1} - 1 \right], \quad \text{by the binomial theorem}$$

$$= 1 + x + \cdots + x^n.$$

Hence, we have shown

$$f(B_n, x) = 1 + x + \cdots + x^{n-1}, \quad n \geq 1,$$

$$g(B_n, x) = 1, \quad n \geq 0.$$

Now suppose that P is Eulerian of rank $n+1$ and $P - \{\hat{1}\}$ is simplicial. Since $g(B_n, x) = 1$ we get from equation (3.76) that

$$f(P, x) = \sum_{Q \neq P} (x-1)^{n-\rho(Q)}$$

$$= \sum_{i=0}^{n} W_i (x-1)^{n-i}, \tag{3.77}$$

where P has W_i elements of rank i.

We come to the main result of this section.

3.16.9 Theorem. *Let P be Eulerian of rank $n+1$. Then $f(P,x) = x^n f(P, 1/x)$. Equivalently, if $f(P,x) = \sum_{i=0}^{n} h_i x^{n-i}$, then $h_i = h_{n-i}$.*

Proof. We write $f(P)$ for $f(P,x)$, $g(P)$ for $g(P,x)$, and so on. Set $y = x - 1$. Multiply equation (3.76) by y and add $g(P)$ to obtain

$$g(P) + y f(P) = \sum_{Q \in \tilde{P}} g(Q) y^{\rho(P) - \rho(Q)}$$

$$\Rightarrow y^{-\rho(P)} (g(P) + y f(P)) = \sum_{Q} g(Q) y^{-\rho(Q)}.$$

By Möbius inversion, we obtain

$$g(P) y^{-\rho(P)} = \sum_{Q} (g(Q) + y f(Q)) y^{-\rho(Q)} \mu_{\tilde{P}}(Q, P).$$

Since $\widetilde{P}$ is Eulerian, we get $\mu_{\widetilde{P}}(Q,P) = (-1)^{\ell(Q,P)}$, so

$$g(P) = \sum_{Q}(g(Q) + yf(Q))(-y)^{\ell(Q,P)}. \tag{3.78}$$

Let $f(Q) = a_0 + a_1 x + \cdots + a_r x^r$, where $\rho(Q) = r+1$. Then

$$g(Q) + yf(Q) = (a_s - a_{s+1})x^{s+1} + (a_{s+1} - a_{s+2})x^{s+2} + \cdots,$$

where $s = \lfloor r/2 \rfloor$. By induction on $\rho(Q)$ we may assume that $a_i = a_{r-i}$ for $r < n$. In this case,

$$g(Q) + yf(Q) = \begin{cases} (a_s - a_{s+1})x^{s+1} + (a_{s-1} - a_{s-2})x^{s+2} + \cdots, & r \text{ even} \\ (a_s - a_{s+1})x^{s+2} + (a_{s-1} - a_{s-2})x^{s+3} + \cdots, & r \text{ odd} \end{cases}$$

$$= x^{\rho(Q)} g(Q, 1/x). \tag{3.79}$$

Now subtract $yf(P) + g(P)$ from both sides of equation (3.78) and use (3.79) to obtain

$$-yf(P) = \sum_{Q < \hat{1}} x^{\rho(Q)} g(Q, 1/x)(-y)^{\ell(Q,P)}$$

$$\Rightarrow f(P) = \sum_{Q < \hat{1}} x^{\rho(Q)} g(Q, 1/x)(-y)^{\ell(Q,P)-1}$$

$$= x^n f(P, 1/x), \text{ by equation (3.76)},$$

and the proof is complete. $\qquad\square$

Equation (3.77) gives a direct combinatorial interpretation of the polynomial $f(P,x)$ provided $P - \{\hat{1}\}$ is simplicial, and in this case Theorem 3.16.9 is equivalent to equation (3.73). In general, however, $f(P,x)$ seems to be an exceedingly subtle invariant of P. See Exercises 176–177 and 179 for further information.

3.17 The *cd*-Index of an Eulerian Poset

In Corollary 3.16.6 we showed that $\beta_P(S) = \beta_P(\overline{S})$ for any Eulerian poset P of rank n, where $S \subseteq [n-1]$ and $\overline{S} = [n-1] - S$. We can ask whether the flag h-vector β_P satisfies additional relations, valid for all Eulerian posets of rank n. For instance, writing $\beta(i)$ as short for $\beta(\{i\})$, we always have the relation

$$\beta_P(1) - \beta_P(2) + \cdots + (-1)^n \beta_P(n-1) = \begin{cases} 1, & n \text{ even}, \\ 0, & n \text{ odd}, \end{cases}$$

an immediate consequence of the Eulerian property. However, there can be additional relations, such as

$$\beta_P(1,2) - \beta_P(1,3) + \beta_P(2,3) = 1$$

when $n = 5$.

In this section, we will find all the linear relations satisfied by the flag h-vector (or equivalently, flag f-vector, since the two flag vectors are linearly related) of an Eulerian poset of rank n. This information is best codified by a certain noncommutative polynomial $\Phi_P(c,d)$, called the *cd-index* of P. When $P = B_n$, the *cd*-index $\Phi_{B_n}(c,d)$ coincides with the *cd*-index $\Phi_n(c,d)$ of $\mathfrak{S}_n$ defined in Section 1.6. The equality of $\Phi_n(c,d)$ and $\Phi_{B_n}(c,d)$ is a consequence of Corollary 3.13.2, which states that $\beta_{B_n}(S)$ is the number $\beta_n(S)$ of permutations $w \in \mathfrak{S}_n$ with descent set S.

Recall from equation (1.60) that we defined the *characteristic monomial* u_S of $S \subseteq [n-1]$ by

$$u_S = e_1 e_2 \cdots e_{n-1},$$

where

$$e_i = \begin{cases} a, & \text{if } i \notin S, \\ b, & \text{if } i \in S. \end{cases}$$

Given any graded poset P of rank n with $\hat{0}$ and $\hat{1}$, define the noncommutative polynomial $\Psi_P(a,b)$, called the *ab-index* of P, by

$$\Psi_P(a,b) = \sum_{S \subseteq [n-1]} \beta_P(S) u_S.$$

Thus, $\Psi_P(a,b)$ is a noncommutative generating function for the flag h-vector β_P. Note that it is an immediate consequence of the definition (3.52) (or equivalently, equation (3.53)) that

$$\Psi_P(a+b,b) = \sum_{S \subseteq [n-1]} \alpha_P(S) u_S.$$

The main result of this section is the following.

3.17.1 Theorem. *Let P be an Eulerian poset of rank n. Then there exists a polynomial $\Phi_P(c,d)$ in the noncommutative variables c and d such that*

$$\Psi_P(a,b) = \Phi_P(a+b, ab+ba).$$

The polynomial $\Phi_P(c,d)$ is called the *cd-index* of P. For instance, let P be the face poset of the regular cell complex of Figure 3.34 (a decomposition of the 2-sphere). Then

$$\Psi_P(a+b,b) = aaa + 4baa + 7aba + 5aab + 14bba + 14bab + 14abb + 28bbb,$$

whence

$$\Psi_P(a,b) = aaa + 3baa + 6aba + 4aab + 4bba + 6bab + 3abb + bbb$$
$$= (a+b)^3 + 3(a+b)(ab+ba) + 2(ab+ba)(a+b).$$

It follows that $\Phi_P(c,d) = c^3 + 3cd + 2dc$.

Figure 3.34 A regular cell complex with four vertices, seven edges, and five faces.

Proof of Theorem 3.17.1. The proof is by induction on rank(P). The result is clearly true for rank(P) = 1, where $\Phi_P(c,d) = 1$. Assume for posets of rank less than n, and let rank(P) = n. We have

$$\Psi_P(a+b,b) = \sum_{S\subseteq[n-1]} \alpha_P(S)u_S$$

$$= \sum_{\hat{0}=t_0<t_1<\cdots<t_k=\hat{1}} a^{\rho(t_0,t_1)-1}ba^{\rho(t_1,t_2)-1}b\cdots a^{\rho(t_{k-1},t_k)-1}.$$

Hence,

$$\Psi_P(a,b) = \sum_{\hat{0}=t_0<t_1<\cdots<t_k=\hat{1}} (a-b)^{\rho(t_0,t_1)-1}b(a-b)^{\rho(t_1,t_2)-1}b\cdots(a-b)^{\rho(t_{k-1},t_k)-1}.$$

$$(3.80)$$

This formula remains true when P is replaced by any interval $[s,u]$ of P (since intervals of Eulerian posets are Eulerian). Write $\Psi_Q = \Psi_Q(a,b)$. Let (s,u) denote as usual the open interval $\{v \in P : s < v < u\}$, and similarly for the half-open interval $(s,u] = \{v \in P : s < v \leq u\}$. Replacing P by $[s,u]$ in equation (3.80) and breaking up the right-hand side according to the value of $t = t_{k-1}$ gives

$$\Psi_{[s,u]} = (a-b)^{\rho(s,u)-1} + \sum_{t\in(s,u)} \Psi_{[s,t]}b(a-b)^{\rho(t,u)-1}, \qquad (3.81)$$

where the first term on the right-hand side comes from the chain $s = t_0 < t_1 = u$. Multiply on the right by $a - b$ and add $\Psi_{[s,u]}b$ to both sides:

$$\Psi_{[s,u]}a = (a-b)^{\rho(s,u)} + \sum_{t\in(s,u]} \Psi_{[s,t]}b(a-b)^{\rho(t,u)}. \qquad (3.82)$$

Let $I(P)$ denote the incidence algebra of P over the noncommutative polynomial ring $\mathbb{Q}\langle a,b\rangle$. Define $f,g,h \in I(P)$ by

$$f(s,t) = \begin{cases} \Psi_{[s,t]}a, & s<t, \\ 1, & s=t, \end{cases}$$

$$g(s,t) = \begin{cases} \Psi_{[s,t]}b, & s<t, \\ 1, & s=t, \end{cases}$$

$$h(s,t) = (a-b)^{\rho(s,t)}.$$

Thus, equation (3.82) is equivalent to $f = gh$.

Note that h^{-1} exists (since $h(t,t) = 1$). Since P is Eulerian and h has the multiplicative property $h(s,t)h(t,u) = h(s,u)$, we have

$$h^{-1}(s,t) = \mu(s,t)(a-b)^{\rho(s,t)} = (-1)^{\rho(s,t)}(a-b)^{\rho(s,t)}.$$

Therefore, the relation $g = fh^{-1}$ can be rewritten

$$\Psi_{[s,u]}b = (-1)^{\rho(s,u)}(a-b)^{\rho(s,u)} + \sum_{t\in(s,u]} \Psi_{[s,t]}a(-1)^{\rho(t,u)}(a-b)^{\rho(t,u)}.$$

Move $\Psi_{[s,u]}a$ to the left-hand side and cancel the factor $b - a$:

$$\Psi_{[s,u]} = -(-1)^{\rho(s,u)}(a-b)^{\rho(s,u)-1} - \sum_{t\in(s,u)} \Psi_{[s,t]}a(-1)^{\rho(t,u)}(a-b)^{\rho(t,u)-1}.$$

$$(3.83)$$

Add equations (3.81) and (3.83). We obtain an expression for $2\Psi_{[s,u]}$ as a function of $\Psi_{[s,t]}$ for $s < t < u$ and of $a+b = c$ and $(a-b)^{2m} = (c^2 - 2d)^m$. By the induction hypothesis, $\Psi_{[s,t]}$ is a polynomial in c and d, so the proof follows by induction. $\square$

It is easy to recover the result $\beta_P(S) = \beta_P(\overline{S})$ (Corollary 3.16.6) from Theorem 3.17.1. For this result is equivalent to $\Psi_P(a,b) = \Psi_P(b,a)$, which is an immediate consequence of $\Psi_P(a,b) = \Phi_P(a+b, ab+ba)$.

If P is an Eulerian poset of rank n, then $\Phi_P(c,d)$ is a homogeneous polynomial in c,d of degree $n - 1$, where we define $\deg(c) = 1$ and $\deg(d) = 2$. The number of monomials of degree $n - 1$ in c and d is the number of compositions of $n - 1$ into parts equal to 1 and 2, which by Exercise 1.35(c) is the Fibonacci number F_n. The coefficients of Φ_P are linear combinations of the 2^{n-1} numbers $\beta_P(S)$, $S \subseteq [n - 1]$, where the coefficients in the linear combination depend only on n, not on P. The coefficients of Φ_P are also linear combinations of the 2^{n-1} numbers $\alpha_P(S)$, $S \subseteq [n - 1]$, since the $\beta_P(S)$'s are linear combinations of the $\alpha_P(T)$'s (equation (3.52)). Let $\mathcal{F}_n$ denote the linear span of all flag h-vectors β_P (or flag f-vectors α_P) of Eulerian posets of rank n in the 2^{n-1}-dimensional vector space of all functions $2^{[n-1]} \to \mathbb{R}$. This argument shows that

$$\dim \mathcal{F}_n \leq F_n. \tag{3.84}$$

We can also ask for the dimension of the *affine* subspace spanned by flag h-vectors of Eulerian posets of rank n. There is at least one additional affine relation, namely, $\beta(\emptyset) = 1$. Equivalently, the coefficient of c^{n-1} in $\Phi_P(c,d)$ is 1. Let $\mathcal{G}_n$ denote the affine span of all flag h-vectors of Eulerian posets of rank n. Thus,

$$\dim \mathcal{G}_n \leq F_n - 1. \tag{3.85}$$

We now show that the bounds in equations (3.84) and (3.85) are tight.

First, we need to define an operation on posets P and Q with $\hat{0}$ and $\hat{1}$. The *join* $P * Q$ of P and Q is the poset

$$P * Q = (P - \{\hat{1}\}) \oplus (Q - \{\hat{0}\}), \tag{3.86}$$

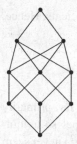

Figure 3.35 The join $B_3 * B_2$.

where $\oplus$ denotes ordinal sum. For instance, Figure 3.35 shows the join of the boolean algebras B_3 and B_2. It is easy to see (Exercise 3.191) that the join of Eulerian posets is Eulerian. Moreover, in this case we have

$$\Phi_{P*Q}(c,d) = \Phi_P(c,d)\Phi_Q(c,d). \tag{3.87}$$

Let Q_m denote the face lattice of a polygon with m vertices, so Q_m is Eulerian with rank-generating function $F(Q_m,q) = 1 + mq + mq^2 + q^3$. Moreover,

$$\Phi_{Q_m}(c,d) = c^2 + (m-2)d.$$

3.17.2 Theorem. *We have* $\dim \mathcal{F}_n = F_n$ *and* $\dim \mathcal{G}_n = F_n - 1$.

Proof. By the previous discussion, we need to show that F_n and $F_n - 1$ are lower bounds for the dimensions of $\mathcal{F}$ and $\mathcal{G}$. Given a cd-monomial $\rho = e_1 e_2 \cdots e_r$, let $P_\rho = Q_1 * Q_2 * \cdots * Q_r$, where

$$Q_i = \begin{cases} B_2, & e_i = c, \\ Q_m, & e_i = d. \end{cases}$$

By equations (3.86) and (3.87), P_ρ is Eulerian, and $\Phi_{P_\rho}(c,d) = h_1 h_2 \cdots h_r$, where

$$h_i = \begin{cases} c, & e_i = c, \\ c^2 + (m-2)d, & e_i = d. \end{cases}$$

For instance, if $\rho = dc^2 d^2$, then

$$\Phi_{P_\rho}(c,d) = (c^2 + (m-2)d)c^2(c^2 + (m-2)d)^2$$

$$= m^3 dc^2 d^2 + O(m^2).$$

In general, $\Phi_{P_\rho}(c,d) = m^k \rho + O(m^{k-1})$, where k is the number of d's in ρ. Hence, unless $\rho = c^n$, we can make the coefficient of ρ arbitrarily large compared to the other coefficients, so it is affinely (and hence linearly) independent from the other coefficients. Hence, $\dim \mathcal{G}_n = F_n - 1$. Since all the coefficients other than that of c^n are affinely independent, while the coefficient of c^n is 1, it follows that the coefficient of c^n is linear independent from the other coefficients. Hence, $\dim \mathcal{F}_n = F_n$. $\square$

3.18 Binomial Posets and Generating Functions

We have encountered many examples of generating functions so far, primarily of the form $\sum_{n\geq 0} f(n)x^n$ or $\sum_{n\geq 0} f(n)x^n/n!$. Why are these types so ubiquitous, and why do generating functions such as $\sum_{n\geq 0} f(n)x^n/(1+n^2)$ never seem to occur? Are there additional classes of generating functions besides the previous two that are useful in combinatorics? The theory of binomial posets seeks to answer these questions. It allows a unified treatment of many of the different types of generating functions that occur in combinatorics. This section and the next will be devoted to this topic. Most of the material in subsequent chapters of this book will be devoted to more sophisticated aspects of generating functions that are not really appropriate to the theory of binomial posets. We should mention that there are several alternative approaches to unifying the theory of generating functions. We have chosen binomial posets for two reasons: (a) We have already developed much of the relevant background information concerning posets, and (b) of all the existing theories, binomial posets give the most explicit combinatorial interpretation of the numbers $B(n)$ appearing in generating functions of the form $\sum_{n\geq 0} f(n)x^n/B(n)$. (Do not confuse these $B(n)$'s with the Bell numbers.)

Let us first consider some of the kinds of generating functions $F(x) \in \mathbb{C}[[x]]$ that have actually arisen in combinatorics. These generating functions should be regarded as "representing" the function $f: \mathbb{N} \to \mathbb{C}$ by the power series $F(x) = \sum_{n\geq 0} f(n)x^n/B(n)$, where the $B(n)$'s are certain complex numbers (which turn out in the theory of binomial posets always to be positive integers). The field $\mathbb{C}$ can be replaced with any field K of characteristic 0 throughout.

3.18.1 Example. a. (ordinary generating functions). These are generating functions of the form $F(x) = \sum_{n\geq 0} f(n)x^n$. (More precisely, we say that F is the *ordinary generating function* of f.) Of course, we have seen many examples of such generating functions, such as

$$\sum_{n\geq 0} \binom{t}{n} x^n = (1+x)^t,$$

$$\sum_{n\geq 0} \left(\binom{t}{n} \right) x^n = (1-x)^{-t},$$

$$\sum_{n\geq 0} p(n)x^n = \prod_{i\geq 1} (1-x^i)^{-1}.$$

b. (exponential generating functions). Here $F(x) = \sum_{n\geq 0} f(n)x^n/n!$. Again, we have many examples, such as

$$\sum_{n\geq 0} B(n)\frac{x^n}{n!} = e^{e^x-1},$$

$$\sum_{n\geq 0} D(n)\frac{x^n}{n!} = \frac{e^{-x}}{1-x}.$$

c. (Eulerian generating functions). Let q be a fixed positive integer (almost always taken in practice to be a prime power corresponding to the field $\mathbb{F}_q$). Sometimes it is advantageous to regard q as an indeterminate, rather than an integer. The corresponding generating function is

$$F(x) = \sum_{n\geq 0} f(n)\frac{x^n}{(n)!},$$

where $(n)! = (1+q)(1+q+q^2)\cdots(1+q+\cdots+q^{n-1})$ is in Section 1.3. Note that $(n)!$ reduces to $n!$ upon setting $q = 1$. As discussed in Section 1.8, sometimes in the literature one sees the denominator replaced with $[n]! = (1-q)(1-q^2)\cdots(1-q^n)$; this amounts to the transformation $x \mapsto x/(1-q)$. We will see that our choice of denominator is the natural one insofar as binomial posets are concerned. One immediate advantage is that an Eulerian generating function reduces to an exponential generating function upon setting $q = 1$. An example of an Eulerian generating function is

$$\sum_{n\geq 0} f(n)\frac{x^n}{(n)!} = \left(\sum_{n\geq 0}\frac{x^n}{(n)!}\right)^2,$$

where $f(n)$ is the total number of subspaces of $\mathbb{F}_q^n$, that is, $f(n) = \sum_{k=0}^n \binom{n}{k}$.

d. (doubly exponential generating functions). These have the form $F(x) = \sum_{n\geq 0} f(n)x^n/n!^2$. For instance, if $f(n)$ is the number of $n \times n$ matrices of non-negative integers such that every row and column sum equals two, then $F(x) = e^{x/2}(1-x)^{-1/2}$ (see Corollary 5.5.11). Sometimes one has occasion to deal with the more general *r-exponential generating function* $F(x) = \sum_{n\geq 0} f(n)x^n/n!^r$, where r is any positive integer.

e. (chromatic generating functions). Fix $q \in \mathbb{P}$. Then

$$F(x) = \sum_{n\geq 0} f(n)\frac{x^n}{q^{\binom{n}{2}}n!}.$$

Sometimes one sees $q^{\binom{n}{2}}$ replaced with $q^{n^2/2}$, amounting to the transformation $x \to xq^{-1/2}$. An example is

$$\sum_{n \geq 0} f(n) \frac{x^n}{2^{\binom{n}{2}} n!} = \left(\sum_{n \geq 0} (-1)^n \frac{x^n}{2^{\binom{n}{2}} n!} \right)^{-1}, \qquad (3.88)$$

where $f(n)$ is the number of *acyclic digraphs* on n vertices; that is, the number of subsets of $[n] \times [n]$ not containing a sequence of elements $(i_0, i_1), (i_1, i_2), (i_2, i_3), \ldots, (i_{j-1}, i_j), (i_j, i_0)$. For instance, $f(3) = 25$, corresponding to the empty set, the six 1-subsets $\{(i, j) : i \neq j\}$, the twelve 2-subsets $\{(i, j), (k, \ell) : i \neq j, k \neq \ell, (i, j) \neq (\ell, k)\}$, and the six 3-subsets obtained from $\{(1, 2), (2, 3), (1, 3)\}$ by permuting 1,2,3. See the solution to Exercise 3.200.

The basic concept that will be used to unify these three above examples follows.

3.18.2 Definition. A poset P is called a *binomial poset* if it satisfies the three conditions:

a. P is locally finite with $\hat{0}$ and contains an infinite chain.
b. Every interval $[s, t]$ of P is graded. If $\ell(s, t) = n$, then we call $[s, t]$ an *n-interval*.
c. For all $n \in \mathbb{N}$, any two n-intervals contain the same number $B(n)$ of maximal chains. We call $B(n)$ the *factorial function* of P.

NOTE. Condition (a) is basically a matter of convenience, and several alternative conditions are possible.

Note that from the definition of binomial poset we have $B(0) = B(1) = 1$, $B(2) = \#[s, t] - 2$, where $[s, t]$ is any 2-interval, and $B(0) \leq B(1) \leq B(2) \leq \cdots$.

3.18.3 Example. The posets below are all binomial posets.

a. Let $P = \mathbb{N}$ with the usual linear order. Then $B(n) = 1$.
b. Let P be the lattice of all finite subsets of $\mathbb{P}$ (or any infinite set), ordered by inclusion. Then P is a distributive lattice and $B(n) = n!$. We will denote this poset as $\mathbb{B}$.
c. Let P be the lattice of all finite-dimensional subspaces of a vector space of infinite dimension over $\mathbb{F}_q$, ordered by inclusion. Then $B(n) = (n)!$. We denote this poset by $\mathbb{B}(q)$.
d. Let P be the set of all ordered pairs (S, T) of finite subsets S, T of $\mathbb{P}$ satisfying $\#S = \#T$, ordered componentwise (i.e., $(S, T) \leq (S', T')$ if $S \subseteq S'$ and $T \subseteq T'$). Then $B(n) = n!^2$. This poset will be denoted $\mathbb{B}_2$. More generally, let $P_1, \ldots, P_k$ be binomial posets with factorial functions $B_1, \ldots, B_k$. Let P be the subposet of $P_1 \times \cdots \times P_k$ consisting of all k-tuples $(t_1, \ldots, t_k)$ such that $\ell(\hat{0}, t_1) = \cdots = \ell(\hat{0}, t_k)$. Then P is binomial with factorial function $B(n) = B_1(n) \cdots B_k(n)$. We write $P = P_1 * \cdots * P_k$, the *Segre product* of $P_1, \ldots, P_k$. Thus $\mathbb{B}_2 = \mathbb{B} * \mathbb{B}$. More generally, we set $\mathbb{B}_r = \mathbb{B} * \cdots * \mathbb{B}$ (r times).

e. Let V be an infinite vertex set, let $q \in \mathbb{P}$ be fixed, and let P be the set of all pairs (G, σ), where G is a function from all 2-sets $\{u, v\} \in \binom{V}{2}$ into $\{0, 1, \ldots, q-1\}$ such that all but finitely many values of G are 0 (think of G as a graph with finitely many edges labeled $1, 2, \ldots, q-1$), and where $\sigma : V \to \{0, 1\}$ is a map satisfying the two conditions:
   1. If $G(\{u, v\}) \neq 0$ then $\sigma(u) \neq \sigma(v)$, and
   2. $\sum_{v \in V} \sigma(v) < \infty$.
   If $(G, \sigma), (H, \tau) \in P$, then define $(G, \sigma) \leq (H, \tau)$ if
   1. $\sigma(v) \leq \tau(v)$ for all $v \in V$, and
   2. If $\sigma(u) = \tau(u)$ and $\sigma(v) = \tau(v)$, then $G(\{u, v\}) = H(\{u, v\})$.
   Then P is a binomial poset with $B(n) = n! q^{\binom{n}{2}}$. We leave to the reader the task of finding a binomial poset Q with factorial function $B(n) = q^{\binom{n}{2}}$ such that $P = Q * \mathbb{B}$, where $\mathbb{B}$ is the binomial poset of Example 3.18.3(b).

f. Let P be a binomial poset with factorial function $B(n)$, and let $k \in \mathbb{P}$. Define the rank-selected subposet (called the *k-th Veronese subposet*)

$$P^{(k)} = \{t \in P : \ell(\hat{0}, t) \text{ is divisible by } k\}.$$

Then $P^{(k)}$ is binomial with factorial function

$$B_k(n) = B(nk)/B(k)^n.$$

Observe that the numbers $B(n)$ considered in Example 3.18.3(a)–(e) appear in the power series generating functions of Example 3.18.1. If we can somehow associate a binomial poset with generating functions of the form $\sum f(n) x^n / B(n)$, the we will have "explained" the form of the generating functions of Example 3.18.3. We also will have provided some justification of the heuristic principle that ordinary generating functions are associated with the nonnegative integers, exponential generating functions with sets, Eulerian generating functions with vector spaces over $\mathbb{F}_q$, and so on.

To begin our study of binomial posets P, choose $i, n \in \mathbb{N}$ and let $\binom{n}{i}_P$ denote the number of elements u of rank i in an n-interval $[s, t]$. Note that since $B(i) B(n-i)$ maximal chains of $[s, t]$ pass through a given element u of rank i, we have

$$\binom{n}{i}_P = \frac{B(n)}{B(i) B(n-i)}, \tag{3.89}$$

so $\binom{n}{i}_P$ depends only on n and i, not on the choice of the n-interval $[s, t]$. When $P = \mathbb{B}$ as in Example 3.18.3(b), then $B(n) = n!$ and $\binom{n}{i}_P = \binom{n}{i}$, explaining our terms "binomial poset" and "factorial function." The analogy with factorials is strengthened further by observing that

$$B(n) = A(n) A(n-1) \cdots A(1),$$

where $A(i) = \binom{i}{1}_P$, the number of atoms in an i-interval.

We can now state the main result concerning binomial posets. Let P be a binomial poset with factorial function $B(n)$ and incidence algebra $I(P)$ over $\mathbb{C}$. Define

$$R(P) = \{f \in I(P): f(s,t) = f(s',t') \text{ if } \ell(s,t) = \ell(s',t')\}.$$

If $f \in R(P)$, then write $f(n)$ for $f(s,t)$ when $\ell(s,t) = n$. Clearly $R(P)$ is a vector subspace of $I(P)$.

3.18.4 Theorem. *The space $R(P)$ is a subalgebra of $I(P)$, and we have an algebra isomorphism $\phi: R(P) \to \mathbb{C}[[x]]$ given by*

$$\phi(f) = \sum_{n \geq 0} f(n) \frac{x^n}{B(n)}.$$

The subalgebra $R(P)$ is called the reduced incidence algebra *of $I(P)$.*

Proof. Let $f, g \in R(P)$. We need to show that $fg \in R(P)$. By definition of $\binom{n}{i}_P$ we have for an n-interval $[s,t]$

$$fg(s,t) = \sum_{u \in [s,t]} f(s,u)g(u,t)$$

$$= \sum_{i=0}^{n} \binom{n}{i}_P f(i)g(n-i). \tag{3.90}$$

Hence, $fg(s,t)$ depends only on $\ell(s,t)$, so $R(P)$ is a subalgebra of $I(P)$. Moreover, the right-hand side of equation (3.90) is just the coefficient of $x^n/B(n)$ in $\phi(f)\phi(g)$, so the proof follows. $\square$

Let us note a useful property of the algebra $R(P)$ that follows directly from Theorem 3.18.4 (and that can also be easily proved without recourse to Theorem 3.18.4).

3.18.5 Proposition. *Let P be a binomial poset and $f \in R(P)$. Suppose that f^{-1} exists in $I(P)$, i.e., $f(t,t) \neq 0$ for all $t \in P$. Then $f^{-1} \in R(P)$.*

Proof. The constant term of the power series $F = \phi(f)$ is equal to $f(t,t) \neq 0$ for any $t \in P$, so F^{-1} exists in $\mathbb{C}[[x]]$. Let $g = \phi^{-1}(F^{-1})$. Since $FF^{-1} = 1$ in $\mathbb{C}[[x]]$, we have $fg = 1$ in $I(P)$. Hence, $f^{-1} = g \in R(P)$. $\square$

We now turn to some examples of the unifying power of binomial posets. We make no attempt to be systematic or as general as possible, but simply try to convey some of the flavor of the subject.

3.18.6 Example. Let $f(n)$ be the cardinality of an n-interval $[s,t]$ of P, that is, $f(n) = \sum_{i=0}^{n} \binom{n}{i}_P$. Clearly by definition the zeta function ζ is in $R(P)$ and

$\phi(\zeta) = \sum_{n\geq 0} x^n/B(n)$. Since $R(P)$ is a subalgebra of $I(P)$, we have $\zeta^2 \in R(P)$. Since $\zeta^2(s,t) = \#[s,t]$, it follows that

$$\sum_{n\geq 0} f(n)\frac{x^n}{B(n)} = \left(\sum_{n\geq 0} \frac{x^n}{B(n)}\right)^2.$$

Thus from Example 3.18.3(a), we have the the cardinality of a chain of length n (or the number of integers in the interval $[0,n]$) satisfies

$$\sum_{n\geq 0} f(n)x^n = \left(\sum_{n\geq 0} x^n\right)^2 = \frac{1}{(1-x)^2} = \sum_{n\geq 0} (n+1)x^n,$$

whence $f(n) = n+1$ (not exactly the deepest result in enumerative combinatorics). Similarly from Example 3.18.3(b), the number $f(n)$ of subsets of an n-set satisfies

$$\sum_{n\geq 0} f(n)\frac{x^n}{n!} = \left(\sum_{n\geq 0} \frac{x^n}{n!}\right)^2 = e^{2x} = \sum_{n\geq 0} 2^n\frac{x^n}{n!},$$

whence $f(n) = 2^n$. The analogous formula for Eulerian generating functions was given in Example 3.18.1(c).

3.18.7 Example. If $\mu(n)$ denotes the Möbius function $\mu(s,t)$ of an n-interval $[s,t]$ of P (which depends only on n, by Proposition 3.18.5), then from Theorem 3.18.4 we have

$$\sum_{n\geq 0} \mu(n)\frac{x^n}{B(n)} = \left(\sum_{n\geq 0} \frac{x^n}{B(n)}\right)^{-1}. \qquad (3.91)$$

Thus with P as in Example 3.18.3(a),

$$\sum_{n\geq 0} \mu(n)x^n = \left(\sum_{n\geq 0} x^n\right)^{-1} = 1-x,$$

agreeing, of course, with Example 3.8.1. Similarly for Example 3.18.3(b),

$$\sum_{n\geq 0} \mu(n)\frac{x^n}{n!} = \left(\sum_{n\geq 0} \frac{x^n}{n!}\right)^{-1} = e^{-x} = \sum_{n\geq 0} (-1)^n\frac{x^n}{n!},$$

giving yet another determination of the Möbius function of a boolean algebra. Thus, formally the Principle of Inclusion-Exclusion is equivalent to the identity $(e^x)^{-1} = e^{-x}$.

3.18.8 Example. The previous two examples can be generalized as follows. Let $Z_n(\lambda)$ denote the zeta polynomial (in the variable λ) of an n-interval $[s,t]$ of P.

Then since $Z_n(\lambda) = \zeta^\lambda(s,t)$, we have

$$\sum_{n \geq 0} Z_n(\lambda) \frac{x^n}{B(n)} = \left(\sum_{n \geq 0} \frac{x^n}{B(n)} \right)^\lambda.$$

This formula is valid for any complex number (or indeterminate) λ.

3.18.9 Example. As a variant of the previous example, fix $k \in \mathbb{P}$ and let $c_k(n)$ denote the number of chains $s = s_0 < s_1 < \cdots < s_k = t$ of length k between s and t in an n-interval $[s,t]$. Since $c_k(n) = (\zeta - 1)^k[s,t]$, we have

$$\sum_{n \geq 0} c_k(n) \frac{x^n}{B(n)} = \left(\sum_{n \geq 1} \frac{x^n}{B(n)} \right)^k.$$

The case $P = \mathbb{B}$ is particularly interesting. Here $c_k(n)$ is the number of chains $\emptyset = S_0 \subset S_1 \subset \cdots \subset S_k = [n]$, or alternatively the number of ordered partitions $(S_1, S_2 - S_1, S_3 - S_2, \ldots, [n] - S_{k-1})$ of $[n]$ into k (nonempty) blocks. Since there are $k!$ ways of ordering a partition with k blocks, we have $c_k(n) = k! S(n,k)$. Hence,

$$\sum_{n \geq 0} S(n,k) \frac{x^n}{n!} = \frac{1}{k!} \left(e^x - 1 \right)^k.$$

Thus, the theory of binomial posets "explains" the simple form of the generating function from equation (1.94b).

3.18.10 Example. Let $c(n)$ be the *total* number of chains from s to t in the n-interval $[s,t]$; that is, $c(n) = \sum_k c_k(n)$. We have seen (Section 3.6) that $c(n) = (2 - \zeta)^{-1}(s,t)$. Hence,

$$\sum_{n \geq 0} c(n) \frac{x^n}{B(n)} = \left(2 - \sum_{n \geq 0} \frac{x^n}{B(n)} \right)^{-1}.$$

For instance, if $P = \mathbb{N}$, then

$$\sum_{n \geq 0} c(n) x^n = \left(2 - \frac{1}{1-x} \right)^{-1} = 1 + \sum_{n \geq 1} 2^{n-1} x^n.$$

Thus, $c(n) = 2^{n-1}, n \geq 1$. Indeed, in the n-interval $[0,n]$, a chain $0 = t_0 < t_1 < \cdots < t_k = n$ can be identified with the composition $(t_1, t_2 - t_1, \ldots, n - t_{k-1})$, so we recover the result in Section 1.1 that there are 2^{n-1} compositions of n. If instead $P = \mathbb{B}$, then

$$\sum_{n \geq 0} c(n) \frac{x^n}{n!} = \frac{1}{2 - e^x}.$$

As seen from Example 3.18.9, $c(n)$ is the total number of ordered partitions of the set $[n]$; that is, $c(n) = \sum_k k! S(n,k)$. One sometimes calls an ordered partition of a set S a *preferential arrangement*, since it corresponds to ranking the elements of S in linear order where ties are allowed.

3.18.11 Example. Let $f(n)$ be the total number of chains $s = s_0 < s_1 < \cdots < s_k = t$ in an n-interval $[s,t]$ of P such that $\ell(s_{i-1}, s_i) \geq 2$ for all $1 \leq i \leq k$, where k is allowed to vary. By now it should be obvious to the reader that

$$\sum_{n \geq 0} f(n) \frac{x^n}{B(n)} = \sum_{k \geq 0} \left(\sum_{n \geq 0} \frac{x^n}{B(n)} - 1 - x \right)^k$$

$$= \left(1 - \sum_{n \geq 2} \frac{x^n}{B(n)} \right)^{-1}. \tag{3.92}$$

For instance, when $P = \mathbb{N}$, we are enumerating subsets of $[0,n]$ that contain 0 and n, and that don't contain two consecutive integers. Equivalently, we are counting compositions $(t_1 - t_0, t_2 - t_1, \ldots, n - t_{k-1})$ of n with no part equal to 1. From equation (3.92), we have

$$\sum_{n \geq 0} f(n) x^n = \left(1 - \frac{x^2}{1-x} \right)^{-1}$$

$$= \frac{1-x}{1-x-x^2} = 1 + \sum_{n \geq 2} F_{n-1} x^n,$$

where F_{n-1} denotes a Fibonacci number, in agreement with Exercise 1.35(b). Similarly when $P = \mathbb{B}$, we get $(2 + x - e^x)^{-1}$ as the exponential generating function for the number of ordered partitions of an n-set with no singleton blocks.

3.19 An Application to Permutation Enumeration

In Section 3.13, we related Möbius functions to the counting of permutations with certain properties. Using the theory of binomial posets, we can obtain generating functions for counting some of these permutations.

Throughout this section P denotes a binomial poset with factorial function $B(n)$. Let $S \subseteq \mathbb{P}$. If $[s,t]$ is an n-interval of P, then denote by $[s,t]_S$ the S-rank selected subposet of $[s,t]$ with s and t adjoined; that is,

$$[s,t]_S = \{u \in [s,t] : u = s, u = t, \text{ or } \ell(s,u) \in S\}. \tag{3.93}$$

Let μ_S denote the Möbius function of $[s,t]_S$, and set $\mu_S(n) = \mu_S(s,t)$. (It is easy to see that $\mu_S(n)$ depends only on n, not on the choice of the n-interval $[s,t]$.)

3.19.1 Lemma. *We have*

$$- \sum_{n \geq 1} \mu_S(n) \frac{x^n}{B(n)} = \left[\sum_{n \geq 1} \frac{x^n}{B(n)} \right] \left[1 + \sum_{n \in S} \mu_S(n) \frac{x^n}{B(n)} \right]. \tag{3.94}$$

Proof. Define a function $\chi : \mathbb{N} \to \{0,1\}$ by $\chi(n) = 1$ if $n = 0$ or $n \in S$, and $\chi(n) = 0$ otherwise. Then the defining recurrence (3.15) for Möbius functions yields $\mu_S(0) = 1$ and

$$\mu_S(n) = -\sum_{i=0}^{n-1} \binom{n}{i}_P \mu_S(i)\chi(i), \quad n \geq 1,$$

where $\binom{n}{i}_P = B(n)/B(i)B(n-i)$ as usual. Hence,

$$-\mu_S(n)(1 - \chi(n)) = \sum_{i=0}^{n} \binom{n}{i}_P \mu_S(i)\chi(i), \quad n \geq 1,$$

which translates into the generating function identity

$$-\sum_{n \geq 0} \mu_S(n)\frac{x^n}{B(n)} + \sum_{n \geq 0} \mu_S(n)\chi(n)\frac{x^n}{B(n)}$$

$$= \left[\sum_{n \geq 0} \frac{x^n}{B(n)} \right] \left[\sum_{n \geq 0} \mu_S(n)\chi(n)\frac{x^n}{B(n)} \right] - 1.$$

This formula is clearly equivalent to equation (3.94). $\square$

We now consider a set S for which the power series $1 + \sum_{n \in S} \mu_S(n)x^n/B(n)$ can be explicitly evaluated.

3.19.2 Lemma. *Let $k \in \mathbb{P}$, and let $S = k\mathbb{P} = \{kn : n \in \mathbb{P}\}$. Then*

$$1 + \sum_{n \in S} \mu_S(n)\frac{x^n}{B(n)} = \left[\sum_{n \geq 0} \frac{x^{kn}}{B(kn)} \right]^{-1}. \tag{3.95}$$

Proof. Let $P^{(k)}$ be the binomial poset of Example 3.18.3(f), with factorial function $B_k(n) = B(kn)/B(k)^n$. If $\mu^{(k)}$ is the Möbius function of $P^{(k)}$, then it follow from equation (3.91) that

$$\sum_{n \geq 0} \mu^{(k)}(n)\frac{x^n}{B_k(n)} = \left[\sum_{n \geq 0} \frac{x^n}{B_k(n)} \right]^{-1}. \tag{3.96}$$

But $\mu^{(k)}(n) = \mu_S(kn)$. Putting $B_k(n) = B(kn)/B(k)^n$ in equation (3.96), we obtain

$$\sum_{n \geq 0} \mu_S(kn)\frac{(B(k)x)^n}{B(kn)} = \left[\sum_{n \geq 0} \frac{(B(k)x)^n}{B(kn)} \right]^{-1}.$$

If we put x^k for $B(k)x$, then we get equation (3.95). $\square$

Combining Lemmas 3.19.1 and 3.19.2 we obtain:

3.19.3 Corollary. *Let $k \in \mathbb{P}$ and $S = k\mathbb{P}$. Then*

$$-\sum_{n \geq 1} \mu_S(n) \frac{x^n}{B(n)} = \left[\sum_{n \geq 1} \frac{x^n}{B(n)} \right] \left[\sum_{n \geq 0} \frac{x^{kn}}{B(kn)} \right]^{-1}.$$

Now specialize to the case $P = \mathbb{B}(q)$ of Example 3.18.3(c). For any $S \subseteq \mathbb{P}$, it follows from Theorem 3.13.3 that

$$(-1)^{\#(S \cap [n-1])-1} \mu_S(n) = \sum_w q^{\mathrm{inv}(w)},$$

where the sum is over all permutations $w \in \mathfrak{S}_n$ with descent set S. If $S = k\mathbb{P}$, then $\#(S \cap [n-1]) = \lfloor (n-1)/k \rfloor$. Hence, we conclude:

3.19.4 Proposition. *Let $k \in \mathbb{P}$, and let $f_{n,k}(q) = \sum_w q^{\mathrm{inv}(w)}$, where the sum is over all permutations $w = a_1 a_2 \cdots a_n \in \mathfrak{S}_n$ such that $a_i > a_{i+1}$ if and only if $k \mid i$. Then*

$$\sum_{n \geq 1} (-1)^{\lfloor (n-1)/k \rfloor} f_{n,k}(q) \frac{x^n}{(n)!} = \left[\sum_{n \geq 1} \frac{x^n}{(n)!} \right] \left[\sum_{n \geq 0} \frac{x^{kn}}{(kn)!} \right]^{-1}. \tag{3.97}$$

Although Proposition 3.19.4 can be proved without the use of binomial posets, our approach yields additional insight as to why equation (3.97) has such a simple form. In particular, the simple denominator $\sum_{n \geq 0} x^{kn}/(kn)!$ arises from dealing with the Möbius function of the poset $P^{(k)}$ where $P = \mathbb{B}(q)$.

We can eliminate the unsightly factor $(-1)^{\lfloor (n-1)/k \rfloor}$ in equation (3.97) by treating each congruence class of n modulo k separately. Fix $1 \leq j \leq k$, substitute $x^k \to -x^k$, and extract from (3.97) only those terms whose exponent is congruent to j modulo k to obtain the elegant formula

$$\sum_{\substack{m \geq 0 \\ n = mk+j}} f_{n,k}(q) \frac{x^n}{(n)!} = \left[\sum_{\substack{m \geq 0 \\ n = mk+j}} (-1)^m \frac{x^n}{(n)!} \right] \left[\sum_{n \geq 0} (-1)^n \frac{x^{nk}}{(nk)!} \right]^{-1} \tag{3.98}$$

In particular, when $j = k$, we can add 1 to both sides of equation (3.98) to obtain

$$\sum_{m \geq 0} f_{mk,k}(q) \frac{x^{mk}}{(mk)!} = \left[\sum_{n \geq 0} (-1)^n \frac{x^{nk}}{(nk)!} \right]^{-1}. \tag{3.99}$$

Equation (3.99) is also a direct consequence of Lemma 3.19.2.

One special case of equation (3.98) deserves special mention. Recall (Sections 1.4 and 1.6) that a permutation $a_1 a_2 \cdots a_n \in \mathfrak{S}_n$ is *alternating* if $a_1 > a_2 < a_3 > \cdots$. It is clear from the definition of $f_{n,k}(q)$ that $f_{n,2}(1)$ is the number E_n of alternating permutations in $\mathfrak{S}_n$. Substituting $k = 2$, $q = 1$, and $j = 1, k$ in equation (3.98) recovers Proposition 1.6.1, namely,

$$\sum_{n \geq 0} E_n \frac{x^n}{n!} = \sec x + \tan x. \tag{3.100}$$

Thus, we have a poset-theoretic explanation for the remarkable elegance of equation (3.100). Moreover, equation (1.59) and Exercise 1.147 can be proved in exactly the same way using $f_{n,k}(1)$.

3.20 Promotion and Evacuation

This section is independent of the rest of this book (except for a few exercises) and can be omitted without loss of continuity. Promotion and evacuation are certain bijections on the set of linear extensions of a finite poset P. They have some remarkable properties and arise in a variety of unexpected situations. See for instance Section A1.2 of Appendix 1 of Chapter 7, as well as Exercises 3.79–3.80.

Let $\mathcal{L}(P)$ denote the set of linear extensions of P. For now we regard a linear extension as an order-preserving bijection $f : P \to \boldsymbol{p}$, where $\#P = p$. Think of the element $t \in P$ as being labeled by $f(t)$. Now define a bijection $\partial : \mathcal{L}(P) \to \mathcal{L}(P)$, as follows. Remove the label 1 from P. Let $t_1 \in P$ satisfy $f(t_1) = 1$. Among the elements of P covering t_1, let t_2 be the one with the smallest label $f(t_2)$. Remove this label from t_2 and place it at t_1. (Think of "sliding" the label $f(t_2)$ down from t_2 to t_1.) Now among the elements of P covering t_2, let t_3 be the one with the smallest label $f(t_3)$. Slide the label from t_3 to t_2. Continue this process until eventually reaching a maximal element t_k of P. After we slide $f(t_k)$ to t_{k-1}, label t_k with $p + 1$. Now subtract 1 from every label. We obtain a new linear extension $f \partial \in \mathcal{L}(P)$, called the *promotion* of f. Note that we let ∂ operate on the *right*. Note also that $t_1 \lessdot t_2 \lessdot \cdots \lessdot t_k$ is a maximal chain of P, called the *promotion chain* of f. Figure 3.36(a) shows a poset P and a linear extension f. The promotion chain is indicated by circled dots and arrows. Figure 3.36(b) shows the labeling after the sliding operations and the labeling of the last element of the promotion chain by $p + 1 = 10$. Figure 3.36(c) shows the linear extension $f \partial$ obtained by subtracting 1 from the labels in Figure 3.36(b).

It should be obvious that $\partial : \mathcal{L}(P) \to \mathcal{L}(P)$ is a bijection. In fact, let ∂^* denote *dual promotion*, (i.e., we remove the *largest* label p from some element $u_1 \in P$, then slide the *largest* label of an element covered by u_1 up to u_1, etc.). After reaching a minimal element u_k, we label it by 0 and then add 1 to each label, obtaining $f \partial^*$. It is easy to check that

$$\partial^{-1} = \partial^*.$$

We next define a variant of promotion called *evacuation*. The evacuation of a linear extension $f \in \mathcal{L}(P)$ is denoted $f \epsilon$ and is another linear extension of P. First,

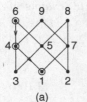

(a)

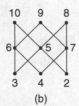

(b)

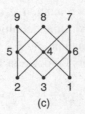

(c)

Figure 3.36 The promotion operator ∂ applied to a linear extension.

compute $f \partial$. Then "freeze" the label p into place and apply ∂ to what remains. In other words, let P_1 consist of those elements of P labelled $1, 2, \ldots, p - 1$ by $f \partial$, and apply ∂ to the restriction of $f \partial$ to P_1. Then freeze the label $p - 1$ and apply ∂ to the $p - 2$ elements that remain. Continue in this way until every element has been frozen. Let $f \epsilon$ be the linear extension, called the *evacuation* of f, defined by the frozen labels.

Figure 3.37 illustrates the evacuation of a linear extension f. The promotion paths are shown by arrows, and the frozen elements are circled. For ease of understanding we don't subtract 1 from the unfrozen labels since they all eventually disappear. The labels are always frozen in descending order $p, p - 1, \ldots, 1$. Figure 3.38 shows the evacuation of $f \epsilon$, where f is the linear extension of Figure 3.37. Note that (seemingly) miraculously, we have $f \epsilon^2 = f$. This example illustrates a fundamental property of evacuation given by Theorem 3.20.1(a).

We can define *dual evacuation ϵ^** analogously to dual promotion (i.e., evacuate from the top of P rather than from the bottom). In symbols, if $f \in \mathcal{L}(P)$, then define $f^* \in \mathcal{L}(P^*)$ by $f^*(t) = p + 1 - f(t)$. Then ϵ^* is given by

$$f \epsilon^* = (f^* \epsilon)^*.$$

We can now state the main result of this section.

3.20.1 Theorem. *Let P be a p-element poset. Then the operators ϵ, ϵ^*, and ∂ satisfy the following properties.*

(a) Evacuation is an involution, that is, $\epsilon^2 = 1$ (the identity operator).
(b) $\partial^p = \epsilon \epsilon^*$.
(c) $\partial \epsilon = \epsilon \partial^{-1}$.

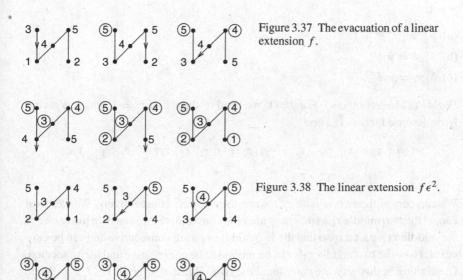

Figure 3.37 The evacuation of a linear extension f.

Figure 3.38 The linear extension $f \epsilon^2$.

Theorem 3.20.1 can be interpreted algebraically as follows. The bijections ϵ and ϵ^* generate a subgroup of the symmetric group $\mathfrak{S}_{\mathcal{L}(P)}$ on all the linear extensions of P. Since ϵ and (by duality) ϵ^* are involutions, the group they generate is a dihedral group (possibly degenerate, i.e., isomorphic to $\{1\}$, $\mathbb{Z}/2\mathbb{Z}$, or $\mathbb{Z}/2\mathbb{Z} \times \mathbb{Z}/2\mathbb{Z}$) of order 1 or $2m$ for some $m \geq 1$. If ϵ and ϵ^* are not both trivial, so they generate a group of order $2m$, then m is the order of ∂^P. In general the value of m, or more generally the cycle structure of ∂^P, is mysterious. For a few cases in which further information is known, see Exercise 3.80.

The main idea for proving Theorem 3.20.1 is to write linear extensions as *words* rather than functions and then to describe the actions of ∂ and ϵ on these words. The proof then becomes a routine algebraic computation. Let us first develop the necessary algebra in a more abstract context.

Let G be the group generated by elements $\tau_1, \ldots, \tau_{p-1}$ satisfying

$$\tau_i^2 = 1, \quad 1 \leq i \leq p-1$$
$$\tau_i \tau_j = \tau_j \tau_i, \quad \text{if } |i-j| > 1. \tag{3.101}$$

Some readers will recognize these relations as a subset of the Coxeter relations defining the symmetric group $\mathfrak{S}_p$. Define the following elements of G for $1 \leq j \leq p-1$:

$$\delta_j = \tau_1 \tau_2 \cdots \tau_j,$$
$$\gamma_j = \delta_j \delta_{j-1} \cdots \delta_1,$$
$$\gamma_j^* = \tau_j \tau_{j-1} \cdots \tau_1 \cdot \tau_j \tau_{j-1} \cdots \tau_2 \cdots \tau_j \tau_{j-1} \cdot \tau_j.$$

3.20.2 Lemma. *In the group G, we have the following identities for $1 \leq j \leq p-1$:*

(a) $\gamma_j^2 = (\gamma_j^*)^2 = 1$.
(b) $\delta_j^{j+1} = \gamma_j \gamma_j^*$.
(c) $\delta_j \gamma_j = \gamma_j \delta_j^{-1}$.

Proof. (a) Induction on j. For $j = 1$, we need to show that $\tau_1^2 = 1$, which is given. Now assume for $j - 1$. Then

$$\gamma_j^2 = \tau_1 \tau_2 \cdots \tau_j \cdot \tau_1 \cdots \tau_{j-1} \cdots \tau_1 \tau_2 \tau_3 \cdot \tau_1 \tau_2 \cdot \tau_1 \cdot \tau_1 \tau_2 \cdots \tau_j \cdot \tau_1 \cdots \tau_{j-1}$$
$$\cdots \tau_1 \tau_2 \tau_3 \cdot \tau_1 \tau_2 \cdot \tau_1.$$

We can cancel the two middle τ_1's since they appear consecutively. We can then cancel the two middle τ_2's since they are now consecutive. We can then move one of the middle τ_3's past a τ_1 so that the two middle τ_3's are consecutive and can be canceled. Now the two middle τ_4's can be moved to be consecutive and then canceled. Continuing in this way, we can cancel the two middle τ_i's for all $1 \leq i \leq j$. When this cancellation is done, what remains is the element γ_{j-1}, which is 1 by induction.

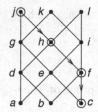

Figure 3.39 The promotion chain of the linear extension *cabdfeghjilk.*

(b,c) Analogous to (a). Details are omitted. □

Proof of Theorem 3.20.1. A glance at Theorem 3.20.1 and Lemma 3.20.2 makes it obvious that they should be connected. To see this connection, regard the linear extension $f \in \mathcal{L}(P)$ as the word (or permutation of P) $f^{-1}(1), \ldots, f^{-1}(p)$. For $1 \le i \le p-1$ define operators $\tau_i : \mathcal{L}(P) \to \mathcal{L}(P)$ by

$$(u_1 u_2 \cdots u_p)\tau_i = \begin{cases} u_1 u_2 \cdots u_p, & \text{if } u_i \text{ and } u_{i+1} \text{ are comparable in } P, \\ u_1 u_2 \cdots u_{i+1} u_i \cdots u_p, & \text{if } u_i \parallel u_{i+1}. \end{cases}$$

(3.102)

Clearly, τ_i is a bijection, and the τ_i's satisfy the relations (3.101). By Lemma 3.20.2, the proof of Theorem 3.20.1 follows from showing that

$$\partial = \delta_{p-1} := \tau_1 \tau_2 \cdots \tau_{p-1}.$$

Note that if $f = u_1 u_2 \cdots u_p$, then $f\delta_{p-1}$ is obtained as follows. Let j be the least integer such that $j > 1$ and $u_1 < u_j$. Since f is a linear extension, the elements $u_2, u_3, \ldots, u_{j-1}$ are incomparable with u_1. Move u_1 so it is between u_{j-1} and u_j. (Equivalently, cyclically shift the sequence $u_1 u_2 \cdots u_{j-1}$ one unit to the left.) Now let k be the least integer such that $k > j$ and $u_j < u_k$. Move u_j so it is between u_{k-1} and u_k. Continue in this way reaching the end. For example, let z be the linear extension *cabdfeghjilk* of the poset in Figure 3.39 (which also shows the evacuation chain for this linear extension). (We denote the linear extension for this one example by z instead of f since we are denoting one of the elements of P by f.) We factor z from left to right into the longest factors for which the first element of each factor is incomparable with the other elements of the factor:

$$z = (cabd)(feg)(h)(jilk).$$

Cyclically shift each factor one unit to the left to obtain $z\delta_{p-1}$:

$$z\delta_{p-1} = (abdc)(egf)(h)(ilkj) = abdcegfhkilj.$$

Now consider the process of promoting the linear extension f of the previous paragraph, given as a function by $f(u_i) = i$ and as a word by $u_1 u_2 \cdots u_p$. The elements $u_2, \ldots, u_{j-1}$ are incomparable with u_1 and thus will have their labels reduced by 1 after promotion. The label j of u_j (the least element in the linear extension f greater than u_1) will slide down to u_1 and be reduced to $j-1$. Hence,

$f\partial = u_2u_3\cdots u_{j-1}u_1\cdots$. Exactly analogous reasoning applies to the next step of the promotion process, when we slide the label k of u_k down to u_j. Hence,

$$f\partial = u_2u_3\cdots u_{j-1}u_1\cdot u_{j+1}u_{j+2}\cdots u_{k-1}u_j\cdots.$$

Continuing in this manner shows that $z\delta = z\partial$, completing the proof of Theorem 3.20.1. $\qquad\square$

3.21 Differential Posets

Differential posets form a class of posets with many explicit enumerative properties that can be proved by linear algebraic techniques. Their combinatorics is closely connected with the combinatorics of the relation $DU - UD = I$.

3.21.1 Definition. Let $r \in \mathbb{P}$. An *r-differential poset* is a poset P satisfying the following three axioms.

(D1) P has a $\hat{0}$ and is locally finite and graded (i.e., every interval $[\hat{0}, t]$ is graded).

(D2) If $t \in P$ covers exactly k elements, then t is covered by exactly $k+r$ elements.

(D3) Let $s, t \in P$, $s \neq t$. If exactly j elements are covered by both s and t, then exactly j elements cover both s and t.

NOTE.

(a) It is easy to see that in axiom (D3) we must have either $j = 0$ or $j = 1$. For suppose $j > 1$ elements $u_1, u_2, \ldots$ are covered by both s and t, where the rank m of s and t is minimal with respect to this property. Then u_1 and u_2 are covered by both s and t, so by (D3) u_1 and u_2 cover at least two elements, contradicting the minimality of m.

(b) Suppose that L is an r-differential lattice. Then by Proposition 3.3.2, axiom (D3) is equivalent to the statement that L is modular.

Let us first give some examples of differential posets.

3.21.2 Example. 1. It is easy to see that if P is r-differential and Q is s-differential, then $P \times Q$ is $(r + s)$-differential.

2. Young's lattice $Y = J_f(\mathbb{N} \times \mathbb{N})$, defined in Example 3.4.4(b), is a 1-differential poset. Axiom (D1) of Definition 3.21.1 is clear, while (D3) follows from the distributivity (and hence modularity) of L. For (D2), note that the positions where we can add a square to the Young diagram of a partition λ (marked $\times$ in the diagram at the top of the next page) alternate along the boundary with the positions where we can remove a square (marked $\bullet$), with $\times$ at the beginning and end.

3. Let P be r-differential "up to rank n." This means that (a) P is graded of rank n with $\hat{0}$, (b) P satisfies axiom (D2) of Definition 3.21.1 if $\text{rank}(t) < n$, and (c) P satisfies axiom (D3) of Definition 3.21.1 if $\text{rank}(s) = \text{rank}(t) < n$. Define $\Omega_r P$, the *reflection extension* of P, as follows. Let P_i denote the set of elements of P

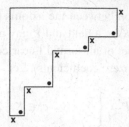

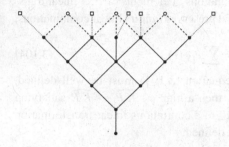

Figure 3.40 The Reflection-Extension construction.

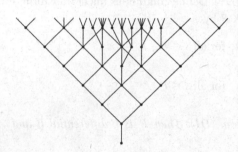

Figure 3.41 The Fibonacci differential poset Z_1.

of rank i, and suppose that $P_{n-1} = \{s_1, \ldots, s_k\}$. First, place $t_1, \ldots, t_k \in P_{n+1}$ with $t_i > u \in P_n$ if and only if $s_i \lessdot u$ (the "reflection of P_{n-1} through P_n"). Then for each $u \in P_n$, adjoin r new elements covering u. Figure 3.40 gives an example for $r = 1$ and $n = 4$. The open circles and dashed lines show the reflection of rank 3 to rank 5, while the open squares and dotted lines indicate the extension of each element of rank 4.

It is easy to see that $\Omega_r P$ is r-differential up to rank $n + 1$, and that

$$p_{n+1} = r p_n + p_{n-1}, \tag{3.103}$$

where $p_i = \#P_i$. In particular, if we infinitely iterate the operation Ω_r on P, then we obtain an r-differential poset $\Omega_r^\infty P$.

4. Let P be the one-element poset, and write $Z_r = \Omega_r^\infty P$. It is easy to check (e.g., by induction) that Z_r is a (modular) lattice, called the r-*Fibonacci* differential poset. If $r = 1$ then it follows from equation (3.103) that $\#(Z_1)_n = F_{n+1}$, a Fibonacci number. See Figure 3.41 for Z_1 up to rank 6.

We now turn to the connection between the axioms (D1)–(D3) of a differential poset and linear algebra. Let K be a field and P any poset. Let KP be the vector space with basis P (i.e., the set of all formal linear combinations of elements of P with only finitely many nonzero coefficients). Let

$$\widehat{KP} = \left\{ \sum_{t \in P} c_t t : c_t \in K \right\},$$

the vector space of all *infinite* linear combinations of elements of P. A linear transformation $\varphi : \widehat{KP} \to \widehat{KP}$ is *continuous* if it preserves *infinite* linear combinations, that is,

$$\varphi\left(\sum c_t t \right) = \sum c_t \varphi(t), \tag{3.104}$$

so in particular the right-hand side of equation (3.104) must be well defined. For instance, if P is infinite and $u \in P$, then a map $\varphi : \widehat{KP} \to \widehat{KP}$ satisfying $\varphi(t) = u$ for all $t \in P$ cannot be extended to a continuous linear transformation since $\varphi\left(\sum_{t \in P} t \right) = \sum_{t \in P} u$, which is not defined.

Now assume that P satisfies axiom (D1). Define continuous linear transformations $U : \widehat{KP} \to \widehat{KP}$ and $D : \widehat{KP} \to \widehat{KP}$ by

$$U(s) = \sum_{t \gtrdot s} t, \text{ for all } s \in P,$$

$$D(s) = \sum_{t \lessdot s} t, \text{ for all } s \in P.$$

3.21.3 Proposition. *Let P satisfy axiom (D1). Then P is r-differential if and only if*

$$DU - UD = rI,$$

where I is the identity linear transformation on $\widehat{KP}$.

Proof. The proof is of the type "follow your nose." By the definition of the product of linear transformations, axiom (D2) is equivalent to the statement that the coefficient of t (when we expand as an infinite linear combination of elements of P) in $(DU - UD)t$ is r. If $s \neq t$, then axiom (D3) is equivalent to the statement that the coefficient of s in $(DU - UD)t$ is 0. $\qquad\square$

If $X \subseteq P$, then write $X = \sum_{t \in X} t \in \widehat{KP}$. Define a pairing $\widehat{KP} \times KP \to K$ by

$$\langle s, t \rangle = \delta_{st}, \quad s, t \in P.$$

Thus,

$$\left\langle \sum_{t \in P} a_t t, \sum_{t \in P} b_t t \right\rangle = \sum a_t b_t,$$

a finite sum since $\sum_{t \in P} b_t t \in KP$. If $f \in \widehat{KP}$ and $t \in P$, then $\langle f, t \rangle$ is just the coefficient of t in f.

3.21.4 Proposition. *Let P be an r-differential poset. Then*

$$DP = (U + r)P.$$

Proof. Equivalent to axiom (D2). In more detail, for $t \in P$ let

$$C^-(t) = \{s \in P : s \lessdot t\}.$$

Then $\langle (U + r)P, t \rangle = r + \#C^-(t)$, and the proof follows. $\qquad\square$

3.21.5 Example. Let $t \in P_n$ (the set of elements of P of rank n), and let $e(t)$ denote the number of saturated chains $\hat{0} = t_0 \lessdot t_1 \lessdot \cdots \lessdot t_n = t$ from $\hat{0}$ to t. Then

$$e(t) = \langle U^n \hat{0}, t \rangle.$$

Similarly [why?],

$$\sum_{t \in P_n} e(t) = \langle U^n \hat{0}, P_n \rangle,$$

$$\sum_{t \in P_n} e(t)^2 = \langle D^n U^n \hat{0}, \hat{0} \rangle.$$

Let us now consider some relations between U and D that are formal consequences of $DU - UD = rI$. The basic goal is to "push" to the right the D's in an expression (power series) involving U and D by using $DU = UD + rI$. A useful way to understand such results is the following. We can "represent" $U = z$ (i.e., multiplication by the indeterminate z) and $D = r\frac{d}{dz}$ since $(r\frac{d}{dz})z - z(r\frac{d}{dz}) = r$ as operators. Thus, familiar identities involving differentiation can be transferred to identities involving U and D. For the algebraically minded, a more precise statement is that we have an isomorphism

$$K\langle\langle U, D \rangle\rangle / (DU - UD - r) \cong K\left\langle\!\!\left\langle z, r\frac{d}{dz} \right\rangle\!\!\right\rangle,$$

where $K\langle\langle \cdot \rangle\rangle$ denotes noncommutative formal power series.

As an example, let $f(U)$ be any power series in U. Since $r\frac{d}{dz} f(z)g(z) = rf'(z)g(z) + f(z)r\frac{d}{dz}g(z)$, we have

$$Df(U) = rf'(U) + f(U)D.$$

This identity can be verified directly by proving by induction on n that $DU^n = rnU^{n-1} + U^n D$ and using linearity and continuity.

3.21.6 Theorem. *(a) Suppose that $DU - UD = rI$. Then*

$$e^{(U+D)x} = e^{\frac{1}{2}rx^2 + Ux}e^{Dx}. \tag{3.105}$$

(b) Let $f(U)$ be a power series in y whose coefficients are polynomials (independent of another variable x) in U. Thus, the action of $f(U)$ on $\widehat{KP}$ is well defined. Then

$$e^{Dx} f(U) = f(U + rx)e^{Dx}.$$

In particular, setting $f(U) = e^{Uy}$ and then $x = y$, we get

$$e^{Dx}e^{Ux} = e^{rx^2 + Ux}e^{Dx}. \tag{3.106}$$

NOTE. Care must be taken in interpreting Theorem 3.21.6 since U and D don't commute. For instance, equation (3.105) asserts that

$$\sum_{n \geq 0}(U+D)^n \frac{x^n}{n!} = \left(\sum_{n \geq 0}\frac{r^n x^{2n}}{2^n n!}\right)\left(\sum_{n \geq 0}U^n \frac{x^n}{n!}\right)\left(\sum_{n \geq 0}D^n \frac{x^n}{n!}\right).$$

Thus, when we equate coefficients of $x^3/3!$ on both sides, we obtain

$$(U+D)^3 = 3!\left[\left(\frac{U^3}{6} + \frac{U^2 D}{2} + \frac{UD^2}{2} + \frac{D^3}{6}\right) + \frac{r}{2}(U+D)\right].$$

Proof of Theorem 3.21.6. Both equations (3.105) and (3.106) can be proved straightforwardly by verifying them for the coefficient of x^n using induction on n. We give other proofs based on the representation $U = z$, $D = r\frac{d}{dz}$ discussed above.

(a) Let $H(x) = e^{(D+U)x} = \sum_{n \geq 0}(D+U)^n \frac{x^n}{n!}$. Then $H(x)$ is uniquely determined by

$$(D+U)H(x) = \frac{d}{dx}H(x), \quad H(0) = 1.$$

Let

$$J(x) = e^{\frac{1}{2}rx^2 + Ux}e^{Dx}.$$

Clearly $J(0) = 1$. Regarding $D = r\frac{d}{dU}$, we have

$$(D+U)J(x) = (rxe^{\frac{1}{2}rx^2 + Ux}e^{Dx} + e^{\frac{1}{2}rx^2 + Ux}De^{Dx}) + UJ(x),$$

$$\frac{d}{dx}J(x) = (rx + U)J(x) + e^{\frac{1}{2}rx^2 + Ux}De^{Dx}$$

$$= (D+U)J(x).$$

Hence, $H(x) = J(x)$, proving (3.105).

(b) The Taylor series expansion of $f(z+rx)$ at z is given by

$$f(z+rx) = \sum_{n \geq 0}\left(r\frac{d}{dz}\right)^n f(z)\frac{x^n}{n!}$$

$$= e^{x\left(r\frac{d}{dz}\right)}f(z),$$

and the proof follows from the representation $D = r\frac{d}{dU}$.

$\square$

We are now ready to give some enumerative applications. A *Hasse walk* of length ℓ from s to t in a poset P is a sequence

$$s = t_0, t_1, \ldots, t_\ell = t, \quad t_i \in P,$$

such that either $t_{i-1} \lessdot t_i$ or $t_{i-1} \gtrdot t_i$ for $1 \le i \le \ell$. Note that in a graded poset, all closed Hasse walks (i.e., those with $s = t$) have even length.

3.21.7 Theorem. *Let P be an r-differential poset, and let κ_ℓ be the number of Hasse walks of length ℓ from $\hat{0}$ to $\hat{0}$, so $\kappa_\ell = 0$ if ℓ is odd. Then*

$$\kappa_{2n} = (2n-1)!! \, r^n = 1 \cdot 3 \cdot 5 \cdots (2n-1) r^n.$$

Proof. Note that $\kappa_{2n} = \langle (U+D)^{2n} \hat{0}, \hat{0} \rangle$. Hence using equation (3.105) and $D^n \hat{0} = 0$ for $n \ge 1$, we get

$$\sum_{n \ge 0} \kappa_{2n} \frac{x^{2n}}{(2n)!} = \left\langle \sum_{n \ge 0} (U+D)^n \frac{x^n}{n!} \hat{0}, \hat{0} \right\rangle$$

$$= \left\langle e^{(U+D)x} \hat{0}, \hat{0} \right\rangle$$

$$= \left\langle e^{\frac{1}{2} r x^2 + U x} e^{D x} \hat{0}, \hat{0} \right\rangle$$

$$= \left\langle e^{\frac{1}{2} r x^2 + U x} \hat{0}, \hat{0} \right\rangle$$

$$= e^{\frac{1}{2} r x^2}$$

$$= \sum_{n \ge 0} r^n (2n-1)!! \frac{x^{2n}}{(2n)!},$$

and the proof follows from equating coefficients of $x^{2n}/(2n)!$. $\qquad \square$

3.21.8 Theorem. *For any r-differential poset P we have*

$$\sum_{t \in P_n} e(t)^2 = r^n n!.$$

Proof. Completely analogous to the previous proof, using equation (3.106) and

$$\sum_{n \ge 0} \left(\sum_{t \in P_n} e(t)^2 \right) \frac{x^{2n}}{n!^2} = \left\langle e^{Dx} e^{Ux} \hat{0}, \hat{0} \right\rangle.$$

$\qquad \square$

We next give some enumerative applications of the identity $DP = (U + r)P$ (Proposition 3.21.4).

3.21.9 Theorem. *Let P be an r-differential poset. Then*

$$e^{Dx}P = e^{rx + \frac{1}{2}rx^2 + Ux}P, \tag{3.107}$$

$$e^{(U+D)x}P = e^{rx + rx^2 + 2Ux}P, \tag{3.108}$$

$$e^{Dx}e^{Ux}P = e^{rx + \frac{3}{2}rx^2 + 2Ux}P. \tag{3.109}$$

Proof. Let $H(x) = e^{Dx}$. Then $H(x)P$ is uniquely determined by the conditions

$$DH(x)P = \frac{d}{dx}H(x)P, \quad H(0)P = P.$$

Let $L(x) = e^{rx + \frac{1}{2}rx^2 + Ux}P$. Then

$$DL(x)P = (rxL(x) + L(x)D)P \quad (\text{since } Df(U) = rf'(U) + f(U)D)$$

$$= (rxL(x) + L(x)(U + r))P$$

$$= (rx + U + r)L(x)P$$

$$= \frac{d}{dx}L(x)P.$$

Clearly, $L(0)P = P$, so $L(x) = H(x)$, proving (3.107).

Now

$$e^{(U+D)x}P = e^{\frac{1}{2}rx^2 + Ux}e^{Dx}P$$

$$= e^{\frac{1}{2}rx^2 + Ux}e^{rx + \frac{1}{2}rx^2 + Ux}P$$

$$= e^{rx + rx^2 + 2Ux}P,$$

and

$$e^{Dx}e^{Ux}P = e^{rx^2 + Ux}e^{Dx}P$$

$$= e^{rx^2 + Ux}e^{rx + \frac{1}{2}rx^2 + Ux}P$$

$$= e^{rx + \frac{3}{2}rx^2 + 2Ux}P,$$

completing the proof. $\qquad \square$

Write $\alpha(0 \to n) = \sum_{t \in P_n} e(t)$, and let δ_n be the number of Hasse walks of length n from $\hat{0}$ (with any ending element).

3.21.10 Theorem. *For any r-differential poset P we have*

$$\sum_{n \geq 0} \alpha(0 \to n)\frac{x^n}{n!} = e^{rx + \frac{1}{2}rx^2}$$

$$\sum_{n \geq 0} \delta_n \frac{x^n}{n!} = e^{rx + rx^2}.$$

NOTE. It follows from Theorem 1.3.3 or from Section 5.1 that Theorem 3.21.10 can be restated in the form

$$\alpha(0 \to n) = \sum_{\substack{w \in \mathfrak{S}_n \\ w^2 = 1}} r^{c(w)}$$

$$\sum_{n \geq 0} \delta_n = \sum_{\substack{w \in \mathfrak{S}_n \\ w^2 = 1}} r^{c(w)} 2^{c_2(w)},$$

where $c(w)$ denotes the number of cycles of w and $c_2(w)$ the number of 2-cycles. For instance, $\alpha(0 \to 3) = r^3 + 3r^2$, and $\delta_3 = r^3 + 6r^2$. Note that in particular if $r = 1$, then $\alpha(0 \to n)$ is just the number of involutions in $\mathfrak{S}_n$. For the case where P is Young's lattice Y, this result is equivalent to Corollary 7.13.9.

Proof of Theorem 3.21.10. Clearly,

$$\alpha(0 \to n) = \langle D^n P, \hat{0} \rangle, \quad \delta_n = \langle (U+D)^n P, \hat{0} \rangle.$$

Now for any $f(U)$ we have $\langle f(U)P, \hat{0} \rangle = f(0)$, so

$$\sum_{n \geq 0} \alpha(0 \to n) \frac{x^n}{n!} = \sum_{n \geq 0} \langle D^n P, \hat{0} \rangle \frac{x^n}{n!}$$

$$= \langle e^{Dx} P, \hat{0} \rangle$$

$$= \langle e^{rx + \frac{1}{2}rx^2 + Ux} P, \hat{0} \rangle$$

$$= e^{rx + \frac{1}{2}rx^2}.$$

An analogous argument works for $\sum_{n \geq 0} \delta_n x^n / n!$. $\qquad \square$

Let us now generalize this result for $\alpha(0 \to n)$ by considering increasing Hasse walks from rank n to rank $n + k$. More precisely, let $\alpha(n \to n + k)$ be the number of such walks, that is,

$$\alpha(n \to n + k) = \#\{t_n \lessdot t_{n+1} \lessdot \cdots \lessdot t_{n+k} : \rho(t_i) = i\},$$

where ρ denotes the rank function of P. In particular, $\alpha(n \to n) = \#P_n$. This special case suggests that the rank-generating function

$$F(P, q) = \sum_{t \in P} q^{\rho(t)} = \sum_{n \geq 0} (\#P_n) q^n$$

will be relevant, so let us note that

$$F(Y^r, q) = \prod_{i \geq 1} \frac{1}{(1 - q^i)^r},$$

$$F(Z_r, q) = \frac{1}{1 - rq - q^2}.$$

3.21.11 Theorem. *For any r-differential poset P we have*

$$\sum_{n\geq0}\sum_{k\geq0}\alpha(n\to n+k)q^n\frac{x^k}{k!} = F(P,q)\exp\left(\frac{rx}{1-q}+\frac{rx^2}{2(1-q^2)}\right).$$

NOTE. Taking the coefficient of $x^k/k!$ for $0\leq k\leq2$ yields

$$\sum_{n\geq0}\alpha(n\to n)q^n = F(P,q),$$

$$\sum_{n\geq0}\alpha(n\to n+1)q^n = \frac{r}{1-q}F(P,q),$$

$$\sum_{n\geq0}\alpha(n\to n+2)q^n = \frac{r(r+1)+r(r-1)q}{(1-q)^2(1-q^2)}F(P,q).$$

In general, it is immediate from Theorem 3.21.11 that for fixed k we have

$$\sum_{n\geq0}\alpha(n\to n+k)q^n = A_k(q)F(P,q),$$

where $A_k(q)$ is a rational function of q (and r) satisfying

$$\sum_{k\geq0}A_k(q)\frac{x^k}{k!} = \exp\left(\frac{rx}{1-q}+\frac{rx^2}{2(1-q^2)}\right). \tag{3.110}$$

Proof of Theorem 3.21.11. Let $\gamma\colon \widehat{KP}\to K[[q]]$ be the continuous linear transformation defined by $\gamma(t) = q^{\rho(t)}$ for all $t\in P$, so $\gamma(P) = F(P,q)$. Now

$$\gamma(e^{Dx}P) = \sum_{k\geq0}\gamma(D^kP)\frac{x^k}{k!}$$

$$= \sum_{n\geq0}\sum_{k\geq0}\alpha(n\to n+k)q^n\frac{x^k}{k!} := G(q,x).$$

But also

$$\gamma(e^{Dx}P) = e^{rx+\frac{1}{2}rx^2}\gamma(e^{Ux}P)$$

$$= e^{rx+\frac{1}{2}rx^2}\sum_{n\geq0}\sum_{k\geq0}\alpha(n-k\to n)q^n\frac{x^k}{k!}$$

$$= e^{rx+\frac{1}{2}rx^2}\sum_{n\geq0}\sum_{k\geq0}\alpha(n-k\to n)q^{n-k}\frac{(qx)^k}{k!}$$

$$= e^{rx+\frac{1}{2}rx^2}G(q,qx).$$

Hence, we have shown that

$$G(q,x) = e^{rx+\frac{1}{2}rx^2}G(q,qx). \tag{3.111}$$

Moreover, it is clear that

$$G(q,0) = F(P,q). \tag{3.112}$$

We claim that equations (3.111) and (3.112) determine $G(q,x)$ uniquely. For if $G(q,x) = \sum a_k(q)x^k$, then

$$\sum a_k(q)x^k = e^{rx+\frac{1}{2}rx^2} \sum a_k(q)(qx)^k.$$

We can equate coefficients of x^k and solve for $a_k(q)$ in terms of $a_0(q), a_1(q), \ldots, a_{k-1}(q)$, with $a_0(q) = F(P,q)$, proving the claim.

Now note that $F(P,q)\exp\left(\frac{rx}{1-q} + \frac{rx^2}{2(1-q^2)}\right)$ satisfies equations (3.111) and (3.112), completing the proof. $\square$

A further aspect of differential posets is the computation of eigenvalues and eigenvectors of certain linear transformations. We illustrate this technique here by computing the eigenvalues of the adjacency matrix of the graph obtained by restricting the Hasse diagram of a differential poset to two consecutive levels. In general, if G is a finite graph, say with no multiple edges, then the *adjacency matrix* of G is the (symmetric) matrix A, say over $\mathbb{C}$, with rows and columns indexed by the vertices of G (in some order), with

$$A_{uv} = \begin{cases} 1, & \text{if } uv \text{ is an edge of } G, \\ 0, & \text{otherwise.} \end{cases}$$

Let P be an r-differential poset, and let $P_{j-1,j}$ denote the restriction of P to $P_{j-1} \cup P_j$. Identify $P_{j-1,j}$ with its Hasse diagram, regarded as an undirected (bipartite) graph. Let A denote the adjacency matrix of $P_{j-1,j}$. We are interested in computing the eigenvalues (or characteristic polynomial) of A. By definition of matrix multiplication, the matrix entry $(A^n)_{uv}$ is the number of walks of length n from u to v. On the other hand, $(A^n)_{uv}$ is closely related to the eigenvalues of A, as discussed in Section 4.7. This suggests that differential poset techniques might be useful in computing the eigenvalues.

3.21.12 Theorem. *Let $p_i = \#P_i$. Then the eigenvalues of A (over $\mathbb{C}$) are as follows:*

- *0 with multiplicity $p_j - p_{j-1}$,*
- *$\pm\sqrt{rs}$ with multiplicity $p_{j-s} - p_{j-s-1}$, $1 \le s \le j$.*

NOTE. The total number of eigenvalues is

$$p_j - p_{j-1} + 2\sum_{s=1}^{j}(p_{j-s} - p_{j-s-1}) = p_{j-1} + p_j,$$

the number of elements of the poset $P_{j-1,j}$.

Proof of Theorem 3.21.12. For any set S write $\mathbb{C}S$ for the complex vector space with basis S. Since $\mathbb{C}P_{j-1,j} = \mathbb{C}P_{j-1} \oplus \mathbb{C}P_j$, any $v \in \mathbb{C}P_{j-1,j}$ can be uniquely written

$$v = v_{j-1} + v_j, \quad v_i \in \mathbb{C}P_i.$$

Then A acts on $\mathbb{C}P_{j-1,j}$ by

$$A(v) = D(v_j) + U(v_{j-1}).$$

Write U_i for the restriction of U to $\mathbb{C}P_i$, and similarly for D_i and I_i. Thus, the identity $DU - UD = rI$ takes the form

$$D_{i+1}U_i - U_{i-1}D_i = rI_i.$$

Now U_{i-1} and D_i are adjoint linear transformations with respect to the bases P_{i-1} and P_i (in other words, their matrices are transposes of one another). Thus, by standard results in linear algebra, the linear transformation $U_{i-1}D_i$ is (positive) semidefinite and hence has nonnegative real eigenvalues. Now

$$D_{i+1}U_i = U_{i-1}D_i + rI_i.$$

The eigenvalues of $U_{i-1}D_i + rI_i$ are obtained by adding r to the eigenvalues of the semidefinite transformation $U_{i-1}D_i$. Hence, $D_{i+1}U_i$ has positive eigenvalues and is therefore invertible. In particular, U_i is injective, so its adjoint D_{i+1} is surjective. Therefore,

$$\dim(\ker D_i) = \dim \mathbb{C}P_i - \dim \mathbb{C}P_{i-1}$$

$$= p_i - p_{i-1}.$$

Case 1. Let $v \in \ker(D_j)$, so $v \in \mathbb{C}P_j$, i.e., $v = v_j$. Hence $Av = Dv = 0$, so $\ker(D_j)$ is an eigenspace of A with eigenvalue 0. Thus 0 is an eigenvalue of A with multiplicity at least $p_j - p_{j-1}$.

Case 2. Let $w \in \ker(D_s)$ for some $0 \le s \le j - 1$. Let

$$w^* = \underbrace{\sqrt{r(j-s)}U^{j-1-s}(w)}_{w^*_{j-1}} + \underbrace{U^{j-s}(w)}_{w^*_j} \in \mathbb{C}P_{j-1,j}.$$

We can choose either sign for the square root. Then

$$\begin{aligned}
A(w^*) &= U(w^*_{j-1}) + D(w^*_j) \\
&= \sqrt{r(j-s)}U^{j-s}(w) + DU^{j-s}(w) \\
&= \sqrt{r(j-s)}U^{j-s}(w) + U^{j-s}\underbrace{D(w)}_{0} + r(j-s)U^{j-s-1}(w) \\
&= \sqrt{r(j-s)}U^{j-s}(w) + r(j-s)U^{j-s-1}(w) \\
&= \sqrt{r(j-s)}w^*.
\end{aligned}$$

If $w_1, \ldots, w_t$ is a basis for $\ker(D_s)$ then $w_1^*, \ldots, w_t^*$ are linearly independent (since U is injective). Hence, $\pm\sqrt{r(j-s)}$ is an eigenvalue of A with multiplicity at least $t = \dim(\ker D_s) = p_s - p_{s-1}$. We have found a total of

$$p_j - p_{j-1} + 2\sum_{s=0}^{j-1}(p_s - p_{s-1}) = p_{s-1} + p_s$$

eigenvalues, so we have them all. $\qquad\qquad\qquad\qquad\qquad\qquad\qquad\qquad\qquad\square$

3.21.13 Corollary. *Fix $j \geq 1$. The number of closed walks of length $2m > 0$ in $P_{j-1,j}$ beginning and ending at some $t \in P_j$ is given by*

$$\sum_{s=1}^{j}(p_{j-s} - p_{j-s-1})(rs)^m.$$

Proof. By the definition of matrix multiplication, the total number of closed walks of length $2m$ in $P_{j-1,j}$ is equal to $\operatorname{tr}A^{2m} = \sum \theta_i^{2m}$, where the θ_i's are the eigenvalues of A. (See Theorem 4.7.1.) Exactly half these walks start at P_j. By Theorem 3.21.12 we have

$$\frac{1}{2}\operatorname{tr}A^{2m} = \frac{1}{2}\sum_{s=1}^{j}(p_{j-s} - p_{j-s-1})\left((\sqrt{rs})^{2m} + (-\sqrt{rs})^{2m}\right)$$

$$= \frac{1}{2}\sum_{s=1}^{j}(p_{j-s} - p_{j-s-1})2(rs)^m,$$

and the proof follows. $\qquad\qquad\qquad\qquad\qquad\qquad\qquad\qquad\qquad\qquad\qquad\square$

Notes

The subject of partially ordered sets and lattices has its origins in the work of G. Boole, C. S. Peirce, E. Schröder, and R. Dedekind during the nineteenth century. However, it was not until the work of Garrett Birkhoff in the 1930s that the development of poset theory and lattice theory as subjects in their own right really began. In particular, the appearance in 1940 of the first edition of Birkhoff's famous book [3.13] played a seminal role in the development of the subject. It is interesting to note that the three successive editions of this book used the terms "partly ordered set," "partially ordered set," and "poset," respectively. More explicit references to the development of posets and lattices can be found in [3.13]. Another important impetus to lattice theory was the work of John von Neumann on continuous geometries, also in the 1930s. For two surveys of this work see Birkhoff [3.12] and Halperin [3.40].

A bibliography of around 1400 items dealing with posets (but not lattices!) appears in Rival [3.57]. This latter reference contains many valuable surveys of the status of poset theory up to 1982. In particular, we mention the survey [3.35] of C. Greene on Möbius functions. An extensive bibliography of lattice theory appears in Grätzer [3.33].

Matroid theory was mentioned at the end of Section 3.3. Some books on this subject are by Oxley [3.54], Welsh [3.89], and White (ed.) [3.90][3.91][3.92].

The idea of incidence algebras can be traced back to Dedekind and E. T. Bell, whereas the Möbius inversion formula for posets is essentially due to L. Weisner in 1935. It was rediscovered shortly thereafter by P. Hall, and stated in its full generality by M. Ward in 1939. Hall proved the basic Proposition 3.8.5 (therefore known as "Philip Hall's theorem"), and Weisner, the equally important Corollary 3.9.3 ("Weisner's theorem"). However, it was not until 1964 that the seminal paper [3.58] of G.-C. Rota that began the systematic development of posets and lattices within combinatorics appeared. Reference to earlier work in this area cited earlier appear in [3.58]. Much additional material on incidence algebras appears in the book of E. Spiegel and C. O'Donnell [3.66].

We now turn to more specific citations, beginning with Section 3.4. Theorem 3.4.1 (the fundamental theorem for finite distributive lattices) was proved by Birkhoff [3.11, Thm. 17.3]. Generalizations to arbitrary distributive lattices were given by M. H. Stone [3.86, Thm. 4] and H. A. Priestley [3.55][3.56]. A nice survey is given by Davey and Priestley [3.24]. The connection between chains in distributive lattices $J(P)$ and order-preserving maps $\sigma : P \to \mathbb{N}$ (Section 3.5) was first explicitly observed by Stanley in [3.67] and [3.68]. The notion of a "generalized Pascal triangle" appears in Stanley [3.73].

The development of a homology theory for posets was considered by Deheuvels, Dowker, Farmer, Nöbeling, Okamoto, and others (see Farmer [3.27] for references), but the combinatorial ramifications of such a theory, including the connection with Möbius functions, was not perceived until Rota [3.58, pp. 355–356]. Some early work along these lines was done by Farmer, Folkman, Lakser, Mather, and others (see Walker [3.87][3.88] for references). In particular, Folkman proved a result equivalent to the statement that geometric lattices are Cohen-Macaulay. The systematic development of the relationship between combinatorial and topological properties of posets was begun by K. Baclawski and A. Björner and continued by J. Walker, followed by many others. A nice survey of topological combinatorics up to 1995 was given by Björner [3.18], while Kozlov [3.48] has written an extensive text. The connection between regular cell complexes and posets is discussed by Björner [3.15]. Cohen-Macaulay complexes were discovered independently by Baclawski [3.3] and Stanley [3.75, §8]. A survey of Cohen-Macaulay posets, including their connection with commutative algebra, appears in Björner–Garsia–Stanley [3.16]. For further information on the subject of "combinatorial commutative algebra" see the books by Stanley [3.78] and by E. Miller and B. Sturmfels [3.52]. The statement preceding Proposition 3.8.9 that $\operatorname{lk} F$ need not be simply connected and $|\operatorname{lk} F|$ need not be a manifold when $|\Delta|$ is a manifold is a consequence of a deep result of R. D. Edwards. See for instance [3.23, II.12].

The Möbius algebra of a poset P (generalizing our definition in Section 3.10 when P is a lattice) was introduced by L. Solomon [3.65] and first systematically investigated by C. Greene [3.34], who showed how it could be used to derive many apparently unrelated properties of Möbius functions.

Proposition 3.10.1 (stated for geometric lattices) is due to Rota [3.58, Thm. 4, p. 357]. The formula (3.34) for the Möbius function of $B_n(q)$ is due to P. Hall [3.39, (2.7)], while the generalization (3.38) appeared in Stanley [3.74, Thm. 3.1] (with $r = 1$). The formula (3.37) for Π_n is due independently to Schützenberger and to Frucht and Rota (see [3.58, p. 359]).

Theorem 3.11.7 is perhaps the first significant result in the theory of hyperplane arrangements. It was obtained by T. Zaslavsky [3.95][3.96] in 1975, though some special cases were known earlier. In particular, Proposition 3.11.8 goes back to L. Schläfli [3.59] (written in 1850–1852 and published in 1901).

Proposition 3.11.3, known as *Whitney's theorem*, was proved by H. Whitney [3.93, §6] for graphs (equivalent to graphical arrangements). Whitney considered this formula further in [3.94]. Many aspects of hyperplane arrangments are best understood *via* matroid theory, as explained for example in Stanley [3.83]. The finite field method had its origins in the work of Crapo and Rota [3.22, §16] but was not applied systematically to computing characteristic polynomials until the work of C. A. Athanasiadis [3.1][3.2]. The theory of hyperplane arrangements has developed into a highly sophisticated subject with deep connections with topology, algebraic geometry, and so on. For a good overview, see the text of P. Orlik and H. Terao [3.53]. For an introduction to the combinatorial aspects of hyperplane arrangements, see the lecture notes of Stanley [3.83]. The Shi arrangement was first defined by Jian-Yi Shi (时偍益) [3.60][3.61], who computed the number of regions in connection with determining the left cells of the affine Weyl group of type $\tilde{A}_{n-1}$. The characteristic polynomial was first computed by P. Headley [3.41].

Zeta polynomials were introduced by Stanley [3.72, §3] and further developed by P. Edelman [3.26].

The idea of rank-selected subposets and the corresponding functions $\alpha_P(S)$ and $\beta_P(S)$ was considered for successively more general classes of posets by Stanley in [3.68, Ch. II][3.69][3.71], finally culminating in [3.76, §5]. Theorem 3.13.1 appeared (in a somewhat more general form) in [3.68, Thm. 9.1], while Theorem 3.13.3 appeared in [3.74, Thm. 3.1] (with $r = 1$).

R-labelings had a development parallel to that of rank-selection. The concept was successively generalized in [3.68][3.69][3.71], culminating this time in Björner [3.14] (from which the term "R-labeling" is taken) and Björner and Wachs [3.17]. Example 3.14.4 comes from [3.69], while Example 3.14.5 is found in [3.71]. A more stringent type of labeling than R-labeling, originally called *L-labeling* and now called *EL-labeling*, was introduced by Björner [3.14] and generalized to *CL-labeling* by Björner and Wachs [3.17]. (The definition of CL-labeling implicitly generalizes the notion of R-labeling to what logically should be called "CR-labeling.") A poset with a CL-labeling (originally, just with an EL-labeling) is called *lexicographically shellable*. While R-labelings are used (as in Section 3.14) to compute Euler characteristics (i.e., Möbius functions), CL-labelings allow one to compute the actual homology groups. From the many important examples, beginning with [3.17][3.19], of posets that can be proved to have a CL-labeling but not an EL-labeling, it seems clear that CL-labeling is the right level of generality for this subject. Note, however, that an even more general concept is due to D. Kozlov [3.47]. We have treated only R-labelings here for ease of presentation and because we are focusing on enumeration, not topology.

The theory of (P, ω)-partitions was foreshadowed by the work of MacMahon (see, for example, [1.55, §§439, 441]) and more explicitly Knuth [3.46], but the first general development appeared in Stanley [3.67][3.68]. Our treatment closely follows [3.68].

Eulerian posets were first explicitly defined in Stanley [3.77, p. 136], though they had certainly been considered earlier. A survey was given by Stanley [3.82]. In particular, Proposition 3.16.1 appears in [3.72, Prop. 3.3] (though stated less generally), while our approach to the Dehn-Sommerville equations (Theorem 3.16.9 in the case when $P - \{\hat{1}\}$ is Eulerian) appears in [3.72, p. 204]. Classically the Dehn-Sommerville equations were stated for face lattices of simplicial convex polytopes or triangulations of spheres (see [3.37, Ch. 9.8]); Klee [3.45] gives a treatment equivalent in generality to ours. A good general reference on polytopes is the book of Ziegler [3.97].

Lemma 3.16.3 and its generalization Lemma 3.16.4 are due independently to Baclawski [3.4, Lem. 4.6] and Stečkin [3.85]. A more general formula is given by Björner and Walker [3.20]. Proposition 3.16.5 and Corollary 3.16.6 appear in [3.77, Prop. 2.2]. Theorem 3.16.9 has an interesting history. It first arose when P is the lattice of faces of a rational convex polytope $\mathcal{P}$ as a byproduct of the computation of the intersection homology $IH(X(\mathcal{P}); \mathbb{C})$ of the toric variety $X(\mathcal{P})$ associated with $\mathcal{P}$. Specifically, setting $\beta_i = \dim IH_i(X(\mathcal{P}); \mathbb{C})$ one has

$$\sum_{i \geq 0} \beta_i q^i = f(P, q^2).$$

But intersection homology satisfies Poincaré duality, which implies $\beta_i = \beta_{2n-i}$. For references and further information, see Exercise 3.179. It was then natural to ask for a more elementary proof in the greatest possible generality, from which Theorem 3.16.9 arose. For further developments in this area, see Exercise 3.179.

The *cd*-index arose from the work of M. Bayer and L. Billera [3.5] on flag f-vectors of Eulerian posets. J. Fine (unpublished) observed that the linear relations obtained by Bayer and Billera are equivalent to the existence of the *cd*-index. Bayer and A. Klapper [3.6] wrote up the details of Fine's argument and developed some further properties of the *cd*-index. Stanley [1.70] gave the proof of the existence of the *cd*-index appearing here as the proof of Theorem 3.17.1 (with a slight

improvement due to G. Hetyei; see L. Billera and R. Ehrenborg, [3.10, p. 427])
and gave some additional results. An important breakthrough was later given by
K. Karu (Exercise 3.192).

The theory of binomial posets was developed by Doubilet, Rota, and Stanley
[3.25, §8]. Virtually all the material of Section 3.18 (some of it in a more general
form) can be found in this reference, with the exception of chromatic generating functions [3.70]. The generating function $(2 - e^x)^{-1}$ of Example 3.18.10 was
first considered by A. Cayley [3.21] in connection with his investigation of trees.
See also O. A. Gross [3.36]. The application of binomial posets to permutation
enumeration (Section 3.19) was developed by Stanley [3.74].

Among the many alternative theories to binomial posets for unifying various
aspects of enumerative combinatorics and generating functions, we mention the
theory of prefabs of Bender and Goldman [3.7], dissects of Henle [3.42], linked
sets of Gessel [3.31], and species of Joyal [3.44]. The most powerful of these theories is perhaps that of species, which is based on category theory. An exposition
is given by F. Bergeron, G. Labelle, and P. Leroux [3.8]. We should also mention
the book of Goulden and Jackson [3.32], which gives a fairly unified treatment
of a large part of enumerative combinatorics related to the counting of sequences
and paths.

Evacuation (Section 3.20) first arose in the theory of the RSK algorithm.
See pages 425–429 of Chapter 7, Appendix I, for this connection. Evacuation
was described by M.-P. Schützenberger [3.62] in a direct way not involving the
RSK algorithm. In two follow-up papers [3.63][3.64], Schützenberger extended
the definition of evacuation to linear extensions of any finite poset and developed the connection with promotion. Schützenberger's work was simplified by
Haiman [3.38] and Malvenuto and Reutenauer [3.51]. A survey of promotion and
evacuation with many additional references was given by Stanley [3.84].

Differential posets were discovered independently by S. Fomin [3.29][3.30]
and Stanley [3.80] [3.81]. Fomin's work goes back to his M.S. thesis [3.28] and
is done in the more general context of "dual graded graphs" (essentially where
the U and D operators act on different posets). Our exposition follows [3.80],
where many further results may be found. Theorem 3.21.6(b) is based on a suggestion of Yan Zhang. Generalizations of differential posets in addition to dual
graded graphs include sequentially differential posets [3.81, §2], weighted differential posets (an example appearing in [3.81, §3]), down-up algebras [3.9], signed
differential posets [3.49], quantized dual graded graphs [3.50], and the updown
categories of M. E. Hoffman [3.43].

Bibliography

[1] C. A. Athanasiadis, Algebraic combinatorics of graph spectra, subspace arrangements,
    and Tutte polynomials, Ph.D. thesis, M.I.T., 1996.
[2] C. A. Athanasiadis, Characteristic polynomials of subspace arrangements and finite
    fields, *Advances in Math.* **122** (1996), 193–233.

[3] K. Baclawski, Cohen-Macaulay ordered sets, *J. Algebra* **63** (1980), 226–258.

[4] K. Baclawski, Cohen-Macaulay connectivity and geometric lattices, *Europ. J. Combinatorics* **3** (1984), 293–305.

[5] M. M. Bayer and L. J. Billera, Generalized Dehn-Sommerville relations for polytopes, spheres and Eulerian partially ordered sets, *Invent. Math.* **79** (1985), 143–157.

[6] M. M. Bayer and A. Klapper, A new index for polytopes, *Discrete Comput. Geom.* **6** (1991), 33–47.

[7] E. A. Bender and J. R. Goldman, Enumerative uses of generating functions, *Indiana Univ. Math. J.* **20** (1971), 753–765.

[8] F. Bergeron, G. Labelle, and P. Leroux, *Combinatorial Species and Tree-like Structures*, Encyclopedia of Mathematics and Its Applications **67**, Cambridge University Press, Cambridge, 1998.

[9] G. Benkart and T. Roby, Down-up algebras, *J. Algebra* **209** (1998), 305–344.

[10] L. J. Billera and R. Ehrenborg, Monotonicity properties of the *cd*-index for polytopes, *Math. Z.* **233** (2000), 421–441.

[11] G. Birkhoff, On the combination of subalgebras, *Proc. Camb. Phil. Soc.* **29** (1933), 441–464.

[12] G. Birkhoff, Von Neumann and latttice theory, *Bull. Amer. Math.* **64** (1958), 50–56.

[13] G. Birkhoff, *Lattice Theory*, 3rd ed., American Mathematical Society, Providence, R.I., 1967.

[14] A. Björner, Shellable and Cohen-Macaulay partially ordered sets, *Trans. Amer. Math. Soc.* **260** (1980), 159–183.

[15] A. Björner, Posets, regular CW complexes and Bruhat order, *Europ. J. Combinatorics* **5** (1984), 7–16.

[16] A. Björner, A. Garsia, and R. Stanley, An introduction to the theory of Cohen-Macaulay partially ordered sets, in *Ordered Sets* (I. Rival, ed.), Reidel, Dordrecht/Boston/-London, 1982, pp. 583–615.

[17] A. Björner and M. L. Wachs, Bruhat order of Coxeter groups and shellability, *Advances in Math.* **43** (1982), 87–100.

[18] A. Björner, Topological methods, in *Handbook of Combinatorics* (R. Graham, M. Grötschel, and L. Lovász, eds.), North-Holland, Amsterdam, 1995, pp. 1819–1872.

[19] A. Björner and M. L. Wachs, On lexicographically shellable posets, *Trans. Amer. Math. Soc.* **277** (1983), 323–341.

[20] A. Björner and J. W. Walker, A homotopy complementation formula for partially ordered sets, *Europ. J. Combinatorics* **4** (1983), 11–19.

[21] A. Cayley, On the theory of the analytical forms called trees, *Phil. Mag.* **18** (1859), 374–378.

[22] H. Crapo and G.-C. Rota, *On the Foundations of Combinatorial Theory: Combinatorial Geometries*, preliminary edition, M.I.T. Press, Cambridge, Mass., 1970.

[23] R. J. Daverman, *Decompositions of Manifolds*, American Mathematical Society, Providence, R.I. 2007.

[24] B. A. Davey and H. A. Priestly, Distributive lattices and duality, Appendix B of [3.33], pp. 499–517.

[25] P. Doubilet, G.-C. Rota, and R. Stanley, On the foundations of combinatorial theory (VI): The idea of generating function, in *Sixth Berkeley Symposium on Mathematical Statistics and Probability*, Vol. II: *Probability Theory*, University of California, Berkeley and Los Angeles, 1972, pp. 267–318.

[26] P. H. Edelman, Zeta polynomials and the Möbius function, *Europ. J. Combinatorics* **1** (1980), 335–340.

[27] F. D. Farmer, Cellular homology for posets, *Math. Japonica* **23** (1979), 607–613.

[28] S. V. Fomin, Two-dimensional growth in Dedekind lattices, M.S. thesis, Leningrad State University, 1979.

[29] S. Fomin, Duality of graded graphs, *J. Algebraic Combinatorics* **3** (1994), 357–404.

[30] S. Fomin, Schensted algorithms for dual graded graphs, *J. Algebraic Combinatorics* **4** (1995), 5–45.

[31] I. M. Gessel, Generating functions and enumeration of sequences, thesis, M.I.T., 1977.

[32] I. P. Goulden and D. M. Jackson, *Combinatorial Enumeration*, John Wiley, New York, 1983; reissued by Dover, New York, 2004.

[33] G. Grätzer, *General Lattice Theory: Vol. 1: The Foundation*, 2nd ed., Birkhäuser, Basel/Boston/Berlin, 2003.

[34] C. Greene, On the Möbius algebra of a partially ordered set, *Advances in Math.* **10** (1973), 177–187.

[35] C. Greene, The Möbius function of a partially ordered set, in [3.57], pp. 555–581.

[36] O. A. Gross, Preferential arrangements, *Amer. Math. Monthly* **69** (1962), 4–8.

[37] B. Grünbaum, *Convex Polytopes*, 2nd ed., Springer-Verlag, New York, 2003.

[38] M. D. Haiman, Dual equivalence with applications, including a conjecture of Proctor, *Discrete Math.* **99** (1992), 79–113.

[39] P. Hall, The Eulerian functions of a group, *Quart. J. Math.* **7** (1936), 134–151.

[40] I. Halperin, A survey of John von Neumann's books on continuous geometry, *Order* **1** (1985), 301–305.

[41] P. Headley, On a family of hyperplane arrangements related to affine Weyl groups, *J. Algebraic Combinatorics* **6** (1997), 331–338.

[42] M. Henle, Dissection of generating functions, *Studies in Applied Math.* **51** (1972), 397–410.

[43] M. E. Hoffman, Updown categories, preprint; math.CO/0402450.

[44] A. Joyal, Une théorie combinatoire des séries formelles, *Advances in Math.* **42** (1981), 1–82.

[45] V. Klee, A combinatorial analogue of Poincaré's duality theorem, *Canad. J. Math.* **16** (1964), 517–531.

[46] D. E. Knuth, A note on solid partitions, *Math. Comp.* **24** (1970), 955–962.

[47] D. N. Kozlov, General lexicographic shellability and orbit arrangements, *Ann. Comb.* **1** (1997), 67–90.

[48] D. N. Kozlov, *Combinatorial Algebraic Topology*, Springer, Berlin, 2008.

[49] T. Lam, Signed differential posets and sign-imbalance, *J. Combinatorial Theory Ser. A* **115** (2008), 466–484.

[50] T. Lam, Quantized dual graded graphs, *Electronic J. Combinatorics* **17**(1) (2010), R88.

[51] C. Malvenuto and C. Reutenauer, Evacuation of labelled graphs, *Discrete Math.* **132** (1994), 137–143.

[52] E. Miller and B. Sturmfels, *Combinatorial Commutative Algebra*, Springer, New York, 2005.

[53] P. Orlik and H. Terao, *Arrangements of Hyperplanes*, Springer-Verlag, Berlin/Heidelberg, 1992.

[54] J. G. Oxley, *Matroid Theory*, Oxford University Press, New York, 1992.

[55] H. A. Priestley, Representation of distributive lattices by means of ordered Stone spaces, *Bull. London Math. Soc.* **2** (1970), 186–190.

[56] H. A. Priestley, Ordered topological spaces and the representation of distributive lattices, *Proc. London Math. Soc. (3)* **24** (1972), 507–530.

[57] I. Rival (ed.), *Ordered Sets*, Reidel, Dordrecht/Boston, 1982.

[58] G.-C. Rota, On the foundations of combinatorial theory I. Theory of Möbius functions, *Z. Wahrscheinlichkeitstheorie* **2** (1964), 340–368.

[59] L. Schläfli, Theorie der vielfachen Kontinuität, in *Neue Denkschrifter den allgemeinen schweizerischen Gesellschaft für die gesamten Naturwissenschaften*, vol. 38, IV, Zürich, 1901; *Ges. Math. Abh.*, vol. 1, Birkhaäuser, Basel, 1950, p. 209.

[60] J.-Y. Shi, *The Kazhdan-Lusztig Cells in Certain Affine Weyl Groups*, Lecture Notes in Mathematics, no. 1179, Springer-Verlag, Berlin/Heidelberg/New York, 1986.

[61] J.-Y. Shi, Sign types corresponding to an affine Weyl group, *J. London Math. Soc.* **35** (1987), 56–74.

[62] M.-P. Schützenberger, Quelques remarques sur une construction de Schensted, *Canad. J. Math.* **13** (1961), 117–128.

[63] M.-P. Schützenberger, Promotions des morphismes d'ensembles ordonnés, *Discrete Math.* **2** (1972), 73–94.

[64] M.-P. Schützenberger, Evacuations, in *Colloquio Internazionale sulle Teorie Combinatorie (Rome, 1973)*, Tomo I, Atti dei Convegni Lincei, No. 17, Accad. Naz. Lincei, Rome, 1976, pp. 257–264.

[65] L. Solomon, The Burnside algebra of a finite group, *J. Combinatorial Theory* **2** (1967), 603–615.

[66] E. Spiegel and C. O'Donnell, *Incidence Algebras*, Marcel Dekker, New York, 1997.

[67] R. Stanley, *Ordered structures and partitions*, thesis, Harvard Univ., 1971.

[68] R. Stanley, Ordered structures and partitions, *Memoirs Amer. Math. Soc.*, no. 119 (1972).

[69] R. Stanley, Supersolvable lattices, *Alg. Univ.* **2** (1972), 197–217.

[70] R. Stanley, Acyclic orientations of graphs, *Discrete Math.* **5** (1973), 171–178.

[71] R. Stanley, Finite lattices and Jordan-Hölder sets, *Alg. Univ.* **4** (1974), 361–371.

[72] R. Stanley, Combinatorial reciprocity theorems, *Advances in Math.* **14** (1974), 194–253.

[73] R. Stanley, The Fibonacci lattice, *Fib. Quart.* **13** (1975), 215–232.

[74] R. Stanley, Binomial posets, Möbius inversion, and permutation enumeration, *J. Combinatorial Theory Ser. A* **20** (1976), 336–356.

[75] R. Stanley, Cohen-Macaulay complexes, in *Higher Combinatorics* (M. Aigner, ed.), Reidel, Dordrecht/Boston, 1977, pp. 51–62.

[76] R. Stanley, Balanced Cohen-Macaulay complexes, *Trans. Amer. Math. Soc.* **249** (1979), 139–157.

[77] R. Stanley, Some aspects of groups acting on finite posets, *J. Combinatorial Theory Ser. A* **32** (1982), 132–161.

[78] R. Stanley, *Combinatorics and Commutative Algebra*, Progress in Mathematics, vol. 41, Birkhäuser, Boston/Basel/Stuttgart, 1983; 2nd ed., Boston/Basel/Berlin, 1996.

[79] R. Stanley, Generalized h-vectors, intersection cohomology of toric varieties, and related results, in *Commutative Algebra and Combinatorics* (M. Nagata and H. Matsumura, eds.), Advanced Studies in Pure Mathematics **11**, Kinokuniya, Tokyo, and North-Holland, Amsterdam/New York, 1987, pp. 187–213.

[80] R. Stanley, Differential posets, *J. Amer. Math. Soc.* **1** (1988), 919–961.

[81] R. Stanley, Variations on differential posets, in *Invariant Theory and Tableaux* (D. Stanton, ed.), The IMA Volumes in Mathematics and Its Applications, vol. 19, Springer-Verlag, New York, 1990, pp. 145–165.

[82] R. Stanley, A survey of Eulerian posets, in *Polytopes: Abstract, Convex, and Computational* (T. Bisztriczky, P. McMullen, R. Schneider, A. I. Weiss, eds.), NATO ASI Series C, vol. 440, Kluwer Academic Publishers, Dordrecht/Boston/London, 1994, pp. 301–333.

[83] R. Stanley, An introduction to hyperplane arrangements, in *Geometric Combinatorics* (E. Miller, V. Reiner, and B. Sturmfels, eds.), IAS/Park City Mathematics Series, vol. 13, American Mathematical Society, Providence, R.I., 2007, pp. 389–496.

[84] R. Stanley, Promotion and evacuation, *Electronic J. Combinatorics* **15**(2) (2008–2009), R9.

[85] B. S. Stečkin, Imbedding theorems for Möbius functions, *Soviet Math. Dokl.* **24** (1981), 232–235 (translated from **260** (1981)).

[86] M. H. Stone, Applications of the theory of boolean rings to general topology, *Trans. Amer. Math. Soc.* **41** (1937), 375–481.

[87] J. Walker, Homotopy type and Euler characteristic of partially ordered sets, *European J. Combinatorics* **2** (1981), 373–384.

[88] J. Walker, Topology and combinatorics of ordered sets, Ph.D. thesis, M.I.T., 1981.

[89] D. J. A. Welsh, *Matroid Theory*, Academic Press, London, 1976; reprinted by Dover, New York, 2010.

[90] N. White (ed.), *Theory of Matroids*, Cambridge University Press, Cambridge, 1986.

[91] N. White (ed.), *Combinatorial Geometries*, Cambridge University Press, Cambridge, 1987.

[92] N. White (ed.), *Matroid Applications*, Cambridge University Press, Cambridge, 1992.

[93] H. Whitney, A logical expansion in mathematics, *Bull. Amer. Math. Soc.* **38** (1932), 572–579.

[94] H. Whitney, The coloring of graphs, *Ann. Math.* **33**(2) (1932), 688–718.

[95] T. Zaslavsky, Counting the faces of cut-up spaces, *Bull. Amer. Math. Soc.* **81** (1975), 916–918.

[96] T. Zaslavsky, Facing up to arrangements: Face-count formulas for partitions of space by hyperplanes, *Mem. Amer. Math. Soc.* **1** (1975), no. 154, vii+102 pages.

[97] G. M. Ziegler, *Lectures on Polytopes*, Springer, New York, 1995.

Exercises for Chapter 3

1.   [3] What is the connection between a partially ordered set and itinerant salespersons who take revenge on customers who don't pay their bills?

2.   **a.** [1+] A *preposet* (or *quasi-ordered set*) is a set P with a binary relation $\leq$ satisfying reflexivity and transitivity (but not necessary antisymmetry). Given a preposet P and $s, t \in P$, define $s \sim t$ if $s \leq t$ and $t \leq s$. Show that $\sim$ is an equivalence relation.

  b. [1+] Let $\widetilde{P}$ denote the set of equivalence classes under $\sim$. If $S, T \in \widetilde{P}$, then define $S \leq T$ if there is an $s \in S$ and $t \in T$ for which $s \leq t$ in P. Show that this definition of $\leq$ makes $\widetilde{P}$ into a poset.

  c. [2–] Let Q be a poset and $f \colon P \to Q$ order-preserving. Show that there is a unique order-preserving map $g \colon \widetilde{P} \to Q$ such that the following diagram commutes:

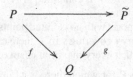

$$P \longrightarrow \widetilde{P}$$

Here the map $P \to \widetilde{P}$ is the canonical map taking t into the equivalence class containing t.

3.   **a.** [1+] Let P be a finite preposet (as defined in Exercise 3.2). Define a subset U of P to be *open* if U is an order ideal (defined in an obvious way for preposets) of P. Show that P becomes a finite topological space, denoted P_{top}.

  b. [2–] Given a finite topological space X, show that there is a unique preposet P (up to isomorphism) for which $P_{\text{top}} = X$. Hence, the correspondence $P \to P_{\text{top}}$ is a bijection between finite preposets and finite topologies.

  c. [2–] Show that the preposet P is a poset if and only if P_{top} is a T_0-space (i.e., distinct points have distinct sets of neighborhoods).

  d. [2–] Show that a map $f \colon P \to Q$ of preposets is order-preserving if and only if f is continuous when regarded as a map $P_{\text{top}} \to Q_{\text{top}}$.

4.   [2–] Let P be a poset. Show that there exists a collection S of sets such that if we partially order S by defining $S \leq T$ if $S \subseteq T$, then $S \cong P$.

5.   **a.** [2] Draw diagrams of the 63 five-element posets (up to isomorphism), 318 six-element posets, and 2045 seven-element posets (straightforward, but time consuming). For readers with a lot of spare time on their hands, continue with eight-element posets, nine-element posets, and so on, obtaining

the numbers 16999, 183231, 2567284, 46749427, 1104891746, 33823827452, 1338193159771, 68275077901156, and 4483130665195087.

b. [5] Let $f(n)$ be the number of nonisomorphic n-element posets. Find a "reasonable" formula for $f(n)$ (probably impossible, and similarly for the case of *labeled posets*, i.e., posets on the vertex set $[n]$).

c. [5] With $f(n)$ as previously, let $\mathcal{P}$ denote the statement that infinitely many values of $f(n)$ are palindromes when written in base 10. Show that $\mathcal{P}$ cannot be proved or disproved in Zermelo–Fraenkel set theory.

d. [3] Show that
$$\log f(n) \sim (n^2/4)\log 2.$$

e. [3+] Improve (d) by showing
$$f(n) \sim C 2^{n^2/4+3n/2} e^n n^{-n-1},$$
where C is a constant given by
$$C = \frac{2}{\pi}\sum_{i\geq 0} 2^{-i(i+1)} \approx 0.80587793 \quad (n \text{ even}),$$
and similarly for n odd.

6.   **a.** [2] Let P be a finite poset and $f\colon P \to P$ an order-preserving bijection. Show that f is an automorphism of P (i.e., f^{-1} is order-preserving).

   b. [2] Show that (a) fails for infinite P.

7.   **a.** [1+] Give an example of a finite poset P such that if ℓ is the length of the longest chain of P, then every $t \in P$ is contained in a chain of length ℓ, yet P has a maximal chain of length less than ℓ.

   b. [2] Let P be a finite poset with no isolated points and with longest chain of length ℓ. Assume that for every t covering s in P there exists a chain of length ℓ containing both s and t. Show that every maximal chain of P has length ℓ.

8.   [3–] Find a finite poset P for which there is a bijection $f\colon P \to P$ such that $s \leq t$ if and only if $f(s) \geq f(t)$ (i.e., P is self-dual), but for which there is *no* such bijection f satisfying $f(f(t)) = t$ for all $t \in P$.

9.   [2–] True or false: The number of nonisomorphic 8-element posets that are not self-dual is 16507.

10. **a.** [2–]* If P is a poset, then let $\mathrm{Int}(P)$ denote the poset of (nonempty) intervals of P, ordered by inclusion. Show that for any posets A and B, we have $\mathrm{Int}(A \times B) \cong \mathrm{Int}(A \times B^*)$.

   b. [2+] Let P and Q be posets. If P has a $\hat{0}$ and $\mathrm{Int}(P) \cong \mathrm{Int}(Q)$, show that $P \cong A \times B$ and $Q \cong A \times B^*$ for some posets A and B.

   c. [3] Find finite posets P, Q such that $\mathrm{Int}(P) \cong \mathrm{Int}(Q)$, yet the conclusion of (b) fails.

11. **a.** [2] Let A be the set of all isomorphism classes of finite posets. Let $[P]$ denote the class of the poset P. Then A has defined on it the operations $+$ and $\cdot$ given by $[P]+[Q] = [P+Q]$ and $P \cdot Q = [P \times Q]$. Show that these operations make A into a commutative semiring (i.e., A satisfies all the axioms of a commutative ring except the existence of additive inverses).

   b. [3–] We can formally adjoin additive inverses to A in an obvious way to obtain a ring B (exactly the same way as one obtains $\mathbb{Z}$ from $\mathbb{N}$). Define a poset to be *irreducible* if it cannot be written in a nontrivial way as a direct product. Show that B is just the polynomial ring $\mathbb{Z}[[P_1],[P_2],\dots]$ where the $[P_i]$'s are the classes of irreducible connected finite posets with more than one element. (The additive identity of B is given by the class of the empty poset, and the multiplicative identity by the class of the one-element poset.)

c. [3–] Find irreducible finite posets P_i satisfying $P_1 \times P_2 \cong P_3 \times P_4$, yet $P_1 \not\cong P_3$ and $P_1 \not\cong P_4$. Why does this not contradict the known fact that $\mathbb{Z}[x_1, x_2, \ldots]$ is a unique factorization domain?

12. [2+] True or false: If every chain and every antichain of a poset P is finite, then P is finite.

13. **a.** [3] Let P be a poset for which every antichain is finite. Show that every antichain of $J_f(P)$ is finite.

   b. [2] Show that if every antichain of P is finite, it need not be the case that every antichain of $J(P)$ is finite.

14. A finite poset P is a *series-parallel poset* if it can be built up from a one-element poset using the operations of disjoint union and ordinal sum. There is a unique four-element poset (up to isomorphism) that is not series-parallel, namely, the zigzag poset Z_4 of Exercise 3.66.

   a. [2+]* Show that a finite poset P is series-parallel if and only if it contains no induced subposet isomorphic to Z_4. Such posets are sometimes called *N-free posets*.

   b. [2+]* Let P_w be the inversion poset of the permutation $w \in \mathfrak{S}_n$, as defined in the solution to Exercise 3.20. Show that P_w is N-free if and only if w is 3142-avoiding and 2413-avoiding. Such permutations are also called *separable*. NOTE. The number of separable permutations in $\mathfrak{S}_n$ is the Schröder number r_{n-1}, as defined in Vol. II, Section 6.2.

15. An *interval order* is a poset P isomorphic to a set of closed intervals of $\mathbb{R}$, with $[a, b] < [c, d]$ if $b < c$.

   a. [3–] Show that a finite poset P is an interval order if and only if it is $(\mathbf{2} + \mathbf{2})$-free (i.e., has no induced subposet isomorphic to $\mathbf{2} + \mathbf{2}$).

   b. [3–] A poset P is a *semiorder* (or *unit interval order*) if it is an interval order corresponding to a set of intervals all of length one. Show that an interval order P is a semiorder if and only if P is $(\mathbf{3} + \mathbf{1})$-free. (For the enumeration of labeled and unlabeled semiorders, see Exercises 6.30 and 6.19(ddd), Vol. II, respectively.)

   c. [3] Let $t(n)$ be the number of nonisomorphic interval orders with n elements. Show that

$$\sum_{n \geq 0} t(n) x^n = \sum_{n \geq 0} \prod_{k=1}^{n} (1 - (1-x)^k)$$

$$= 1 + x + 2x^2 + 5x^3 + 15x^4 + 53x^5 + 217x^6 + \cdots.$$

   d. [3] Let $u(n)$ be the number of labeled n-element interval orders (i.e., interval orders on the set $[n]$). Show that

$$\sum_{n \geq 0} u(n) \frac{x^n}{n!} = \sum_{n \geq 0} \prod_{k=1}^{n} (1 - e^{-kx})$$

$$= 1 + x + 3\frac{x^2}{2!} + 19\frac{x^3}{3!} + 207\frac{x^4}{4!} + 3451\frac{x^5}{5!} + 81663\frac{x^6}{6!} + \cdots.$$

   e. [2+] Show that the number of nonisomorphic n-element interval orders that are also series-parallel posets (defined in Exercise 3.14) is the Catalan number C_n.

f. [2]* Let $\ell_1,\ldots,\ell_n$ be positive real numbers. Let $g(\ell_1,\ldots,\ell_n)$ be the number of interval orders P that can be formed from intervals $I_1,\ldots,I_n$, where I_i has length ℓ_i, such that the element t of P corresponding to I_i is labeled i. For instance, if $n=4$ and P is isomorphic to $\mathbf{3}+\mathbf{1}$, then $\mathbf{3}$ can be labeled a,b,c from bottom to top and $\mathbf{1}$ labeled d if and only if $\ell_d \geq \ell_b$ (12 labelings in all if l_a, l_b, l_c, l_d are distinct). Show that $g(\ell_1,\ldots,\ell_n)$ is equal to the number of regions of the real hyperplane arrangement

$$x_i - x_j = \ell_i, \ \ i \neq j$$

$(n(n-1)$ hyperplanes in all).

g. [3] Suppose that $\ell_1,\ldots,\ell_n$ in (f) are linearly independent over $\mathbb{Q}$. Define a power series

$$y = 1 + x + 5\frac{x^2}{2!} + 46\frac{x^3}{3!} + 631\frac{x^4}{4!} + 9655\frac{x^5}{5!} + \cdots$$

by the equation

$$1 = y(2 - e^{xy}),$$

or equivalently

$$y - 1 = \left(\frac{1}{1+x}\log\frac{1+2x}{1+x}\right)^{\langle-1\rangle},$$

where $\langle-1\rangle$ denotes compositional inverse. Let

$$z = \sum_{n\geq 0} g(\ell_1,\ldots,\ell_n)\frac{x^n}{n!}$$

$$= 1 + x + 3\frac{x^2}{2!} + 19\frac{x^3}{3!} + 195\frac{x^4}{4!} + 2831\frac{x^5}{5!} + 53703\frac{x^6}{6!} + \cdots.$$

Show that z is the unique power series satisfying

$$\frac{z'}{z} = y^2, \ \ z(0) = 1.$$

Note that it is by no means a priori obvious that $g(\ell_1,\ldots,\ell_n)$ is independent of $\ell_1,\ldots,\ell_n$ (provided they are linearly independent over $\mathbb{Q}$).

16. a. [3] Let $f(n)$ be the number of graded $(\mathbf{3}+\mathbf{1})$-free partial orderings of an n-element set. Set

$$G(x) = \sum_{m,n\geq 0} 2^{mn}\frac{x^{m+n}}{m!n!}.$$

Show that

$$\sum_{n\geq 0} f(n)\frac{x^n}{n!} = \frac{e^{2x}(2e^x - 3) + e^x(e^x - 2)^2 G(x)}{e^x(1+2e^x) + (e^{2x} - 2e^x - 1)G(x)}$$

$$= 1 + x + 3\frac{x^2}{2!} + 13\frac{x^3}{3!} + 111\frac{x^4}{4!} + 1381\frac{x^5}{5!} + 22383\frac{x^6}{6!} + \cdots.$$

b. [5] What can be said about the *total* number of $(\mathbf{3}+\mathbf{1})$-free posets on an n-element set?

17. a. [3–] Let $\mathcal{S}$ be a collection of finite posets, all of whose automorphism groups are trivial. Let $\mathcal{T}$ be the set of all nonisomorphic posets that can be obtained by replacing each element t of some $P \in \mathcal{S}$ with a finite nonempty antichain A_t. (Thus if t

covers s in P, then each $t' \in A_t$ covers each $s' \in A_s$.) Let $f(n)$ be the number of nonisomorphic n-element posets in $\mathcal{T}$. Let $g(n)$ be the number of posets on the set $[n]$ that are isomorphic to some poset in $\mathcal{T}$. Set

$$F(x) = \sum_{n \geq 0} f(n)x^n, \quad G(x) = \sum_{n \geq 0} g(n)\frac{x^n}{n!}.$$

Show that $G(x) = F(1 - e^{-x})$.

b. [2] What are $F(x)$ and $G(x)$ when $\mathcal{S} = \{\mathbf{1}, \mathbf{2}, \dots\}$, where i denotes an i-element chain?

c. [2+] Show that we can take $\mathcal{S}$ to consist of all interval orders (respectively, all semiorders) with no nontrivial automorphisms. Then $\mathcal{T}$ consists of all nonisomorphic interval orders (respectively, semiorders). Formulas for $F(x)$ and $G(x)$ appear in Exercises 3.15(c,d): in Vol. II, Exercises 6.30, and 6.19(ddd).

d. [3–]* Show that the number of nonisomorphic n-element graded semiorders is $1 + F_{2n-2}$, where F_{2n-2} is a Fibonacci number.

18. [3] Let P be a finite $(\mathbf{3}+\mathbf{1})$-free poset. Let c_i denote the number of i-element chains of P (with $c_0 = 1$). Show that all the zeros of the polynomial $C(P, x) = \sum_i c_i x^i$ are real.

19. **a.** [3–] An element t of a finite poset P is called *irreducible* if t covers exactly one element or is covered by exactly one element of P. A subposet Q of P is called a *core* of P, written $Q = \text{core}\, P$, if
  i. one can write $P = Q \cup \{t_1, \dots, t_k\}$ such that t_i is an irreducible element of $Q \cup \{t_1, \dots, t_i\}$ for $1 \leq i \leq k$, and
  ii. Q has no irreducible elements.
  Show that any two cores of P are isomorphic (though they need not be equal). Hence, the notation $\text{core}\, P$ determines a unique poset up to isomorphism.

b. [1+]* If P has a $\hat{0}$ or $\hat{1}$, then show that $\text{core}\, P$ consists of a single element.

c. [3–] Show that $\#(\text{core}\, P) = 1$ if and only if the poset P^P of order-preserving maps $f : P \to P$ is connected. (Such posets are called *dismantlable*.)

d. [5–] Is it possible to enumerate nonisomorphic n-element dismantlable posets or dismantlable posets on an n-element set?

20. [2+]
a. Let $d \in \mathbb{P}$. Show that the following two conditions on a finite poset P are equivalent:
  i. P is the intersection of d linear orderings of $[n]$, where $\#P = n$.
  ii. P is isomorphic to a subposet of $\mathbb{N}^d$.
b. Moreover, show that when $d = 2$ the two conditions are also equivalent to:
  iii. There exists a poset Q on $[n]$ such that $s < t$ or $s > t$ in Q if and only if s and t are incomparable in P.

21. [3+] A finite poset $P = \{t_1, \dots, t_n\}$ is a *sphere order* if for some $d \geq 1$ there exist $(d-1)$-dimensional spheres $S_1, \dots, S_n$ in $\mathbb{R}^d$ such that S_i is inside S_j if and only if $t_i < t_j$. *Prove or disprove:* Every finite poset is a sphere order.

22. [2+] Let P be a poset with elements $t_1, \dots, t_p$, which we regard as indeterminates. Define a $p \times p$ matrix A by

$$A_{ij} = \begin{cases} 0, & \text{if } t_i < t_j, \\ 1, & \text{otherwise.} \end{cases}$$

Define the diagonal matrix $D = \text{diag}(t_1, \dots, t_p)$, and let I denote the $p \times p$ identity matrix. Show that

$$\det(I + DA) = \sum_C \prod_{t_i \in C} t_i,$$

where C ranges over all chains in P.

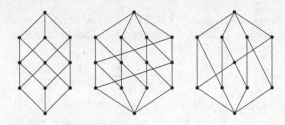

Figure 3.42 Which are lattices?

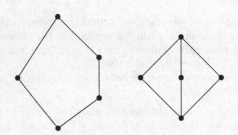

Figure 3.43 Obstructions to distributivity and modularity.

23. [2+] Show that the boolean algebra $B_{\mathbb{P}}$ of *all* subsets of $\mathbb{P}$, ordered by inclusion, contains both countable and uncountable maximal chains.

24. [2+] Let $n \geq 5$. Show that up to isomorphism there is one n-element poset with 2^n order ideals, one with $(3/4)2^n$ order ideals, two with $(5/8)2^n$ order ideals, three with $(9/16)2^n$, and two with $(17/32)2^n$. Show also that all other n-element posets have at most $(33/64)2^n$ order ideals.

25. [2–] Which of the posets of Figure 3.42 are lattices?

26. [2] Give an example of a meet-semilattice with $\hat{1}$ (necessary infinite) that is not a lattice.

27. [3–] Let L be a finite lattice, and define the subposet $\mathrm{Irr}(L)$ of irreducibles of L by

$$\mathrm{Irr}(L) = \{x \in L : x \text{ is join-irreducible or meet-irreducible (or both)}\}.$$

Show that L can be uniquely recovered from the poset $\mathrm{Irr}(L)$.

28. [3–] Give an example of a finite atomic and coatomic lattice that is not complemented.

29. [5–] A finite lattice L has n join-irreducibles. What is the most number $f(n)$ of meet-irreducible elements L can have?

30. a. [2+] Show that a lattice is distributive if and only if it does not contain a sublattice isomorphic to either of the two lattices of Figure 3.43.
   b. [2+] Show that a lattice is modular if and only if it does not contain a sublattice isomorphic to the first lattice of Figure 3.43.

31. a. [2+]* A poset is called *locally connected* if every nonempty open interval (s,t) is either an antichain or is connected. Show that a finite locally connected poset with $\hat{0}$ and $\hat{1}$ is graded.
   b. [3] Let L be a finite locally connected lattice for which every interval of rank 3 is a distributive lattice. Show that L is a distributive lattice.
   c. [2–] Deduce from (b) that if L is a finite locally connected lattice for which every interval of rank 3 is a product of chains, then L is a product of chains.
   d. [2–] Deduce from (b) that if L is a finite locally connected lattice for which every interval of rank 3 is a boolean algebra, then L is a boolean algebra.

32. **a.** [3–] For a finite graded poset P with $\hat{0}$ and with rank function ρ, let $f(P)$ be the largest integer d such that there exists a partition of $P - \{\hat{0}\}$ into (pairwise disjoint) closed intervals $[s,t]$ satisfying $\rho(t) \geq d$. Find $f(B_n)$. For instance, $f(B_3) = 2$, corresponding to the partition

$$\pi = \{[1,12],[2,23],[3,13],[123,123]\}.$$

b. [3–] Show that $f(k^n) = (k-1)f(B_n)$.
c. [2] Let $a \leq b$. Show that $f(a \times b) = b$.
d. [2+] Let $a \leq b \leq c$. Show that $f(a \times b \times c) = \max\{a+b,c\}$.
e. [3–] Let $a \leq b \leq c \leq d$. Show that

$$f(a \times b \times c \times d) = \max\{d, \min\{b+d, a+b+c\}\}.$$

f. [3] Find $f(a \times b \times c \times d \times e)$.

33. [2+] Characterize all positive integers n for which there exists a connected poset with exactly n chains (including the empty chain). The empty poset is not considered to be connected.

34. [2]* Find all nonisomorphic posets P such that

$$F(J(P),x) = (1+x)(1+x^2)(1+x+x^2).$$

35. **a.** [2] Let $f_k(n)$ be the number of nonisomorphic n-element posets P such that if $1 \leq i \leq n-1$, then P has exactly k order ideals of cardinality i. Show that $f_2(n) = 2^{n-3}$, $n \geq 3$.
b. [2+] Let $g(n)$ be the number of those posets enumerated by $f_3(n)$ with the additional property that the only 3-element antichains of P consist of the three minimal elements and three maximal elements of P. Show that $g(n) = 2^{n-7}$, $n \geq 7$.
c. [3] Show that

$$\sum_{n\geq 0} f_3(n)x^n = \frac{x^3 - x^4 - x^5 - xg(x) - x^2 g(x)}{1 - 2x - 3x^2}$$

$$= x^3 + x^5 + x^6 + 3x^7 + 6x^8 + 16x^9 + 39x^{10} + \cdots,$$

where

$$g(x) = \frac{x^3 - 2x^4 - x^5 - x^6}{1 - 2x - 2x^2 + 2x^4 + 3x^5}.$$

d. [1+] Find $f_k(n)$ for $k > 3$.

36. **a.** [2] Let L be a finite semimodular lattice. Let L' be the subposet of L consisting of elements of L that are joins of atoms of L (including $\hat{0}$ as the empty join). Show that L' is a geometric lattice.
b. [3–] Is L' a sublattice of L?

37. **a.** [3–] Let W be a subspace of the vector space K^n, where K is a field of characteristic 0. The *support* of a vector $v = (v_1, \ldots, v_n) \in K^n$ is given by $\mathrm{supp}(v) = \{i : v_i \neq 0\}$. Let L denote the set of supports of all vectors in W, ordered by reverse inclusion. Show that L is a geometric lattice.
b. [2+] An *isthmus* of a graph H is an edge e of H whose removal disconnects the component to which e belongs. Let G be a finite graph, allowing loops and multiple

edges. Let D_G be the set of all spanning subgraphs of G that do not have an isthmus, ordered by reverse edge inclusion. Use (a) to show that D_G is a geometric lattice.

38. [2+] Let $k \in \mathbb{N}$. In a finite distributive lattice L, let P_k be the subposet of elements that cover k elements, and let R_k be the subposet of elements that are covered by k elements. Show that $P_k \cong R_k$, and describe in terms of the structure of L an explicit isomorphism $\phi \colon P_k \to R_k$.

39. [2+]* Find all finitary distributive lattices L (up to isomorphism) such that $L \cong V_t$ for all $t \in L$. If we only require that L is a locally finite distributive lattice with $\hat{0}$, are there other examples?

40. **a.** [3−] Let L be a finite distributive lattice of length kr that contains k join-irreducibles of rank i for $1 \le i \le r$ (and therefore no other join-irreducibles). What is the most number of elements that L can have? Show that the lattice L achieving this number of elements is unique (up to isomorphism).

b. [2+] Let L be a finitary distributive lattice with exactly two join-irreducible elements at each rank $n \in \mathbb{P}$, and let L_i denote the set of elements of L at rank i. Show that $\#L_i \le F_{i+2}$ (a Fibonacci number), with equality for all i if and only if $L \cong J_f(P+1)$, where P is the poset of Exercise 3.62(b).

c. [5−] Suppose that L is a finitary distributive lattice with an infinite antichain $t_1, t_2, \ldots$ such that t_i has rank i. Does it follow that $\#L_i \ge F_{i+2}$?

41. **a.** [2] Let L be a finite lattice. Given $f \colon L \to \mathbb{N}$, choose s, t incomparable in L such that $f(s) > 0$ and $f(t) > 0$. Define $\Gamma f \colon L \to \mathbb{N}$ by

$$\Gamma f(s) = f(s) - 1,$$
$$\Gamma f(t) = f(t) - 1,$$
$$\Gamma f(s \wedge t) = f(s \wedge t) + 1,$$
$$\Gamma f(s \vee t) = f(s \vee t) + 1,$$
$$\Gamma f(u) = f(u), \text{ otherwise.}$$

Show that for some $n > 0$ we have $\Gamma^n f = \Gamma^{n+1} f$ (i.e., $\Gamma^n f$ is supported on a chain).

b. [3−] Show the limiting function $\tilde{f} = \Gamma^n f$ (where $\Gamma^n f$ is supported on a chain) does not depend on the way in which we choose the pairs u, v (though the number of steps n may depend on these choices) if and only if L is distributive.

c. [2+] Show from (b) that $\tilde{f}$ has the following description: It is the unique function $\tilde{f} \colon L \to \mathbb{N}$ supported on a chain, such that for all join-irreducibles $t \in L$ and for $t = \hat{0}$, we have

$$\sum_{s \ge t} f(s) = \sum_{s \ge t} \tilde{f}(s).$$

d. [2] For $t \in L$ let $g(t) = \sum_{s \ge t} f(s)$. Order the join-irreducibles $t_1, \ldots, t_m$ of L such that $g(t_1) \ge g(t_2) \ge \cdots \ge g(t_m)$, and let $u_i = t_1 \vee t_2 \vee \cdots \vee t_i$. Deduce from (c) that $\tilde{f}(u_i) = g(t_i) - g(t_{i+1})$ (where we set $g(t_0) = g(\hat{0})$ and $g(t_{m+1}) = 0$), and that $\tilde{f}(u) = 0$ for all other $u \in L$.

e. [2+] Let $L = B_n$, the boolean algebra of subsets of $[n]$. Define $f \colon B_n \to \mathbb{N}$ by $f(S) = \#\{w \in \mathfrak{S}_{n+1} : \text{Exc}(w) = S\}$, where

$$\text{Exc}(w) = \{i : w(i) > i\},$$

the *excedance set* of w. Show that for all $0 \leq i \leq n$, we have

$$\tilde{f}(\{n-i+1, n-i+2, \ldots, n\}) = n!,$$

so all other $\tilde{f}(S) = 0$ (since $\tilde{f}$ is supported on a chain).

42.   **a.** [2–]* Regard Young's lattice Y as the lattice of all partitions of all integers $n \geq 0$, ordered componentwise. Let Z be the subposet of Y consisting of all partitions with odd parts. Show that Z is a sublattice of Y.

    b. [2]* Since sublattices of distributive lattices are distributive, it follows that Z is a finitary distributive lattice. (This fact is also easy to see directly.) For what poset P do we have $Z \cong J_f(P)$?

43.   [3–] Let P be the poset with elements s_i and t_i for $i \geq 1$, and cover relations

$$s_1 \lessdot s_2 \lessdot \cdots, \quad t_1 \lessdot t_2 \lessdot \cdots, \quad s_{2i} \lessdot t_i \text{ for } i \geq 1.$$

Find a nice product formula for the rank-generating function $F_{J_f(J_f(P))}(q)$.

44.   **a.** [3–]* Let $w = a_1 a_2 \cdots a_n \in \mathfrak{S}_n$. Let $P_w = \{(i, a_i) : i \in [n]\}$, regarded as a subposet of $\mathbb{P} \times \mathbb{P}$. In other words, define $(i, a_i) \leq (k, a_k)$ if $i \leq k$ and $a_i \leq a_k$. Let $j(P)$ denote the number of order ideals of the poset P. Show that

$$\sum_{w \in \mathfrak{S}_n} j(P_w) = \sum_{i=0}^{n} \frac{n!}{i!} \binom{n}{i}.$$

    b. [3]* Let w be as in (a), and let $Q_w = \{(i, j) : 1 \leq i < j \leq n, \ a_i < a_j\}$. Partially order Q_w by $(i, j) \leq (r, s)$ if $r \leq i < j \leq s$. Show that

$$\sum_{w \in \mathfrak{S}_n} j(Q_w) = (n+1)^{n-1}.$$

45.   **a.** [2]* Let $L_k(n)$ denote the number of k-element order ideals of the boolean algebra B_n. Show that for fixed k, $L_k(n)$ is a polynomial function of n of degree $k-1$ and leading coefficient $1/(k-1)!$. Moreover, the differences $\Delta^i L_k(0)$ are all nonnegative integers.

    b. [3–]* Show that

$$L_0(n) = L_1(n) = 1,$$

$$L_2(n) = \binom{n}{1},$$

$$L_3(n) = \binom{n}{2},$$

$$L_4(n) = \binom{n}{2} + \binom{n}{3},$$

$$L_5(n) = 3\binom{n}{3} + \binom{n}{4},$$

$$L_6(n) = 3\binom{n}{3} + 6\binom{n}{4} + \binom{n}{5},$$

$$L_7(n) = \binom{n}{3} + 15\binom{n}{4} + 10\binom{n}{5} + \binom{n}{6},$$

$$L_8(n) = \binom{n}{3} + 20\binom{n}{4} + 45\binom{n}{5} + 15\binom{n}{6} + \binom{n}{7},$$

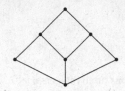

Figure 3.44 A meet-distributive lattice that is not distributive.

$$L_9(n) = 19\binom{n}{4} + 120\binom{n}{5} + 105\binom{n}{6} + 21\binom{n}{7} + \binom{n}{8},$$

$$L_{10}(n) = 18\binom{n}{4} + 220\binom{n}{5} + 455\binom{n}{6} + 210\binom{n}{7} + 28\binom{n}{8} + \binom{n}{9},$$

$$L_{11}(n) = 13\binom{n}{4} + 322\binom{n}{5} + 1385\binom{n}{6} + 1330\binom{n}{7} + 378\binom{n}{8}$$

$$+ 36\binom{n}{9} + \binom{n}{10}.$$

NOTE. It was conjectured that $L_k(n)$ has only real zeros. This conjecture fails, however, for $k = 11$.

46. **a.** [2]* Let $f(n)$ be the number of sublattices of rank n of the boolean algebra B_n. Show that $f(n)$ is also the number of partial orders P on $[n]$.

b. [2+]* Let $g(n)$ be the number of sublattices of B_n that contain $\emptyset$ and $[n]$ (the $\hat{0}$ and $\hat{1}$ of B_n). Write

$$F(x) = \sum_{n \geq 0} f(n) \frac{x^n}{n!},$$

$$G(x) = \sum_{n \geq 0} g(n) \frac{x^n}{n!}.$$

Show that $G(x) = F(e^x - 1)$.

c. [2]* Let $h(n)$ be the number of nonempty sublattices of B_n. Write

$$H(x) = \sum_{n \geq 0} h(n) \frac{x^n}{n!}.$$

Using (b), show that $H(x) = e^{2x} G(x)$.

47. A finite meet-semilattice is *meet-distributive* if for any interval $[s,t]$ of L such that s is the meet of the elements of $[s,t]$ covered by t, we have that $[s,t]$ is a boolean algebra. For example, distributive lattices are meet-distributive, while the lattice of Figure 3.44 is meet-distributive but not distributive.

a. [2-]* Show that a meet-distributive lattice is lower semimodular and hence graded.

b. [2] Let L be a meet-distributive meet-semilattice, and let $f_k = f_k(L)$ be the number of intervals of L isomorphic to the boolean algebra B_k. Also let $g_k = g_k(L)$ denote the number of elements of L that cover exactly k elements. Show that

$$\sum_{k \geq 0} g_k (1+x)^k = \sum_{k \geq 0} f_k x^k.$$

c. [1] Deduce from (b) that

$$\sum_{k \geq 0} (-1)^k f_k = 1. \qquad (3.113)$$

d. [2+] Let $L = J(m \times n)$ in (a). Explicitly compute f_k and g_k.

e. [3−] Given $m \le n$, let Q_{mn} be the subposet of $\mathbb{P} \times \mathbb{P}$ defined by

$$Q_{mn} = \{(i, j) \in P \times P : 1 \le i \le j \le m + n - i,\ 1 \le i \le m\},$$

and set $P_{mn} = m \times n$. Show that P_{mn} and Q_{mn} have the same zeta polynomial.

f. [3+] Show that P_{mn} and Q_{mn} have the same order polynomial.

g. [3−] Show that $J(P_{mn})$ and $J(Q_{mn})$ have the same values of f_k and g_k.

48. [2+] Let L be a meet-distributive lattice, as defined in Exercise 3.47, and let $t \in L$. Show that the number of join-irreducibles s of L satisfying $s \le t$ is equal to the rank $\rho(t)$ of t.

49. [2] Let L_p denote the set of all natural partial orders P of $[p]$ (that is, $i <_P j \Rightarrow i <_\mathbb{Z} j$), ordered by refinement. The bottom element is an antichain, and the top element is the chain p. Figure 3.6 shows a poset isomorphic to L_3. Show that L_p is meet-distributive of rank $\binom{p}{2}$.

50. [2+] Let L be a finitary distributive lattice with finitely many elements of each rank. Let $u(i, j)$ be the number of elements of L of rank i that cover exactly j elements, and let $v(i, j)$ be the number of elements of rank i that are covered by exactly j elements. Show that for all $i \le j \le 0$,

$$\sum_{k \ge 0} u(i, k) \binom{k}{j} = \sum_{k \ge 0} v(i - j, k) \binom{k}{j}. \tag{3.114}$$

(Each sum has finitely many nonzero terms.)

51. Let $f : \mathbb{N} \to \mathbb{N}$. A finitary distributive lattice L is said to have the *cover function* f if whenever $t \in L$ covers i elements, then t is covered by $f(i)$ elements.

a. [2+] Show that there is at most one (up to isomorphism) finitary distributive lattice with a given cover function f.

b. [2+] Show that if L is a *finite* distributive lattice with a cover function f, then L is a boolean algebra.

c. [2+] Let $k \in \mathbb{P}$. Show that there exist finitary distributive lattices with cover functions $f(n) = k$ and $f(n) = n + k$.

d. [2+] Let $a, k \in \mathbb{P}$ with $a \ge 2$. Show that there does not exist a finitary distributive lattice L with cover function $f(n) = an + k$.

e. [3] Show in fact that $f(n)$ is the cover function of a finitary distributive lattice L if and only if it belongs to one of the following seven classes. (Omitted values of f have no effect on L.)

- Let $k \ge 1$. Then $f(n) = k$ for $0 \le n \le k$.
- Let $k \ge 1$. Then $f(n) = n + k$.
- Let $k \ge 2$. Then $f(0) = 1$, and $f(n) = k$ for $1 \le n \le k$.
- $f(0) = 2$, and $f(n) = n + 1$ for $n \ge 1$.
- Let $k \ge 0$. Then $f(n) = k - n$ for $0 \le n \le k$.
- Let $k \ge 2$. Then $f(n) = k - n$ for $0 \le n < k$, and $f(k) = k$.
- $f(0) = 2$, $f(1) = f(2) = 1$.

52. [2+]* What is the maximum possible value of $e(P)$ for a connected n-element poset P?

53. [2]* Let P be a finite n-element poset. Simplify the two sums

$$f(P) = \sum_{I \in J(P)} e(I)e(\bar{I}),$$

$$g(P) = \sum_{I \in J(P)} \binom{n}{\#I} e(I)e(\bar{I}),$$

where $\bar{I}$ denotes the complement $P - I$ of the order ideal I.

54. [2+]* Let P be a finite poset. Simplify the sum

$$f(P) = \sum_{t_1 < \cdots < t_n} \frac{1}{(\#V_{t_1} - 1) \cdots (\#V_{t_{n-1}} - 1)},$$

where the sum ranges over all nonempty chains of P for which t_n is a maximal element of P. Generalize.

55. **a.** [3] Generalize Corollary 1.6.5 as follows. Let T be a tree on the vertex set $[n]$. Given an orientation $\mathfrak{o}$ of the edges of T, let $P(T, \mathfrak{o})$ be the reflexive and transitive closure of $\mathfrak{o}$, so $P(T, \mathfrak{o})$ is a poset. Clearly for fixed T, exactly two of these posets (one the dual of the other) have no 3-element chains. Let us call the corresponding orientations *bipartite*. Show that for fixed T, the number $e(P(T, \mathfrak{o}))$ of linear extensions of $(P, \mathfrak{o})$ is maximized when $\mathfrak{o}$ is bipartite. (Corollary 1.6.5 is the case when T is a path.)

b. [5–] Does (a) continue to hold when T is replaced with any finite bipartite graph?

56. [3–] Let P be a finite poset, and let $f(P)$ denote the number of ways to partition the elements of P into (nonempty) disjoint saturated chains. For instance, $f(n) = 2^{n-1}$. Suppose that every element of P covers at most two elements and is covered by at most two elements. Show that $f(P)$ is a product of Fibonacci and Lucas numbers. In particular, compute $f(m \times n)$.

57. **a.** [2]* Let P be an n-element poset. If $t \in P$, then set $\lambda_t = \#\{s \in P : s \leq t\}$. Show that

$$e(P) \geq \frac{n!}{\prod_{t \in P} \lambda_t}. \tag{3.115}$$

b. [2+]* Show that equality holds in equation (3.115) if and only if every component of P is a rooted tree (where the root as usual is the maximum element of the tree).

58. [3–]* Let P be a finite poset. Let A be an antichain of P which intersects every maximal chain. Show that

$$e(P) = \sum_{t \in A} e(P - t).$$

Try to give an elegant bijective proof.

59. [2+] Let P be a finite p-element poset. Choose two incomparable elements $s, t \in P$. Define $P_{s<t}$ to be the poset obtained from P by adjoining the relation $s < t$ (and all those implied by transitivity). Similarly define $P_{s>t}$. Define $P_{s=t}$ to be the poset obtained from P by identifying s and t. Hence $\#P_{s=t} = p - 1$. Write formally

$$P \to P_{s<t} + P_{s>t} + P_{s=t}.$$

Now choose two incomparable elements (if they exist) of each summand and apply the same decomposition to them. Continue until P is formally written as a linear

Figure 3.45 The composition poset $\mathcal{C}$.

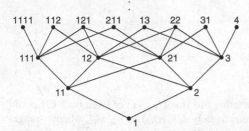

combination of chains:

$$P \to \sum_{i=1}^{p} a_i i.$$

Show that the numbers a_i are independent of the way in which the decomposition was obtained, and find a combinatorial interpretation of a_i.

60. Let P be a p-element poset. A bijection $f \colon P \to [p]$ is called a *dropless labeling* if we never have $s < t$ and $f(s) = f(t) + 1$.

   a. [1]* Show that every linear extension of P is a dropless labeling.

   b. [3–] Let G be an (undirected) graph, say with no loops or multiple edges. An *acyclic orientation* of G is an assignment of a direction $u \to v$ or $v \to u$ to each edge uv of G so that no directed cycles $u_1 \to u_2 \to \cdots \to u_k \to u_1$ result. Show that the number of dropless labelings of P is equal to the number of acyclic orientations of the incomparability graph $\mathrm{inc}(P)$. (For further information on the number of acyclic orientations of a graph, see Exercise 3.109.)

   c. [2+]* Give a bijective proof that the number of dropless labelings of P is equal to the number of bijections $g \colon P \to P$ such that we never have $g(t) < t$. *Hint.* Use Proposition 1.3.1(a).

61. [2+] Let $\mathcal{C}$ be the set of all compositions of all positive integers. Define a partial ordering on $\mathcal{C}$ by letting τ cover $\sigma = (\sigma_1, \ldots, \sigma_k)$ if τ can be obtained from σ either by adding 1 to a part, or adding 1 to a part and then splitting this part into two parts. More precisely, for some j we have either

$$\tau = (\sigma_1, \ldots, \sigma_{j-1}, \sigma_j + 1, \sigma_{j+1}, \ldots, \sigma_k)$$

or

$$\tau = (\sigma_1, \ldots, \sigma_{j-1}, h, \sigma_j + 1 - h, \sigma_{j+1}, \ldots, \sigma_k)$$

for some $1 \le h \le \sigma_j$. See Figure 3.45. For each $\sigma \in \mathcal{C}$, find in terms of a "familiar" number the number of saturated chains from the composition 1 (the bottom element of $\mathcal{C}$) to σ. What is the total number of saturated chains from 1 to some composition of n?

62. **a.** [2]* Let P_n be the poset with elements s_i, t_i for $i \in [n]$, and cover relations $s_1 < s_2 < \cdots < s_n$ and $t_i > s_i$ for all $i \in [n]$. For example, P_3 has the Hasse diagram of Figure 3.46.

   Find a "nice" expression for the rank-generating function $F(J(P_n), x)$.

   b. [2–]* Let $P = \lim_{n \to \infty} P_n$. Find the rank-generating function $F(J_f(P), x)$.

   c. [2]* Find a simple formula for $e(P_n)$.

   d. [2]* Let $\Omega_{P_n}(m)$ denote the order polynomial of P_n (naturally labeled). For $m \in \mathbb{P}$ express $\Omega_{P_n}(m)$ in terms of Stirling numbers of the second kind.

   e. [2+]* For $m \in \mathbb{P}$ express $\Omega_{P_n}(-m)$ in terms of Stirling numbers of the first kind.

Figure 3.46 The poset P_3 of Exercise 3.62.

f. [2+] Let P be as in (b). The generating function $U_{P,m}(x)$ of Exercise 3.171 is still well-defined although P is infinite, namely, $U_{P,m}(x)) = \sum_\sigma x^{|\sigma|}$, where σ ranges over all order-reversing maps $\sigma: P \to [0,m]$ such that $|\sigma| := \sum_{t \in P} \sigma(t) < \infty$. Such a map σ is called a *protruded partition* of $n = |\sigma|$ (with largest part at most m). Thus, we can regard a protruded partition of n as a pair (λ,μ), where $\lambda = (\lambda_1, \lambda_2, \dots)$ is a partition, $\mu = (\mu_1, \mu_2, \dots)$ is a sequence of nonnegative integers satisfying $\mu_i \le \lambda_i$, and $\sum(\lambda_i + \mu_i) = n$. For instance, there are six protruded partitions of 3, given by

$$(3,0),\ (21,00),\ (111,000),\ (2,1),\ (11,10),\ (11,01).$$

Show that

$$U_{P,m}(x) = \prod_{i=1}^{m}(1 - x^i - x^{i+1} - \cdots - x^{2i})^{-1}.$$

g. [2+] Show that

$$\sum_{m \ge 0} U_{P_n}(x)q^n = P(q,x) \sum_{j \ge 0} \frac{x^{j(j+1)}q^j}{[j]!(1 - x - x^2)(1 - x - x^3)\cdots(1 - x - x^{j+1})},$$

where $[j]! = (1 - x)(1 - x^2)\cdots(1 - x^j)$ and $P(q,x) = 1/(1 - q)(1 - qx)$ $(1 - qx^2)\cdots$.

63. **a.** [2]* Let P be a p-element poset, with every maximal chain of length ℓ. Let e_s (respectively, $\bar{e}_s$) denote the number of surjective (respectively, strict surjective) order-preserving maps $f: P \to s$. (The order-preserving map f is *strict* if $s < t$ in P implies $f(s) < f(t)$.) Use Corollary 3.15.18 to show that
  (i) $2e_{p-1} = (p + \ell - 1)e(P)$.
  (ii) $2\bar{e}_{p-1} = (p - \ell - 1)e(P)$.
  (iii) $\sum_{s=1}^{p} e_s = 2^\ell \sum_{s=1}^{p} \bar{e}_s$.
  b. [1+]* With P as in (a). show that if $p \equiv \ell \pmod 2$, then $e(P)$ is even.
  c. [2]* With P as in (a), suppose that $\ell = p - 4$. Let $j(P)$ denote the number of order ideals of P. Show that $e(P) = 2(j(P) - p)$.

64. **a.** [2+]* Let $\varphi: \mathbb{Q}[n] \to \mathbb{Q}[x]$ be the $\mathbb{Q}$-linear function on polynomials with rational coefficients that takes n^k to $\sum_j c_j(k)x^j$, where $c_j(k)$ is the number of ordered partitions of $[k]$ into j blocks (entry 3 of the Twelvefold Way or Example 3.18.9). Let (P,ω) be a labeled poset. Show that

$$\varphi\Omega_{P,\omega}(n) = \sum_j a_j(P,\omega)x^j,$$

where $a_j(P,\omega)$ is the number of chains $\emptyset = I_0 < I_1 < \cdots < I_j = P$ in $J(P)$ for which the restriction of ω to every set $I_i - I_{i-1}$ is order-preserving.
  b. [1+]* Let $c(k)$ denote the total number of ordered set partitions of $[k]$. Deduce from (a) that when we substitute $c(k)$ for n^k in $\Omega_P(n)$ (so P is naturally labeled), then we obtain the total number of chains from $\hat{0}$ to $\hat{1}$ in $J(P)$.

c. [2+]* Now let $\sigma: \mathbb{Q}[n] \to \mathbb{Q}[x]$ be defined by $\sigma(n^k) = (-1)^k \sum_j c_j(k) x^j$. Let $\#P = p$. Show that

$$\sigma \Omega_{P,\omega}(n) = (-1)^p \sum_j b_j(P,\omega) x^j,$$

where $b_j(P,\omega)$ is the number of chains $\emptyset = I_0 < I_1 < \cdots < I_j = P$ in $J(P)$ for which the restriction of ω to every set $I_i - I_{i-1}$ is order-reversing.

d. [1+]* Deduce from (c) that when we substitute $(-1)^k c(k)$ for n^k in $\Omega(P,n)$, then we obtain $(-1)^p$ times the number of chains $\emptyset = I_0 < I_1 < \cdots < I_j = P$ in $J(P)$ for which every interval $I_i - I_{i-1}$ is a boolean algebra.

65. [2] Let $n \in \mathbb{P}$ and $r,s,t \in \mathbb{N}$. Let $P(r,s,2t,n)$ be the poset with elements x_i ($1 \le i \le n$), y_{ij} ($1 \le i \le r, 1 \le j \le n$), z_{ij} ($1 \le i \le s, 1 \le j \le n$), and a_{ijk} ($1 \le j < k \le n, 1 \le i \le 2t$), and cover relations

$$x_1 \lessdot x_2 \lessdot \cdots \lessdot x_n,$$

$$y_{1j} \lessdot y_{2j} \lessdot \cdots \lessdot y_{rj} \lessdot x_j, \quad 1 \le j \le n,$$

$$x_j \lessdot z_{1j} \lessdot z_{2j} \lessdot \cdots \lessdot z_{sj}. \quad 1 \le j \le n,$$

$$x_j \lessdot a_{1jk} \lessdot a_{2jk} \lessdot \cdots \lessdot a_{2t,j,k} \lessdot x_k, \quad 1 \le j < k \le n.$$

Use Exercise 1.11(b) to show that

$$e(P,r,s,2t,n) = \frac{[(r+s+1)n + tn(n-1)]!}{n! \, r!^n \, s!^n \, t!^n \, (2t)!^{\binom{n}{2}}}$$

$$\cdot \prod_{j=1}^{n} \frac{(r+(j-1)t)!(s+(j-1)t)!(jt)!}{(r+s+1+(n+j-2)t)!}. \tag{3.116}$$

Figure 3.47 shows the poset $p = P(2,1,2,3)$ for which $e(P) = 4725864 = 2^3 \cdot 3^5 \cdot 11 \cdot 13 \cdot 17$.

66. Let Z_n denote the n-element "zigzag poset" or *fence*, with elements $t_1, \ldots, t_n$ and cover relations $t_{2i-1} < t_{2i}$ and $t_{2i} > t_{2i+1}$.

a. [2] How many order ideals does Z_n have?

b. [2+] Let $W_n(q)$ denote the rank-generating function of $J(Z_n)$, so $W_0(q) = 1$, $W_1(q) = 1+q$, $W_2(q) = 1+q+q^2$, $W_3(q) = 1+2q+q^2+q^3$, etc. Find a simple explicit formula for the generating function

$$F(x) := \sum_{n \ge 0} W_n(q) x^n.$$

c. [2] Find the number $e(Z_n)$ of linear extensions of Z_n.

d. [3−] Let $\Omega_{Z_n}(m)$ be the order polynomial of Z_n. Set

$$G_m(x) = 1 + \sum_{n \ge 0} \Omega_{Z_n}(m) x^{n+1}, \quad m \ge 1.$$

Find a recurrence relation expressing $G_m(x)$ in terms of $G_{m-2}(x)$, and give the initial conditions $G_1(x)$ and $G_2(x)$.

67. [3] For $p \le q$ define a poset P_{pq} to consist of three chains $s_1 > \cdots > s_p, t_1 > \cdots > t_q$, and $u_1 > \cdots > u_q$, with $s_i < u_i$ and $t_i < u_i$. Show that the number of linear extensions of P_{pq} is given by

$$e(P_{pq}) = \frac{2^{2p}(p+2q)!(2q-2p+2)!}{p!(2q+2)!(q-p)!(q-p+1)!}.$$

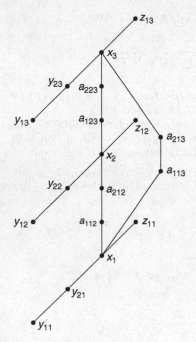

Figure 3.47 The "Selberg poset" $P(2,1,2,3)$.

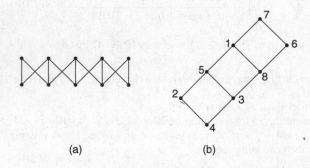

Figure 3.48 A garland and an alternating labeling of 2×4.

(a)   (b)

68. The *garland* or *double fence* G_n is the poset with vertices $s_1, \ldots, s_n, t_1, \ldots, t_n$ and cover relations $s_i < t_i$ $(1 \leq i \leq n)$, $s_i < t_{i-1}$ $(2 \leq i \leq n)$, and $s_i < t_{i+1}$ $(1 \leq i \leq n-1)$. Figure 3.48(a) shows the garland G_5. An *alternating labeling* of an m-element poset P is a bijection $f : P \to [m]$ such that every maximal chain $t_1 < t_2 < \cdots < t_k$ has alternating labels, (i.e., $f(t_1) > f(t_2) < f(t_3) > f(t_4) < \cdots$). Figure 3.48(b) shows an alternating labeling of the poset 2×4.

a. [2+] Let $j(n,k)$ denote the number of k-element order ideals of G_n. Show that

$$\sum_{n \geq 0} \sum_{k \geq 0} j(n,k) x^k y^n = \frac{1 - x^2 y^2}{1 - (1 + x + x^2)y + x^2 y^2 + x^3 y^3}.$$

b. [2]* Show that $e(G_n)$ is the number of alternating labelings of $2 \times n$.

c. [5–] Find a nice formula or generating function for $e(G_n)$. The values $e(G_n)$ for $1 \le n \le 6$ are

$$1, 4, 44, 896, 29392, 1413792.$$

69. a. [3] Fix an element t of a p-element poset P, and let $\mathcal{L}(P)$ denote the set of all linear extensions $f: P \to [p]$. Show that the polynomial

$$P_t(x) = \sum_{f \in \mathcal{L}(P)} x^{f(t)}$$

is log-concave, as defined in Exercise 1.50.

b. [3+] Suppose that the finite poset P is not a chain. Show that there exist elements $s, t \in P$ such that $f(s) < f(t)$ in more than a fraction $\frac{5-\sqrt{5}}{10} = 0.276\cdots$ and less than a fraction $\frac{5+\sqrt{5}}{10} = 0.723\cdots$ of the linear extensions f of P.

70. a. [2–]* Let E_n denote the poset of all subsets of $[n]$ whose elements have even sum, ordered by inclusion. Find $\#E_n$.

b. [2+]* Compute $\mu(S,T)$ for all $S \le T$ in E_n.

c. [3–] Generalize (b) as follows. Let $k \ge 3$, and let P_k denote the poset of all subsets of $\mathbb{P}$ whose elements have sum divisible by k. Given $T \le S$ in P_k, let

$$i_j = \#\{n \in T - S : n \equiv j \,(\mathrm{mod}\, k)\}.$$

Clearly $\mu(S,T)$ depends only on the k-tuple $(i_0, i_1, \ldots, i_{k-1})$, so write $\mu(i_0, \ldots, i_{k-1})$ for $\mu(S,T)$. Show that

$$\sum_{i_0, \ldots, i_{k-1} \ge 0} \mu(i_0, \ldots, i_{k-1}) \frac{x_0^{i_0} \cdots x_{k-1}^{i_{k-1}}}{i_0! \cdots i_{k-1}!}$$

$$= k \left[\sum_{j=0}^{k-1} \exp\left(x_0 + \zeta^j x_1 + \zeta^{2j} x_2 + \cdots + \zeta^{(k-1)j} x_{k-1} \right) \right]^{-1},$$

where ζ is a primitive kth root of unity.

71. [3–] Let P be a finite poset. The *free distributive lattice* $\mathrm{FD}(P)$ generated by P is, intuitively, the largest distributive lattice containing P as a subposet and generated (as a lattice) by P. More precisely, if L is any distributive lattice containing P and generated by P, then there is a (surjective) lattice homomorphism $f: \mathrm{FD}(P) \to L$ that is the identity on P. Show that $\mathrm{FD}(P) \cong J(J(P)) - \{\hat{0}, \hat{1}\}$. In particular, $\mathrm{FD}(P)$ is finite. When $P = n\mathbf{1}$ (an n-element antichain), we write $\mathrm{FD}(P) = \mathrm{FD}(n)$, the free distributive lattice with n generators, so that $\mathrm{FD}(n) \cong J(B_n) - \{\hat{0}, \hat{1}\}$.

NOTE. Sometimes one defines $\mathrm{FD}(P)$ to be the free *bounded* distributive lattice generated by P. In this case, we need to add an extra $\hat{0}$ and $\hat{1}$ to $\mathrm{FD}(P)$, so one sometimes sees the statement $\mathrm{FD}(P) \cong J(J(P))$ and $\mathrm{FD}(n) \cong J(B_n)$.

72. a. [2] Let P be a finite poset with largest antichain of cardinality k. Every antichain A of P corresponds to an order ideal

$$\langle A \rangle = \{s : s \le t \text{ for some } t \in A\} \in J(P).$$

Show that the set of all order ideals $\langle A \rangle$ of P with $\#A = k$ forms a sublattice $M(P)$ of $J(P)$.

b. [3–] Show that every finite distributive lattice is isomorphic to $M(P)$ for some P.

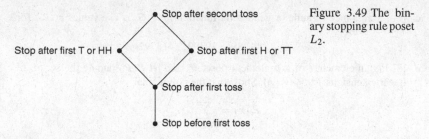

Figure 3.49 The binary stopping rule poset L_2.

73. **a.** [2+] Let P be a finite poset, and define $G_P(q,x) = \sum_I q^{\#I} x^{m(I)}$, where I ranges over all order ideals of P and where $m(I)$ denotes the number of maximal elements of I. (Thus, $G_P(q,1)$ is the rank-generating function of $J(P)$.) Let Q be an n-element poset. Show that

$$G_{P \otimes Q}(q,x) = G_P(q^n, q^{-n}(G_Q(q,x) - 1)),$$

where $P \otimes Q$ denotes ordinal product.
   b. [2+] Show that if $\#P = p$, then

$$G_P\left(q, \frac{q-1}{q}\right) = q^p.$$

74. [2+] A *binary stopping rule* of length n is (informally) a rule for telling a person when to stop tossing a coin, so that he is guaranteed to stop within n tosses. Two rules are considered the same if they result in the same outcome. For instance, "toss until you get three consecutive heads or four consecutive tails, or else after n tosses" is a stopping rule of length n. Partially order the stopping rules of length n by $A \le B$ if the tosser would never stop later using rule A rather than rule B. Let L_n be the resulting poset. For example, L_2 is shown in Figure 3.49. Show that L_n is a distributive lattice, and compute its poset of join-irreducibles. Find a simple recurrence for the rank-generating function $F(L_n, q)$ in terms of $F(L_{n-1}, q)$.

75. Let G be a finite connected graph, allowing multiple edges but not loops. Fix a vertex v of G, and let $\mathrm{ao}(G, v)$ be the set of acyclic orientations of G such that v is a sink. If $\mathfrak{o}, \mathfrak{o}' \in \mathrm{ao}(G, v)$, then define $\mathfrak{o} \le \mathfrak{o}'$ if we can obtain $\mathfrak{o}'$ from $\mathfrak{o}$ by a sequence of operations that consist of choosing a *source* vertex w not adjacent to v and orienting all edges of G incident to w toward w, keeping the rest of $\mathfrak{o}$ unchanged.
   a. [2+] Show that $(\mathrm{ao}(G, v), \le)$ is a poset.
   b. [2−] Let G be a 6-cycle. By symmetry the choice of v is irrelevant. Show that

$$\mathrm{ao}(G, v) \cong J(2 \times 2) + 4 + 4 + 1 + 1.$$

   c. [3−] Show that every connected component of $(\mathrm{ao}(G, v), \le)$ is a distributive lattice.

76. In this exercise, P and Q denote locally finite posets and $I(P)$, $I(Q)$ their incidence algebras over a field K.
   a. [2] Show that the (Jacobson) radical of $I(P)$ is $\{f : f(t,t) = 0 \text{ for all } t \in P\}$. The Jabcobson radical can be defined as the intersection of all maximal right ideals of $I(P)$.
   b. [2+] Show that the lattice of two-sided ideals of $I(P)$ is isomorphic to the set of all order ideals A of $\mathrm{Int}(P)$ (the poset of intervals of P), ordered by reverse inclusion.
   c. [3−] Show that if $I(P)$ and $I(Q)$ are isomorphic as K-algebras, then P and Q are isomorphic.

Figure 3.50 A poset for Exercise 3.77.

d. [3] Describe the group of K-automorphisms and the space of K-derivations of $I(P)$.

77. **a.** [3] Let P be a p-element poset, and define nonnegative integers λ_i by setting $\lambda_1 + \cdots + \lambda_i$ equal to the maximum size of a union of i chains in P. For instance, the poset P of Figure 3.50 satisfies $\lambda_1 = 5$, $\lambda_2 = 3$, $\lambda_3 = 1$, and $\lambda_i = 0$ for $i \geq 4$. Note that the largest chain has five elements, but that the largest union of two chains does not contain a five-element chain. Show that $\lambda_1 \geq \lambda_2 \geq \cdots$ (i.e., if we set $\lambda = (\lambda_1, \lambda_2, \ldots)$ then $\lambda \vdash p$).
   b. [3] Define μ_i's analogously by letting $\mu_1 + \cdots + \mu_i$ be the maximum size of a union of i antichains. For the poset of Figure 3.50 we have $\mu = (\mu_1, \mu_2, \ldots) = (3, 2, 2, 1, 1, 0, 0, \ldots)$. Show that $\mu \vdash p$.
   c. [3] Show that $\mu = \lambda'$, the conjugate partition to λ.
   d. [2–] Deduce from (c) *Dilworth's theorem*: The minimum k for which P is a union of k chains is equal to the size of the largest antichain of P.
   e. [2] Prove directly that $\lambda_1 = \mu_1'$ (i.e., the size of the longest chain of P is equal to the minimum k for which P is a union of k antichains).
   f. [3] Let A be the matrix whose rows and columns are indexed by P, with

   $$A_{st} = \begin{cases} x_{st}, & \text{if } s < t, \\ 0, & \text{otherwise,} \end{cases}$$

   where the x_{st}'s are independent indeterminates. It is clear that A is nilpotent (i.e., every eigenvalue is 0). Show that the Jordan block sizes of A are the numbers $\lambda_i > 0$ of (a).

78. **a.** [3–] Find the partitions λ and μ of Exercise 1.77 for the boolean algebra B_n.
   b. [5] Do the same for the partition lattice Π_n.
   c. [5] Do the same for the lattice Par(n) of partitions of n ordered by dominance (defined in Exercise 3.136).

79. [3–] Let P be a finite poset on the set $[p]$, such that if $s < t$ in P then $s < t$ in $\mathbb{Z}$. A linear extension of P can therefore be regarded as a permutation $w = a_1 a_2 \cdots a_p \in \mathfrak{S}_p$ such that if $a_i < a_j$ in P, then $i < j$ in $\mathbb{Z}$. Define the *comajor index* comaj$(w) = \sum_{i \in D(w)} (p - i)$, where $D(w)$ denotes the descent set of w. A *P-domino tableau* is a chain $\emptyset = I_0 \subset I_1 \subset \cdots \subset I_r = P$ of order ideals of P such that $I_i - I_{i-1}$ is a two-element chain for $2 \leq i \leq r$, while I_1 is either a two-element or one-element chain (depending on whether p is even or odd). In particular, $r = \lceil p/2 \rceil$. Show that the following three quantities are equal.
   i.   The sum $w(P) = \sum_{w \in \mathcal{L}(P)} (-1)^{\text{comaj}(w)}$. NOTE. If p is even, then comaj$(w) \equiv$ maj$(w) \pmod 2$. In this case $w(P) = W_P(-1)$ in the notation of Section 3.15.
   ii. The number of P-domino tableaux.
   iii. The number of self-evacuating linear extensions of P (i.e., linear extensions f satisfying $f\epsilon = f$, where ϵ denotes evacuation).

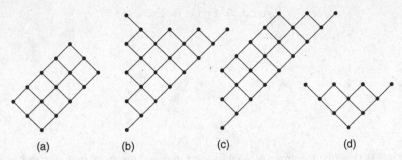

(a)   (b)   (c)   (d)

Figure 3.51 Four posets with nice promotion properties.

80. **a.** [3] Show that for the following p-element posets P we have $f\partial^p = f$, where ∂ is
the promotion operator and f is a linear extension of P. (We give an example of
each type of poset, from which the general definition should be clear.)
   i.   Rectangles: Figure 3.51(a).
   ii. Shifted double staircases: Figure 3.51(b).
   iii. Shifted trapezoids: Figure 3.51(c).
   b. [3] Show that if P is a staircase (illustrated in Figure 3.51(d)), then $f\partial^p$ is obtained
by reflecting P (labeled by f) about a vertical line. Thus, $f\partial^{2p} = f$.

81. An n-element poset P is *sign-balanced* if the set $\mathcal{E}_P$ of linear extensions of P (regarded
as a permutation of the elements of P with respect to some fixed ordering of the ele-
ments) contains the same number of even permutations as odd permutations. (This
definition does not depend on the fixed ordering of the elements of P, since changing
the ordering simply multiplies the elements of $\mathcal{E}_P$ by a fixed permutation in $\mathfrak{S}_n$).
   a. [2−] Suppose that $n \geq 2$. Show that if every nonminimal element of P is greater
than at least two minimal elements, then P is sign-balanced. For instance, atomic
lattices with at least three elements are sign-balanced (since we can clearly remove
$\hat{0}$ without affecting the property of being sign-balanced).
   b. [2+] Suppose that the length $\ell(C)$ of every maximal chain C of P satisfies
$\ell(C) \equiv n \,(\mathrm{mod}\,2)$. Show that P is sign-balanced.

82. [3] Show that a product $\boldsymbol{p} \times \boldsymbol{q}$ of two chains is sign-balanced if and only if $p, q > 1$
and $p \equiv q \,(\mathrm{mod}\,2)$.

83. [2] Show that P is sign-balanced if $\#P$ is even and there does not exist a P-domino
tableau, as defined in Exercise 3.79.

84. [2+] A mapping $t \mapsto \bar{t}$ on a poset P is called a *closure operator* (or *closure*) if for all
$s, t \in P$,

$$t \leq \bar{t},$$
$$s \leq t \Rightarrow \bar{s} \leq \bar{t},$$
$$\bar{\bar{t}} = \bar{t}.$$

An element t of P is *closed* if $t = \bar{t}$. The set of closed elements of P is denoted $\overline{P}$,
called the *quotient* of P relative to the closure $\bar{}$. If $s \leq t$ in P, then define $\bar{s} \leq \bar{t}$ in $\overline{P}$.
It is easy to see that $\overline{P}$ is a poset.
Let P be a locally finite poset with closure $t \mapsto \bar{t}$ and quotient $\overline{P}$. Show that for all
$s, t \in P$,

$$\sum_{\substack{u \in P \\ \bar{u} = \bar{t}}} \mu(s, u) = \begin{cases} \mu_{\overline{P}}(\bar{s}, \bar{t}), & \text{if } s = \bar{s} \\ 0, & \text{if } s < \bar{s}. \end{cases}$$

85. [2+]* Let P be a finite poset. Show that the following two conditions are equivalent:
i. For all $s < t$, the interval $[s,t]$ has an odd number of atoms.
ii. For all $s < t$, the interval $[s,t]$ has an odd number of coatoms.
Hint. Consider $\mu(s,t)$ modulo 2.

86. [2+] Let f and g be functions on a finite lattice L, with values in a field of characteristic 0, satisfying

$$f(s) = \sum_{\substack{t \\ s \wedge t = \hat{0}}} g(t). \tag{3.117}$$

Show that if $\mu(\hat{0}, u) \neq 0$ for all $u \in L$, then equation (3.117) can be inverted to yield

$$g(s) = \sum_t \alpha(s,t) f(t),$$

where

$$\alpha(s,t) = \sum_u \frac{\mu(s,u)\mu(t,u)}{\mu(\hat{0},u)}.$$

87. **a.** [2+] Let P be a finite poset with $\hat{0}$ and $\hat{1}$, and let μ be its Möbius function. Let $f : P \to \mathbb{C}$. Show that

$$\sum (f(t_1) - 1)(f(t_2) - 1) \cdots (f(t_k) - 1)$$

$$= \sum (-1)^{k+1} \mu(\hat{0}, t_1) \mu(t_1, t_2) \cdots \mu(t_k, \hat{1}) f(t_1) f(t_2) \cdots f(t_k),$$

where both sums range over all chains $\hat{0} < t_1 < \cdots < t_k < \hat{1}$ of P.
b. [1+] Deduce that

$$\sum_{\hat{0} = t_0 < t_1 < \cdots < t_k = \hat{1}} (-1)^k \mu(t_0, t_1) \mu(t_1, t_2) \cdots \mu(t_{k-1}, t_k) = 1.$$

c. [2] Give a proof of (b) using incidence algebras.
d. [2−] Deduce equation (3.113) from (a) when L is a meet-distributive lattice.

88. [2] Let P be a finite poset with $\hat{0}$ and $\hat{1}$, and with Möbius function μ. Show that

$$\sum_{s \le t} \mu(s,t) = 1.$$

89. [2]* For a finite lattice L, let $f_L(m)$ be the number of m-tuples $(t_1, \ldots, t_m) \in L^m$ such that $t_1 \wedge t_2 \wedge \cdots \wedge t_m = \hat{0}$. Give two proofs that

$$f_L(m) = \sum_{t \in L} \mu(\hat{0}, t)(\#V_t)^m.$$

The first proof should be by direct Möbius inversion, and the second by considering $\left(\sum_{t \in L} t \right)^m$ in the Möbius algebra $A(L, \mathbb{R})$.

90. [2]* Let P be a finite graded poset, and let $m(s,t)$ denote the number of maximal chains of the interval $[s,t]$. Define $f \in I(P, \mathbb{C})$ by

$$f(s,t) = \frac{m(s,t)}{\ell(s,t)!}.$$

Show that

$$f^{-1}(s,t) = (-1)^{\ell(s,t)} f(s,t).$$

91. **a.** [3–]* Let L be a finite lattice with n atoms. Show that

$$|\mu(\hat{0},\hat{1})| \leq \binom{n-1}{\lfloor (n-1)/2 \rfloor},$$

and that this result is best possible.

b. [3+]* Assume also that the longest chain of L has length at most ℓ. show that

$$|\mu(\hat{0},\hat{1})| \leq \binom{n-1}{k},$$

where $k = \min(\ell - 1, \lfloor (n-1)/2 \rfloor)$, and that this result is best possible.

92. [3–] Assume that L is a finite lattice and fix $t \in L$. Show that

$$\mu(\hat{0},\hat{1}) = \sum_{u,v} \mu(\hat{0},u)\zeta(u,v)\mu(v,\hat{1}),$$

where u,v range over all pairs of complements of t. Deduce that if $\mu(\hat{0},\hat{1}) \neq 0$, then L is complemented.

93. **a.** [2] Let L be a finite lattice such that for every $t > \hat{0}$, the interval $[\hat{0},t]$ has even cardinality. Use Exercise 3.92 to show that L is complemented.

b. [3–] Find a simple proof that avoids Möbius functions.

94. [2+] Let $L = J(P)$ be a finite distributive lattice. A function $v\colon L \to \mathbb{C}$ is called a *valuation* (over $\mathbb{C}$) if $v(\hat{0}) = 0$ and $v(s) + v(t) = v(s \wedge t) + v(s \vee t)$ for all $s,t \in L$. Prove that v is uniquely determined by its values on the join-irreducibles of L (which we may identify with P). More precisely, show that if I is an order ideal of P, then

$$v(I) = -\sum_{t \in I} v(t)\mu(t,\hat{1}),$$

where μ denotes the Möbius function of I (considered as a subposet of P) with a $\hat{1}$ adjoined.

95. [3–] Let L be a finite lattice and fix $z \in L$. Show that the following identity holds in the Möbius algebra of L (over some field):

$$\sum_{t \in L} \mu(\hat{0},t)t = \left(\sum_{u \leq z} \mu(\hat{0},u)u \right) \cdot \left(\sum_{v \wedge z = \hat{0}} \mu(\hat{0},v)v \right).$$

96. **a.** [3–] Let L be a finite lattice (or meet-semilattice), and let $f(s,v)$ be a function (say with values in a commutative ring) defined for all $s,v \in L$. Set $F(s,v) = \sum_{u \leq s} f(u,v)$. Show that

$$\det [F(s \wedge t,s)]_{s,t \in L} = \prod_{s \in L} f(s,s).$$

b. [2] Deduce that

$$\det [\gcd(i,j)]_{i,j=1}^n = \prod_{k=1}^n \phi(k),$$

where ϕ is the Euler totient (or φ) function.

c. [2] Choose $f(s,v) = \mu(\hat{0},s)$ to deduce that if L is a finite meet-semilattice such that $\mu(\hat{0},s) \neq 0$ for all $s \in L$, then there exists a permutation $w \colon L \to L$ satisfying $s \wedge w(s) = \hat{0}$ for all $s \in L$.

d. [2] Let L be a finite geometric lattice of rank n with W_i elements of rank i. Deduce from (c) (more precisely, the dualized form of (c)) that for $k \leq n/2$,

$$W_1 + \cdots + W_k \leq W_{n-k} + \cdots + W_{n-1}. \tag{3.118}$$

In particular, $W_1 \leq W_{n-1}$.

e. [3–] If equality holds in equation (3.118) for any one value of k, then show that L is modular.

f. [5] With L as in (d), show that $W_k \leq W_{n-k}$ for all $k \leq n/2$.

97. [3–] Let L be a finite lattice such that $\mu(t,\hat{1}) \neq 0$ and $\mu(\hat{0},t) \neq 0$ for all $t \in L$. Prove that there is a permutation $w \colon L \to L$ such that for all $t \in L$, t and $w(t)$ are complements. Show that this conclusion is false if one merely assumes that $\mu(\hat{0},t) \neq 0$ for all $t \in L$.

98. [2+]* Let L be a finite geometric lattice, and let t be a coatom of L. Let $\eta(t)$ be the number of atoms $s \in L$ satisfying $s \not\leq t$. Show that

$$|\mu(\hat{0},\hat{1})| \leq |\mu(\hat{0},t)| \cdot \eta(t).$$

99. [2+]* Let L be a finite geometric lattice of rank n. Let $L' = L - \{\hat{0}\}$, and let $f \colon L' \to A$ be a function from L' to the set A of atoms of L satisfying $f(t) \leq t$ for all $t \in L'$. Let $\alpha(L,f)$ be the number of maximal chains $\hat{0} = t_0 < t_1 < \cdots < t_n = \hat{1}$ of L such that

$$f(t_1) \vee \cdots \vee f(t_n) = \hat{1}.$$

Show that $\alpha(L,f) = (-1)^n \mu(\hat{0},\hat{1})$.

100. Let L be a finite geometric lattice.

a. [2] Show that every element of L is a meet of coatoms (where we regard $\hat{1}$ as being the meet of the empty set of coatoms).

b. [2] Show that Proposition 3.10.1 has the following improvement for geometric lattices: the Möbius function of L *strictly* alternates in sign. In other words, if $s \leq t$ in L then $(-1)^{\rho(t) - \rho(s)} \mu(s,t) > 0$.

c. [2+] Show that if $\mu(s,t) = \pm 1$, then the interval $[s,t]$ is a boolean algebra.

d. [3–]* Let $n \in \mathbb{P}$. Show that there exist finitely many geometric lattices $L_1, \ldots, L_k$ such that if L is any finite geometric lattice satisfying $|\mu(\hat{0},\hat{1})| = n$, then $L \cong L_i \times B_d$ for some i and d.

101. a. [3] Let L be a finite lattice and A, B subsets of L. Suppose that for all $t \notin A$ there exists $t^* > t$ such that $\mu(t,t^*) \neq 0$ and $t^* \neq t \vee u$ whenever $u \in B$. (Thus, $\hat{1} \in A$.) Show that there exists an injective map $\phi \colon B \to A$ satisfying $\phi(s) \geq s$ for all $s \in B$.

b. [2+] Let K be a finite modular lattice. Show the following: (i) If $\hat{1}$ is a join of atoms of K, then K is a geometric lattice and hence $\mu(\hat{0},\hat{1}) \neq 0$. (ii) With K as in (i), K has the same number of atoms as coatoms. (iii) For any $a, b \in K$, the map $\psi_b \colon [a \wedge b, a] \to [b, a \vee b]$ defined by $\psi_b(t) = t \vee b$ is a lattice (or poset) isomorphism.

c. [2+] Let L be a finite modular lattice, and let J_k (respectively, M_k) be the set of elements of L that cover (respectively, are covered by) at most k elements. (Thus, $J_0 = \{\hat{0}\}$ and $M_0 = \{\hat{1}\}$.) Deduce from (a) and (b) the existence of an injective map $\phi \colon J_k \to M_k$ satisfying $\phi(s) \geq s$ for all $s \in J_k$.

d. [2–] Deduce from (c) that the number of elements in L covering exactly k elements equals the number of elements covered by exactly k elements.

e. [2] Let P_k be the subposet of elements of L that cover k elements, and let R_k be the subposet of elements that are covered by k elements. Show by example that we need not have $P_k \cong R_k$, unlike the situation for distributive lattices (Exercise 3.38).

f. Deduce Exercise 3.96(d) from (a).

102. a. [5] Let L be a finite lattice with n elements. Does there exist a join-irreducible t of L such that the principal dual order ideal $V_t := \{s \in L : s \geq t\}$ has at most $n/2$ elements?

b. [2+] Let L be any finite lattice with n elements. Suppose that there is a $t \neq \hat{0}$ in L such that $\#V_t > n/2$. Show that $\mu(\hat{0}, s) = 0$ for some $s \in L$.

103. [3] Let L be a finite lattice, and suppose that L contains a subset S of cardinality n such that (i) any two elements of S are incomparable (i.e., S is an antichain), and (ii) every maximal chain of L meets S. Find, as a function of n, the smallest and largest possible values of $\mu(\hat{0}, \hat{1})$. For instance, if $n = 2$, then $0 \leq \mu(\hat{0}, \hat{1}) \leq 1$, while if $n = 3$ then $-1 \leq \mu(\hat{0}, \hat{1}) \leq 2$.

104. a. [3–] Let P be an $(n+2)$-element poset with $\hat{0}$ and $\hat{1}$. What is the largest possible value of $|\mu(\hat{0}, \hat{1})|$?

b. [5] Same as (a) for n-element lattices L.

105. [5–] Let $k, \ell \in \mathbb{P}$. Find $\max_P |\mu(\hat{0}, \hat{1})|$, where P ranges over all finite posets with $\hat{0}$ and $\hat{1}$ and longest chain of length ℓ, such that every element of P is covered by at most k elements.

106. [2+] Let L be a finite lattice for which $|\mu_L(\hat{0}, \hat{1})| \geq 2$. Does it follow that L contains a sublattice isomorphic to the 5-element lattice $\mathbf{1} \oplus (\mathbf{1}+\mathbf{1}+\mathbf{1}) \oplus \mathbf{1}$?

107. [3–] Let $k \geq 0$, and let I be an order ideal of the boolean algebra B_n. Suppose that for any $t \in I$ of rank at most k, we have $\sum_{\substack{u \in I \\ u \geq t}} \mu(t, u) = 0$. Show that $\#I$ is divisible by 2^{k+1}.

108. Let G be a (simple) graph with finite vertex set V and edge set $E \subseteq \binom{V}{2}$. Write $p = \#V$. An n-*coloring* of G (sometimes called a *proper n-coloring*) is a function $f : V \to [n]$ such that $f(a) \neq f(b)$ if $\{a,b\} \in E$. Let $\chi_G(n)$ be the number of n-colorings of G. The function $\chi_G : \mathbb{N} \to \mathbb{N}$ is called the *chromatic polynomial* of G.

a. [2–]* A *stable partition* of V is a partition π of V such that every block B of π is stable (or independent), that is, no two vertices of B are adjacent. Let $S_G(j)$ be the number of stable partitions of V with k blocks. Show that

$$\chi_G(n) = \sum_j S_G(j)(n)_j.$$

Deduce that $\chi_G(n)$ is a monic polynomial in n of degree p with integer coefficients. Moreover, the coefficient of n^{p-1} is $-(\#E)$.

b. [2+] A set $A \subseteq V$ is *connected* if the induced subgraph on A is connected (i.e., for any two vertices $v, v' \in A$ there is a path from v to v' using only vertices in A). Let L_G be the poset (actually a geometric lattice) of all partitions π of V ordered by refinement, such that every block of V is connected. Show that

$$\chi_G(n) = \sum_{\pi \in L_G} \mu(\hat{0}, \pi) n^{\#\pi},$$

where $\#\pi$ is the number of blocks of π and μ is the Möbius function of L_G. It follows that the chromatic polynomial $\chi_G(n)$ and characteristic polynomial $\chi_{L_G}(n)$ are related by $\chi_G(n) = n^c \chi_{L_G}(n)$, where c is the number of connected components of G. Note that when G is the complete graph K_p (i.e., $E = \binom{V}{2}$), then we obtain equation (3.38).

c. [2+] Let $\mathcal{B}_G$ be the hyperplane arrangement in $\mathbb{R}^p$ with hyperplanes $x_i = x_j$ whenever $\{i, j\} \in E$. We call $\mathcal{B}_G$ a *graphical arrangement*. Show that $L_G \cong L(\mathcal{B}_G)$ (the intersection poset of $\mathcal{B}$). Deduce that $\chi_G = \chi_{\mathcal{B}_G}$.

d. [2+] Let e be an edge of G. Let $G - e$ (also denoted $G \backslash e$) denote G with e deleted, and let G/e denote G with e contracted to a point, and all resulting multiple edges replaced by a single edge (so that G/e is simple). Deduce from (c) and Proposition 3.11.5 that

$$\chi_G(n) = \chi_{G-e}(n) - \chi_{G/e}(n). \tag{3.119}$$

Give also a direct combinatorial proof.

e. [2+] Let $\varphi \colon \mathbb{Q}[n] \to \mathbb{Q}[x]$ be the $\mathbb{Q}$-linear function defined by $\varphi(n^k) = \sum_j S(k, j)x^j$, where $S(k, j)$ denotes a Stirling number of the second kind. Show that

$$\varphi(\chi_G(n)) = \sum_j S_G(j)x^j. \tag{3.120}$$

In particular, if B_G denotes the total number of stable partitions of G (a G-analogue of the Bell number $B(n)$), then we have the "umbral" formula $\chi_G(B) = B_G$. That is, expand $\chi_G(B)$ as a polynomial in B (regarding B as an indeterminate), and then replace B^k by $B(k)$.

109. Preserve the notation of the previous exercise. Let $\mathrm{ao}(G)$ denote the number of acyclic orientations of G, as defined in Exercise 3.60.

a. [2+] Use equation (3.119) to prove that

$$\mathrm{ao}(G) = (-1)^p \chi_G(-1). \tag{3.121}$$

b. [2+] Give another proof of equation (3.121) using Theorem 3.11.7.

110. [3] Let $w \in \mathfrak{S}_n$, and let $\mathcal{A}_w$ be the arrangement in $\mathbb{R}^n$ determined by the equations $x_i = x_j$ for all inversions (i, j) of w.

a. Show that $r(\mathcal{A}_w) \geq \#\Lambda_w$, where Λ_w is the principal order ideal generated by w in the Bruhat order on $\mathfrak{S}_n$ (as defined in Exercise 3.183).

b. Show that equality holds in (a) if and only if w avoids all the patterns 4231, 35142, 42513, and 351624.

111. [3–] Give a bijective proof that the number of regions of the Shi arrangement $\mathcal{S}_n$ is $(n + 1)^{n-1}$ (Corollary 3.11.14).

112. A sequence $\mathfrak{A} = (\mathcal{A}_1, \mathcal{A}_2, \dots)$ of arrangements is called an *exponential sequence of arrangements* (ESA) if it satisfies the following three conditions.

- $\mathcal{A}_n$ is in K^n for some field K (independent of n).
- Every $H \in \mathcal{A}_n$ is parallel to some hyperplane H' in the braid arrangement $\mathcal{B}_n$ (over K).
- Let S be a k-element subset of $[n]$, and define

$$\mathcal{A}_n^S = \{H \in \mathcal{A}_n : H \text{ is parallel to } x_i - x_j = 0 \text{ for some } i, j \in S\}.$$

Then $L(\mathcal{A}_n^S) \cong L(\mathcal{A}_k)$.

a. [1+]* Show that the braid arrangements $(\mathcal{B}_1, \mathcal{B}_2, \dots)$ and Shi arrangements $(\mathcal{S}_1, \mathcal{S}_2, \dots)$ form ESAs.

b. [3–] Let $\mathfrak{A} = (\mathcal{A}_1, \mathcal{A}_2, \ldots)$ be an ESA. Show that

$$\sum_{n \geq 0} \chi_{\mathcal{A}_n}(x) \frac{z^n}{n!} = \left(\sum_{n \geq 0} (-1)^n r(\mathcal{A}_n) \frac{z^n}{n!} \right)^{-x}.$$

c. [3–] Generalize (b) as follows. For $n \geq 1$ let $\mathcal{A}_n$ be an arrangement in $\mathbb{R}^n$ such that every $H \in \mathcal{A}_n$ is parallel to a hyperplane of the form $x_i = cx_j$, where $c \in \mathbb{R}$. Just as in (b), define for every subset S of $[n]$ the arrangement

$$\mathcal{A}_n^S = \{H \in \mathcal{A}_n : H \text{ is parallel to some } x_i = cx_j, \text{ where } i, j \in S\}.$$

Suppose that for every such S we have $L_{\mathcal{A}_n^S} \cong L_{\mathcal{A}_k}$, where $k = \#S$. Let

$$F(z) = \sum_{n \geq 0} (-1)^n r(\mathcal{A}_n) \frac{z^n}{n!}$$

$$G(z) = \sum_{n \geq 0} (-1)^{\text{rank}(\mathcal{A}_n)} b(\mathcal{A}_n) \frac{z^n}{n!}.$$

Show that

$$\sum_{n \geq 0} \chi_{\mathcal{A}_n}(x) \frac{z^n}{n!} = \frac{G(z)^{(x+1)/2}}{F(z)^{(x-1)/2}}.$$

113. [2] Use the finite field method (Theorem 3.11.10) to give a proof of the Deletion-Restriction recurrence (Proposition 3.11.5) for arrangements defined over $\mathbb{Q}$.

114. For the arrangements $\mathcal{A}$ below (all in $\mathbb{R}^n$), show that the characteristic polynomials are as indicated.
 a. [2–]* $x_i = x_j$ for $1 \leq i < j \leq n$ and $x_i = 0$ for $1 \leq i \leq n$. Then

$$\chi_{\mathcal{A}}(x) = (x-1)^2 (x-2)(x-3) \cdots (x-n+1).$$

 b. [2+]* $x_i = x_j$ for $1 \leq i < j \leq n$ and $x_1 + x_2 + \cdots + x_n = 0$. Then

$$\chi_{\mathcal{A}}(x) = (x-1)^2 (x-2)(x-3) \cdots (x-n+1).$$

 c. [3–]* $x_i = 2x_j$ and $x_i = x_j$ for $1 \leq i < j \leq n$, and $x_i = 0$ for $1 \leq i \leq n$. Then

$$\chi_{\mathcal{A}}(x) = (x-1)(x-n-1)^{n-1}.$$

115. For the arrangements $\mathcal{A}$ below (all in $\mathbb{R}^n$), show that the characteristic polynomials are as indicated.
 a. [2+] The *Catalan arrangement* $\mathcal{C}_n$: $x_i - x_j = -1, 0, 1,\ 1 \leq i < j \leq n$. Then

$$\chi_{\mathcal{C}_n}(x) = x(x-n-1)(x-n-2)(x-n-3) \cdots (x-2n+1).$$

 b. [3] The *Linial arrangement* $\mathcal{L}_n$: $x_i - x_j = 1,\ 1 \leq i < j \leq n$. Then

$$\chi_{\mathcal{L}_n}(x) = \frac{1}{2^n} \sum_{k=0}^{n} \binom{n}{k} (x-k)^{n-1}. \tag{3.122}$$

c. [3–] The *threshold arrangement* $\mathcal{T}_n$: $x_i + x_j = 0, 1$, $1 \le i < j \le n$. Then

$$\sum_{n \ge 0} \chi_{\mathcal{T}_n}(x) \frac{z^n}{n!} = (1+z)(2e^z - 1)^{(x-1)/2}.$$

d. [2] The *type B braid arrangement* $\mathcal{B}_n^B$: $x_i - x_j = 0$, $x_i + x_j = 0$, $1 \le i < j \le n$, and $x_i = 0$, $1 \le i \le n$. Then

$$\chi_{\mathcal{B}_n^B} = (x-1)(x-3)(x-5)\cdots(x-2n+1).$$

116. [3–] Let $v_1, \ldots, v_k$ be "generic" points in $\mathbb{R}^n$. Let $\mathcal{C} = \mathcal{C}(v_1, \ldots, v_k)$ be the arrangement consisting of the perpendicular bisectors of all pairs of the points. Thus, $\#\mathcal{C} = \binom{k}{2}$. Find the characteristic polynomial $\chi_{\mathcal{C}}(x)$ and number of regions $r(\mathcal{C})$.

117. [3–] Let $(t_1, x_1), \ldots, (t_k, x_k)$ be "generic" events (points) in $(n+1)$-dimensional Minkowski space $\mathbb{R} \times \mathbb{R}^n$ with respect to some reference frame. Assume that the events are spacelike with respect to each other (i.e., there can be no causal connection among them). Suppose that $t_1 < \cdots < t_k$ (i.e., the events occur in the order $1, 2, \ldots, k$). In another reference frame moving at a constant velocity v with respect to the first, the events may occur in a different order $a_1 a_2 \cdots a_k \in \mathfrak{S}_k$. What is the number of different orders in which observers can see the events? Express your answer in terms of the signless Stirling numbers $c(n, i)$ of the first kind.
NOTE. Write $v = \tanh(\rho)u$, where u is a unit vector in $\mathbb{R}^n$ and $\tanh(\rho)$ is the speed (with the speed of light $c = 1$). Part of the Lorentz transformation states that the coordinates (t, x) and (t', x') of the two frames are related by

$$t' = \cosh(\rho)t - \sinh(\rho)x \cdot u. \tag{3.123}$$

118. a. [3+] Let $\mathcal{A} = \{H_1, \ldots, H_\nu\}$ be a linear arrangement of hyperplanes in $\mathbb{R}^d$ with intersection lattice $L(\mathcal{A})$. Let $r = d - \dim(H_1 \cap \cdots \cap H_\nu) = \operatorname{rank} L(\mathcal{A})$. Define

$$\Omega = \Omega(\mathcal{A}) = \{p = (p_1, \ldots, p_d) : p_i \in \mathbb{R}[x_1, \ldots, x_d], \text{ and for all } i \in [\nu]$$

$$\text{and } \alpha \in H_i \text{ we have } p(\alpha) \in H_i\}.$$

Clearly, Ω is a module over the ring $R = \mathbb{R}[x_1, \ldots, x_d]$, that is, if $p \in \Omega$ and $q \in R$, then $qp \in \Omega$. One easily shows that Ω has rank r, that is, Ω contains r (and no more) elements linearly independent over R. Suppose that Ω is a *free* R-module—that is, we can find $p_1, \ldots, p_r \in \Omega$ such that $\Omega = p_1 R \oplus \cdots \oplus p_r R$. (The additional condition that $p_i R \cong R$ as R-modules is automatic here.) We then call $\mathcal{A}$ a *free* arrangement. It is easy to see that we can choose each p_i so that all its components are homogeneous of the same degree e_i. Show that the characteristic polynomial of $L(\mathcal{A})$ is given by

$$\chi_{L(\mathcal{A})}(x) = \prod_{i=1}^{r} (x - e_i).$$

b. [3] Show that Ω is free if L is supersolvable, and find a free Ω for which L is not supersolvable.
c. [3] For $n \ge 3$ let $H_1, \ldots, H_\nu$ ($\nu = \binom{n}{2} + \binom{n}{3}$) be defined by the equations

$$x_i = x_j, \quad 1 \le i < j \le n,$$

$$x_i + x_j + x_k = 0, \quad 1 \le i < j < k \le n.$$

Is Ω free?

d. [5] Suppose that $\mathcal{A}$ and $\mathcal{A}'$ are two linear hyperplane arrangements in $\mathbb{R}^d$ with corresponding modules $\boldsymbol{\Omega}$ and $\boldsymbol{\Omega}'$. If $L(\mathcal{A}) \cong L(\mathcal{A}')$ and $\boldsymbol{\Omega}$ is free, does it follow that $\boldsymbol{\Omega}'$ is free? In other words, is freeness a property of $L(\mathcal{A})$ alone, or does it depend on the actual position of the hyperplanes?

e. [3] Let $\mathcal{A} = \{H_1, \dots, H_v\}$ as in (a), and let $t \in L(\mathcal{A})$. With $\mathcal{A}_t$ as in equation (3.43), show that if $\boldsymbol{\Omega}(\mathcal{A})$ is free, then $\boldsymbol{\Omega}(\mathcal{A}_t)$ is free.

f. [3] Continuing (e), let $\mathcal{A}^t$ be as in equation (3.44). Give an example where $\boldsymbol{\Omega}(\mathcal{A})$ is free but $\boldsymbol{\Omega}(\mathcal{A}^t)$ is not free.

119. [2] Let V be an n-dimensional vector space over $\mathbb{F}_q$, and let L be the lattice of subspaces of V. Let X be a vector space over $\mathbb{F}_q$ with x vectors. By counting the number of injective linear transformations $V \to X$ in two ways (first way – direct, second way – Möbius inversion on L) show that

$$\prod_{k=0}^{n-1} (x - q^k) = \sum_{k=0}^{n} \binom{n}{k} (-1)^k q^{\binom{k}{2}} x^{n-k}.$$

This is an identity valid for infinitely many x and hence valid as a polynomial identity (with x an indeterminate). Note that if we substitute $-1/x$ for x then we obtain equation (1.87) (the q-binomial theorem).

120. a. [3–] Let P be a finite graded poset of rank n, and let $q \geq 2$. Show that the following two conditions are equivalent:

- For every interval $[s,t]$ of length k we have $\mu(s,t) = (-1)^k q^{\binom{k}{2}}$.
- For every interval $[s,t]$ of length k and all $0 \leq i \leq k$, the number of elements of $[s,t]$ of rank i (where the rank is computed in $[s,t]$, not in P) is equal to the q-binomial coefficient $\binom{k}{i}$ (evaluated at the positive integer q).

b. [5–] Is it true that for n sufficiently large, such posets P must be isomorphic to $B_n(q)$ (the lattice of subspaces of $\mathbb{F}_q^n$)?

121. [3–] Fix $k \geq 2$. Let L_n' be the poset of all subsets S of $[n]$, ordered by inclusion, such that S contains no k consecutive integers. Let L_n be L_n' with a $\hat{1}$ adjoined. Let μ_n denote the Möbius function of L_n. Find $\mu_n(\emptyset, \hat{1})$. Your answer should depend only on the congruence class of n modulo $2k+2$.

122. [2] A positive integer d is a *unitary divisor* of n if $d|n$ and $(d, n/d) = 1$. Let L be the poset of all positive integers with $a \leq b$ if a is a unitary divisor of b. Describe the Möbius function of L. State a unitary analogue of the classical Möbius inversion formula of number theory.

123. a. [2+] Let M be a monoid (semigroup with identity ε) with generators $g_1, \dots, g_n$ subject only to relations of the form $g_i g_j = g_j g_i$ for certain pairs $i \neq j$. Order the elements of M by $s \leq t$ if there is a u such that $su = t$. For instance, suppose that M has generators $1, 2, 3, 4$ (short for $g_1, \dots, g_4$) with relations

$$13 = 31, \ 14 = 41, \ 24 = 42.$$

Then the interval $[\varepsilon, 11324]$ is shown in Figure 3.52. Show that any interval $[\varepsilon, w]$ in M is a distributive lattice L_w, and describe the poset P_w for which $L_w = J(P_w)$.

b. [1+] Deduce from (a) that the number of factorizations $w = g_{i_1} \cdots g_{i_\ell}$ is equal to the number $e(P_w)$ of linear extensions of P_w.

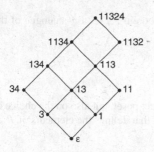

Figure 3.52 The distributive lattice L_{11324} when $13 = 31$, $14 = 41$, $24 = 42$.

c. [2–] Deduce from (a) that the Möbius function of M is given by

$$\mu(s,su) = \begin{cases} (-1)^r, & \text{if } u \text{ is a product of } r \text{ distinct} \\ & \text{pairwise commuting } g_i, \\ 0, & \text{otherwise.} \end{cases}$$

d. [2] Let $N(a_1,a_2,\ldots,a_n)$ denote the number of *distinct* elements of M of degree a_i in g_i. (E.g., $g_1^2 g_2 g_1 g_4^2$ has $a_1 = 3$, $a_2 = 1$, $a_3 = 0$, $a_4 = 2$.) Let $x_1,\ldots,x_n$ be independent (commuting) indeterminates. Deduce from (c) that

$$\sum_{a_1 \geq 0} \cdots \sum_{a_n \geq 0} N(a_1,\ldots,a_n) x_1^{a_1} \cdots x_n^{a_n} = \left(\sum (-1)^r x_{i_1} x_{i_2} \cdots x_{i_r} \right)^{-1},$$

where the last sum is over all $(i_1,i_2,\ldots,i_r)$ such that $1 \leq i_1 < i_2 < \cdots < i_r \leq n$ and $g_{i_1}, g_{i_2}, \ldots, g_{i_r}$ pairwise commute.
e. [2–] What identities result in (d) when no g_i and g_j commute ($i \neq j$), or when all g_i and g_j commute?

124. Let L be a finite supersolvable semimodular lattice, with M-chain $C : \hat{0} = t_0 < t_1 < \cdots < t_n = \hat{1}$.
a. [3–] Let a_i be the number of atoms s of L such that $s \leq t_i$ but $s \nleq t_{i-1}$. Show that

$$\chi_L(q) = (q - a_1)(q - a_2) \cdots (q - a_n).$$

b. [3–] If $t \in L$ then define

$$\Lambda(t) = \{i : t \vee t_{i-1} = t \vee t_i\} \subseteq [n].$$

One easily sees that $\#\Lambda(t) = \rho(t)$ and that if u covers t then (in the notation of equation (3.56)) $\Lambda(u) - \Lambda(t) = \{\lambda(t,u)\}$. Now let P be any natural partial ordering of $[n]$ (i.e., if $i < j$ in P, then $i < j$ in $\mathbb{Z}$), and define

$$L_P = \{t \in L : \Lambda(t) \in J(P)\}.$$

Show that L_P is an R-labelable poset satisfying

$$\beta_{L_P}(S) = \sum_{\substack{w \in \mathcal{L}(P) \\ D(w) = S}} \beta_L(S),$$

where $\mathcal{L}(P)$ denotes the Jordan-Hölder set of P (defined in Section 3.13).

In particular, taking $L = B_n(q)$ yields from Theorem 3.13.3 a q-analogue of the distributive lattice $J(P)$, satisfying

$$\beta_{L_P}(S) = \sum_{\substack{w \in \mathcal{L}(P) \\ D(w) = S}} q^{\text{inv}(w)}.$$

Note that L_P depends not only on P as an abstract poset, but also on the choice of linear extension P (or maximal chain of $J(P)$) that defines the elements of P as elements of $[n]$.

125. [3–] Let L be a finite graded lattice of rank n. Show that the following two conditions are equivalent:
   - L is supersolvable.
   - L has an R-labeling for which the label of every maximal chain is a permutation of $1, 2, \ldots, n$.

126. Fix a prime p and integer $k \geq 1$, and define posets $L_k^{(1)}(p)$, $L_k^{(2)}(p)$, and $L_k^{(3)}(p)$ as follows:
   - $L_k^{(1)}(p)$ consists of all subgroups of the free abelian group $\mathbb{Z}^k$ that have finite index p^m for some $m \geq 0$, ordered by reverse inclusion.
   - $L_k^{(2)}(p)$ consists of all finite subgroups of $(\mathbb{Z}/p^\infty\mathbb{Z})^k$ ordered by inclusion, where

$$\mathbb{Z}/p^\infty\mathbb{Z} = \mathbb{Z}[1/p]/\mathbb{Z},$$

$$\mathbb{Z}[1/p] = \{\alpha \in \mathbb{Q} : p^m\alpha \in \mathbb{Z} \text{ for some } m \geq 0\}.$$

   - $L_k^{(3)}(p) = \bigcup_n L_{n,k}(p)$, where $L_{n,k}(p)$ denotes the lattice of subgroups of the abelian group $(\mathbb{Z}/p^n/\mathbb{Z})^k$, and where we regard

$$L_{n,k}(p) \subset L_{n+1,k}(p)$$

via the embedding

$$\left(\mathbb{Z}/p^n\mathbb{Z}\right)^k \hookrightarrow \left(\mathbb{Z}/p^{n+1}/\mathbb{Z}\right)^k$$

defined by

$$(a_1, \ldots, a_k) \mapsto (pa_1, \ldots, pa_k).$$

a. [2+] Show that $L_k^{(1)}(p) \cong L_k^{(2)}(p) \cong L_k^{(3)}(p)$. Calling this poset $L_k(p)$, show that $L_k(p)$ is a locally finite modular lattice with $\hat{0}$ such that each element is covered by finitely many elements (and hence $L_k(p)$ has a rank function $\rho : L_k(p) \to \mathbb{N}$).
b. [2–] Show that for any $t \in L_k(p)$, the principal dual order ideal V_t is isomorphic to $L_k(p)$.
c. [3–] Show that $L_k(p)$ has $\binom{n+k-1}{k-1}$ elements of rank n, and hence has rank-generating function

$$F(L_k(p), x) = \frac{1}{(1-x)(1-px)\cdots(1-p^{k-1}x)}.$$

All q-binomial coefficients in this exercise are in the variable p.

d. [1+] Deduce from (b) and (c) that if $S = \{s_1, s_2, \ldots, s_j\}_< \subset \mathbb{P}$, then

$$\alpha_{L_k(p)}(S) = \binom{s_1+k-1}{k-1}\binom{s_2-s_1+k-1}{k-1}$$

$$\ldots \binom{s_j-s_{j-1}+k-1}{k-1}.$$

e. [2+] Let N_k denote the set of all infinite words $w = e_1 e_2 \cdots$ such that $e_i \in [0, k-1]$ and $e_i = 0$ for i sufficiently large. Define $\sigma(w) = e_1 + e_2 + \cdots$, and as usual define the descent set

$$D(w) = \{i : e_i > e_{i+1}\}.$$

Use (d) to show that for any finite $S \subset \mathbb{P}$,

$$\alpha_{L_k(p)}(S) = \sum_{\substack{w \in N_k \\ D(w) \subseteq S}} p^{\sigma(w)}$$

$$\beta_{L_k(p)}(S) = \sum_{\substack{w \in N_k \\ D(w) = S}} p^{\sigma(w)}.$$

127. a. [2−]* How many maximal chains does Π_n have?

b. [2+]* The symmetric group $\mathfrak{S}_n$ acts on the partition lattice Π_n in an obvious way. This action induces an action on the set $\mathcal{M}$ of maximal chains of Π_n. Show that the number $\#\mathcal{M}/\mathfrak{S}_n$ of $\mathfrak{S}_n$-orbits on $\mathcal{M}$ is equal to the Euler number E_{n-1}. For instance, when $n = 5$, a set of orbit representatives is given by (omitting $\hat{0}$ and $\hat{1}$ from each chain, and writing e.g. 12-34 for the partition whose non-singleton blocks are $\{1,2\}$ and $\{3,4\}$): $12 < 123 < 1234$, $12 < 123 < 123$-45, $12 < 12$-$34 < 125$-34, $12 < 12$-$34 < 12$-345, $12 < 12$-$34 < 1234$. Hint. Use Proposition 1.6.2.

c. [2]* Let Λ_n denote the subposet of Π_n consisting of all partitions of $[n]$ satisfying (i) if i is the least element of a nonsingleton block B, then $i + 1 \in B$, and (ii) if $i < n$ and $\{i\}$ is a singleton block, then $\{i + 1\}$ is also a singleton block. Figure 3.53 shows Λ_6, where we have omitted singleton blocks from the labels. Show that the number of maximal chains of Λ_n is E_{n-1}.

d. [2+]* Show that Λ_n is a supersolvable lattice of rank $n-1$. Hence for all $S \subseteq [n-2]$ we have by Example 3.14.4 that $\beta_{\Lambda_n}(S) \geq 0$.

e. [5−] Find an *elegant* combinatorial interpretation of $\beta_{\Lambda_n}(S)$ as the number of alternating permutations in $\mathfrak{S}_{n-1}$ with some property depending on S.

f. [2]* Show that the number of elements of Λ_n whose nonsingleton block sizes are $\lambda_1 + 1, \ldots, \lambda_\ell + 1$ is the number of partitions of a set of cardinality $m = \sum \lambda_i$ whose block sizes are $\lambda_1, \ldots, \lambda_\ell$ (given explicitly by equation (3.36)) provided that $m + \ell \leq n$, and is 0 if $m + \ell > n$. As a corollary, the number of elements of Λ_n of rank k is $\sum_{j=0}^{\min\{k,n-k\}} S(k, j)$, while the total number of elements of Λ_n is

$$\#\Lambda_n = \sum_{\substack{j+k \leq n \\ j \leq k}} S(k, j),$$

including the term $S(0,0) = 1$.

g. [2]* We can identify Λ_n with a subposet of Λ_{n+1} by adjoining a single block $\{n + 1\}$ to each $\pi \in \Lambda_n$. Hence we can define $\Lambda = \lim_{n \to \infty} \Lambda_n$. Show that Λ has $B(n)$ elements of rank n, where $B(n)$ denotes a Bell number.

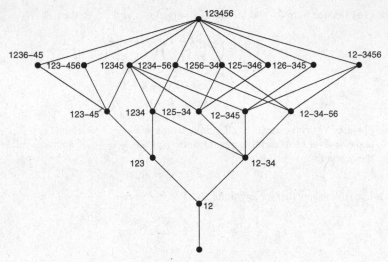

Figure 3.53 The poset Λ_6.

h. [2+]* Write

$$\exp \sum_{i \geq 1} E_i t_i \frac{x^i}{i!} = \sum_{k \geq 0} P_k(t_1, t_2, \ldots, t_k) \frac{x^k}{k!}.$$

Show the the coefficient of $t_1^{\alpha_1} \cdots t_k^{\alpha_k}$ in $P_k(t_1, \ldots, t_k)$ is the number of saturated chains in Λ from $\hat{0}$ to a partition with $\alpha_i + 1$ nonsingleton blocks of cardinality i. Thus, if $M(k, j)$ denotes the number of saturated chains in Λ from $\hat{0}$ to some element of rank k with j nonsingleton blocks, then

$$\sum_{j,k \geq 0} M(k, j) t^j \frac{x^k}{k!} = e^{t(\tan x + \sec x - 1)}.$$

128. [3–]* Let $\pi \in \Pi_n$ have type $(a_1, a_2, \ldots)$, with $\#\pi = \sum a_i = m$. Let $f(\pi)$ be the number of $\sigma \in \Pi_n$ satisfying $\pi \vee \sigma = \hat{1}$, $\pi \wedge \sigma = \hat{0}$, and $\#\sigma = n + 1 - \#\pi$. Show that

$$f(\pi) = 1^{a_1} 2^{a_2} \cdots n^{a_n} (n - m + 1)^{m-2}.$$

129. [2+]* Let P be a finite poset, and let μ be the Möbius function of $\widehat{P} = P \cup \{\hat{0}, \hat{1}\}$. Suppose that P has a fixed-point free automorphism $\sigma : P \to P$ of prime order p (i.e., $\sigma(t) \neq t$ and $\sigma^p(t) = t$ for all $t \in P$). Show that

$$\mu(\hat{0}, \hat{1}) \equiv -1 \pmod{p}.$$

What does this say in the case $\widehat{P} = \Pi_p$?

130. Let P be a finite poset satisfying: (i) P is graded of rank n and has a $\hat{0}$ and $\hat{1}$, and (ii) for $0 \leq i \leq n$, there is a poset Q_i such that $[t, \hat{1}] \cong Q_i$ whenever $n - \rho(t) = i$. In particular, $P \cong Q_n$. We call the poset P *uniform*.
a. [2+] Let $V(i, j)$ be the number of elements of Q_i that have rank $i - j$, and let

$$v(i, j) = \sum_t \mu(\hat{0}, t),$$

where t ranges over all $t \in Q_i$ of rank $i - j$. (Thus, $V(i, j) = W_{i-j}$ and $v(i, j) = w_{i-j}$, where w and W denote the Whitney numbers of Q_i of the first and second kinds, as defined in Section 3.10.) Show that the matrices $[V(i, j)]_{0 \le i, j \le n}$ and $[v(i, j)]_{0 \le i, j \le n}$ are inverses of one another. (Note that Proposition 1.9.1 corresponds to the case $Q_i = \Pi_{i+1}$.)

b. [5] Find interesting uniform posets. Can all uniform geometric lattices be classified? (See Exercise 3.131(d).)

131. Let X be an n-element set and G a finite group of order m. A *partial partition* of X is a collection $\{A_1, \ldots, A_r\}$ of nonempty, pairwise-disjoint subsets of X. A *partial G-partition* of X is a family $\alpha = \{a_1, \ldots, a_r\}$ of functions $a_j \colon A_j \to G$, where $\{A_1, \ldots, A_r\}$ is a partial partition of X. Define two partial G-partitions $\alpha = \{a_1, \ldots, a_r\}$ and $\beta = \{b_1, \ldots, b_s\}$ to be *equivalent* if their underlying partial partitions are the same (so $r = s$), say $\{A_1, \ldots, A_r\}$, and if for each $1 \le j \le r$, there is some $w \in G$ (depending on j) such that $a_j(t) = w \cdot b_j(t)$ for all $t \in A_j$. Define a poset $Q_n(G)$ as follows. The elements of $Q_n(G)$ are equivalence classes of partial G-partitions. Representing a class by one of its elements, define $\alpha = \{a_1, \ldots, a_r\} \le \beta = \{b_1, \ldots, b_s\}$ in $Q_n(G)$ if every block A_i of the underlying partial partition $\{A_1, \ldots, A_r\}$ of α is either (1) contained in a block B_j of the underlying partial partition σ of β, in which case there is a $w \in G$ for which $a_i(t) = w \cdot b_j(t)$ for all $t \in A_i$, or else (2) every block of σ is disjoint from A_i. (Thus, $Q_n(G)$ has a top element consisting of the empty set.)

a. [2−] Show that if $m = 1$, then $Q_n(G) \cong \Pi_{n+1}$.

b. [3−] Show that $Q_n(G)$ is a supersolvable geometric lattice of rank n.

c. [2] Use (b) and Exercise 3.124 to show that the characteristic polynomial of $Q_n(G)$ is given by

$$\chi_{Q_n(G)}(t) = \prod_{i=1}^{n-1} (t - 1 - mi).$$

d. [2] Show that $Q_n(G)$ is uniform in the sense of Exercise 3.130.

132. [2+] Let P_n be the set of all sets $\{i_1, \ldots, i_{2k}\} \subset \mathbb{P}$ where

$$0 < i_1 < i_2 < \cdots < i_{2k} < 2n + 1,$$

and $i_1, i_2 - i_1, \ldots, i_{2k} - i_{2k-1}, 2n + 1 - i_{2k}$ are all odd. Order the elements of P_n by inclusion. Then P_n is graded of rank n, with $\hat{0}$ and $\hat{1}$. Compute the number of elements of P_n of rank k, the total number of elements of P_n, the Möbius function $\mu(\hat{0}, \hat{1})$, and the number of maximal chains of P_n. Show that if $\rho(t) = k$ then $[\hat{0}, t] \cong P_k$ while $[t, \hat{1}]$ is isomorphic to a product of P_i's. (Thus P_n^* is uniform in the sense of Exercise 3.130.)

133. Let L_n denote the lattice of all subgroups of the symmetric group $\mathfrak{S}_n$, ordered by inclusion. Let μ_n denote the Möbius function of L_n.

a. [2+] Show that

$$\sum \mu_n(\hat{0}, G) = (-1)^{n-1}(n-1)!,$$

where G ranges over all transitive subgroups of $\mathfrak{S}_n$.

b. [3] Show that $\mu_n(\hat{0}, \hat{1})$ is divisible by $n!/2$.

c. [3] Let C_n denote the collection of transitive proper subgroups of $\mathfrak{S}_n$ that contain an odd involution (i.e., an involution with an odd number of 2-cycles). Show that

$$\mu_n(\hat{0}, \hat{1}) = (-1)^{n-1} \frac{n!}{2} - \sum_{H \in C_n} \mu_n(\hat{0}, H).$$

d. [3–] Let p be prime. Deduce from (c) that

$$\mu_p(\hat{0},\hat{1}) = (-1)^{p-1}\frac{p!}{2}.$$

e. [3–] Let $n = 2^a$ for some positive integer a. Deduce from (c) that

$$\mu_n(\hat{0},\hat{1}) = -\frac{n!}{2}.$$

f. [3] Let p be an odd prime and $n = 2p$. Deduce from (c) that

$$\mu_n(\hat{0},\hat{1}) = \begin{cases} -n!, & \text{if } n-1 \text{ is prime and } p \equiv 3 \,(\text{mod}\,4), \\ n!/2, & \text{if } n = 22, \\ -n!/2, & \text{otherwise.} \end{cases}$$

134. a. [3–]* Let A be a finite alphabet and A^* the free monoid generated by A. If $w = a_1 a_2 \cdots a_n$ is a word in the free monoid A^* with each $a_i \in A$, then a *subword* of w is a word $v = a_{i_1} a_{i_2} \cdots a_{i_k}$ where $1 \le i_1 < i_2 < \cdots < i_k \le n$. Partially order A^* by $u \le v$ if u is a subword of v. We call this partial ordering the *subword order* on A^*. Let μ be the Möbius function of A^*. Given $v = a_1 a_2 \cdots a_n$ where $a_i \in A$, call the letter a_i *special* if $a_i = a_{i-1}$. Show that

$$\mu(u,v) = (-1)^{\ell(v)-\ell(u)} s(u,v),$$

where $s(u,v)$ is the number of subwords of v isomorphic to u which use every special letter of v. For instance,

$$\mu(aba, aba\underline{a}ba) = -2$$

(where we have underlined the only special letter.) There is a simple proof using Philip Hall's theorem (Proposition 3.8.5), based on a sign-reversing involution acting on a subset of the set of chains of the open interval (u,v).

b. [3–]* Now define $u \le v$ if u is a factor of v, as defined in Example 4.7.7. We call this partial ordering the *factor order* on A^*. Given a word $w = a_1 a_2 \cdots a_n \in A^*$, $n \ge 2$, define $\iota w = a_2 a_3 \cdots a_{n-1}$ and φw to be the longest $v \ne w$ (possibly the empty word 1) which is both a left factor and right factor of w (so one can write $w = vw' = w''v$). The word w is *trivial* if $a_1 = a_2 = \cdots = a_n$. Show that the Möbius function of the factor order is determined recursively by

$$\mu(u,v) = \begin{cases} \mu(u,\varphi v), & \text{if } \ell(u,v) > 2 \text{ and } u \le \varphi v \nleq \iota v, \\ 1, & \text{if } \ell(u,v) = 2, \ v \text{ is nontrivial and } u = \iota v \text{ or } u = \varphi v, \\ (-1)^{\ell(u,v)}, & \text{if } \ell(u,v) < 2, \\ 0, & \text{in all other cases.} \end{cases}$$

In particular $\mu(u,v) \in \{0, +1, -1\}$.

135. Let Λ_n denote the set of all $p(n)$ partitions of the integer $n \ge 0$. Order Λ_n by refinement. This means that $\lambda \le \rho$ if the parts of λ can be partitioned into blocks so that the parts of ρ are precisely the sum of the elements in each block of λ. For instance, $(4,4,3,2,2,1,1) \le (9,4,4,2)$, corresponding to $9 = 4+2+2+1$, $4 = 4$, $4 = 3+1$, $2 = 2$.

a. [2–]* Show that Λ_n is graded of rank $n-1$.

b. [5] Determine the Möbius function $\mu(\lambda,\rho)$ of Λ_n. (This is trivial when $\lambda = \langle 1^n \rangle$ and easy when $\lambda = \langle 1^{n-2} 2^1 \rangle$.)

c. [3] Does the Möbius function μ of Λ_n alternate in sign; that is, $(-1)^\ell \mu(\lambda, \rho) \geq 0$ if $[\lambda, \rho]$ is an interval of length ℓ? Is Λ_n a Cohen-Macaulay poset?

136. [3] Let Λ_n be as in Exercise 3.135, but now order Λ_n by *dominance*. This means that $(\lambda_1, \lambda_2, \ldots) \leq (\rho_1, \rho_2, \ldots)$ if $\lambda_1 + \lambda_2 + \cdots + \lambda_i \leq \rho_1 + \rho_2 + \cdots + \rho_i$ for all $i \geq 1$. Find μ for this ordering.

137. [2] Let P and Q be finite posets. Express the zeta polynomial values $Z(P + Q, m)$, $Z(P \oplus Q, m)$, and $Z(P \times Q, m)$ in terms of $Z(P, j)$ and $Z(Q, j)$ for suitable values of j.

138. a. [2] Let P be a finite poset and $\text{Int}(P)$ the poset of (nonempty) intervals of P, ordered by inclusion. How are the zeta polynomials $Z(P, n)$ and $Z(\text{Int}(P), n)$ related?

b. [2] Suppose that P has a $\hat{0}$ and $\hat{1}$. Let Q denote $\text{Int}(P)$ with a $\hat{0}$ adjoined. How are $\mu_P(\hat{0}, \hat{1})$ and $\mu_Q(\hat{0}, \hat{1})$ related?

139. [2+]* Let U_k denote the ordinal sum of k 2-element antichains, so $\#U_k = 2k$. Show that

$$\sum_{k \geq 0} Z(U_k, n) x^k = \frac{1}{2} \left(\frac{1+x}{1-x} \right)^n - \frac{1}{2}.$$

140. [2+] Let $\varphi : \mathbb{Q}[n] \to \mathbb{Q}[x]$ be the $\mathbb{Q}$-linear function on polynomials with rational coefficients that takes n^k to $\sum_j c_j(k) x^j$, where $c_j(k) = j! S(k, j)$, the number of ordered partitions of $[k]$ into j blocks, or equivalently, the number of surjective functions $[k] \to [j]$ (Example 3.18.9). (Set $\varphi(1) = 1$.) Let $Z(P, n)$ denote the zeta polynomial of the poset P. Show that

$$\varphi Z(P, n+2) = \sum_{j \geq 1} c_j(P) x^{j-1},$$

where $c_j(P)$ is the number of j-element chains of P.

141. a. [2] Let P be a finite poset, and let $Q = \text{ch}(P)$ denote the poset of nonempty chains of P, ordered by inclusion. Let Q_0 denote Q with a $\hat{0}$ (the empty chain of P) adjoined. Show that if $Z(P, m+1) = \sum_{i \geq 1} a_i \binom{m-1}{i}$, then $Z(Q_0, m+1) = 1 + \sum_{i \geq 1} a_i m^i$.

b. [2] Let $\widehat{P}$ and $\widehat{Q}$ denote P and Q, respectively, with a $\hat{0}$ and $\hat{1}$ adjoined. Express $\mu_{\widehat{Q}}(\hat{0}, \hat{1})$ in terms of $\mu_{\widehat{P}}(\hat{0}, \hat{1})$.

c. [2−] Let P be an Eulerian poset with $\hat{0}$ and $\hat{1}$ removed. Show that $\widehat{Q}$ is Eulerian.

d. [2+] Define $F_n(x) = \sum_{k=1}^n k! S(n, k) x^{k-1}$, where $S(n, k)$ denotes a Stirling number of the second kind. By letting $E = B_n$ in (c), deduce that

$$F_n(x) = (-1)^{n-1} F_n(-x - 1). \tag{3.124}$$

142. a. [2] We say that a finite graded poset P of rank n is *chain-partitionable*, or just *partitionable*, if for every maximal chain K of P there is a chain $r(K) \subseteq K$ (the *restriction* of K) such that every chain (including $\emptyset$) of P lies in exactly one of the intervals $[r(K), K]$ of Q_0. Given a chain C of P, define its *rank set* $\rho(C) = \{\rho(t) : t \in C\} \subseteq [0, n]$. Show that if P is partitionable, then $\beta(P, S)$ is equal to the number of maximal chains K of P for which $\rho(r(K)) = S$. Thus, a necessary condition that P is partitionable is that $\beta(P, S) \geq 0$ for all $S \subseteq [0, n]$.

b. [2+] Show that if P is a poset for which $\widehat{P} := P \cup \{\hat{0}, \hat{1}\}$ is R-labelable, then P is partitionable.

c. [5] Is every Cohen-Macaulay poset partitionable?

143. a. [3–] If P is a poset, then the *comparability graph* Com(P) is the graph whose vertices are the elements of P, and two vertices s and t are connected by an (undirected) edge if $s < t$ or $t < s$. Show that the order polynomial $\Omega_P(m)$ of a finite poset P depends only on Com(P).

   b. [2] Give an example of two finite posets P, Q for which Com(P) $\not\cong$ Com(Q) but $\Omega_P(m) = \Omega_Q(m)$.

144. [2]* Let B_k denote a boolean algebra of rank k, and $\Omega_{B_k}(m)$ its order polynomial. Show that $\Omega_{B_{n+1}}(2) = \Omega_{B_n}(3)$.

145. [2+] Let $\Omega_P(n)$ denote the order polynomial of the finite poset P, so from Section 3.12 we have $\Omega_P(n) = Z(J(P), n)$. Let $p = \#P$. Use Example 3.9.6 to give another proof of the reciprocity theorem for order polynomials (Theorem 3.15.10) in the case of natural labelings, that is, for $n \in \mathbb{P}$, $(-1)^p \Omega_P(-n)$ is equal to the number of strict order-preserving maps $\tau \colon P \to \mathbf{n}$.

146. [1+] Compute $\Omega_P(n)$ and $(-1)^p \Omega_P(-n)$ explicitly when (i) P is a p-element chain, and (ii) P is a p-element antichain.

147. [1+] Compute $Z(L, n)$ when L is the lattice of faces of each of the five Platonic solids.

148. [2]* Let P be a p-element poset. Find a simple expression for $\sum_\omega \Omega_{P,\omega}(n)$, where ω ranges over all $p!$ labelings of P.

149. [3] Let Y be Young's lattice (defined in Section 3.4). Fix $\mu \le \lambda$ in Y, and let $Z(n) = \zeta^n(\mu, \lambda)$ be the zeta polynomial of the interval $[\mu, \lambda]$. Choose r so that $\lambda_{r+1} = 0$, and set $\binom{a}{b} = 0$ if $a < 0$ (in contravention to the usual definition). Show that

$$Z(n+1) = \det \left[\binom{\lambda_i - \mu_j + n}{i - j + n} \right]_{1 \le i, k \le r}.$$

150. a. [3] Let $S = \{a_1, \ldots, a_j\}_< \subset \mathbb{P}$. Define $f_S(n)$ to be the number of chains $\lambda^0 < \lambda^1 < \cdots < \lambda^j$ of partitions λ^i in Young's lattice Y such that $\lambda^0 \vdash n$ and $\lambda^i \vdash n + a_i$ for $i \in [j]$. Thus in the notation of Section 3.13, we have $f_S(n) = \alpha_Y(T)$, where $T = \{n, n + a_1, \ldots, n + a_j\}$. Set

$$\sum_{n \ge 0} f_S(n) q^n = P(q) A_S(q),$$

where $P(q) = \prod_{i \ge 1}(1 - q^i)^{-1}$. For instance, $A_\emptyset(q) = 1$. Show that $A_S(q)$ is a rational function whose denominator can be taken as

$$\phi_{a_j}(q) = (1 - q)(1 - q^2) \cdots (1 - q^{a_j}).$$

   b. [2+] Compute $A_S(q)$ for $S \subseteq [3]$.

   c. [3–] Show that for $k \in \mathbb{P}$,

$$\sum_{S \subseteq [k]} (-1)^{k - \#S} A_S(q) = q^{\binom{k+1}{2}} \phi_k(q)^{-1}. \tag{3.125}$$

   d. [2+] Deduce from (c) that if $\beta_Y(S)$ is defined as in Section 3.13, then for $k \in \mathbb{N}$ we have

$$\sum_{n \ge 0} \beta_Y([n, n+k]) q^{n+k} = P(q) \sum_{i=0}^{k} \left(\frac{q^{i(i+3)/2}(-1)^{k-i}}{\phi_i(q)} \right) - \frac{(-1)^k}{1 - q}.$$

e. [2] Give a simple combinatorial proof that $A_{\{1\}}(q) = (1-q)^{-1}$.

151. a. [2+] Let P be a p-element poset, and let $S \subseteq [p-1]$ such that $\beta_{J(P)}(S) \neq 0$. Show that if $T \subseteq S$, then $\beta_{J(P)}(T) \neq 0$.

b. [5−] Find a "nice" characterization of the collections Δ of subsets of $[p-1]$ for which there exists a p-element poset P satisfying

$$\beta_{J(P)}(S) \neq 0 \Leftrightarrow S \in \Delta.$$

c. [2+] Show that (a) continues to hold if we replace $J(P)$ with any finite supersolvable lattice L of rank p.

152. a. [2+] Let P be a finite naturally labeled poset. Construct explicitly a simplicial complex Δ_P whose faces are the linear extensions of P, such that the dimension of the face w is $\mathrm{des}(w) - 1$. (In particular, the empty face $\emptyset \in \Delta_P$ is the linear extension $12 \cdots p$, where $p = \#P$.)

b. [2] Draw a picture (geometric realization) of Δ_P when P is a four-element antichain.

c. [3] Show that when $P = p\mathbf{1}$ (a p-element antichain), we have $\widetilde{H}_i(\Delta_P; \mathbb{Z}) \neq 0$ if and only if

$$\frac{p-4}{3} \leq i \leq \frac{2p-5}{3},$$

where $\widetilde{H}$ denotes reduced homology.

153. [2+] Let $p \in \mathbb{P}$ and $S \subseteq [p-1]$. What is the least number of linear extensions a p-element poset P can have if $\beta_{J(P)}(S) > 0$?

154. [3−] If L and L' are distributive lattices of rank n such that $\beta_L(S) = \beta_{L'}(S)$ for all $S \subseteq [n-1]$ (or equivalently $\alpha_L(S) = \alpha_{L'}(S)$ for all $S \subseteq [n-1]$), then are L and L' isomorphic?

155. a. [2+] Let P be a finite graded poset of rank n with $\hat{0}$ and $\hat{1}$, and suppose that every interval of P is self-dual. Let $S = \{n_1, n_2, \ldots, n_s\}_< \subseteq [n-1]$. Show that $\alpha_P(S)$ depends only on the multiset of numbers $n_1, n_2 - n_1, n_3 - n_2, \ldots, n_s - n_{s-1}, n - n_s$ (not on their order).

b. [2]* Let L be a finite distributive lattice for which every interval is self-dual. Show that L is a product of chains. (For a stronger result, see Exercise 3.166.)

c. [5] Find all finite modular lattices for which every interval is self-dual.

156. [2+] Let $P = \mathbb{N} \times \mathbb{N}$. For any finite $S \subset \mathbb{P}$ we can define $\alpha_P(S)$ and $\beta_P(S)$ exactly as in Section 3.13 (even though P is infinite). Show that if $S = \{m_1, m_2, \ldots, m_s\}_< \subset \mathbb{N}$, then

$$\beta_{\mathbb{N} \times \mathbb{N}}(S) = m_1(m_2 - m_1 - 1) \cdots (m_s - m_{s-1} - 1).$$

157. Let P be a finite graded poset of rank n with $\hat{0}$ and $\hat{1}$.

a. [2] Show that

$$\Delta^{k+1} Z(P, 0) = \sum_{\substack{S \subseteq [n-1] \\ \#S = k}} \alpha_P(S).$$

b. [2+] Show that

$$(1-x)^{n+1} \sum_{m \geq 0} Z(P, m) x^m = \sum_{k \geq 0} \beta_k x^{k+1},$$

where

$$\beta_k = \sum_{\substack{S \subseteq [n-1] \\ \#S = k}} \beta_P(S).$$

c. [2] Show that the characteristic polynomial of P is given by $\chi_P(q) = \sum_{k \geq 0} w_k q^{n-k}$, where

$$(-1)^k w_k = \beta_P([k-1]) + \beta_P([k]).$$

(Set $\beta_P([n]) = \beta_P([-1]) = 0$.)

158. a. [3–] Let $k, t \in \mathbb{P}$. Let $P_{k,t}$ denote the poset of all partitions π of the set $[kt] = \{1, 2, \ldots, kt\}$, ordered by refinement (i.e., $P_{k,t}$ is a subposet of Π_{kt}), satisfying the two conditions:
   a. Every block of π has cardinality divisible by k.
   b. If $a < b < c < d$ and if B and B' are blocks of π such that $a, c \in B$ and $b, d \in B'$, then $B = B'$.
   Show combinatorially that the zeta polynomial of $P_{k,t}$ is given by

$$Z(P_{k,t}, n+1) = \frac{((kn+1)t)_{t-1}}{t!}.$$

b. [1+] Note that $P_{k,t}$ always has a $\hat{1}$, and that $P_{1,t}$ has a $\hat{0}$. Use (a) to show that $P_{1,t}$ has C_t elements and that $\mu_{P_{1,t}}(\hat{0}, \hat{1}) = (-1)^{t-1} C_{t-1}$, where C_r denotes a Catalan number.

c. [3–] Show that $P_{2,t} \cong \mathrm{Int}(P_{1,t})$, the poset of intervals of $P_{1,t}$..

d. [3] Note that $P_{k,t}$ is graded of rank $t - 1$. If $S = \{m_1, \ldots, m_s\}_< \subseteq [0, t-2]$, then show thats

$$\alpha_{P_{k,t}}(S) = \frac{1}{t} \binom{t}{m_1} \binom{kt}{m_2 - m_1} \cdots \binom{kt}{m_s - m_{s-1}} \binom{kt}{t-1-m_s}.$$

e. [1] Deduce that $P_{k,t}$ has $\frac{1}{t} \binom{t}{m} \binom{kt}{m-1}$ elements of rank $t - m$ and has $k(kt)!^{t-2}$ maximal chains.

f. [3–] Let $\lambda \vdash t$. Show that the number N_λ of $\pi \in P_{1,t}$ of type λ (i.e., with blocks sizes $\lambda_1, \lambda_2, \ldots$) is given by

$$N_\lambda = \frac{(n)_{\ell(\lambda)-1}}{m_1(\lambda)! \cdots m_n(\lambda)!},$$

where λ has $m_i(\lambda)$ parts equal to i.

g. [2+] Use Exercise 3.125 to show that $P_{1,t}$ is a supersolvable lattice (though not semimodular for $t \geq 4$).

159. [2+] Define a partial order $A(\mathfrak{S}_n)$ on the symmetric group $\mathfrak{S}_n$, called the *absolute order*, as follows. We say that $u \lessdot v$ in $A(\mathfrak{S}_n)$ if $v = (i, j)u$ for some transposition (i, j), and if v has fewer cycles (necessarily exactly one less) than u. See Figure 3.54 for the case $n = 4$. Clearly, the maximal elements of $A(\mathfrak{S}_n)$ are the n-cycles, while there is a unique minimal element $\hat{0}$ (the identity permutation). Show that if w is an n-cycle then $[\hat{0}, w] \cong P_{1,n}$, where $P_{1,n}$ is defined in Exercise 3.158.

160. [2] Let P be a p-element poset. Define two labelings $\omega, \omega' : P \to [p]$ to be *equivalent* if $\mathcal{A}(P, \omega) = \mathcal{A}(P, \omega')$. Clearly, this definition of equivalence is an equivalence relation. For instance, one equivalence class consists of the natural labelings. Show that the number of equivalence classes is equal to the number of acyclic orientations

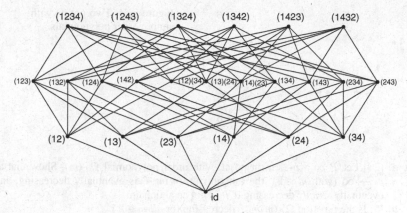

Figure 3.54 The absolute order on $\mathfrak{S}_4$.

of the Hasse diagram $\mathcal{H}$ of P, considered as an undirected graph. (See Exercise 3.109 for further information on the number of acyclic orientations of a graph.)

161. [2] Fix $j,k \geq 1$. Given two permutations $u = u_1 \cdots u_j$ and $v = v_1 \cdots v_k$ of disjoint finite sets U and V of integers, a *shuffle* of u and v is a permutation $w = w_1 \cdots w_{j+k}$ of $U \cup V$ such that u and v are subsequences of w. Let $\mathrm{Sh}(u,v)$ denote the set of shuffles of u and v. For instance, $\mathrm{Sh}(14,26) = \{1426, 1246, 1264, 2146, 2164, 2614\}$. In general, $\#\mathrm{Sh}(u,v) = \binom{j+k}{j}$. Let $S \subseteq [j+k-1]$. Show that the number of permutations in $\mathrm{Sh}(u,v)$ with descent set S depends only on $D(u)$ and $D(v)$ (the descent sets of u and v). (Use the theory of (P,ω)-partitions.)

162. **a.** [2+] Let P_1, P_2 be disjoint posets with $p_i = \#P_i$. Let ω be a labeling of $P_1 + P_2$ (disjoint union). Let ω_i be the labeling of P_i whose labels are in the same relative order as they are in the restriction of ω to P_i. Show that

$$W_{P_1+P_2,\omega}(x) = \binom{p_1+p_2}{p_1}_x W_{P_1,\omega_1}(x)W_{P_2,\omega_2}(x),$$

where $\binom{p_1+p_2}{p_1}_x$ indicates that the q-binomial coefficient should be taken in the variable x.

b. [2] Let $\{B_1,\ldots,B_k\} \in \Pi_n$, and let w^i be a permutation of B_i. Extending to k permutations the definition of shuffle in Exercise 3.161, define a *shuffle* of $w^1,\ldots,w^k$ to be a permutation $w = a_1 \cdots a_n$ of $[n]$ such that the subword of w consisting of letter from B_i is w^i. For instance 469381752 is a shuffle of 4812, 67, and 935. Let $\mathrm{sh}(w^1,\ldots,w^k)$ denote the set of all shuffles of $w^1,\ldots,w^k$, so if $\#B_i = b_i$ then

$$\#\mathrm{sh}(w^1,\ldots,w^k) = \binom{n}{b_1,\ldots,b_k}.$$

Show that

$$\sum_{w\in\mathrm{sh}(w^1,\ldots,w^k)} x^{\mathrm{maj}(w)} = x^\alpha \binom{n}{b_1,\ldots,b_k}_x, \tag{3.126}$$

where $\alpha = \sum_i \mathrm{maj}(w^i)$.

c. [2+]* Deduce from (b) that the minimum value of $\mathrm{maj}(w)$ for $w \in \mathrm{sh}(w^1,\ldots,w^k)$ is equal to $\sum \mathrm{maj}(w^i)$, and that this value is achieved for a unique w. Find an explicit description of this extremal permutation w.

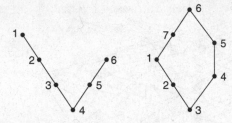

Figure 3.55 Two posets with simple order polynomials.

163. a. [2+] Let P be a p-element poset, with order polynomial $\Omega_P(m)$. Show that as $m \to \infty$ (with $m \in \mathbb{P}$), the function $\Omega_P(m)m^{-p}$ is eventually decreasing, and eventually *strictly* decreasing if P is not an antichain.
   b. [5] Is the function $\Omega_P(m)m^{-p}$ decreasing for *all* $m \in \mathbb{P}$?

164. [2+] Let P be a finite poset. Does the order polynomial $\Omega_P(m)$ always have nonnegative coefficients?

165. [3+] Let (P, ω) be a finite labeled poset. Does the (P, ω)-Eulerian polynomial $A_{P,\omega}(x)$ have only real zeros? What if ω is a natural labeling?

166. Let $(P, \omega) = \{t_1, \ldots, t_p\}$ be a labeled p-element poset. Define the formal power series $G_{P,\omega}(x)$ in the variables $x = (x_0, x_1, \ldots)$ by

$$G_{P,\omega}(x) = \sum_\sigma x_{\sigma(t_1)} \cdots x_{\sigma(t_p)} = \sum_\sigma x_0^{\#\sigma^{-1}(0)} x_1^{\#\sigma^{-1}(1)} \cdots,$$

where the sums range over all (P, ω)-partitions $\sigma : P \to \mathbb{N}$.
   a. [3] Suppose that ω is natural, and write $G_P(x)$ for $G_{P,\omega}(x)$. Show that $G_P(x)$ is a symmetric function (i.e., $G_P(x) = G_P(wx)$ for any permutation w of $\mathbb{N}$, where $wx = (x_{w(0)}, x_{w(1)}, \ldots)$) if and only if P is a disjoint union of chains.

NOTE. It is easily seen that $G_P(x)$ is symmetric if and only if for

$$S = \{n_1, n_2, \ldots, n_s\}_< \subseteq [p-1],$$

the number $\alpha_{J(P)}(S)$ depends only on the multiset of numbers $n_1, n_2 - n_1, \cdots, n_s - n_{s-1}, p - n_s$ (not on their order). See Exercise 3.155.
   b. [5] Show that $G_{P,\omega}(x)$ is symmetric if and only if P is isomorphic to a (finite) convex subset of $\mathbb{N} \times \mathbb{N}$, labeled so that $\omega(i, j) > \omega(i+1, j)$ and $\omega(i, j) < \omega(i, j+1)$.

167. a. [2+]* Let P be a finite poset that is a disjoint union of two chains, with a $\hat{0}$ or $\hat{1}$ or both added. Label P so that the labels i and $i + 1$ always occur on two elements that form an edge of the Hasse diagram. This gives a labeled poset (P, ω). Two examples are shown in Figure 3.55.
   Show that all $w \in \mathcal{L}(P, \omega)$ have the same number of descents, and as a consequence give an explicit formula for the (P, ω)-order polynomial $\Omega_{P,\omega}(m)$.
   b. [3–] Show that if (P, ω) is a labeled poset such that all $w \in \mathcal{L}(P, \omega)$ have the same number of descents, then P is an ordinal sum of the posets of (a), and describe the possible labelings ω.

168. [3]* Let P be a finite naturally labeled poset. Suppose that every connected order ideal I of P is either principal, or else there is a *unique* way to write $I = I_1 \cup I_2$ (up to order) where I_1 and I_2 are connected order ideals properly contained in I. The poset P of Figure 3.56 is an example. Write as usual $(i) = 1 + q + \cdots + q^{i-1}$ and $(n)! = (1)(2) \cdots (n)$. Show that

Figure 3.56 A poset P for which $W_P(x)$ has a nice product formula.

$$W_P(x) = (n)! \frac{\prod_{\{I_1,I_2\}}(\#I_1 + \#I_2)}{\prod_I(\#I)},$$

where I runs over all *connected* order ideals of P, and $\{I_1, I_2\}$ runs over all pairs of incomparable (in $J(P)$) connected order ideals such that $I_1 \cap I_2 \neq \emptyset$. For the poset P of Figure 3.56 we get

$$W_P(x) = (5)! \frac{(6)}{(1)(1)(2)(2)(4)(5)} = 1 + x + 2x^2 + x^3 + 2x^4 + x^5 + x^6.$$

169. a. [2] Let $M = \{1^{r_1}, 2^{r_2}, \ldots, m^{r_m}\}$ be a finite multiset on $[m]$, and let $\mathfrak{S}_M$ be the set of all $\binom{r_1 + \cdots + r_m}{r_1, \ldots, r_m}$ permutations $w = (a_1, a_2, \ldots a_r)$ of M, where $r = r_1 + \cdots + r_m = \#M$. Let $\mathrm{des}(w)$ be the number of descents of w, and set

$$A_M(x) = \sum_{w \in \mathfrak{S}_M} x^{1 + \mathrm{des}(w)},$$

$$\overline{A}_M(x) = \sum_{w \in \mathfrak{S}_M} x^{r - \mathrm{des}(w)}.$$

Show that

$$\sum_{n \geq 0} \left(\!\!\binom{n}{r_1}\!\!\right)\left(\!\!\binom{n}{r_2}\!\!\right) \cdots \left(\!\!\binom{n}{r_m}\!\!\right) x^n = \frac{A_M(x)}{(1 - x)^{r+1}},$$

$$\sum_{n \geq 0} \binom{n}{r_1}\binom{n}{r_2} \cdots \binom{n}{r_m} x^n = \frac{\overline{A}_M(x)}{(1 - x)^{r+1}}.$$

b. [2+] Find the coefficients of $A_M(x)$ explicitly in the case $m = 2$.

170. Let us call a finite graded poset P (with rank function ρ) *pleasant* if the rank-generating function $F(L, q)$ of $L = J(P)$ is given by

$$F(L, q) = \prod_{t \in P} \frac{1 - q^{\rho(t)+2}}{1 - q^{\rho(t)+1}}.$$

In (a)–(g) show that the given posets P are pleasant. (Note that (a) is a special case of (b), and (c) is a special case of (d).)

a. [2] $P = m \times n$, where $m, n \in \mathbb{P}$.
b. [3] $P = l \times m \times n$, where $l, m, n \in \mathbb{P}$.
c. [2] $P = J(2 \times n)$, where $n \in \mathbb{P}$.
d. [3+] $P = m \times J(2 \times n)$, where $m, n \in \mathbb{P}$.
e. [3+] $P = J(3 \times n)$, where $n \in \mathbb{P}$.
f. [2+] $P = m \times (n \oplus (1+1) \oplus n)$, where $m, n \in \mathbb{P}$.
g. [3–] $P = m \times J(J(2 \times 3))$ and $P = m \times J(J(J(2 \times 3)))$, where $m \in \mathbb{P}$.
h. [5] Find a reasonable expression for $F(J(P))$, where $P = n_1 \times n_2 \times n_3 \times n_4$ or $P = J(4 \times n)$. (In general, these posets P are not pleasant.)

i. [5] Are there any other "nice" classes of connected pleasant posets? Can all pleasant posets be classified?

171. a. [2–] Let (P, ω) be a finite labeled poset and $m \in \mathbb{N}$. Define a polynomial

$$U_{P,\omega,m}(q) = \sum_{\sigma} q^{|\sigma|},$$

where σ ranges over all (P, ω)-partitions $\sigma : P \to [0, m]$. In particular, $U_{P,0}(q) = 1$ (as usual, the suppression of ω from the notation indicates that ω is natural) and $U_{P,\omega,m}(1) = \Omega_{P,\omega}(m + 1)$. Show that $U_{P,m}(q) = F(J(m \times P), q)$, the rank-generating function of $J(m \times P)$.

b. [2+] If $\#P = p$ and $0 \le i \le p - 1$, then define

$$W_{P,\omega,i}(q) = \sum_{w} q^{\mathrm{maj}(w)}, \tag{3.127}$$

where w ranges over all permutations in $\mathcal{L}(P, \omega)$ with exactly i descents. Note that $W_{P,\omega}(q) = \sum_i W_{P,\omega,i}(q)$. Show that for all $m \in \mathbb{N}$,

$$U_{P,\omega,m}(q) = \sum_{i=0}^{p-1} \binom{p+m-i}{p} W_{P,\omega,i}(q). \tag{3.128}$$

c. [2] Let ω^* be the labeling of the dual P^* defined by $\omega^*(t) = p + 1 - \omega(t)$. (Note that ω^* and $\overline{\omega}$ have the same values. However, ω^* is a labeling of P^*, while $\overline{\omega}$ is a labeling of P.) Show that

$$W_{P^*,\omega^*,i}(q) = q^{pi} W_{P,\omega,i}(1/q), \tag{3.129}$$

$$U_{P^*,\omega^*,m}(q) = q^{pm} U_{P,\omega,m}(1/q). \tag{3.130}$$

d. [1+]* The formula

$$\binom{a}{b} = \frac{(1 - q^a)(1 - q^{a-1}) \cdots (1 - q^{a-b+1})}{(1 - q^b)(1 - q^{b-1}) \cdots (1 - q)}$$

allows us to define $\binom{a}{b}$ for any $a \in \mathbb{Z}$ and $b \in \mathbb{N}$. Show that

$$\binom{-a}{b} = (-1)^b q^{-b(2a+b-1)/2} \binom{a+b-1}{b}$$

$$= (-1)^b q^{\binom{b+1}{2}} \binom{a+b-1}{b}_{1/q}.$$

e. [2+] Equation (3.96) and part (d) above allow us to define $U_{P,\omega,m}(q)$ for any $m \in \mathbb{Z}$. Show that for $m \in \mathbb{P}$,

$$U_{P,\omega,-m}(q) = (-1)^p \sum_{\tau} q^{-|\tau|},$$

where τ ranges over all $(P, \overline{\omega})$-partitions $\tau : P \to [m - 1]$.

f. [3–] If $t \in P$, then let $\delta(t)$ and δ_i, $0 \le i \le \ell = \ell(P)$, be as in Section 3.15.4. Define

$$\Delta_r = \delta_r + \delta_{r+1} + \cdots + \delta_\ell, \quad 1 \le r \le \ell,$$

and set
$$M(P) = [p-1] - \{\Delta_1, \Delta_2, \ldots, \Delta_\ell\}.$$

Show that the degree of $W_{P,i}(q)$ is equal to the sum of the largest i elements of $M(P)$. Note also that if P is graded of rank ℓ, then

$$\Delta_r = \#\{t \in P : \rho(t) \leq \ell - r\}.$$

172. Let $P = \{t_1, \ldots, t_p\}$ be a finite poset. We say that P is *Gaussian* if there exist integers $h_1, \ldots, h_p > 0$ such that for all $m \in \mathbb{N}$,

$$U_{P,m}(q) = \prod_{i=1}^{p} \frac{1 - q^{m+h_i}}{1 - q^{h_i}}, \tag{3.131}$$

where $U_{P,m}(q)$ is given by Exercise 3.171.
a. [3–] Show that P is Gaussian if and only if every connected component of P is Gaussian.
b. [3–] If P is connected and Gaussian, then show that every maximal chain of P has the same length ℓ. (Thus, P is graded of rank ℓ.)
c. [3] Let P be connected and Gaussian, with rank function ρ (which exists by (b)). Show that the multisets $\{h_1, \ldots, h_p\}$ and $\{1 + \rho(t) : t \in P\}$ coincide.

NOTE. It follows easily from (c) that a finite connected poset P is Gaussian if and only if $P \times m$ is pleasant (as defined in Exercise 3.170) for all $m \in \mathbb{P}$.
d. [2+] Suppose that P is connected and Gaussian, with $h_1, \ldots, h_p$ labeled so that $h_1 \leq h_2 \leq \cdots \leq h_p$. Show that $h_i + h_{p+1-i} = \ell(P) + 2$ for $1 \leq i \leq p$.
e. [2+] Let P be connected and Gaussian. Show that every element of P of rank one covers exactly one minimal element of P.
f. [3+] Show that the following posets are Gaussian:
   i. $r \times s$, for all $r, s \in \mathbb{P}$,
   ii. $J(2 \times r)$, for all $r \in \mathbb{P}$,
   iii. the ordinal sum $r \oplus (1+1) \oplus r$, for all $r \in \mathbb{P}$,
   iv. $J(J(2 \times 3))$,
   v. $J(J(J(2 \times 3)))$.
g. [5] Are there any other connected Gaussian posets? In particular, must a connected Gaussian poset be a distributive lattice?

173. Let (P, ω) be a labeled poset, and set

$$\mathcal{E}(P) = \{(s,t) : s \lessdot t\}.$$

Define $\epsilon = \epsilon_\omega : \mathcal{E}(P) \to \{-1, 1\}$ by

$$\epsilon(s,t) = \begin{cases} 1, & \omega(s) < \omega(t), \\ -1, & \omega(s) > \omega(t). \end{cases}$$

We say that ϵ is a *sign-grading* if for all maximal chains $t_0 \lessdot t_1 \lessdot \cdots \lessdot t_\ell$ in P the quantity $\sum_{i=1}^{\ell} \epsilon(t_{i-1}, t_i)$ is the same, denoted $r(\epsilon)$ and called the *rank* of ϵ. A labeled poset (P, ω) with a sign-grading ϵ is called a *sign-graded* poset. In that case, we have a *rank function* $\rho = \rho_\epsilon$ given by

$$\rho(t) = \sum_{i=1}^{m} \epsilon(t_{i-1}, t_i),$$

where $t_0 \lessdot t_1 \lessdot \cdots \lessdot t_m = t$ is a saturated chain from a minimal element t_0 to t. (The definition of sign-grading insures that $\rho(t)$ is well-defined.)

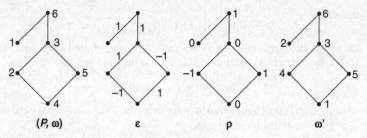

Figure 3.57 A labeling, sign-grading, rank function, and canonical labeling.

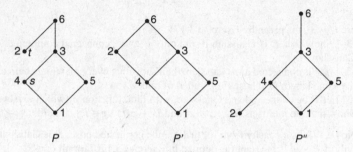

Figure 3.58 The procedure $P \to (P', P'')$.

a. [1+]* Suppose that ω is natural. Show that ϵ is a sign-grading if and only if P is graded.

b. [2]* Show that a finite poset P has a labeling ω for which (P, ω) is a sign-graded poset if and only if the lengths of all maximal chains of P have the same parity.

c. [2+]* Suppose that ω and ω' are labelings of P which both give rise to sign-gradings ϵ and ϵ'. Show that the (P, ω) and (P, ω')-Eulerian polynomials are related by

$$x^{r(\epsilon)/2} A_{P,\omega}(x) = x^{r(\epsilon')/2} A_{P,\omega'}(x).$$

d. [2]* Suppose that (P, ω) is a sign-graded poset whose corresponding rank function ρ takes on only the values 0 and 1, and $\rho(s) < \rho(t)$ implies $\omega(s) < \omega(t)$. We then call ω a *canonical* labeling of P. Show that every sign-graded poset (P, ω) has a canonical labeling. Figure 3.57 shows a poset (P, ω), the sign-grading ϵ_ω, the rank function ρ, and a canonical labeling ω'.

e. [2+]* Let (P, ω) be sign-graded, where ω is a canonical labeling. Let $s \parallel t$ in P, with $\rho(t) = \rho(s) + 1$. Define

$$P' = P \text{ with } s < t \text{ adjoined,}$$

$$P'' = P \text{ with } s > t \text{ adjoined.}$$

Thus ω continues to be a labeling of P' and P''. Show that P' and P'' are sign-graded, and that ω is a canonical labeling for both. Figure 3.58 shows an example of this decomposition for the poset of Figure 3.57. We take s to be the element labeled 4 (of rank 1) and t to be the element labeled 2 (of rank 1).

f. [2–]* Show that $\mathcal{L}(P, \omega) = \mathcal{L}(P', \omega) \,\dot\cup\, \mathcal{L}(P'', \omega)$.

g. [2+] Write $A'_j(x) = A_j(x)/x$, where $A_j(x)$ is an Eulerian polynomial. Iterate the procedure $P \to (P', P'')$ as long as possible. Deduce that if (P, ω) is sign-graded,

Figure 3.59 A poset for Exercise 3.173(i).

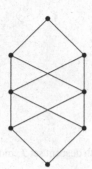

Figure 3.60 The poset $\mathbf{1} \oplus 21 \oplus 21 \oplus 21 \oplus \mathbf{1}$.

then we can write the (P, ω)-Eulerian polynomial $A_{P,\omega}(x)$ as a sum of terms of the form $x^b A'_{a_1}(x) \cdots A'_{a_k}(x)$, where $b \in \mathbb{N}$. Moreover, all these terms are symmetric (in the sense of Exercise 3.50) with the same center of symmetry.

    h. [2] Deduce from (g) that the coefficients of $A_{P,\omega}(x)$ are symmetric and unimodal.

    i. [2–] Carry out the procedure of (g) for the poset of Figure 3.59, naturally labeled.

174. a. [2–] Show that a finite graded poset with $\hat{0}$ and $\hat{1}$ is semi-Eulerian if and only if for all $s < t$ in P except possibly $(s,t) = (\hat{0}, \hat{1})$, the interval $[s,t]$ has as many elements of odd rank as of even rank. Show that P is Eulerian if in addition P has as many elements of odd rank as of even rank.

    b. [2] Show that if P is semi-Eulerian of rank n, then

$$(-1)^n Z(P, -m) = Z(P, m) + m((-1)^n \mu_P(\hat{0}, \hat{1}) - 1).$$

    c. [2] Show that a semi-Eulerian poset of odd rank n is Eulerian.

175. [2+] Suppose that P and Q are Eulerian, and let $P' = P - \{\hat{0}\}$, $Q' = Q - \{\hat{0}\}$, $R = (P' \times Q') \cup \{\hat{0}\}$. Show that R is Eulerian.

176. a. [2] Let P_n denote the ordinal sum $\mathbf{1} \oplus 21 \oplus 21 \oplus \cdots \oplus 21 \oplus \mathbf{1}$ (n copies of 21). For example, P_3 is shown in Figure 3.60. Compute $\beta_{P_n}(S)$ for all $S \subseteq [n]$.

    b. [1+] Use (a) and Exercise 3.157(b) to compute $\sum_{m \geq 0} Z(P_n, m) x^m$.

    c. [2+] It is easily seen that P_n is Eulerian. Compute the polynomials $f(P_n, x)$ and $g(P_n, x)$ of Section 3.16.

177. a. [2] Let L_n denote the lattice of faces of an n-dimensional cube, ordered by inclusion. Show that L_n is isomorphic to the poset $\mathrm{Int}(B_n)$ with a $\hat{0}$ adjoined, where B_n denotes a boolean algebra of rank n.

    b. [2] Show that L_n is isomorphic to Λ^n with a $\hat{0}$ adjoined, where Λ is the three-element poset $\bigwedge$.

    c. [2] Let P_n be the poset of Exercise 3.176. Show that L_n is isomorphic to the poset of chains of P_n that don't contain $\hat{0}$ and $\hat{1}$ (including the empty chain), ordered by reverse inclusion, with a $\hat{0}$ adjoined.

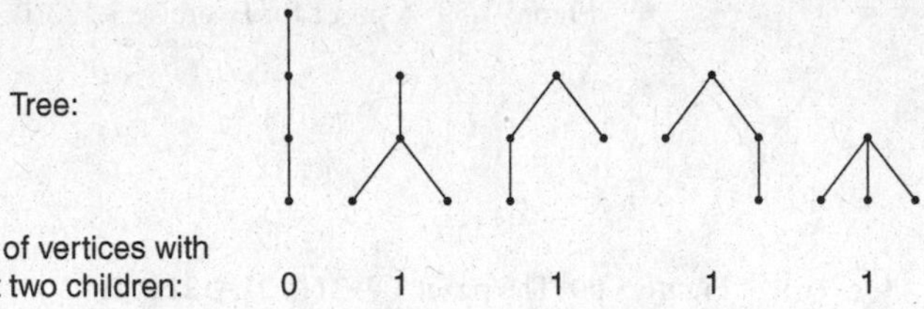

Tree:

Number of vertices with
at least two children: 0   1   1   1   1

Figure 3.61 Plane trees with four vertices.

d. [3–] Let $S \subseteq [n]$. Show that

$$\beta_{L_n}(S) = \sum_{i=0}^{n} \binom{n}{i} D_{n+1}(\overline{S}, i+1),$$

where $D_m(T, j)$ denotes the number of permutations of $[m]$ with descent set T and last element j, and where $\overline{S} = [n] - S$.

e. [2+] Compute $Z(L_n, m)$.

f. [3] Since L_n is the lattice of faces of a convex polytope, it is Eulerian by Proposition 3.8.9. Compute the polynomial $g(L_n, x)$ of Section 3.16. Show in particular that

$$g(L_n, 1) = \frac{1}{n+1}\binom{2n}{n} \quad \text{and} \quad f(L_n, 1) = 2\binom{2(n-1)}{n-1}.$$

g. [3–] Use (f) to show that $g(L_n, x) = \sum a_i x^i$, where a_i is the number of plane trees with $n+1$ vertices such that exactly i vertices have at least two children. For example, see Figure 3.61 for $n = 3$, which shows that $g(L_3, x) = 1 + 4x$.

178. [2+]* Let $f(n)$ be the total number of chains containing $\hat{0}$ and $\hat{1}$ in the lattice L_n of the previous exercise. Show that

$$\sum_{n \geq 0} f(n)\frac{x^n}{n!} = \frac{e^x}{2 - e^{2x}}.$$

179. [4] Let L be the face lattice of a convex polytope $\mathcal{P}$. Show that the coefficients of $g(L, x)$ are nonnegative. Equivalently (since $f(L, 0) = g(L, 0) = 1$), the coefficients of $f(L, x)$ are nonnegative and *unimodal* (i.e., weakly increase to a maximum and then decrease).

180. a. [2+] Let $n, d \in \mathbb{P}$ with $n \geq d + 1$. Define L'_{nd} to be the poset of all subsets S of $[n]$, ordered by inclusion, satisfying the following condition: S is contained in a d-subset T of $[n]$ such that whenever $1 \leq i \notin T$, $[i+1, i+k] \subseteq T$, and $n \geq i+k+1 \notin T$, then k is even. Let L_{nd} be L'_{nd} with a $\hat{1}$ adjoined. Show that L_{nd} is an Eulerian lattice of rank $d + 1$. The lattice L_{42} is shown in Figure 3.62.

b. [2]* Show that L_{nd} has $\binom{n}{k}$ elements of rank k for $0 \leq k \leq \lfloor d/2 \rfloor$.

181. a. [3–] Let $L = L_0 \cup L_1 \cup \cdots \cup L_{d+1}$ be an Eulerian lattice of rank $d + 1$. Suppose that the truncation $L_0 \cup L_1 \cup \cdots \cup L_{\lceil (d+1)/2 \rceil}$ is isomorphic to the truncation $M = M_0 \cup M_1 \cup \cdots \cup M_{\lceil (d+1)/2 \rceil}$, where M is a boolean algebra B_n of rank $n = \#L_1$. Does it follow that $n = d + 1$ and that $L \cong M$? Note that by Exercise 3.180 this result is best possible (i.e., $\lceil (d+1)/2 \rceil$ cannot be replaced with $\lceil (d-1)/2 \rceil$).

b. [2] What if we only assume that L is an Eulerian poset?

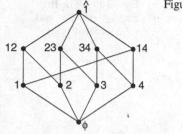

Figure 3.62 The Eulerian lattice L_{42}.

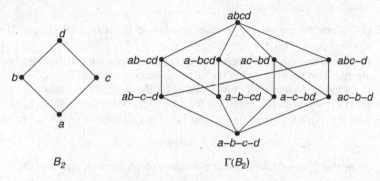

B_2 $\Gamma(B_2)$

Figure 3.63 The Eulerian lattice Γ_{B_2}.

182. [3–] Let P be a finite poset, and let π be a partition of the elements of P such that every block of π is connected (as a subposet of P). Define a relation $\leq$ on the blocks of π as follows: $B \leq B'$ if for some $t \in B$ and $t' \in B'$ we have $t \leq t'$ in P. If this relation is a partial order, then we say that π is P-*compatible*. Let Γ_P be the set of all P-compatible partitions of P, ordered by refinement (so $\Gamma(P)$ is a subposet of Π_P). See Figure 3.63 for an example. Show that Γ_P is an Eulerian lattice.

183. a. [2–]* Define a partial order, called the *(strong) Bruhat order* on the symmetric group $\mathfrak{S}_n$, by defining its cover relations as follows. We say that w covers v if $w = (i, j)v$ for some transposition (i, j) and if $\mathrm{inv}(w) = 1 + \mathrm{inv}(v)$. For instance, 75618324 covers 73618524; here $(i, j) = (2, 6)$. We always let the "default" partial ordering of $\mathfrak{S}_n$ be the Bruhat order, so any statement about the poset structure of $\mathfrak{S}_n$ refers to the Bruhat order. The poset $\mathfrak{S}_3$ is shown in Figure 3.64(a), while the solid and broken lines of Figure 3.65 show $\mathfrak{S}_4$. Show that $\mathfrak{S}_n$ is a graded poset with $\rho(w) = \mathrm{inv}(w)$, so that the rank-generating function is given by $F(\mathfrak{S}_n, q) = (n)!$.

    b. [3–] Given $w = a_1 a_2 \cdots a_n \in \mathfrak{S}_n$, define a left-justified triangular array T_w whose ith row consists of $a_1, \ldots, a_i$ written in increasing order. For instance, if $w = 31524$, then

$$
T_w = \begin{array}{l} 3 \\ 1\,3 \\ 1\,3\,5 \\ 1\,2\,3\,5 \\ 1\,2\,3\,4\,5. \end{array}
$$

Show that $v \leq w$ if and only if $T_v \leq T_w$ (component-wise ordering).

    c. [3] Show that $\mathfrak{S}_n$ is Eulerian.

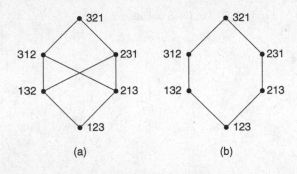

Figure 3.64 The Bruhat order $\mathfrak{S}_3$ and weak order $W(\mathfrak{S}_3)$.

(a)   (b)

d. [2+] Show that the number of cover relations in $\mathfrak{S}_n$ is $(n+1)!(H_{n+1}-2)+n!$, where $H_{n+1}=1+\frac{1}{2}+\frac{1}{3}+\cdots+\frac{1}{n+1}$.

e. [2+]* Find the number of elements $w \in \mathfrak{S}_n$ for which the interval $[\hat{0},w]$ is a boolean algebra. Your answer shouldn't involve any sums or products.

f. [5–] Find the total number of intervals of $\mathfrak{S}_n$ that are boolean algebras.

g. [3+] Let $v \lessdot w$ in $\mathfrak{S}_n$, so $w=(i,j)v$ for some $i<j$. Define the weight $\omega(v,w)=j-i$. Set $r=\binom{n}{2}$. If $C: \hat{0}=v_0 \lessdot v_1 \lessdot \cdots \lessdot v_r = \hat{1}$ is a maximal chain of $\mathfrak{S}_n$, then define

$$\omega(C) = \omega(v_0,v_1)\omega(v_1,v_2)\cdots\omega(v_{r-1},v_r).$$

Show that $\sum_C \omega(C) = r!$, where C ranges over all maximal chains of $\mathfrak{S}_n$.

184. [3] Let $\mathfrak{I}_n$ denote the subposet of $\mathfrak{S}_n$ (under Bruhat order) consisting of the involutions in $\mathfrak{S}_n$. Show that $\mathfrak{I}_n$ is Eulerian.

185. a. [2–]* Define a partial order $W(\mathfrak{S}_n)$, called the *weak (Bruhat) order* on $\mathfrak{S}_n$, by defining its cover relations as follows. We say that w covers v if $w=(i,i+1)v$ for some *adjacent* transposition $(i,i+1)$ and if $\mathrm{inv}(w)=1+\mathrm{inv}(v)$. For instance, 75618325 covers 75613825. The poset $W(\mathfrak{S}_3)$ is shown in Figure 3.64(b), while the solid lines of Figure 3.65 show $W(\mathfrak{S}_4)$. Show that $W(\mathfrak{S}_n)$ is a graded poset with $\rho(w) = \mathrm{inv}(w)$, so that the rank-generating function is given by $F(W(\mathfrak{S}_n),q)=(n)!$.

b. [2+] Show that $W(\mathfrak{S}_n)$ is a lattice.

c. [2] Show that the number of cover relations in $W(\mathfrak{S}_n)$ is $(n-1)n!/2$.

d. [3–] Let μ denote the Möbius function of $W(\mathfrak{S}_n)$. Show that

$$\mu(v,w) = \begin{cases} (-1)^k, & \text{if } w \text{ can be obtained from } v \text{ by reversing the elements} \\ & \text{in each of } k+1 \text{ disjoint increasing factors of } v, \\ 0, & \text{otherwise.} \end{cases}$$

e. [3] Show that the zeta polynomial of $W(\mathfrak{S}_n)$ satisfies

$$Z(W(\mathfrak{S}_n),-j)=(-1)^{n-1}j, \quad 1 \le j \le n-1.$$

f. [2]* Characterize permutations $w \in W(\mathfrak{S}_n)$ for which the interval $[\hat{0},w]$ is a boolean algebra in terms of pattern avoidance.

g. [2+]* Find the number of elements $w \in W(\mathfrak{S}_n)$ for which the interval $[\hat{0},w]$ is a boolean algebra. Your answer shouldn't involve any sums or products. More generally, find a simple formula for the generating function $\sum_{n \ge 0} \sum_w q^{\mathrm{rank}(w)} x^n$, where w ranges over all elements of $W(\mathfrak{S}_n)$ for which the interval $[\hat{0},w]$ is a boolean algebra.

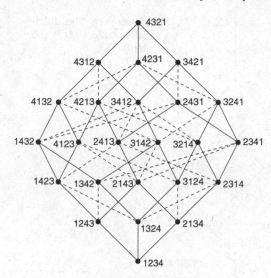

Figure 3.65 The Bruhat order $\mathfrak{S}_4$ and weak order $W(\mathfrak{S}_4)$.

h. [3–]* Let $f(n,i)$ denote the total number of intervals of $W(\mathfrak{S}_n)$ that are isomorphic to the boolean algebra B_i. Show that

$$\sum_{n\geq 0}\sum_{i\geq 0} f(n,i)q^i \frac{x^n}{n!} = \frac{1}{1-x-\frac{qx^2}{2}}$$

$$= 1 + x + (q+2)\frac{x^2}{2!} + (6q+6)\frac{x^3}{3!}$$

$$+ (6q^2+36q+24)\frac{x^4}{4!} + (90q^2+240q+120)\frac{x^5}{5!}$$

$$(90q^3+1080q^2+1800q+720)\frac{x^6}{6!} + \cdots.$$

i. [3–]* Find the number of elements $w \in W(\mathfrak{S}_n)$ for which the interval $[\hat{0}, w]$ is a distributive lattice. Your answer shouldn't involve any sums or products.

j. [5–] Find the total number of intervals of $W(\mathfrak{S}_n)$ that are distributive lattices. The values for $1 \leq n \leq 8$ are 1, 2, 16, 124, 1262, 15898, 238572, 4152172.

k. [3+] Show that the number M_n of maximal chains of $W(\mathfrak{S}_n)$ is given by

$$M_n = \frac{\binom{n}{2}!}{1^{n-1}\, 3^{n-2}\, 5^{n-3}\cdots(2n-3)^1}.$$

l. [3+] Let $v \lessdot w$ in $W(\mathfrak{S}_n)$, so $w = (i,i+1)v$ for some i. Define the weight $\sigma(v,w) = i$. Set $r = \binom{n}{2}$. If $C: \hat{0} = v_0 \lessdot v_1 \lessdot \cdots \lessdot v_r = \hat{1}$ is a maximal chain of $W(\mathfrak{S}_n)$, then define

$$\sigma(C) = \sigma(v_0,v_1)\sigma(v_1,v_2)\cdots\sigma(v_{r-1},v_r).$$

Show that $\sum_C \sigma(C) = r!$, where C ranges over all maximal chains of $\mathfrak{S}_n$.

m. [5–] Is the similarity between (l) and Exercise 3.183(g) just a coincidence?

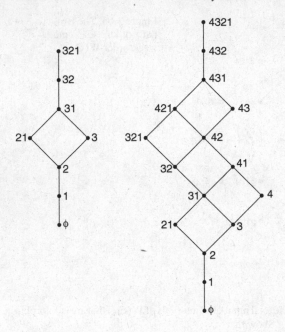

Figure 3.66 The distributive lattices $M(3)$ and $M(4)$.

186. a. [3–] Let $w \in \mathfrak{S}_n$ be separable, as defined in Exercise 3.14(b). Show that the rank-generating functions of the intervals $\Lambda_w = [\hat{0}, w]$ and $V_w = [w, \hat{1}]$ in $W(\mathfrak{S}_n)$ (where $\mathrm{rank}(w) = 0$ in V_w) satisfy

$$F(\Lambda_w, q) F(V_w, q) = (n)!.$$

b. [3–] Show that the polynomials $F(\Lambda_w, q)$ and $F(V_w, q)$ are symmetric and unimodal (as defined in Exercise 1.50).

c. [3–] Let $w = a_1 a_2 \cdots a_n \in \mathfrak{S}_n$ be 231-avoiding. Set $a_{n+1} = n + 1$. Show that

$$F(\Lambda_w, q) = \prod_{i=1}^{n} (c_i),$$

where c_i is the least positive integer for which $a_{i+c_i} > a_i$.

d. [5–] What can be said about other permutations $w \in \mathfrak{S}_n$ for which $F(\Lambda_w, q)$ is symmetric or, more strongly, is a divisor of $(n)!$?

187. a. [2]* For $n \geq 0$, define a partial order $M(n)$ on the set $2^{[n]}$ of all subsets of $[n]$ as follows. If $S = \{a_1, \ldots, a_s\}_> \in M(n)$ and $T = \{b_1, \ldots, b_t\}_> \in M(n)$, then $S \geq T$ if $s \geq t$ and $a_i \geq b_i$ for $1 \leq i \leq t$. The posets $M(3)$ and $M(4)$ are shown in Figure 3.66. Show that $M(n) \cong J(J(\mathbf{2} \times \mathbf{n}))$ and that the rank-generating function of $M(n)$ is given by

$$F(M(n), q) = (1 + q)(1 + q^2) \cdots (1 + q^n).$$

b. [2+] Consider the following variation $\mathcal{G}_n$ of the weak order on $\mathfrak{S}_n$, which we call the *greedy weak order*. It is a partial order on a certain subset (also denoted $\mathcal{G}_n$) of $\mathfrak{S}_n$. First, we let $12 \cdots n \in \mathcal{G}_n$. Suppose now that $w = a_1 a_2 \cdots a_n \in \mathcal{G}_n$ and that the permutations that cover w in the weak order $W(\mathfrak{S}_n)$ are obtained from w by transposing adjacent elements $(a_{i_1}, a_{i_1+1}), \ldots, (a_{i_k}, a_{i_k+1})$. In other words, the ascent

Figure   3.67 An   Eulerian
poset with a simple *cd*-index.

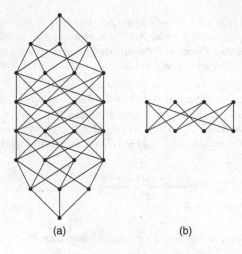

(a)   (b)

set of w is $\{i_1,\ldots,i_k\}$. Then the permutations that cover w in $\mathcal{G}_n$ are obtained by transposing one of the pairs (a_{i_j},a_{i_j+1}) for which (a_{i_j},a_{i_j+1}) is *minimal* in the poset $\mathbb{P}\times\mathbb{P}$ among all the pairs $(a_{i_1},a_{i_1+1}),\ldots,(a_{i_k},a_{i_k+1})$. For instance, the elements that cover 342561 in weak order are obtained by transposing the pairs $(3,4)$, $(2,5)$, and $(5,6)$. The minimal pairs are $(3,4)$ and $(2,5)$. Hence in $\mathcal{G}_6$, 342561 is covered by 432561 and 345261. Show that $\mathcal{G}_n \cong M(n-1)$.

   c. [2+] Describe the elements of the set $\mathcal{G}_n$.

188. [3–] Let $a = (a_1,a_2,\ldots,a_n)$ be a finite sequence of integers with no two consecutive elements equal. Let $P = P(a)$ be the set of all subsequences $a' = (a_{i_1},a_{i_2},\ldots,a_{i_m})$ (so $1 \le i_1 < i_2 < \cdots < i_m \le n$) of a such that no two consecutive elements of a' are equal. Order P by the rule $b \le c$ if b is a subsequence of c. Show that P is Eulerian.

189. [2+]* Let P be an Eulerian poset of rank $d+1$ with d atoms, such that $P - \{\hat{1}\}$ is a simplicial poset. Show that if d is even, then P has an even number of coatoms.

190. **a.** [3–] Let P_n be the poset of rank $n+1$ illustrated in Figure 3.67(a) for $n = 6$. The restriction of P_n to ranks i and $i+1$, $2 \le i \le n-2$, is the poset of Figure 3.67(b). Show that P_n is Eulerian with cd-index equal to the sum of all cd monomials of degree n (where $\deg c = 1$, $\deg d = 2$).

   b. [2]* Let $M(n)$ denote the number of maximal chains of P_n. Show that

$$\sum_{n \ge 0} M(n)x^n = \frac{1}{1-2x-2x^2}.$$

191. **a.** [2]* Let P and Q be Eulerian posets. Show that $P * Q$ is Eulerian, where $P * Q$ is the join of P and Q as defined by equation (3.86).

   b. [2]* Show that $\Phi_{P*Q}(c,d) = \Phi_P(c,d)\Phi_Q(c,d)$, where Φ denotes the cd-index.

192. [4–] Let P be a Cohen–Macaulay and Eulerian poset. Such posets are also called *Gorenstein* posets*, as in the topological digression of Section 3.16. Show that the cd-index $\Phi_P(c,d)$ has nonnegative coefficients.

193. **a.** [3–] Give an example of an Eulerian poset whose cd-index has a negative coefficient.

   b. [3] Strengthening (a), give an example of an Eulerian poset P whose flag h-vector β_P has a negative value.

194. [5–] Let P be a finite graded poset of rank n, with p_i elements of rank i. We say that P is *rank-symmetric* if $p_i = p_{n-i}$ for $0 \leq i \leq n$. Find the dimension $d(n)$ of the space spanned, say over $\mathbb{Q}$, by the flag f-vectors (or equivalently flag h-vectors) of all Eulerian posets P of rank n with the additional condition that every interval of P is rank-symmetric.

195. [3] Fix $k \geq 1$, and let $\mathcal{F}_n(k)$ be the space spanned over $\mathbb{Q}$ by the flag f-vectors of all graded posets of rank n with $\hat{0}$ and $\hat{1}$ such that every interval of rank k is Eulerian. Such posets are called *k-Eulerian*. Since $\mathcal{F}_n(k) = \mathcal{F}_n(k+1)$ for k even by Exercise 3.174(c), we may assume that k is odd, say $k = 2j + 1$. Let $d_n(k) = \dim \mathcal{F}_n(k)$. Show that

$$\sum_{n \geq 0} d_n(2j+1)x^n = 1 + \frac{x(1+x)(1+x^{2j})}{1-x-x^2-x^{2j+2}}.$$

196. a. [2] Show that if $B(n)$ is the factorial function of a binomial poset, then $B(n)^2 \leq B(n-1)B(n+1)$.

b. [5] What functions $B(n)$ are factorial functions of binomial posets? In particular, can one have $B(n) = F_1 F_2 \cdots F_n$, where F_i is the ith Fibonacci number ($F_1 = F_2 = 1$, $F_{n+1} = F_n + F_{n-1}$)?

197. [3–] Show that there exist an uncountable number of pairwise nonisomorphic binomial posets P_α such that (a) they all have the same factorial function $B(n)$, and (b) each P_α has a maximal chain $\hat{0} = t_0 \lessdot t_1 \lessdot t_2 \lessdot \cdots$ such that $P_\alpha = \bigcup_{n \geq 0}[\hat{0}, t_n]$.

198. [3–] Let P be an Eulerian binomial poset (i.e., a binomial poset for which every interval is Eulerian). Show that either every n-interval of P is a boolean algebra B_n, or else every n-interval is a "butterfly poset" or ladder $1 \oplus A_2 \oplus A_2 \oplus \cdots \oplus A_2 \oplus 1$, where $A_2 = 1 + 1$, a two-element antichain.

199. [2–] Find all finite distributive lattices L that are binomial posets, except for the axiom of containing an infinite chain.

200. [2+] Let P_n be an n-interval of the $q = 2$ case of the binomial poset of Example 3.18.3(e), so $B(n) = 2^{\binom{n}{2}}n!$. Show that the zeta polynomial of P_n is given by

$$Z(P_n, m) = \sum_G \chi_G(m), \tag{3.132}$$

where G ranges over all simple graphs on the vertex set $[n]$, and where χ_G is the chromatic polynomial of G. (Note that Example 3.18.9 gives a generating function for $Z(P_n, m)$.)

201. a. [2] Let P be a locally finite poset with $\hat{0}$ for which every maximal chain is infinite and every interval $[s, t]$ is graded. Thus, P has a rank function ρ. Call P a *triangular poset* if there exists a function $B: \{(i, j) \in \mathbb{N} \times \mathbb{N} : i \leq j\} \to \mathbb{P}$ such that any interval $[s, t]$ of P with $\rho(s) = m$ and $\rho(t) = n$ has $B(m, n)$ maximal chains. Define a subset T of the incidence algebra $I(P) = I(P, K)$, where $\text{char}(K) = 0$, by

$$T(P) = \{f \in I(P) : f(s, t) = f(s', t') \text{ if } \rho(s) = \rho(s') \text{ and } \rho(t) = \rho(t')\}.$$

If $f \in T(P)$, then write $f(m, n)$ for $f(s, t)$ when $\rho(s) = m$ and $\rho(t) = n$. Show that $T(P)$ is isomorphic to the algebra of all infinite upper-triangular matrices $[a_{ij}]_{i,j \geq 0}$

over K, the isomorphism being given by

$$f \mapsto \begin{bmatrix} \dfrac{f(0,0)}{B(0,0)} & \dfrac{f(0,1)}{B(0,1)} & \dfrac{f(0,2)}{B(0,2)} & \cdots \\[2ex] 0 & \dfrac{f(1,1)}{B(1,1)} & \dfrac{f(1,2)}{B(1,2)} & \cdots \\[2ex] 0 & 0 & \dfrac{f(2,2)}{B(2,2)} & \cdots \\[1ex] \vdots & \vdots & \vdots & \end{bmatrix},$$

where $f \in T(P)$.

b. [3–] Let L be a triangular lattice. Set $D(n) = B(n, n+1) - 1$. Show that L is (upper) semimodular if and only if for all $n \geq m + 2$,

$$\frac{B(m,n)}{B(m+1,n)} = 1 + \sum_{i=0}^{n-m-2} D(m) D(m+1) \cdots D(m+i).$$

c. [2] Let L be a triangular lattice. If $D(n) \neq 0$ for all $n \geq 0$ then show that L is atomic. Use (b) to show that the converse is true if L is semimodular.

202. [3] The *shuffle poset* W_{mn} with respect to alphabets A and B is defined in Exercise 7.48(g). Let $[u, v]$ be an interval of W_{mn}, where $u = u_1 \cdots u_r$ and $v = v_1 \cdots v_s$. Let $u_{i_1} \cdots u_{i_t}$ and $v_{j_1} \cdots v_{j_t}$ be the subwords of u and v, respectively, formed by the letters in common to both words. Because $u \leq v$, the shuffle property implies $u_{i_p} = v_{i_p}$ for each $p = 1, \ldots, t$. Moreover, the remaining letters of u belong to A, and the remaining letters of v belong to B. Therefore the interval $[u, v]$ is isomorphic to the product of shuffle posets $W_{i_p - i_{p-1} - 1, j_p - j_{p-1} - 1}$ for $p = 1, 2, \ldots, t+1$, where we set $i_0 = j_0 = 0$, $i_{t+1} = r + 1$ and $j_{t+1} = s + 1$. We write

$$[u, v] \simeq_c \prod_p W_{i_p - i_{p-1} - 1, j_p - j_{p-1} - 1}, \tag{3.133}$$

the *canonical isomorphism type* of the interval $[u, v]$. (The reason for this terminology is that some of the factors in equation (3.133) can be one-element posets, any of which could be omitted without affecting the isomorphism type.) Consider now the poset $W_{\infty\infty}$ whose elements are shuffles of finite words using the lower alphabet $A = \{a_i : i \in \mathbb{P}\}$ and the upper alphabet $B = \{b_i : i \in \mathbb{P}\}$, with the same definition of $\leq$ as for finite alphabets. A *multiplicative function* on $W_{\infty\infty}$ is a function f in the incidence algebra $I(W_{\infty\infty}, \mathbb{C})$ for which $f_{00} = 1$ and which has the following two properties:

- If $[u, v]$ and $[u', v']$ are two intervals both canonically isomorphic to W_{ij}, then $f(u, v) = f(u', v')$. We denote this value by f_{ij}.
- If $[u, v] \simeq_c \prod_{i,j} W_{ij}^{c_{ij}}$, then $f(u, v) = \prod_{ij} f_{ij}^{c_{ij}}$.

Let f and g be two multiplicative functions on $W_{\infty\infty}$, and let

$$F = F(x, y) = \sum_{i,j \geq 0} f_{ij} x^i y^j,$$

$$G = G(x, y) = \sum_{i,j \geq 0} g_{ij} x^i y^j,$$

$$F * G = (F * G)(x, y) = \sum_{i,j \geq 0} (f * g)_{ij} x^i y^j,$$

where $*$ denotes convolution in the incidence algebra $I(W_{\infty\infty}, \mathbb{C})$. Let $F_0 = F(x, 0)$, $G_0 = G(0, y)$, and

$$\widetilde{F}(x, y) = F(x, G_0 y),$$
$$\widetilde{G}(x, y) = G(F_0 x, y).$$

Show that

$$\frac{1}{F * G} = \frac{1}{\widetilde{F}G_0} + \frac{1}{F_0\widetilde{G}} - \frac{1}{F_0 G_0}.$$

203. [3–] Fix an integer sequence $0 \le a_1 < a_2 < \cdots < a_r < m$. For $k \in [r]$, let $f_k(n)$ denote the number of permutations $b_1 b_2 \cdots b_{mn+a_k}$ of $[mn + a_k]$ such that $b_j > b_{j+1}$ if and only if $j \equiv a_1, \ldots, a_r \pmod{m}$. Let

$$F_k = F_k(x) = \sum_{n \ge 0} (-1)^{nr+k} f_k(n) \frac{x^{mn+a_k}}{(mn + a_k)!},$$

$$\Phi_j(x) = \sum_{n \ge 0} \frac{x^{mn+j}}{(mn + j)!}.$$

Let $\bar{a}$ denote the least nonnegative residue of $a \pmod{m}$, and set $\psi_{ij} = \Phi_{\overline{a_i - a_j}}(x)$. Show that

$$
\begin{array}{ccccccccc}
F_1\psi_{11} & + & F_2\psi_{12} & + & \cdots & + & F_r\psi_{1r} & = & 1 \\
F_1\psi_{21} & + & F_2\psi_{22} & + & \cdots & + & F_r\psi_{2r} & = & 0 \\
& & & & & & & \vdots & \\
F_1\psi_{r1} & + & F_2\psi_{r2} & + & \cdots & + & F_r\psi_{rr} & = & 0.
\end{array}
$$

Solve these equations to obtain an explicit expression for $F_k(x)$ as a quotient of two determinants.

204. a. [2+] Let P be a locally finite poset for which every interval is graded. For any $S \subseteq \mathbb{P}$ and $s \le t$ in P, define $[s, t]_S$ as in equation (3.93), and let $\mu_S(s, t)$ denote the Möbius function of the poset $[s, t]_S$ evaluated at the interval $[s, t]$. Let z be an indeterminate, and define $g, h \in I(P)$ by

$$
g(s, t) = \begin{cases} 1, & \text{if } s = t, \\ (1+z)^{n-1}, & \text{if } \ell(s, t) = n \ge 1, \end{cases}
$$

$$
h(s, t) = \begin{cases} 1, & \text{if } s = t, \\ \displaystyle\sum_S \mu_S(s, t) z^{n-1-\#S}, & \text{if } s < t, \text{ where } \ell(s, t) = n \ge 1 \text{ and} \\ & S \text{ ranges over all subsets of } [n-1]. \end{cases}
$$

Show that $h = g^{-1}$ in $I(P)$.

b. [1+] For a binomial poset P write $h(n)$ for $h(s, t)$ when $\ell(s, t) = n$, where h is defined in (a). Show that

$$1 + \sum_{n \ge 1} h(n) \frac{x^n}{B(n)} = \left[1 + \sum_{n \ge 1} (1+z)^{n-1} \frac{x^n}{B(n)} \right]^{-1}.$$

c. [2] Define

$$G_n(q, z) = \sum_{w \in \mathfrak{S}_n} z^{\text{des}(w)} q^{\text{inv}(w)},$$

where des(w) and inv(w) denote the number of descents and inversions of w, respectively. Show that

$$1+z\sum_{n\geq 1}G_n(q,z)\frac{x^n}{(n)!}=\left[1-z\sum_{n\geq 1}(z-1)^{n-1}\frac{x^n}{(n)!}\right]^{-1}.$$

In particular, setting $q=1$ we obtain Proposition 1.4.5:

$$1+\sum_{n\geq 1}z^{-1}A_n(z)\frac{x^n}{n!}=\left[1-\sum_{n\geq 1}(x-1)^{n-1}\frac{x^n}{n!}\right]^{-1}$$

$$=\frac{1-z}{e^{x(z-1)}-z},$$

where $A_n(z)$ denotes an Eulerian polynomial.

205. **a.** [2+] Give an example of a 1-differential poset that is not isomorphic to Young's lattice Y nor to $\Omega_1^\infty Y[n]$ for any n, where $Y[n]$ denotes the rank n truncation of Y (i.e., the subposet of Y consisting of all elements of rank at most n).
   b. [3] Show that there are two nonisomorphic 1-differential posets up to rank 5, five up to rank 6, 35 up to rank 7, 643 up to rank 8, and 44605 up to rank 9.
   c. [3–] Give an example of a 1-differential poset that is not isomorphic to Y nor to a poset $\Omega_1^\infty P$, where P is 1-differential up to some rank n.

206. [3] Show that the only 1-differential lattices are Y and Z_1.

207. [2+] Let P be an r-differential poset, and let $A_k(q)$ be as in equation (3.110). Write $\alpha(n-2\to n\to n-1\to n)$ for the number of Hasse walks $t_0 \lessdot t_1 \lessdot t_2 \gtrdot t_3 \lessdot t_4$ in P, where $\rho(t_0)=n-2$. Show that

$$\sum_{n\geq 0}\alpha(n-2\to n\to n-1\to n)q^n=F(P,q)(2rq^2A_2(q)+rq^3A_3(q)+q^4A_4(q)).$$

208. **a.** [2+] Let P be an r-differential poset, and let $t\in P$. Define a word (noncommutative monomial) $w=w(U,D)$ in the letters U and D to be a *valid t-word* if $\langle w(U,D)\hat{0},t\rangle \neq 0$. Note that if $s\in P$, then a valid t-word is also a valid s-word if and only if $\rho(s)=\rho(t)$. Let $w=w_1\cdots w_l$ be a valid t-word. Let $S=\{i : w_i=D\}$. For each $i\in S$, let a_i be the number of D's in w to the right of w_i, and let b_i be the number of U's in w to the right of w_i. Show that

$$\langle w\hat{0},t\rangle=e(t)r^{\#S}\prod_{i\in S}(b_i-a_i),$$

where $e(t)$ is defined in Example 3.21.5.
   b. [2–]* Deduce from (a) that if $n=\rho(t)$ then

$$\langle w\hat{0},P\rangle=\alpha(0\to n)r^{\#S}\prod_{i\in S}(b_i-a_i).$$

   c. [2–]* Deduce the special case $\langle UDUU\hat{0},P\rangle=2r^2(r+1)$. Also deduce this result from Exercise 3.207.

209. [2] Let U and D be operators (or indeterminates) satisfying $DU-UD=1$. Show that

$$(UD)^n=\sum_{k=0}^n S(n,k)U^kD^k,\qquad(3.134)$$

where $S(n,k)$ denotes a Stirling number of the second kind.

210. [2+] A word w in U and D is *balanced* if it contains the same number of U's as D's. Show that if $DU - UD = 1$, then any two balanced words in U and D commute.

211. [3–]* Let P be an r-differential poset. Let $c(t)$ denote the number of elements covering $t \in P$, and set $f(n) = \sum_{t \in P_n} c(t)^2$. Show that

$$\sum_{n \geq 0} f(n)q^n = \frac{r^2 + (r+1)q - q^2}{(1-q)(1-q^2)} F(P,q).$$

212. Let P be an r-differential poset, and fix $k \in \mathbb{N}$. Let $\kappa(n \to n+k \to n)$ denote the number of *closed* Hasse walks in P of the form $t_0 \lessdot t_1 \lessdot \cdots \lessdot t_k \gtrdot t_{k+1} \gtrdot \cdots \gtrdot t_{2k}$ (so $t_0 = t_{2k}$) such that $\rho(t_0) = n$.
a. [2–]* Show that

$$\kappa(n \to n+k \to n) = \sum_{t \in P_n} \langle D^k U^k t, t \rangle$$

$$= \sum_{s \in P_n} \sum_{t \in P_{n+k}} e(s,t)^2,$$

where $e(s,t)$ denotes the number of saturated chains $s = s_0 \lessdot s_1 \lessdot \cdots \lessdot s_k = t$.
b. [2+]* Show that

$$\sum_{n \geq 0} \kappa(n \to n+k \to n)q^n = r^k k! (1-q)^{-k} F(P,q).$$

213. [2+] Let P be an r-differential poset, and let $\kappa_{2k}(n)$ denote the total number of closed Hasse walks of length $2k$ starting at some element of P_n. Show that for fixed k,

$$\sum_{n \geq 0} \kappa_{2k}(n)q^n = \frac{(2k)! r^k}{2^k k!} \left(\frac{1+q}{1-q}\right)^k F(P,q).$$

214. a. [3–] Show that the "fattest" r-differential poset is Z_r (i.e., has at least as many elements of any rank i as any r-differential poset).
b. [5] Show that the "thinnest" r-differential poset is Y^r.

215. a. [2] Let P be an r-differential poset, and let $p_i = \#P_i$. Show that $p_0 \leq p_1 \leq \cdots$. Hint. Use linear algebra.
b. [2+]* Show that $\lim_{i \to \infty} p_i = \infty$.
c. [5] Show that $p_i < p_{i+1}$ except for the case $i = 0$ and $r = 1$.

Solutions to Exercises

1. Itinerant salespersons who take revenge on customers who don't pay their bills are retaliatory peddlers, and "retaliatory peddlers" is an anagram of "partially ordered set" (i.e., they have the same multiset of letters).

2. Routine. See [3.13], Lemma 1 on page 21.

3. The correspondence between finite posets and finite topologies (or more generally arbitrary posets and topologies for which any intersection of open sets is open) seems first to have been considered by P. S. Alexandroff, *Mat. Sb. (N.S.)* **2** (1937), 510–518, and has been rediscovered many times.

4. Let $\mathcal{S} = \{\Lambda_t : t \in P\}$, where $\Lambda_t = \{s \in P : s \leq t\}$. This exercise is the poset analogue of Cayley's theorem that every group is isomorphic to a group of permutations of a set.

5. **a.** The enumeration of n-element posets for $1 \leq n \leq 7$ appears in John A. Wright, thesis, Univ. of Rochester, 1972. Naturally computers have allowed the values of n to be considerably extended. At the time of this writing, the most recent paper on this topic is G. Brinkmann and B. D. McKay, *Order* **19** (2002), 147–179.
   c. The purpose of this seemingly frivolous exercise is to point out that some simply stated facts about posets may be forever unknowable.
   d. D. J. Kleitman and B. L. Rothschild, *Proc. Amer. Math. Soc.* **25** (1970), 276–282. The lower bound for this estimate is obtained by considering posets of rank one with $\lfloor n/2 \rfloor$ elements of rank 0 and $\lceil n/2 \rceil$ elements of rank 1.
   e. D. J. Kleitman and B. L. Rothschild, *Trans. Amer. Math. Soc.* **205** (1975), 205–220. The asymptotic formula given there is more complicated but can be simplified to that given here. It follows from the proof that almost all posets have longest chain of length two.

6. **a.** The function f is a permutation of a finite set, so $f^n = 1$ for some $n \in \mathbb{P}$. But then $f^{-1} = f^{n-1}$, which is order-preserving.
   b. Let $P = \mathbb{Z} \cup \{t\}$, with $t < 0$ and t incomparable with all $n < 0$. Let $f(t) = t$ and $f(n) = n+1$ for $n \in \mathbb{Z}$.

7. **a.** An example is shown in Figure 3.68. There are four other 6-element examples, and none smaller. For the significance of this exercise, see Corollary 3.15.18(a).
   b. Use induction on ℓ, removing all minimal elements from P. This proof is due to D. West. The result (with a more complicated proof) first appeared in [2.19, pp. 19–20].

8. The poset Q of Figure 3.69 was found by G. Ziegler, and with a $\hat{0}$ and $\hat{1}$ adjoined is a lattice. Ziegler also has an example of length one with 24 elements, and an example which is a graded lattice of length three with 26 elements. Another example is presumably the one referred to by Birkhoff in [3.13, Exer. 10, p. 54].

9. False. Non-self-dual posets come in pairs P, P^*, so the number of each order is even. The actual number is 16506. The number of self-dual 8-element posets is 493.

10. **b.** Suppose that $f : \text{Int}(P) \to \text{Int}(Q)$ is an isomorphism. Let $f([\hat{0}, \hat{0}]) = [s, s]$, where $\hat{0} \in P$ and $s \in Q$. Define A to be the subposet of Q of all elements $t \geq s$, and define B to be all elements $t \leq s$. Check that $P \cong A \times B^*$, $Q \cong A \times B$.
   This result is due independently to A. Gleason (unpublished) and M. Aigner and G. Prins, *Trans. Amer. Math. Soc.* **166** (1972), 351–360.
   c. (A. Gleason, unpublished) See Figure 3.70. The poset P may be regarded as a "twisted" direct product (not defined here) of the posets A and B of Figure 3.71, and Q a twisted direct product of A and C. These twisted direct products exist since the poset A is, in a suitable sense, not simply-connected but has the covering poset $\tilde{A}$ of Figure 3.71. A general theory was presented by A. Gleason at an M.I.T. seminar in December 1969.

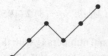

Figure 3.68 A solution to Exercise 3.7.

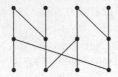

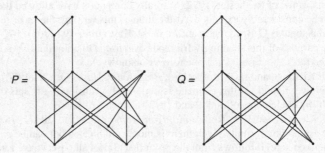

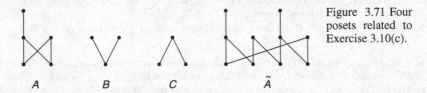

Figure 3.69 A self-dual poset with no involutive antiautomorphism.

Figure 3.70 A solution to Exercise 3.10(c).

Figure 3.71 Four posets related to Exercise 3.10(c).

A B C $\tilde{A}$

For the determination of which posets have isomorphic posets of *convex* subposets, see G. Birkhoff and M. K. Bennett, *Order* **2** (1985), 223–242 (Theorem 13).

11.   **a.** See Birkhoff [3.13, Thm. 2, p. 57].

   b. See [3.13, Thm. 2, pp. 68–69].

   c. See [3.13, p. 69]. If P is any connected poset with more than one element, then we can take $P_1 = 1 + P^3$, $P_2 = 1 + P + P^2$, $P_3 = 1 + P^2 + P^4$, $P_4 = 1 + P$, where **1** denotes the one-element poset. There is no contradiction, because although $\mathbb{Z}[x_1, x_2, \ldots]$ is a UFD, this does not mean that $\mathbb{N}[x_1, x_2, \ldots]$ is a unique factorization semiring. In the ring B, we have (writing Q for $[Q]$)

$$P_1 P_2 = P_3 P_4 = (1+P)(1-P+P^2)(1+P+P^2).$$

12.   True. If the number of maximal chains of P is finite, then P is clearly finite, so assume that P has infinitely many maximal chains. These chains are all finite, so in particular every maximal chain containing a nonmaximal element t of P must contain an element covering t. The set $C^+(t)$ of elements covering $t \in P$ is an antichain, so $C^+(t)$ is finite. Since P has only finitely many minimal elements (since they form an antichain), infinitely many maximal chains C contain the same minimal element t_0. Since $C^+(t_0)$ is finite, infinitely many of the chains C contain the same element $t_1 \in C^+(t_0)$. Continuing in this way, we obtain an infinite chain $t_0 < t_1 < \cdots$, a contradiction.

13.   **a.** This result is a consequence of G. Higman, *Proc. London Math. Soc. (3)* **2** (1952), 326–336.

   b. Let $P = P_1 + P_2$, where each P_i is isomorphic to the rational numbers $\mathbb{Q}$ with their usual linear order. Every antichain of P has at most two elements. For any real

$\alpha > 0$, let

$$I_\alpha = \{a \in P_1 : a < -\alpha\} \cup \{b \in P_2 : b < \alpha\}.$$

Then the I_α's form an infinite (in fact, uncountable) antichain in $J(P)$.

15. **a.** Straightforward proof by induction on $\#P$. This result is implicit in work of A. Ghouila-Houri *C. R. Acad. Sci. Paris* **254** (1962), 1370–1371, and P. C. Gilmore and A. J. Hoffman *Canad. J. Math.* **16** (1964), 539–548. The first explicit statement was given by P. C. Fishburn, *J. Math. Psych.* **7** (1970), 144–149. Two references with much further information on interval orders are P. C. Fishburn, *Interval Orders and Interval Graphs*, Wiley-Interscience, New York, 1985, and W. T. Trotter, *Combinatorics and Partially Ordered Sets*, The Johns Hopkins University Press, Baltimore, 1992. In particular, Fishburn (pp. 19–22) discusses the history of interval orders and their applications to such areas as psychology.

   b. This result is due to D. Scott and P. Suppes, *J. Symbolic Logic* **23** (1958), 113–128. Much more information on semiorders may be found in the books of Fishburn and Trotter cited in (a).

   c. See M. Bousquet-Mélou, A. Claesson, M. Dukes, and S. Kitaev, *J. Combinatorial Theory Ser. A* **117** (2010), 884–909 (Theorem 13). This paper mentions several other objects (some of which were already known) counted by $t(n)$, in particular, the *regular linearized chord diagrams* (RLCD) (not defined here) of D. Zagier, *Topology* **40** (2001), 945–960.

   The paper of Zagier also contains the remarkable result that if we set $F(x) = \sum_{n \geq 0} t(n) x^n$, then

$$F(1 - e^{-24x}) = e^x \sum_{n \geq 0} u(n) \frac{x^n}{n!},$$

   where

$$\sum_{n \geq 0} u(n) \frac{x^{2n+1}}{(2n+1)!} = \frac{\sin 2x}{2 \cos 3x}.$$

   d. This result is a consequence of equation (4) of Zagier, ibid., and the bijection between interval orders and RLCD's given by Bousquet-Mélou et al., op. cit. Note that if we set $F(x) = \sum_{n \geq 0} t(n) x^n$ and $G(x) = \sum_{n \geq 0} u(n) \frac{x^n}{n!}$, then $G(x) = F(1 - e^{-x})$. This phenomenon also occurs for semiorders (see equation (6.57)) and other objects (see Exercise 3.17 and the solution to Exercise 6.30).

   e. A series-parallel interval order can clearly be represented by intervals such that for any two of these intervals, they are either disjoint or one is contained in the other. Conversely any such finite set of intervals represents a series-parallel interval order. Now use Exercise 6.19(o). For a refinement of this result see J. Berman and P. Dwinger, *J. Combin. Math. Combin. Comput.* **16** (1994), 75–85. An interesting characterization of series-parallel interval orders was given by M. S. Rhee and J. G. Lee, *J. Korean Math. Soc.* **32** (1995), 1–5.

   f. (sketch) Let $\mathcal{G}_n$ denote the arrangement in part (e). Putting $x = -1$ in Proposition 3.11.3 gives

$$r(\mathcal{G}_n) = \sum_{\substack{\mathcal{B} \subseteq \mathcal{G}_n \\ \mathcal{B} \text{ central}}} (-1)^{\#\mathcal{B} - \text{rank}(\mathcal{B})}. \tag{3.135}$$

   Given a central subarrangement $\mathcal{B} \subseteq \mathcal{G}_n$, define a digraph $G_\mathcal{B}$ on $[n]$ by letting $i \to j$ be a (directed) edge if the hyperplane $x_i - x_j = \ell_i$ belongs to $\mathcal{B}$. One then shows that as an undirected graph $G_\mathcal{B}$ is bipartite. Moreover, if B is a block of $G_\mathcal{B}$ (as defined in Exercise 5.20), say with vertex bipartition (U_B, V_B), then either all edges of B are directed from U_B to V_B, or all edges are directed from V_B to U_B.

It can also be seen that all such directed bipartite graphs can arise in this way. It follows that equation (3.135) can be rewritten

$$r(\mathcal{G}_n) = (-1)^n \sum_G (-1)^{e(G)+c(G)} 2^{b(G)},\qquad(3.136)$$

where G ranges over all (undirected) bipartite graphs on $[n]$, $e(G)$ denotes the number of edges of G, and $b(G)$ denotes the number of blocks of G.

Equation (3.136) reduces the problem of determining $r(\mathcal{G}_n)$ to a (rather difficult) problem in enumeration, whose solution may be found in A. Postnikov and R. Stanley, *J. Combinatorial Theory Ser. A* **91** (2000), 544–597 (§6).

16. This result is due to J. Lewis and Y. Zhang, Enumeration of graded (3+1)-avoiding posets, `arXiv:1106.5480`.

17. **a.** Hint. First show that the formula $G(x) = F(1 - e^{-x})$ is equivalent to

$$n! f(n) = \sum_{k=1}^n c(n,k) g(k),$$

where $c(n,k)$ denotes a signless Stirling number of the first kind.

The special case where $\mathcal{T}$ consists of the nonisomorphic finite semiorders is due to J. L. Chandon, J. Lemaire, and J. Pouget, *Math. et Sciences Humaines* **62** (1978), 61–80, 83. (See Vol. II, Exercise 6.30.) The generalization to the present exercise (and beyond) is due to Y. Zhang, in preparation (2011).

b. $F(x) = (1-x)/(1-2x)$, the ordinary generating function for the number of compositions of n, and $G(x) = 1/(2-e^x)$, the exponential generating function for the number of ordered partitions of $[n]$. See Example 3.18.10.

c. These results follow from two properties of interval orders and semiorders P: (i) any automorphism of P is obtained by permuting elements in the same autonomous subset (as defined in the solution to Exercise 3.143), and (ii) replacing elements in an interval order (respectively, semiorder) by antichains preserves the property of being an interval order (respectively, semiorder).

18. Originally this result was proved using symmetric functions (R. Stanley, *Discrete Math.* **193** (1998), 267–286). Later M. Skandera, *J. Combinatorial Theory (A)* **93** (2001), 231–241, showed that for a certain ordering of the elements of P, the *square* of the anti-incidence matrix of Exercise 3.22, with each $t_i = 1$, is totally nonnegative (i.e., every minor is nonnegative). The result then follows easily from Exercise 3.22 and the standard fact that totally nonnegative square matrices have real eigenvalues. Note that if $P = \mathbf{3}+\mathbf{1}$ then $C_P(x) = x^3 + 3x^2 + 4x + 1$, which has the approximate nonreal zeros $-1.34116 \pm 1.16154i$.

19. **a,c.** These results (in the context of finite topological spaces) are due to R. E. Stong, *Trans. Amer. Math. Soc.* **123** (1966), 325–340 (see page 330). For (a), see also D. Duffus and I. Rival, in *Colloq. Math. Soc. János Bolyai* (A. Hajnal and V. T. Sós, eds.), vol. 1, North-Holland, New York, pp. 271–292 (page 272), and J. D. Farley, *Order* **10** (1993), 129–131. For (c), see also D. Duffus and I. Rival, *Discrete Math.* **35** (1981), 53–118 (Theorem 6.13). Part (c) is generalized to infinite posets by K. Baclawski and A. Björner, *Advances in Math.* **31** (1979), 263–287 (Thm. 4.5). For a general approach to results such as (a) where any way of carrying out a procedure leads to the same outcome, see K. Eriksson, *Discrete Math.* **153** (1996), 105–122; *Europ. J. Combinatorics* **17** (1996), 379–390; and *Discrete Math.* **139** (1995), 155–166.

20. **a,b.** The least d for which (i) or (ii) holds is called the *dimension* of P. For a survey of this topic, see D. Kelly and W. T. Trotter, in [3.57, pp. 171–211]. In particular, the

equivalence of (i) and (ii) is due to Ore, while (iii) is an observation of Dushnik and Miller. Note that a 2-dimensional poset P on $[n]$ which is compatible with the usual ordering of $[n]$ (i.e., if $s < t$ in P, then $s < t$ in $\mathbb{Z}$) is determined by the permutation $w = a_1 \cdots a_n \in \mathfrak{S}_n$ for which P is the intersection of the linear orders $1 < 2 < \cdots < n$ and $a_1 < a_2 < \cdots < a_n$. We call $P = P_w$ the *inversion poset* of the permutation w. In terms of w, we have that $a_i < a_j$ in P if and only if $i < j$ and $a_i < a_j$ in $\mathbb{Z}$. For further results on posets of dimension 2, see K. A. Baker, P. C. Fishburn, and F. S. Roberts, *Networks* **2** (1972), 11–28. Much additional information appears in P. C. Fishburn, *Interval Orders and Interval Graphs*, John Wiley, New York, 1985, and in W. T. Trotter, *Combinatorics and Partially Ordered Sets: Dimension Theory*, The John Hopkins University Press, Baltimore, 1992.

21. The statement is false. It was shown by S. Felsner, W. T. Trotter, and P. C. Fishburn, *Discrete Math.* **201** (1999), 101–132, that the poset n^3, for n sufficiently large, is not a sphere order.

22. This result is an implicit special case of a theorem of D. M. Jackson and I. P. Goulden, *Studies Appl Math.* **61** (1979), 141–178 (Lemma 3.12). It was first stated explictly by R. Stanley, *J. Combinatorial Theory Ser. A* **74** (1996), 169–172 (in the more general context of acyclic digraphs). To prove it directly, use the fact that the coefficient of x^j in $\det(I + xDA)$ is the sum of the principal $j \times j$ minors of DA. Let $DA[W]$ denote the principal submatrix of DA whose rows and columns are indexed by $W \subseteq [p]$. It is not difficult to show that

$$\det DA[W] = \begin{cases} \prod_{i \in W} t_i, & \text{if } W \text{ is the set of vertices of a chain,} \\ 0, & \text{otherwise,} \end{cases}$$

and the proof follows.

23. Of course $\emptyset < [1] < [2] < \cdots$ is a countable maximal chain. Now clearly $B_{\mathbb{P}} \cong B_{\mathbb{Q}}$ since $\mathbb{P}$ and $\mathbb{Q}$ are both countable infinite sets. For each $\alpha \in \mathbb{R}$, define $t_\alpha \in B_{\mathbb{Q}}$ by $t_\alpha = \{s \in \mathbb{Q} : s < \alpha\}$. Then the elements t_α, together with $\hat{0}$ and $\hat{1}$, form an uncountable maximal chain.

24. For an extension to all n-element posets having at least $(7/16)2^n$ order ideals, see R. Stanley, *J. Combinatorial Theory Ser. A* **10** (1971), 74–79. For further work on the number of n-element posets with k order ideals, see M. Benoumhani, *J. Integer Seq.* **9** (2006), 06.2.6 (electronic), and K. Ragnarsson and B. Tenner, *J. Combinatorial Theory Ser. A* **117** (2010), 138–151.

25. None.

26. Perhaps the simplest example is $\mathbf{1} \oplus (\mathbf{1} + \mathbf{1}) \oplus \mathbb{N}^*$, where $\mathbb{N}^*$ denotes the dual of $\mathbb{N}$ with the usual linear order. We could replace $\mathbb{N}^*$ with $\mathbb{Z}$.

27. Let B be the boolean algebra of all subsets of $\mathrm{Irr}(L)$, and let L' be the meet-semilattice of B generated by the principal order ideals of $\mathrm{Irr}(L)$. One can show that L is isomorphic to L' with a $\hat{1}$ adjoined.
In fact, L is the *MacNeille completion* (e.g., [3.13, Ch. V.9]) of $\mathrm{Irr}(L)$, and this exercise is a result of B. Banaschewski, *Z. Math. Logik* **2** (1956), 117–130. An example is shown in Figure 3.72.

28. Let L be the sub-meet-semilattice of the boolean algebra B_6 generated by the subsets 1234, 1236, 1345, 2346, 1245, 1256, 1356, 2456, with a $\hat{1}$ adjoined. By definition L is coatomic. One checks that each singleton subset $\{i\}$ belongs to L, $1 \leq i \leq 6$, so L is atomic. However, the subset $\{1, 2\}$ has no complement.
This example was given by I. Rival (personal communication) in February 1978. See *Discrete Math.* **29** (1980), 245–250 (Fig. 5).

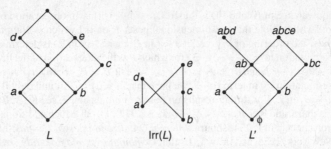

Figure 3.72 The MacNeille completion of Irr(L).

29. D. Kleitman has shown (unpublished) that

$$\binom{n}{\lfloor n/2 \rfloor}\left(1 + \frac{1}{n}\right) < f(n) < \binom{n}{\lfloor n/2 \rfloor}\left(1 + \frac{1}{\sqrt{n}}\right),$$

and conjectures that the lower bound is closer to the truth.

30. a,b. Since sublattices of distributive (respectively, modular) lattices are distributive (respectively, modular), the "only if" part is immediate from the nonmodularity of the first lattice in Figure 3.43 and the nondistributivity of both lattices. For the "if" part, it is not hard to check that the failure of the distributive law (3.8) for a triple (s,t,u) forces the sublattice generated by s,t,u to contain (as a sublattice) one of the two lattices of Figure 3.43. Similarly the failure of the modular law (3.7) forces the first lattice of Figure 3.43. This result goes back to R. Dedekind, *Festschrift Techn. Hoch. Braunschweig* (1897), 1–40; reprinted in *Ges. Werke*, vol. 2, 103–148, and *Math. Ann.* **53** (1900), 371–403; reprinted in *Ges. Werke*, vol. 2, 236–271.

31. b. See J. D. Farley and S. E. Schmidt, *J. Combinatorial Theory Ser. A* **92** (2000), 119–137.

    c. This result was originally conjectured by R. Stanley (unpublished) and proved by D. J. Grabiner, *Discrete Math.* **199** (1999), 77–84.

    d. This result is originally due to R. Stanley (unpublished).

32. a. *Answer.* $f(B_n) = \lceil n/2 \rceil$. See C. Biró, D. M. Howard, M. T. Keller, W. T. Trotter, and S. J. Young, *J. Combinatorial Theory Ser. A* **117** (2010), 475–482. See Exercise 3.142 for a more general context to this topic.

    b–e. See Y. H. Wang, The new Stanley depth of some power sets of multisets, `arXiv:0908.3699`.

33. *Answer* (in collaboration with J. Shareshian). $n \neq 1,3,5,7$.

35. a. By Theorem 3.4.1, $f_2(n)$ is equal to the number of distributive lattices L of rank n with exactly two elements of every rank $1, 2, \ldots, n-1$. We build L from the bottom up. Ranks $0, 1, 2$ must look (up to isomorphism) like the diagram in Figure 3.73(a), where we have also included $u = s \vee t$ of rank 3. We have two choices for the remaining element v of rank 3—place it above s or above t, as shown in Figure 3.73(b). Again we have two choices for the remaining element of rank 4—place it above u or above v. Continuing this line of reasoning, we have two independent choices a total of $n-3$ times, yielding the result. When, for example, $n = 5$, the four posets are shown in Figure 3.74.

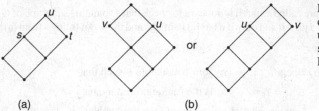

Figure 3.73 A construction used in the solution to Exercise 3.35.

(a)   (b)

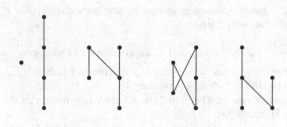

Figure 3.74 The four posets enumerated by $f_2(5)$.

b. Similar to (a).

c. See J. D. Farley and R. Klippenstine, *J. Combinatorial Theory Ser. A* **116** (2009), 1097–1119.

d. (suggested by P. Edelman) $f_k(n) = 0$ for $k > 3$ since $\binom{k}{2} > k$.

36. **a.** Clearly, L' is a join-semilattice of L with $\hat{0}$; hence by Proposition 3.3.1, L' is a lattice. By definition L' is atomic. Suppose t covers s in L'. Then $t = s \vee a$ for some atom a of L. The semimodularity property of Proposition 3.3.2(ii) is inherited from L by L'. Thus L' is geometric.

b. No. Let K be the boolean algebra B_5 of all subsets of $[5]$, with all four-element subsets removed. Let L consist of K with an additional element t adjoined such that t covers $\{1\}$ and is covered by $\{1,2,3\}$ and $\{1,4,5\}$. Then $t \notin L'$ but t belongs to the sublattice of L generated by L'.

37. **a.** This result is an immediate consequence of a much more general result of W. T. Tutte, *J. Res. Natl. Bur. Stand., Sect. B* **69** (1965), 1–47. For readers who know some matroid theory we provide some more details. Tutte shows (working in the broader context of "chain groups") that the set of minimal nonempty supports are the set of circuits of a matroid. Since char$(K) = 0$ the support sets coincide with the unions of minimal nonempty supports. This means that the supports coincide with the sets of unions of circuits. The complements of circuits are hyperplanes of the dual matroid. The proper flats of a matroid coincide with the intersections of hyperplanes so their complements are unions of circuits of the dual, and the present exercise follows.

b. Let K^E denote the vector space of all functions $E \to K$, and let V denote the vertex set of G. Choose an orientation o of the edges of G. For each vertex v, let v^+ denote the set of edges pointing out of v, and v^- the set of edges pointing in (with respect to o). Let

$$W = \left\{ f \in K^E : \forall v \in V \ \sum_{e \in v^+} f(e) = \sum_{e \in v^-} f(e) \right\}.$$

Elements of W are called *flows*. It is not hard to check that a spanning subgraph of G is the support of a flow if and only if it has no isthmus, and the proof follows from (a).

38. If $t \in P_k$, then define

$$\phi(t) = \sup\{u : u \not\geq \text{ any join-irreducible } t_i \text{ such that}$$

$$t = t_1 \vee \cdots \vee t_n \text{ is the (unique) irredundant}$$

$$\text{expression of } t \text{ as a join of join-irreducibles}\}. \qquad (3.137)$$

In particular, if $t \in P_1$ then $\phi(t) = \sup\{u : u \not\geq t\}$.
It is fairly easy to see that ϕ has the desired properties by dealing with the poset P for which $L = J(P)$, rather than with L itself.

40. **a.** *Answer.* $\dfrac{2^{k-1}(2^{r(k-1)} - 1)}{2^{k-1} - 1} + 2^{r(k-1)}$. This result is a special case of a more general result, where the number of elements of every rank is specified, of R. Stanley, *J. Combinatorial Theory* **14** (1973), 209–214 (Corollary 1).

   b. See R. Stanley, ibid. (special case of Theorem 2). For further information on the extremal lattice $J(P+1)$, see [3.73].

   c. This inequality, if true, is best possible, as seen by taking $L = J_f(P+1)$ as in (b). Note that $J_f(P+1)$ is *maximal* with respect to having two join-irreducibles at each positive rank, and is conjectured to be *minimal* with respect to having an antichain passing through each positive rank.

41. **a.** Let $t_1, \ldots, t_p$ be a linear extension of L, regarded as a permutation of the elements of L. Let $\sigma_i = (\Gamma^i(t_1), \ldots, \Gamma^i(t_p))$. All the sequences σ_i have the same sum of their terms. Moreover, if $\sigma_i \neq \sigma_{i+1}$ then $\sigma_i < \sigma_{i+1}$ in dominance order. It follows that eventually we must have $\sigma_n = \sigma_{n+1}$.

   b. The "if" part of the statement is equivalent to Problem A3 on the 69th William Lowell Putnam Mathematical Competition (2008). The "only if" part follows easily from Exercise 3.30(a). The "only if" part was observed by T. Belulovich and is discussed at the Putnam Archive,

   $\langle$`www.unl.edu/amc/a-activities/a7-problems/`

   `putnamindex.shtml`$\rangle$.

   c. This result was shown by F. Liu and R. Stanley, October 2009.

   d. This observation is due to R. Ehrenborg, October 2009.

   e. Use (d) and the fact that $\#\{w \in \mathfrak{S}_{n+1} : w(i) > i\} = (n+1-i)n!$. This result is due to R. Ehrenborg, October 2009.

43. First show that $J_f(P)$ can be identified with the subposet of $\mathbb{N} \times \mathbb{N}$ consisting of all (i, j) for which $0 \leq j \leq \lfloor i/2 \rfloor$. Then show that $J_f(J_f(P))$ can be identified with the subposet (actually a sublattice) of Young's lattice consisting of all partitions whose parts differ by at least 2. It follows from Exercise 1.88 that

$$F_{J_f(J_f(P))}(q) = \frac{1}{\displaystyle\prod_{k \geq 0}(1 - q^{5k+1})(1 - q^{5k+4})}.$$

47. **b.** Induction on $\#L$. Trivial for $\#L = 1$. Now let $\#L \geq 2$, and let t be a maximal element of L. Suppose that t covers j elements of L, and set $L' = L - \{t\}$. The meet-distributivity hypothesis implies that the number of $s \leq t$ for which $[s, t] \cong B_k$

is equal to $\binom{j}{k}$. Hence,

$$\sum_{k\geq 0} g_k(L)x^k = x^j + \sum_{k\geq 0} g_k(L')(1+x)^k, \text{ and}$$

$$\sum_{k\geq 0} f_k(L)x^k = \sum_{k=0}^{j} \binom{j}{k} x^k + \sum_{k\geq 0} f_k(L')x^k$$

$$= (1+x)^j + \sum_{k\geq 0} f_k(L')x^k,$$

and the proof follows by induction since L' is meet-distributive.

Note that in the special case $L = J(P)$, $g_k(L)$ is equal to the number of k-element antichains of P.

c. Let $x = -1$ in (b). This result was first proved (in a different way) for distributive lattices by S. K. Das, *J. Combinatorial Theory Ser. B* **26** (1979), 295–299. It can also be proved using the identity $\zeta \mu \zeta = \zeta$ in the incidence algebra of the lattice $L \cup \{\hat{1}\}$.
TOPOLOGICAL REMARK. This exercise has an interesting topological generalization (done in collaboration with G. Kalai). Given L, define an abstract cubical complex $\Omega = \Omega(L)$ as follows: the vertices of Ω are the elements of L, and the faces of Ω consist of intervals $[s,t]$ of L isomorphic to boolean algebras. (It follows from Exercise 3.177(a) that Ω is indeed a cubical complex.)
Proposition. *The geometric realization* $|\Omega|$ *is contractible.* In fact, Ω is collapsible.
Sketch of Proof. Let t be a maximal element of L, let $L' = L - \{t\}$, and let s be the meet of elements that t covers, so $[s,t] \cong B_k$ for some $k \in \mathbb{P}$. Then $|\Omega(L')|$ is obtained from $|\Omega(L)|$ by collapsing the cube $|[s,t]|$ onto its boundary faces that don't contain t. Thus by induction, $\Omega(L)$ is collapsible, so $|\Omega(L)|$ is contractible. $\square$

The formula $\sum (-1)^k f_k = 1$ asserts merely that the Euler characteristic of $\Omega(L)$ or $|\Omega(L)|$ is equal to 1; the statement that $|\Omega(L)|$ is contractible is much stronger. For some further results along these lines involving homotopy type, see P. H. Edelman, V. Reiner, and V. Welker, *Discrete & Computational Geometry* **27**(1) (2002), 99–116.

d. A k-element antichain A of $m \times n$ has the form

$$A = \{(a_1,b_1),(a_2,b_2),\ldots,(a_k,b_k)\},$$

where $1 \leq a_1 < a_2 < \cdots < a_k \leq m$ and $n \geq b_1 > b_2 > \cdots > b_k \geq 1$. Hence, $g_k = \binom{m}{k}\binom{n}{k}$.
It is easy to compute, either by a direct combinatorial argument or by (b) and Vandermonde's convolution (Example 1.1.17), that $f_k = \binom{m}{k}\binom{m+n-k}{m}$.

e. This result was proved independently by J. R. Stembridge (unpublished) and R. A. Proctor, *Proc. Amer. Math. Soc.* **89** (1983), 553–559 (Theorem 2). Later Stembridge gave another proof in *Europ. J. Combinatorics* **7** (1986), 377–387 (Corollary 2.2).

f. R. A. Proctor, op. cit., Theorem 1.

g. This result was conjectured by P. H. Edelman for $m = n$, and first proved in general by R. Stanley and J. Stembridge using the theory of "jeu de taquin" (see Chapter 7, Vol. II, Appendix A1.2). An elementary proof was given by M. Haiman (unpublished). See J. R. Stembridge, *Europ. J. Combinatorics* **7** (1986), 377–387, for details and additional results (see in particular Corollary 2.4).

48. Induction on $\rho(t)$. Clearly true for $\rho(t) \leq 1$. Assume true for $\rho(t) < k$, and let $\rho(t) = k$. If t is join-irreducible, then the conclusion is clear. Otherwise t covers $r > 1$ elements. By the Principle of Inclusion-Exclusion and the induction hypothesis, the number of join-irreducibles $s \leq t$ is

$$r(k-1) - \binom{r}{2}(k-2) + \binom{r}{3}(k-3) - \cdots \pm \binom{r}{k-1} = k.$$

For further information on this result and on meet-distributive lattices in general, see B. Monjardet, *Order* **1** (1985), 415–417, and P. H. Edelman, *Contemporary Math.* **57** (1986), 127–150. Other references include C. Greene and D. J. Kleitman, *J. Combinatorial Theory Ser. A* **20** (1976), 41–68 (Thm. 2.31); P. H. Edelman, *Alg. Universalis* **10** (1980), 290–299; and P. H. Edelman and R. F. Jamison, *Geometriae Ded.* **19** (1985), 247–270.

49. Routine. For more information on the posets L_p, see R. A. Dean and G. Keller, *Canad. J. Math.* **20** (1968), 535–554.

50. The left-hand side of equation (3.114) counts the number of pairs (s, S) where s is an element of L of rank i and S is a set of j elements that s covers. Similarly the right-hand side is equal to the number of pairs (t, T) where $\rho(t) = i - j$ and T is a set of j elements that cover t. We set up a bijection between the pairs (s, S) and (t, T) as follows. Given (s, S), let $t = \bigwedge_{w \in S} w$, and define T to be set of all elements in the interval $[t, s]$ that cover t.

51. **a.** Let L be a finitary distributive lattice with cover function f. Let L_k denote the sublattice of L generated by all join-irreducibles of rank at most k. We prove by induction on k that L_k is unique (if it exists). Since $L = \bigcup L_k$, the proof will follow.

True for $k = 0$, since L_0 is a point. Assume for k. Now L_k contains all elements of L of rank at most k. Suppose that t is an element of L_k of rank k covering n elements, and suppose that t is covered by c_t elements in L_k. Let $d_t = f(n) - c_t$. If $d_t < 0$ then L does not exist, so assume $d_t \geq 0$. Then the d_t elements of $L - L_k$ that cover t in L must be join-irreducibles of L. Thus for each $t \in L_k$ of rank k, attach d_t join-irreducibles covering t, yielding a meet-semilattice L'_k. Let P_{k+1} denote the poset of join-irreducibles of L'_k. Then P'_{k+1} must coincide with the poset of join-irreducibles of L_{k+1}. Hence, $L_{k+1} = J(P_{k+1})$, so L_{k+1} is uniquely determined.

b. See Proposition 2 on page 226 of [3.73].

c. If $f(n) = k$ then $L = \mathbb{N}^k$. If $f(n) = n + k$, then $L = J_f(\mathbb{N}^2)^k$.

d. Use Exercise 3.50 to show that

$$u(5, 1) = -(k/3)(2a^3 - 2a^2 - 3).$$

Hence, $u(5, 1) < 0$ if $a \geq 2$ and $k \geq 1$, so L does not exist.

e. See J. D. Farley, *Graphs and Combinatorics* **19** (2003), 475–491 (Theorem 11.1).

55. **a.** See K. Saito, *Advances in Math.* **212** (2007), 645–688 (Theorem 3.2).

b. Essentially this question was raised by Saito, ibid. (Remark 4).

56. Let E be the set of all (undirected) edges of the Hasse diagram of P. Define $e, f \in E$ to be *equivalent* if e has vertices s, u and f has vertices t, u, such that either both $s < u$ and $t < u$, or both $s > u$ and $t > u$. Extend this equivalence to an equivalence relation using reflexivity and transitivity. The condition on P implies that the equivalence classes are paths and cycles. We obtain a partition of P into disjoint saturated chains by choosing a set of edges, no two consecutive, from each equivalence class. If an element t of P does not lie on one of the chosen edges, then it forms a one-element saturated chain. The number of ways to choose a set of edges, no two consecutive, from a path of length ℓ is the Fibonacci number $F_{\ell+2}$. The number of ways to choose a set of edges, no two consecutive, from a cycle of length ℓ is the Lucas number L_ℓ (see Exercise 1.40), and the proof follows. This result is due to R. Stanley, *Amer. Math. Monthly* **99** (1992); published solution by W. Y. C. Chen, **101** (1994), 278–279.

For $P = m \times n$ the equivalence classes consist of all cover relations between two consecutive ranks. Assuming $m \le n$, we obtain

$$f(m \times n) = F_{2m+3}^{n-m} \prod_{i=1}^{m} F_{2i+2}^2.$$

59. It is straightforward to prove by induction on n that a_i is the number of strict surjective maps $\tau : P \to i$, i.e., τ is surjective, and if $s < t$ in P then $\tau(s) < \tau(t)$. See R. Stanley, *Discrete Math.* **4** (1973), 77–82.

60. **b.** This result is implicit in J. R. Goldman, J. T. Joichi, and D. E. White, *J. Combinatorial Theory Ser. B* **25** (1978), 135–142 (put $x = -1$ in Theorem 2) and J. P. Buhler and R. L. Graham, *J. Combinatorial Theory Ser. A* **66** (1994), 321–326 (put $\lambda = -1$ and use our equation (3.121) in the theorem on page 322), and explicit in E. Steingrímsson, Ph.D. thesis, M.I.T., 1991 (Theorem 4.12). For an application see R. Stanley, *J. Combinatorial Theory Ser. A* **100** (2002), 349–375 (Theorem 4.8).

Sketch of Proof. Given the dropless labeling $f : P \to [p]$, define an acyclic orientation $\mathfrak{o} = \mathfrak{o}(f)$ as follows. If st is an edge of $\mathrm{inc}(P)$, then let $s \to t$ in $\mathfrak{o}$ if $f(s) < f(t)$. Clearly $\mathfrak{o}$ is an acyclic orientation of $\mathrm{inc}(P)$. Conversely, let $\mathfrak{o}$ be an acyclic orientation of $\mathrm{inc}(P)$. The set of sources (i.e., vertices with no arrows into them) form a chain in P since otherwise two are incomparable, and there is an arrow between them that must point into one of them. Let s be the minimal element of this chain (i.e., the unique minimal source). If f is a dropless labeling of P with $\mathfrak{o} = \mathfrak{o}(f)$, then we claim $f(s) = 1$. Suppose to the contrary that $f(s) = i > 1$. Let j be the largest integer satisfying $j < i$ and $t := f^{-1}(j) \not< s$. Note that j exists since $f^{-1}(1) > s$. We must have $t > s$ since s is a source. But then $f^{-1}(j+1) \le s < t = f^{-1}(j)$, contradicting the fact that f is dropless. Thus, we can set $f(s) = 1$, remove s from $\mathrm{inc}(P)$, and proceed inductively to construct a unique f satisfying $\mathfrak{o} = \mathfrak{o}(f)$.

61. Write $\mathrm{Comp}(n)$ for the set of compositions of n. Regarding n as given, and given a set $S = \{i_1, i_2, \ldots, i_j\}_< \subseteq [n-1]$, define the composition

$$\sigma_S = (i_1, i_2 - i_1, \ldots, i_j - i_{j-1}, n - i_j) \in \mathrm{Comp}(n).$$

Given a sequence $u = b_1 \cdots b_k$ of distinct integers, let $D(u) = \{i_1, i_2, \ldots, i_j\}_< \subseteq [k-1]$ be its descent set. Now given a permutation $w = a_1 \cdots a_n \in \mathfrak{S}_n$, let $w[k] = a_1 \cdots a_k$. It can be checked that

$$\sigma_{D(w[1])} \lessdot \sigma_{D(w[2])} \lessdot \cdots \lessdot \sigma_{D(w[n])}$$

is a saturated chain $\mathfrak{m}$ in $\mathcal{C}$ from 1 to $\sigma = \sigma(w[n])$, and that the map $w \mapsto \mathfrak{m}$ is a bijection from $\mathfrak{S}_n$ to saturated chains in $\mathcal{C}$ from 1 to a composition of n. Hence, the number of saturated chains from 1 to $\sigma \in \mathrm{Comp}(n)$ is $\beta_n(S)$, the number of $w \in \mathfrak{S}_n$ with descent set S, where $\sigma = \sigma_S$. In particular, the total number of saturated chains from 1 to some composition of n is $\sum_S \beta_n(S) = n!$. This latter fact also follows from the fact that every $\alpha \in \mathrm{Comp}(n)$ is covered in $\mathcal{C}$ by exactly $n+1$ elements.

The poset $\mathcal{C}$ was first defined explicitly in terms of compositions by Björner and Stanley (unpublished). It was pointed out by S. Fomin that $\mathcal{C}$ is isomorphic to the subword order on all words in a two-letter alphabet (see Exercise 3.134). A generalization was given by B. Drake and T. K. Petersen, *Electronic J. Combinatorics* **14**(1) (2007), #R23.

62. **f.** Let k_i be the number of λ_j's that are equal to i in a protruded partition (λ, μ). If some $a_j = i$, then μ_j can be any of $0, 1, \ldots, i$, so $a_j + b_j$ is one of $i, i+1, \ldots, 2i$. Hence,

$$U_{P_n}(x) = \prod_{i=1}^{n} \left(\sum_{k \geq 0} (x^i + x^{i+1} + \cdots + x^{2i})^k \right)$$

$$= \prod_{i=1}^{n} (1 - x^i - x^{i+1} - \cdots - x^{2i})^{-1}.$$

g. Write

$$\sum_{n \geq 0} U_{P_n}(x) q^n = P(q, x) \sum_{j \geq 0} W_j(x) q^j.$$

The poset P satisfies $P \cong \mathbf{1} \oplus (\mathbf{1} + P)$. This leads to the recurrence

$$W_j(x) = \frac{x^{2j}}{1-x} W_{j-1}(x) + x^j \frac{1 - x^{j+1}}{1-x} W_j(x), \quad W_0(x) = 1.$$

Hence $W_j(x) = x^{2j} W_{j-1}(x)/(1-x^j)(1-x-x^{j+1})$, from which the proof follows. Protruded partitions are due to Stanley [2.19, Ch. 5.4][2.20, §24], where more details of the above argument can be found. For a less combinatorial approach, see Andrews [1.2, Exam. 18, p. 51].

65. The fact that Exercise 1.11 can be interpreted in terms of linear extensions is an observation of I. M. Pak (private communication). NOTE. Equation (3.116) continues to hold if $2t$ is an odd integer, provided we replace any factorial $m!$ with the corresponding Gamma function value $\Gamma(m+1)$.

66. **a.** The Fibonacci number F_{n+2}—a direct consequence of Exercise 1.35(e).
   b. Simple combinatorial proofs can be given of the recurrences

$$W_{2n} = W_{2n-1} + q^2 W_{2n-2}, \ n \geq 1,$$

$$W_{2n+1} = q W_{2n} + W_{2n-1}, \ n \geq 1.$$

It follows easily from multiplying these recurrences by x^{2n} and x^{2n+1}, respectively, and summing on n, that

$$F(x) = \frac{1 + (1+q)x - q^2 x^3}{1 - (1+q+q^2)x^2 + q^2 x^4}.$$

c. A bijection $\sigma: Z_n \to [n]$ is a linear extension if and only if the sequence $n + 1 - \sigma(t_1), \ldots, n + 1 - \sigma(t_n)$ is an alternating permutation of $[n]$ (as defined in Section 1.4). Hence $e(Z_n)$ is the Euler number E_n, and by Proposition 1.6.1 we have

$$\sum_{n \geq 0} e(Z_n) \frac{x^n}{n!} = \tan x + \sec x.$$

d. Adjoin an extra element t_{n+1} to Z_n to create Z_{n+1}. We can obtain an order-preserving map $f: Z_n \to \mathbf{m+2}$ as follows. Choose a composition $a_1 + \cdots + a_k = n+1$, and associate with it the partition $\{t_1, \ldots, t_{a_1}\}, \{t_{a_1+1}, \ldots, t_{a_1+a_2}\}, \ldots$ of Z_{n+1}. For example, choosing $n = 17$ and $3 + 1 + 2 + 4 + 1 + 2 + 2 + 3 = 18$ gives the partition shown in Figure 3.75. Label the last element t of each block by 1 or $m+2$, depending on whether t is a minimal or maximal element of Z_{n+1}, as shown in

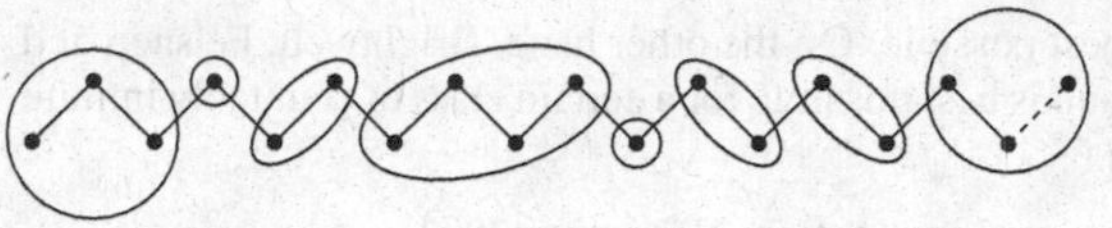

Figure 3.75 Illustration of the solution to Exercise 3.66(d).

$m+2$ $m+2$ $m+2$ $m+2$

1   1   1   1

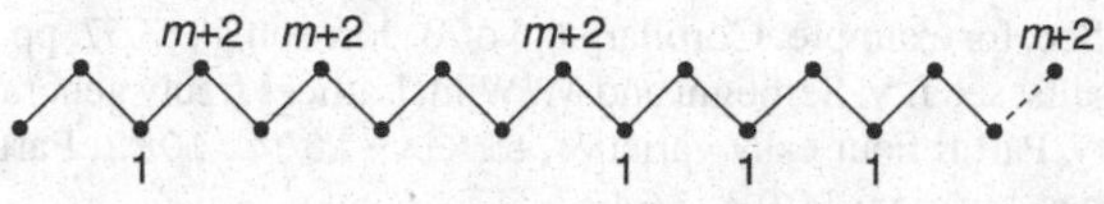

Figure 3.76 Continuing the solution to Exercise 3.66(d).

Figure 3.76. Removing these labeled elements from Z_{n+1} yields a disjoint union $Y_1 + \cdots + Y_k$, where Y_i is isomorphic to Z_{a_i-1} or $Z^*_{a_i-1}$ (where $*$ denotes dual). For each i choose an order-preserving map $Y_i \to [2, m+1]$ in $\Omega_{Z_{a_i-1}}(m)$ ways. There is one additional possibility. If some $a_i = 2$, then we can also assign the unique element t of Y_i the same label (1 or $m+2$) as the remaining element s in the block containing t (so t is labeled 1 if it is a maximal element of Z_{n+1} and $m+2$ if it is minimal). This procedure yields each order-preserving map $f: Z_n \to m+2$ exactly once. Hence,

$$\Omega_{Z_n}(m+2) = \sum_{a_1+\cdots+a_k=n+1} \prod_{i=1}^{k} (\Omega_{Z_{a_i-1}}(m) + \delta_{2,a_i})$$

$$\Rightarrow G_{m+2}(x) = \sum_{k \geq 0} (G_m(x) - 1 + x^2)^k$$

$$= (2 - x^2 - G_m(x))^{-1}.$$

The initial conditions are $G_1(x) = 1/(1-x)$ and $G_2(x) = 1/(1-x-x^2)$. An equivalent result was stated without proof (with an error in notation) in Example 3.2 of R. Stanley, *Annals of Discrete Math.* **6** (1980), 333–342. Moreover, G. Ziegler has shown (unpublished) that

$$G_{m+1}(x) = \frac{1 + G_m(x)}{3 - x^2 - G_m(x)}.$$

67. A complicated proof was first given by G. Kreweras, *Cahiers Bur. Univ. Rech. Operationnelle*, no. 6, 1965 (eqn. (85)). Subsequent proofs were given by H. Niederhausen, Proc. West Coast Conf. on Combinatorics, Graph Theory, and Computing (Arcata, Calif., 1979), *Utilitas Math.*, Winnipeg, Man., 1980, pp. 281–294, and Kreweras and Niederhausen, *Europ. J. Combinatorics* **4** (1983), 161–167.

68. **a.** See E. Munarini, *Ars Combin.* **76** (2005), 185–192. For further properties of order ideals and antichains of garlands, see E. Munarini, *Integers* **9** (2009), 353–374.

69. **a.** This result is due to R. Stanley, *J. Combinatorial Theory Ser. A* **31** (1981), 56–65 (see Theorem 3.1). The proof uses the Aleksandrov-Fenchel inequalities from the theory of mixed volumes.

    b. This result was proved by J. N. Kahn and M. Saks, *Order* **1** (1984), 113–126, with $\frac{5\pm\sqrt{5}}{10}$ replaced with $\frac{3}{11}$ and $\frac{8}{11}$. The improvement to $\frac{5\pm\sqrt{5}}{10}$ is due to G. R. Brightwell, S. Felsner, and W. T. Trotter, *Order* **12** (1995), 327–349. Both proofs use (a). It is conjectured that there exist s, t in P such that $f(s) < f(t)$ in no fewer than $\frac{1}{3}$ and no more than $\frac{2}{3}$ of the linear extensions of P. The poset $\mathbf{2} + \mathbf{1}$ shows that this result, if true, would be best possible. On the other hand, Brightwell, Felsner, and Trotter show that their result is best possible for a certain class of countably infinite posets, called *thin* posets.

70. **c.** Due to Ethan Fenn, private communication, November 2002.

71. The result for FD(n) is due to Dedekind. See [3.13, Ch. III, §4]. The result for FD(P) is proved the same way. See, for example, Corollary 6.3 of B. Jónsson, in [3.57, pp. 3–41]. For some related results, see J. V. Semegni and M. Wild, Lattices freely generated by posets within a variety. Part I: Four easy varieties, arXiv:1004.4082; Part II: Finitely generated varieties, arXiv:1007.1643.

72. **a.** The proof easily reduces to the following statement: if A and B are k-element antichains of P, then $A \cup B$ has k maximal elements. Let C and D be the set of maximal and minimal elements, respectively, of $A \cup B$. Since $t \in A \cap B$ if and only if $t \in C \cap D$, it follows that $\#C + \#D = 2k$. If $\#C < k$, then D would be an antichain of P with more than k elements, a contradiction.

This result is due to R. P. Dilworth, in *Proc. Symp. Appl. Math.* (R. Bellman and M. Hall, Jr., eds.), American Mathematical Society, Providence, R.I., 1960, pp. 85–90. An interesting application appears in §2 of C. Greene and D. J. Kleitman, in *Studies in Combinatorics* (G.-C. Rota, ed.), Mathematical Association of America, 1978, pp. 22–79.

b. R. M. Koh, *Alg. Univ.* **17** (1983), 73–86, and **20** (1985), 217–218.

73. **a.** Let $p \colon P \otimes Q \to P$ be the projection map onto P (i.e., $p(s,t) = s$), and let I be an order ideal of $P \otimes Q$. Then $p(I)$ is an order ideal of P, say with m maximal elements $t_1, \ldots, t_m$ and k nonmaximal elements $s_1, \ldots, s_k$. Then I is obtained by taking $p^{-1}(s_1) \cup \cdots \cup p^{-1}(s_k)$ together with a *nonempty* order ideal I_i of each $p^{-1}(t_i) \cong Q$. We then have $\#I = kn + \sum \#I_i$ and $m(I) = \sum m(I_i)$. Hence,

$$\sum_{I \in J(P \otimes Q)} q^{\#I} x^{m(I)} = \sum_{T \in J(P)} q^{n(\#T - m(T))} (G_Q(q,x) - 1)^{m(T)}$$

$$= G_P(q^n, q^{-n}(G_Q(q,x) - 1)).$$

b. Let t be a maximal element of P, and let $\Lambda_t = \{s \in P : s \le t\}$. Set $P_1 = P - t$ and $P_2 = P - \Lambda_t$. write $G(P) = G_P(q, (q-1)/q)$. One sees easily that

$$G(P) = G(P_1) + (q-1)q^{\#\Lambda_t - 1} G(P_2),$$

by considering for each $I \in J(P)$ whether $t \in I$ or $t \notin I$. By induction we have $G(P_1) = q^{p-1}$ and $G(P_2) = q^{\#(P - \Lambda_t)}$, so the proof follows.

This exercise is due to M. D. Haiman.

74. If $L_n = J(P_n)$, then P_n is the complete dual binary tree of height n, as illustrated in Figure 3.77. An order ideal I of P_n defines a stopping rule as follows: start at $\hat{0}$, and move up one step left (respectively, right) after tossing a tail (respectively, head). Stop as soon as you leave I or have reached a maximal element of P_n.

Since $P_n = \mathbf{1} \oplus (P_{n-1} + P_{n-1})$, it follows easily that

$$F(L_n, q) = 1 + q F(L_{n-1}, q)^2.$$

75. This result is due to J. Propp, Lattice structure for orientations of graphs, preprint, 1993, and was given another proof using hyperplane arrangements by R. Ehrenborg and M. Slone, *Order* **26** (2009), 283–288.

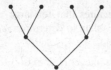

Figure 3.77 The complete dual binary tree P_3.

76. See R. Stanley, *Bull. Amer. Math. Soc.* **76** (1972), 1236–1239; [3.25, §3]; K. Baclawski, *Proc. Amer. Math. Soc.* **36** (1972), 351–356; R. B. Feinberg, *Pacific J. Math.* **65** (1976), 35–45; R. B. Feinberg, *Discrete Math.* **17** (1977), 47–70; M. Wild, *Linear Algebra Appl.* **430** (2009), 1007–1016; Y. Drozd and P. Kolesnik, *Comm. Alg.* **35** (2007), 3851–3854. A wealth of additional material can be found in Spiegel and O'Donnell [3.66].

77. a–c. These results are part of a beautiful theory of chains and antichains developed originally by C. Greene and D. J. Kleitman, *J. Combinatorial Theory Ser. A* **20** (1976), 41–68, and C. Greene, *J. Combinatorial Theory Ser. A* **20** (1976), 69–70. They were rediscovered by S. Fomin, *Soviet Math. Dokl.* **19** (1978), 1510–1514. Subsequently, two other elegant approaches were discovered, the first based on linear algebra by E. R. Gansner, *SIAM J. Algebraic Discrete Methods* **2** (1981), 429–440, and the second based on network flows by A. Frank, *J. Combinatorial Theory, Ser. B* **29** (1980), 176–184. A survey of this latter method (with much additional information) appears in T. Britz and S. Fomin, *Advances in Math.* **158** (2001), 86–127.

  d. Clearly, $k = \ell(\lambda)$, and by (c) we have $\ell(\lambda) = \mu_1$. This famous result, which can be regarded as a special case of the duality theorems of network flows and linear programming, is due to R. P. Dilworth, *Ann. Math.* **51** (1950), 161–166.

  e. Clearly $\mu_1' \geq \lambda_1$, since an antichain intersects a chain in at most one element. On the other hand, we have $P = P_1 \cup \cdots \cup P_{\lambda_1}$, where P_1 is the set of minimal elements of P, P_2 is the set of minimal elements of $P - P_1$, and so on. Each P_i is an antichain, so $\mu_1' \leq \lambda_1$. Note that this "dual" version of Dilworth's theorem is much easier to prove than Dilworth's theorem itself.

  f. See M. Saks, *SIAM J. Algebraic Discrete Methods* **1** (1980), 211–215, and *Discrete Math.* **59** (1986), 135–166, and E. R. Gansner, op. cit. An erroneous determination of the Jordan block sizes of A was earlier given by A. C. Aitken, *Proc. London Math. Soc. (2)* **38** (1934), 354–376, and D. E. Littlewood *Proc. London Math. Soc. (2)* **40** (1936), 370–381, and Vol. II, [7.88, §10.2].

78. a. The lattice B_n has the property, known as the *strong Sperner property*, that a maximum size union of k antichains consists of the union of the k largest ranks. Hence μ_i is just the ith largest binomial coefficient $\binom{n}{j}$. Some other posets with the strong Sperner property are any finite product of chains, $B_n(q)$, $J(m \times n \times r)$ for any $m, n, r \geq 1$, and $J(J(2 \times n))$. On the other hand, it is unknown whether $J(m \times n \times r \times s)$ has the strong Sperner property. For further information see K. Engel, *Sperner Theory*, Cambridge University Press, Cambridge, 1997.

  b. G.-C. Rota, *J. Combinatorial Theory* **2** (1967), 104, conjectured that the size of the largest antichain in Π_n was the maximum Stirling number $S(n,k)$ (i.e., the largest rank in Π_n was a maximum size antichain). This conjecture was disproved by E. R. Canfield, *Bull. Amer. Math. Soc.* **84** (1978), 164. For additional information, see E. R. Canfield, *J. Combinatorial Theory Ser. A* **83** (1998), 188–201.

  c. By Vol. II, Exercise 7.2(f) we have $\lambda_1 = \frac{1}{3}m(m^2 + 3r - 1)$, where $n = \binom{m+1}{2} + r$, $0 \leq r \leq m$. E. Early, Ph.D. thesis, M.I.T., 2004 (§2), showed that $\lambda_2 = \lambda_1 - 6$

for $n > 16$, and $\lambda_3 = \lambda_2 - 6$ for $n > 135$. Early conjectures that for large n, $\lambda_i - \lambda_{i+1}$ depends only on i. It is an interesting open problem to determine μ_1. Some observations on this problem are given by Early, ibid.

79. **i=ii.** Let $w = a_1 \cdots a_p \in \mathcal{L}(P)$. Let i be the least nonnegative integer (if it exists) for which

$$w' := a_1 \cdots a_{p-2i-2} a_{p-2i} a_{p-2i-1} a_{p-2i+1} \cdots a_p \in \mathcal{L}(P).$$

Note that $w'' = w$. Now exactly one of w and w' has the descent $p - 2i - 1$. The only other differences in the descent sets of w and w' occur (possibly) for the numbers $p - 2i - 2$ and $p - 2i$. Hence, $(-1)^{\text{comaj}(w)} + (-1)^{\text{comaj}(w')} = 0$. The surviving permutations $w = b_1 \cdots b_p$ in $\mathcal{L}(P)$ (those for which i does not exist) are exactly those for which the chain of order ideals

$$\emptyset \subset \cdots \subset \{b_1, b_2, \ldots, b_{p-4}\} \subset \{b_1, b_2, \ldots, b_{p-2}\} \subset \{b_1, b_2, \ldots, b_p\} = P$$

is a P-domino tableau. We call w a *domino linear extension*; they are in bijection with domino tableaux. Such permutations w can only have descents in positions $p - j$ where j is even, so $(-1)^{\text{comaj}(w)} = 1$. Hence, (i) and (ii) are equal. This result, stated in a dual form, appears in R. Stanley, *Advances Appl. Math.* **34** (2005), 880–902 (Theorem 5.1(a)).

ii=iii. Let τ_i be the operator on $\mathcal{L}(P)$ defined by equation (3.102). Thus, w is self-evacuating if and only if

$$w = w \tau_1 \tau_2 \cdots \tau_{p-1} \cdot \tau_1 \cdots \tau_{p-2} \cdots \tau_1 \tau_2 \tau_3 \cdot \tau_1 \tau_2 \cdot \tau_1.$$

On the other hand, note that w is a domino linear extension if and only if

$$w \tau_{p-1} \tau_{p-3} \tau_{p-5} \cdots \tau_h = w,$$

where $h = 1$ if p is even, and $h = 2$ if p is odd. We claim that w is a domino linear extension if and only if

$$\widetilde{w} := w \tau_1 \cdot \tau_3 \tau_2 \tau_1 \cdot \tau_5 \tau_4 \tau_3 \tau_2 \tau_1 \cdots \tau_m \tau_{m-1} \cdots \tau_1$$

is self-evacuating, where $m = p - 1$ if p is even, and $m = p - 2$ if p is odd. The proof follows from this claim since the map $w \mapsto \widetilde{w}$ is then a bijection between domino linear extensions and self-evacuating linear extensions of P.

The claim is proved by an elementary argument analogous to the proof of Theorem 3.20.1. The cases p even and p odd need to be treated separately. We won't give the details here but will prove the case $p = 6$ as an example. For notational simplicity we write simply i for τ_i. We need to show that the two conditions

$$w = w135 \tag{3.138}$$

$$w132154321 = w132154321 \cdot 123451234123121 \tag{3.139}$$

are equivalent. (The first condition says that w is a domino linear extension, and the second that $w132154321$ is self-evacuating.) The internal factor $32154321 \cdot 12345123$ cancels out of the right-hand side of equation (3.139). We can also cancel the rightmost 21 on both sides of (3.139). Thus, (3.139) is equivalent to $w1321543 = w141231$. Now $w1321543 = w1352143$ and $w141231 = w112143 = w2143$. Cancelling 2143 from the right of both sides yields $w135 = w$. Since all steps are reversible, the claim is proved for $p = 6$.

The equality of (ii) and (iii) was first proved by J. R. Stembridge, *Duke Math. J.* **82** (1996), 585–606, for the special case of standard Young tableaux (i.e., when P is a finite order ideal of $\mathbb{N} \times \mathbb{N}$). Stembridge's proof was based on representation theory.

He actually proved a more general result involving *semistandard* tableaux that does not seem to extend to other posets. The bijective argument given here, again for the case of semistandard tableaux, is due to A. Berenstein and A. N. Kirillov, *Discrete Math.* **225** (2000), 15–24.

The equivalence of (i) and (iii) is an instance of Stembridge's "$q = -1$ phenomenon." Namely, suppose that an involution ι acts on a finite set S. Let $f : S \to \mathbb{Z}$. (Usually f will be a "natural" combinatorial or algebraic statistic on S.) Then we say that the triple (S, ι, f) exhibits the $q = -1$ phenomenon if the number of fixed points of ι is given by $\sum_{t \in S} (-1)^{f(t)}$. See J. R. Stembridge, *J. Combinatorial Theory Ser. A* **68** (1994), 373–409; *Duke Math. J.* **73** (1994), 469–490; and *Duke Math. J.* **82** (1996), 585–606. The $q = -1$ phenomenon has been generalized to the action of cyclic groups by V. Reiner, D. Stanton, and D. E. White, *J. Combinatorial Theory Ser. A* **108** (2004), 17–50, where it is called the "cyclic sieving phenomenon." For further examples of the cyclic sieving phenomenon, see C. Bessis and V. Reiner, *Ann. Combinatorics*, submitted, arXiv:math/0701792; H. Barcelo, D. Stanton, and V. Reiner, *J. London Math. Soc. (2)* **77** (2009), 627–646; and B. Rhoades, Cyclic sieving and promotion, preprint.

80. Part (a)(i) follows easily from work of Schützenberger, whereas (a)(ii)–(a)(iii) are due to Haiman and (b) to Edelman and Greene. For further details, see R. Stanley, *Electronic J. Combinatorics* **15**(2) (2008–2009), #R9 (§4):

81. **a.** If $a_1 a_2 a_3 \cdots a_n \in \mathcal{L}(P)$, then $a_2 a_1 a_3 \cdots a_n \in \mathcal{L}(P)$.
    b. Hint. Show that the promotion operator $\partial : \mathcal{L}(P) \to \mathcal{L}(P)$ always reverses the parity of the linear extension f to which it is applied. See R. Stanley, *Advances in Appl. Math.* **34** (2005), 880–902 (Corollary 2.2). Corollary 2.4 of this reference gives another result of a similar nature.

82. This result is due to D. White, *J. Combinatorial Theory Ser. A* **95** (2001), 1–38 (Corollary 20 and §8). White also computes the "sign imbalance" $\left| \sum_{w \in \mathcal{E}_{p \times q}} \operatorname{sgn}(w) \right|$ when $p \times q$ in not sign-balanced. A conjectured generalization for any finite-order ideal of $\mathbb{N} \times \mathbb{N}$ appears in R. Stanley, *Advances in Appl. Math.* **34** (2005), 880–902 (Conjecture 3.6).

83. Let $\#P = 2m$, and suppose that there does not exist a P-domino tableau. Let $w = a_1 a_2 \cdots a_{2m} \in \mathcal{E}_P$. Since there does not exist a P-domino tableau, there is a least i for which a_{2i-1} and a_{2i} are incomparable. Let w' be the permutation obtained from w by transposing a_{2i-1} and a_{2i}. Then the map $w \mapsto w'$ is an involution on $\mathcal{E}_P$ that reverses parity, and the proof follows. This result appears in R. Stanley, ibid. (Corollary 4.2), with an analogous result for $\#P$ odd.

84. We have

$$\sum_{\substack{u \in P \\ \bar{u} = \bar{t}}} \mu(s, u) = \sum_{u} \mu(s, u) \delta_{\overline{P}}(\bar{u}, \bar{t})$$

$$= \sum_{u, \bar{v}} \mu(s, u) \zeta_{\overline{P}}(\bar{u}, \bar{v}) \mu_{\overline{P}}(\bar{v}, \bar{t})$$

$$= \sum_{u, \bar{v}} \mu(s, u) \zeta(u, \bar{v}) \mu_{\overline{P}}(\bar{v}, \bar{t}) \quad (\text{since } u \leq \bar{v} \Leftrightarrow \bar{u} \leq \bar{v})$$

$$= \sum_{\bar{v} \in \overline{P}} \delta(s, \bar{v}) \mu_{\overline{P}}(\bar{v}, \bar{t}).$$

This fundamental result was first given by H. Crapo, *Archiv der Math.* **19** (1968), 595–607 (Thm. 1), simplifying some earlier work of G.-C. Rota in [3.58]. For an exposition of the theory of Möbius functions based on closure operators, see Ch. IV.3 of M. Aigner, *Combinatorial Theory*, Springer-Verlag, Berlin/Heidelberg/New York, 1979.

86. Let $G(s) = \sum_{t \geq s} g(t)$. It is easy to show that

$$\sum_{\hat{0} \leq u \leq s} \mu(\hat{0}, u) G(u) = \sum_{\substack{t \\ s \wedge t = \hat{0}}} g(t) = f(s).$$

Now use Möbius inversion to obtain

$$\mu(\hat{0}, t) G(t) = \sum_{u \leq t} \mu(u, t) f(u). \tag{3.140}$$

On the other hand, Möbius inversion also yields

$$g(s) = \sum_{t \geq s} \mu(s, t) G(t). \tag{3.141}$$

Substituting the value of $G(t)$ from equation (3.140) into (3.141) yields the desired result.

This formula is a result of P. Doubilet, *Studies in Appl. Math.* **51** (1972), 377–395 (lemma on page 380).

87. a. Given $C: \hat{0} < t_1 < \cdots < t_k < \hat{1}$, the coefficient of $f(t_1) \cdots f(t_k)$ on the left-hand side is

$$\sum_{C' \supseteq C} (-1)^{\#(C'-C)} = (-1)^{k+1} \mu(\hat{0}, t_1) \mu(t_1, t_2) \cdots \mu(t_k, \hat{1}),$$

by Proposition 3.8.5. Here C' ranges over all chains of $P - \{\hat{0}, \hat{1}\}$ containing C. Essentially the same result appears in Ch. II, Lemma 3.2, of [3.67].

b. Put each $f(t) = 1$. All terms on the left-hand side are 0 except for the term indexed by the chain $\hat{0} < \hat{1}$ (an empty product is equal to 1).

c. We have

$$\sum_{\hat{0} = t_0 < t_1 < \cdots < t_k = \hat{1}} (-1)^k \mu(t_0, t_1) \mu(t_1, t_2) \cdots \mu(t_{k-1}, t_k)$$

$$= (1 - (\mu - 1) + (\mu - 1)^2 - (\mu - 1)^3 + \cdots)(\hat{0}, \hat{1})$$

$$= (1 + (\mu - 1))^{-1}(\hat{0}, \hat{1})$$

$$= \zeta(\hat{0}, \hat{1})$$

$$= 1.$$

d. By Example 3.9.6 we have

$$\mu(t_0, t_1) \mu(t_1, t_2) \cdots \mu(t_{k-1}, t_k)$$

$$= \begin{cases} (-1)^{\ell}, & \text{if the chain } t_0 < t_1 < \cdots < t_k \text{ is boolean,} \\ 0, & \text{otherwise,} \end{cases}$$

and the proof follows easily from (a).

88. Consider $\zeta \mu \zeta(\hat{0}, \hat{1})$ in the incidence algebra $I(P, \mathbb{C})$. For a similar trick, see the solution to Exercise 3.47(b). A solution can also be given based on Philip Hall's theorem (Proposition 3.8.5).

92. This result is known as the "Crapo complementation theorem." See H. H. Crapo, *J. Combinatorial Theory* **1** (1966), 126–131 (Thm. 3). For topological aspects of this result, see A. Björner, *J. Combinatorial Theory Ser. A* **30** (1981), 90–100.

93. **a.** By the inductive definition (3.15) of the Möbius function, it follows that $\mu_L(\hat{0}, t)$ is *odd* (and therefore nonzero) for all $t \in L$. Now use Exercise 3.92.

b. See R. Freese and Univ. of Wyoming Problem Group, *Amer. Math. Monthly* **86** (1979), 310–311.

94. This result (stated slightly differently) is due to G.-C. Rota, in *Studies in Pure Mathematics* (L. Mirsky, ed.), Academic Press, London, 1971, pp. 221–233 (Thm. 2). Related papers include G.-C. Rota, in *Proc. Univ. Houston Lattice Theory Conf.*, 1973, pp. 575–628; L. D. Geissinger, *Arch. Math. (Basel)* **24** (1973), 230–239, 337–345, and in *Proc. Third Caribbean Conference on Combinatorics and Computing*, University of the West Indies, Cave Hill, Barbados, pp. 125–133; R. L. Davis, *Bull. Amer. Math. Soc.* **76** (1970), 83–87; H. Dobbertin, *Order* **2** (1985), 193–198. See also Exercise 4.58.

95. See Greene [3.34, Thm. 5].

96. Our exposition for this entire exercise is based on Greene [3.35].

a. Define a matrix $M = [M(s,t)]$ by setting $M(s,t) = \zeta(s,t)f(s,t)$. Clearly if we order the rows and columns of M by some linear extension of L, then M is triangular and $\det M = \prod_s f(s,s)$. On the other hand (writing ζ for the matrix of the ζ-function of L with respect to the basis L, that is, ζ is the incidence matrix of the relation L),

$$M^t \zeta = \left[\sum_u f(u,s)\zeta(u,s)\zeta(u,t) \right]_{s,t \in L}$$

$$= \left[\sum_{u \le s \wedge t} f(u,s) \right]_{s,t \in L} = [F(s \wedge t), s].$$

Thus, $\det[F(s \wedge t, s)] = \det M^t \zeta = \det M$.

This formula is a result of B. Lindström, *Proc. Amer. Math. Soc.* **20** (1969), 207–208, and (in the case where $F(s,v)$ depends only on s) H. S. Wilf, *Bull. Amer. Math. Soc.* **74** (1968), 960–964.

b. Take L to be the set $[n]$ ordered by divisibility, and let $f(s,v) = \phi(s)$ (so $F(s,v) = s$). For a proof from scratch, see G. Pólya and G. Szegö, *Problems and Theorems in Analysis II*, Springer-Verlag, Berlin/Heidelberg/New York, 1976 (Part VIII, Ch. 1, no. 33).

c. When $f(s,v) = \mu(\hat{0}, s)$, we have (suppressing v)

$$F(s \wedge t) = \sum_{u \le s \wedge t} \mu(\hat{0}, u) = \delta(\hat{0}, s \wedge t).$$

Hence the matrix $R = [F(s \wedge t)]$ is just the incidence matrix of the relation $s \wedge t = \hat{0}$. By (a), $\det R \ne 0$. Hence some term in the expansion of $\det R$ must be nonzero, and this term yields the desired permutation w.

This result is due to T. A. Dowling and R. M. Wilson, *Proc. Amer. Math. Soc.* **47** (1975), 504–512 (Thm. 2*).

d. By Exercise 3.100(b) we have that $\mu(s,t) \ne 0$ for all $s \le t$ in a geometric lattice L. Apply (c) to the dual L^*. We get a permutation $w : L \to L$ such that $s \vee w(s) = \hat{1}$ for all $s \in L$. Semimodularity implies $\rho(s) + \rho(w(s)) \ge n$, so w maps elements of rank at most k injectively into elements of rank at least $n - k$.

Figure 3.78 A lattice with no complementing permutation.

This result is also due to T. A. Dowling and R. M. Wilson, ibid. (Thm. 1). The case $k = 1$ was first proved by C. Greene, *J. Combinatorial Theory* **2** (1970), 357–364.

  e. T. A. Dowling and R. M. Wilson, op. cit. (Thm. 1).

97. See T. A. Dowling, *J. Combinatorial Theory Ser. B* **23** (1977), 223–226. The following elegant proof is due to R. M. Wilson (unpublished). Let ζ be the matrix in the solution to Exercise 3.96(a), and let

$$\Delta_0 = \operatorname{diag}(\mu(\hat{0}, t) : t \in L),$$

$$\Delta_1 = \operatorname{diag}(\mu(t, \hat{1}) : t \in L).$$

By the solution to Exercise 3.96(c) (and its dual), we have that

$$\left[\zeta^t \Delta_0 \zeta\right]_{uv} = \delta(\hat{0}, u \wedge v),$$

$$\left[\zeta \Delta_1 \zeta^t\right]_{uv} = \delta(u \vee v, \hat{1}).$$

Let $C = \zeta \Delta_1 \zeta^t \Delta_0 \zeta$. Since $C = (\zeta \Delta_1 \zeta^t) \Delta_0 \zeta = \zeta \Delta_1 (\zeta^t \Delta_0 \zeta)$, it follows that $C_{uv} = 0$ unless u and v are complements. But the hypothesis on L implies that $\det C \neq 0$, and so a nonzero term in the expansion of $\det C$ gives the desired permutation w.

For an example where $\mu(\hat{0}, t) \neq 0$ for all $t \in L$, yet a "complementing permutation" does not exist, see Figure 3.78.

100. a. Suppose that s is an element of L of maximal rank that is not a meet of coatoms. Thus, s is covered by a unique element t [why?]. Clearly, $t \neq \hat{1}$; else s would be a coatom. Since L is atomic, there is an atom a of L such that $a \not\leq t$; else $\hat{1}$ would not be a join of atoms. Since L is semimodular, $s \vee a$ covers a. Since $a \not\leq t$ we have $s \vee a \neq t$, a contradiction.

   b. Consider equation (3.33). By (a) the sum is not empty, so the proof follows by induction on the rank of L (the case $\operatorname{rank}(L) = 1$ being trivial).

   c. Induction on $n = \operatorname{rank}(L)$, the case $n = 1$ being trivial. Now assume for $\operatorname{rank}(L) = n - 1$, and let $\operatorname{rank}(L) = n$. Let a be an atom of L. By (b), the sum over t in equation (3.33) has exactly one term. Thus, there is exactly one coatom t of L not lying above a, and moreover by the induction hypothesis the interval $[\hat{0}, t]$ is isomorphic to B_{n-1}. Since this result holds for all atoms a, it is easy to see that all subsets of the atoms have different joins. Hence, $L \cong B_n$.

101. a. Let $f : L \to \mathbb{Q}$, and define $\widehat{f} : L \to \mathbb{Q}$ by

$$\widehat{f}(t) = \sum_{s \leq t} f(s).$$

For any $t \le t^*$ in L, we have

$$\sum_{t \le s \le t^*} \widehat{f}(s)\mu(s,t^*) = \sum_{t \le s \le t^*} \mu(s,t^*) \sum_{u \le s} f(u)$$

$$= \sum_{u} f(u) \sum_{\substack{t \le s \le t^* \\ u \le s}} \mu(s,t^*)$$

$$= \sum_{u} f(u) \sum_{t \vee u \le s \le t^*} \mu(s,t^*)$$

$$= \sum_{\substack{u \\ t \vee u = t^*}} f(u).$$

Now suppose that $f(t) = 0$ unless $t \in B$. We claim that the restriction $\widehat{f}_A$ of $\widehat{f}$ to A determines $\widehat{f}$ (and hence f since $f(t) = \sum_{s \le t} \widehat{f}(s)\mu(s,t)$ by Möbius inversion). We prove the claim by induction on the length $\ell(t, \hat{1})$ of the interval $[t, \hat{1}]$. If $t = \hat{1}$ that $\hat{1} \in A$ by hypothesis, so $\widehat{f}(\hat{1}) = \widehat{f}_A(\hat{1})$. Now let $t < \hat{1}$. If $t \in A$, then there is nothing to prove, since $\widehat{f}(t) = \widehat{f}_A(t)$. Thus, assume $t \notin A$. Let t^* be as in the hypothesis. Then

$$\sum_{\substack{u \\ t \vee u = t^*}} f(u) = 0 \quad \text{(empty sum)},$$

so

$$\sum_{t \le s \le t^*} \widehat{f}(s)\mu(s,t^*) = 0.$$

By induction, we know $\widehat{f}(s)$ for $t < s$. Since $\mu(t,t^*) \ne 0$, we can then solve for $\widehat{f}(t)$. Hence, the claim is proved.

It follows that the matrix $[\zeta(s,t)]_{\substack{s \in B \\ t \in A}}$ has rank $b = \#B$. Thus, some $b \times b$ submatrix has nonzero determinant. A nonzero term in the expansion of this determinant defines an injective function $\phi: B \to A$ with $\phi(s) \ge s$, and the proof follows.

This result and the following applications are due to J. P. S. Kung, *Order* **2** (1985), 105–112; *Math. Proc. Cambridge Phil. Soc.* **101** (1987), 221–231. The solution given here was suggested by C. Greene.

b. These are standard results in lattice theory; for example [3.13], Theorem 13 on page 13 and §IV.6–IV.7.

c. Choose $A = M_k$ and $B = J_k$ in (a). Given $t \in L$, let t^* be the join of elements covering t. By (i) from (b) we have $\mu(t,t^*) \ne 0$. Moreover by (ii) if t is covered by j elements of L then t^* covers j elements of $[t,t^*]$. Thus, if $t \notin A = M_k$ then T^* covers more than k elements of $[t,t^*]$. Let $u \in B = J_k$. Then by (iii), $[t \wedge u, u] \cong [t, t \vee u]$. Hence, $t \vee u \ne t^*$, so the hypotheses of (a) are satisfied, and the result follows.

d. By (c), $\#J_k \le \#M_k$. Since the dual of a modular lattice is modular, we also have $\#M_k \le \#J_k$, and the result follows. This result was first proved (in a more complicated way) by R. P. Dilworth, *Ann. Math.* (2) **60** (1954), 359–364, and later by B. Ganter and I. Rival, *Alg. Universalis* **3** (1973), 348–350.

e. See Figure 3.79.

f. With L as in Exercise 3.96(d), choose

$$A = \{t \in L : \rho(t) \ge n - k\},$$

$$B = \{t \in L : \rho(t) \le k\}.$$

Define $t^* = \hat{1}$ for all $t \in L$. The hypotheses of (a) are easily checked, so in particular $\#B \le \#A$ as desired.

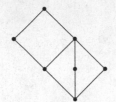

Figure 3.79 A modular lattice for which $P_1 \not\cong Q_1$.

102. a. This diabolical problem is equivalent to a conjecture of P. Frankl. See page 525 of *Graphs and Order* (I. Rival, ed.), Reidel, Dordrecht/Boston, 1985. For some further work on this conjecture, see R. Morris, *Europ. J. Combinatorics* **27** (2006), 269–282, and the references therein.

b. If not, then by Exercise 3.96(c) there is a permutation $w \colon L \to L$ for which $s \wedge w(s) = \hat{0}$ for all $s \in L$. But if $\#V_t > n/2$, then $V_t \cap w(V_t) \neq \emptyset$, and any $u \in V_t \cap w(V_t)$ satisfies $u \wedge w(u) \geq t$. One can also give a simple direct proof (avoiding Möbius inversion) of the following stronger result. Let L be a finite lattice with n elements, such that for all $\hat{0} < s \leq t$ in L, there exists $u \neq t$ for which $s \vee u = t$. Then every $t > \hat{0}$ satisfies $\#V_t \leq n/2$.

103. *Answer.* If $n \geq 3$ then

$$-\binom{n-1}{2\lfloor (n-1)/4 \rfloor} \leq \mu(\hat{0}, \hat{1}) \leq \binom{n-1}{2\lfloor (n-1)/4 \rfloor + 1}.$$

See H. Scheid, *J. Combinatorial Theory* **13** (1972), 315–331 (Satz 5).

104. a. E. E. Maranich, *Mat. Zametki* **44** (1988), 469–487, 557; translation in *Math. Notes* **44** (1988), 736–447 (1989), and independently G. M. Ziegler, *J. Combinatorial Theory Ser. A* **56** (1991), 203–222, have shown by induction on the length of P that the answer is

$$\max (a_1 - 1) \cdots (a_k - 1),$$

where the maximum is taken over all partitions $a_1 + \cdots + a_j = n$. (One can show that the maximum is obtained by taking at most four of the a_i's not equal to five.) This bound is achieved by taking P to be the ordinal sum $\mathbf{1} \oplus a_1 \mathbf{1} \oplus \cdots \oplus a_k \mathbf{1} \oplus \mathbf{1}$. For some additional results, see D. N. Kozlov, *Combinatorica* **19** (1999), 533–548.

b. One can achieve $n^{2-\epsilon}$ (for any $\epsilon > 0$ and sufficiently large n) by taking L to be the lattice of subspaces of a suitable finite-dimensional vector space over a finite field. It seems plausible that $n^{2-\epsilon}$ is best possible. This problem was suggested by L. Lovász. A subexponential upper bound is given by Ziegler, op. cit.

105. This problem was suggested by P. H. Edelman. It is plausible to conjecture that the maximum is obtained by taking P to be the ordinal sum $\mathbf{1} \oplus k\mathbf{1} \oplus k\mathbf{1} \oplus \cdots \oplus k\mathbf{1} \oplus \mathbf{1}$ ($\ell - 1$ copies of $k\mathbf{1}$ in all), yielding $|\mu(\hat{0}, \hat{1})| = (k-1)^{\ell-1}$, but this conjecture is false. The first counterexample was given by Edelman; and G. M. Ziegler, op. cit., attained $|\mu(\hat{0}, \hat{1})| = (k-1)(k^{\ell-1} - 1)$, together with some related results.

106. No, an example being given in Figure 3.80. The first such example (somewhat more complicated) was given by C. Greene (private communication, 1972).

107. This result is due to R. Stanley (proposer), Problem 11453, *Amer. Math. Monthly* **116** (2009), 746. The following solution is due to R. Ehrenborg.

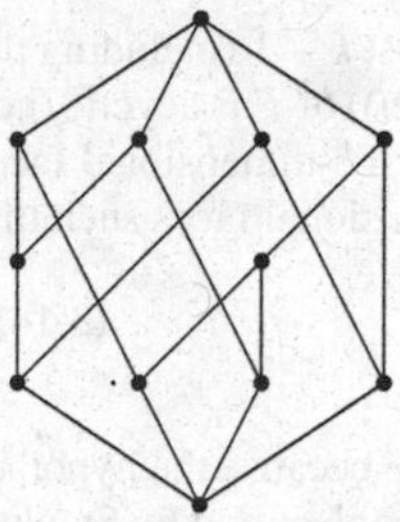

Figure 3.80 A lattice with no sublattice $\mathbf{1} \oplus (\mathbf{1}+\mathbf{1}+\mathbf{1}) \oplus \mathbf{1}$.

We can rewrite the identity (regarding elements of I as subsets of $[n]$) as

$$\sum_{\substack{u \in I \\ u \geq t}} (-1)^{\#u} = 0.$$

Sum the given identity over all sets v in I of cardinality j, where $0 \leq j \leq k$:

$$0 = \sum_{\substack{v \in I \\ \#v=j}} \sum_{\substack{u \in I \\ u \geq v}} (-1)^{\#u}$$

$$= \sum_{u \in I} \sum_{\substack{v \geq u \\ \#v=j}} (-1)^{\#u}$$

$$= \sum_{u \in I} \binom{\#u}{j} (-1)^{\#u}.$$

Multiply this equation by $(-2)^j$ and take modulo 2^{k+1}, to obtain

$$0 \equiv \sum_{u \in I} \binom{\#u}{j} (-1)^{\#u-j} 2^j \bmod 2^{k+1}.$$

Observe that this congruence is also true for $j > k$, that is, it holds for all nonnegative integers j. Now summing over all j and using the binomial theorem, we have modulo 2^{k+1} that

$$0 \equiv \sum_{j \geq 0} \sum_{u \in I} \binom{\#u}{j} (-1)^{\#u-j} 2^j$$

$$\equiv \sum_{u \in I} \sum_{j \geq 0} \binom{\#u}{j} (-1)^{\#u-j} 2^j$$

$$\equiv \sum_{u \in I} (-1+2)^{\#u}$$

$$\equiv \sum_{u \in I} 1$$

$$\equiv \#I.$$

This result is the combinatorial analogue of a much deeper topological result of G. Kalai, in *Computational Commutative Algebra and Combinatorics*, Advanced Studies in Pure Mathematics **23** (2002), 121–163 (Theorem 4.2), a special case of which can be stated as follows. Let Δ be a finite simplicial complex, or equivalently, an order ideal I of B_n. Suppose that for any face F of dimension at most $k-1$ (including the empty face of dimension -1), the link (defined in equation (3.26)) of F is acyclic (i.e., has vanishing reduced homology). Let f_i denote the number of i-dimensional faces of Δ. Then there exists a simplicial complex Γ with g_i i-dimensional faces such that

$$\sum_{i \geq -1} f_i x^i = (1+x)^{k+1} \sum_{i \geq -1} g_i x^i. \tag{3.142}$$

(Note that equation (3.142) does not imply the present exercise because the hypothesis on Δ is stronger for (3.142).) An even stronger result was conjectured by Stanley, *Discrete Math.* **120** (1993), 175–182 (Conjecture 2.4), as follows. Let L be the poset (or meet-semilattice) of faces of Δ. Then there exists a partitioning of L into intervals $[s,t]$ of rank $k+1$ such that the bottom elements s of the intervals form an order ideal of L. The case $k=0$ was proved by Stanley, *Discrete Math.* **120** (1993), 175–182, and some generalizations by A. M. Duval, *Israel J. Math.* **87** (1994), 77–87, and A. M. Duval and P. Zhang, *Israel J. Math.* **121** (2001), 313–331. A stronger conjecture than the one just stated is due to Kalai, op. cit. (Conjecture 22).

108. b. If σ is a partition of V, then let $\chi_\sigma(n)$ be the number of maps $f: V \to [n]$ such that (i) if a and b are in the same block of σ then $f(a) = f(b)$, and (ii) if a and b are in different blocks and $\{a,b\} \in E$, then $f(a) \neq f(b)$. Given *any* $f: V \to [n]$, there is a unique $\sigma \in L_G$ such that f is one of the maps enumerated by $\chi_\sigma(n)$. It follows that for any $\pi \in L_G$, we have $n^{\#\pi} = \sum_{\sigma \geq \pi} \chi_\sigma(n)$. By Möbius inversion $\chi_\pi(n) = \sum_{\sigma \geq \pi} n^{\#\sigma} \mu(\pi,\sigma)$. But $\chi_{\hat{0}}(n) = \chi_G(n)$, so the proof follows. This interpretation of $\chi_G(n)$ in terms of Möbius functions is due to G.-C. Rota [3.58, §9].

c. Denote the hyperplane with defining equation $x_i - x_j$ by H_e, where e is the edge with vertices i and j. Let i_T be an intersection of some set T of hyperplanes of the arrangement $\mathcal{B}_G$. Let G_T be the spanning subgraph of G with edge set $\{e : H_e \in T\}$. If e' is an edge of G such that its vertices belong to the same connected component of G_T, then it is easy to see that $i_T = i_{T \cup \{e'\}}$. From this observation, it follows that $L_{\mathcal{B}_G}$ is isomorphic to the set of connected partitions of G ordered by refinement, as desired. It follows from (b) that χ_G and $\chi_{\mathcal{B}_G}$ differ at most by a power of q. Equality then follows, for example, from the fact that both have degree equal to $\#V$.

d. It is routine to verify equation (3.119) from (c) and Proposition 3.11.5. To give a direct combinatorial proof, let $e = \{u,v\}$. Show that $\chi_G(n)$ is the number of proper colorings f of $G - e$ such that $f(u) \neq f(v)$, while $\chi_{G/e}(n)$ is the number of proper colorings f of $G - e$ such that $f(u) = f(v)$.

e. It follows from equation (1.96) and Proposition 1.9.1(a) that $\varphi((n)_k) = x^k$. Now use (a). Chung Chan has pointed out that this result can also be proved from (d) by first showing that if we set $g_G = \sum_j S_G(j) x^j$, then $g_G = g_{G-e} - g_{G/e}$ for any edge e of G. Equation (3.120) is equivalent to an unpublished result of Rhodes Peele.

109. a. We need to prove that

$$\mathrm{ao}(G) = \mathrm{ao}(G-e) + \mathrm{ao}(G/e),$$

together with the initial condition $\mathrm{ao}(G) = 1$ if G has no edges. Let $\mathfrak{o}$ be an acyclic orientation of $G - e$, where $e = \{u,v\}$. Let $\mathfrak{o}_1$ be $\mathfrak{o}$ with $u \to v$ adjoined, and $\mathfrak{o}_2$ be $\mathfrak{o}$ with $v \to u$ adjoined, so $\mathfrak{o}_1$ and $\mathfrak{o}_2$ are orientations of G. The key step is to show the following: Exactly one of $\mathfrak{o}_1$ and $\mathfrak{o}_2$ is acyclic, except for $\mathrm{ao}(G/e)$ cases for which both $\mathfrak{o}_1$ and $\mathfrak{o}_2$ are acyclic. See R. Stanley, *Discrete Math.* **5** (1973), 171–178.

b. A region of the graphical arrangement $\mathcal{B}_G$ is obtained by specifying for each edge $\{i,j\}$ of G whether $x_i < x_j$ or $x_i > x_j$. Such a specification is consistent if and only if the following condition is satisfied: Let $\mathfrak{o}$ be the orientation obtained by letting $i \rightarrow j$ whenever we choose $x_i < x_j$. Then $\mathfrak{o}$ is acyclic. Hence, the number of regions of $\mathcal{B}_G$ is $\mathrm{ao}(G)$. Now use Exercise 3.108(c) and Theorem 3.11.7. This proof is due to G. Greene and T. Zaslavsky, *Trans. Amer. Math. Soc.* **280** (1983), 97–126.

110. Part (a) was proved and (b) was conjectured by A. E. Postnikov, Total positivity, Grassmannians, and networks, `arXiv:math/0609764` (Conjecture 24.4(1)). Postnikov's conjecture was proved by A. Hultman, S. Linusson, J. Shareshian, and J. Sjöstrand, *J. Combinatorial Theory Ser. A* **116** (2009), 564–580.

111. See C. A. Athanasiadis and S. Linusson, *Discrete Math.* **204** (1999), 27–39; and R. Stanley, in *Mathematical Essays in Honor of Gian-Carlo Rota* (B. Sagan and R. Stanley, eds.), Birkhäuser, Boston/Basel/Berlin, 1998, pp. 359–375.

112. **b.** By Whitney's theorem (Proposition 3.11.3) we have for any arrangement $\mathcal{A}$ in K^n that

$$\chi_{\mathcal{A}}(x) = \sum_{\substack{\mathcal{B} \subseteq \mathcal{A} \\ \mathcal{B} \text{ central}}} (-1)^{\#\mathcal{B}} x^{n - \mathrm{rank}(\mathcal{B})}.$$

Let $\mathfrak{A} = (\mathcal{A}_1, \mathcal{A}_2, \dots)$, and let $\mathcal{B} \subseteq \mathcal{A}_n$ for some n. Define $\pi(\mathcal{B}) \in \Pi_n$ to have blocks that are the vertex sets of the connected components of the graph G on $[n]$ with edges

$$E(G) = \{ij : \exists\, x_i - x_j = c \text{ in } \mathcal{B}\}. \tag{3.143}$$

Define

$$\tilde{\chi}_{\mathcal{A}_n}(x) = \sum_{\substack{\mathcal{B} \subseteq \mathcal{A} \\ \mathcal{B} \text{ central} \\ \pi(\mathcal{B}) = \{[n]\}}} (-1)^{\#\mathcal{B}} x^{n - \mathrm{rank}(\mathcal{B})}.$$

Then

$$\chi_{\mathcal{A}_n}(x) = \sum_{\pi = \{B_1, \dots, B_k\} \in \Pi_n} \sum_{\substack{\mathcal{B} \subseteq \mathcal{A} \\ \mathcal{B} \text{ central} \\ \pi(\mathcal{B}) = \pi}} (-1)^{\#\mathcal{B}} x^{n - \mathrm{rank}(\mathcal{B})}$$

$$= \sum_{\pi = \{B_1, \dots, B_k\} \in \Pi_n} \tilde{\chi}_{\mathcal{A}_{\#B_1}}(x) \tilde{\chi}_{\mathcal{A}_{\#B_2}}(x) \cdots \tilde{\chi}_{\mathcal{A}_{\#B_k}}(x).$$

Thus by the exponential formula (Corollary 5.1.6),

$$\sum_{n \geq 0} \chi_{\mathcal{A}_n}(x) \frac{z^n}{n!} = \exp \sum_{n \geq 1} \tilde{\chi}_{\mathcal{A}_n}(x) \frac{z^n}{n!}.$$

But $\pi(\mathcal{B}) = \{[n]\}$ if and only if $\mathrm{rank}(\mathcal{B}) = n - 1$, so $\tilde{\chi}_{\mathcal{A}_n}(x) = c_n x$ for some $c_n \in \mathbb{Z}$. We therefore get

$$\sum_{n \geq 0} \chi_{\mathcal{A}_n}(x) \frac{z^n}{n!} = \exp x \sum_{n \geq 1} c_n \frac{z^n}{n!} \tag{3.144}$$

$$= \left(\sum_{n \geq 0} b_n \frac{z^n}{n!} \right)^x,$$

where $\exp \sum_{n \geq 1} c_n \frac{z^n}{n!} = \sum_{n \geq 0} b_n \frac{z^n}{n!}$. Put $x = -1$ to get

$$\sum_{n \geq 0} (-1)^n r(\mathcal{A}_n) \frac{z^n}{n!} = \left(\sum_{n \geq 0} b_n \frac{z^n}{n!} \right)^{-1},$$

from which it follows that

$$\sum_{n\geq 0}\chi_{\mathcal{A}_n}(x)\frac{z^n}{n!} = \left(\sum_{n\geq 0}(-1)^n r(\mathcal{A}_n)\frac{z^n}{n!}\right)^{-x}.$$

This result was stated without proof by R. Stanley, *Proc. Nat. Acad. Sci.* **93** (1996), 2620–2625 (Theorem 1.2), and proved in [3.83, Thm. 5.17].

c. Similarly to equation (3.144), we get

$$\sum_{n\geq 0}\chi_{\mathcal{A}_n}(x)\frac{z^n}{n!} = \exp\sum_{n\geq 1}(c_n x + d_n)\frac{z^n}{n!}$$

$$= A(z)^x B(z),$$

say, where $A(z)$ and $B(z)$ are independent of x. Put $x = -1$ and $x = 1$, and solve for $A(z)$ and $B(z)$ to complete the proof. This result appears without proof in [3.83, Exer. 5.10].

113. Let $\mathcal{A}_q$ lie in $\mathbb{F}_q^n$. Suppose that $\mathcal{A}_q' = \mathcal{A}_q - \{H_0\}$. The points of $\mathbb{F}_q^n$ that do not lie in any $H \in \mathcal{A}'$ are a disjoint union of those points that do not lie on any $H \in \mathcal{A}_q$, together with the points $\alpha \in H_0$ that do not lie on any $H \in \mathcal{A}_q$. These points α are just those points in H_0 that do not lie on $\mathcal{A}_q''$, so the proof follows. This proof was suggested by A. Postnikov, private communication, February 2010.

115. a. Let p be a large prime. By Theorem 3.11.10 we want the number of ways to choose an n-tuple $(a_1,\ldots,a_n) \in \mathbb{F}_p^n$ such that no $a_i - a_j = 0, \pm 1$ $(i \neq j)$. Once we choose a_1 in p ways, we need to choose $n-1$ points (in order) from $[p-3]$ so that no two are consecutive. Now use Exercise 1.34 for $j = 2$. This arrangement is called the "Catalan arrangement" because the number of regions is $n! C_n$. Perhaps the first explicit appearance of this arrangement and determination of the number of regions is R. Stanley, *Proc. Nat. Acad. Sci.* **93** (1996), 2620–2625 (special case of Theorem 2.2). The evaluation of $\chi_{\mathcal{C}_n}(x)$ appears in C. A. Athanasiadis, *Advances in Math.* **122** (1996), 193–233 (special case of Theorem 5.1).

b. The case $x = -1$ (i.e., the number of regions of $\mathcal{L}_n$) was raised by N. Linial. Equation (3.122) was first proved by C. Athanasiadis, ibid. (Theorem 5.2), generalized further in *J. Alg. Comb.* **10** (1999), 207–225 (§3), using the finite field method (Theorem 3.11.10). A proof based on Whitney's theorem (Proposition 3.11.3) was given by A. E. Postnikov, Ph.D. thesis, M.I.T., 1997. Numerous generalizations appear in A. E. Postnikov and R. Stanley, *J. Combinatorial Theory, Ser. A* **91** (2000), 544–597. See also Exercise 5.41 for some combinatorial interpretations of $r(\mathcal{L}_n)$.

c. This result is a special case of Exercise 3.112(c). It first appeared (without proof) as Exercise 5.25 of [3.83]. The arrangement $\mathcal{T}_n$ is called the "threshold arrangement" because the number of regions is equal to the number of threshold graphs with vertex set $[n]$ (see Exercise 5.4).

d. Let p be a large prime ($p > 2$ will do). Choose $a_1 \neq 0$ in $p-1$ ways. Since p is odd, we can choose $a_2 \neq 0, \pm a_1$ in $p-3$ ways. We can then choose $a_3 \neq 0, \pm a_1, \pm a_2$ in $p-5$ ways, and so on, giving

$$\chi_{\mathcal{B}_n^B}(x) = (x-1)(x-3)(x-5)\cdots(x-2n+1).$$

A nice introduction to the combinatorics of hyperplane arrangements related to root systems is T. Zaslavsky, *Amer. Math. Monthly* **88** (1981), 88–105.

116. It is not so difficult to show that the intersection poset $L(\mathcal{C})$ is isomorphic to the rank k truncation of the partition lattice Π_n (i.e., the order ideal of Π_n consisting of all partitions with at least $n-k$ blocks). It follows from Proposition 1.3.7 and equations (3.38) and (3.46) that

$$\chi_C(x) = \sum_{i=0}^{k} (-1)^i c(n, n-i) x^{n-i},$$

$$r(C) = c(n,n) + c(n,n-1) + \cdots + c(n,n-k).$$

This problem was first considered by I. J. Good and T. N. Tideman, *J. Combinatorial Theory Ser. A* **23** (1977), 34–45, in connection with voting theory. They obtained the formula for $r(C)$ by a rather complicated induction argument. Later Zaslavsky, *Discrete Comput. Geom.* **27** (2002), 303–351, corrected an oversight in the proof of Good and Tideman and reproved their result by using standard techniques from the theory of arrangements (working in a more general context than here). H. Kamiya, P. Orlik, A. Takemura, and H. Terao, *Ann. Combinatorics* **10** (2006), 219–235, considered additional aspects of this topic in an analysis of ranking patterns.

117. It follows from equation (3.123) that in a reference frame at velocity v, the events $p_i = (t_i, x_i)$ and $p_j = (t_j, x_j)$ occur at the same time if and only if

$$t_1 - t_2 = (x_1 - x_2) \cdot v.$$

The set of all such $v \in \mathbb{R}^n$ forms a hyperplane. The set of all such $\binom{k}{2}$ hyperplanes forms an arrangement $\mathcal{E} = \mathcal{E}(p_1, \ldots, p_k)$, which we call the *Einstein arrangement*. The number of different orders in which the events can be observed is therefore $r(\mathcal{E})$. As in the previous exercise, the intersection poset $L(\mathcal{E})$ is isomorphic to the rank k truncation of Π_n, so we obtain as earlier that

$$r(\mathcal{E}) = c(n,n) + c(n,n-1) + \cdots + c(n,n-k).$$

For instance, when $n = 3$, we get

$$r(\mathcal{E}) = \frac{1}{48} \left(k^6 - 7k^5 + 23k^4 - 37k^3 + 48k^2 - 28k + 48 \right).$$

For further details, see R. Stanley, *Advances in Appl. Math.* **37** (2006), 514–525. Some additional results are due to M. I. Heiligman, Sequentiality restrictions in special relativity, preprint dated February 4, 2010.

118. **a.** This remarkable result is equivalent to the main theorem of H. Terao, *Invent. Math.* **63** (1981), 159–179. For an exposition, see Orlik and Terao [3.53, Thm. 4.6.21].

   b. The result that Ω is free when L is supersolvable (due independently to R. Stanley and to M. Jambu and H. Terao, *Advances in Math.* **52** (1984), 248–258) can be proved by induction on v using the Removal Theorem of H. Terao, *J. Fac. Sci. Tokyo (IA)* **27** (1980), 293–312, and the fact that if $L = L(H_1, \ldots, H_v)$ is supersolvable, then for some $i \in [v]$ we have that $L(H_1, \ldots, H_{i-1}, H_{i+1}, \ldots, H_v)$ is also supersolvable. Examples of free Ω when L is not supersolvable appear in the previous reference and in H. Terao, *Proc. Japan Acad. (A)* **56** (1980), 389–392.

   c. This question was raised by Orlik–Solomon–Terao, who verified it for $n \le 7$. The numbers $(e_1, \ldots, e_n)$ for $3 \le n \le 7$ are given by $(1,1,2)$, $(1,2,3,4)$, $(1,3,4,5,7)$, $(1,4,5,7,8,10)$, and $(1,5,7,9,10,11,13)$. However, G. M. Ziegler showed in *Advances in Math.* **101** (1993), 50–58, that the arrangement is not free for $n \ge 9$. The case $n = 8$ remains open.

   d. This question is alluded to on page 293 of H. Terao, *F. Fac. Sci. Tokyo (IA)* **27** (1980), 293–312. It is a central open problem in the theory of free arrangements, though most likely the answer is negative.

e. See H. Terao, *Invent. Math.* **63** (1981), 159–179 (Prop. 5.5), and Orlik and Terao [3.53, Thm. 4.2.23]. Is there a more elementary proof?

f. The question of the freeness of $\mathcal{A}^t$ was raised by P. Orlik. A counterexample was discovered by P. H. Edelman and V. Reiner, *Proc. Amer. Math. Soc.* **118** (1993), 927–929.

119. Let $N(V,X)$ be the number of injective linear transformations $V \to X$. It is easy to see that $N(V,X) = \prod_{k=0}^{n-1}(x - q^k)$. On the other hand, let W be a subspace of V and let $F_=(W)$ be the number of linear $\theta: V \to X$ with kernel (null space) W. Let $F_\geq(W)$ be the number with kernel containing W. Thus $F_\geq(W) = \sum_{W' \geq W} F_=(W')$, so by Möbius inversion we get

$$N(V,X) = F_=(\{0\}) = \sum_{W'} F_\geq(W')\mu(\hat{0}, W').$$

Clearly, $F_\geq(W') = x^{n-\dim W'}$, whereas by equation (3.34) $\mu(\hat{0}, W') = (-1)^k q^{\binom{k}{2}}$, where $k = \dim W'$. Since there are $\binom{n}{k}$ subspaces W' of dimension k, we get

$$N(V,X) = \sum_{k=0}^{n}(-1)^k q^{\binom{k}{2}}\binom{n}{k}x^{n-k}.$$

120. See R. Stanley, *J. Amer. Math. Soc.* **5** (1992), 805–851 (Proposition 9.1). This exercise suggests that there is no good q-analogue of an Eulerian poset.

121. First Solution. Let $f(i,n)$ be the number of i-subsets of $[n]$ with no k consecutive integers. Since the interval $[\emptyset, S]$ is a boolean algebra for $S \in L_n'$, it follows that $\mu(\emptyset, S) = (-1)^{\#S}$. Hence, setting $a_n = \mu_n(\emptyset, \hat{1})$,

$$-a_n = \sum_{i=0}^{n}(-1)^i f(i,n).$$

Define $F(x,y) = \sum_{i\geq 0}\sum_{n\geq 0} f(i,n)x^i y^n$. The recurrence

$$f(i,n) = f(i,n-1) + f(i-1,n-2) + \cdots + f(i-k+1,n-k)$$

(obtained by considering the largest element of $[n]$ omitted from $S \in L_n'$) yields

$$F(x,y) = \frac{1 + xy + x^2y^2 + \cdots + x^{k-1}y^{k-1}}{1 - y(1 + xy + \cdots + x^{k-1}y^{k-1})}.$$

Since $-F(-1,y) = \sum_{n\geq 0} a_n y^n$, we get

$$\sum_{n\geq 0} a_n y^n = \frac{-(1 - y + y^2 - \cdots \pm y^{k-1})}{1 - y(1 - y + y^2 - \cdots \pm y^{k-1})}$$

$$= \frac{1 + (-1)^{k-1}y^k}{1 + (-1)^k y^{k+1}}$$

$$= -(1 + (-1)^{k-1}y^k)\sum_{i\geq 0}(-1)^i(-1)^{ki}y^{i(k+1)}$$

$$\Rightarrow a_n = \begin{cases} -1, & \text{if } n \equiv 0, -1 \pmod{2k+2}, \\ (-1)^k, & \text{if } n \equiv k, k+1 \pmod{2k+2}, \\ 0, & \text{otherwise.} \end{cases}$$

Figure 3.81 The poset P_{11324} when $13 = 31$, $14 = 41$, $24 = 42$.

Second Solution (E. Grimson and J. B. Shearer, independently). Let $\emptyset \neq a \in L_n'$. The dual form of Corollary 3.9.3 asserts that

$$\sum_{t \vee a = \hat{1}} \mu(\emptyset, t) = 0.$$

Now $t \vee a = \hat{1} \Rightarrow t = \hat{1}$ or $t = \{2, 3, \ldots, k\} \cup A$ where $A \subseteq \{k+2, \ldots, n\}$. It follows easily that

$$a_n - (-1)^{k-1} a_{n-k-1} = 0.$$

This recurrence, together with the initial conditions $a_0 = -1$, $a_i = 0$ if $i \in [k-1]$, and $a_k = (-1)^k$ determine a_n uniquely.

122. An interval $[d, n]$ of L is isomorphic to the boolean algebra $B_{\nu(n/d)}$, where $\nu(m)$ denotes the number of distinct prime divisors of m. Hence $\mu(d, n) = (-1)^{\nu(n/d)}$. Write $d \parallel n$ if $d \leq n$ in L. Given $f, g : \mathbb{P} \to \mathbb{C}$, we have

$$g(n) = \sum_{d \parallel n} f(d), \quad \text{for all } n \in \mathbb{P},$$

if and only if

$$f(n) = \sum_{d \parallel n} (-1)^{\nu(n/d)} g(d), \quad \text{for all } n \in \mathbb{P}.$$

123. **a,b.** Choose a factorization $w = g_{i_1} \cdots g_{i_\ell}$. Define P_w to be the multiset $\{i_1, \ldots, i_\ell\}$ partially ordered by letting $i_r < i_s$ if $r < s$ and $g_{i_r} g_{i_s} \neq g_{i_s} g_{i_r}$, or if $r < s$ and $i_r = i_s$. For instance, with $w = 11324$ as in Figure 3.52, we have P_w as in Figure 3.81. One can show that I is an order ideal of P_w if and only if for some (or any) linear extension $g_{i_1}, \ldots, g_{i_k}$ of I, we have $w = g_{i_1} \cdots g_{i_k} z$ for some $z \in M$. It follows readily that $L_w = J(P_w)$, and (b) is then immediate.

The monoid M was introduced and extensively studied by P. Cartier and D. Foata, *Lecture Notes in Math.*, no. 85, Springer-Verlag, Berlin/Heidelberg/New York, 1969. It is known as a *free partially commutative monoid* or *trace monoid*. The first explicit statement that $L_w = J(P_w)$ seems to have been made by I. M. Gessel in a letter dated February 8, 1978. This result is implicit, however, in Exercise 5.1.2.11 of D. E. Knuth [1.48]. This exercise of Knuth is essentially the same as our (b), though Knuth deals with a certain representation of elements of M as multiset permutations. An equivalent approach to this subject is the theory of *heaps*, developed by X. G. Viennot [4.60] after a suggestion of A. M. Garsia. For the connection between factorization and heaps, see C. Krattenthaler, appendix to electronic edition of Cartier-Foata, ⟨www.mat.univie.ac.at/~slc/books/cartfoa.pdf⟩.

c. The intervals $[v, vw]$ and $[\varepsilon, w]$ are clearly isomorphic (*via* the map $x \mapsto vx$), and it follows from (a) that P_w is an antichain (and hence $[\varepsilon, w]$ is a boolean algebra) if and only if w is a product of r distinct pairwise commuting g_i. The proof follows from Example 3.9.6.

A different proof appears in P. Cartier and D. Foata, op. cit, Ch. II.3.

d. If $w \in M$, then let x^w denote the (commutative) monomial obtained by replacing in w each g_i by x_i. By (c) we want to show that

$$\left(\sum_{w \in M} x^w \right) \left(\sum_{v \in M} \mu(\varepsilon, v) x^v \right) = 1. \qquad (3.145)$$

Expand the left-hand side of equation (3.145), take the coefficient of a monomial x^u, and use the defining recurrence (3.15) for μ to complete the proof.

(e) We have

$$\sum_{a_1 \geq 0} \cdots \sum_{a_n \geq 0} \binom{a_1 + \cdots + a_n}{a_1, \ldots, a_n} x_1^{a_1} \cdots x_n^{a_n} = \frac{1}{1 - (x_1 + \cdots + x_n)}$$

and

$$\sum_{a_1 \geq 0} \cdots \sum_{a_n \geq 0} x_1^{a_1} \cdots x_n^{a_n} = \frac{1}{(1 - x_1) \cdots (1 - x_n)},$$

respectively.

124. a. See [3.68, Thm. 4.1].

b. This exercise is jointly due to A. Björner and R. Stanley. Given $t \in L$, let $D_t = J(Q_t)$ be the distributive sublattice of L generated by C and t. The M-chain C defines a linear extension of Q_t and hence defines Q_t as a natural partial ordering of $[n]$. One sees easily that $L_P \cap D_t = J(P \cap Q_t)$. From this, all statements follow readily. Let us mention that it is not always the case that L_P is a lattice.

125. See P. McNamara, *J. Combinatorial Theory, Ser. A* **101** (2003), 69–89. McNamara shows that there is a third equivalent condition: L admits a good local action of the 0-Hecke algebra $\mathcal{H}_n(0)$. This condition is too technical to be explained here.

126. a. The isomorphism $L_k^{(2)}(p) \cong L_k^{(3)}(p)$ is straightfoward, whereas $L_k^{(1)}(p) \cong L_k^{(2)}(p)$ follows from standard duality results in the theory of abelian groups (or more generally abelian categories). A good elementary reference is Chapter 2 of P. J. Hilton and Y.-C. Wu, *A Course in Modern Algebra*, Wiley, New York, 1974. In particular, the functor taking G to $\mathrm{Hom}_{\mathbb{Z}}(G, \mathbb{Z}/p^\infty \mathbb{Z})$ is an order-reversing bijection between subgroups G of index p^m (for some $m \geq 0$) in $\mathbb{Z}^k$ and subgroups of order p^m in $(\mathbb{Z}/p^\infty \mathbb{Z})^k \cong \mathrm{Hom}_{\mathbb{Z}}(G, \mathbb{Z}/p^\infty \mathbb{Z})$.
The remainder of (a) is routine.

b. Follows, for example, from the fact that every subgroup of $\mathbb{Z}^k$ of finite index is isomorphic to $\mathbb{Z}^k$.

c. This result goes back to Eisenstein, Königl. Preuss. Akad. Wiss. Berlin (1852), 350–359, and Hermite, *J. Reine u. angewandte Mathematik* **41** (1851), 191–216. The proof follows directly from the theory of Hermite normal form (see, e.g., §6 of M. Newman, *Integral Matrices*, Academic Press, New York, 1972), which implies that every subgroup G of $\mathbb{Z}^k$ of index p^n has a unique $\mathbb{Z}$-basis $y_1, \ldots, y_k$ of the form

$$y_i = (a_{i1}, a_{i2}, \ldots, a_{ii}, 0, \ldots, 0),$$

where $a_{ii} > 0$, $0 \leq a_{ij} < a_{ii}$ if $j < i$, and $a_{11} a_{22} \cdots a_{kk} = p^n$. Hence, the number of such subgroups is

$$\sum_{b_1 + \cdots + b_k = n} p^{b_2 + 2b_3 + \cdots + (k-1)b_k} = \binom{n + k - 1}{k - 1}.$$

For some generalizations, see L. Solomon, *Advances in Math.* **26** (1977), 306–326, and L. Solomon, in *Relations between Combinatorics and Other Parts of Mathematics* (D.-K. Ray-Chaudhuri, ed.), Proc. Symp. Pure Math., vol 34, American Mathematical Society, Providence, R.I., 1979, pp. 309–329.

d. If $t_1 < \cdots < t_j$ in $L_k(p)$ with $\rho(t_i) = s_i$, then t_1 can be chosen in $\binom{s_1+k-1}{k-1}$ ways, next t_2 in $\binom{s_2-s_1+k-1}{k-1}$ ways, and so on.

e. A word $w = e_1 e_2 \cdots \in N_k$ satisfies $D(w) \subseteq S = \{s_1, \ldots, s_j\}_<$ if and only if $e_1 \leq e_2 \leq \cdots \leq e_{s_1}, e_{s_1+1} \leq \cdots \leq e_{s_2}, \ldots, e_{s_{j-1}+1} \leq \cdots \leq e_{s_j}, e_{s_j+1} = e_{s_j+2} = \cdots = 0$. Now for fixed i and k,

$$\sum_{0 \leq d_1 \leq \cdots \leq d_i \leq k-1} p^{d_1 + \cdots + d_i} = \binom{i+k-1}{k-1},$$

and the proof follows easily.

The problem of computing $\alpha_{L_\lambda}(S)$ and $\beta_{L_\lambda}(S)$, where L_λ is the lattice of subgroups of a *finite* abelian group of type $\lambda = (\lambda_1, \ldots, \lambda_k)$ (or more generally, a *q-primary lattice* as defined in R. Stanley, *Electronic J. Combinatorics* **3**(2) (1996), #R6 (page 9)) is more difficult. (The present exercise deals with the "stable" case $\lambda_i \to \infty$, $1 \leq i \leq k$.) One can show fairly easily that $\beta_{L_\lambda}(S)$ is a polynomial in p, and the theory of symmetric functions can be used to give a combinatorial interpretation of its coefficients that shows they are nonnegative. An independent proof of this fact is due to L. M. Butler, Ph.D. thesis, M.I.T., 1986, and *Memoirs Amer. Math. Soc.* **112**, no. 539 (1994) (Theorem 1.5.5).

130. a. For any fixed $t \in Q_i$ we have

$$0 = \sum_{s \leq t} \mu(\hat{0}, s) = \sum_j \left(\sum_{\substack{s \leq t \\ \rho(t) = i-j}} \mu(\hat{0}, s) \right).$$

Sum on all $t \in Q_i$ of fixed rank $i - k > 0$ to get (since $[x, \hat{1}] \cong Q_j$)

$$0 = \sum_j \left(\sum_{\substack{s \in Q_i \\ \rho(s) = i-j}} \mu(\hat{0}, s) \right) \left(\sum_{\substack{t \in Q_j \\ \rho(t) = j-k}} 1 \right)$$

$$= \sum_j v(i, j) V(j, k).$$

On the other hand, it is clear that $\sum_j v(i, j) V(j, i) = 1$, and the proof follows. This result (for geometric lattices) is due to T. A. Dowling, *J. Combinatorial Theory Ser. B* **14** (1973), 61–86 (Thm. 6).

b. See M. Aigner, *Math. Ann.* **207** (1974), 1–22; M. Aigner, *Aeq. Math.* **16** (1977), 37–50; and J. R. Stonesifer, *Discrete Math.* **32** (1980), 85–88. For some related results, see J. N. Kahn and J. P. S. Kung, *Trans. Amer. Math. Soc.* **271** (1982), 485–489, and J. P. S. Kung, *Geom. Dedicata* **21** (1986), 85–105.

131. See T. A. Dowling, *J. Combinatorial Theory Ser. B* **14** (1973), 61–86. Erratum, same journal **15** (1973), 211.

A far-reaching extension of these remarkable "Dowling lattices" appears in the work of Zaslavsky on signed graphs (corresponding to the case $\#G = 2$) and gain graphs (arbitrary G). Zaslavsky's work on the calculation of characteristic polynomials and related invariants appears in *Quart. J. Math. Oxford (2)* **33** (1982), 493–511. A general reference for enumerative results on gain graphs is T. Zaslavsky, *J. Combinatorial Theory Ser. B* **64** (1995), 17–88.

132. Number of elements of rank k is $\binom{n+k}{2k}$:

$\#P_n = F_{2n+1}$ (Fibonacci number),

$(-1)^n \mu(\hat{0},\hat{1}) = \frac{1}{n+1}\binom{2n}{n}$ (Catalan number),

number of maximal chains is $1 \cdot 3 \cdot 5 \cdots (2n-1)$.
This exercise is due to K. Baclawski and P. H. Edelman.

133. a. Define a closure operator (as defined in Exercise 3.84) on L_n by setting $\overline{G} = \mathfrak{S}(\mathcal{O}_1) \times \cdots \times \mathfrak{S}(\mathcal{O}_k)$, where $\mathcal{O}_1,\ldots,\mathcal{O}_k$ are the orbits of G and $\mathfrak{S}(\mathcal{O}_i)$ denotes the symmetric group on $\mathcal{O}_i$. Then $\overline{L}_n \cong \Pi_n$. In Exercise 3.84 choose $s = \hat{0}$ and $t = \hat{1}$, and the result follows from equation (3.37).

    b. A generalization valid for any finite group G is given in Theorem 3.1 of C. Kratzer and J. Thévenaz, *Comment. Math. Helvetici* **59** (1984), 425–438.

    c–f. See J. Shareshian, *J. Combinatorial Theory Ser. A* **78** (1997), 236–267. For a topological refinement, see J. Shareshian, *J. Combinatorial Theory Ser. A* **104** (2003), 137–155.

135. b. The poset Λ_n is defined in Birkhoff [3.12, Ch. I.8, Ex. 10]. The problem of computing the Möbius function is raised in Exercise 13 on p. 104 of the same reference. (In this exercise, 0 should be replaced with the partition $\langle 1^{n-2}2 \rangle$.)

    c. It was shown by G. M. Ziegler, *J. Combinatorial Theory Ser. A* **42** (1986), 215–222, that Λ_n is not Cohen–Macaulay for $n \geq 19$, and that the Möbius function does not alternate in sign for $n \geq 111$. (These bounds are not necessarily tight.) For some further information on Λ_n, see F. Bédard and A. Goupil, *Canad. Math. Bull.* **35** (1992), 152–160.

136. See T. H. Brylawski, *Discrete Math.* **6** (1973), 201–219 (Prop. 3.10), and C. Greene, *Europ. J. Combinatorics* **9** (1988), 225–240. For further information on this poset, see Exercises 3.78(c) and 7.2, as well as A. Björner and M. L. Wachs, *Trans. Amer. Math. Soc.* **349** (1997), 3945–3975 (§8); J. N. Kahn, *Discrete and Comput. Geometry* **2** (1987), 1–8; and S. Linusson, *Europ. J. Combinatorics* **20** (1999), 239–257.

137. *Answer.* $Z(P+Q,m) = Z(P,m) + Z(Q,m)$,

$$Z(P \oplus Q,m) = \sum_{j=2}^{m-1} Z(P,j)Z(Q,m+1-j) + Z(P,m) + Z(Q,m),$$

$$m \geq 2,$$

$$Z(P \times Q,m) = Z(P,m)Z(Q,m).$$

138. a. By definition, $Z(\mathrm{Int}(P),n)$ is equal to the number of multichains

$$[s_1,t_1] \leq [s_2,t_2] \leq \cdots \leq [s_{n-1},t_{n-1}]$$

of intervals of P. Equivalently,

$$s_{n-1} \leq s_{n-2} \leq \cdots \leq s_1 \leq t_1 \leq t_2 \leq t_{n-1}.$$

Hence, $Z(\mathrm{Int}(P),n) = Z(P,2n-1)$.

    b. It is easily seen that

$$Z(Q,n) - Z(Q,n-1) = Z(\mathrm{Int}(P),n).$$

Put $n = 0$ and use Proposition 3.12.1(c) together with (a) to obtain $\mu_Q(\hat{0}, \hat{1}) = -Z(P, -1) = -\mu_P(\hat{0}, \hat{1})$. When P is the face lattice of a convex polytope, much more can be said about Q. This is unpublished work of A. Björner, though an abstract appears in the Oberwolfach Tagungsbericht 41/1997, pp. 7–8, and a shorter version in Abstract 918-05-688, *Abstracts Amer. Math. Soc.* **18:1** (1997), 19.

140. Since $n^k = \sum_j j! S(k, j) \binom{n}{j}$ by equation (1.94d), and since the polynomials $\binom{n}{j}$ are linearly independent over $\mathbb{Q}$, it follows that $\varphi(\binom{n}{j}) = x^j$. But by equation (3.51) we have

$$Z(P, n+2) = \sum_{j \geq 1} c_j(P) \binom{n}{j-1}.$$

Applying φ to both sides completes the proof. Note the similarity to Exercise 3.108(e).

141. **a.** For any chain C of P, let $Z_C(Q_0, m+1)$ be the number of multichains $C_1 \leq C_2 \leq \cdots \leq C_m = C$ in Q_0. Since the interval $[\emptyset, C]$ in Q_0 is a boolean algebra, we have by Example 3.12.2 that $Z_C(Q_0, m+1) = m^{\#C}$. Hence, $Z(Q_0, m+1) = \sum_{C \in Q_0} m^{\#C} = \sum a_i m^i$, where P has a_i i-chains, and the proof follows from Proposition 3.12.1(a).

b. *Answer.* $\mu_{\widehat{P}}(\hat{0}, \hat{1}) = \mu_{\widehat{Q}}(\hat{0}, \hat{1})$. Topologically, this identity reflects the fact that a finite simplicial complex and its first barycentric subdivision have homeomorphic geometric realizations and therefore equal Euler characteristics.

c. Follows easily from (b).

d. It is easy to see that the number of elements of $\widehat{Q}$ of rank $k - 1$ is $k! S(n, k)$, $1 \leq k \leq n$. It is not hard to see that equation (3.124) is then a consequence of Theorem 3.16.9. The formula (3.124) was first observed empirically by M. Bóna (private communication, dated 27 October 2009). NOTE. The dual poset $\widehat{Q}^*$ is the face lattice of the *permutohedron*, the polytope of Exercise 4.64(a).

142. **a.** Let $\gamma_P(S)$ denote the number of intervals $[r(K), K)]$ for which $\rho(r(K)) = S$. If C is any chain of P with $\rho(C) = S$, then C is contained in a unique interval $[r(K), K]$ such that $\rho(r(K)) \subseteq S$; and conversely an interval $[r(K), K]$ such that $\rho(r(K)) \subseteq S$ contains a unique chain C of P such that $\rho(C) = S$. Hence,

$$\sum_{T \subseteq S} \gamma_P(T) = \alpha_P(S),$$

and the proof follows from equation (3.53).

The concept of chain-partitionable posets is due independently to J. S. Provan, thesis, Cornell Univ., 1977 (Appendix 4); R. Stanley [3.76, p. 149]; and A. M. Garsia, *Advances in Math.* **38** (1980), 229–266 (§4). The first two of these references work in the more general context of simplicial complexes, whereas the third uses the term "ER-poset" for our (chain-)partitionable poset.

b. Let $\lambda: \mathcal{H}(\widehat{P}) \to \mathbb{Z}$ be an R-labeling and $K: t_1 < \cdots < t_{n-1}$ a maximal chain of P, so $\hat{0} = t_0 < t_1 < \cdots < t_{n-1} < t_n = \hat{1}$ is a maximal chain of $\widehat{P}$. Define

$$r(K) = \{t_i : \lambda(t_{i-1}, t_i) > \lambda(t_i, t_{i+1})\}.$$

Given any chain $C: s_1 < \cdots < s_k$ of P, define K to be the (unique) maximal chain of P that consists of increasing chains of the intervals $[\hat{0}, s_1], [s_1, s_2], \ldots, [s_k, \hat{1}]$, with $\hat{0}$ and $\hat{1}$ removed. It is easily seen that $C \in [r(K), K]$, and that K is the only maximal chain of P for which $C \in [r(K), K]$. Hence, P is partitionable.

c. The posets in a special class of Cohen–Macaulay posets called "shellable" are proved to be partitionable in the three references given in (a). It is not known whether all Cohen–Macaulay shellable posets (or in

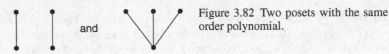

Figure 3.82 Two posets with the same order polynomial.

fact all Cohen–Macaulay posets) are R-labelable. On the other hand, it seems quite likely that there exist Cohen–Macaulay R-labelable posets that are not shellable, though this fact is also unproved. (Two candidates are Figures 18 and 19 of Björner–Garsia–Stanley [3.16].) A very general ring-theoretic conjecture that would imply that Cohen–Macaulay posets are partitionable appears in R. Stanley, *Invent. Math.* **68** (1982), 175–193 (Conjecture 5.1). For some progress on this conjecture, see for instance I. Anwar and D. Popescu, *J. Algebra* **318** (2007), 1027–1031; Y. H. Shen, *J. Algebra* **321** (2009), 1285–1292; D. Popescu, *J. Algebra* **321** (2009), 2782-2797; M. Cimpoeaş, *Matematiche (Catania)* **63** (2008), 165–171; J. Herzog, M. Vladoiu, and X. Zheng, *J. Algebra 322* (2009), 3151-3169; D. Popescu and M. I. Qureshi, *J. Algebra* **323** (2010), 2943–2959; and the two surveys D. Popescu, Stanley depth, ⟨www.univ-ovidius.ro/math/sna/17/PDF/17_Lectures.pdf⟩ and S. A. Seyed, M. Tousi, and S. Yassemi, *Notices Amer. Math. Soc.* **56** (2009), 1106–1108.

143. **a.** *First Proof.* It is implicit in the work of several persons (e.g., Faigle-Schrader, Gallai, Golumbic, Habib, Kelly, Wille) that two finite posets P and Q have the same comparability graph if and only if there is a sequence $P = P_0, P_1, \ldots, P_k = Q$ such that P_{i+1} is obtained from P_i by "turning upside-down" (dualizing) a subset $T \subseteq P_i$ such that every element $t \in P_i - T$ satisfies either (a) $t < s$ for all $s \in T$, or (b) $t > s$ for all $s \in T$, or (c) $s \parallel t$ for all $s \in T$. (Such subsets T are called *autonomous subsets*.) The first explicit statement and proof seem to be in B. Dreesen, W. Poguntke, and P. M. Winkler, *Order* **2** (1985), 269–274 (Thm. 1). A further proof appears in D. A. Kelly, *Order* **3** (1986), 155–158. It is easy to see that P_i and P_{i+1} have the same order polynomial, so the proof of the present exercise follows.

Second Proof. Let $\Gamma_P(m)$ be the number of maps $g \colon P \to [0, m-1]$ satisfying $g(t_1) + \cdots + g(t_k) \leq m - 1$ for every chain $t_1 < \cdots < t_k$ of P. We claim that $\Omega_P(m) = \Gamma_P(m)$. To prove this claim, given g as above define for $t \in P$

$$f(t) = 1 + \max\{g(t_1) + \cdots + g(t_k) : t_1 < \cdots < t_k = t\}.$$

Then $f \colon P \to [m]$ is order-preserving. Conversely, given f then

$$g(t) = \min\{f(t) - f(s) : t \text{ covers } s\}.$$

Thus, $\Omega_P(m) = \Gamma_P(m)$. But by definition $\Gamma_P(m)$ depends only on $\mathrm{Com}(P)$. This proof appears in R. Stanley, *Discrete Comput. Geom.* **1** (1986), 9–23 (Cor. 4.4).

b. See Figure 3.82.

For a general survey of comparability graphs of posets, see D. A. Kelly, in *Graphs and Order* (I. Rival, ed.), Reidel, Dordrecht/Boston, 1985, pp. 3–40.

145. We have $\Omega_P(-n) = Z(J(P), -n) = \mu_{J(P)}^n(\hat{0}, \hat{1})$. By Example 3.9.6.

$$\mu^n(\hat{0}, \hat{1}) = \sum (-1)^{\#(I_1 - I_0) + \cdots + \#(I_n - I_{n-1})},$$

summed over all multichains $\emptyset = I_0 \subseteq I_1 \subseteq \cdots \subseteq I_n = P$ of order ideals of P such that each $I_i - I_{i-1}$ is an antichain of P. Since $\#(I_1 - I_0) + \cdots + \#(I_n - I_{n-1}) = p$, we have that $(-1)^p \mu^n(\hat{0}, \hat{1})$ is equal to the number of such multichains. But such a multichain

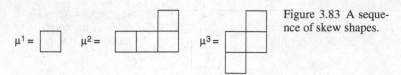

Figure 3.83 A sequence of skew shapes.

$\mu^1 =$ $\mu^2 =$ $\mu^3 =$

corresponds to the strict order-preserving map $\tau : P \to n$ defined by $\tau(t) = i$ if $t \in I_i - I_{i-1}$, and the proof follows. This proof appeared in Stanley [3.67, Thm. 4.2].

146.

$$\Omega_{p1}(n) = (-1)^p \Omega_{p1}(-n) = n^p,$$

$$\Omega_p(n) = \left(\binom{n}{p} \right) = \binom{n+p-1}{p},$$

$$(-1)^p \Omega_p(-n) = \binom{n}{p}.$$

147. Tetrahedron: $Z(L,n) = n^4$.
cube or octahedron: $Z(L,n) = 2n^4 - n^2$. icosahedron or dodecahedron: $Z(L,n) = 5n^4 - 4n^2$.
Note that in all cases $Z(L,n) = Z(L,-n)$, a consequence of Proposition 3.16.1.

149. The case $\mu = \emptyset$ is equivalent to a result of P. A. MacMahon [1.55] (put $x = 1$ in the implied formula for $GF(p_1, p_2, \ldots, p_m; n)$ on page 243) and has been frequently rediscovered in various guises. The general case is due to G. Kreweras, *Cahiers du BURO*, no. 6, Institut de Statistique de L'Univ. Paris, 1965 (Section 2.3.7) and is also a special case (after a simple preliminary bijection) of Theorem 2.7.1. When $\mu = \emptyset$ and λ has the form $(M - d, M - 2d, \ldots, M - \ell d)$ the determinant can be explicitly evaluated; see Vol. II, Exercise 7.101(b). A different approach to these results was given by I. M. Gessel, *J. Stat. Planning and Inference* **14** (1986), 49–58, and by R. A. Pemantle and H. S. Wilf, *Electronic J. Combinatorics* **16** (2009), #R60. For an extensive survey of the evaluation of combinatorial determinants, see C. Krattenthaler, *Sém. Lotharingien Combin.* **42** (1999), article B42q, and *Linear Algebra Appl.* **411** (2005), 68-166.

150. a. When the Young diagram λ^0 is removed from λ^j, there results an ordered disjoint union (the order being from lower left to upper right) of rookwise connected skew diagrams (or skew shapes, as defined in Section 7.10) $\mu^1, \ldots, \mu^r$. For example, if $\lambda^0 = (5,4,4,3,1)$ and $\lambda^j = (6,6,5,4,4,4,1)$, then we obtain the sequence of skew diagrams shown in Figure 3.83. Since $|\mu^1| + \cdots + |\mu^r| = a_j$, there are only finitely many possible sequences $\mu = (\mu^1, \ldots, \mu^k)$ for fixed S. Thus, if we let $f_S(\mu, n)$ be the number of chains $\lambda^0 < \lambda^1 < \cdots < \lambda^j$ under consideration yielding the sequence μ, then it suffices to show that the power series $A_S(\mu, q)$ defined by

$$\sum_{n \geq 0} f_S(\mu, n) q^n = P(q) A_S(\mu, q) \tag{3.146}$$

is rational with numerator $\phi_{a_j}(q)$.
We illustrate the computation of $A_S(\mu, q)$ for μ given by Figure 3.83 and leave the reader the task of seeing that the argument works for arbitrary μ. First, it is easy to see that there is a constant $c_S(\mu) \in \mathbb{P}$ for which $A_S(\mu, q) = c_S(\mu) A_{\{a_j\}}(\mu, q)$, so we may assume that $S = \{a_j\} = \{9\}$. Consider a typical λ^j, as shown in Figure 3.84. Here a, b, c mark the lengths of the indicated rows, so $c \geq b + 2 \geq a + 5$. When the

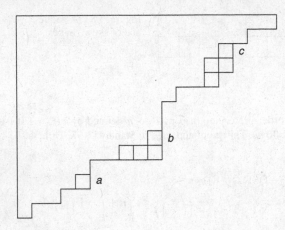

Figure 3.84 An example of the computation of $A_S(\mu, q)$.

rows intersecting some μ^i are removed from λ^j, there results a partition ν with no parts equal to $b-1$, $b-2$, or $c-1$, and every such ν occurs exactly once. Hence,

$$\sum_{n \geq 0} f_{\{9\}}(\mu, n) q^{n+9}$$

$$= P(q) \sum_{c \geq b+2 \geq a+5 \geq 6} q^{a+2b+(3c-1)}(1 - q^{b-1})(1 - q^{b-2})(1 - q^{c-1}).$$

To evaluate this sum, expand the summand into eight terms, and sum on c, b, a in that order. Each sum will be a geometric series, introducing a factor $1 - q^i$ in the denominator and a monomial in the numerator. Since among the eight terms the maximum sum of coefficients of a, b, c in the exponent of q is $a_j = 9$ (coming from $q^{a+4b+4c-5}$), it follows that the eight denominators will consist of distinct factors $1 - q^i$, $1 \leq i \leq 9$. Hence, they have a common denominator $\phi_9(q)$, as desired. Is there a simpler proof?

b. Let $A_S(q) = B_S(q)/\phi_{a_j}(q)$. Then

$$B_\emptyset = 1, \quad B_1 = 1, \quad B_2 = 2 - q, \quad B_3 = 3 - q - q^2, \quad B_{1,2} = 2,$$

$$B_{1,3} = 3 + 2q - q^2 - q^3, \quad B_{2,3} = 4 - q + 2q^2 - 2q^3, \text{ and}$$

$$B_{1,2,3}(q) = 2(2 - q)(1 + q + q^2).$$

Is there a simple formula for $B_{[n]}(q)$?

c. (with assistance from L. M. Butler) First check that the coefficient $g(n)$ of q^n, in the product of the left-hand side of equation (3.125) with $P(q)$, is equal to $\beta_Y([n, n+k]) + \beta_Y([n+1, n+k])$. We now want to apply Theorem 3.13.1. Regard $\mathbb{N}^2$ with the usual product order as a coarsening of the total (lexicographic) order

$$(i, j) \leq (i', j') \text{ if } i < i' \text{ or if } i = i', \ j \leq j'.$$

By Theorem 3.13.1, $g(n)$ is equal to the number of chains $\boldsymbol{\nu}: \nu^0 < \nu^1 < \cdots < \nu^k$ of partitions ν^i such that (1) $\nu_i \vdash n + i$; (2) ν^{i+1} is obtained from ν^i by adding a square (in the Young diagram) strictly above the square that was added in obtaining ν^i from ν^{i-1}; and (3) the square added in ν^k from ν^{k-1} is not in the top row. (This last condition guarantees a descent at $n + k$.) Here ν^0 can be arbitrary and ν^1 can be obtained by adding any square to ν^0. (If the square added to ν^0 starts a new row or is in the bottom row of v^0, then the chain $\boldsymbol{v}$ contributes to $\beta_Y([n+1, n+k])$; otherwise, it contributes to $\beta_Y([n, n+k])$. We can now argue as in the solution to (a); namely, the added k squares belong to columns of length $2 \le i_1 < i_2 < \cdots < i_k$, and when these rows are removed, any partition can be left. Hence,

$$\sum_{n \ge 0} g(n) q^{n+k} = P(q) \sum_{2 \le i_1 < i_2 \le \cdots \le i_k} q^{i_1 + \cdots + i_k}$$

$$= q^{k + \binom{k+1}{2}} P(q) \phi_k(q),$$

and the proof follows. Is there a simple proof avoiding Theorem 3.13.1?

d. Follows readily from the first sentence of the solution to (c), upon noting that

$$\beta_Y([n, n+k]) = \sum_{i=0}^{k} (-1)^i \left(\beta_Y([n+i, n+k]) \right.$$

$$\left. + \beta_Y([n+i+1, n+k]) \right) - (-1)^k.$$

(The term $-(-1)^k$ is needed to cancel the term $(-1)^k \beta_Y[n+k+1, n+k]) = (-1)^k \beta_Y(\emptyset) = (-1)^k$ arising in the summand with $i = k$.)

e. We want to show that the number $f(n)$ of chains $\lambda < \mu$ with $\lambda \vdash n$ and $\mu \vdash n+1$ is equal to $p(0) + p(1) + \cdots + p(n)$, where $p(j)$ is the number of partitions of j. Now see Exercise 1.71. (This bijection is implicit in the proof of (a) or (c).)

151. a. By Theorem 3.13.1 there exists a permutation $w = a_1 a_2 \cdots a_n \in \mathcal{L}(P)$ with descent set $S = \{i_1, \ldots, i_k\}_<$. Set $i_0 = 0$, $i_{k+1} = p$, and choose $1 \le r \le k$. Rearrange all the elements $a_{i_{r-1}+1}, a_{i_{r-1}+1}, \ldots, a_{i_{r+1}}$ in increasing order, obtaining a permutation w'. Since P is naturally labeled we have $w' \in \mathcal{L}(P)$. Moreover, $D(w') = D(w) - \{i_r\}$, from which the proof is immediate. This result is due to Stanley [3.67, Ch. III, Cor. 1.2][3.68, Cor. 15.2].

b. Some necessary conditions on Δ are given in [3.68, §16].

c. We use the characterization of supersolvable lattices given by Exercise 3.125. The proof then parallels that of (a) with linear extensions replaced with labels of maximal chains. Specifically, let $\mathfrak{m}: \hat{0} = t_0 \lessdot t_1 \lessdot \cdots \lessdot t_p = \hat{1}$ be a maximal chain with label $\lambda(\mathfrak{m}) = (\lambda_1, \ldots, \lambda_p) \in \mathfrak{S}_p$ such that $D(\lambda(\mathfrak{m})) = S$. Using the notation of (a), replace $t_{i_{r-1}+1}, t_{i_{r-1}+2}, \ldots, t_{i_{r+1}}$ with the unique increasing chain between $t_{i_{r-1}+1}$ and $t_{i_{r+1}}$, obtaining a new maximal chain $\mathfrak{m}'$. Because the labels of maximal chains are permutations of $1, 2, \ldots, p$ it follows that $D(\lambda(\mathfrak{m}')) = D(\lambda(\mathfrak{m})) - \{i_r\}$, and the proof is immediate as in (a).

NOTE. The result we have just proved is true under the even more general hypothesis that L is a finite Cohen–Macaulay poset, but the proof now involves algebraic techniques. See R. Stanley [3.76, Cor. 4.5] and Björner-Garsia-Stanley [3.16, p. 24].

152. a. Let $w = a_1 \cdots a_n$ be a linear extension of P. Define the vertices of w to be the linear extensions obtained by choosing $i \in D(w)$ and writing the elements $a_1, \ldots, a_i$ in increasing order, followed by writing the elements $a_{i+1}, \ldots, a_n$ in increasing order. It is easy to check that we obtain a simplicial complex Δ_P with the desired properties.

Example. If $w = 3642175$, then the vertices of w are 3612457, 3461257, 2346157, and 1234675. The simplicial complex Δ_P was investigated (in a more general context) by P. H. Edelman and V. Reiner, *Advances in Math.* **106** (1994), 36–62.

b. See Figure 3.85.

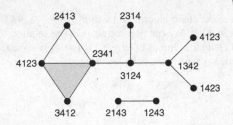

Figure 3.85 A simplicial complex whose faces are the permutations $w \in \mathfrak{S}_4$.

Figure 3.86 Posets with equal values of $\beta_{J(P)}(S)$.

c. See P. L. Hersh, *J. Combinatorial Theory, Ser. A* **105** (2004), 111–126.

153. *Answer.* Let $[p-1] - S = \{i_1, \ldots, i_k\}_<$. Then the minimum value of $e(P)$ is

$$\min e(P) = i_1!(i_2 - i_1)!(i_3 - i_2)! \cdots (p - i_k)!,$$

achieved uniquely by $P = i_1 \oplus (i_2 - i_1) \oplus \cdots \oplus (p - i_k)$.

154. The answer is affirmative for $n \le 6$. Of the 2045 nonisomorphic seven-element posets, it was checked by J. R. Stembridge that there is a unique pair P, Q that satisfy $\beta_{J(P)}(S) = \beta_{J(Q)}(S)$ for all $S \subseteq [6]$. The Hasse diagrams of P and Q are given in Figure 3.86.

155. a. Let $1 \le k \le n$, and define in the incidence algebra $I(P)$ (over $\mathbb{R}$, say) a function η_k by

$$\eta_k(s,t) = \begin{cases} 1, & \text{if } \rho(t) - \rho(s) = k, \\ 0, & \text{otherwise.} \end{cases}$$

The self-duality of $[s,t]$ implies that $\eta_j \eta_k(s,t) = \eta_k \eta_j(s,t)$ for all j and k, so η_j and η_k commute. But

$$\alpha_P(S) = \eta_{n_1} \eta_{n_2 - n_1} \cdots \eta_{n - n_k}(\hat{0}, \hat{1}),$$

and the proof follows since the various η_j's can be permuted arbitrarily.

c. It follows from a result of F. Regonati, *J. Combinatorial Theory, Ser. A* **60** (1992), 34–49 (theorem on page 45) that such lattices are products of certain modular lattices known as *q-primary* (though not conversely). See also Theorem 3.4 of R. Stanley, *Electronic J. Combinatorics* **3**, #R6 (1996); reprinted in *The Foata Festschrift* (J. Désarménien, A. Kerber, and V. Strehl, eds.), Imprimerie Louis-Jean, Gap, 1996, pp. 165–186. There is an almost complete classification of primary modular lattices (which includes the q-primary modular lattices) by Baer, Inaba, and Jónsson-Monk; see B. Jónsson and G. S. Monk, *Pacific J. Math.* **30** (1969), 95–139. A complete classification of finite modular lattices for which every interval is self-dual (or the more general products of q-primary lattices) seems hopeless since it involves such problems as the classification of finite projective planes. For some further work related to primary modular lattices, see F. Regonati and S. D. Sarti, *Ann. Combinatorics* **4** (2000), 109–124.

156. We have $\mathbb{N} \times \mathbb{N} = J_f(Q)$, where the elements of Q are $s_1 < s_2 < \cdots$ and $t_1 < t_2 < \cdots$. Regard Q as being contained in the total order where $s_i < t_j$ for all i, j. By Theorem 3.13.1 (extended in an obvious way to finitary distributive lattices), we have that

$\beta_{\mathbb{N} \times \mathbb{N}}(S)$ is equal to the number of linear orderings $u_1, u_2, \ldots$ of Q such that the s_i's appear in increasing order, the t_i's appear in increasing order, and a t_i is immediately followed by an s_j if and only if $t_i = u_k$ where $k \in S$. Thus, $u_1, \ldots, u_{m_1}$ can be chosen as $s_1, \ldots, s_i, t_1, \ldots, t_{m_1 - i}$ $(0 \le i \le m_1 - 1)$ in m_1 ways. Then $u_{m_1 + 1} = s_{i+1}$, whereas $u_{m_1 + 2}, \ldots, u_{m_2}$ can be chosen in $m_2 - m_1 - 1$ ways, and so on, giving the desired result. A less combinatorial proof appears in [3.68, Prop. 23.7].

157. a. Let $\alpha_k = \sum_{\#S=k} \alpha_P(S)$. Now $Z(P, m)$ is equal to the number of multichains $\hat{0} = t_0 \le t_1 \le \cdots \le t_m = \hat{1}$. Such a multichain K is obtained by first choosing a chain $C : \hat{0} < u_1 < \cdots < u_k < \hat{1}$ in α_k ways, and then choosing K whose support (underlying set) is C in $\left(\binom{k+2}{m-1-k} \right) = \binom{m}{k+1}$ ways. Hence, $Z(P, m) = \sum_k \binom{m}{k+1} \alpha_k$; that is, $\Delta^{k+1} Z(P, 0) = \alpha_k$.

b. Divide both sides of the desired equality by $(1 - x)^{n+1}$ and take the coefficient of x^m. Then we need to show that

$$Z(P, m) = \sum_j \beta_j (-1)^{m-j-1} \binom{-n-1}{m-j-1}$$

$$= \sum_j \beta_j \binom{n+m-j-1}{n}.$$

Now

$$\alpha_k = \sum_{\#S = k} \sum_{T \subseteq S} \beta_P(T)$$

$$= \sum_j \sum_{\#T = j} \binom{n-1-j}{n-1-k} \beta_P(T)$$

$$= \sum_j \binom{n-1-j}{n-1-k} \beta_j.$$

Hence from (a),

$$Z(P, m) = \sum_k \binom{m}{k+1} \alpha_k$$

$$= \sum_{j,k} \binom{m}{k+1} \binom{n-1-j}{n-1-k} \beta_j.$$

But

$$\sum_k \binom{m}{k+1} \binom{n-1-j}{n-1-k} = \binom{n+m-j-1}{n}$$

(e.g., by Example 1.1.17), and the proof follows.

A more elegant proof can be given along the following lines. Introduce variables $x_1, \ldots, x_{n-1}$, and for $S \subseteq [n-1]$ write $x_S = \prod_{i \in S} x_i$. Moreover, for a multichain $K : t_1 \le \cdots \le t_m$ of $P - \{\hat{0}, \hat{1}\}$, write $x_K = \prod_{i=1}^m x_{\rho(t_i)}$. One easily sees that

$$\sum_K x_K = \sum_S \alpha_P(S) \left(\prod_{i \in S} \frac{x_i}{1 - x_i} \right)$$

$$= \frac{\sum_S \beta_P(S) x_S}{(1 - x_1)(1 - x_2) \cdots (1 - x_{n-1})}.$$

Set each $x_i = x$ and multiply by $(1 - x)^{-2}$ (corresponding to adjoining $\hat{0}$ and $\hat{1}$) to obtain (a) and (b).

NOTE. If $f(m)$ is any polynomial of degree n, then Section 4.3 discusses the generating function $\sum_{m \geq 0} f(m)x^m$, in particular, its representation in the form $P(x)(1 - x)^{-n-1}$. Hence, the present exercise may be regarded as "determining" $P(x)$ when $f(m) = Z(P, m)$.

c. By definition of $\chi_P(q)$, we have

$$w_k = \sum_{\substack{t \in P \\ \rho(t) = k}} \mu(\hat{0}, t)$$

$$= \sum_{\rho(t) \leq k} \mu(\hat{0}, t) - \sum_{\rho(t) \leq k-1} \mu(\hat{0}, t).$$

Letting μ_S denote the Möbius function of the S-rank-selected subposet P_S of P as in Section 3.13, then by the defining recurrence (3.15) for μ we get

$$w_k = -\mu_{[k]}(\hat{0}, \hat{1}) + \mu_{[k-1]}(\hat{0}, \hat{1}).$$

The proof follows from equation (3.54).

158. In the case $k = 1$, a noncombinatorial proof of (a) was first given by G. Kreweras, *Discrete Math.* **1** (1972), 333–350, followed by a combinatorial proof by Y. Poupard, *Discrete Math.* **2** (1972), 279–288. The case of general k, as well as (c) and (d), is due to P. H. Edelman, *Discrete Math.* **31** (1980), 171–180. See also P. H. Edelman, *Discrete Math.* **40** (1982), 171–179. Of course (b) follows from (a) by taking $n = 1$ and $n = -2$, whereas (e) follows from (d) by taking $S = \{t - m\}$ and $S = [0, t - 2]$. Part (f) is due to Kreweras, op. cit. (Thm. 4), while (g) first appeared in P. L. Hersh, Ph.D. thesis, M.I.T., 1999 (Theorem 4.3.2), and *J. Combinatorial Theory Ser. A* **103** (2003), 27–52 (Theorem 6.3). To solve (g) using Exercise 3.125, define a labeling $\lambda \colon \mathcal{H}(P_{1,t}) \to \mathbb{Z}$ as follows. If $\pi \lessdot \sigma$ in $P_{1,t}$, then σ is obtained from π by merging two blocks B, B'. Define

$$\lambda(\pi, \sigma) = \max(\min B, \min B') - 1.$$

It is routine to check that λ has the necessary properties. This labeling is due to P. H. Edelman and A. Björner, and appears in A. Björner, *Trans. Amer. Math. Soc.* **260** (1980), 159–183 (page 165). A different edge labeling related to parking functions appears in R. Stanley, *Electronic J. Combinatorics* **4** (1997), #R20. Note that $P_{1,t}$ is a lattice by Proposition 3.3.1 because it is a meet-semilattice of Π_t with $\hat{1}$.

Partitions π satisfying (ii) are called *noncrossing partitions* and have received much attention. For some additional information and references, see Vol. II, Exercises 5.35, 6.19(pp), and 7.48(f).

159. By symmetry it suffices to take $w = (1, 2, \ldots, n)$. In this case, it can be checked that an isomorphism $\varphi \colon [\hat{0}, w] \to P_{1,n}$ is obtained by taking the set of elements of each cycle of $u \in [\hat{0}, w]$ to be the blocks of $\varphi(w)$. This result is due to P. Biane, *Discrete Math.* **175** (1997), 41–53 (Theorem 1). For further information on the absolute order, see C. A. Athanasiadis and M. Kallipoliti, *J. Combinatorial Theory Ser. A* **115** (2008), 1286–1295.

160. Given a labeling ω, define an orientation $\mathfrak{o}_\omega$ of $\mathcal{H}$ by directing an edge ij from i to j if $i < j$. Clearly, $\mathfrak{o}_\omega$ is acyclic, and it is easy to check that ω and ω' are equivalent if and only if $\mathfrak{o}_\omega = \mathfrak{o}_{\omega'}$. The problem of counting the number of equivalence classes was raised by Stanley [3.67, p. 25], with the answer stated without proof in [3.68, p. 7].

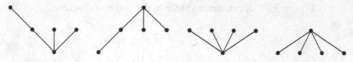

Figure 3.87 Posets for which $\Omega_P(m)$ has a negative coefficient.

161. Without loss of generality we may assume $U \cup V = [j+k]$. Let P be the j-element chain with elements labeled $u_1,\ldots,u_j$ from bottom to top, and similarly Q with elements labeled $v_1,\ldots,v_k$. Denote this labeling of $P+Q$ by ω. Note that $\mathcal{A}(P+Q,\omega)$ depends only on $D(u)$ and $D(v)$, and that $\mathcal{L}(P,\omega) = \mathrm{sh}(u,v)$. The proof follows easily from Lemmas 3.15.3 and 3.15.4. Exercises 7.93 and 7.95 are related.

162. **a.** Although this problem can be done using the formula (3.13) for $W_{P,\omega}(x)$, it is easier to observe that

$$G_{P_1+P_2,\omega}(x) = G_{P_1,\omega_1}(x)G_{P_2,\omega_2}(x)$$

and then use equation (3.63).

   b. Let $P = P_1 + \cdots + P_k$, where P_k is a chain labeled by the word w^i from bottom to top. Then $\mathcal{L}(P,\omega) = \mathrm{sh}(w^1,\ldots,w^k)$. Moreover, $W_{P_i,\omega_i} = q^{\mathrm{maj}(w^i)}$, so the proof follows from iterating (a).

163. **a.** If P is an antichain then $\Omega_P(m) = m^p$, and the conclusion is clear. It thus suffices to show that when P is not an antichain, the coefficient of m^{p-1} in $\Omega_P(m)$ is positive. The coefficient is equal to $2e_P(p-1) - (p-1)e(p)$ (as defined preceding Theorem 3.15.8). Let A be the set of all ordered pairs (σ,i), where $\sigma: P \to p$ is a linear extension and $i \in [p-1]$. Let B be the set of all ordered pairs (τ,j), where $\tau: P \to p-1$ is a surjective order-preserving map and $j = 1$ or 2. Since $\#A = (p-1)e(p)$ and $\#B = 2e(p-1)$, it suffices to find an injection $\phi: A \to B$ that is not surjective. Choose an indexing $\{t_1,\ldots,t_p\}$ of the elements of P. Given $(\sigma,i) \in A$, define $\phi(\sigma,i) = (\tau,j)$, where

$$\tau(t) = \begin{cases} \sigma(t), & \text{if } \sigma(t) \le i, \\ \sigma(t) - 1, & \text{if } \sigma(t) > i, \end{cases}$$

$$j = \begin{cases} 1, & \text{if } \sigma(t_r) = i, \ \sigma(t_s) = i+1, \text{ and } r < s, \\ 2, & \text{if } \sigma(t_r) = i, \ \sigma(t_s) = i+1, \text{ and } r > s. \end{cases}$$

   It is easily seen that ϕ is injective. If t covers s in P and $\tau: P \to p-1$ is an order-preserving surjection for which $\tau(s) = \tau(t)$ (such a τ always exists), then one of $(\tau,1)$ and $(\tau,2)$ cannot be in the image of ϕ. Hence, ϕ is not surjective.

   b. This problem was raised by J. N. Kahn and M. Saks, who found the foregoing proof of (a) independently from this writer.

164. No. There are four 5-element posets for which $\Omega_P(m)$ has a negative coefficient, and none smaller. These four posets are shown in Figure 3.87.

165. J. Neggers, *J. Combin. Inform. System Sci.* **3** (1978), 113–133, made a conjecture equivalent to $A_P(x)$ having only real zeros (the naturally labeled case). In 1986 Stanley (unpublished) suggested that this conjecture could be extended to arbitrary labelings. The first published reference seems to be F. Brenti, *Mem. Amer. Math. Soc.*, no. 413 (1989). These conjectures became known as the *poset conjecture* or the *Neggers–Stanley conjecture*. Counterexamples to the conjecture of Stanley were obtained by P. Brändén, *Electron. Res. Announc. Amer. Math. Soc.* **10** (2004), 155–158. Finally, J.

Figure 3.88 A counterexample to the poset conjecture.

R. Stembridge, *Trans. Amer. Math. Soc.* **359** (2007), 1115–1128, produced counterexamples to the original conjecture of Neggers. Stembridge's smallest counterexample has 17 elements. One such poset P is given by Figure 3.88, for which

$$A_P(x) = x + 32x^2 + 336x^3 + 1420x^4 + 2534x^5 + 1946x^6 + 658x^7 + 86x^8 + 3x^9,$$

which has zeros near $-1.858844 \pm 0.149768i$. It is still open whether every *graded* natural poset satisfies the poset conjecture.

166. a. The "if" part is easy; we sketch a proof of the "only if" part. Let P be the smallest poset for which $G_P(x)$ is symmetric and P is not a disjoint union of chains. Define $\overline{G}_P(x) = \sum_{\tau} x_{\tau(t_1)} \cdots x_{\tau(t_p)}$, where τ ranges over all *strict P-partitions* $\tau \colon P \to \mathbb{N}$. The technique used to prove Theorem 3.15.10 shows that $G_P(x)$ is symmetric if and only if $\overline{G}_P(x)$ is symmetric. Let M be the set of minimal elements of P. Set $m = \#M$ and $P_1 = P - M$. The coefficient of x_0^m in $\overline{G}_P(x)$ is $\overline{G}_{P_1}(x')$, where $x' = (x_1, x_2, \ldots)$. Hence, $\overline{G}_{P_1}(x)$ is symmetric, so P_1 is a disjoint union of chains. Similarly, if M' denotes the set of maximal elements of P, then $P - M'$ is a disjoint union of chains.

Now note that m is the largest power of x_0 that can appear in a monomial in $\overline{G}_P(x)$. Hence, m is the largest power of *any* x_i that can appear in a monomial in $\overline{G}_P(x)$. Let A be an antichain of P. We can easily find a strict P-partition that is constant on A, so $\#A \leq m$. Hence, the largest antichain of P has size m. By Dilworth's theorem (Exercise 3.77(d)), P is a union of m chains. Each such chain intersects M and M'. It is easy to conclude that P is a disjoint union of chains $C_1, \ldots, C_k$, together with relations $s < t$, where s is a minimal element of some C_i and t a maximal element of some C_j, $i \neq j$.

Next note that the coefficient of $x_0^m x_1 x_2 \cdots x_{p-m}$ in $\overline{G}_P(x)$ is equal to $e(P_1)$, the number of linear extensions of P_1, so the coefficient of $x_0 x_1 \cdots x_{i-1} x_i^m x_{i+1} \cdots x_{p-m}$ is also $e(P_1)$ for any $0 \leq i \leq p - m$. Let $Q = C_1 + \cdots + C_k$. Then the coefficient of $x_0^m x_1 x_2 \cdots x_{p-m}$ in $\overline{G}_Q(x)$ is again equal to $e(P_1)$, since $P_1 \cong Q - \{\text{minimal elements of } Q\}$. Thus, the coefficient of $x_0 x_1 \cdots x_{i-1} x_i^m x_{i+1} \cdots x_{p-m}$ in $\overline{G}_Q(x)$ is $e(P_1)$. Since P is a refinement of Q it follows that if $\tau \colon P \to [0, p-m]$

is a strict Q-partition such that $\tau^{-1}(j)$ has one element for all $j \in [0, p-m]$ with a single exception $\#\tau^{-1}(i) = m$, then (regarding P as a refinement of Q) $\tau: P \to [0, p-m]$ is a strict P-partition. Now let $s < t$ in P but $s \parallel t$ in Q. One can easily find a strict Q-partition $\tau: Q \to [0, p-m]$ with $\tau(s) = \tau(t) = i$, say, and with $\#\tau^{-1}(i) = m, \#\tau^{-1}(j) = 1$ if $j \neq i$. Then $\tau: P \to [0, p-m]$ is not a strict P-partition, a contradiction.

b. This conjecture is due to R. Stanley, [3.68, p. 81]. For a proof of the "if" part, see Vol. II, Theorem 7.10.2. An interesting special case (different from (a)) is due to C. Malvenuto, *Graphs and Combinatorics* **9** (1993), 63–73.

167. b. The idea is to rule out subposets of P until the only P that remain have the desired form. For instance, P cannot have a three-element antichain A. For let i, j, k be the labels of the elements of A. Then there are linear extensions of P of the form $\sigma i' j' k' \tau$ for fixed σ and τ, where $i' j' k'$ is any permutation of ijk. One can check that these six linear extensions cannot all have the same number of descents.

169. a. Apply Theorem 3.15.8 to the case $P = r_1 + \cdots + r_m$, naturally labeled.

b. Suppose that $w \in \mathfrak{S}_M$ with $\mathrm{des}(w) = k-1$. Then w consists of x_1 1's, then y_1 2's, and so on, where $x_1 + \cdots + x_k = r_1$, $y_1 + \cdots + y_k = r_2$, and $x_1 \in \mathbb{N}$, $x_i \in \mathbb{P}$ for $2 \le i \le k$, $y_i \in \mathbb{P}$ for $1 \le i \le k-1$, $y_k \in \mathbb{N}$. Conversely, any such x_i's and y_i's yield a $w \in \mathfrak{S}_M$ with $\mathrm{des}(w) = k-1$. There are $\binom{r_1}{k-1}$ ways of choosing the x_i's and $\binom{r_2}{k-1}$ ways of choosing the y_i's. Hence,

$$A_M(x) = \sum_{k=0}^{r_1+r_2} \binom{r_1}{k} \binom{r_2}{k} x^{k+1}.$$

A q-analogue of this result appears in [3.68, Cor. 12.8]. Exercise 3.156 is related.

170. a. An order ideal of $J(m \times n)$ of rank r can easily be identified with a partition of r into at most m parts, with largest part at most n. Now use Proposition 1.7.3 to show that $F(L, q) = \binom{m+n}{m}$, which is equivalent to pleasantness.

b. Equivalent to a famous result of MacMahon. See Vol. II, Theorem 7.21.7 and the discussion following it. A further reference is R. Stanley, *Studies in Applied Math.* **50** (1971), 167–188, 259–279.

c. An order ideal of $J(2 \times n)$ of rank r can easily be identified with a partition of r into *distinct parts*, with largest part at most n, whence $F(L, q) = (1+q)(1+q^2) \cdots (1+q^n)$.

d. This result is equivalent to a conjecture of Bender and Knuth, shown by G. E. Andrews, *Pacific J. Math.* **72** (1977), 283–291, to follow from a much earlier conjecture of MacMahon. MacMahon's conjecture was proved independently by G. E. Andrews, *Adv. Math. Suppl. Studies* **1** (1978), 131–150; B. Gordon, *Pacific J. Math.* **108** (1983), 99–113; and I. G. Macdonald, *Symmetric Functions and Hall Polynomials*, Oxford Univ. Press, Oxford, 1979 (Ex. 19 on p. 53), 2nd ed., 1995 (Ex. 19 on p. 86). MacMahon's conjecture and similar results can be unified by the theory of minuscule representations of finite-dimensional complex semisimple Lie algebras; see R. A. Proctor, *Europ. J. Combinatorics* **5** (1984), 331–350.

e. This result is equivalent to the conjectured "q-enumeration of totally symmetric plane partitions," alluded to by G. E. Andrews, *Abstracts Amer. Math. Soc.* **1** (1980), 415, and D. P. Robbins (unpublished), and stated more explicitly in R. Stanley, *J. Combinatorial Theory, Ser. A* **43** (1986), 103–113 (equation (2)). The $q = 1$ case was first proved by J. R. Stembridge, *Advances in Math.* **111** (1995),

227–243, and later by G. E. Andrews, P. Paule, and C. Schneider, *Advances in Appl. Math.* **34** (2005), 709–739. A proof of the general case was finally given by C. Koutschan, M. Kauers, and D. Zeilberger, arXiv:1002.4384, 23 February 2010. Several persons have shown that $F(L,q)$ is also equal to $\sum_A (\det A)$, where A ranges over all square submatrices (including the empty matrix $\emptyset$, with $\det \emptyset = 1$) of the $(n+1) \times (n+1)$ matrix

$$\left[q^{i+1+\binom{j+1}{2}} \binom{i}{j} \right]_{i,j=0}^{n}.$$

f, g. Follows from either Theorem 6 or the proof of Theorem 8 of R. A. Proctor, *Europ. J. Combinatorics* **5** (1984), 331–350. (It is not difficult to give a direct proof of (f).) The proof of Proctor's Theorem 8 involves the application of the techniques of our Section 3.15 to these posets.

171. a. Follows from the bijection given in the proof of Proposition 3.5.1.
   b. This result appears in [3.68, Prop. 8.2] and is proved in the same way as Theorems 3.15.7 or 3.15.16.
   c. Equation (3.129) follows directly from the definition (3.127); see [3.68, Prop. 12.1]. Equation (3.130) is then a consequence of (3.96) and (3.97). Alternatively, (3.130) follows directly from (a).
   (e) Analogous to the proof of Theorem 3.15.10.
   (f) See [3.68, Prop. 17.3(ii)].

172. a. First note that

$$\binom{p+m-i}{p} = \frac{(1-q^{p-i}y)(1-q^{p-i-1}y)\cdots(1-q^{-i+1}y)}{(1-q^p)(1-q^{p-1})\cdots(1-q)},$$

where $y = q^m$. It follows from Exercise 3.171(b) that there is a polynomial $V_P(y)$ of degree p in y, whose coefficients are rational functions of q, such that

$$U_{P,m}(q) = V_P(q^m).$$

The polynomial $V_P(y)$ is unique since it is determined by its values on the infinite set $\{1, q, q^2, \dots\}$.
Since $U_{P_1+P_2,m}(q) = U_{P_1,m}(q) U_{P_2,m}(q)$, it follows that if each component of P is Gaussian, then so is P. Conversely, suppose that $P_1 + P_2$ is Gaussian. Thus,

$$V_{P_1+P_2}(y) = R(q) \prod_{i=1}^{p} \left(1 - yq^{h_i}\right),$$

where $R(q)$ depends only on q (not on y). But clearly $V_{P_1+P_2}(y) = V_{P_1}(y) V_{P_2}(y)$. Since each factor $1 - yq^{h_i}$ is irreducible (as a polynomial in y) and since $\deg V_{P_i}(y) = \#P_i$, we must have

$$V_{P_i}(y) = R_i(q) \prod_{j \in S_i} \left(1 - yq^{h_i}\right),$$

where j ranges over some subset S_i of $[p]$. Since $U_{P_i,0}(q) = V_{P_i}(1) = 1$, it follows that $R_i(q) = \prod_{j \in S_i} \left(1 - q^{h_i}\right)^{-1}$, so P_i is Gaussian.
   b. Clearly for any finite poset P, we have

$$\lim_{m \to \infty} U_{P,m}(q) = G_p(q),$$

as defined by equation (3.62). Hence, if P is Gaussian, we get

$$G_P(q) = \frac{W_P(q)}{(1-q)\cdots(1-q^p)} = \prod_{i=1}^{p} (1-q^{h_i})^{-1}. \qquad (3.147)$$

Hence, $W_P(q) = q^{d(P)} W_P(1/q)$, where $d(P) = \deg W_P(q)$, so by Theorem 3.15.16 P satisfies the δ-chain condition.

Now by equation (3.130), we have

$$U_{m,P}(q) = q^{pm} U_{P,m}(1/q) = U_{P,m}(q).$$

It follows that P^* is also Gaussian, and hence P^* satisfies the δ-chain condition. But if P is connected, then both P and P^* satisfy the δ-chain condition if and only if P is graded, and the proof follows.

c. Suppose that a_i of the h_j's are equal to i. Then by equation (3.128) we have

$$\frac{(p)!}{(1)^{a_1}\cdots(p)^{a_p}} (1-qy)^{a_1}(1-q^2y)^{a_2}\cdots(1-q^py)^{a_p}$$

$$= \sum_{i=0}^{p-1}(1-q^{p-i}y)(1-q^{p-i-1}y)\cdots(1-q^{-i+1}y)W_{P,i}(q). \qquad (3.148)$$

Pick $1 \le j \le p+1$, and let $b_i = a_i$ if $i \ne j$, and $b_j = a_j + 1$ (where we set $a_{p+1} = 0$). Set

$$\frac{(p+1)!}{(1)^{b_1}\cdots(p+1)^{b_{p+1}}}(1-qy)^{b_1}\cdots(1-q^{p+1}y)^{b_{p+1}}$$

$$= \sum_{i=0}^{p}(1-q^{p+1-i}y)\cdots(1-q^{-i+1}y)X_i(P,q).$$

This equation uniquely determines each $X_i(P,q)$.

Now we note the identity

$$(1-q^{p+1}y)(1-q^j y) = (1-q^{i+j})(1-q^{p+1-i}y)$$
$$+ (q^{i+j} - q^{p+1})(1-q^{-i}y). \qquad (3.149)$$

Multiply equation (3.148) by (3.149) to obtain

$$(1-q^j)\sum_{i=0}^{p}(1-q^{p+1-i}y)\cdots(1-q^{-i+1}y)X_i(P,q)$$

$$= \sum_{i=0}^{p-1}[(1-q^{i+j})(1-q^{p+1-i}y)\cdots(1-q^{-i+1}y)$$

$$+ (q^{i+j}-q^{p+1})(1-q^{p-i}y)\cdots(1-q^{-i}y)]W_{P,i}(q).$$

It follows that

$$(1-q^j)X_i = (1-q^{i+j})W_i + (q^{i+j-1}-q^{p+1})W_{i-1}. \qquad (3.150)$$

Next define

$$[p-1] - \{a_1 + a_2 + \cdots + a_i : i \ge 1\} = \{c_1,\ldots,c_k\}_>.$$

If we assume by induction that if we know deg W_{i-1} and deg W_i in equation (3.150), we can compute deg X_i. It then follows by induction that

$$\deg W_i = c_1 + \cdots + c_i, \; 0 \le i \le k.$$

Comparing with Exercise 3.171(f) completes the proof.

d. If $U_{P,m}(q)$ is given by equation (3.131), then

$$q^{pm} U_{P,m}(1/q) = U_{P,m}(q).$$

Comparing with equation (3.130) shows that $U_{P,m}(q) = U_{P*,m}(q)$. Let ρ^* denote the rank function of P^*. It follows from (c) that

$$\{1 + \rho(t) : t \in P\} = \{1 + \rho^*(t) : t \in P\}$$
$$= \{\ell(P) + 1 - \rho(t) : t \in P\}$$

(as multisets). Hence by (c), the multisets $\{h_1, \ldots, h_p\}$ and $\{\ell(P) + 2 - h_1, \ldots, \ell(P) + 2 - h_p\}$ coincide, and the proof follows. (This result was independently obtained by P. J. Hanlon.)

e. Let P have W_i elements of rank i. Using equation (3.131) and (c), one computes that the coefficient of q^2 in $U_{P,1}(q)$ is $\binom{W_0}{2} + W_1$. By Exercise 3.171(a), this number is equal to the number of two-element order ideals of P. Any of the $\binom{W_0}{2}$ two-element subsets of minimal elements forms such an order ideal. The remaining W_1 two-element order ideals must consist of an element of rank one and the unique element that it covers, completing the proof.

f. A uniform proof of (i)-(v), using the representation theory of semisimple Lie algebras, is due to R. A. Proctor, *Europ. J. Combinatorics* **5** (1984), 313–321. For ad hoc proofs (using the fact that a connected poset P is Gaussian if and only if $P \times m$ is pleasant for all $m \in \mathbb{P}$). see the solution to Exercise 3.170(b,d,f,g).

NOTE. Posets P satisfying equation (3.147) are called *hook length posets*. R. A. Proctor and D. Peterson found many interesting classes of such posets. See Proctor, *J. Algebra* **213** (1999), 272–303 (§1). Proctor discusses a uniform proof based on representation theory and calls these posets *d-complete*. For a classification of d-complete posets, see Proctor, *J. Algebraic Combinatorics* **9** (1999), 61–94. For a further important property of d-complete posets, see Proctor, preprint, arXiv:0905.3716.

173. This beautiful theory is due to P. Brändén, *Electronic J. Combinatorics* **11(2)** (2004), #R9. Note that as a special case of (h), $A_{P,\omega}(x)$ has symmetric unimodal coefficients if P is graded and ω is natural. (Symmetry of the coefficients also follows from Corollary 3.15.18 and Corollary 4.2.4 (iii).) In this special case unimodality was shown by V. Reiner and V. Welker, *J. Combinatorial Theory Ser. A* **109** (2005), 247–280 (Corollary 3.8 and Theorem 3.14), and later as part of more general results by C. A. Athanasiadis, *J. reine angew. Math.* **583** (2005), 163–174 (Lemma 3.8), and *Electronic J. Combinatorics* **11** (2004), #R6 (special case of Theorem 4.1), by using deep results on toric varieties. A combinatorial proof using a complicated recursion argument was given by J. D. Farley, *Advances in Applied Math.* **34** (2005), 295–312.

g. For a canonical labeling ω the procedure will end when each poset is an ordinal sum $Q_1 \oplus \cdots \oplus Q_k$ of antichains, labeled so that every label of elements of Q_i is either less than or greater than every label of elements in Q_{i+1}, depending on whether i is odd or even. From this observation the proof follows easily (using (c) to extend the result to any labeling ω for which (P, ω) is sign-graded).

h. Use Exercise 1.50(c,e).

i. We obtain

$$A_P(x) = (1+x)(1+4x+x^2) + 4(x+x^2) = 1 + 9x + 9x^2 + x^3.$$

174. a. The statement that the interval $[s,t]$ has as many elements of odd rank as of even rank is equivalent to $\sum_{u \in [s,t]} (-1)^{\rho(u)-\rho(s)} = 0$. The proof now follows easily from the defining recurrence (3.15) for μ.

b. Analogous to Proposition 3.16.1.

c. If n is odd, then by (b),

$$Z(P,m) + Z(P,-m) = -m((-1)^n \mu_P(\hat{0},\hat{1}) - 1).$$

The left-hand side is an even function of m, whereas the right-hand side is even if and only if $\mu_P(\hat{0},\hat{1}) = (-1)^n$. (There are many other proofs.)

175. By Proposition 3.8.2, $P \times Q$ is Eulerian. Hence, every interval $[z',z]$ of R with $z' \neq \hat{0}_R$ is Eulerian. Thus by Exercise 3.174(a), it suffices to show that for every $z = (s,t) > \hat{0}_R$ in R, we have

$$\sum_{z' \leq z} (-1)^{\rho_R(z')} = 0,$$

where ρ_R denotes the rank function in R. Since for any $v \neq \hat{0}_R$, we have $\rho_R(v) = \rho_{P \times Q}(v) - 1$, there follows

$$\sum_{\substack{z \leq z' \\ \text{in } R}} (-1)^{\rho_R(z')} = \sum_{\substack{u \leq z \\ \text{in } P \times Q}} (-1)^{\rho_{P \times Q}(u)-1}$$

$$- \sum_{\substack{\hat{0}_P \neq s' \leq s \\ \text{in } P}} (-1)^{\rho_P(s')-1} - \sum_{\substack{\hat{0}_Q \neq t' \leq t \\ \text{in } Q}} (-1)^{\rho_Q(t')-1}$$

$$+ (-1)^{\rho_{P \times Q}(\hat{0}_P \times Q)-1} + (-1)^{\rho_R(\hat{0}_R)}$$

$$= 0 - 1 - 1 + 1 + 1 = 0.$$

For further information related to the poset R, see M. K. Bennett, *Discrete Math.* **79** (1990), 235–249.

176. a. *Answer.* $\beta_{P_n}(S) = 1$ for all $S \subseteq [n]$.

b. By Exercise 3.157(b),

$$\sum_{m \geq 0} Z(P_n, m)x^m = \frac{x(1+x)^n}{(1-x)^{n+2}}.$$

(One could also appeal to Exercise 3.137.)

c. Write $f_n = f(P_n, x)$, $g_n = g(P_n, x)$. The recurrence (3.76) yields

$$f_n = (x-1)^n + 2\sum_{i=0}^{n-1} g_i(x-1)^{n-1-i}. \tag{3.151}$$

Equations (3.75) and (3.151), together with the initial conditions $f_0 = g_0 = 1$, completely determine f_n and g_n. Calculating some small cases leads to the guess

$$g_n = \sum_{k=0}^{\lfloor n/2 \rfloor} (-1)^k \left[\binom{n-1}{k} - \binom{n-1}{k-2} \right] x^k, \qquad (3.152)$$

$$f_n = \sum_{k=0}^{\lfloor n/2 \rfloor} (-1)^k \left[\binom{n-1}{k} - \binom{n-1}{k-1} \right] (x^k + x^{n-k}).$$

It is not difficult to check that these polynomials satisfy the necessary recurrences. Note also that $g_{2m} = (1-x)g_{2m-1}$ and $f_{2m+1} = (1-x)^{2m}(1+x)$.

177. a. Let $C_n = \{(x_1,\ldots,x_n) \in \mathbb{R}^n : 0 \le x_i \le 1\}$, an n-dimensional cube. A nonempty face F of C_n is obtained by choosing a subset $T \subseteq [n]$ and a function $\phi: T \to \{0,1\}$, and setting

$$F = \{(x_1,\ldots,x_n) \in C_n : x_i = \phi(i) \text{ if } i \in T\}.$$

Let F correspond to the interval $[\phi^{-1}(1), \phi^{-1}(1) \cup ([n] - T)]$ of B_n. This yields the desired (order-preserving) bijection.

b. Denote the elements of Λ as follows:

Let F be as in (a), and correspond to F the n-tuple $(y_1,\ldots,y_n) \in \Lambda^n$ where $y_i = \phi(i)$ if $i \in T$ and $y_i = u$ if $i \notin T$. This yields the desired (order-preserving) bijection.

c. Denote the two elements of P_n of rank i by a_i and b_i, $1 \le i \le n$. Associate with the chain $z_1 < z_2 < \cdots < z_k$ of $P_n - \{\hat{0},\hat{1}\}$ the n-tuple $(y_1,\ldots,y_n) \in \Lambda^n$ as follows:

$$y_i = \begin{cases} 0, & \text{if some } z_j = a_i, \\ 1, & \text{if some } z_j = b_i, \\ u, & \text{otherwise.} \end{cases}$$

This yields the desired bijection.

d. Follows from (c), Exercise 3.176(a), and [3.77, Thm. 8.3].

e. With Λ as in (b), we have $Z(\Lambda,m) = 2m - 1$, so $Z(\Lambda^n,m) = (2m-1)^n$ by Exercise 3.137. It follows easily that

$$Z(L_n,m) = 1^n + 3^n + 5^n + \cdots + (2m-1)^n.$$

f. *Answer.* $g(L_n,x) = \sum_{k \ge 0} \frac{1}{n-k+1} \binom{n}{k} \binom{2n-2k}{n} (x-1)^k$ (obtained in collaboration with I. M. Gessel). A generating function for $g(L_n,x)$ was given in R. Stanley, *J. Amer. Math. Soc.* **5** (1992), 805–851 (Proposition 8.6), namely,

$$\sum_{n \ge 0} g(L_n,x) \frac{y^n}{n!} = e^{2y} \sum_{n \ge 0} (-1)^n g_n \frac{y^n}{n!},$$

where g_n is given by equation (3.152).

g. This result was deduced from (f) by L. W. Shapiro (private communication). For further work in this area, see G. Hetyei, A second look at the toric h-polynomial of a cubical complex, `arXiv:1002.3601`.

179. For *rational* polytopes (i.e., those whose vertices have rational coordinates), this result follows from the hard Lefschetz theorem for the intersection homology of projective toric varieties; see Stanley [3.79]. For arbitrary convex polytopes the notion of intersection homology needs to be defined despite the absence of a corresponding variety, and the hard Lefschetz theorem must be proved in this context. The theory of "combinatorial intersection homology" was developed by G. Barthel, J.-P. Brasselet, K.-H. Fiesler, and L. Kaup, *Tohoku Math. J.* **54** (2002), 1–41, and independently by P. Bressler and V. A. Lunts, *Compositio Math.* **135**:3 (2003), 245–278. K. Karu, *Invent. math.* **157** (2004), 419–447, showed that the hard Lefschetz theorem held for this theory, thereby proving the nonnegativity of the coefficients of $g(L,x)$. An improvement to Karu's result was given by Bressler and Lunts, *Indiana Univ. Math. J.* **54** (2005), 263–307. A more direct approach to the work of Bressler and Lunts was given by Barthel, Brasselet, Fiesler, and Kaup, *Tohoku Math. J.* **57** (2005), 273–292. It remains open to prove the nonnegativity of the coefficients of $g(P,x)$ (or even $f(P,x)$) when P is both Cohen–Macaulay and Eulerian.

180. L_{nd} is in fact the lattice of faces of a certain d-dimensional convex polytope $C(n,d)$ called a *cyclic polytope*. Hence by Proposition 3.8.9, L_{nd} is an Eulerian lattice of rank $d + 1$. The combinatorial description of L_{nd} given in the problem is called "Gale's evenness condition." See, for example, page 85 of P. McMullen and G. C. Shephard, *Convex Polytopes and the Upper Bound Conjecture*, Cambridge Univ. Press, London, 1971, or [3.37, p. 62], or G. M. Ziegler, *Lectures on Polytopes*, Springer-Verlag, New York, 1995 (Theorem 0.7).

181. **a.** If L is the face lattice of a convex d-polytope $\mathcal{P}$, then the result goes back to Carathéodory. For a direct proof, see B. Grünbaum, *Convex Polytopes*, 2nd ed., Springer-Verlag, New York, 2003 (item 4 on page 123). The extension to Eulerian lattices is due to H. Bidkhori, Ph.D. thesis, M.I.T., 2010 (Section 3.5). The proof first shows by induction on d that $L - \{\hat{1}\}$ is simplicial (as defined in Section 3.16). It then follows from equation (3.73) that L has $\binom{d+1}{k}$ elements of rank k for all k. Since L is atomic (e.g., by Corollary 3.9.5) it must be a boolean algebra.
   b. In Exercise 3.191 let $P = B_d$ and $Q = B_2$. Then $P * Q$ (defined by equation (3.86)) is Eulerian of rank $d + 1$ whose truncation $(P * Q)_0 \cup (P * Q)_1 \cup \cdots \cup (P * Q)_{d-1}$ is a truncated boolean algebra, yet $P * Q$ itself is not a boolean algebra.

182. Let $P = \{t_1, \ldots, t_n\}$, and define

$$\mathcal{P} = \{(\alpha_1, \ldots, \alpha_n) \in \mathbb{R}^n : 0 \le \alpha_i \le 1, \text{ and } t_i \le t_j \Rightarrow \alpha_i \le \alpha_j\}.$$

Then $\mathcal{P}$ is a convex polytope, and it is not difficult to show (as first noted by L. D. Geissinger, in *Proc. Third Carribean Conf. on Combinatorics*, 1981, pp. 125–133) that Γ_P is isomorphic to the dual of the lattice of faces of $\mathcal{P}$ and hence is an Eulerian lattice. For further information on the polytope $\mathcal{P}$, see R. Stanley, *J. Disc. and Comp. Geom.* **1** (1986), 9–23.

183. **b.** This description of the Bruhat order goes back to C. Ehresmann, *Ann. Math.* **35** (1934), 396–443, who was the first person to define the order. For an exposition, see A. Björner and F. Brenti, *Combinatorics of Coxeter Groups*, Springer, New York, 2005 (Chapter 2). This book is the standard reference on the combinatorics of Coxeter groups, which we will refer to as B-B for the remainder of this exercise and in Exercise 3.185.
   c. The Bruhat order can be generalized to arbitrary Coxeter groups. In this context, $\mathfrak{S}_n$ was shown to be Eulerian by D.-N. Verma, *Ann. Sci. Éc. Norm. Sup.* **4** (1971), 393–398, and V. V. Deodhar, *Invent. Math.* **39** (1977), 187–198. See B-B, Corollary 2.7.10. More recent proofs were given by J. R. Stembridge, *J. Algebraic Combinatorics* **25** (2007), 141–148, B. C. Jones, *Order* **26** (2009), 319–330, and M. Marietti, *J. Algebraic Combinatorics* **26** (2007), 363–382. This last paper introduces a new class of Eulerian posets called *zircons*, which give a combinatorial

generalization of Bruhat order. A far-reaching topological generalization of the present exercise is due to A. Björner and M. L. Wachs [3.17]. A survey of Bruhat orders is given by A. Björner, *Contemp. Math.* **34** (1984), pp. 175–195.

d. First show that for fixed $i < j$, the number of permutations v for which $v < (i, j)v$ is $n!/(j - i + 1)$. Then sum on $1 \le i < j \le n$. This argument is due to D. Callan, as reported in *The On-Line Encyclopedia of Integer Sequences*, A002538.

g. This result goes back to Chevalley in 1958 (for arbitrary finite Coxeter groups), but the first explicit statement seems to be due to J. R. Stembridge, *J. Algebraic Combinatorics* **15** (2002), 291–301. For additional information see A. Postnikov and R. Stanley, *J. Algebraic Combinatorics* **29** (2009), 133–174.

184. See F. Incitti, *J. Algebraic Combinatorics* **20** (2004), 243–261. For further work on this poset, see A. Hultman and K. Vorwerk, *J. Algebraic Combinatorics* **30** (2009), 87–102.

185. b. Given $w = a_1 a_2 \cdots a_n \in \mathfrak{S}_n$, let $I_w = \{(a_i, a_j) : i < j, a_i > a_j\}$, the *inversion set* of w. It is easy to see that $v \le w$ in $W(\mathfrak{S}_n)$ if and only if $I_v \subseteq I_w$. From this observation it follows readily that $v \vee w$ is defined by $I_{v \vee w} = \overline{I_v \cup I_w}$, where the overline denotes transitive closure. Hence, $W(\mathfrak{S}_n)$ is a join-semilattice. Since it has a $\hat{0}$ (or, in fact, since it is self-dual *via* the anti-automorphism $a_1 a_2 \cdots a_n \mapsto a_n \cdots a_2 a_1$), it follows that $W(\mathfrak{S}_n)$ is a lattice. This argument appears in C. Berge, *Principles of Combinatorics*, Academic Press, New York, 1971 (§4.4, Prop. 3). For further information see A. Hammett and B. G. Pittel, Meet and join in the weak order lattice, preprint, 2006. An exposition of the weak order for arbitrary Coxeter groups appears in B-B, Chapter 3.

c. Every vertex in the Hasse diagram of $W(\mathfrak{S}_n)$ has degree $n - 1$, from which the result is immediate.

d. Follows from Corollary 3 on page 185 of A. Björner, *Contemp. Math.* **34** (1984), pp. 175–195. A topological generalization appears in B-B, Corollary 3.2.8.

e. This result was shown by P. H. Edelman, Geometry and the Möbius function of the weak Bruhat order of the symmetric group, unpublished.

f. This result was first proved by R. Stanley, *Europ. J. Combinatorics* **5** (1984), 359–372 (Corollary 4.3). Subsequent proofs were announced in P. H. Edelman and C. Greene, *Contemporary Math.* **34** (1984), 155–162, and A. Lascoux and M.-P. Schützenberger, *C. R. Acad. Sc. Paris* **295**, Série I (1982), 629–633. The proof of Edelman and Greene appears in *Advances in Math.* **63** (1987), 42–99. An interesting exposition was given by A. M. Garsia, Publications du LaCIM, Université du Québec á Montréal, Montréal, vol. 29, 2002. The number M_n is just the number of standard Young tableaux of the staircase shape $(n - 1, n - 2, \ldots, 1)$; see Exercise 7.22.

g. This result was first proved by I. G. Macdonald, *Notes on Schubert polynomials*, Publications du LaCIM, Université du Québec à Montréal, Montréal, vol. 6, 1991 (equation (6.11)). A simpler proof, as well as a proof of a q-analogue conjectured by Macdonald, was given by S. Fomin and R. Stanley, *Advances in Math.* **103** (1994), 196–207 (§2).

186. a–c. Fan Wei, The weak Bruhat order and separable permutations, arXiv:1009:5740.
     d.   It has been checked for $n \le 8$ that if $w \in \mathfrak{S}_n$ and $F(\Lambda_w, q)$ is symmetric, then every zero of $F(\Lambda_w, q)$ is a root of unity.

187. b. For every set $S \subseteq [n - 1]$, there exists a unique permutation $w \in \mathcal{G}_n$ with descent set $D(w) = S$. The map $w \mapsto D(w)$ is an isomorphism from $\mathcal{G}_n$ to $M(n)$.

c. These permutations $w = a_1 \cdots a_n$ are just those of Exercise 1.114(b) (i.e., for all $1 \le i \le n$, the set $\{a_1, a_2, \ldots, a_i\}$ consists of consecutive integers (in some order)). Another characterization of such permutations w is the following. For $1 \le i \le n$,

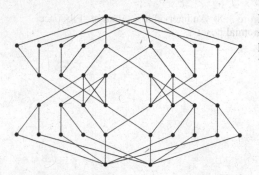

Figure 3.89 A poset Q for which $Q \otimes (1+1) \cup \{\hat{0}, \hat{1}\}$ has a nonpositive flag h-vector.

let μ_i be the number of terms of w that lie to the left of i and that are greater than i (a variation of the inversion table of w). Then $\mu = (\mu_1, \mu_2, \ldots, \mu_n)$ is a partition into distinct parts (i.e., for some k we have $\mu_1 > \mu_2 > \cdots > \mu_k = \mu_{k+1} = \cdots = \mu_n = 0$). Note that we also have $D(w) = \{\mu_1, \cdots, \mu_{k-1}\}$ and $\mathrm{maj}(w) = \mathrm{inv}(w)$. These permutations are also the possible *ranking patterns* as defined by H. Kamiya, P. Orlik, A. Takemura, and H. Terao, *Ann. Combinatorics* **10** (2006), 219–235.

188. The poset P is an interval of the poset of *normal words* introduced by F. D. Farmer, *Math. Japonica* **23** (1979), 607–613. It was observed by A. Björner and M. L. Wachs [3.19, §6] that the poset of all normal words on a finite alphabet $S = \{s_1, \ldots, s_n\}$ is just the Bruhat order of the Coxeter group $W = \langle S : s_i^2 = 1 \rangle$. Hence, P is Eulerian by the Verma-Deodhar result mentioned in the solution to Exercise 3.183. A direct proof can also be given.

190. **a.** This result is due to R. Ehrenborg, G. Hetyei, and M. A. Readdy, Level Eulerian posets, preprint dated June 12, 2010 (Corollary 8.3), as a special case of a much more general situation.

192. This deep result was proved by K. Karu, *Compositio Math.* **142** (2006), 701–708. Karu gives another proof for a special case (complete fans) in Lefschetz decomposition and the cd-index of fans, preprint, math.AG/0509220. Ehrenborg and Karu, *J. Algebraic Combin.* **26** (2007), 225–251, continue this work, proving in particular a conjecture of Stanley that the cd-index of a Gorenstein* lattice is minimized on boolean algebras.

193. **a.** The simplest example is obtained by taking two butterfly posets (as defined in Exercise 3.198) of rank 5 and identifying their top and bottom elements. For this poset P, we have
$$\Phi_P(c, d) = c^4 + 2c^2 d + 2cd^2 - 4d^2.$$
For further information on negative coefficients of the cd-index, see M. M. Bayer, *Proc. Amer. Math. Soc.* **129** (2001), 2219–2225.

   b. It follows from the work of M. M. Bayer and G. Hetyei, *Europ. J. Combinatorics* **22** (2001), 5–26, that such a poset must have rank at least seven. For an example of rank 7, let Q be the poset of Figure 3.89, and let $P = Q \otimes (1 + 1)$, with a $\hat{0}$ and $\hat{1}$ adjoined (where $\otimes$ denotes ordinal product). Then it can be checked that P is Eulerian, with $\beta_P(4, 5, 6) = -2$. The poset Q appears as Figure 2 in Bayer and Hetyei, ibid.

194. If $S = \{a_1, a_2, \ldots, a_k\}_\leq \subseteq [n-1]$, then let $\rho = (a_1, a_2 - a_1, a_3 - a_2, \ldots, n - a_k)$, a composition of n. We write $\alpha_P(\rho)$ for $\alpha_P(S)$. By Exercise 3.155(a), we have $\alpha_P(\rho) = \alpha_P(\sigma)$ if ρ and σ have the same multiset of parts. By a result of Bayer and Billera [3.5, Prop. 2.2], α_P is determined by its values on those ρ with no part

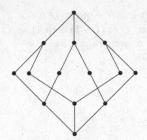

Figure 3.90 An interval in a putative "Fibonacci binomial poset".

equal to 1. From this we get $d(n) \leq p(n) - p(n-1)$, where $p(n)$ denotes the number of partitions of n. On the other hand, the work of T. Bisztriczky, *Mathematika* **43** (1996), 274–285, gives a lower bound for $d(n)$, though it is far from the upper bound $p(n) - p(n-1)$. For the lower bound see also M. M. Bayer, A. Bruening, and J. Stewart, *Discrete Comput. Geom.* **27** (2002), 49–63.

195. See R. Ehrenborg, *Order* **18** (2001), 227–236 (Prop. 4.6).

196. a. Let $[s,t]$ be an $(n+1)$-interval of P, and let u be a coatom (element covered by t) of $[s,t]$ Then $[s,t]$ has $A(n+1) = B(n+1)/B(n)$ atoms, while $[s,u]$ has $A(n) = B(n)/B(n-1)$ atoms. Since every atom of $[s,u]$ is an atom of $[s,t]$ we have $A(n+1) \geq A(n)$, and the proof follows.

b. The poset of Figure 3.90 could be a 4-interval in a binomial poset where $B(n) = F_1 F_2 \cdots F_n$. It is known that the *Fibonomial coefficient*

$$\binom{n}{k}_F = \frac{F_n F_{n-1} \cdots F_{n-k+1}}{F_k F_{k-1} \cdots F_1}$$

is an integer, a necessary condition for the existence of a binomial poset with $B(n) = F_1 F_2 \cdots F_n$. For a combinatorial interpretation of $\binom{n}{k}_F$, see A. T. Benjamin and S. S. Plott, *Fib. Quart.* **46/47** (2008/2009), 7–9.

197. See J. Backelin, Binomial posets with non-isomorphic intervals, `arXiv:math/0508397`. Backelin's posets have factorial function $B(1) = 1$ and $B(n) = 2^{n-2}$ for $n \geq 2$.

198. See R. Ehrenborg and M. A. Readdy, *J. Combinatorial Theory Ser. A* **114** (2007), 339–359. For further work in this area, see H. Bidkhori, Ph.D. thesis, M.I.T., 2010, and Finite Eulerian posets which are binomial, Sheffer or triangular, `arXiv:1001.3175`.

199. *Answer: L* is a chain or a boolean algebra.

200. Equation (3.132) is equivalent to a result of R. C. Read, *Canad. J. Math.* **12** (1960), 410–414 (also obtained by E. A. Bender and J. R. Goldman [3.7]). The connection with binomial posets was pointed out by Stanley, *Discrete Math.* **5** (1973), 171–178 (§3). Note that equation (3.88) (the chromatic generating function for the number of acyclic digraphs on $[n]$) follows immediately from equations (3.132) and (3.121).

201. a. See [3.25, Prop. 9.1]. This result is proved in exact analogy with Theorem 3.18.4.

b,c. See [3.25, Prop. 9.3].

202. See Theorem 5.2 of R. Simion and R. Stanley, *Discrete Math.* **204** (1999), 369–396.

203. As the notation becomes rather messy, let us illustrate the proof with the example $a_1 = 0, a_2 = 3, a_3 = 4, m = 6$. Let

$$S_n = \{6i, 6i + 3, 6i + 4 : 0 \le i \le n\},$$
$$S_n' = S_n \cup \{6n\},$$
$$S_n'' = S_n \cup \{6n, 6n + 3\}.$$

Let P be the binomial poset $\mathbb{B}$ of all finite subsets of $\mathbb{N}$, ordered by inclusion, and let $\mu_S(n)$ be as in Section 3.19. Then by Theorem 3.13.1, we have

$$(-1)^n f_1(n) = \mu_{S_n}(6n) := g_1(n),$$
$$(-1)^{n+1} f_2(n) = \mu_{S_n'}(6n + 3) := g_2(n),$$
$$(-1)^{n+2} f_3(n) = \mu_{S_n''}(6n + 4) := g_3(n).$$

By the defining recurrence (3.15) we have

$$g_1(n) = -\sum_{i=0}^{n-1} \left[\binom{6n}{6i} g_1(i) + \binom{6n}{6i+3} g_2(i) + \binom{6n}{6i+4} g_3(i) \right], \quad n > 0,$$

$$g_2(n) = -\sum_{i=0}^{n} \binom{6n+3}{6i} g_1(i) - \sum_{i=0}^{n-1} \binom{6n+3}{6i+3} g_2(i) - \sum_{i=0}^{n-1} \binom{6n+3}{6i+4} g_3(i),$$

$$g_3(n) = -\sum_{i=0}^{n} \binom{6n+4}{6i} g_1(i) - \sum_{i=0}^{n} \binom{6n+4}{6i+3} g_2(i) - \sum_{i=0}^{n-1} \binom{6n+4}{6i+4} g_3(i).$$

These formulas may be rewritten (incorporating also $g_1(0) = 1$)

$$\delta_{0n} = \sum_{i=0}^{n} \left[\binom{6n}{6i} g_1(i) + \binom{6n}{6i+3} g_2(i) + \binom{6n}{6i+4} g_3(i) \right],$$

$$0 = \sum_{i=0}^{n} \left[\binom{6n+3}{6i} g_1(i) + \binom{6n+3}{6i+3} g_2(i) + \binom{6n+3}{6i+4} g_3(i) \right],$$

$$0 = \sum_{i=0}^{n} \left[\binom{6n+4}{6i} g_1(i) + \binom{6n+4}{6i+3} g_2(i) + \binom{6n+4}{6i+4} g_3(i) \right].$$

Multiplying the three equations by $x^{6n}/(6n)!$, $x^{6n+3}/(6n+3)!$, and $x^{6n+4}/(6n+4)!$, respectively, and summing on $n \ge 0$ yields

$$F_1 \Phi_0 + F_2 \Phi_3 + F_3 \Phi_2 = 1,$$
$$F_1 \Phi_3 + F_2 \Phi_0 + F_3 \Phi_5 = 0,$$
$$F_1 \Phi_4 + F_2 \Phi_1 + F_3 \Phi_0 = 0,$$

as desired. We leave to the reader to see that the general case works out in the same way. Note that we can replace $f_k(n)$ by the more refined $\sum_w q^{\text{inv}(w)}$, where w ranges over all permutations enumerated by $f_k(n)$, simply by replacing $\mathbb{B}$ by $\mathbb{B}_q$ and thus $a!$ by $(a)!$ and $\binom{a}{b}$ by $\binom{a}{b}$ throughout.

An alternative approach to this problem is given by D. M. Jackson and I. P. Goulden, *Advances in Math.* **42** (1981), 113–135.

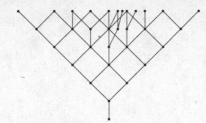

Figure 3.91 A 1-differential poset up to rank 6.

204. a. See [3.74, Lemma 2.5].

b. Apply Theorem 3.18.4 to (a). See [3.74, Cor. 2.6].

c. Specialize (b) to $P = \mathbb{B}(q)$, and note that by Theorem 3.13.3 we have $G_n(q,z) = (-1)^n h(n)|_{z \to -z}$. A more general result is given in [3.74, Cor. 3.6].

205. a. See Figure 3.91 for a 1-differential poset P up to rank 6 that is not isomorphic to $\Omega^i Y[6-i]$ for any $0 \le i \le 6$. Then $\Omega^\infty P$ is the desired example.

b. These results were computed by Patrick Byrnes, private communication, dated 7 March 2008.

c. Examples of this nature appear in J. B. Lewis, On differential posets, undergraduate thesis, Harvard University, 2007;

$$\langle \texttt{http://math.mit.edu/~jblewis/}$$

$$\texttt{JBLHarvardSeniorThesis.pdf} \rangle.$$

206. This result was proved by P. Byrnes, preprint, 2011, based on work of Y. Qing, Master's thesis, M.I.T., 2008. It is reasonable to conjecture that the only r-differential lattices are direct products of a suitable number of copies of Y and Z_k's, $k \ge 1$.

207. With γ as in the proof of Theorem 3.21.11, we have

$$\gamma(UDUUP) = \sum_{n \ge 0} \alpha(n-2 \to n \to n-1 \to n)q^n.$$

Repeated applications of $DU = UD + rI$ gives $UDUU = 2rU^2 + U^3 D$. Then use $DP = (U+r)P$ (Proposition 3.21.3) to get

$$UUDUP = (2rU^2 + rU^3 + U^4)P.$$

The proof follows easily from Theorem 3.21.11. This result appeared in [3.80, Exam. 3.5] as an illustration of a more general result, where $UUDU$ is replaced by any word in U and D.

208. a. Use the relation $DU = UD + rI$ to put w in the form

$$w = \sum_{i,j} c_{ij}(w)U^i D^j, \tag{3.153}$$

where $c_{ij}(w)$ is a polynomial in r, and where if $c_{ij}(w) \ne 0$ then $i - j = \rho(t)$. It is easily seen that this representation of w is unique. Apply U on the left to equation (3.153). By uniqueness of the c_{ij}'s there follows [why?]

$$c_{ij}(Uw) = c_{i-1,j}(w). \tag{3.154}$$

Now apply D on the left to equation (3.153). Using $DU^i = U^i D + riU^{i-1}$ we get [why?]

$$c_{ij}(Dw) = c_{i,j-1}(w) + r(i+1)c_{i+1,j}(w). \tag{3.155}$$

Setting $j = 0$ in equations (3.154) and (3.155) yields

$$c_{i0}(Uw) = c_{i-1,0}(w), \tag{3.156}$$

$$c_{i0}(Dw) = r(i+1)c_{i+1,0}. \tag{3.157}$$

Now let (3.153) operate on $\hat{0}$. We get $w(\hat{0}) = c_{n0}(w)U^n(\hat{0})$. Thus, the coefficient of t in $w(\hat{0})$ is given by

$$\langle w(\hat{0}), t \rangle = c_{n0}(w)e(t).$$

It is easy to see from equations (3.156) and (3.157) that

$$c_{n0}(w) = r^{\#S} \prod_{i \in S}(b_i - a_i),$$

and the proof follows.

209. Easily proved by induction on n. In particular, assume for n, multiply by UD on the left, and use the identity $DU^k = kU^{k-1} + U^k D$. See [3.80, Prop. 4.9].

210. Using $DU = UD + 1$, we can write a balanced word $w = w(U, D)$ as a linear combination of words $U^k D^k$. By Proposition 1.9.1, we can invert equation (3.134) to get

$$U^n D^n = \sum_{k=0}^{n} s(n,k)(UD)^k = UD(UD - 1) \cdots (UD - n + 1).$$

It follows that every balanced word is a polynomial in UD. Since any two polynomials in UD commute, the proof follows. This result appeared in [3.80, Cor. 4.11(a)].

213. We have (using equation (3.105))

$$\sum_{n \geq 0} \sum_{k \geq 0} \kappa_{2k}(n) \frac{q^n x^{2k}}{(2k)!} = \sum_{t \in P} \langle e^{(D+U)x}t, t \rangle q^{\rho(t)}$$

$$= e^{rx^2/2} \sum_{t \in P} \langle e^{Ux} e^{Dx} t, t \rangle q^{\rho(t)}.$$

From Exercise 3.212(b), it is easy to obtain

$$\sum_{n \geq 0} \sum_{k \geq 0} \kappa_{2k}(n) \frac{q^n x^{2k}}{(2k)!} = F(P, q) \exp\left(\frac{1}{2}rx^2 + \frac{rqx^2}{1 - q}\right).$$

Extracting the coefficient of $x^{2k}/(2k)!$ on both sides completes the proof. This result first appeared in [3.80, Cor. 3.14].

214. This is a result of P. Byrnes, preprint, 2011.

215. a. We showed in the proof of Theorem 3.21.12 that the linear transformation $U_i: \mathbb{C}P_i \to \mathbb{C}P_{i+1}$ is injective. Hence,

$$p_i = \dim \mathbb{C}P_i \leq \dim \mathbb{C}P_{i+1} = p_{i+1}.$$

4

Rational Generating Functions

4.1 Rational Power Series in One Variable

The theory of binomial posets developed in the previous chapter sheds considerable light on the "meaning" of generating functions and reduces certain types of enumerative problems to a routine computation. However, it does not seem worthwhile to attack more complicated problems from this point of view. The remainder of this book will for the most part be concerned with other techniques for obtaining and analyzing generating functions. We first consider the simplest general class of generating functions, namely, the *rational* generating functions. In this chapter we will concern ourselves primarily with rational generating functions in one variable; that is, generating functions of the form $F(x) = \sum_{n \geq 0} f(n)x^n$ that are rational functions in the ring $K[[x]]$, where K is a field. This means that there exist polynomials $P(x), Q(x) \in K[x]$ such that $F(x) = P(x)Q(x)^{-1}$ in $K[[x]]$. Here it is assumed that $Q(0) \neq 0$, so that $Q(x)^{-1}$ exists in $K[[x]]$. The field of all rational functions in x over K is denoted $K(x)$, so the ring of rational power series is given by $K[[x]] \cap K(x)$. For our purposes here it suffices to take $K = \mathbb{C}$ or sometimes $\mathbb{C}$ with some indeterminates adjoined.

The fundamental property of rational functions in $\mathbb{C}[[x]]$ from the viewpoint of enumeration is the following.

4.1.1 Theorem. *Let* $\alpha_1, \alpha_2, \ldots, \alpha_d$ *be a fixed sequence of complex numbers, $d \geq 1$ and $\alpha_d \neq 0$. The following conditions on a function $f : \mathbb{N} \to \mathbb{C}$ are equivalent:*

i.

$$\sum_{n \geq 0} f(n)x^n = \frac{P(x)}{Q(x)}, \qquad (4.1)$$

where $Q(x) = 1 + \alpha_1 x + \alpha_2 x^2 + \cdots + \alpha_d x^d$ *and $P(x)$ is a polynomial in x of degree less than d.*

ii. For all $n \geq 0$,

$$f(n+d) + \alpha_1 f(n+d-1) + \alpha_2 f(n+d-2) + \cdots + \alpha_d f(n) = 0. \quad (4.2)$$

iii. For all $n \geq 0$,

$$f(n) = \sum_{i=1}^{k} P_i(n)\gamma_i^n, \tag{4.3}$$

where $1 + \alpha_1 x + \alpha_2 x^2 + \cdots + \alpha_d x^d = \prod_{i=1}^{k}(1 - \gamma_i x)^{d_i}$, the γ_i's are distinct and nonzero, and $P_i(n)$ is a polynomial of degree less than d_i.

Proof. Fix $Q(x) = 1 + \alpha_1 x + \cdots + \alpha_d x^d$. Define four complex vector spaces as follows:

$V_1 = \{f : \mathbb{N} \to \mathbb{C}$ such that (i) holds$\}$,

$V_2 = \{f : \mathbb{N} \to \mathbb{C}$ such that (ii) holds$\}$,

$V_3 = \{f : \mathbb{N} \to \mathbb{C}$ such that (iii) holds$\}$,

$V_4 = \{f : \mathbb{N} \to \mathbb{C}$ such that $\sum_{n \geq 0} f(n)x^n = \sum_{i=1}^{k} \sum_{j=1}^{d_i} \beta_{ij}(1 - \gamma_i x)^{-j}$,

for some $\beta_{ij} \in \mathbb{C}$, where γ_i and d_i have the same meaning as in (iii)$\}$.

We first claim that $\dim V_4 = d$ (all dimensions are taken over $\mathbb{C}$). Now V_4 is spanned over $\mathbb{C}$ by the rational functions $R_{ij}(x) = (1 - \gamma_i x)^{-j}$, where $1 \leq i \leq k$ and $1 \leq j \leq d_i$. There are $\sum d_i = d$ such functions, so $\dim V_4 \leq d$. It remains to show that the $R_{ij}(x)$'s are linearly independent. Suppose to the contrary that we have a linear relation

$$\sum c_{ij} R_{ij}(x) = 0, \tag{4.4}$$

where $c_{ij} \in \mathbb{C}$ and not all $c_{ij} = 0$. Let i be such that some $c_{ij} \neq 0$, and then let j be the largest integer for which $c_{ij} \neq 0$. Multiply equation (4.4) by $(1 - \gamma_i x)^j$ and set $x = 1/\gamma_i$. We obtain $c_{ij} = 0$, a contradiction, proving that $\dim V_4 = d$.

Now in (i) we may choose the d coefficients of $P(x)$ arbitrarily. Hence, $\dim V_1 = d$. In (ii) we may choose $f(0), f(1), \ldots, f(d-1)$ and then the other $f(n)$'s are uniquely determined. Hence, $\dim V_2 = d$. In (iii) we see that $f(n)$ is determined by the d coefficients of the $P_i(n)$'s, so $\dim V_3 \leq d$. (It is not so apparent, as it was for (i) and (ii), that different choices of $P_i(n)$'s will produce different $f(n)$'s.) Now for $j \geq 0$ we have

$$\frac{1}{(1 - \gamma x)^j} = \sum_{n \geq 0} (-\gamma)^n \binom{-j}{n} x^n = \sum_{n \geq 0} x^n \gamma^n \binom{j+n-1}{j-1}.$$

Since $\binom{j+n-1}{j-1}$ is a polynomial in n of degree j, we get $V_4 \subseteq V_3$. Since $\dim V_4 = d \geq \dim V_3$, we have $V_3 = V_4$.

If $f \in V_1$, then equate coefficients of x^n in the identity $Q(x) \sum_{n \geq 0} f(n)x^n = P(x)$ to get $f \in V_2$. Since $\dim V_1 = \dim V_2$, there follows $V_1 = V_2$.

By putting the sum $\sum_{i=1}^{k} \sum_{j=1}^{d_i} \beta_{ij}(1 - \gamma_i x)^{-j}$ over a common denominator, we see that $V_4 \subseteq V_1$. Since $\dim V_1 = \dim V_4$, there follows $V_1 = V_2 = V_3 (= V_4)$, so the proof is complete. $\qquad\square$

Before turning to some interesting variations and special cases of Theorem 4.1.1, we first give a couple of examples of how a rational generating function arises in combinatorics.

4.1.2 Example. The prototypical example of a function $f(n)$ satisfying the conditions of Theorem 4.1.1 is given by $f(n) = F_n$, a Fibonacci number. The recurrence $F_{n+2} = F_{n+1} + F_n$ yields the generating function $\sum_{n \geq 0} F_n x^n = P(x)/(1 - x - x^2)$ for some polynomial $P(x) = a + bx$. The initial conditions $F_0 = 0, F_1 = 1$ imply that $P(x) = x$. Hence,

$$\sum_{n \geq 0} F_n x^n = \frac{x}{1 - x - x^2}.$$

Now $1 - x - x^2 = (1 - \varphi x)(1 - \bar{\varphi} x)$, where

$$\varphi = \frac{1 + \sqrt{5}}{2}, \quad \bar{\varphi} = \frac{1 - \sqrt{5}}{2} = 1 - \varphi = -\frac{1}{\varphi}.$$

Hence, $F_n = \alpha \varphi^n + \beta \bar{\varphi}^n$. Setting $n = 0, 1$ yields the linear equations

$$\alpha + \beta = 0, \quad \varphi \alpha + \bar{\varphi} \beta = 1,$$

with solution $\alpha = 1/\sqrt{5}$ and $\beta = -1/\sqrt{5}$. Hence,

$$F_n = \frac{\varphi^n - \bar{\varphi}^n}{\sqrt{5}}. \tag{4.5}$$

Although equation (4.5) has no direct combinatorial meaning, it still has many uses. For instance, since $-1 < \bar{\varphi} < 0$, it is easy to deduce that F_n is the nearest integer to $\varphi^n/\sqrt{5}$. Thus, we have a very accurate expression for the rate of growth of F_n. Moreover, the explicit formula (4.5) often gives a routine method for proving various identities and formulas involving F_n, though sometimes there are more enlightening combinatorial or algebraic proofs. An instance is mentioned in Example 4.7.16.

4.1.3 Example. Let $f(n)$ be the number of paths with n steps starting from $(0,0)$, with steps of the type $(1,0)$, $(-1,0)$, or $(0,1)$, and never intersecting themselves. For instance, $f(2) = 7$, as shown in Figure 4.1 (with the initial point at $(0,0)$ circled). Equivalently, letting $E = (1,0)$, $W = (-1,0)$, $N = (0,1)$, we want the number of words $A_1 A_2 \cdots A_n$ ($A_i = E, W$, or N) such that EW and WE never appear as factors. Let $n \geq 2$. There are $f(n-1)$ words of length n ending in N. There are $f(n-1)$ words of length n ending in EE, WW, or NE. There are $f(n-2)$ words of length n ending in NW. Every word of length at least 2 ends in exactly one of N, EE, WW, NE, or NW. Hence,

$$f(n) = 2f(n-1) + f(n-2), \quad f(0) = 1, \quad f(1) = 3.$$

By Theorem 4.1.1, there are numbers A and B for which $\sum_{n \geq 0} f(n) x^n = (A + Bx)/(1 - 2x - x^2)$. By, for example, comparing coefficients of 1 and x, we obtain

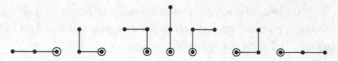

Figure 4.1 Some non-self-intersecting lattice paths.

$A = B = 1$, so

$$\sum_{n \geq 0} f(n)x^n = \frac{1+x}{1-2x-x^2}.$$

We have $1 - 2x - x^2 = (1 - (1 + \sqrt{2})x)(1 - (1 - \sqrt{2})x)$. Again by Theorem 4.1.1 we have $f(n) = a(1 + \sqrt{2})^n + b(1 - \sqrt{2})^n$ for some numbers a and b. By, for example, setting $n = 0, 1$, we obtain $a = \frac{1}{2}(1 + \sqrt{2})$ and $b = \frac{1}{2}(1 - \sqrt{2})$. Hence,

$$f(n) = \frac{1}{2}\left((1 + \sqrt{2})^{n+1} + (1 - \sqrt{2})^{n+1}\right). \tag{4.6}$$

Note that without the restriction that the path doesn't self-intersect, there are 3^n paths with n steps. With the restriction, the number has been reduced from 3^n to roughly $(1 + \sqrt{2})^n = (2.414 \cdots)^n$. Note also that since $-1 < 1 - \sqrt{2} < 0$, it follows from equation (4.6) that

$$f(n) = \begin{cases} \left\lfloor \frac{1}{2}(1 + \sqrt{2})^{n+1} \right\rfloor, & n \text{ even}, \\ \left\lceil \frac{1}{2}(1 + \sqrt{2})^{n+1} \right\rceil, & n \text{ odd}. \end{cases}$$

4.2 Further Ramifications

In this section, we will consider additional information that can be gleaned from Theorem 4.1.1. First, we give an immediate corollary that is concerned with the possibilities of "simplifying" the formulas (4.1), (4.2), (4.3).

4.2.1 Corollary. *Suppose that $f : \mathbb{N} \to \mathbb{C}$ satisfies any (or all) of the three equivalent conditions of Theorem 4.1.1, and preserve the notation of that theorem. The following conditions are equivalent.*

i. *$P(x)$ and $Q(x)$ are relatively prime. In other words, there is no way to write $P(x)/Q(x) = P_1(x)/Q_1(x)$, where P_1, Q_1 are polynomials and $\deg Q_1 < \deg Q = d$.*

ii. *There does not exist an integer $1 \leq c < d$ and complex numbers $\beta_1, \ldots, \beta_c$ such that*

$$f(n+c) + \beta_1 f(n+c-1) + \cdots + \beta_c f(n) = 0$$

for all $n \geq 0$. In other words, equation (4.2) is the homogeneous linear recurrence with constant coefficients of least degree satisfied by $f(n)$.

iii. $\deg P_i(n) = d_i - 1$ *for* $1 \leq i \leq k$.

Next we consider the coefficients of *any* rational function $P(x)/Q(x)$, where $P, Q \in \mathbb{C}[x]$, not just those with $\deg P < \deg Q$. Write $\mathbb{C}^* = \mathbb{C} - \{0\}$.

4.2.2 Proposition. *Let* $f: \mathbb{N} \to \mathbb{C}$ *and suppose that* $\sum_{n \geq 0} f(n)x^n = P(x)/Q(x)$, *where* $P, Q \in \mathbb{C}[x]$. *Then there is a unique finite set* $E_f \subset \mathbb{N}$ *(called the* exceptional set *of* f*) and a unique function* $f_1: E_f \to \mathbb{C}^*$ *such that the function* $g: \mathbb{N} \to \mathbb{C}$ *defined by*

$$g(n) = \begin{cases} f(n), & \text{if } n \notin E_f, \\ f(n) + f_1(n), & \text{if } n \in E_f. \end{cases}$$

satisfies $\sum_{n \geq 0} g(n)x^n = R(x)/Q(x)$ *where* $R \in \mathbb{C}[x]$ *and* $\deg R < \deg Q$. *Moreover, assuming* $E_f \neq \emptyset$ *(i.e.,* $\deg P \geq \deg Q$*), define* $m(f) = \max\{i : i \in E_f\}$. *Then*

i. $m(f) = \deg P - \deg Q$.
ii. $m(f)$ *is the largest integer n for which equation (4.2) fails to hold.*
iii. Writing $Q(x) = \prod_{i=1}^{k}(1 - \gamma_i x)^{d_i}$ *as in Theorem 4.1.1(iii), there are unique polynomials* $P_1, \ldots, P_k$ *for which equation (4.3) holds for all n sufficiently large. Then* $m(f)$ *is the largest integer n for which (4.3) fails.*

Proof. By the division algorithm for polynomials in one variable, there are unique polynomials $L(x)$ and $R(x)$ with $\deg R < \deg Q$ such that

$$\frac{P(x)}{Q(x)} = L(x) + \frac{R(x)}{Q(x)}. \tag{4.7}$$

Thus, we must define E_f, $g(n)$, and $f_1(n)$ by

$$\sum_{n \geq 0} g(n)x^n = \frac{R(x)}{Q(x)}, \quad E_f = \{i : [x^i]L(x) \neq 0\}, \quad \sum_{n \in E_f} f_1(n)x^n = -L(x).$$

The rest of the proof is then immediate. $\square$

We next describe a fast method for computing the coefficients of a rational function $P(x)/Q(x) = \sum_{n \geq 0} f(n)x^n$ by inspection. Suppose (without loss of generality) that $Q(x) = 1 + \alpha_1 x + \cdots + \alpha_d x^d$, and let $P(x) = \beta_0 + \beta_1 x + \cdots + \beta_e x^e$ (possibly $e \geq d$). Equating coefficients of x^n in

$$Q(x) \sum_{n \geq 0} f(n)x^n = P(x)$$

yields

$$f(n) = -\alpha_1 f(n-1) - \cdots - \alpha_d f(n-d) + \beta_n. \tag{4.8}$$

where we set $f(k) = 0$ for $k < 0$ and $\beta_k = 0$ for $k > e$. The recurrence (4.8) can easily be implemented by inspection (at least for reasonably small values of d and α_i). For instance, let

$$\frac{P(x)}{Q(x)} = \frac{1 - 2x + 4x^2 - x^3}{1 - 3x + 3x^3 - x^3}.$$

Then

$$f(0) = \beta_0 = 1,$$

$$f(1) = 3f(0) + \beta_1 = 3 - 2 = 1,$$

$$f(2) = 3f(1) - 3f(0) + \beta_2 = 3 - 3 + 4 = 4,$$

$$f(3) = 3f(2) - 3f(1) + f(0) + \beta_3 = 12 - 3 + 1 - 1 = 9,$$

$$f(4) = 3f(3) - 3f(2) + f(1) = 27 - 12 + 1 = 16,$$

$$f(5) = 3f(4) - 3f(3) + f(2) = 48 - 27 + 4 = 25,$$

and so on. The sequence of values $1, 1, 4, 9, 16, 25, \ldots$ looks suspiciously like $f(n) = n^2$, except for $f(0) = 1$. Indeed, the exceptional set $E_f = \{0\}$, and

$$\frac{P(x)}{Q(x)} = 1 + \frac{x + x^2}{(1-x)^3} = 1 + \sum_{n \geq 0} n^2 x^n.$$

We will discuss in Section 4.3 the situation when $f(n)$ is a polynomial, and in particular the case $f(n) = n^k$.

Proposition 4.2.2(i) explains the significance of the number $\deg P - \deg Q$ when $\deg P \geq \deg Q$. What about the case $\deg P < \deg Q$? This is best explained in the context of a kind of duality theorem. If $\sum_{n \geq 0} f(n)x^n = P(x)/Q(x)$ with $\deg P < \deg Q$, then the formulas (4.2) and (4.3) are valid. Either of them may be used to extend the domain of f to *negative* integers. In (4.2) we can just run the recurrence backwards (since by assumption $\alpha_d \neq 0$) by successively substituting $n = -1, -2, \ldots$. It follows that there is a *unique* extension of f to all of $\mathbb{Z}$ satisfying (4.2) for all $n \in \mathbb{Z}$. In (4.3) we can let n be a negative integer on the right-hand side. It is easy to see that these two extensions of f to $\mathbb{Z}$ agree.

4.2.3 Proposition. *Let $d \in \mathbb{N}$ and $\alpha_1, \ldots, \alpha_d \in \mathbb{C}$ with $\alpha_d \neq 0$. Suppose that $f : \mathbb{Z} \to \mathbb{C}$ satisfies*

$$f(n+d) + \alpha_1 f(n+d-1) + \cdots + \alpha_d f(n) = 0 \quad \text{for all } n \in \mathbb{Z}.$$

Thus, $\sum_{n \geq 0} f(n) = F(x)$ is a rational function, as is $\sum_{n \geq 1} f(-n)x^n = \overline{F}(x)$. We then have

$$\overline{F}(x) = -F(1/x),$$

as rational functions.

NOTE. It is important to realize that Proposition 4.2.3 is a statement about the equality of *rational functions*, not power series. For instance, suppose that $f(n) = 1$

for all $n \in \mathbb{Z}$. Then $F(x) = \sum_{n \geq 0} x^n = 1/(1-x)$ and $\overline{F}(x) = \sum_{n \geq 1} x^n = x/(1-x)$. Then as rational functions we have

$$-F(1/x) = -\frac{1}{1 - 1/x} = -\frac{x}{x-1} = \frac{x}{1-x} = \overline{F}(x).$$

Proof. Let $F(x) = P(x)/Q(x)$, where $Q(x) = 1 + \alpha_1 x + \cdots + \alpha_d x^d$. Let $\mathcal{L}$ denote the complex vector space of all formal Laurent series $\sum_{n \in \mathbb{Z}} a_n x^n$, $a_n \in \mathbb{C}$. Although two such Laurent series cannot be formally multiplied in a meaningful way, we *can* multiply such a Laurent series by the polynomial $Q(x)$. The map $\mathcal{L} \overset{Q}{\to} \mathcal{L}$ given by multiplication by $Q(x)$ is a linear transformation. The hypothesis on f implies that

$$Q(x) \sum_{n \in \mathbb{Z}} f(n) x^n = 0.$$

Since multiplication by $Q(x)$ is linear, we have

$$Q(x) \sum_{n \geq 1} f(-n) x^{-n} = -Q(x) \sum_{n \geq 0} f(n) x^n = -P(x).$$

Substituting $1/x$ for x yields

$$\sum_{n \geq 1} f(-n) x^n = -\frac{P(1/x)}{Q(1/x)} = -F(1/x),$$

as desired. (The reader suspicious of this argument should check carefully that all steps are formally justified. Note in particular that the vector space $\mathcal{L}$ contains the two rings $\mathbb{C}[[x]]$ and $\left\{ \sum_{n \leq n_0} a_n x^n \right\}$, whose intersection is $\mathbb{C}[x]$.) $\qquad\square$

Proposition 4.2.3 allows us to explain the significance of certain properties of the rational function $P(x)/Q(x)$.

4.2.4 Corollary. *Let* $d \in \mathbb{P}$ *and* $\alpha_1, \ldots, \alpha_d \in \mathbb{C}$ *with* $\alpha_d \neq 0$. *Suppose that* $f : \mathbb{Z} \to \mathbb{C}$ *satisfies*

$$f(n+d) + \alpha_1 f(n+d-1) + \cdots + \alpha_d f(n) = 0$$

for all $n \in \mathbb{Z}$. *Thus,* $\sum_{n \geq 0} f(n) x^n = P(x)/Q(x)$ *where* $Q(x) = 1 + \alpha_1 x + \cdots + \alpha_d x^d$ *and* $\deg P < \deg Q$. *Say* $P(x) = \beta_0 + \beta_1 x + \cdots + \beta_{d-1} x^{d-1}$.

i. $\min\{n \in \mathbb{N} : f(n) \neq 0\} = \min\{j \in \mathbb{N} : \beta_j \neq 0\}$.
   Moreover, if r *denotes the value of the above minimum, then* $f(r) = \beta_r$.
ii. $\min\{n \in \mathbb{P} : f(-n) \neq 0\} = \min\{j \in \mathbb{P} : \beta_{d-j} \neq 0\} = \deg Q - \deg P$.
   Moreover, if s *denotes the value of the above minimum, then* $f(-s) = -\alpha_d^{-1} \beta_s$.
iii. *Let* $F(x) = P(x)/Q(x)$, *and let* r *and* s *be as above. Then* $F(x) = \pm x^{r-s} F(1/x)$ *if and only if* $f(n) = \mp f(-n+r-s)$ *for all* $n \in \mathbb{Z}$.

Proof. If

$$P(x) = \beta^r x^r + \beta_{r+1} x^{r+1} + \cdots + \beta_{d-1} x^{d-1},$$

then $P(x)/Q(x) = \beta_r x^r + \cdots$, so (i) is clear. If

$$P(x) = \beta_{d-s} x^{d-s} + \beta_{d-s-1} x^{d-s-1} + \cdots + \beta_0,$$

then by Proposition 4.2.3 we have

$$\sum_{n \geq 1} f(-n)x^n = -\frac{P(1/x)}{Q(1/x)} = -\frac{\beta_{d-s} x^{-(d-s)} + \cdots + \beta_0}{1 + \alpha_1 x^{-1} + \cdots + \alpha_d x^{-d}}$$

$$= \frac{-\alpha_d^{-1}(\beta_{d-s} x^s + \cdots + \beta_0 x^d)}{1 + \alpha_{d-1} \alpha_d^{-1} x + \cdots + \alpha_d^{-1} x^d} = -\alpha_d^{-1} \beta_{d-s} x^s + \cdots,$$

from which (ii) follows. Finally, (iii) is immediate from Proposition 4.2.3. $\qquad \square$

Corollary 4.2.4(ii) answers the question raised earlier regarding the significance of $\deg Q - \deg P$ when $\deg Q > \deg P$. A situation to which this result applies is Corollary 3.15.13.

It is clear that if $F(x)$ and $G(x)$ are rational power series belonging to $\mathbb{C}[[x]]$, then $\alpha F(x) + \beta G(x)$ $(\alpha, \beta, \in \mathbb{C})$ and $F(x)G(x)$ are also rational. Moreover, if $F(x)/G(x) \in \mathbb{C}[[x]]$, then $F(x)/G(x)$ is rational. Perhaps somewhat less obvious is the closure of rational power series under the operation of *Hadamard product*. The Hadamard product $F * G$ of the power series $F(x) = \sum_{n \geq 0} f(n)x^n$ and $G(x) = \sum_{n \geq 0} g(n)x^n$ is defined by

$$F(x) * G(x) = \sum_{n \geq 0} f(n)g(n)x^n.$$

4.2.5 Proposition. *If $F(x)$ and $G(x)$ are rational power series, then so is the Hadamard product $F * G$.*

Proof. By Theorem 4.1.1 and Proposition 4.2.2, the power series $H(x) = \sum_{n \geq 0} h(n)x^n$ is rational if and only if $h(n) = \sum_{i=1}^m R_i(n)\zeta_i^n$, where $\zeta_1, \ldots, \zeta_m$ are fixed nonzero complex numbers, and $R_1, \ldots, R_m$ are fixed polynomials in n. Thus, if $F(x) = \sum_{n \geq 0} f(n)x^n$ and $G(x) = \sum_{n \geq 0} g(n)x^n$, then $f(n) = \sum_{i=1}^k P_i(n)\gamma_i^n$ and $g(n) = \sum_{j=1}^l Q_i(n)\delta_j^n$ for n large. Then

$$f(n)g(n) = \sum_{i,j} P_i(n)Q_j(n)(\gamma_i \delta_j)^n$$

for n large, so $F * G$ is rational. $\qquad \square$

4.3 Polynomials

An important special class of functions $f : \mathbb{N} \to \mathbb{C}$ whose generating function $\sum_{n \geq 0} f(n)x^n$ is rational are the *polynomials*. Indeed, the following result is an immediate corollary of Theorem 4.1.1.

4.3.1 Corollary. *Let* $f: \mathbb{N} \to \mathbb{C}$, *and let* $d \in \mathbb{N}$. *The following three conditions are equivalent:*

i. $\displaystyle\sum_{n \geq 0} f(n)x^n = \frac{P(x)}{(1-x)^{d+1}}$, *where* $P(x) \in \mathbb{C}[x]$ *and* $\deg P \leq d$.

ii. *For all* $n \geq 0$,

$$\sum_{i=0}^{d+1} (-1)^{d+1-i} \binom{d+1}{i} f(n+i) = 0.$$

In other words, $\Delta^{d+1} f(n) = 0$.

iii. $f(n)$ *is a polynomial function of* n *of degree at most* d. *(Moreover,* $f(n)$ *has degree exactly* d *if and only if* $P(1) \neq 0$.)

Note that the equivalence of (ii) and (iii) is just Proposition 1.9.2(a). Also note that when $P(1) \neq 0$, so that $\deg f = d$, then the leading coefficient of $f(n)$ is $P(1)/d!$. This may be seen, for example, by considering the coefficient of $(1 - x)^{d+1}$ is the Laurent expansion of $\sum_{n \geq 0} f(n)x^n$ about $x = 1$.

The set of all polyonomials $f: \mathbb{N} \to \mathbb{C}$ (or $f: \mathbb{Z} \to \mathbb{C}$) of degree at most d is a vector space P_d of dimension $d + 1$ over $\mathbb{C}$. This vector space has many natural choices of a basis. A description of these bases and the transition matrices among them would occupy a book in itself. Here we list what are perhaps the four most important bases, with a brief discussion of their significance. Note that any set of polynomials $p_0(n), p_1(n), \ldots, p_d(n)$ with $\deg p_i = i$ is a basis for P_d [why?].

a. n^i, $0 \leq i \leq d$. When a polynomial $f(n)$ is expanded in terms of this basis, then we of course obtain the usual coefficients of $f(n)$.

b. $\binom{n}{i}$, $0 \leq i \leq d$. (Alternatively, we could use $(n)_i = i!\binom{n}{i}$.) By Proposition 1.9.2(b) we have the expansion $f(n) = \sum_{i=0}^{d} (\Delta^i f(0))\binom{n}{i}$, the discrete analogue of the Taylor series (still assuming that $f(n) \in P_d$) $f(x) = \sum_{i=0}^{d} D^i f(0)\frac{x^i}{i!}$, where $Df(t) = \frac{d}{dt}f(t)$. By Proposition 1.9.2(c), the transition matrices between the bases n^i and $\binom{n}{i}$ are essentially the Stirling numbers of the first and second kind, that is

$$n^j = \sum_{i=0}^{j} i!S(j,i)\binom{n}{i} = \sum_{i=0}^{j} S(j,i)(n)_i,$$

$$\binom{n}{j} = \frac{1}{j!}\sum_{i=0}^{j} s(j,i)n^i, \text{ or } (n)_j = \sum_{i=0}^{j} s(j,i)n^i.$$

c. $\left(\!\binom{n}{i}\!\right) = (-1)^i \binom{-n}{i}$, $0 \leq i \leq d$. (Alternatively, we could use the rising factorial $n(n+1)\cdots(n+i-1) = i! \left(\!\binom{n}{i}\!\right)$.) We thus have

$$f(n) = \sum_{i=0}^{d} (-1)^i (\Delta^i f(-n))_{n=0} \left(\!\binom{n}{i}\!\right).$$

Equivalently, if one forms the difference table of $f(n)$ then the coefficients of $\left(\binom{n}{i}\right)$ in the expansion $f(n) = \sum c_i \left(\binom{n}{i}\right)$ are the elements of the diagonal beginning with $f(0)$ and moving southwest. For instance, if $f(n) = n^3 + n + 1$ then we get the difference table

$$
\begin{array}{cccc}
-29 & & -9 & & -1 & & 1 = f(0) \\
& 20 & & 8 & & 2 \\
& & -12 & & -6 \\
& & & 6,
\end{array}
$$

so $n^3 + n + 1 = 1 + 2\left(\binom{n}{1}\right) - 6\left(\binom{n}{2}\right) + 6\left(\binom{n}{3}\right)$. The transition matrices with n^i and with $\binom{n}{i}$ are given by

$$
n^j = \sum_{i=0}^{j} (-1)^{j-i} i! S(j,i) \left(\binom{n}{i}\right),
$$

$$
\left(\binom{n}{j}\right) = \frac{1}{j!} \sum_{i=0}^{j} c(j,i) n^i, \quad \text{where } c(j,i) = (-1)^{j-i} s(j,i),
$$

$$
\binom{n}{j} = \sum_{i=1}^{j} (-1)^{j-i} \binom{j-1}{i-1} \left(\binom{n}{i}\right),
$$

$$
\left(\binom{n}{j}\right) = \sum_{i=1}^{j} \binom{j-1}{i-1} \binom{n}{i}.
$$

d. $\binom{n+d-i}{d}$, $0 \le i \le d$. There are (at least) two quick ways to see that this is a basis for P_d. Given that $f(n) = \sum_{i=0}^{d} c_i \binom{n+d-i}{d}$, set $n = 0$ to obtain c_0 uniquely. Then set $n = 1$ to obtain c_1 uniquely, and so on. Thus, the $d+1$ polynomials $\binom{n+d-i}{d}$ are linearly independent and therefore form a basis for P_d. Alternatively, observe that

$$
\sum_{n \ge 0} \binom{n+d-i}{d} x^n = \frac{x^i}{(1-x)^{d+1}}.
$$

Hence, the statement that the polynomials $\binom{n+d-i}{d}$ form a basis for P_d is equivalent (in view of Corollary 4.3.1) to the obvious fact that the rational functions $x^i/(1-x)^{d+1}$, $0 \le i \le d$, form a basis for all rational functions $P(x)/(1-x)^{d+1}$, where $P(x)$ is a polynomial of degree at most d. If

$$
\sum_{n \ge 0} f(n) x^n = \frac{w_0 + w_1 x + \cdots + w_d x^d}{(1-x)^{d+1}},
$$

then the numbers $w_0, w_1, \ldots, w_d$ are called the f-*Eulerian numbers*, and the polynomial $P(x) = w_0 + w_1 x + \cdots + w_d x^d$ is called the f-*Eulerian polynomial*. If in particular $f(n) = n^d$, then it follows from Proposition 1.4.4 that the f-Eulerian numbers are simply the *Eulerian numbers* $A(d,i)$, whereas the

f-Eulerian polynomial is the *Eulerian polynomial* $A_d(x)$. Just as for ordinary Eulerian numbers, the f-Eulerian numbers frequently have combinatorial significance. A salient example are order polynomials $\Omega_{P,\omega}(m)$ of labeled posets (Theorem 3.15.8). We could discuss the transition matrices between the basis $\binom{n+d-i}{d}$ and the other three bases considered earlier, but this is not a particularly fruitful endeavor and will be omitted.

4.4 Quasipolynomials

A *quasipolynomial* (known by many other names, such as *pseudopolynomial* and *polynomial on residue classes* (PORC)) of *degree d* is a function $f : \mathbb{N} \to \mathbb{C}$ (or $f : \mathbb{Z} \to \mathbb{C}$) of the form

$$f(n) = c_d(n)n^d + c_{d-1}(n)n^{d-1} + \cdots + c_0(n),$$

where each $c_i(n)$ is a *periodic function* (with integer period), and where $c_d(n)$ is not identically zero. Equivalently, f is a quasipolynomial if there exists an integer $N > 0$ (namely, a common period of $c_0, c_1, \ldots, c_d$) and polynomials $f_0, f_1, \ldots, f_{N-1}$ such that

$$f(n) = f_i(n) \text{ if } n \equiv i \pmod{N}.$$

The integer N (which is not unique) will be called a *quasiperiod* of f.

4.4.1 Proposition. *The following conditions on a function $f : \mathbb{N} \to \mathbb{C}$ and integer $N > 0$ are equivalent.*

i. *f is a quasipolynomial of quasiperiod N.*

ii. $\sum_{n \geq 0} f(n)x^n = \dfrac{P(x)}{Q(x)}$, *where $P(x), Q(x) \in \mathbb{C}[x]$, every zero α of $Q(x)$ satisfies $\alpha^N = 1$ (provided $P(x)/Q(x)$ has been reduced to lowest terms), and $\deg P < \deg Q$.*

iii. *For all $n \geq 0$,*

$$f(n) = \sum_{i=1}^{k} P_i(n)\gamma_i^n, \tag{4.9}$$

where each P_i is a polynomial function of n and each γ_i satisfies $\gamma_i^N = 1$.

Moreover, the degree of $P_i(n)$ in equation (4.9) is equal to one less than the multiplicity of the root γ_i^{-1} in $Q(x)$, provided $P(x)/Q(x)$ has been reduced to lowest terms.

Proof. The proof is a simple consequence of Theorem 4.1.1; the details are omitted. $\square$

4.4.2 Example. Let $\overline{p}_k(n)$ denote the number of partitions of n into at most k parts. Thus from equation (1.76), we have

$$\sum_{n\geq 0}\overline{p}_k(n)x^n = \frac{1}{(1-x)(1-x^2)\cdots(1-x^k)}.$$

Hence $\overline{p}_k(n)$ is a quasipolynomial. Its minimum quasiperiod in equal to the least common multiple of $1,2,\ldots,k$, and its degree is $k-1$. Much more precise statements are possible; consider for instance the case $k=6$. Then

$$\overline{p}_6(n) = c_5 n^5 + c_4 n^4 + c_3 n^3 + c_2(n)n^2 + c_1(n)n + c_0(n),$$

where $c_3, c_4, c_5 \in \mathbb{Q}$ (and in fact $c_5 = 1/5!6!$, as may be seen by considering the coefficient of $(1-x)^{-6}$ in the Laurent expansion of $1/(1-x)(1-x^2)\cdots(1-x^6)$ about $x = 1$), $c_2(n)$ has period 2, $c_1(n)$ has period 6, and $c_0(n)$ has period 60. (These need not be the minimum periods.) Moreover, $c_1(n)$ is in fact the sum of periodic functions of periods 2 and 3. The reader should be able to read these facts off from the generating function $1/(1-x)(1-x^2)\cdots(1-x^6)$.

The case $k=3$ is particularly elegant. Let us write $[a_0, a_1, \ldots, a_{p-1}]_p$ for the periodic function $c(n)$ of period p satisfying $c(n) = a_i$ if $n \equiv i \pmod{p}$. A rather tedious computation yields

$$\overline{p}_3(n) = \frac{1}{12}n^2 + \frac{1}{2}n + \left[1, \frac{5}{12}, \frac{2}{3}, \frac{3}{4}, \frac{2}{3}, \frac{5}{12}\right]_6.$$

It is essentially an "accident" that this expression for $\overline{p}_3(n)$ can be written in the concise form $\| \frac{1}{12}(n+3)^2 \|$, where $\| t \|$ denotes the nearest integer to the real number t, that is, $\| t \| = \lfloor t + \frac{1}{2} \rfloor$.

4.5 Linear Homogeneous Diophantine Equations

The remainder of this chapter will be devoted to two general areas in which rational generating functions play a prominent role. Another such area is the theory of (P,ω)-partitions developed in Section 3.15.

Let Φ be an $r \times m$ matrix with integer entries (or $\mathbb{Z}$-*matrix*). Many combinatorial problems turn out to be equivalent to finding all (column) vectors $\alpha \in \mathbb{N}^m$ satisfying

$$\Phi\alpha = 0, \tag{4.10}$$

where $0 = (0,0,\ldots,0) \in \mathbb{N}^r$. (For convenience of notation we will write column vectors as row vectors.) Equation (4.10) is equivalent to a system of r homogeneous linear equations with integer coefficients in the m unknowns $\alpha = (\alpha_1,\ldots,\alpha_m)$. Note that if we were searching for solutions $\alpha \in \mathbb{Z}^m$ (rather than $\alpha \in \mathbb{N}^m$), then there would be little problem. The solutions in $\mathbb{Z}^m$ (or $\mathbb{Z}$-*solutions*) form a subgroup G of $\mathbb{Z}^m$ and hence by the theory of finitely generated abelian groups, G is a finitely generated free abelian group. The minimal number of generators (or *rank*) of G is equal to the nullity of the matrix Φ, and there are well-known algorithms for finding the generators of G explicitly. The situation for solutions in $\mathbb{N}^m$ (or $\mathbb{N}$-*solutions*) is not so clear. The set of solutions forms not a group but rather a (commutative) *monoid* (semigroup with identity) $E = E_\Phi$. It certainly is not the case that E is a free commutative monoid; that is, there exist $\alpha_1, \ldots, \alpha_s \in E$ such that every $\alpha \in E$ can be written uniquely as $\sum_{i=1}^s a_i \alpha_i$, where $a_i \in \mathbb{N}$. For instance, take $\Phi = [1, 1, -1, -1]$. Then in E there is the nontrivial relation $(1, 0, 1, 0) + (0, 1, 0, 1) = (1, 0, 0, 1) + (0, 1, 1, 0)$.

Without loss of generality, we may assume that the rows of Φ are linearly independent; that is, rank $\Phi = r$. If now $E \cap \mathbb{P}^m = \emptyset$ (i.e., equation (4.10) has no $\mathbb{P}$-solution), then for some $i \in [m]$, every $(\alpha_1, \ldots, \alpha_m) \in E$ satisfies $\alpha_i = 0$. It costs nothing to ignore this entry α_i. Hence, we may assume from now on that $E \cap \mathbb{P}^m \neq \emptyset$. We then call E a *positive* monoid.

We will analyze the structure of the monoid E to the extent of being able to write down a formula for the generating function

$$E(x) = E(x_1, \ldots, x_m) = \sum_{\alpha \in E} x^\alpha, \tag{4.11}$$

where if $\alpha = \{\alpha_1, \ldots, \alpha_m\}$ then $x^\alpha = x_1^{\alpha_1} \cdots x_m^{\alpha_m}$. We will also consider the closely related generating function

$$\overline{E}(x) = \sum_{\alpha \in \overline{E}} x^\alpha, \tag{4.12}$$

where $\overline{E} = E \cap \mathbb{P}^m$. Since we are assuming that $\overline{E} \neq \emptyset$, it follows that $\overline{E}(x) \neq 0$. In general throughout this section, if G is any subset of $\mathbb{N}^m$, then we write

$$G(x) = \sum_{\alpha \in G} x^\alpha.$$

First, let us note that there is no real gain in generality by also allowing *inequalities* of the form $\Psi \alpha \geq 0$ for some $s \times m$ $\mathbb{Z}$-matrix Ψ. This is because we can introduce slack variables $\gamma = (\gamma_1, \ldots, \gamma_s)$ and replace the inequality $\Psi \alpha \geq 0$ by the equality $\Psi \alpha - \gamma = 0$. An $\mathbb{N}$-solution to the latter equality is equivalent to an $\mathbb{N}$-solution to the original inequality. In particular, the theory of P-partitions (where the labeling ω is natural) of Section 3.15 can be subsumed by the general theory of $\mathbb{N}$-solutions to equation (4.10), though P-partitions have many additional special features. Specifically, introduce variables α_t for all $t \in P$ and α_{st} for all pairs $s < t$ (or in fact just for $s \lessdot t$). Then an $\mathbb{N}$-solution α to the system

$$\alpha_s - \alpha_t - \alpha_{st} = 0, \quad \text{for all } s < t \text{ in } P \text{ (or just for all } s \lessdot t) \tag{4.13}$$

is equivalent to the P-partition $\sigma : P \to \mathbb{N}$ given by $\sigma(t) = \alpha_t$. Moreover, a $\mathbb{P}$-solution to equation (4.13) is equivalent to a strict P-partition τ with positive parts. If we merely subtract one from each part, then we obtain an arbitrary strict P-partition. Hence by Theorem 3.15.10, the generating functions $E(x)$ and

$\overline{E}(x)$ of (4.11) and (4.12), for the system (4.13), are related by

$$\overline{E}(x) = (-1)^p E(1/x), \tag{4.14}$$

where $1/x$ denotes the substitution of $1/x_i$ for x_i in the rational function $E(x)$. This suggests a reciprocity theorem for the general case (4.10), and one of our goals will be to prove such a theorem. (We do not even know yet whether $E(x)$ and $\overline{E}(x)$ are rational functions; otherwise, equation (4.14) makes no sense.) The theory of P-partitions provides clues about obtaining a formula for $E(x)$. Ideally, we would like to partition in an explicit and canonical way the monoid E into finitely many easily understood parts. Unfortunately, we will have to settle for somewhat less. We will express E as a union of nicely behaved parts (called "simplicial monoids"), but these parts will not be disjoint, and it will be necessary to analyze how they intersect. Moreover, the simplicial monoids themselves will be obtained by a rather arbitrary construction (not nearly as elegant as associating a P-partition to a unique $w \in \mathcal{L}(P)$), and it will require some work to analyze the simplicial monoids themselves. But the reward for all this effort will be an extremely general theory with a host of interesting and significant applications.

Although the theory we are about to derive can be developed purely algebraically, it is more convenient and intuitive to proceed geometrically. To this end we will briefly review some of the basic theory of convex polyhedral cones. A *linear half-space* $\mathcal{H}$ of $\mathbb{R}^m$ is a subset of $\mathbb{R}^m$ of the form $\mathcal{H} = \{v : v \cdot w \geq 0\}$ for some fixed nonzero vector $w \in \mathbb{R}^m$. A *convex polyhedral cone* $\mathcal{C}$ in $\mathbb{R}^m$ is defined to be the intersection of finitely many half-spaces. (Some authorities would require that $\mathcal{C}$ contain a vector $v \neq 0$.) We say that $\mathcal{C}$ is *pointed* if it doesn't contain a line; or equivalently, whenever $0 \neq v \in \mathcal{C}$ then $-v \notin \mathcal{C}$. A *supporting hyperplane* $\mathcal{H}$ of $\mathcal{C}$ is a linear hyperplane of which $\mathcal{C}$ lies entirely on one side. In other words, $\mathcal{H}$ divides $\mathbb{R}^m$ into two closed half-spaces $\mathcal{H}^+$ and $\mathcal{H}^-$ (whose intersection is $\mathcal{H}$), such that either $\mathcal{C} \subseteq \mathcal{H}^+$ or $\mathcal{C} \subseteq \mathcal{H}^-$. A *face* of $\mathcal{C}$ is a subset $\mathcal{C} \cap \mathcal{H}$ of $\mathcal{C}$, where $\mathcal{H}$ is a supporting hyperplane. Every face $\mathcal{F}$ of $\mathcal{C}$ is itself a convex polyhedral cone, including the degenerate face $\{0\}$. The *dimension* of $\mathcal{F}$, denoted $\dim \mathcal{F}$, is the dimension of the subspace of $\mathbb{R}^m$ spanned by $\mathcal{F}$. If $\dim \mathcal{F} = i$, then $\mathcal{F}$ is called an *i-face*. In particular, $\{0\}$ and $\mathcal{C}$ are faces of $\mathcal{C}$, called *improper*, and $\dim\{0\} = 0$. A 1-face is called an *extreme ray*, and if $\dim \mathcal{C} = d$ then a $(d-1)$-face is called a *facet*. We will assume the standard result that a pointed polyhedral cone $\mathcal{C}$ has only finitely many extreme rays, and that $\mathcal{C}$ is the convex hull of its extreme rays. A *simplicial cone* σ is an e-dimensional pointed convex polyhedral cone with e extreme rays (the minimum possible). Equivalently, σ is simplicial if there exist *linearly independent* vectors $\beta_1, \ldots, \beta_e$ for which $\sigma = \{a_1\beta_1 + \cdots + a_e\beta_e : a_i \in \mathbb{R}_+\}$. A *triangulation* of $\mathcal{C}$ consists of a finite collection $\Gamma = \{\sigma_1, \ldots, \sigma_t\}$ of simplicial cones satisfying: (i) $\cup \sigma_i = \mathcal{C}$, (ii) if $\sigma \in \Gamma$, then every face of σ is in Γ, and (iii) $\sigma_i \cap \sigma_j$ is a common face of σ_i and σ_j. An element of Γ is called a *face* of Γ.

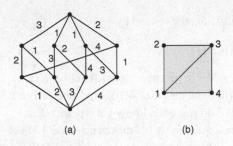

Figure 4.2 An edge-labeled face lattice
and corresponding triangulation.

(a)   (b)

4.5.1 Lemma. *A pointed polyhedral cone C possesses a triangulation Γ whose 1-faces (= 1-dimensional faces of Γ) are the extreme rays of C.*

Proof. Let L denote the lattice of faces of C, so the face $\{0\}$ is the unique minimal element of L. The extreme rays of C are the atoms of L. Choose an ordering $\mathcal{R}_1,\ldots,\mathcal{R}_m$ of the extreme rays. Given an edge $e = uv$ of the Hasse diagram of L (so $u \lessdot v$ in L), define $\lambda(e)$ to be the least integer i for which $v = u \vee \mathcal{R}_i$ in L. Let $\mathfrak{m}$ be a maximal chain of L, say $\hat{0} = t_0 \lessdot t_1 \lessdot \cdots \lessdot t_d = C$, for which $\lambda(t_0,t_1) > \lambda(t_1,t_2) > \cdots > \lambda(t_{d-1},t_d)$. Suppose that $\lambda(t_{i-1},t_i) = j_i$. Let $\Delta_{\mathfrak{m}}$ be the convex hull of the extreme rays $\mathcal{R}_{j_1},\ldots,\mathcal{R}_{j_d}$. We leave it to the reader to check that the $\Delta_{\mathfrak{m}}$'s are the facets of a triangulation Γ whose 1-faces are the extreme rays of C. (Is the similarity to Example 3.14.5 just a coincidence?)   □

As an illustration of the preceding proof, consider a 3-dimensional cone C whose cross-section is a quadrilateral Q. Let $\mathcal{R}_1,\ldots,\mathcal{R}_4$ be the extreme rays of C in cyclic order. Figure 4.2(a) shows the edge-labeled face lattice of C (or the face lattice of Q). There are two decreasing chains, labeled 321 and 431. The corresponding triangulation of Q (a cross-section of the triangulation Γ of C) is shown in Figure 4.2(b).

The *boundary* of C, denoted ∂C, is the union of all facets of C. (This definition coincides with the usual topological notion of boundary.) If Γ is a triangulation of C, define the *boundary* $\partial \Gamma = \{\sigma \in \Gamma : \sigma \subseteq \partial C\}$, and define the *interior* $\Gamma^\circ = \Gamma - \partial \Gamma$.

4.5.2 Lemma. *Let Γ be any triangulation of C. Let $\widehat{\Gamma}$ denote the poset (actually a lattice) of elements of Γ, ordered by inclusion, with a $\hat{1}$ adjoined. Let μ denote the Möbius function of $\widehat{\Gamma}$. Then $\widehat{\Gamma}$ is graded of rank $d + 1$, where $d = \dim C$, and*

$$\mu(\sigma,\tau) = \begin{cases} (-1)^{\dim\tau - \dim\sigma}, & \text{if } \sigma \leq \tau < \hat{1}, \\ (-1)^{d - \dim\sigma + 1}, & \text{if } \sigma \in \Gamma^\circ \text{ and } \tau = \hat{1}, \\ 0, & \text{if } \sigma \in \partial\Gamma \text{ and } \tau = \hat{1}. \end{cases}$$

Proof. This result is a special case of Proposition 3.8.9.   □

Let us now return to the system of equations (4.10). Let C denote the set of solutions $\boldsymbol{\alpha}$ in nonnegative *real* numbers. Then C is a pointed convex polyhedral cone. We will always denote $\dim C$ by the letter d. Since we are assuming that rank $\Phi = r$ and that E is positive, it follows that $d = m - r$ [why?]. Although we

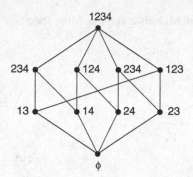

Figure 4.3 A support lattice $L(E)$.

don't require it here, it is natural to describe the faces of $\mathcal{C}$ directly in terms of E. We will simply state the relevant facts without proof. If $\alpha = (\alpha_1, \ldots, \alpha_m) \in \mathbb{R}^m$, then define the *support* of α, denoted $\operatorname{supp}\alpha$, by $\operatorname{supp}\alpha = \{i : \alpha_i \neq 0\}$. If X is any subset of $\mathbb{R}^m$, then define

$$\operatorname{supp} X = \bigcup_{\alpha \in X} (\operatorname{supp}\alpha).$$

Let $L(\mathcal{C})$ be the lattice of faces of $\mathcal{C}$, and let $L(E) = \{\operatorname{supp}\alpha : \alpha \in E\}$, ordered by inclusion. Define a map $f : L(\mathcal{C}) \to B_m$ (the boolean algebra on $[m]$) by $f(\mathcal{F}) = \operatorname{supp}\mathcal{F}$. Then f is an isomorphism of $L(\mathcal{C})$ onto $L(E)$.

4.5.3 Example. Let $\Phi = [1, 1, -1, -1]$. The poset $L(E)$ is given by Figure 4.3. Thus, $\mathcal{C}$ has four extreme rays and four 2-faces. The four extreme rays are the rays from $(0, 0, 0, 0)$ passing through $(1, 0, 1, 0)$, $(1, 0, 0, 1)$, $(0, 1, 1, 0)$, and $(0, 1, 0, 1)$.

Now let Γ be a triangulation of $\mathcal{C}$ whose extreme rays are the extreme rays of $\mathcal{C}$. Such a triangulation exists by Lemma 4.5.1. If $\sigma \in \Gamma$, then let

$$E_\sigma = \sigma \cap \mathbb{N}^m. \tag{4.15}$$

Then each E_σ is a submonoid of E, and $E = \bigcup_{\sigma \in \Gamma} E_\sigma$. Moreover, if we set

$$\overline{E}_\sigma = \{u \in E_\sigma : u \notin E_\tau \text{ for any } \tau \subset \sigma\}, \tag{4.16}$$

then $\overline{E} = \bigcup_{\sigma \in \Gamma} \overline{E}_\sigma$ (disjoint union). This provides the basic decomposition of E and $\overline{E}$ into "nice" subsets, just as Lemma 3.15.3 did for (P, ω)-partitions.

The "triangulation" $\{E_\sigma : \sigma \in \Gamma\}$ of E and $\{\overline{E}_\sigma : \sigma \in \Gamma^\circ\}$ of $\overline{E}$ yield the following result about generating functions.

4.5.4 Lemma. *The generating functions $E(x)$, $\overline{E}(x)$ and $E_\sigma(x)$, $\overline{E}_\sigma(x)$ are related by*

$$E(x) = -\sum_{\sigma \in \Gamma} \mu(\sigma, \hat{1}) E_\sigma(x), \tag{4.17}$$

$$\overline{E}(x) = \sum_{\sigma \in \Gamma^\circ} \overline{E}_\sigma(x). \tag{4.18}$$

Proof. Equation (4.17) follows immediately from Möbius inversion. More specifically, set $\overline{E}_{\hat{1}}(x) = 0$ and define

$$H_\sigma(x) = \sum_{\tau \le \sigma} \overline{E}_\tau(x), \quad \sigma \in \widehat{\Gamma}.$$

Clearly,

$$H_\sigma(x) = E_\sigma(x), \quad \sigma \in \Gamma,$$

$$H_{\hat{1}}(x) = E(x). \tag{4.19}$$

By Möbius inversion,

$$0 = \overline{E}_{\hat{1}}(x) = \sum_{\sigma \le \hat{1}} H_\sigma(x)\mu(\sigma,\hat{1}),$$

so equation (4.17) follows from (4.19).

Equation (4.18) follows immediately from the fact that the union $\overline{E} = \bigcup_{\sigma \in \Gamma^\circ} \overline{E}_\sigma$ is disjoint. $\qquad\square$

4.5.5 Example. Let E be the monoid of Example 4.5.3. Triangulate $\mathcal{C}$ as shown in Figure 4.4, where $\operatorname{supp} a = \{1,3\}$, $\operatorname{supp} b = \{1,4\}$, $\operatorname{supp} c = \{2,4\}$, $\operatorname{supp} d = \{2,3\}$. Then the poset $\widehat{\Gamma}$ is given by Figure 4.5. Note also that $\Gamma^\circ = \{bd, abd, bcd\}$. Lemma 4.5.4 states that

$$E(x) = E_{abd}(x) + E_{bcd}(x) - E_{bd}x$$

$$\overline{E}(x) = \overline{E}_{abd}(x) + \overline{E}_{bcd}(x) + \overline{E}_{bd}(x). \tag{4.20}$$

Figure 4.4 Triangulation of a cross-section of a cone $\mathcal{C}$.

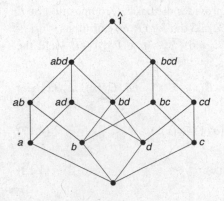

Figure 4.5 Face poset of the triangulation of Figure 4.4.

Our next step is the evaluation of the generating functions $E_\sigma(x)$ and $\overline{E}_\sigma(x)$ appearing in equations (4.17) and (4.18). Let us call a submonoid F of $\mathbb{N}^m$ (or even $\mathbb{Z}^m$) *simplicial* if there exist linearly independent vectors $\alpha_1, \ldots, \alpha_t \in F$ (called *quasigenerators* of F) such that

$$F = \{\gamma \in \mathbb{N}^m : n\gamma = a_1\alpha_1 + \cdots + a_t\alpha_t \text{ for some } n \in \mathbb{P} \text{ and } a_i \in \mathbb{N}\}.$$

The quasigenerators $\alpha_1, \ldots, \alpha_t$ are not quite unique. If $\alpha_1', \ldots, \alpha_s'$ is another set of quasigenerators, then $s = t$ and with suitable choice of subscripts $\alpha_1' = q_i\alpha_i'$ where $q_i \in \mathbb{Q}$, $q_i > 0$. Define the *interior* $\overline{F}$ of F by

$$\overline{F} = \{\alpha \in \mathbb{N}^m : n\alpha = a_1\alpha_1 + \cdots + a_t\alpha_t \text{ for some } n \in \mathbb{P} \text{ and } a_i \in \mathbb{P}\}. \quad (4.21)$$

Note that $\overline{F}$ depends only on F, not on $\alpha_1, \ldots, \alpha_t$.

4.5.6 Lemma. *The submonoids E_σ of E defined by equation (4.15) are simplicial. If $\mathcal{R}_1, \ldots, \mathcal{R}_t$ are the extreme rays of σ, then we can pick as quasigenerators of E_σ any nonzero integer vectors in $\mathcal{R}_1, \ldots, \mathcal{R}_t$ (one vector from each $\mathcal{R}_i$). Moreover, the interior of E_σ, as defined by equation (4.21), coincides with the definition (4.16) of $\overline{E}_\sigma$.*

Proof. This is an easy consequence of the fact that σ is a simplicial cone. The details are left to the reader. $\qquad\square$

If $F \subseteq \mathbb{N}^m$ is a simplicial monoid with quasigenerators $Q = \{\alpha_1, \ldots, \alpha_t\}$, then define two subsets D_F and $\overline{D}_F$ (which depend on the choice of Q) as follows:

$$D_F = \{\gamma \in F : \gamma = a_1\alpha_1 + \cdots + a_t\alpha_t, \, 0 \leq a_i < 1\}, \quad (4.22)$$

$$\overline{D}_F = \{\gamma \in F : \gamma = a_1\alpha_1 + \cdots + a_t\alpha_t, \, 0 < a_i \leq 1\}. \quad (4.23)$$

Note that D_F and $\overline{D}_F$ are finite sets, since they are contained in the intersection of the discrete set F (or $\mathbb{N}^m$) with the bounded set of all vectors $a_1\alpha_1 + \cdots + a_t\alpha_t \in \mathbb{R}^m$ with $0 \leq a_i \leq 1$.

4.5.7 Lemma. *Let $F \subseteq \mathbb{N}^m$ be a simplicial monoid with quasigenerators $\alpha_1, \ldots, \alpha_t$.*

i. Every element $\gamma \in F$ can be written uniquely in the form

$$\gamma = \beta + a_1\alpha_1 + \cdots + a_t\alpha_t,$$

where $\beta \in D_F$ and $a_i \in \mathbb{N}$. Conversely, any such vector belongs to F.
ii. Every element $\gamma \in \overline{F}$ can be written uniquely in the form

$$\gamma = \overline{\beta} + a_1\alpha_1 + \cdots + a_t\alpha_t,$$

where $\overline{\beta} \in \overline{D}_F$ and $a_i \in \mathbb{N}$. Conversely, any such vector belongs to $\overline{F}$.

Proof. i. Let $\gamma \in F$, and write (uniquely) $\gamma = b_1\alpha_1 + \cdots + b_t\alpha_t$, $b_i \in \mathbb{Q}$. Let $a_i = \lfloor b_i \rfloor$, and let $\beta = \gamma - a_1\alpha_1 - \cdots - a_t\alpha_t$. Then $\beta \in F$, and since $0 \le b_i - a_i < 1$, in fact $\beta \in D_F$. If $\gamma = \beta' + a_1'\alpha_1 + \cdots + a_t'\alpha_t$ were another such representation, then $0 = \beta - \beta' = (a_1 - a_1')\alpha_1 + \cdots + (a_t - a_t')\alpha_t$. Each $a_i - a_i' \in \mathbb{Z}$, whereas if $\beta - \beta' = c_1\alpha_1 + \cdots + c_t\alpha_t$, then $-1 < c_i < 1$. Hence $c_i = 0$ and the two representations agree. The converse statement is clear.

ii. The proof is analogous to (i). Instead of $a_i = \lfloor b_i \rfloor$ we take $a_i = \lceil b_i - 1 \rceil$, and so on.

$\square$

4.5.8 Corollary. *The generating functions*

$$F(x) = \sum_{\alpha \in F} x^\alpha, \ \overline{F}(x) = \sum_{\alpha \in \overline{F}} x^\alpha$$

are given by

$$F(x) = \left(\sum_{\beta \in D_F} x^\beta \right) \prod_{i=1}^t \left(1 - x^{\alpha_i} \right)^{-1}, \tag{4.24}$$

$$\overline{F}(x) = \left(\sum_{\beta \in \overline{D}_F} x^\beta \right) \prod_{i=1}^t \left(1 - x^{\alpha_i} \right)^{-1}. \tag{4.25}$$

Proof. Immediate from Lemma 4.5.7. $\square$

NOTE. For the algebraic-minded, we mention the algebraic significance of the sets D_F and $\overline{D}_F$. Let G be the subgroup of $\mathbb{Z}^m$ generated by F, and let H be the subgroup of G generated by the quasigenerators $\alpha_1, \ldots, \alpha_t$. Then each of D_F and $\overline{D}_F$ is a set of coset representatives for H in G. Moreover, D_F (respectively, $\overline{D}_F$) consists of those coset representatives that belong to F (respectively, $\overline{F}$) and are closest to the origin. It follows from general facts about finitely generated abelian groups that the index $[G : H]$ (i.e., the cardinalities of D_F and $\overline{D}_F$) is equal to the greatest common divisor of the determinants of the $t \times t$ submatrices of the matrix whose rows are $\alpha_1, \ldots, \alpha_t$.

4.5.9 Example. Let $\alpha_1 = (1,3,0)$ and $\alpha_2 = (1,0,3)$. The greatest common divisor of the determinants

$$\begin{vmatrix} 1 & 3 \\ 1 & 0 \end{vmatrix}, \begin{vmatrix} 1 & 0 \\ 1 & 3 \end{vmatrix}, \begin{vmatrix} 3 & 0 \\ 0 & 3 \end{vmatrix}$$

is $3 = \#D_F = \#\overline{D}_F$. Indeed, $D_F = \{(0,0,0),(1,1,2),\ (1,2,1)\}$ and $\overline{D}_F = \{(1,1,2),(1,2,1),(2,3,3)\}$. Hence,

$$F(x) = \frac{1 + x_1 x_2 x_3^2 + x_1 x_2^2 x_3}{(1 - x_1 x_2^3)(1 - x_1 x_3^3)},$$

$$\overline{F}(x) = \frac{x_1 x_2 x_3^2 + x_1 x_2^2 x_3 + x_1^2 x_2^3 x_3^3}{(1 - x_1 x_2^3)(1 - x_1 x_3^3)}.$$

We mentioned earlier that if the simplicial monoid $F \subseteq \mathbb{N}^m$ has quasigenerators $\alpha_1,\ldots,\alpha_t$, then any nonzero rational multiples of $\alpha_1,\ldots,\alpha_t$ (provided they lie in $\mathbb{N}^m$) can be taken as the quasigenerators. Thus there is a unique set $\beta_1,\ldots,\beta_t$ of quasigenerators such that any other set has the form $a_1\beta_1,\ldots,a_t\beta_t$, where $a_i \in \mathbb{P}$. We call $\beta_1,\ldots,\beta_t$ the *completely fundamental* elements of F and write $\mathrm{CF}(F) = \{\beta_1,\ldots,\beta_t\}$. Now suppose that E is the monoid of all $\mathbb{N}$-solutions to equation (4.10). Define $\beta \in E$ to be *completely fundamental* if for all $n \in \mathbb{P}$ and $\alpha,\alpha' \in E$ for which $n\beta = \alpha + \alpha'$, we have $\alpha = i\beta$ and $\alpha' = (n-i)\beta$ for some $i \in \mathbb{P}$, $0 \le i \le n$. Denote the set of completely fundamental elements of E by $\mathrm{CF}(E)$.

4.5.10 Proposition. *Let Γ be a triangulation of $\mathcal{C}$ whose extreme rays coincide with those of $\mathcal{C}$, and let $E = \bigcup_{\sigma \in \Gamma} E_\sigma$ be the corresponding decomposition of E into simplicial monoids E_σ. Then the following sets are identical:*

    i. $\mathrm{CF}(E)$,

    ii. $\bigcup_{\sigma \in \Gamma} \mathrm{CF}(E_\sigma)$,

    iii. $\{\beta \in E : \beta$ *lies on an extreme ray of* $\mathcal{C}$*, and* $\beta \ne n\beta'$ *for some* $n \ge 1$*,* $\beta' \in E\}$,

    iv. *The nonzero elements* β *of* E *of minimal support that are not of the form* $n\beta'$ *for some* $n > 1$*,* $\beta' \in E$.

Proof. Suppose that $0 \ne \beta \in E$ and $\mathrm{supp}\,\beta$ is not minimal. Then some $\alpha \in E$ satisfies $\mathrm{supp}\,\alpha \subset \mathrm{supp}\,\beta$. Hence for $n \in \mathbb{P}$ sufficiently large, $n\beta - \alpha \ge 0$ and so $n\beta - \alpha \in E$. Setting $\alpha' = n\beta - \alpha$, we have $n\beta = \alpha + \alpha'$ but $\alpha \ne i\beta$ for any $i \in \mathbb{N}$. Thus, $\beta \notin \mathrm{CF}(E)$.

Suppose that $\beta \in E$ belongs to set (iv), and let $n\beta = \alpha + \alpha'$, where $n \in \mathbb{P}$ and $\alpha,\alpha' \in E$. Since $\mathrm{supp}\,\beta$ is minimal, either $\alpha = 0$ or $\mathrm{supp}\,\alpha = \mathrm{supp}\,\beta$. In the latter case, let p/q be the largest rational number where $q \in \mathbb{P}$, for which $\beta - (p/q)\alpha \ge 0$. Then $q\beta - p\alpha \in E$ and $\mathrm{supp}\,(q\beta - p\alpha) \subset \mathrm{supp}\,\beta$. By the minimality of $\mathrm{supp}\,\beta$, we conclude $q\beta = p\alpha$. Since $\beta \ne \beta'$ for $n > 1$ and $\beta' \in E$, it follows that $p = 1$ and therefore $\beta \in \mathrm{CF}(E)$. Thus, the sets (i) and (iv) coincide.

Now let $\mathcal{R}$ be an extreme ray of $\mathcal{C}$, and suppose that $\alpha \in \mathcal{R}, \alpha = \alpha_1 + \alpha_2, \alpha_i \in \mathcal{C}$. By definition of extreme ray, it follows that $\alpha_1 = a\alpha_2, 0 \le a \le 1$. (Otherwise, α_1 and α_2 lie on different sides of the hyperplane $\mathcal{H}$ supporting $\mathcal{R}$.) From this observation, it is easy to deduce that the sets (i) and (iii) coincide.

Since the extreme rays of Γ and $\mathcal{C}$ coincide, an element β of $\mathrm{CF}(E_\sigma)$ lies on some extreme ray $\mathcal{R}$ of $\mathcal{C}$ and hence in set (iii). Conversely, if $\sigma \in \Gamma$ contains the extreme ray $\mathcal{R}$ of $\mathcal{C}$ and if $\mathcal{H}$ supports $\mathcal{R}$ in $\mathcal{C}$, then $\mathcal{H}$ supports $\mathcal{R}$ in σ. Thus, $\mathcal{R}$ is an extreme ray of σ. Since $E = \bigcup_{\sigma \in \Gamma} E_\sigma$, it follows that set (iii) is contained in set (ii). $\qquad\qquad\square$

We finally come to the first of the two main theorems of this section.

4.5.11 Theorem. *The generating functions $E(x)$ and $\overline{E}(x)$ represent rational functions of $x = (x_1, \ldots, x_m)$. When written in lowest terms, both these rational functions have denominator*

$$D(x) = \prod_{\beta \in \mathrm{CF}(E)} \left(1 - x^\beta\right).$$

Proof. Let Γ be a triangulation of $\mathcal{C}$ whose extreme rays coincide with those of $\mathcal{C}$ (existence guaranteed by Lemma 4.5.1). Let $E = \bigcup_{\sigma \in \Gamma} E_\sigma$ be the corresponding decomposition of E. Since $\mathrm{CF}(E_\sigma)$ is a set of quasigenerators for the simplicial monoid E_σ, it follows from Corollary 4.5.8 that $E_\sigma(x)$ and $\overline{E}_\sigma(x)$ can be written as rational functions with denominator

$$D(x) = \prod_{\beta \in \mathrm{CF}(E)} \left(1 - x^\beta\right).$$

By Proposition 4.5.10, $\mathrm{CF}(E_\sigma) \subseteq \mathrm{CF}(E)$. Hence by Lemma 4.5.4, we can put the expressions (4.17) and (4.18) for $E(x)$ and $\overline{E}(x)$ over the common denominator $D(x)$.

It remains to prove that $D(x)$ is the *least* possible denominator. We will consider only $E(x)$, the proof being essentially the same (and also following from Theorem 4.5.14) for $\overline{E}(x)$. Write $E(x) = N(x)/D(x)$. Suppose that this fraction is not in lowest terms. Then some factor $T(x)$ divides both $N(x)$ and $D(x)$. By the unique factorization theorem for the polynomial ring $\mathbb{C}[x_1, \ldots, x_m]$, we may assume that $T(x)$ divides $1 - x^\gamma$ for some $\gamma \in \mathrm{CF}(E)$. Since $\gamma \neq n\gamma'$ for any integer $n > 1$ and any $\gamma' \in \mathbb{N}^m$, the polynomial $1 - x^\gamma$ is irreducible. Hence, we may assume that $T(x) = 1 - x^\gamma$. Thus, we can write

$$F(x) = \frac{N'(x)}{\displaystyle\prod_{\substack{\beta \in \mathrm{CF}(E) \\ \beta \neq \gamma}} \left(1 - x^\beta\right)}, \qquad\qquad (4.26)$$

where $N'(x) \in \mathbb{C}[x_1, \ldots, x_m]$. Since, for any $n \in \mathbb{P}$ and $a_\beta \in \mathbb{N}$ $(\beta \neq \gamma)$, we have

$$n\gamma \neq \sum_{\substack{\beta \in \mathrm{CF}(E) \\ \beta \neq \gamma}} a_\beta \cdot \beta,$$

it follows that only finitely many terms of the form $x^{n\gamma}$ can appear in the expansion of the right-hand side of equation (4.26). This contradicts the fact that each $n\gamma \in E$, and completes the proof. $\qquad\qquad\square$

Our next goal is the reciprocity theorem that connects $E(x)$ and $\overline{E}(x)$. As a preliminary lemma we need to prove a reciprocity theorem for simplicial monoids.

4.5.12 Lemma. *Let $F \subseteq \mathbb{N}^m$ be a simplicial monoid with quasigenerators $\alpha_1, \ldots, \alpha_t$, and suppose that $D_F = \{\beta_1, \ldots, \beta_s\}$. Then*

$$\overline{D}_F = \{\alpha - \beta_1, \ldots, \alpha - \beta_s\},$$

where $\alpha = \alpha_1 + \cdots + \alpha_t$.

Proof. Let $\gamma = a_1\alpha_1 + \cdots + a_t\alpha_t \in F$. Since $0 \le a_i < 1$ if and only if $0 < 1 - a_i \le 1$, the proof follows from the definitions (4.22) and (4.23) of D_F and $\overline{D}_F$. $\square$

Recall that if $R(x) = R(x_1, \ldots, x_m)$ is a rational function, then $R(1/x)$ denotes the rational function $R(1/x_1, \ldots, 1/x_m)$.

4.5.13 Lemma. *Let $F \subseteq \mathbb{N}^m$ be a simplicial monoid of dimension t. Then*

$$\overline{F}(x) = (-1)^t F(1/x).$$

Proof. By equation (4.24), we have

$$F(1/x) = \left(\sum_{\beta \in D_S} x^{-\beta} \right) \prod_{i=1}^{t} \left(1 - x^{-\alpha_i}\right)^{-1}$$

$$= (-1)^t \left(\sum_{\beta \in D_S} x^{\alpha-\beta} \right) \prod_{i=1}^{t} \left(1 - x^{\alpha_i}\right)^{-1},$$

where α is as in Lemma 4.5.12. By Lemma 4.5.12,

$$\sum_{\beta \in D_S} x^{\alpha-\beta} = \sum_{\beta \in \overline{D}_S} x^{\beta}.$$

The proof follows from equation (4.25). $\square$

We now have all the necessary tools to deduce the second main theorem of this section.

4.5.14 Theorem (the reciprocity theorem for linear homogeneous diophantine equations). *Assume (as always) that the monoid E of $\mathbb{N}$-solutions to equation (4.10) is positive, and let $d = \dim \mathcal{C}$. Then*

$$\overline{E}(x) = (-1)^d E(1/x).$$

Proof. By Lemma 4.5.2 and equation (4.17), we have

$$E(1/x) = -\sum_{\sigma \in \Gamma^\circ} (-1)^{d - \dim \sigma + 1} E_\sigma(1/x).$$

Thus by Lemma 4.5.13,

$$E(1/x) = (-1)^d \sum_{\sigma \in \Gamma^\circ} \overline{E}_\sigma(x).$$

Comparing with equation (4.18) completes the proof. □

We now give some examples and applications of the preceding theory. First, we dispose of the equation $\alpha_1 + \alpha_2 - \alpha_3 - \alpha_4 = 0$ discussed in Examples 4.5.3 and 4.5.5.

4.5.15 Example. Let $E \subset \mathbb{N}^4$ be the monoid of $\mathbb{N}$-solutions to $\alpha_1 + \alpha_2 - \alpha_3 - \alpha_4 = 0$. According to equation (4.20), we need to compute $E_{abd}(x)$, $E_{bcd}(x)$, and $E_{bd}(x)$. Now $\mathrm{CF}(E) = \{\beta_1, \beta_2, \beta_3, \beta_4\}$, where $\beta_1 = (1,0,1,0)$, $\beta_2 = (1,0,0,1)$, $\beta_3 = (0,1,0,1)$, $\beta_4 = (0,1,1,0)$. A simple computation reveals that $D_{abd} = D_{bcd} = D_{bd} = \{(0,0,0,0)\}$ (the reason for this being that each of the sets $\{\beta_1, \beta_2, \beta_4\}$, $\{\beta_2, \beta_3, \beta_4\}$, and $\{\beta_2, \beta_4\}$ can be extended to a set of free generators of the group $\mathbb{Z}^4$). Hence by Lemma 4.5.12, we have $\overline{D}_{abd} = \{\beta_1 + \beta_2 + \beta_4\} = \{(2,1,2,1)\}$, $\overline{D}_{bcd} = \{\beta_2 + \beta_3 + \beta_4\} = \{(1,2,1,2)\}$, $\overline{D}_{bd} = \{\beta_2 + \beta_4\} = \{(1,1,1,1)\}$. There follows

$$E(x) = \frac{1}{(1 - x_1 x_3)(1 - x_1 x_4)(1 - x_2 x_3)}$$

$$+ \frac{1}{(1 - x_1 x_4)(1 - x_2 x_4)(1 - x_2 x_3)}$$

$$- \frac{1}{(1 - x_1 x_4)(1 - x_2 x_3)}$$

$$= \frac{1 - x_1 x_2 x_3 x_4}{(1 - x_1 x_3)(1 - x_1 x_4)(1 - x_2 x_3)(1 - x_2 x_4)},$$

$$\overline{E}(x) = \frac{x_1^2 x_2 x_3^2 x_4}{(1 - x_1 x_3)(1 - x_1 x_4)(1 - x_2 x_3)}$$

$$+ \frac{x_1 x_2^2 x_3 x_4^2}{(1 - x_1 x_4)(1 - x_2 x_4)(1 - x_2 x_3)}$$

$$+ \frac{x_1 x_2 x_3 x_4}{(1 - x_1 x_4)(1 - x_2 x_3)}$$

$$= \frac{x_1 x_2 x_3 x_4 (1 - x_1 x_2 x_3 x_4)}{(1 - x_1 x_3)(1 - x_1 x_4)(1 - x_2 x_3)(1 - x_2 x_4)}.$$

Note that indeed $\overline{E}(x) = -E(1/x)$. Note also that $\overline{E}(x) = x_1 x_2 x_3 x_4 E(x)$. This is because $\alpha \in E$ if and only if $\alpha + (1,1,1,1) \in \overline{E}$. More generally, we have the following result.

4.5.16 Corollary. *Let E be the monoid of $\mathbb{N}$-solutions to equation (4.10), and let $\gamma \in \mathbb{Z}^m$. The following two conditions are equivalent.*

i. $E(1/x) = (-1)^d x^\gamma E(x)$,
ii. $\overline{E} = \gamma + E$ *(i.e., $\alpha \in E$ if and only if $\alpha + \gamma \in \overline{E}$).*

Proof. Condition (ii) is clearly equivalent to $\overline{E}(x) = x^\gamma E(x)$. The proof follows from Theorem 4.5.14. $\qquad\square$

NOTE. There is another approach toward computing the generating function $E(x)$ of Example 4.5.15. Namely, the monoid E is generated by the vectors β_1, β_2, β_3, β_4, subject to the single relation $\beta_1 + \beta_3 = \beta_2 + \beta_4$. Hence, the number of representations of a vector δ in the form $\sum a_i \beta_i$, $a_i \in \mathbb{N}$, is one more than the number of representations of $\delta - (1,1,1,1)$ in this form. It follows that

$$E(x) = \frac{1 - x_1 x_2 x_3 x_4}{(1 - x_1 x_3)(1 - x_1 x_4)(1 - x_2 x_3)(1 - x_2 x_4)}.$$

The relation $\beta_1 + \beta_3 = \beta_2 + \beta_4$ is called a *syzygy of the first kind*. In general, there can be relations among the relations, called *syzygies of the second kind*, and so on. In order to develop a "syzygetic proof" of Theorem 4.5.11, techniques from commutative algebra are necessary but which will not be pursued here.

Only in the simplest cases is it practical to compute $E(x)$ by brute force, such as was done in Example 4.5.15. However, even if we can't compute $E(x)$ explicitly, we can still draw some interesting conclusions, as we now discuss. First, we need a preliminary result concerning specializations of the generating function $E(x)$.

4.5.17 Lemma. *Let E be the monoid of $\mathbb{N}$-solutions to equation (4.10). Let $a_1, \ldots, a_m \in \mathbb{Z}$ such that for each $r \in \mathbb{N}$, the number $g(r)$ of solutions $\alpha = (\alpha_1, \ldots, \alpha_m) \in E$ satisfying $L(\alpha) := a_1 \alpha_1 + \cdots + a_m \alpha_m = r$ is finite. Assume that $g(r) > 0$ for at least one $r > 0$. Let $G(\lambda) = \sum_{r \geq 0} g(r) \lambda^r$. Then*

i. $G(\lambda) = E(\lambda^{a_1}, \ldots, \lambda^{a_m}) \in \mathbb{C}(\lambda)$, *where $E(x) = \sum_{\gamma \in E} x^\gamma$ as usual.*
ii. $\deg G(\lambda) < 0$.

Proof. i. We first claim that $g(s) = 0$ for all $s < 0$. Let $\alpha \in E$ satisfy $L(\alpha) = r > 0$, and suppose that there exists $\beta \in E$ with $L(\beta) = s < 0$. Then for all $t \in \mathbb{N}$, the vectors $-ts\alpha + tr\beta$ are distinct elements of E, contradicting $g(0) < \infty$. Hence, the claim is proved, from which it is immediate that $G(\lambda) = E(\lambda^{a_1}, \ldots, \lambda^{a_m})$. Since $E(x) \in \mathbb{C}(x)$, we have $G(\lambda) \in \mathbb{C}(\lambda)$.
ii. By equation (4.17) and Lemma 4.5.2, it suffices to show that $\deg E_\sigma(\lambda^{a_1}, \ldots, \lambda^{a_m}) < 0$ for all $\sigma \in \Gamma^\circ$. Consider the expression (4.24) for $E_\sigma(x)$ (where $F = E_\sigma$), and let $\beta \in D_S$. Thus by equation (4.22), $\beta = b_1 \alpha_1 + \cdots + b_t \alpha_t$, $0 \leq b_i < 1$. Hence, $L(\beta) \leq L(\alpha_1) + \cdots + L(\alpha_t)$ with equality if and only if $t = 0$ (so $\sigma = \{0\}$). But $\{0\} \notin \Gamma^\circ$, so $L(\beta) < L(\alpha_1) + \cdots + L(\alpha_t)$. Since the

monomial x^β evaluated at $x = (x^{a_1}, \ldots, x^{a_m})$ has degree $L(\beta)$, it follows that each term of the numerator of $E_\sigma(\lambda^{a_1}, \ldots, \lambda^{a_m})$ has degree less than the degree $L(\alpha_1) + \cdots + L(\alpha_t)$ of the denominator.

$$\square$$

Note that in the preceding proof we did not need Lemma 4.5.2 to show that $G(\lambda) \leq 0$. We only required this result to show that the constant term $G(0)$ of $G(\lambda)$ was "correct" (in the sense of Proposition 4.2.2).

4.6 Applications

4.6.1 Magic Squares

We now come to our first real application of the preceding theory. Let $H_n(r)$ be the number of $n \times n$ $\mathbb{N}$-matrices such that every row and column sums to r. We call such matrices *magic squares*, though our definition is far less stringent than the classical one. For instance, $H_1(r) = 1$ (corresponding the the 1×1 matrix $[r]$), $H_2(r) = r + 1$ (corresponding to $\begin{bmatrix} i & r-i \\ r-i & i \end{bmatrix}$, $0 \leq i \leq r$), and $H_n(1) = n!$ (corresponding to all $n \times n$ permutation matrices). Introduce n^2 variables α_{ij} for $(i,j) \in [n] \times [n]$. Then an $n \times n$ $\mathbb{N}$-matrix with every row and column sum r corresponds to an $\mathbb{N}$-solution to the system of equations

$$\sum_{i=1}^{n} \alpha_{ij} = \sum_{i=1}^{n} \alpha_{ki}, \quad 1 \leq j \leq n, \ 1 \leq k \leq n, \tag{4.27}$$

with $\alpha_{11} + \alpha_{12} + \cdots + \alpha_{1n} = r$. It follows from Lemma 4.5.17(i) that if E denotes the monoid of $\mathbb{N}$-solutions to equation (4.27), then

$$E(x_{ij}) \Big|_{\substack{x_{1j}=\lambda \\ x_{ij}=1, i>1}} = \sum_{r \geq 0} H_n(r) \lambda^r. \tag{4.28}$$

In particular, $H_n(r)$ is a quasipolynomial in r. To proceed further, we must find the set $\mathrm{CF}(E)$.

4.6.1 Lemma. *The set* $\mathrm{CF}(E)$ *consists of the* $n!$ $n \times n$ *permutation matrices.*

Proof. Let π be a permutation matrix, and suppose that $k\pi = \alpha_1 + \alpha_2$, where $\alpha_1, \alpha_2 \in E$. Then α_1 and α_2 have at most one nonzero entry in every row and column (since $\mathrm{supp}\,\alpha_i \subseteq \mathrm{supp}\,\pi$) and hence are multiples of π. Thus, $\pi \in \mathrm{CF}(E)$.

Conversely, suppose that $\pi = (\pi_{ij}) \in E$ is not a permutation matrix. If π is a proper multiple of a permutation matrix, then clearly $\pi \notin \mathrm{CF}(E)$. Hence, we may assume that some row, say i_1, has at least two nonzero entries $\pi_{i_1 j_1}$ and $\pi_{i_1 j_1'}$. Since column j_1 has the same sum as row i_1, there is another nonzero entry in column j_1, say $\pi_{i_2 j_1}$. Since row i_2 has the same sum as column j_1, there is another nonzero entry in row i_2, say $\pi_{i_2 j_2}$. If we continue in this manner, we eventually must reach some entry twice. Thus, we have a sequence of at least four nonzero entries indexed by (i_r, j_r), (i_{r+1}, j_r), $(i_{r+1}, j_{r+1}), \ldots, (i_s, j_{s-1})$, where $i_s = i_r$ (or possibly beginning (i_{r+1}, j_r)—this is irrelevant). Let $\boldsymbol{\alpha_1}$ (respectively, $\boldsymbol{\alpha_2}$) be the matrix obtained from π by adding 1 to (respectively, subtracting 1 from) the entries in positions (i_r, j_r), $(i_{r+1}, j_{r+1}), \ldots, (i_{s-1}, j_{s-1})$ and subtracting 1 from (respectively, adding 1 to) the entries in positions (i_{r+1}, j_r), $(i_{r+2}, j_{r+1}), \ldots,$ (i_s, j_{s-1}). Then $\boldsymbol{\alpha_1}, \boldsymbol{\alpha_2} \in E$ and $2\pi = \boldsymbol{\alpha_1} + \boldsymbol{\alpha_2}$. But neither $\boldsymbol{\alpha_1}$ nor $\boldsymbol{\alpha_2}$ is a multiple of π, so $\pi \notin \mathrm{CF}(E)$. $\qquad \square$

We now come to the main result concerning the function $H_n(r)$.

4.6.2 Proposition. *For fixed $n \in \mathbb{P}$ the function $H_n(r)$ is a polynomial in r of degree $(n-1)^2$. Since it is a polynomial, it can be evaluated at any $r \in \mathbb{Z}$, and we have*

$$H_n(-1) = H_n(-2) = \cdots = H_n(-n+1) = 0,$$

$$(-1)^{n-1} H_n(-n-r) = H_n(r). \qquad (4.29)$$

Proof. By Lemma 4.6.1, any $\pi = (\pi_{ij}) \in \mathrm{CF}(E)$ satisfies $\pi_{11} + \pi_{12} + \cdots + \pi_{1n} = 1$. Hence if we set $x_{ij} = \lambda$ and $x_{ij} = 1$ for $i \geq 2$ in $1 - x^\pi$ (where $x^\pi = \prod_{i,j} x_{ij}^{\pi_{ij}}$), then we obtain $1 - \lambda$. Let

$$F_n(\lambda) = \sum_{r \geq 0} H_n(r) \lambda^r.$$

Then by Theorem 4.5.11 and Lemma 4.5.17, $F_n(\lambda)$ is a rational function of degree less than 0 and with denominator $(1-\lambda)^{t+1}$ for some $t \in \mathbb{N}$. Thus by Corollary 4.3.1, $H_n(r)$ is a polynomial function of r.

Now $\boldsymbol{\alpha}$ is an $\mathbb{N}$-solution to equation (4.27) if and only if $\boldsymbol{\alpha} + \kappa$ is a $\mathbb{P}$-solution, where κ is the $n \times n$ matrix of all 1's. Thus by Corollary 4.6.16,

$$E(1/\boldsymbol{x}) = \pm \left(\prod_{i,j} x_{ij} \right) E(\boldsymbol{x}).$$

Substituting $x_{1j} = \lambda$ and $x_{ij} = 1$ if $j > 1$, we obtain

$$F_n(1/\lambda) = \pm \lambda^n F_n(\lambda) = \pm \sum_{r \geq 0} \overline{H}_n(r) \lambda^r,$$

where $\overline{H}_n(r)$ is the number of $n \times n$ $\mathbb{P}$-matrices with every row and column sum equal to r. Hence by Proposition 4.2.3,

$$H_n(-n-r) = \pm H_n(r)$$

(the sign being $(-1)^{\deg H_n(r)}$). Since $\overline{H}_n(1) = \cdots = \overline{H}_n(n-1) = 0$, we also get $H_n(-1) = \cdots = H_n(n-1) = 0$.

There remains to show that $\deg H_n(r) = (n-1)^2$. We will give two proofs, one analytic and one algebraic. First, we give the analytic proof. If $\boldsymbol{\alpha} = (\alpha_{ij})$ is an

$\mathbb{N}$-matrix with every row and column sum equal to r, then (a) $0 \le \alpha_{ij} \le r$, and (b) if α_{ij} is given for $(i, j) \in [n-1] \times [n-1]$, then the remaining entries are uniquely determined. Hence,

$$H_n(r) \le (r+1)^{(n-1)^2}, \text{ so } \deg H_n(r) \le (n-1)^2.$$

On the other hand, if we arbitrarily choose

$$\frac{(n-2)r}{(n-1)^2} \le \alpha_{ij} \le \frac{r}{n-1}$$

for $(i, j) \in [n-1] \times [n-1]$, then when we fill in the rest of α to have row and column sums equal to r, every entry will be in $\mathbb{N}$. Thus,

$$H_n(r) \ge \left(\frac{r}{n-1} - \frac{(n-2)r}{(n-1)^2} \right)^{(n-1)^2}$$

$$= \left(\frac{r}{(n-1)!^2} \right)^{(n-1)^2},$$

so $\deg H_n(r) \ge (n-1)^2$. Hence, $\deg H_n(r) = (n-1)^2$.

For the algebraic proof that $\deg H_n(r) = (n-1)^2$, we compute the dimension of the cone $\mathcal{C}$ of all solutions to equation (4.27) in nonnegative real numbers. The n^2 equations appearing in (4.27) are highly redundant; we need for instance only

$$\sum_{j=1}^{n} \alpha_{1j} = \sum_{j=1}^{n} \alpha_{ij}, \ 2 \le i \le n,$$

and

$$\sum_{i=1}^{n} \alpha_{i1} = \sum_{i=1}^{n} \alpha_{ij}, \ 2 \le j \le n.$$

Thus, $\mathcal{C}$ is defined by $2n-2$ linearly independent equations in $\mathbb{R}^{n^2}$, so $\dim \mathcal{C} = n^2 - 2n + 2$. Hence, the denominator of the rational generating function $\sum_{r \ge 0} H_n(r) \lambda^r$, when reduced to lowest terms, is $(1-\lambda)^{n^2-2n+2}$, so $\deg H_n(r) = n^2 - 2n + 1 = (n-1)^2$. $\qquad\square$

One immediate use of Proposition 4.6.2 is for the actual computation of the values $H_n(r)$. Since $H_n(r)$ is a polynomial of degree $(n-1)^2$, we need to compute $(n-1)^2 + 1$ values to determine it completely. Since $H_n(-1) = \cdots = H_n(-n+1) = 0$ and $H_n(-n-r) = (-1)^{n-1} H_n(r)$, once we compute $H_n(0), H_n(1), \dots, H_n(i)$ we know $2i + n + 1$ values. Hence it suffices to take $i = \binom{n-1}{2}$ in order to determine $H_n(r)$. For instance, to compute $H_3(r)$, we only need the trivially computed values $H_3(0) = 1$ and $H_3(1) = 3! = 6$. To compute $H_4(r)$, we need only $H_4(0) = 1$, $H_4(1) = 24$, $H_4(2) = 282$, $H_4(3) = 2008$. Some small values of $F_n(\lambda)$ are given by

$$F_3(\lambda) = \frac{1+\lambda+\lambda^2}{(1-\lambda)^5},$$

$$F_4(\lambda) = \frac{1+14\lambda+87\lambda^2+148\lambda^3+87\lambda^4+14\lambda^5+\lambda^6}{(1-\lambda)^{10}},$$

$$F_5(\lambda) = \frac{P_5(\lambda)}{(1-\lambda)^{17}},$$

where

$$P_5(\lambda) = 1 + 103\lambda + 4306\lambda^2 + 63110\lambda^3$$
$$+ 388615\lambda^4 + 1115068\lambda^5 + 1575669\lambda^6 + 1115068\lambda^7$$
$$388615\lambda^8 + 63110\lambda^9 + 4306\lambda^{10} + 103\lambda^{11} + \lambda^{12}.$$

NOTE. We can apply the method discussed in the Note following Corollary 4.5.16 to the computation of $H_n(r)$. When $n = 3$ the computation can easily be done without recourse to commutative algebra. This approach is the subject of Exercise 2.15, which we now further explicate. Let P_w be the permutation matrix corresponding to the permutation $w \in \mathfrak{S}_3$. Any five of these matrices are linearly independent, and all six of them satisfy the unique linear dependence (up to multiplication by a nonzero scalar)

$$P_{123} + P_{231} + P_{312} = P_{213} + P_{132} + P_{321}. \tag{4.30}$$

Let E be the monoid of all 3×3 $\mathbb{N}$-matrices with equal row and column sums. For $A = (a_{ij}) \in E$, write

$$x^A = \prod_{i,j=1}^{3} x_{ij}^{a_{ij}}.$$

In particular,

$$x^{P_w} = \prod_{i=1}^{3} x_{i,w(i)}.$$

It follows easily from equation (4.30) that

$$\sum_{A \in E} x^A = \frac{1 - x^{P_{123}} x^{P_{231}} x^{P_{312}}}{\prod_{w \in \mathfrak{S}_3}(1 - x^{P_w})}. \tag{4.31}$$

Hence,

$$\sum_{r \geq 0} H_3(r)\lambda^r = \frac{1-\lambda^3}{(1-\lambda)^6}$$

$$= \frac{1+\lambda+\lambda^2}{(1-\lambda)^5}.$$

Moreover, we can write the numerator of the right-hand side of equation (4.31) as

$$(1 - x^{P_{123}}) + x^{P_{123}}(1 - x^{P_{231}}) + x^{P_{123}} x^{P_{231}}(1 - x^{P_{312}}).$$

Each expression in parentheses cancels a factor of the denominator. It follows that we can describe a *canonical form* for the elements of E. Namely, every element of E can be uniquely written in exactly one of the forms

$$a P_{132} + b P_{213} + c P_{231} + d P_{312} + e P_{321},$$

$$(a + 1) P_{123} + b P_{132} + c P_{213} + d P_{312} + e P_{321},$$

$$(a + 1) P_{123} + b P_{132} + c P_{213} + (d + 1) P_{231} + e P_{321},$$

where $a, b, c, d, e \in \mathbb{N}$.

As a modification of Proposition 4.6.2, consider the problem of counting the number $S_n(r)$ of *symmetric* $\mathbb{N}$-matrices with every row (and hence every column) sum equal to r. Again the crucial result is the analogue of Lemma 4.6.1.

4.6.3 Lemma. *Let E be the monoid of symmetric $n \times n$ $\mathbb{N}$-matrices with all row (and column) sums equal. Then $\mathrm{CF}(E)$ is contained in the set of matrices of the form π or $\pi + \pi^t$, where π is a permutation matrix and π^t is its transpose (or inverse).*

Proof. Let $\alpha \in E$. Forgetting for the moment that α is symmetric, we have by Lemma 4.6.1 that $\mathrm{supp}\,\alpha$ contains the support of some permutation matrix π. Thus for some $k \in \mathbb{P}$ (actually, $k = 1$ will do, but this is irrelevant), $k\alpha = \pi + \rho$ where ρ is an $\mathbb{N}$-matrix with equal line sums. Therefore, $2k\alpha = k(\alpha + \alpha^t) = (\pi + \pi^t) + (\rho + \rho^t)$. Hence, $\mathrm{supp}(\pi + \pi^t) \subseteq \mathrm{supp}(\alpha)$. It follows that any $\beta \in \mathrm{CF}(E)$ satisfies $j\beta = \pi + \pi^t$ for some $j \in \mathbb{P}$ and permutation matrix π. If $\pi = \pi^t$, then we must have $j = 2$; otherwise, $j = 1$, and the proof follows. $\qquad\square$

4.6.4 Proposition. *For fixed $n \in \mathbb{P}$, there exist polynomials $P_n(r)$ and $Q_n(r)$ such that $\deg P_n(r) = \binom{n}{2}$ and*

$$S_n(r) = P_n(r) + (-1)^r Q_n(r).$$

Moreover,

$$S_n(-1) = S_n(-2) = \cdots = S_n(-n + 1) = 0,$$

$$S_n(-n - r) = (-1)^{\binom{n}{2}} S_n(r).$$

Proof. By Lemma 4.6.3, any $\beta = (\beta_{ij}) \in \mathrm{CF}(E)$ satisfies $\beta_{11} + \beta_{12} + \cdots + \beta_{1n} = 1$ or 2. Hence, if we set $x_{ij} = \lambda$ and $x_{ij} = 1$ for $i \geq 2$ in $1 - x^\beta$, then we obtain either $1 - \lambda$ or $1 - \lambda^2$. Set $G_n(\lambda) = \sum_{r \geq 0} S_n(r)\lambda^r$. Then by Theorem 4.5.11 and Lemma 4.5.17, $G_n(x)$ is a rational function of negative degree and with denominator $(1 - \lambda)^s (1 - \lambda^2)^t$ for some $s, t \in \mathbb{N}$. Hence by Proposition 4.4.1 (or the more general Theorem 4.1.1), $S_n(r) = P_n(r) + (-1)^r Q_n(r)$ for certain polynomials $P_n(r)$ and $Q_n(r)$. The remainder of the proof is analogous to that of Proposition 4.6.2. $\qquad\square$

For the problem of computing $\deg Q_n(r)$, see equation (4.50) and the sentence following.

Some small values of $G_n(\lambda)$ are given by

$$G_1(\lambda) = \frac{1}{1-\lambda}, \quad G_2(\lambda) = \frac{1}{(1-\lambda)^2},$$

$$G_3(\lambda) = \frac{1+\lambda+\lambda^2}{(1-\lambda)^4(1+\lambda)},$$

$$G_4(\lambda) = \frac{1+4\lambda+10\lambda^2+4\lambda^3+\lambda^4}{(1-\lambda)^7(1+\lambda)},$$

$$G_5(\lambda) = \frac{V_5(\lambda)}{(1-\lambda)^{11}(1+\lambda)^6},$$

where

$$V_5(\lambda) = 1 + 21\lambda + 222\lambda^2 + 1082\lambda^3 + 3133\lambda^4$$
$$+ 5722\lambda^5 + 7013\lambda^6 + 5722\lambda^7 + 3133\lambda^8$$
$$+ 1082\lambda^9 + 222\lambda^{10} + 21\lambda^{11} + \lambda^{12}.$$

4.6.2 The Ehrhart Quasipolynomial of a Rational Polytope

An elegant and useful application of the preceding theory concerns a certain function $i(\mathcal{P}, n)$ associated with a convex polytope $\mathcal{P}$. By definition, a *convex polytope* $\mathcal{P}$ is the convex hull of a finite set of points in $\mathbb{R}^m$. Then $\mathcal{P}$ is homeomorphic to a ball $\mathbb{B}^d$. We write $d = \dim \mathcal{P}$ and call $\mathcal{P}$ a *d-polytope*. Equivalently, the affine span $\mathrm{aff}(\mathcal{P})$ of $\mathcal{P}$ is a d-dimensional affine subspace of $\mathbb{R}^m$. By $\partial \mathcal{P}$ and $\mathcal{P}^\circ$ we denote the boundary and interior of $\mathcal{P}$ in the usual topological sense (with respect to the embedding of $\mathcal{P}$ in its affine span). In particular $\partial \mathcal{P}$ is homeomorphic to the $(d-1)$-sphere $\mathbb{S}^{d-1}$.

A point $\alpha \in \mathcal{P}$ is a *vertex* of $\mathcal{P}$ if there exists a closed affine half-space $\mathcal{H} \subset \mathbb{R}^m$ such that $\mathcal{P} \cap \mathcal{H} = \{\alpha\}$. Equivalently, $\alpha \in \mathcal{P}$ is a vertex if it does not lie in the interior of any line segment contained in $\mathcal{P}$. Let V be the set of vertices of $\mathcal{P}$. Then V is finite and $\mathcal{P} = \mathrm{conv}\, V$, the convex hull of V. Moreover, if $S \subset \mathbb{R}^m$ is any set for which $\mathcal{P} = \mathrm{conv}\, V$, then $V \subseteq S$. The (convex) polytope $\mathcal{P}$ is called *rational* if each vertex of $\mathcal{P}$ has rational coordinates.

If $\mathcal{P} \subset \mathbb{R}^m$ is a rational convex polytope and $n \in \mathbb{P}$, then define integers $i(\mathcal{P}, n)$ and $\bar{i}(\mathcal{P}, n)$ by

$$i(\mathcal{P}, n) = \mathrm{card}(n\mathcal{P} \cap \mathbb{Z}^m),$$

$$\bar{i}(\mathcal{P}, n) = \mathrm{card}(n\mathcal{P}^\circ \cap \mathbb{Z}^m),$$

where $n\mathcal{P} = \{n\alpha : \alpha \in \mathcal{P}\}$. Equivalently, $i(\mathcal{P}, n)$ (respectively, $\bar{i}(\mathcal{P}, n)$) is equal to the number of rational points in $\mathcal{P}$ (respectively, $\mathcal{P}^\circ$) all of whose coordinates have least denominator dividing n. We call $i(\mathcal{P}, n)$ (respectively, $\bar{i}(\mathcal{P}, n)$) the

Ehrhart quasipolynomial of $\mathcal{P}$ (respectively, $\mathcal{P}^\circ$). Of course, we have to justify this terminology by showing that $i(\mathcal{P},n)$ and $\bar{i}(\mathcal{P},n)$ are indeed quasipolynomials.

4.6.5 Example. a. Let $\mathcal{P}_m$ be the convex hull of the set $\{(\varepsilon_1,\ldots,\varepsilon_m)\in\mathbb{R}^m : \varepsilon_i = 0 \text{ or } 1\}$. Thus, $\mathcal{P}_m$ is the *unit cube* in $\mathbb{R}^m$. It should be geometrically obvious that $i(\mathcal{P}_m,n)=(n+1)^m$ and $\bar{i}(\mathcal{P}_m,n)=(n-1)^m$.

b. Let $\mathcal{P}$ be the line segment joining 0 and $\alpha > 0$ in $\mathbb{R}$, where $\alpha \in \mathbb{Q}$. Clearly, $i(\mathcal{P},n)=\lfloor n\alpha \rfloor + 1$, which is a quasipolynomial of minimum quasiperiod equal to the denominator of α when written in lowest terms.

In order to prove the fundamental result concerning the Ehrhart quasipolynomials $i(\mathcal{P},n)$ and $\bar{i}(\mathcal{P},n)$, we will need the standard fact that a convex polytope $\mathcal{P}$ may also be defined as a bounded intersection of finitely many half-spaces. In other words, $\mathcal{P}$ is the set of all real solutions $\alpha \in \mathbb{R}^m$ to a finite system of linear inequalities $\alpha \cdot \delta \le a$, provided that this solution set is bounded. (Note that the equality $\alpha \cdot \delta = a$ is equivalent to the two inequalities $\alpha \cdot (-\delta) \le -a$ and $\alpha \cdot \delta \le a$, so we are free to describe $\mathcal{P}$ using inequalities and equalities.) The polytope $\mathcal{P}$ is rational if and only if the inequalities can be chosen to have rational (or integral) coefficients.

Since $i(\mathcal{P},n)$ and $\bar{i}(\mathcal{P},n)$ are not affected by replacing $\mathcal{P}$ with $\mathcal{P}+\gamma$ for $\gamma \in \mathbb{Z}^m$, we may assume that all points in $\mathcal{P}$ have nonnegative coordinates, denoted $\mathcal{P} \ge 0$. We now associate with a rational convex polytope $\mathcal{P} \ge 0$ in $\mathbb{R}^m$ a monoid $E_\mathcal{P} \subseteq \mathbb{N}^{m+1}$ of $\mathbb{N}$-solutions to a system of homogeneous linear inequalities. (Recall that an inequality may be converted to an equality by introducing a slack variable.) Suppose that $\mathcal{P}$ is the set of solutions α to the system

$$\alpha \cdot \delta_i \le a_i, \ \ 1 \le i \le s,$$

where $\delta_i \in \mathbb{Q}^m$, $a_i \in \mathbb{Q}$. Introduce new variables $\gamma = (\gamma_1,\ldots,\gamma_m)$ and t, and define $E_\mathcal{P} \subseteq \mathbb{N}^{m+1}$ to be the set of all $\mathbb{N}$-solutions to the system

$$\gamma \cdot \delta_i \le a_i t, \ \ 1 \le i \le s.$$

4.6.6 Lemma. *A nonzero vector $(\gamma,t)\in\mathbb{N}^{m+1}$ belongs to $E_\mathcal{P}$ if and only if γ/t is a rational point of $\mathcal{P}$.*

Proof. Since $\mathcal{P} \ge 0$, any rational point $\gamma/t \in \mathcal{P}$ with $\gamma \in \mathbb{Z}^m$ and $t \in \mathbb{P}$ satisfies $\gamma \in \mathbb{N}^m$. Hence a nonzero vector $(\gamma,t)\in\mathbb{N}^{m+1}$ with $t > 0$ belongs to $E_\mathcal{P}$ if and only if γ/t is a rational point of $\mathcal{P}$.

It remains to show that if $(\gamma,t)\in E_\mathcal{P}$ and $t = 0$, then $\gamma = 0$. Because $\mathcal{P}$ is bounded, it is easily seen that every vector $\beta \ne 0$ in $\mathbb{R}^m$ satisfies $\beta \cdot \delta_i > 0$ for some $1 \le i \le s$. Hence, the only solution γ to $\gamma \cdot \delta_i \le 0$, $1 \le i \le s$, is $\gamma = 0$, and the proof follows. $\qquad\square$

Our next step is to determine $\mathrm{CF}(E_\mathcal{P})$, the completely fundamental elements of $E_\mathcal{P}$. If $\alpha \in \mathbb{Q}^m$, then define $\mathrm{den}\,\alpha$ (the *denominator* of α) as the least integer

$q \in \mathbb{P}$ such that $q\alpha \in \mathbb{Z}^m$. In particular, if $\alpha \in \mathbb{Q}$, then $\operatorname{den}\alpha$ is the denominator of α when written in lowest terms.

4.6.7 Lemma. *Let $\mathcal{P} \geq 0$ be a rational convex polytope in $\mathbb{R}^m$ with vertex set V. Then*

$$\mathrm{CF}(E_{\mathcal{P}}) = \{((\operatorname{den}\alpha)\alpha, \operatorname{den}\alpha) : \alpha \in V\}.$$

Proof. Let $(\gamma, t) \in E_{\mathcal{P}}$, and suppose that for some $k \in \mathbb{P}$ we have

$$k(\gamma, t) = (\gamma_1, t_1) + (\gamma_2, t_2),$$

where $(\gamma_i, t_i) \in E_{\mathcal{P}}$, $t_i \neq 0$. Then

$$\gamma/t = (t_1/kt)(\gamma_1/t_1) + (t_2/kt)(\gamma_2/t_2),$$

where $(t_1/kt) + (t_2/kt) = 1$. Thus, γ/t lies on the line segment joining γ_1/t_1 and γ_2/t_2. It follows that $(\gamma, t) \in \mathrm{CF}(E_{\mathcal{P}})$ if and only if $\gamma/t \in V$ (so that $\gamma_1/t_1 = \gamma_2/t_2 = \gamma/t$) and $(\gamma, t) \neq j(\gamma', t')$ for $(\gamma', t') \in \mathbb{N}^{m+1}$ and an integer $j > 1$. Thus, we must have $t = \operatorname{den}(\gamma/t)$, and the proof follows. $\square$

It is now easy to establish the two basic facts concerning $i(\mathcal{P}, n)$ and $\bar{i}(\mathcal{P}, n)$.

4.6.8 Theorem. *Let $\mathcal{P}$ be a rational convex polytope of dimension d in $\mathbb{R}^m$ with vertex set V. Let $F(\mathcal{P}, \lambda) = 1 + \sum_{n\geq 1} i(\mathcal{P}, n)\lambda^n$. Then $F(\mathcal{P}, \lambda)$ is a rational function of λ of degree less than 0, which can be written with denominator $\prod_{\alpha \in V}(1 - \lambda^{\operatorname{den}\alpha})$. (Hence, in particular, $i(\mathcal{P}, n)$ is a quasipolynomial whose "correct" value at $n = 0$ is $i(\mathcal{P}, 0) = 1$.) The complex number $\lambda = 1$ is a pole of $F(\mathcal{P}, \lambda)$ of order $d + 1$, while no value of λ is a pole whose order exceeds $d + 1$.*

Proof. Let the variables x_i correspond to γ_i and y to t in the generating function $E_{\mathcal{P}}(x, y)$; that is,

$$E_{\mathcal{P}}(x, y) = \sum_{(\gamma, t) \in E_{\mathcal{P}}} x^\gamma y^t.$$

Lemma 4.6.6, together with the observation $E_{\mathcal{P}}(\mathbf{0}, 0) = 1$, shows that

$$E_{\mathcal{P}}(1, \ldots, 1, \lambda) = F(P, \lambda). \tag{4.32}$$

Hence by Lemma 4.5.17, $F(\mathcal{P}, \lambda)$ is a rational function of degree less than 0. By Theorem 4.5.11 and Lemma 4.6.7, the denominator of $E_{\mathcal{P}}(x, y)$ is equal to

$$\prod_{\alpha \in V}\left(1 - x^{(\operatorname{den}\alpha)\alpha} y^{\operatorname{den}\alpha}\right).$$

Thus by equation (4.32), the denominator of $F(\mathcal{P}, \lambda)$ can be taken as $\prod_{\alpha \in V}\left(1 - \lambda^{\operatorname{den}\alpha}\right)$.

Now $\dim E_{\mathcal{P}}$ is equal to the dimension of the vector space $\langle \mathrm{CF}(E_{\mathcal{P}})\rangle$ spanned by $\mathrm{CF}(E_{\mathcal{P}}) = \{((\operatorname{den}\alpha)\alpha, \operatorname{den}\alpha) : \alpha \in V\}$. Clearly then we also have $\langle \mathrm{CF}(E_{\mathcal{P}})\rangle = \langle (\alpha, 1) : \alpha \in V\rangle$. The dimension of this latter space is just the maximum number of $\alpha \in V$ that are affinely independent in $\mathbb{R}^m$ (i.e., such that no nontrivial linear combination with zero coefficient sum is equal to 0). Since $\mathcal{P}$ spans a d-dimensional affine subspace of $\mathbb{R}^m$ there follows $\dim E_{\mathcal{P}} = d + 1$. Now by Lemmas 4.5.2 and 4.5.4, we have

$$E_{\mathcal{P}}(\boldsymbol{x}, y) = \sum_{\sigma \in \overline{\Gamma}} (-1)^{d+1-\dim \sigma} E_\sigma(\boldsymbol{x}, y),$$

so

$$F(\mathcal{P}, \lambda) = \sum_{\sigma \in \overline{\Gamma}} (-1)^{d+1-\dim \sigma} E_\sigma(1, \ldots, 1, \lambda). \tag{4.33}$$

Looking at the expression (4.24) for $E_\sigma(\boldsymbol{x}, y)$, we see that those terms of equation (4.33) with $\dim \sigma = d + 1$ have a positive coefficient of $(\lambda - 1)^{d+1}$ in the Laurent expansion about $\lambda = 1$, whereas all other terms have a pole of order at most d at $\lambda = 1$. Moreover, no term has a pole of order greater than $d + 1$ at any $\lambda \in \mathbb{C}$. The proof follows. $\qquad\square$

4.6.9 Theorem (the reciprocity theorem for Ehrhart quasipolynomials). *Since* $i(\mathcal{P}, n)$ *is a quasipolynomial, it can be defined for all* $n \in \mathbb{Z}$. *If* $\dim \mathcal{P} = d$, *then* $\bar{i}(\mathcal{P}, n) = (-1)^d i(\mathcal{P}, -n)$.

Proof. A vector $(\boldsymbol{\gamma}, t) \in \mathbb{N}^m$ lies in $\overline{E}_{\mathcal{P}}$ if and only if $\boldsymbol{\gamma}/t \in \mathcal{P}^\circ$. Thus,

$$\overline{E}_{\mathcal{P}}(1, \ldots, 1, \lambda) = \sum_{n \geq 1} \bar{i}(\mathcal{P}, n)\lambda^n.$$

The proof now follows from Theorem 4.5.14, Proposition 4.2.3, and the fact (shown in the proof of the previous theorem) that $\dim E_{\mathcal{P}} = d + 1$. $\qquad\square$

Unlike Theorem 4.5.11, the denominator $D(\lambda) = \prod_{\alpha \in V} \left(1 - \lambda^{\mathrm{den}\,\alpha}\right)$ of $F(\mathcal{P}, \lambda)$ is not in general the *least* denominator of $F(\mathcal{P}, \lambda)$. By Theorem 4.6.8, the least denominator has a factor $(1 - \lambda)^{d+1}$ but not $(1 - \lambda)^{d+2}$, while $D(\lambda)$ has a factor $(1 - \lambda)^{\#V}$. We have $\#V = d + 1$ if and only if $\mathcal{P}$ is a simplex. For roots of unity $\zeta \neq 1$, the problem of finding the highest power of $1 - \zeta\lambda$ dividing the least denominator of $F(\mathcal{P}, \lambda)$ is very delicate and subtle. A result in this direction is given by Exercise 4.66. Here we will content ourselves with one example showing that there is no obvious solution to this problem.

4.6.10 Example. Let $\mathcal{P}$ be the convex 3-polytope in $\mathbb{R}^3$ with vertices $(0, 0, 0)$, $(1, 0, 0)$, $(0, 1, 0)$, $(1, 1, 0)$, and $(\frac{1}{2}, 0, \frac{1}{2})$. An examination of all the preceding theory will produce no theoretical reason why $F(\mathcal{P}, \lambda)$ does not have a factor $1 + \lambda$ in its least denominator, but such is indeed the case. It is just an "accident" that the factor $1 + \lambda$ appearing in $\prod_{\alpha \in V} \left(1 - \lambda^{\mathrm{den}\,\alpha}\right) = (1 - \lambda)^5 (1 + \lambda)$ is eventually canceled, yielding $F(\mathcal{P}, \lambda) = (1 - \lambda)^{-4}$.

One special case of Theorems 4.6.8 and 4.6.9 deserves special mention.

4.6.11 Corollary. *Let* $\mathcal{P} \subset \mathbb{R}^m$ *be an* integral *convex* d-*polytope (i.e., each vertex has integer coordinates). Then* $i(\mathcal{P}, n)$ *and* $\bar{i}(\mathcal{P}, n)$ *are polynomial functions of* n

of degree d, satisfying

$$i(\mathcal{P},0) = 1, \quad i(\mathcal{P},n) = (-1)^d \bar{i}(\mathcal{P},n).$$

Proof. By Theorem 4.6.8, the least denominator of $F(\mathcal{P},\lambda)$ is $(1-\lambda)^{d+1}$. Now apply Corollary 4.3.1. □

If $\mathcal{P} \subset \mathbb{R}^m$ is an integral polytope, then of course we call $i(\mathcal{P},n)$ and $\bar{i}(\mathcal{P},n)$ the *Ehrhart polynomials* of $\mathcal{P}$ and $\mathcal{P}^\circ$. One interesting and unexpected application of Ehrhart polynomials is to the problem of finding the volume of $\mathcal{P}$. Somewhat more generally, we need the concept of the relative volume of an integral d-polytope. If $\mathcal{P} \subset \mathbb{R}^m$ is such a polytope, then the integral points of the affine space $\mathcal{A}$ spanned by $\mathcal{P}$ is a translate (coset) of some d-dimensional sublattice $L \cong \mathbb{Z}^d$ of $\mathbb{Z}^m$. Hence, there exists an invertible affine transformation $\phi : \mathcal{A} \to \mathbb{R}^d$ satisfying $\phi(\mathcal{A} \cap \mathbb{Z}^m) = \mathbb{Z}^d$. The image $\phi(\mathcal{P})$ of $\mathcal{P}$ under ϕ is an integral convex d-polytope in $\mathbb{R}^d$, so $\phi(\mathcal{P})$ has a positive volume (= Jordan content or Lebesgue measure) $v(\mathcal{P})$, called the *relative volume* of $\mathcal{P}$. It is easy to see that $v(\mathcal{P})$ is independent of the choice of ϕ and hence depends on $\mathcal{P}$ alone. If $d = m$ (i.e., $\mathcal{P}$ is an integral d-polytope in $\mathbb{R}^d$), then $v(\mathcal{P})$ is just the usual volume of $\mathcal{P}$ since we can take ϕ to be the identity map.

4.6.12 Example. Let $\mathcal{P} \subset \mathbb{R}^2$ be the line segment joining $(3,2)$ to $(5,6)$. The affine span $\mathcal{A}$ of $\mathcal{P}$ is the line $y = 2x - 4$, and $\mathcal{A} \cap \mathbb{Z}^2 = \{(x, 2x-4) : x \in \mathbb{Z}\}$. For the map $\phi : \mathcal{A} \to \mathbb{R}$, we can take $\phi(x, 2x-4) = x$. The image $\phi(\mathcal{P})$ is the interval $[3,5]$, which has length 2. Hence, $v(\mathcal{P}) = 2$. To visualize this geometrically, draw a picture of $\mathcal{P}$ as in Figure 4.6(a). When "straightened out" $\mathcal{P}$ looks like Figure 4.6(b), which has length 2 when we think of the integer points $(3,2)$, $(4,4)$, $(5,6)$ as consecutive integers on the real line.

4.6.13 Proposition. *Let $\mathcal{P} \subset \mathbb{R}^m$ be an integral convex d-polytope. Then the leading coefficient of $i(\mathcal{P},n)$ is $v(\mathcal{P})$.*

Sketch of proof. The map $\phi : \mathcal{A} \to \mathbb{R}^d$ constructed earlier satisfies $i(\mathcal{P},n) = i(\phi(\mathcal{P}),n)$. Hence, we may assume $m = d$. Given $n \in \mathcal{P}$, for each point $\gamma \in \mathcal{P}$

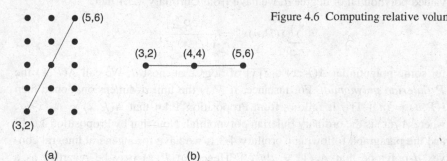

Figure 4.6 Computing relative volume.

(a)   (b)

with $m\gamma \in \mathbb{Z}^d$ construct a d-dimensional hypercube H_γ with center γ and sides of length $1/n$ parallel to the coordinate axes. These hypercubes fit together to fill $\mathcal{P}$ without overlap, except for a small error on the boundary of $\mathcal{P}$. There are $i(\mathcal{P},n)$ hypercubes is all with a volume n^{-d} each, and hence a total volume of $n^{-d}i(\mathcal{P},n)$. As $n \to \infty$, it is geometrically obvious (and not hard to justify rigorously – this is virtually the definition of the Riemann integral) that the volume of these hypercubes will converge to the volume of $\mathcal{P}$. Hence, $\lim_{n\to\infty} n^{-d}i(\mathcal{P},n) = v(\mathcal{P})$, and the proof follows. $\qquad\square$

4.6.14 Corollary. *Let $\mathcal{P} \subset \mathbb{R}^m$ be an integral convex d-polytope. If we know any d of the numbers $i(\mathcal{P},1), \bar{i}(\mathcal{P},1), i(\mathcal{P},2), \bar{i}(\mathcal{P},2),\ldots$, then we can determine $v(\mathcal{P})$.*

Proof. Since $i(\mathcal{P},0) = 1$ and $i(\mathcal{P},-n) = (-1)^d \bar{i}(\mathcal{P},n)$, once we know d of the given numbers we know $d+1$ values of the polynomial $i(\mathcal{P},n)$ of degree d. Hence, we can find $i(\mathcal{P},n)$ and in particular its leading coefficient $v(\mathcal{P})$. $\qquad\square$

4.6.15 Example. a. If $\mathcal{P} \subset \mathbb{R}^m$ is an integral convex 2-polytope, then

$$v(\mathcal{P}) = \frac{1}{2}(i(\mathcal{P},1) + \bar{i}(\mathcal{P},1) - 2).$$

This classical formula (for $m = 2$) is usually stated in the form

$$v(\mathcal{P}) = \frac{1}{2}(2A - B - 2),$$

where $A = \#(\mathbb{Z}^2 \cap \mathcal{P}) = i(\mathcal{P},1)$ and $B = \#(\mathbb{Z}^2 \cap \partial\mathcal{P}) = i(\mathcal{P},1) - \bar{i}(\mathcal{P},1)$.
b. If $\mathcal{P} \subset \mathbb{R}^m$ is an integral convex 3-polytope, then

$$v(\mathcal{P}) = \frac{1}{6}(i(\mathcal{P},2) - 3i(\mathcal{P},1) + \bar{i}(\mathcal{P},1) + 3).$$

c. If $\mathcal{P} \in \mathbb{R}^m$ is an integral convex d-polytope, then

$$v(\mathcal{P}) = \frac{1}{d!}\left((-1)^d + \sum_{k=1}^{d} \binom{d}{k}(-1)^{d-k}i(\mathcal{P},k)\right).$$

Let $\mathcal{P}$ be an integral convex d-polytope in $\mathbb{R}^m$. Because $i(\mathcal{P},n)$ is an integer-valued polynomial of degree d, we have from Corollary 4.3.1 that

$$\sum_{n\geq 0} i(\mathcal{P},n)x^n = \frac{A(\mathcal{P},x)}{(1-x)^{d+1}}$$

for some polynomial $A(\mathcal{P},x) \in \mathbb{Z}[x]$ of degree at most d. We call $A(\mathcal{P},x)$ the *$\mathcal{P}$-Eulerian polynomial*. For instance, if $\mathcal{P}$ is the unit d-dimensional cube then $i(\mathcal{P},n) = (n+1)^d$. It follows from Proposition 1.4.4 that $A(\mathcal{P},x) = A_d(x)/x$, where $A_d(x)$ is the ordinary Eulerian polynomial. Note that by Proposition 4.6.13 and the paragraph following Corollary 4.3.1, we have for a general integral convex d-polytope that $A_d(1) = d!v(\mathcal{P})$. Hence, $A(\mathcal{P},x)$ may be regarded as a refinement of the relative volume $v(\mathcal{P})$. If $A(\mathcal{P}, x) = \sum_{i=0}^{d} h_i^* x^i$, then the vector $h^*(\mathcal{P}) = (h_0^*, \ldots, h_d^*)$ is called the h^*-*vector* or δ-*vector* of $\mathcal{P}$. It can be shown that the h^*-vector is nonnegative (Exercise 4.48).

NOTE. Corollary 4.6.14 extends without difficulty to the case where $\mathcal{P}$ is not necessarily convex. We need only assume that $\mathcal{P} \subset \mathbb{R}^m$ is an integral polyhedral d-manifold with boundary; that is, a union of integral convex d-polytopes in $\mathbb{R}^m$ such that the intersection of any two is a common face of both and such that $\mathcal{P}$, regarded as a topological space, is a manifold with boundary. (In fact, we can replace this last condition with a weaker condition about the Euler characteristic of $\mathcal{P}$ and local Euler characteristic of $\mathcal{P}$ at any point $\alpha \in \mathcal{P}$, but we will not enter into the details here.) Assume for simplicity that $m = d$. Then the only change in the theory is that now $i(\mathcal{P}, 0) = \chi(\mathcal{P})$, the Euler characteristic of $\mathcal{P}$. Details are left to the reader.

We conclude with two more examples.

4.6.16 Example (Propositions 4.6.2 and 4.6.4 revisited). a. Let $\mathcal{P} = \Omega_n \subset \mathbb{R}^{n^2}$, the convex polytope of all $n \times n$ *doubly-stochastic matrices* (i.e., matrices of nonnegative real numbers with every row and column sum equal to one). Clearly, $M \in r\Omega_n \cap \mathbb{Z}^{n^2}$ if and only if M is an N-matrix with every row and column sum equal to r. Hence, $i(\Omega_n, r)$ is just the function $H_n(r)$ of Proposition 4.6.2. Lemma 4.6.1 is equivalent to the statement that $V(\Omega_n)$ consists of the $n \times n$ permutation matrices. Thus, Ω_n is an integral polytope, and the conclusions of Proposition 4.6.2 follow also from Corollary 4.6.11.

b. Let $\mathcal{P} = \Sigma_n \in \mathbb{R}^{n^2}$, the convex polytope of all *symmetric* doubly stochastic matrices. As in (a), we have $i(\Sigma_n, r) = S_n(r)$, where $S_n(r)$ is the function of Proposition 4.6.4. Lemma 4.6.3 is equivalent to the statement that

$$V(\Sigma_n) \subseteq \left\{ \frac{1}{2}(P + P^t) : P \text{ is an } n \times n \text{ permutation matrix} \right\}.$$

Hence, den $M = 1$ or 2 for all $M \in V(\Sigma_n)$, and the conclusions of Proposition 4.6.4 follow also from Theorem 4.6.8.

4.6.17 Example. Let $P = \{t_1, \ldots, t_p\}$ be a finite poset. Let $\mathcal{O} = \mathcal{O}(P)$ be the convex hull of incidence vectors of dual-order ideals K of P; that is, vectors of the form $(\varepsilon_1, \ldots, \varepsilon_p)$, where $\varepsilon_i = 1$ if $t_i \in K$ and $\varepsilon_i = 0$ otherwise. Then

$$\mathcal{O} = \{(a_1, \ldots, a_p) \in \mathbb{R}^p : 0 \leq a_i \leq 1 \text{ and } a_i \leq a_j \text{ if } t_i \leq t_j\}.$$

Thus, $(b_1, \ldots, b_p) \in n\mathcal{O} \cap \mathbb{Z}^p$ if and only if (i) $b_i \in \mathbb{Z}$, (ii) $0 \leq b_i \leq n$, and (iii) $b_i \leq b_j$ if $t_i \leq t_j$. Hence, $i(\mathcal{O}(P), n) = \Omega_P(n+1)$, where Ω_P is the order polynomial of P. The volume of $\mathcal{O}(P)$ is $e(P)/p!$, the leading coefficient of $\Omega_P(n+1)$ or $\Omega_P(n)$. (The volume is the same as the relative volume since dim $\mathcal{O}(P) = p$.) The polytope $\mathcal{O}(P)$ is called the *order polytope* of P.

4.7 The Transfer-Matrix Method

4.7.1 Basic Principles

The transfer-matrix method, like the Principle of Inclusion-Exclusion and the Möbius inversion formula, has simple theoretical underpinnings but a very wide range of applicability. The theoretical background can be divided into two parts – combinatorial and algebraic. First, we discuss the combinatorial part. A (finite) *directed graph* or *digraph* D is a triple (V, E, ϕ), where $V = \{v_1, \ldots, v_p\}$ is a set of *vertices*, E is a finite set of (directed) *edges* or *arcs*, and ϕ is a map from E to $V \times V$. If $\phi(e) = (u, v)$, then e is called an edge *from u to v*, with *initial vertex* u and *final vertex* v. This is denoted $u = \mathrm{init}\, e$ and $v = \mathrm{fin}\, e$. If $u = v$, then e is called a *loop*. A *walk* Γ in D of *length n* from u to v is a sequence $e_1 e_2 \cdots e_n$ of n edges such that $\mathrm{init}\, e_1 = u$, $\mathrm{fin}\, e_n = v$, and $\mathrm{fin}\, e_i = \mathrm{init}\, e_{i+1}$ for $1 \le i < n$. If also $u = v$, then Γ is called a *closed walk based at u*. (Note that if Γ is a closed walk, then $e_i e_{i+1} \cdots e_n e_1 \cdots e_{i-1}$ is in general a different closed walk. In some graph-theoretical contexts this distinction would not be made.)

Now let $w \colon E \to R$ be a *weight function* with values in some commutative ring R. (For our purposes here, we can take $R = \mathbb{C}$ or a polynomial ring over $\mathbb{C}$.) If $\Gamma = e_1 e_2 \cdots e_n$ is a walk, then the *weight* of Γ is defined by $w(\Gamma) = w(e_1) w(e_2) \cdots w(e_n)$. Let $i, j \in [p]$ and $n \in \mathbb{N}$. Since D is finite, we can define

$$A_{ij}(n) = \sum_{\Gamma} w(\Gamma),$$

where the sum is over all walks Γ in D of length n from v_i to v_j. In particular, $A_{ij}(0) = \delta_{ij}$. If all $w(e) = 1$, then we are just counting the *number* of walks of length n from u to v. The fundamental problem treated by the transfer matrix method is the evaluation of $A_{ij}(n)$. The first step is to interpret $A_{ij}(n)$ as an entry in a certain matrix. Define a $p \times p$ matrix $A = (A_{ij})$ by

$$A_{ij} = \sum_{e} w(e),$$

where the sum ranges over all edges e satisfying $\mathrm{init}\, e = v_i$ and $\mathrm{fin}\, e = v_j$. In other words, $A_{ij} = A_{ij}(1)$. The matrix A is called the *adjacency matrix* of D, with respect to the weight function w. The eigenvalues of the adjacency matrix A play a key role in the enumeration of walks. These eigenvalues are also called the *eigenvalues of D* (as a weighted digraph).

4.7.1 Theorem. *Let $n \in \mathbb{N}$. Then the (i, j)-entry of A^n is equal to $A_{ij}(n)$. (Here we define $A^0 = I$ even if A is not invertible.)*

Proof. The proof is immediate from the definition of matrix multiplication. Specifically, we have

$$(A^n)_{ij} = \sum A_{ii_1} A_{i_1 i_2} \cdots A_{i_{n-1} j},$$

where the sum is over all sequences $(i_1, \ldots, i_{n-1}) \in [p]^{n-1}$. The summand is 0 unless there is a walk $e_1 e_2 \cdots e_n$ from v_i to v_j with $\operatorname{fin} e_k = v_{i_k}$ $(1 \le k < n)$ and $\operatorname{init} e_k = v_{i_{k-1}}$ $(1 < k \le n)$. If such a walk exists, then the summand is equal to the sum of the weights of all such walks, and the proof follows. $\qquad \square$

The second step of the transfer-matrix method is the use of linear algebra to analyze the behavior of the function $A_{ij}(n)$. Define the generating function

$$F_{ij}(D, \lambda) = \sum_{n \ge 0} A_{ij}(n) \lambda^n.$$

4.7.2 Theorem. *The generating function $F_{ij}(D, \lambda)$ is given by*

$$F_{ij}(D, \lambda) = \frac{(-1)^{i+j} \det(I - \lambda A : j, i)}{\det(I - \lambda A)}, \tag{4.34}$$

where $(B : j, i)$ denotes the matrix obtained by removing the j-th row and i-th column of B. Thus in particular $F_{ij}(D, \lambda)$ is a rational function of λ whose degree is strictly less than the multiplicity n_0 of 0 as an eigenvalue of A.

Proof. $F_{ij}(D, \lambda)$ is the (i, j)-entry of the matrix $\sum_{n \ge 0} \lambda^n A^n = (I - \lambda A)^{-1}$. If B is any invertible matrix, then it is well known from linear algebra that $(B^{-1})_{ij} = (-1)^{i+j} \det(B : j, i) / \det(B)$, so equation (4.34) follows.

Suppose now that A is a $p \times p$ matrix. Then

$$\det(I - \lambda A) = 1 + \alpha_1 \lambda + \cdots + \alpha_{p-n_0} \lambda^{p-n_0},$$

where

$$(-1)^p \left(\alpha_{p-n_0} \lambda^{n_0} + \cdots + \alpha_1 \lambda^{p-1} + \lambda^p \right)$$

is the characteristic polynomial $\det(A - \lambda I)$ of A. Thus as polynomials in λ, we have $\deg \det(I - \lambda A) = p - n_0$ and $\deg \det(I - \lambda A : j, i) \le p - 1$. Hence,

$$\deg F_{ij} \le p - 1 - (p - n_0) < n_0. \qquad \square$$

One special case of Theorem 4.7.2 is particularly elegant. Let

$$C_D(n) = \sum_{\Gamma} w(\Gamma),$$

where the sum is over all closed walks Γ in D of length n. For instance, $C_D(1) = \operatorname{tr} A$, where tr denotes trace.

4.7.3 Corollary. *Let $Q(\lambda) = \det(I - \lambda A)$. Then*

$$\sum_{n \ge 1} C_D(n) \lambda^n = -\frac{\lambda Q'(\lambda)}{Q(\lambda)}.$$

Proof. By Theorem 4.7.1, we have

$$C_D(n) = \sum_{i=1}^{p} A_{ii}(n) = \operatorname{tr} A^n.$$

Let $\omega_1, \ldots, \omega_q$ be the nonzero eigenvalues of A. Then

$$\operatorname{tr} A^n = \omega_1^n + \cdots + \omega_q^n, \tag{4.35}$$

so

$$\sum_{n \geq 1} C_D(n) \lambda^n = \frac{\omega_1 \lambda}{1 - \omega_1 \lambda} + \cdots + \frac{\omega_q \lambda}{1 - \omega_q \lambda}.$$

When put over the denominator $(1 - \omega_1 \lambda) \cdots (1 - \omega_q \lambda) = Q(\lambda)$, the numerator becomes $-\lambda Q'(\lambda)$. (Alternatively, this result may be deduced directly from Theorem 4.7.2.) $\qquad \square$

4.7.2 Undirected Graphs

The preceding theory applies also to ordinary (undirected) graphs G. If we replace each edge e in G between vertices u and v with the two directed edges e' from u to v and e'' from v to u, then walks in the resulting digraph D_G of length n from u to v correspond exactly to walks in G of length n from u to v, as defined in the Appendix. The same remarks apply to weighted edges and walks. Hence, the counting of walks in undirected graphs G is just a special case of counting walks in digraphs. The undirected case corresponds to a *symmetric* adjacency matrix A. Symmetric matrices enjoy algebraic properties that lead to some additional formulas for the enumeration of walks.

Recall that a real symmetric $p \times p$ matrix A has p linearly independent real eigenvectors, which can in fact be chosen to be orthonormal (i.e., orthogonal and of unit length). Let $u_1^t, \ldots, u_p^t$ (where ^{t} denotes transpose, so u_i is a row vector) be real orthonormal eigenvectors for A, with corresponding eigenvalues $\lambda_1, \ldots, \lambda_p$. Each u_i is a row vector, so u_i^t is a column vector. Thus, the dot (or scalar or inner) product of the vectors u and v is given by uv^t (ordinary matrix multiplication). In particular, $u_i u_j^t = \delta_{ij}$. Let $U = (u_{ij})$ be the matrix whose columns are $u_1^t, \ldots, u_p^t$, denoted $U = [u_1^t, \ldots, u_p^t]$. Thus, U is an orthogonal matrix and

$$U^t = U^{-1} = \begin{bmatrix} u_1 \\ \vdots \\ u_p \end{bmatrix},$$

the matrix whose rows are $u_1, \ldots, u_p$. Recall from linear algebra that the matrix U *diagonalizes* A, that is,

$$U^{-1} A U = \operatorname{diag}(\lambda_1, \ldots, \lambda_p),$$

where $\operatorname{diag}(\lambda_1, \ldots, \lambda_p)$ denotes the diagonal matrix with diagonal entries $\lambda_1, \ldots, \lambda_p$.

4.7.4 Corollary. *Given the graph G as above, fix the two vertices v_i and v_j. Let $\lambda_1, \ldots, \lambda_p$ be the eigenvalues of G, that is, of the adjacency matrix $A = A(G)$.*

Then there exist real numbers $c_1, \ldots, c_p$ such that for all $n \geq 1$, we have

$$(A^n)_{ij} = c_1 \lambda_1^n + \cdots + c_p \lambda_p^n.$$

In fact, if $U = (u_{rs})$ is a real orthogonal matrix such that $U^{-1}AU = \mathrm{diag}(\lambda_1, \ldots, \lambda_p)$, then we have

$$c_k = u_{ik} u_{jk}.$$

Proof. We have [why?]

$$U^{-1} A^n U = \mathrm{diag}(\lambda_1^n, \ldots, \lambda_p^n).$$

Hence,

$$A^n = U \cdot \mathrm{diag}(\lambda_1^n, \ldots, \lambda_p^n) U^{-1}.$$

Taking the (i, j)-entry of both sides (and using $U^{-1} = U^t$) gives

$$(A^n)_{ij} = \sum_k u_{ik} \lambda_k^n u_{jk},$$

as desired. $\qquad\square$

4.7.3 Simple Applications

With the basic theory out of the way, let us look at some applications.

4.7.5 Example. For $p, n \geq 1$, let $f_p(n)$ denote the number of sequences $a_1 a_2 \cdots a_n \in [p]^n$ such that $a_i \neq a_{i+1}$ for $1 \leq i \leq n$, and $a_n \neq a_1$. We are simply counting closed walks of length n in the complete graph K_p; we begin at vertex a_1, then walk to a_2, and so on. Let A be the adjacency matrix of K_p. Then $A + I$ is the all 1's matrix J and, hence, has rank 1. Thus, $p - 1$ eigenvalues of $A + I$ are equal to 0, so $p - 1$ eigenvalues of A are equal to -1. To obtain the remaining eigenvalue of A, note that $\mathrm{tr}\, A = 0$. Since the trace is the sum of the eigenvalues, the remaining eigenvalue of A is $p - 1$. This may also be seen by noting that the column vector $[1, 1, \ldots, 1]^t$ is an eigenvector for A with eigenvalue $p - 1$. We obtain from equation (4.35) that

$$f_p(n) = (p-1)^n + (p-1)(-1)^n. \qquad (4.36)$$

By symmetry, the number of closed walks of length n in K_p that start at a particular vertex, say 1, is given by

$$(A^n)_{11} = \frac{1}{p} f_p(n) = \frac{1}{p} \left((p-1)^n + (p-1)(-1)^n \right).$$

The number of walks of length n between two *unequal* vertices, say 1 and 2, is given by

$$(A^n)_{12} = \frac{1}{p-1} \left((p-1)^n - (A^n)_{11} \right)$$

$$= \frac{1}{p} \left((p-1)^n - (-1)^n \right).$$

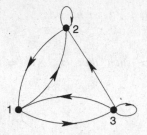

Figure 4.7 A digraph illustrating the transfer-matrix method.

Another way to obtain these results is to note that $J^k = p^{k-1} J$ for $k \geq 1$. Hence,

$$A^n = (J - I)^n$$

$$= (-1)^n I + \sum_{k=1}^{n} (-1)^{n-k} \binom{n}{k} J^k$$

$$= (-1)^n I + \left(\sum_{k=1}^{n} (-1)^{n-k} \binom{n}{k} p^{k-1} \right) J$$

$$= (-1)^n I + \frac{1}{p} \left((p-1)^n - (-1)^n \right) J.$$

It is now easy to extract the $(1,1)$ and $(1,2)$ entries.

4.7.6 Example. Let $f(n)$ be the number of sequences $a_1 a_2 \cdots a_n \in [3]^n$ such that neither 11 nor 23 appear as two consecutive terms $a_i a_{i+1}$. Let D be the digraph on $V = [3]$ with an edge (i, j) if j is allowed to follow i in the sequence. Thus, D is given by Figure 4.7. If we set $w(e) = 1$ for every edge e, then clearly $f(n) = \sum_{i,j=1}^{3} A_{ij}(n-1)$. Setting $Q(\lambda) = \det(I - \lambda A)$ and $Q_{ij}(\lambda) = \det(I - \lambda A : j, i)$, there follows from Theorem 4.7.2 that

$$F(\lambda) := \sum_{n \geq 0} f(n+1) \lambda^n = \frac{\sum_{i,j=1}^{3} (-1)^{i+j} Q_{ij}(\lambda)}{Q(\lambda)}.$$

Now

$$A = \begin{bmatrix} 0 & 1 & 1 \\ 1 & 1 & 0 \\ 1 & 1 & 1 \end{bmatrix},$$

so by direct calculation,

$$(1 - \lambda A)^{-1} = \frac{1}{1 - 2\lambda - \lambda^2 + \lambda^3} \begin{bmatrix} (1-\lambda)^2 & \lambda & \lambda(1-\lambda) \\ \lambda(1-\lambda) & 1-\lambda-\lambda^2 & \lambda^2 \\ \lambda & \lambda(1+\lambda) & 1-\lambda-\lambda^2 \end{bmatrix}.$$

It follows that

$$F(\lambda) = \frac{3 + \lambda - \lambda^2}{1 - 2\lambda - \lambda^2 + \lambda^3}, \tag{4.37}$$

or, equivalently,

$$\sum_{n\geq 0} f(n)\lambda^n = \frac{1+\lambda}{1-2\lambda-\lambda^2+\lambda^3}.$$

In the present situation, we do not actually have to compute $(I-\lambda A)^{-1}$ in order to write down equation (4.37). First, compute $\det(I-\lambda A)=1-2\lambda-\lambda^2+\lambda^3$. Since this polynomial has degree 3, it follows from Theorem 4.7.2 that $\deg F(\lambda)<0$. Hence, the numerator of $F(\lambda)$ is determined by the initial values $f(1)=3$, $f(2)=7$, $f(3)=16$. This approach involves a considerably easier computation than evaluating $(I-\lambda A)^{-1}$.

Now suppose that we impose the additional restriction on the sequence $a_1a_2\cdots a_n$ that $a_na_1 \neq 11$ or 23. Let $g(n)$ be the number of such sequences. Then $g(n)=C_D(n)$, the number of closed walks in D of length n. Hence with no further computation, we obtain

$$\sum_{n\geq 1} g(n)\lambda^n = -\frac{\lambda Q'(\lambda)}{Q(\lambda)} = \frac{\lambda(2+2\lambda-3\lambda^2)}{1-2\lambda-\lambda^2+\lambda^3}. \tag{4.38}$$

It is somewhat magical that, unlike the case for $f(n)$, we did not need to consider any initial conditions. Note that equation (4.38) yields the value $g(1)=2$. The method disallows the sequence 1, since $a_1a_n=11$. This illustrates a common phenomenon in applying Corollary 4.7.3 – for small values of n (never larger than $p-1$) the value of $C_D(n)$ may not conform to our combinatorial expectations.

4.7.7 Example. A *factor* of a word w is a subword of w consisting of consecutive letters. In other words, v is a factor of w if we can write $w=uvy$ for some words u and y. Let $f(n)$ be the number of words (i.e., sequences) $a_1a_2\cdots a_n \in [3]^n$ such that there are no factors of the form $a_ia_{i+1}=12$ or $a_ia_{i+1}a_{i+2}=213, 222, 231$, or 313. At first sight, it may seem as if the transfer-matrix method is inapplicable, since an allowed value of a_i depends on more than just the previous value a_{i-1}. A simple trick, however, circumvents this difficulty – make the digraph D big enough to incorporate the required past history. Here we take $V=[3]^2$, with edges (ab,bc) if abc is allowed as three consecutive terms of the word. Thus, D is given by Figure 4.8. If we now define all weights $w(e)=1$, then

$$f(n) = \sum_{ab,cd\in V} A_{ab,cd}(n-2).$$

Thus, $\sum_{n\geq 0} f(n)\lambda^n$ is a rational function with denominator $Q(\lambda)=\det(I-\lambda A)$ for a certain 8×8 matrix A. (The vertex 12 is never used, so we can take A to be 8×8 rather than 9×9.)

It is clear that this technique applies equally well to prove the following result.

4.7.8 Proposition. *Let S be a finite set, and let $\mathcal{F}$ be a finite set of finite words with terms (letters) from S. Let $f(n)$ be the number of words $a_1a_2\cdots a_n \in S^n$ such that*

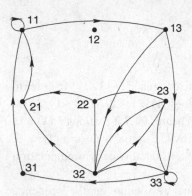

Figure 4.8 The digraph for Example 4.7.7.

no factor $a_i a_{i+1} \cdots a_{i+j}$ appears in $\mathcal{F}$. Then $\sum_{n \geq 0} f(n) \lambda^n \in \mathbb{Q}(\lambda)$. The same is true if we take the subscripts appearing in $a_i a_{i+1} \cdots a_{i+j}$ modulo n. In this case, if $g(n)$ is the number of such words, then $\sum_{n \geq 1} g(n) \lambda^n = -\lambda Q'(\lambda)/Q(\lambda)$ for some $Q(\lambda) \in \mathbb{Q}[\lambda]$, provided that $g(n)$ is suitably interpreted for small n.

Even though there turn out to be special methods for actually computing the generating functions appearing in Proposition 4.7.8 (see for example Exercise 4.40), at least the transfer-matrix method shows transparently that the generating functions are rational.

4.7.9 Example. Let $f(n)$ be the number of permutations $a_1 a_2 \cdots a_n \in \mathfrak{S}_n$ such that $|a_i - i| = 0$ or 1. Again it may first seem that the transfer-matrix method is inapplicable, since the allowed values of a_i depend on *all* the previous values $a_1, \ldots, a_{i-1}$. Observe, however, that there are really only three possible choices for a_i – namely, $i - 1$, i, or $i + 1$. Moreover, none of these values could be used prior to a_{i-2}, so the choices available for a_i depend only on the choices already made for a_{i-2} and a_{i-1}. Thus, the transfer-matrix method is applicable. The vertex set V of the digraph D consists of those pairs $(\alpha, \beta) \in \{-1, 0, 1\}^2$ for which it is possible to have $a_i - i = \alpha$ and $a_{i+1} - i - 1 = \beta$. An edge connects (α, β) to (β, γ) if it is possible to have $a_i - i = \alpha$, $a_{i+1} - i - 1 = \beta$, $a_{i+2} - i - 2 = \gamma$. Thus, $V = \{v_1, \ldots, v_7\}$, where $v_1 = (-1, -1)$, $v_2 = (-1, 0)$, $v_3 = (-1, 1)$, $v_4 = (0, 0)$, $v_5 = (0, 1)$, $v_6 = (1, -1)$, $v_7 = (1, 1)$. (Note, for instance, that $(1, 0)$ cannot be a vertex, since if $a_i - i = 1$ and $a_{i+1} - i - 1 = 0$, then $a_i = a_{i+1}$.) Writing $\alpha_1 \alpha_2$ for the vertex (α_1, α_2), and so on, it follows that a walk $(\alpha_1 \alpha_2, \alpha_2 \alpha_3)$, $(\alpha_2 \alpha_3, \alpha_3 \alpha_4)$, $\ldots, (\alpha_n \alpha_{n+1}, \alpha_{n+1} \alpha_{n+2})$ of length n in D corresponds to the permutation $1 + \alpha_1$, $2 + \alpha_2, \ldots, n + 2 + \alpha_{n+2}$ of $[n + 2]$ of the desired type, provided that $\alpha_1 \neq -1$ and $\alpha_{n+2} \neq 1$. Hence, $f(n + 2)$ is equal to the number of walks of length n in D from one of the vertices v_4, v_5, v_6, v_7 to one of the vertices v_1, v_2, v_4, v_6. Thus, if we set $w(e) = 1$ for all edges e in D, then

$$f(n + 2) = \sum_{i=4,5,6,7} \sum_{j=1,2,4,6} (A^n)_{ij}.$$

The adjacency matrix is given by

$$\begin{bmatrix} 1 & 1 & 1 & 0 & 0 & 0 & 0 \\ 0 & 0 & 0 & 1 & 1 & 0 & 0 \\ 0 & 0 & 0 & 0 & 0 & 1 & 1 \\ 0 & 0 & 0 & 1 & 1 & 0 & 0 \\ 0 & 0 & 0 & 0 & 0 & 1 & 1 \\ 0 & 1 & 1 & 0 & 0 & 0 & 0 \\ 0 & 0 & 0 & 0 & 0 & 0 & 1 \end{bmatrix}$$

and $Q(\lambda) = \det(I - \lambda A) = (1 - \lambda)^2(1 - \lambda - \lambda^2)$. As in Example 4.7.6, we can compute the numerator of $\sum_{n\geq 0} f(n+2)\lambda^n$ using initial values, rather than finding $(I - \lambda A)^{-1}$. According to Theorem 4.7.2, the polynomial $(1 - \lambda^2)(1 - \lambda - \lambda^2)\sum_{n\geq 0} f(n+2)\lambda^n$ may have degree as large as 6, so in order to compute $\sum_{n\geq 0} f(n)\lambda^n$ we need the initial values $f(0), f(1), \ldots, f(6)$. If this work is actually carried out, then we obtain

$$\sum_{n\geq 0} f(n)\lambda^n = \frac{1}{1 - \lambda - \lambda^2}, \tag{4.39}$$

so that $f(n)$ is just the Fibonacci number F_{n+1} (!).

Similarly we may ask for the number $g(n)$ of permutations $a_1 a_2 \cdots a_n \in \mathfrak{S}_n$ such that $a_i - i \equiv 0, \pm 1 \pmod{n}$. This condition has the effect of allowing $a_1 = n$ and $a_n = 1$, so that $g(n)$ is just the number of closed walks $(\alpha_1\alpha_2, \alpha_2\alpha_3)$, $(\alpha_2\alpha_3, \alpha_3\alpha_4), \ldots, (\alpha_{n-1}\alpha_n, \alpha_n\alpha_1), (\alpha_n\alpha_1, \alpha_1\alpha_2)$ in D of length n. Hence,

$$\sum_{n\geq 1} g(n)\lambda^n = -\frac{\lambda Q'(\lambda)}{Q(\lambda)} = \frac{2\lambda}{1 - \lambda} + \frac{\lambda(1 + 2\lambda)}{1 - \lambda - \lambda^2}. \tag{4.40}$$

Hence, $g(n) = 2 + L_n$, where L_n is the nth Lucas number. Note the "spurious" values $g(1) = 3$, $g(2) = 5$.

It is clear that the preceding arguments generalize to the following result.

4.7.10 Proposition. *a. Let S be a finite subset of $\mathbb{Z}$. Let $f_S(n)$ be the number of permutations $a_1 a_2 \cdots a_n \in \mathfrak{S}_n$ such that $a_i - i \in S$ for $i \in [n]$. Then $\sum_{n\geq 0} f_S(n)\lambda^n \in \mathbb{Q}(\lambda)$.*

b. Let $g_S(n)$ be the number of permutations $a_1 a_2 \cdots a_n \in \mathfrak{S}_n$ such that for all $i \in [n]$ there is a $j \in S$ for which $a_i - i \equiv j \pmod{n}$. If we suitably interpret $g_S(n)$ for small n, then there is a polynomial $Q(\lambda) \in \mathbb{Q}[\lambda]$ for which $\sum_{n\geq 1} g(n)\lambda^n = -\lambda Q'(\lambda)/Q(\lambda)$.

4.7.4 Factorization in Free Monoids

The reader is undoubtedly wondering, in view of the simplicity of the generating functions (4.39) and (4.40), whether there is a simpler way of obtaining them. Surely it seems unnecessary to find the characteristic polynomial of a 7×7 matrix

$\mathcal{A}$ when the final answer is $1/(1 - \lambda - \lambda^2)$. The five eigenvalues $0,0,0,1,1$ do not seem relevant to the problem. Actually, the vertices v_5 and v_7 are not needed for computing $f(n)$, but we are still left with a 5×5 matrix. This brings us to an important digresson – the method of factoring words in a free monoid. While this method has limited application, when it does work it is extremely elegant and simple.

Let $\mathcal{A}$ be a finite set, called the *alphabet*. A *word* is a finite sequence $a_1 a_2 \cdots a_n$ of elements of $\mathcal{A}$, including the empty word 1. The set of all words in the alphabet $\mathcal{A}$ is denoted $\mathcal{A}^*$. Define the *product* of two words $u = a_1 \cdots a_n$ and $v = b_1 \cdots b_m$ to be their juxtaposition,

$$uv = a_1 \cdots a_n b_1 \cdots b_m.$$

In particular, $1u = u1 = u$ for all $u \in \mathcal{A}^*$. The set $\mathcal{A}^*$, together with the product just defined, is called the *free monoid* on the set $\mathcal{A}$. (A *monoid* is a set with an associative binary operation and an identity element.) If $u = a_1 \cdots a_n \in \mathcal{A}^*$ with $a_i \in \mathcal{A}$, then define the *length* of u to be $\ell(u) = n$. In particular, $\ell(1) = 0$. If $\mathcal{C}$ is any subset of $\mathcal{A}^*$, then define

$$\mathcal{C}_n = \{u \in \mathcal{C} : \ell(u) = n\}.$$

Let $\mathcal{B}$ be a subset of $\mathcal{A}^*$ (possibly infinite), and let $\mathcal{B}^*$ be the submonoid of $\mathcal{A}^*$ generated by $\mathcal{B}$; that is, $\mathcal{B}^*$ consists of all words $u_1 u_2 \cdots u_n$ where $u_i \in \mathcal{B}$. We say that $\mathcal{B}^*$ is *freely generated* by $\mathcal{B}$ if every word $u \in \mathcal{B}^*$ can be written *uniquely* as $u_1 u_2 \cdots u_n$ where $u_i \in \mathcal{B}$. For instance, if $\mathcal{A} = \{a, b\}$ and $\mathcal{B} = \{a, ab, aab\}$, then $\mathcal{B}^*$ is not freely generated by $\mathcal{B}$ (since $a \cdot ab = aab$), but is freely generated by $\{a, ab\}$. On the other hand, if $\mathcal{B} = \{a, ab, ba\}$, then $\mathcal{B}^*$ is not freely generated by any subset of $\mathcal{A}^*$ (since $ab \cdot a = a \cdot ba$).

Now suppose that we have a *weight function* $w\colon \mathcal{A} \to R$ (where R is a commutative ring), and define $w(u) = w(a_1) \cdots w(a_n)$ if $u = a_1 \cdots a_n, a_i \in \mathcal{A}$. In particular, $w(1) = 1$. For any subset $\mathcal{C}$ of $\mathcal{A}^*$, define the generating function

$$\mathcal{C}(\lambda) = \sum_{u \in \mathcal{C}} w(u) \lambda^{\ell(u)} \in R[[\lambda]].$$

Thus, the coefficient $f(n)$ of λ^n in $\mathcal{C}(\lambda)$ is $\sum_{u \in \mathcal{C}_n} w(u)$. The following proposition is almost self-evident.

4.7.11 Proposition. *Let $\mathcal{B}$ be a subset of $\mathcal{A}^*$ that freely generates $\mathcal{B}^*$. Then*

$$\mathcal{B}^*(\lambda) = (1 - \mathcal{B}(\lambda))^{-1}.$$

Proof. We have

$$f(n) = \sum_{i_1 + \cdots + i_k = n} \prod_{j=1}^{k} \left(\sum_{u \in \mathcal{B}_{i_j}} w(u) \right).$$

Multiplying by λ^n and summing over all $n \in \mathbb{N}$ yields the result. $\square$

As we shall soon see, even the very straightforward Proposition 4.7.11 has interesting applications. But first we seek a result, in the context of the preceding proposition, analogous to Corollary 4.7.3. It turns out that we need the monoid $\mathcal{B}^*$ to satisfy a property stronger than being freely generated by $\mathcal{B}$. This property depends on the way in which $\mathcal{B}^*$ is embedded in $\mathcal{A}^*$, and not just on the abstract structure of $\mathcal{B}^*$. If $\mathcal{B}^*$ is freely generated by $\mathcal{B}$, then we say that $\mathcal{B}^*$ is *very pure* if the following condition, called *unique circular factorization* (UCF), holds:

(UCF) Let $u = a_1 a_2 \cdots a_n \in \mathcal{B}^*$, where $\mathcal{B}^*$ is freely generated by $\mathcal{B}$, with $a_i \in \mathcal{A}$. Thus for unique integers $0 < n_1 < n_2 < \cdots < n_k < n$, we have

$$a_1 a_2 \cdots a_{n_1} \in \mathcal{B}, \ a_{n_1+1} a_{n_1+2} \cdots a_{n_2} \in \mathcal{B},$$

$$a_{n_2+1} a_{n_2+2} \cdots a_{n_3} \in \mathcal{B}, \ldots, a_{n_k+1} a_{n_k+2} \cdots a_n \in \mathcal{B}.$$

Suppose that for some $i \in [n]$ we have $a_i a_{i+1} \cdots a_n a_1 \cdots a_{i-1} \in \mathcal{B}^*$. Then $i = n_j + 1$ for some $0 \le j \le k$, where we set $n_0 = 0$.

In other words, if the letters of u are written in clockwise order around a circle, as in Figure 4.9(a), with the initial letter u_1 *not* specified, then there is a unique way of inserting bars between pairs of consecutive letters such that the letter between any two consecutive bars, read clockwise, form a word in $\mathcal{B}$. See Figure 4.9(b).

For example, if $\mathcal{A} = \{a\}$ and $\mathcal{B} = \{aa\}$, then $\mathcal{B}^*$ fails to have UCF since the word $u = aa$ can be "circularly factored" in the two ways shown in Figure 4.10. Similarly, if $\mathcal{A} = \{a, b, c\}$ and $\mathcal{B} = \{abc, ca, b\}$ then $\mathcal{B}^*$ again fails to have UCF since the word $u = abc$ can be circularly factored as shown in Figure 4.11.

Though not necessary for what follows, for the sake of completeness we state the following characterization of very pure monoids. The proof is left to the reader.

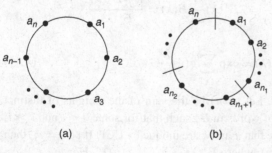

Figure 4.9 Unique circular factorization.

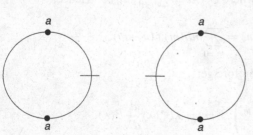

Figure 4.10 Failure of unique circular factorization.

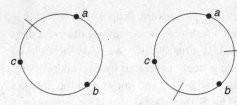

Figure 4.11 Another failure of unique circular factorization.

4.7.12 Proposition. *Suppose that $\mathcal{B}^*$ is freely generated by $\mathcal{B} \subset \mathcal{A}^*$. The following two conditions are equivalent:*

i. $\mathcal{B}^*$ is very pure.
ii. If $u \in \mathcal{A}^*$, $v \in \mathcal{A}^*$, $uv \in \mathcal{B}^*$ and $vu \in \mathcal{B}^*$, then $u \in \mathcal{B}^*$ and $v \in \mathcal{B}^*$.

Suppose now that $\mathcal{B}^*$ has UCF. We always compute the length of a word with respect to the alphabet $\mathcal{A}$, so $\mathcal{B}_n^* = \mathcal{B}^* \cap \mathcal{A}_n^*$. If $a_j \in \mathcal{A}$ and $u = a_1 a_2 \cdots a_n \in \mathcal{B}_n^*$, then an $\mathcal{A}^*$-*conjugate* (or *cyclic shift*) of u is a word $a_i a_{i+1} \cdots a_n a_1 \cdots a_{i-1} \in \mathcal{A}_n^*$. Define $g(n) = \sum w(u)$, where the sum is over all *distinct* $\mathcal{A}^*$-conjugates u of words in $\mathcal{B}_n^*$. For instance, if $\mathcal{A} = \{a, b\}$ and $\mathcal{B} = \{a, ab\}$, then

$$g(4) = w(aaaa) + w(aaab) + w(aaba) + w(abaa) + w(baaa)$$

$$+ w(abab) + w(baba) = w(a)^4 + 4w(a)^3 w(b) + 2w(1)^2 w(b)^2.$$

Define the generating function

$$\widetilde{\mathcal{B}}(\lambda) = \sum_{n \geq 1} g(n) \lambda^n.$$

4.7.13 Proposition. *Assume $\mathcal{B}^*$ is very pure. Then*

$$\widetilde{\mathcal{B}}(\lambda) = \frac{\lambda \frac{d}{d\lambda} \mathcal{B}(\lambda)}{1 - \mathcal{B}(\lambda)} = \lambda \mathcal{B}^*(\lambda) \frac{d}{d\lambda} \mathcal{B}(\lambda) = \frac{\lambda \frac{d}{d\lambda} \mathcal{B}^*(\lambda)}{\mathcal{B}^*(\lambda)}.$$

Equivalently,

$$\mathcal{B}^*(\lambda) = \exp \sum_{n \geq 1} g(n) \frac{\lambda^n}{n}. \tag{4.41}$$

First Proof. Fix a word $v \in \mathcal{B}$. Let $g_v(n)$ be the sum of the weights of distinct $\mathcal{A}^*$-conjugates $a_i a_{i+1} \cdots a_{i-1}$ of words in $\mathcal{B}_n^*$ such that for some $j \leq i$ and $k \geq i$, we have $a_j a_{j+1} \cdots a_k = v$. Note that j and k are unique by UCF. If $\ell(v) = m$, then clearly $g_v(n) = mw(v) f(n - m)$, where $\mathcal{B}^*(\lambda) = \sum_{n \geq 0} f(n) \lambda^n$. Hence,

$$g(n) = \sum_{v \in \mathcal{B}} g_v(n) = \sum_{m=0}^{n} mb(m) f(n - m),$$

where $b(n) = \sum_{v \in \mathcal{B}_n} w(v)$. We therefore get

$$\widetilde{\mathcal{B}}(\lambda) = \left(\sum_{m \geq 0} mb(m) \lambda^m \right) \mathcal{B}^*(\lambda) = \lambda \mathcal{B}^*(\lambda) \frac{d}{d\lambda} \mathcal{B}(\lambda). \qquad \square$$

Our second proof of Proposition 4.7.13 is based on a purely combinatorial lemma involving the relationship between "ordinary" words in $\mathcal{B}^*$ and their $\mathcal{A}^*$-conjugates. This is the general result mentioned after the first proof of Lemma 2.3.4.

4.7.14 Lemma. *Assume that $\mathcal{B}^*$ is very pure. Let $f_k(n) = \sum_u w(u)$, where u ranges over all words in $\mathcal{B}_n^*$ that are a product of k words in $\mathcal{B}$. Let $g_k(n) = \sum_v w(v)$, where v ranges over all distinct $\mathcal{A}^*$-conjugates of the above words u. Then $nf_k(n) = kg_k(n)$.*

Proof. Let A be the set of ordered pairs (u, i), where $u \in \mathcal{B}_n^*$ and u is the product of k words in $\mathcal{B}$, and where $i \in [n]$. Let B be the set of ordered pairs (v, j), where v has the preceding meaning, and where $j \in [k]$. Clearly, $\#A = nf_k(n)$ and $\#B = kg_k(n)$. Define a map $\psi : A \to B$ as follows: Suppose that $u = a_1 a_2 \cdots a_n = b_1 b_2 \cdots b_k \in \mathcal{B}_n^*$, where $a_i \in \mathcal{A}$, $b_i \in \mathcal{B}$. Then let

$$\psi(u, i) = (a_i a_{i+1} \cdots a_{i-1}, j),$$

where a_i is one of the letters of b_j. It is easily seen that ψ is a bijection that preserves the weight of the first component, and the proof follows. $\square$

Second Proof of Proposition 4.7.13. By Lemma 4.7.14,

$$nf(n) = \sum_k nf_k(n) = \sum_k kg_k(n). \tag{4.42}$$

The right-hand side of equation (4.42) counts all pairs (v, b_i), where v is an $\mathcal{A}^*$-conjugate of some word $b_1 b_2 \cdots b_k \in \mathcal{B}_n^*$, with $b_j \in \mathcal{B}$. Thus, v may be written uniquely in the form $b_j' b_{j+1} \cdots b_k b_1 b_2 \cdots b_{j-1} b_j''$. where $b_j'' b_j' = b_j$. Associate with v the ordered pair $(b_i b_{i+1} \cdots b_{i-1}, b_j' b_{j+1} \cdots b_{i-1} b_j'')$. This sets up a bijection between the pairs (v, b_i) above and pairs (y_1, y_2), where $y_1 \in \mathcal{B}^*$, y_2 is an $\mathcal{A}^*$-conjugate of an element of $\mathcal{B}^*$, and $\ell(y_1) + \ell(y_2) = n$. Hence,

$$\sum_k kg_k(n) = \sum_{i=0}^n f(i)g(n-i).$$

By equation (4.42), this says $\lambda \frac{d}{d\lambda} B^*(\lambda) = B^*(\lambda) \widetilde{B}(\lambda)$. $\square$

Note that when $\mathcal{B}$ is finite, $\widetilde{B}(\lambda)$ and $B^*(\lambda)$ are rational. See Exercise 4.8 for further information on this situation.

4.7.15 Example. Let us take another look at Lemma 2.3.4 from the viewpoint of Lemma 4.7.14. Let $\mathcal{A} = \{0, 1\}$ and $\mathcal{B} = \{0, 10\}$. An $\mathcal{A}^*$-conjugate of an element of $\mathcal{B}_m^*$ that is also the product of $m - k$ words in $\mathcal{B}$ corresponds to choosing k points, no two consecutive, from a collection of m points arranged in a circle. (The position of the 1's corresponds to the selected points.) Since there are $\binom{m-k}{k}$ permutations of $m - 2k$ 0's and k 10's, we have $f_{m-k}(m) = \binom{m-k}{k}$. By Lemma 4.7.14, $g_{m-k}(m) = \frac{m}{m-k} \binom{m-k}{k}$, which is Lemma 2.3.4. Note that $f_1(1) = 0$, not 1, since a single point on a circle is regarded as being adjacent to itself.

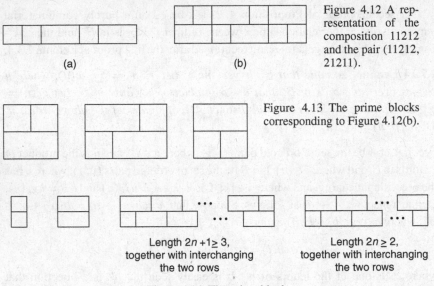

Figure 4.12 A representation of the composition 11212 and the pair (11212, 21211).

(a)   (b)

Figure 4.13 The prime blocks corresponding to Figure 4.12(b).

Length $2n+1 \geq 3$,
together with interchanging
the two rows

Length $2n \geq 2$,
together with interchanging
the two rows

Figure 4.14 The prime blocks.

4.7.16 Example. Recall from Exercise 1.35(c) that the Fibonacci number F_{n+1} counts the number of compositions of n into parts equal to 1 or 2. We may represent such a composition as a row of "bricks" of length 1 or 2; for example, the composition $1+1+2+1+2$ is represented by Figure 4.12(a). An ordered pair (α, β) of such compositions of n is therefore represented by two rows of bricks, as in Figure 4.12(b). The vertical line segments passing from top to bottom serve to "factor" these bricks into blocks of smaller length. For example, Figure 4.13 shows the factorization of Figure 4.12(b). The prime blocks (i.e., those that cannot be factored any further) are given by Figure 4.14. Since there are F_{n+1}^2 pairs (α, β), we conclude that

$$\sum_{n \geq 0} F_{n+1}^2 \lambda^n = \left(1 - \lambda - \lambda^2 - \frac{2\lambda^2}{1-\lambda}\right)^{-1}$$

$$= \frac{1-\lambda}{(1+\lambda)(1-3\lambda+\lambda^2)}.$$

In principle, the same type of reasoning would yield combinatorial evaluations of the generating functions $\sum_{n \geq 0} F_{n+1}^k \lambda^n$, where $k \in \mathbb{P}$. However, it is no longer easy to enumerate the prime blocks when $k \geq 3$. On the contrary, we can reverse the above reasoning to enumerate the prime blocks. For instance, it can be deduced from the explicit formula (4.5) for F_n (or otherwise) that

$$y := \sum_{n \geq 0} F_{n+1}^3 \lambda^n = \frac{1 - 2\lambda - \lambda^2}{1 - 3\lambda - 6\lambda^2 + 3\lambda^3 + \lambda^4}.$$

Let $g_3(n)$ be the number of prime blocks of length n and height 3, and set $z = \sum_{n\geq 1} g_3(n)\lambda^n$. Since $y = 1/(1 - z)$, we get

$$z = 1 - \frac{1}{y}$$

$$= \frac{\lambda + 5\lambda^2 - 3\lambda^3 - \lambda^4}{1 - 2\lambda - \lambda^2}$$

$$= \lambda + 7\lambda^2 + 12\lambda^3 + 30\lambda^4 + 72\lambda^5 + 174\lambda^6 + \cdots.$$

Can the recurrence $g_3(n + 2) = 2g_3(n + 1) + g_3(n)$ for $n \geq 3$ be proved combinatorially?

As a variant of the preceding one-row case, where the generating function is $F(\lambda) = 1/(1 - \lambda - \lambda^2)$, suppose that we have n points on a circle that we cover by bricks of length 1 or 2, where a brick of length i covers i consecutive points. Let $g(n)$ be the number of such coverings. If we choose the second point in clockwise order of each brick of length two, then we obtain a bijection with subsets of the n points, no two consecutive. Hence by Exercise 1.40, we have $g(n) = L_n$. On the other hand, by Proposition 4.7.13 we have

$$\sum_{n\geq 0} g(n)\lambda^n = \frac{\lambda\frac{d}{d\lambda}(\lambda + \lambda^2)}{1 - \lambda - \lambda^2} = \frac{\lambda + 2\lambda^2}{1 - \lambda - \lambda^2}.$$

Moreover, we can build a circular covering by bricks one unit at a time (say in clockwise order) by adding at each step either a brick of length one, the first half of a brick of length two, or the second half of a brick of length two. The rules for specifying what steps can follow what other steps are encoded by the transfer matrix

$$A = \begin{bmatrix} 1 & 1 & 0 \\ 0 & 0 & 1 \\ 1 & 1 & 0 \end{bmatrix}.$$

The eigenvalues of A are 0 and $(1 \pm \sqrt{5})/2$, so by equation (4.35) we get

$$g(n) = \left(\frac{1 + \sqrt{5}}{2}\right)^n + \left(\frac{1 - \sqrt{5}}{2}\right)^n.$$

We therefore have a theoretical explanation of why L_n has the simple form $\alpha^n + \beta^n$.

We now derive equations (4.39) and (4.40) using Propositions 4.7.11 and 4.7.13.

4.7.17 Example. Represent a permutation $a_1a_2 \cdots a_n \in \mathfrak{S}_n$ by drawing n vertices $v_1, \ldots, v_n$ in a line and connecting v_i to v_{a_i} by a directed edge. For instance, the permutation 31542 is represented by Figure 4.15. A permutation $a_1a_2 \cdots a_n \in \mathfrak{S}_n$ for which $|a_i - i| = 0$ or 1 is then represented as a sequence of the "prime" graphs G and H of Figure 4.16. In other words, if we set $\mathcal{A} = \{a, b, c\}$ amd $G = a$, $H = bc$, then the function $f(n)$ of Example 4.7.9 is just the number of words in

Figure 4.15 A representation of the permutation 31542.

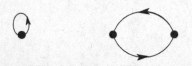

Figure 4.16 The prime graphs for permutations satisfying $|a_i - i| = 0, 1$.

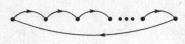

Figure 4.17 Graphs of two exceptional permutations.

and

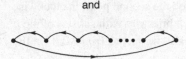

$\mathcal{B}_n^*$, where $\mathcal{B} = \{a, bc\}$. Setting $w(a) = w(b) = w(c) = 1$, we therefore have by Proposition 4.7.11 that

$$\sum_{n \geq 0} f(n) \lambda^n = \mathcal{B}^*(\lambda) = (1 - \mathcal{B}(\lambda))^{-1},$$

where $\mathcal{B}(\lambda) = w(a)\lambda^{\ell(a)} + w(bc)\lambda^{\ell(bc)} = \lambda + \lambda^2$. Consider now the number $g(n)$ of permutations $a_1 a_2 \cdots a_n \in \mathfrak{S}_n$ such that $a_i = 0, \pm 1 \pmod{n}$. Every cyclic shift of a word in $\mathcal{B}^*$ gives rise to one such permutation. There are exactly two other such permutations ($n \geq 3$), namely, $234 \cdots n1$ and $n123 \cdots (n-1)$, as shown in Figure 4.17. Hence,

$$\sum_{n \geq 1} g(n) \lambda^n = \frac{\lambda \frac{d}{d\lambda} \mathcal{B}(\lambda)}{1 - \mathcal{B}(\lambda)} + \sum_{n \geq 1} 2\lambda^n$$

$$= \frac{\lambda(1 + 2\lambda)}{1 - \lambda - \lambda^2} + \frac{2\lambda}{1 - \lambda},$$

provided of course we suitably interpret $g(1)$ and $g(2)$.

4.7.18 Example. Let $f(n)$ be the number of permutations $a_1 a_2 \cdots a_n \in \mathfrak{S}_n$ with $a_i - i = \pm 1$ or ± 2. To use the transfer-matrix method would be quite unwieldy, but the factorization method is very elegant. A permutation enumerated by $f(n)$ is represented by a sequence of graphs of the types shown in Figure 4.18. Hence,

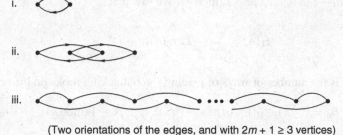

i.

ii.

iii.

(Two orientations of the edges, and with $2m + 1 \geq 3$ vertices)

iv.

(Two orientations of the edges, and with $2m \geq 4$ vertices)

Figure 4.18 The prime graphs for permutations satisfying $a_i - i = \pm 1$ or ± 2.

Figure 4.19 Two additional prime graphs.

v.

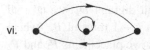

vi.

$\mathcal{B}(\lambda) = \lambda^2 + \lambda^4 + 2 \sum_{m \geq 3} \lambda^m$, and

$$\sum_{n \geq 0} f(n) \lambda^n = \mathcal{B}^*(\lambda) = \left(1 - \lambda^2 - \lambda^4 - \frac{2\lambda^3}{1 - \lambda} \right)^{-1}$$

$$= \frac{1 - \lambda}{1 - \lambda - \lambda^2 - \lambda^3 - \lambda^4 + \lambda^5}.$$

Suppose now that we also allow $a_i - i = 0$. Thus, let $f^*(n)$ be the number of permutations $a_1 a_2 \cdots a_n \in \mathfrak{S}_n$ with $a_i - i = \pm 1, \pm 2$, or 0. There are exactly two new elements of $\mathcal{B}$ introduced by this change, shown in Figure 4.19. Hence,

$$\sum_{n \geq 0} f^*(n) \lambda^n = \left(1 - \lambda - \lambda^2 - \lambda^3 - \lambda^4 - \frac{2\lambda^3}{1 - \lambda} \right)^{-1}$$

$$= \frac{1 - \lambda}{1 - 2\lambda - 3\lambda^3 + \lambda^5}.$$

4.7.19 Example (k-discordant permutations). In Section 2.3, we discussed the problem of counting the number $f_k(n)$ of k-*discordant permutations* $a_1 a_2 \cdots a_n \in$

$\mathfrak{S}_n$, that is, $a_i - i \not\equiv 0, 1, \ldots, k-1 \pmod{n}$. We saw that

$$f_k(n) = \sum_{i=0}^{n} (-1)^i r_i(n)(n-i)!,$$

where $r_i(n)$ is the number of ways of placing i nonattacking rooks on the board

$$B_n = \{(r,s) \in [n] \times [n] : s - r \equiv 0, 1, \ldots, k-1 \pmod{n}\}.$$

The evaluation of $r_i(n)$, or equivalently the rook polynomial $R_n(x) = \sum_i r_i(n)x^i$, can be accomplished by methods analogous to those used to determine $g_S(n)$ in Proposition 4.7.10. The transfer-matrix method will tell us the general form of the generating function $F_k(x, y) = \sum_{n \geq 1} R_n(x)y^n$ (suitably interpreting $R_n(x)$ for $n < k$), whereas the factorization method will enable us to compute $F_k(x, y)$ easily when k is small.

First, we consider the transfer-matrix approach. We begin with the first row of B_n and either place a rook in a square of this row or leave the row empty. We then proceed to the second row, either placing a rook that doesn't attack a previously placed rook or leaving the row empty. If we continue in this manner, then the options available to us at the ith row depend on the configuration of the rooks on the previous $k-1$ rows. Hence, for the vertices of our digraph D_k, we take all possible placements of nonattacking rooks on the first $k-1$ rows of B_n (where $n \geq 2k-1$ to allow all possibilities). An edge connects two placements P_1 and P_2 if the last $k-2$ rows are identical to the first $k-2$ rows of P_2, and if we overlap P_1 and P_2 in this way (yielding a configuration with k rows), then the rooks remain nonattacking. For instance, D_2 is given by Figure 4.20. There is no arrow from v_2 to v_3 since their overlap would be shown in Figure 4.21(a), which is not allowed. Similarly D_3 has 14 vertices, a typical edge being shown in Figure 4.21(b). If we overlap these two vertices, then we obtain the legal configuration shown in Figure 4.21(c). Define the weight $w(P_1, P_2)$ of an edge (P_1, P_2) to be $x^{\nu(P_2)}$, where $\nu(P_2)$ is the number of rooks in the last row of P_2. It is then clear that a closed walk Γ of length n and weight $x^{\nu(\Gamma)}$ in D_k corresponds to a placement of $\nu(\Gamma)$ nonattacking rooks on B_n (provided $n \geq k$). Hence, if A_k is the adjacency matrix of D_k with respect to the weight function w, then

$$R_n(x) = \operatorname{tr} A_k^n, \quad n \geq k.$$

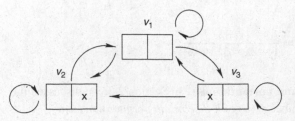

Figure 4.20 The nonattacking rook digraph D_2.

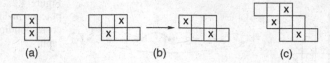

Figure 4.21 Nonedges and edges in the digraphs D_2 and D_3.

Thus, if we set $Q_k(\lambda) = \det(I - \lambda A_k) \in \mathbb{C}[x, \lambda]$, then by Corollary 4.7.3 we conclude

$$\sum_{n\geq 1} R_n(x)\lambda^n = -\frac{\lambda Q'_k(\lambda)}{Q_k(\lambda)}. \tag{4.43}$$

For instance, when $k = 2$ (the "problème des ménages") then with the vertex labeling given by Figure 4.20, we read off from Figure 4.20 that

$$A_k = \begin{bmatrix} 1 & x & x \\ 1 & x & 0 \\ 1 & x & x \end{bmatrix},$$

so that

$$Q_2(\lambda) = \det \begin{bmatrix} 1-\lambda & -\lambda x & -\lambda x \\ -\lambda & 1-\lambda x & 0 \\ -\lambda & -\lambda x & 1-\lambda x \end{bmatrix}$$

$$= 1 - \lambda(1 + 2x) + \lambda^2 x^2.$$

Therefore,

$$\sum_{n\geq 1} R_n(x)\lambda^n = \frac{\lambda(1+2x) - 2\lambda^2 x^2}{1 - \lambda(1+2x) + \lambda^2 x^2} \quad (k=2).$$

The preceding technique, applied to the case $k = 3$, would involve the determinant of a 14×14 matrix. The factorization method yields a much easier derivation. Regard a placement P of nonattacking rooks on B_n (or on any subset of $[n] \times [n]$) as a digraph with vertices $1, 2, \ldots, n$, and with a directed edge from i to j if a rook is placed in row i and column j. For instance, the placement shown in Figure 4.22(a) corresponds to the digraph shown in Figure 4.22(b). In the case $k = 2$, every such digraph is a sequence of the primes shown in Figure 4.23(a), together with the additional digraph shown in Figure 4.23(b). If we weight such a digraph with q edges by x^q, then by Proposition 4.7.13 there follows

$$\sum_{n\geq 1} R_n(x)\lambda^n = \frac{\lambda \frac{d}{d\lambda}\mathcal{B}(\lambda)}{1 - \mathcal{B}(\lambda)} + \sum_{n\geq 2} x^n \lambda^n, \tag{4.44}$$

where

$$\mathcal{B}(\lambda) = x\lambda + \sum_{i\geq 1} x^{i-1}\lambda^i$$

$$= x\lambda + \frac{\lambda}{1 - x\lambda}.$$

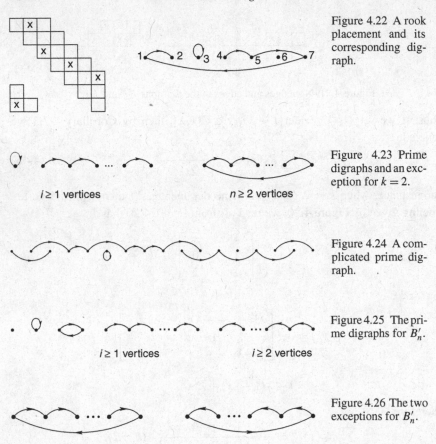

Figure 4.22 A rook placement and its corresponding digraph.

Figure 4.23 Prime digraphs and an exception for $k = 2$.

Figure 4.24 A complicated prime digraph.

Figure 4.25 The prime digraphs for B'_n.

Figure 4.26 The two exceptions for B'_n.

This yields the same answer as before, except that we get the correct value $R_1(x) = 1 + x$ rather than the spurious value $R_1(x) = 1 + 2x$. To obtain $R_1(x) = 1 + 2x$, we would have to replace $\sum_{n \geq 2} x^n \lambda^n$ in equation (4.44) by $\sum_{n \geq 1} x^n \lambda^n$. Thus in effect, we are counting the first digraph of Figure 4.23(a) twice, once as a prime and once as an exception.

When the foregoing method is applied to the case $k = 3$, it first appears extremely difficult because of the complicated set of prime digraphs that can arise, as in Figure 4.24. A simple trick eliminates this problem; namely, instead of using the board $B_n = \{(j,j), (j,j+1), (j,j+2) \,(\mathrm{mod}\, n)\}$, use instead $B'_n = \{(j,j-1), (j,j), (j,j+1) \,(\mathrm{mod}\, n)\}$. Clearly, B_n and B'_n are isomorphic and therefore have the same rook polynomials, but surprisingly B'_n has a much simpler set of prime placements than B_n. The primes for B'_n are given by Figure 4.25. In addition, there are exactly two exceptional placements, shown in Figure 4.26. Hence,

$$\sum_{n \geq 1} R_n(x) \lambda^n = \frac{\lambda \frac{d}{d\lambda} \mathcal{B}(\lambda)}{1 - \mathcal{B}(\lambda)} + 2 \sum_{n \geq 3} x^n \lambda^n, \qquad (4.45)$$

where

$$\mathcal{B}(\lambda) = \lambda + x\lambda + x^2\lambda^2 + 2\sum_{i \ge 2} x^{i-1}\lambda^i$$

$$= \lambda + x\lambda + x^2\lambda^2 + \frac{2x\lambda^2}{1-x\lambda}.$$

If we replace $\sum_{n \ge 3} x^n\lambda^n$ in equation (4.45) by $\sum_{n \ge 1} x^n\lambda^n$ (causing $R_1(x)$ and $R_2(x)$ to be spurious), then after simplification there results

$$\sum_{n \ge 1} R_n(x)\lambda^n = \frac{\lambda(1 + 2x + 2x\lambda - 3x^3\lambda^2)}{1 - (1+2x)\lambda - x\lambda^2 + x^3\lambda^3} + \frac{x\lambda}{1-x\lambda}.$$

4.7.5 Some Sums Over Compositions

Here we will give a more complex use of the transfer-matrix than treated previously.

A *polyomino* is a finite union P of unit squares in the plane such that the vertices of the squares have integer coordinates, and P is connected and has no finite cut set. Two polynominoes will be considered *equivalent* if there is a translation that transforms one into the other (reflections and rotations not allowed). A polynomino P is *horizontally convex* (or HC) if each "row" of P is an unbroken line of squares, that is, if L is any line segment parallel to the x-axis with its two endpoints in P, then $L \subset P$. Let $f(n)$ be the number of HC-polyominoes with n squares. Thus $f(1) = 1$, $f(2) = 2$, $f(3) = 6$, as shown by Figure 4.27. Suppose that we build up an HC-polyomino one row at a time, starting at the bottom. If the ith row has r squares, then we can add an $(i+1)$-st row of s squares in $r + s - 1$ ways. It follows that

$$f(n) = \sum (n_1 + n_2 - 1)(n_2 + n_3 - 1)\cdots(n_s + n_{s+1} - 1), \qquad (4.46)$$

where the sum is over all 2^{n-1} compositions $n_1 + n_2 + \cdots + n_{s+1}$ of n (where the composition with $s = 0$ contributes 1 to the sum). This formula suggests studying the more general sum, over all compositions $n_1 + n_2 + \cdots + n_{s+k-1} = n$ with $s \ge 0$, given by

$$f(n) = \sum (f_1(n_1) + f_2(n_2) + \cdots + f_k(n_k))(f_1(n_2) + f_2(n_3) + \cdots + f_k(n_{k+1}))$$
$$\cdots (f_1(n_s) + f_2(n_{s+1}) + \cdots + f_k(n_{s+k-1})), \qquad (4.47)$$

where $f_1, \ldots, f_k$ are arbitrary functions from $\mathbb{P} \to \mathbb{C}$ (or to any commutative ring R). We make the convention that the term in equation (4.47) with $s = 0$ is 1. The

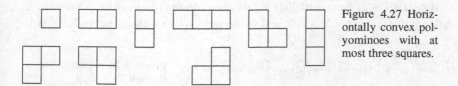

Figure 4.27 Horizontally convex polyominoes with at most three squares.

situation (4.46) corresponds to $f_1(m) = m + \alpha$ and $f_2(m) = m - \alpha - 1$ for any fixed $\alpha \in \mathbb{C}$.

It is surprising that the transfer-matrix method can be used to write down an explicit expression for the generating function $F(x) = \sum_{n \geq 1} f(n) x^n$ in terms of the generating functions $F_i(x) = \sum_{n \geq 1} f_i(n) x^n$. We may compute a typical term of the product appearing in equation (4.47) by first choosing a term $f_{i_1}(n_{i_1})$ from the first factor $\phi_1 = f_1(n_1) + f_2(n_2) + \cdots + f_k(n_k)$, then a term $f_{i_2}(n_{i_2+1})$ from the second factor $\phi_2 = f_1(n_2) + f_2(n_3) + \cdots + f_k(n_{k+1})$, and so on, and finally multiplying these terms together.

Alternatively we could have obtained this term by first deciding from which factors we choose a term of the form $f_{i_1}(n_1)$, then deciding from which factors we choose a term of the form $f_{i_2}(n_2)$, and so on. Once we've chosen the terms $f_{i_j}(n_{i_j})$, the possible choices for $f_{i_{j+1}}(n_{i_{j+1}})$ are determined by which of the $k - 1$ factors $\phi_{j-k+2}, \phi_{j-k+3}, \ldots, \phi_j$ we have already chosen a term from. Hence, define a digraph D_k with vertex set $V = \{(\varepsilon_1, \ldots, \varepsilon_{k-1}) : \varepsilon_i = 0 \text{ or } 1\}$. The vertex $(\varepsilon_1, \ldots, \varepsilon_{k-1})$ indicates that we have already chosen a term from ϕ_{j-k+l} if and only if $\varepsilon_{l-1} = 1$. Draw an edge from $(\varepsilon_1, \ldots, \varepsilon_{k-1})$ to $(\varepsilon'_1, \ldots, \varepsilon'_{k-1})$ if it is possible to choose terms of the form $f_{i_j}(n_{j+1})$ consistent with $(\varepsilon_1, \ldots, \varepsilon_{k-1})$, and then of the form $f_{i_{j+1}}(n_{j+1})$ consistent with $(\varepsilon'_1, \ldots, \varepsilon'_{k-1})$ and our choice of $f_{i_j}(n_j)$'s. Specifically, this means that $(\varepsilon'_1, \ldots, \varepsilon'_{k-1})$ can be obtained from $(\varepsilon_2, \ldots, \varepsilon_{k-1}, 0)$ by changing some 0's to 1's. It now follows that a path in D_k of length $s + k - 1$ that starts at $(1, 1, \ldots, 1)$ (corresponding to the fact that when we first pick out terms of the form $f_{i_1}(n_{i_1})$, we cannot choose from nonexistent factors prior to ϕ_1) and ends at $(0, 0, \ldots, 0)$ (since we cannot have chosen from nonexistent factors following ϕ_s) corresponds to a term in the expansion of $\phi_1 \phi_2 \cdots \phi_s$. For instance, if $k = 3$, then the term $f_3(n_3) f_1(n_2) f_1(n_3) f_2(n_5) f_3(n_7)$ in the expansion of $\phi_1 \phi_2 \cdots \phi_5$ corresponds to the path shown in Figure 4.28. The first edge in the path corresponds to choosing no term $f_{i_1}(n_1)$, the second edge to choosing $f_1(n_2)$, the third to $f_1(n_3) f_3(n_3)$, the fourth to no term $f_{i_4}(n_4)$, the fifth to $f_2(n_5)$, the sixth to no term $f_{i_6}(n_6)$, and the seventh to $f_3(n_7)$.

We now have to consider the problem of weighting the edges of D_k. For definiteness, consider for example the edge e from $v = (0, 0, 1, 0, 0, 1)$ to $v' = (1, 1, 0, 1, 1, 0)$. This means that we have chosen a factor $f_3(m) f_6(m) f_7(m)$, as illustrated schematically by

	7	6	5	4	3	2	1
v	0	0	1	0	0	1	
v'		1	1	0	1	1	0

If $2 \leq i \leq k - 1$, then we include $f_i(m)$ when column i is given by $\begin{smallmatrix} 0 \\ 1 \end{smallmatrix}$. We include $f_k(m)$ if the first entry of v is 0, and we include $f_1(m)$ if the last entry of v' is 1.

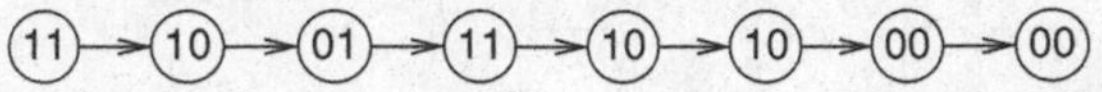

$$\boxed{11} \rightarrow \boxed{10} \rightarrow \boxed{01} \rightarrow \boxed{11} \rightarrow \boxed{10} \rightarrow \boxed{10} \rightarrow \boxed{00} \rightarrow \boxed{00}$$

Figure 4.28 A path in the digraph D_3.

We are free to choose m to be any positive integer. Thus, if we weight edge e with the generating function

$$\sum_{m \geq 1} f_3(m) f_6(m) f_7(m) x^m = F_3 * F_6 * F_7,$$

where $*$ denotes the Hadamard product, then the total weight of a path from $(1, 1, \ldots, 1)$ to $(0, 0, \ldots, 0)$ is precisely the contribution of this path to the generating function $F(x)$. Note that in the case of an edge e where we pick *no* terms of the $f_i(m)$ for fixed m, then we are contributing a factor of 1, so that the edge must be weighted by $\sum_{m \geq 1} x^m = x/(1-x)$, which we will denote as $J(x)$. Since there is no need to keep track of the length of the path, it follows from Theorem 4.7.2 that $F(x) = F_{ij}(D_k, 1)$, where i is the index of $(1, 1, \ldots, 1)$ and j of $(0, 0, \ldots, 0)$. (In general, it is meaningless to set $\lambda = 1$ in $F_{ij}(D, \lambda)$, but here the weight function has been chosen so that $F_{ij}(D_k, 1)$ is a well-defined formal power series. Of course, if we wanted to do so, we could consider the more refined generating function $F_{ij}(D_k, \lambda)$, which keeps track of the number of parts of each composition.)

We can sum up our conclusions in the following result.

4.7.20 Proposition. *Let A_k be the following $2^{k-1} \times 2^{k-1}$ matrix whose rows and columns are indexed by $V = \{0, 1\}^{k-1}$. If $v = (\varepsilon_1, \ldots, \varepsilon_{k-1})$, $v' = (\varepsilon'_1, \ldots, \varepsilon'_{k-1}) \in V$, then define the (v, v')-entry of A_k as follows:*

$$(A_k)_{vv'} = \begin{cases} 0, & \text{if for some } 1 \leq i \leq k-2, \text{ we have } \varepsilon_{i+1} = 1, \\ & \text{and } \varepsilon'_i = 0, \\ F_{i_1} * \cdots * F_{i_r}, & \text{otherwise, where } \{i_1, \ldots, i_r\} = \{i : \varepsilon_{k-i+1} = 0 \text{ and} \\ & \varepsilon'_{k-i} = 1\}, \text{ and where we set } \varepsilon_k = 0, \varepsilon'_0 = 1, \text{ and an} \\ & \text{empty Hadamard product equal to } J = x/(1-x). \end{cases}$$

Let B_k be the matrix obtained by deleting row $(0, 0, \ldots, 0)$ and column $(1, 1, \ldots, 1)$ from $I - A_k$ (where I is the identity matrix) and multiplying by the appropriate sign. Then the generating function $F(x) = \sum_{n \geq 1} f(n) x^n$, as defined by equation (4.47), is given by

$$F(x) = \frac{\det B_k}{\det(I - A_k)}.$$

In particular, if each $F_i(x)$ is rational, then $F(x)$ is rational, by Proposition 4.2.5.

Here are some small examples. When $k = 2$, we have D_2 given by Figure 4.29, while

Figure 4.29 The digraph D_2.

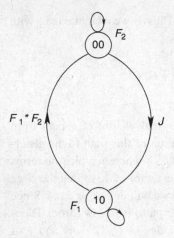

$$A_2 = \begin{bmatrix} F_2 & F_1 * F_2 \\ J & F_1 \end{bmatrix}, \quad B_2 = [J],$$

$$F(x) = \frac{J}{(1 - F_1)(1 - F_2) - J \cdot (F_1 * F_2)}. \tag{4.48}$$

In the original problem of enumerating HC-polyominoes,

$$F_1(x) = \sum_{n \geq 1} n x^n = x/(1 - x)^2,$$

$$F_2(x) = \sum_{n \geq 1} (n - 1) x^n = x^2/(1 - x)^2,$$

$$(F_1 * F_2)(x) = \sum_{n \geq 1} n(n - 1) x^n = 2x^2/(1 - x)^3,$$

yielding

$$F(x) = \frac{x/(1 - x)}{\left(1 - \dfrac{x}{(1 - x)^2}\right)\left(1 - \dfrac{x^2}{(1 - x)^2}\right) - \dfrac{x}{1 - x} \cdot \dfrac{2x^2}{(1 - x)^3}}$$

$$= \frac{x(1 - x)^3}{1 - 5x + 7x^2 - 4x^3}.$$

It is by no means obvious that $f(n)$ satisfies the recurrence

$$f(n + 3) = 5f(n + 2) - 7f(n + 1) + 4f(n), \quad n \geq 2, \tag{4.49}$$

and it is difficult to give a combinatorial proof.

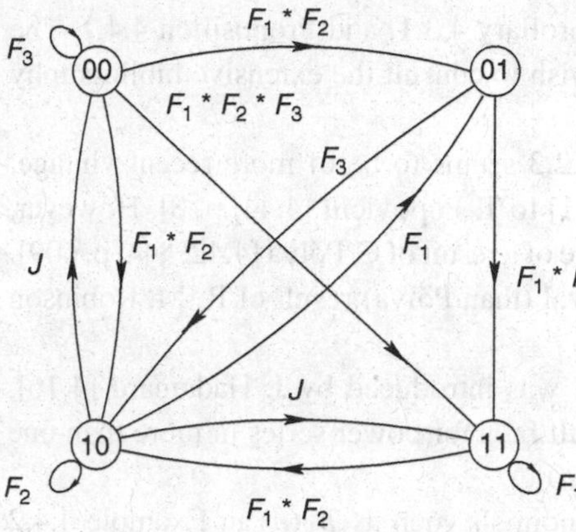

Figure 4.30 The digraph D_3.

Finally, let us consider the case $k = 3$. Figure 4.30 shows D_3, while

$$A_3 = \begin{bmatrix} F_3 & F_1 * F_3 & F_2 * F_3 & F_1 * F_2 * F_3 \\ 0 & 0 & F_3 & F_1 * F_3 \\ J & F_1 & F_2 & F_1 * F_2 \\ 0 & 0 & J & F_1 \end{bmatrix},$$

$$B_3 = \begin{bmatrix} 0 & 1 & -F_3 \\ -J & -F_1 & 1 - F_2 \\ 0 & 0 & J \end{bmatrix},$$

$$F(x) = \frac{J^2}{\det(I - A_3)},$$

where

$$\det(I - A_3) = (1 - F_1)(1 - F_3)(1 - F_2 - F_1 F_3)$$
$$- J(1 - F_1)(F_2 * F_3 + F_3(F_1 * F_3))$$
$$- J(1 - F_3)(F_1 * F_2 + F_1(F_1 * F_3))$$
$$- J^2((F_1 * F_3)^2 + F_1 * F_2 * F_3).$$

Notes

The basic theory of rational generating functions in one variable belongs to the calculus of finite differences. Charles Jordan [4.25] ascribes the origin of this calculus to Brook Taylor in 1717 but states that the real founder was James Stirling in 1730. The first treatise on the subject was written by Euler in 1755, where the notation Δ for the difference operator was introduced. It would probably be an arduous task to ascertain the precise origin of the various parts of Theorem 4.1.1, Corollary 4.2.1,

Proposition 4.2.2, Proposition 4.2.5, Corollary 4.3.1, and Proposition 4.4.1. The reader interested in this question may wish to consult the extensive bibliography in Nörlund [4.39].

The reciprocity result Proposition 4.2.3 seems to be of more recent vintage. It is attributed by E. Ehrhart [4.12, p. 21] to T. Popoviciu [4.44, p. 8]. However, Proposition 4.2.3 is actually a special case of a result of G. Pólya [4.42, §44, p. 609]. It is also a special case of the less general (than Pólya) result of R. M. Robinson [4.47, §3].

The operation of Hadamard product was introduced by J. Hadamard [4.16], who proved Proposition 4.2.5. This result fails for power series in more than one variable, as observed by A. Hurwitz [4.20].

Methods for dealing with quasipolynomials such as $\overline{p}_k(n)$ in Example 4.4.2 were developed by Herschel, Cayley, Sylvester, Glaisher, Bell, and others. For references, see [2.3, §2.6]. Some interesting properties of quasipolynomials are given by I. G. Macdonald as an appendix to the monograph [4.14, pp. 145–155] of Ehrhart and by N. Li and S. Chen [4.30, §3].

The theory of linear homogeneous diophantine equations developed in Section 4.5 was investigated in the weaker context of Ehrhart quasipolynomials by E. Ehrhart beginning around 1955. (It is remarkable that Ehrhart did most of his work as a teacher in a *lycée* and did not receive his Ph.D. until 1966 at the age of 59 or 60.) Ehrhart's work is collected together in his monograph [4.14], which contains detailed references. Some aspects of Ehrhart's work were corrected, streamlined, and expanded by I. G. Macdonald [4.34][4.35].

The extension of Ehrhart's work to linear homogeneous diophantine equations appeared in Stanley [4.52] and is further developed in [4.54][4.57]. In these references, commutative algebra is used as a fundamental tool. The approach given here in Section 4.5 is more in line with Ehrhart's original work. Reference [4.57] is primarily concerned with *inhomogeneous* equations and the extension of Theorem 4.5.14 (reciprocity) to this case. A more elementary but less comprehensive approach to inhomogeneous equations and reciprocity is given in [4.53, §§8–11]; see also Exercises 4.34 and 4.35. For further background information on convex polytopes, see Ziegler [3.97].

Other approaches toward "Ehrhart theory" appear in M. Beck and F. Sottile [4.5], P. McMullen [4.37], S. V. Sam [4.48], S. V. Sam and K. M. Woods [4.49], and R. Stanley [4.56]. A nice exposition of Ehrhart theory and related topics at the undergraduate level is given by M. Beck and S. Robins [4.4].

The triangulation defined in the proof of Lemma 4.5.1 is called the *pulling triangulation* and has several other descriptions. See for instance Beck–Robins [4.4, Appendix] and De Loera–Rambau–Santos Leal [4.11, §4.3.2]. Our description of the pulling triangulation follows Stanley [4.56, Lemma 1.1].

The study of "magic squares" (as defined in Section 4.6) was initiated by MacMahon [4.36][1.55, §404–419]. In the first of these two references MacMahon writes down in Art. 129 a multivariate generating function for all 3×3 magic squares, though he doesn't explictly write down a formula for $H_3(r)$. In the second reference he does give the formula in §407. For MacMahon's proof, see Exercise 2.15. Proposition 4.6.2 was conjectured by H. Anand, V. C. Dumir, and H. Gupta [4.1] and was first proved by Stanley [4.52]. Ehrhart [4.13] also gave a proof of Proposition 4.6.2 using his methods. An elementary proof (essentially an application of the transfer-matrix method) of part of Proposition 4.6.2 was given by J. H. Spencer [4.51]. The fundamental Lemma 4.6.1 on which Proposition 4.6.2 rests is due to Garrett Birkhoff [4.6]. It was rediscovered by J. von Neumann [4.61] and is sometimes called the "Birkhoff–von Neumann theorem." The proof given here is that of von Neumann. There are several papers earlier than that of Birkhoff that are equivalent to or easily imply the Birkhoff–von Neumann theorem. Perhaps the first such results are two nearly identical papers, one in German and one in Hungarian, by D. König [4.27][4.28].

L. Carlitz [4.8, p. 782] conjectured that Proposition 4.6.4 is valid for some *constant* $Q_n(r)$ and proved this fact for $n \leq 4$. The value of $G_5(r)$ given after Proposition 4.6.4 shows that Carlitz's conjecture is false for $n = 5$. Proposition 4.6.4 itself was first proved by Stanley [4.52], and a refinement appears in [4.54, Thm. 5.5]. In particular, it was shown that

$$\deg Q_n(r) \leq \begin{cases} \binom{n-1}{2} - 1, & n \text{ odd}, \\ \binom{n-2}{2} - 1, & n \text{ even}, \end{cases} \tag{4.50}$$

and it was conjectured that equality holds for all n. This conjecture was proved by R.-Q. Jia, [4.23][4.24]. The values of $F_n(\lambda)$ (given for $n \leq 5$ preceding Lemma 4.6.3) were computed for $n \leq 6$ by D. M. Jackson and G. H. J. van Rees [4.21]. They were extended to $n \leq 9$ by M. Beck and D. Pixton [4.3]. The values for $G_n(\lambda)$ for $n \leq 5$ appearing after Proposition 4.6.4 were first given in [4.54].

Example 4.6.15(a) is a classical result of G. A. Pick [4.41]. The extension (b) to three dimensions is due to J. E. Reeve [4.45], while the general case (c) (or even more general Corollary 4.6.14) is due to Macdonald [4.34].

The connection between the powers A^n of the adjacency matrix A of a digraph D and the counting of walks in D (Theorem 4.7.1) is part of the folklore of graph theory. An extensive account of the adjacency matrix A is given by D. M. Cvetković, M. Doob, and H. Sachs [4.10]; see §1.8 and §7.5 in particular for its use in counting walks. We should also mention that the transfer-matrix method is essentially the same as the theory of finite Markov chains in probability theory. For a noncommutative version of the transfer-matrix method, see §6.5 of volume 2 of the present text.

The transfer-matrix method has been used with great success by physicists in the study of phase transitions in statistical mechanics. See for instance Baxter [4.2] and Percus [4.40] for further information.

For more information on Example 4.7.7, see Exercise 4.40 and the references given there. For work related to Examples 4.7.9, 4.7.17, and 4.7.18, see Lagrange [4.29] and Metropolis, Stein, and Stein [4.38], and the references given there. These approaches are less combinatorial than ours.

Our discussion of factorization in free monoids merely scratched the surface of an extensive subject. An excellent overall reference is Lothaire [4.31], from which we have taken most of our terminology and notation. Sequels appear in [4.32][4.33]. Other interesting references include Cohn [4.9] and Fliess [4.15]. The application to summing $\sum F_{n+1}^2 \lambda^n$ (Example 4.7.16) appears in Shapiro [4.50]. For more information on powers of Fibonacci numbers, see Jarden and Motzkin [4.22], Hathaway and Brown [4.17], Riordan [4.46], Carlitz [4.7], and Horadam [4.19].

Two topics with close connections to factorization in monoids are the combinatorial theory of orthogonal polynomials and the theory of heaps. Basic references are two papers [4.59][4.60] of X. G. Viennot.

The first published statement for the generating function $F(x)$ for HC-polyominoes appearing before equation (4.49) seems to be due to H. N. V. Temperley [4.58]. Earlier the recurrence (4.49) was found by Pólya in 1938 but was unpublished by him until 1969 [4.43]. A proof of the more general equation (4.48) is given by Klarner [4.26], while an algebraic version of this proof appears in Stanley [4.55, Ex. 4.2]. The elegant transfer-matrix approach given here was suggested by I. M. Gessel. The combinatorial proof of equation (4.49) alluded to after (4.49) is due to D. R. Hickerson [4.18].

Bibliography

[1] H. Anand, V. C. Dumir, and H. Gupta, A combinatorial distribution problem, *Duke Math. J.* **33** (1966), 757–770.

[2] R. J. Baxter, *Exactly Solved Models in Statistical Mechanics*, Academic Press, London/New York, 1982.

[3] M. Beck and D. Pixton, The Ehrhart polynomial of the Birkhoff polytope, *Discrete Comput. Geom.* **30** (2003), 623–637.

[4] M. Beck and S. Robins, *Computing the Continuous Discretely: Integer-Point Enumeration in Polyhedra*, Springer-Verlag, New York, 2007.

[5] M. Beck and F. Sottile, Irrational proofs for three theorems of Stanley, *European J. Combinatorics* **28** (2007), 403–409.

[6] G. Birkhoff, Tres observaciones sobre el algebra lineal, *Univ. Nac. Tucamán Rev. Ser. (A)* **5** (1946), 147–150.

[7] L. Carlitz, Generating functions for powers of a certain sequence of numbers, *Duke Math. J.* **29** (1962), 521–537.

[8] L. Carlitz, Enumeration of symmetric arrays, *Duke Math. J.* **33** (1966), 771–782.

[9] P. M. Cohn, Algebra and language theory, *Bull. London Math. Soc.* **7** (1975), 1–29.

[10] D. M. Cvetković, M. Doob, and H. Sachs, *Spectra of Graphs*, 3rd ed., Johann Ambrosius Barth, Heidelberg, 1995.

[11] J. A. De Loera, J. Rambau, and F. Santos Leal, *Triangulations*, Springer-Verlag, Berlin, 2010.

[12] E. Ehrhart, Sur les problème de géométrie diophantienne linéaire I, II, *J. Reine Angew. Math.* **226** (1967), 1–29, and **227** (1967), 25–49. Correction, **231** (1968), 220.

[13] E. Ehrhart, Sur les carrés magiques, *C. R. Acad. Sci. Paris* **227 A** (1973), 575–577.

[14] E. Ehrhart, *Polynômes arithmétiques et Méthode des Polyèdres en Combinatoire*, International Series of Numerical Mathematics, vol. 35, Birkhäuser, Basel/Stuttgart, 1977.

[15] M. Fliess, Sur divers produits de séries formelles, *Bull. Soc. Math. France* **102** (1974), 181–191.

[16] J. S. Hadamard, Théorèm sur les séries entières, *Acta Math.* **22** (1899), 55–63.

[17] D. K. Hathaway and S. L. Brown, Fibonacci powers and a fascinating triangle, *College Math. J.* **28** (1997), 124–128.

[18] D. R. Hickerson, Counting horizontally convex polyominoes, *J. Integer Sequences* **2** (1999), article 99.1.8.

[19] A. F. Horadam, Generating functions for powers of a certain generalized sequence of numbers, *Duke Math. J.* **32** (1965), 437–446.

[20] A. Hurwitz, Sur une théorème de M. Hadamard, *C. R. Acad. Sci. Paris* **128** (1899), 350–353.

[21] D. M. Jackson and G. H. J. van Rees, The enumeration of generalized double stochastic nonnegative integer square matrices, *SIAM J. Comput.* **4** (1975), 474–477.

[22] D. Jarden and T. Motzkin, The product of sequences with a common linear recursion formula of order 2 (Hebrew with English summary), *Riveon Lematematika* **3** (1949), 25–27, 38.

[23] R.-Q. Jia, Symmetric magic squares and multivariate splines, *Linear Algebra Appl.* **250** (1997), 69–103.

[24] R.-Q. Jia, Multivariate discrete splines and linear Diophantine equations, *Trans. Amer. Math. Soc.* **340** (1993), 179–198.

[25] C. Jordan, *Calculus of Finite Differences*, 3rd ed., Chelsea, New York, 1965.

[26] D. A. Klarner, A combinatorial formula involving the Fredholm integral equation, *J. Combinatorial Theory* **5** (1968), 59–74.

[27] D. König, Grafok és alkalmaszásuk és a halmazok elméletére, *Mat. Termész. Ért.* **34** (1916), 104–119.

[28] D. König, Über Graphen und ihre Anwendung auf Determinantentheorie and Mengenlehre, *Math. Ann.* **77** (1916), 453–465.

[29] M. R. Lagrange, Quelques résultats dans la métrique des permutations, *Ann. scient. Éc. Norm. Sup.* **79** (1962), 199–241.

[30] N. Li and S. Chen, On Popoviciu type formulas for generalized restricted partition function, arXiv:0709.3571.

[31] M. Lothaire, *Combinatorics on Words*, Addison-Wesley, Reading, Mass., 1983; reprinted by Cambridge University Press, Cambridge, 1997.

[32] M. Lothaire, *Algebraic Combinatorics on Words*, Encyclopedia of Mathematics and Its Applications **90**, Cambridge University Press, Cambridge, 2002.

[33] M. Lothaire, *Applied Combinatorics on Words*, Encyclopedia of Mathematics and Its Applications **105**, Cambridge University Press, Cambridge, 2005.

[34] I. G. Macdonald, The volume of a lattice polyhedron, *Proc. Camb. Phil. Soc.* **59** (1963), 719–726.

[35] I. G. Macdonald, Polynomials associated with finite cell complexes, *J. London Math. Soc. (2)* **4** (1971), 181–192.

[36] P. A. MacMahon, Memoir on the theory of partitions of numbers—Part III, *Phil. Trans.* **205** (1906), 37–58.

[37] P. McMullen, Lattice invariant valuations on rational polytopes, *Arch. Math. (Basel)* **31** (1978/79), 509–516.

[38] N. C. Metropolis, M. L. Stein, and P. R. Stein, Permanents of cyclic (0,1)-matrices, *J. Combinatorial Theory* **7** (1969), 291–321.

[39] N. E. Nörlund, *Vorlesungen über Differenzenrechnung*, Springer-Verlag, Berlin, 1924.

[40] J. K. Percus, *Combinatorial Methods*, Springer-Verlag, Berlin/Heidelberg/New York, 1971.

[41] G. A. Pick, Geometrisches zur Zahlenlehre, *Sitzungsber. Lotos (Prague)* **19** (1899), 311–319.

[42] G. Pólya, Untersuchungen über Lücken and Singularitäten von Potenzreihen, *Math. Zeit.* **29** (1928–1929), 549–640.

[43] G. Pólya, On the number of certain lattice polygons, *J. Combinatorial Theory* **6** (1969), 102–105.

[44] T. Popoviciu, Asupra unei probleme de partitie a numerelor, *Acad. R. P. R., Filiala Cluj*, Studie și certetari șiintifice, 1–2, anul. IV (1953), 7–58.

[45] J. E. Reeve, On the volume of lattice polyhedra, *Proc. London Math. Soc.* **7** (1957), 378–395.

[46] J. Riordan, Generating functions for powers of Fibonacci numbers, *Duke Math. J.* **29** (1962), 5–12.

[47] R. M. Robinson, Integer-valued entire functions, *Trans. Amer. Math. Soc.* **153** (1971), 451–468.

[48] S. V. Sam, A bijective proof for a theorem of Ehrhart, *Amer. Math. Monthly* **116** (2009), 688–701.

[49] S. V. Sam and K. M. Woods, A finite calculus approach to Ehrhart polynomials, *Electronic J. Combin.* **17** (2010), R68.

[50] L. W. Shapiro, A combinatorial proof of a Chebyshev polynomial identity, *Discrete Math.* **34** (1981), 203–206.

[51] J. H. Spencer, Counting magic squares, *Amer. Math. Monthly* **87** (1980), 397–399.

[52] R. Stanley, Linear homogeneous diophantine equations and magic labelings of graphs, *Duke Math. J.* **40** (1973), 607–632.

[53] R. Stanley, Combinatorial reciprocity theorems, *Advances in Math.* **14** (1974), 194–253.

[54] R. Stanley, Magic labelings of graphs, symmetric magic squares, systems of parameters, and Cohen-Macaulay rings, *Duke Math. J.* **43** (1976), 511–531.

[55] R. Stanley, Generating functions, in *Studies in Combinatorics* (G.-C. Rota, ed.), Mathematical Association of America, Washington, DC, 1978, pp. 100–141.

[56] R. Stanley, Decompositions of rational convex polytopes, *Ann. Discrete Math.* **6** (1980), 333–342.

[57] R. Stanley, Linear diophantine equations and local cohomology, *Inventiones Math.* **68** (1982), 175–193.

[58] H. N. V. Temperley, Combinatorial problems suggested by the statistical mechanics of domains and of rubber-like molecules, *Phys. Rev. (2)* **103** (1956), 1–16.

[59] X. G. Viennot, A combinatorial theory for general orthogonal polynomials with extensions and applications, in *Orthogonal polynomials and applications (Bar-le-Duc, 1984)*, Lecture Notes in Math., no. 1171, Springer, Berlin, 1985, pp. 139–157.

[60] X. G. Viennot, Heaps of pieces. I. Basic definitions and combinatorial lemmas, *Graph Theory and Its Applications: East and West (Jinan, 1986)*, Ann. New York Acad. Sci. **576** (1989), 542–570.

[61] J. von Neumann, A certain zero-sum two person game equivalent to the optimal assignment problem, in *Contributions to the Theory of Games*, vol. 2 (H. W. Kuhn and A. W. Tucker, eds.), Annals of Mathematical Studies, no. 28, Princeton University Press, Princeton, 1950, pp. 5–12.

Exercises for Chapter 4

1. [2+] Let $F(x)$ and $G(x)$ be rational functions. Is it true that $F(x) + G(x)$ is also rational?

2. **a.** [3–] Suppose that $f(x) = \sum_{n \geq 0} a_n x^n$ is a rational function with integer coefficients a_n. Show that we can write $f(x) = P(x)/Q(x)$, where P and Q are relatively prime (over $\mathbb{Q}[x]$) polynomials with integer coefficients such that $Q(0) = 1$.

   b. [3–] Suppose that $f(x_1, \ldots, x_n)$ is a formal power series (over $\mathbb{C}$, say) that represents a rational function $P(x_1, \ldots, x_n)/Q(x_1, \ldots, x_n)$, where P and Q are relatively prime polynomials. Show that $Q(0, 0, \ldots, 0) \neq 0$.

3. [3–] Suppose that $f(x) \in \mathbb{Z}[[x]]$, $f(0) \neq 0$, and $f'(x)/f(x) \in \mathbb{Z}[[x]]$. Prove or disprove that $f(x)/f(0) \in \mathbb{Z}[[x]]$. (While this problem has nothing to do with rational functions, it is similar in flavor to Exercise 4.2(a).)

4. **a.** [3+] Suppose that $\sum_{n\geq 0} a_n x^n \in \mathbb{C}[[x]]$ is rational. Define $\chi : \mathbb{C} \to \mathbb{Z}$ by

$$\chi(a) = \begin{cases} 1, & a \neq 0, \\ 0, & a = 0. \end{cases}$$

Show that $\sum_{n\geq 0} \chi(a_n)x^n$ is also rational (and hence its coefficients are eventually periodic, by Exercise 4.46(b)).

b. [2+] Show that the corresponding result is false for $\mathbb{C}[[x,y]]$; that is, we can have $\sum a_{mn} x^m y^n$ rational but $\sum \chi(a_{mn})x^m y^n$ nonrational.

c. [3+] Let $\sum_{n\geq 0} a_n x^n$ and $\sum_{n\geq 0} b_n x^n$ be rational functions with integer coefficients a_n and b_n. Suppose that $c_n := a_n/b_n$ is an integer for all n (so in particular $b_n \neq 0$). Show that $\sum_{n\geq 0} c_n x^n$ is also rational.

5. [5] Given polynomials $P(x), Q(x) \in \mathbb{Q}[x]$ for which $P(x)/Q(x) = \sum_{n\geq 0} a_n x^n$, is it decidable whether there is some n for which $a_n = 0$?

6. [3–] Given a sequence $a = (a_0, a_1, \dots)$ with entries in a field, the *Hankel determinant* $H_n(a)$ is defined by

$$H_n(a) = \det(a_{i+j})_{0\leq i,j \leq n}.$$

Show that the power series $\sum_{n\geq 0} a_n x^n$ is rational if and only if $H_n(a) = 0$ for all sufficiently large n. Equivalently, the infinite matrix (a_{i+j}) has finite rank.

7. **a.** [2+] Let $b_i \in \mathbb{P}$ for $i \geq 1$. Use Exercise 4.4 to show that the formal power series

$$F(x) = \sum_{i\geq 1} (1 - x^{2i-1})^{-b_i}$$

is not a rational function of x.

b. [2+] Find $a_i \in \mathbb{P}$ $(i \geq 1)$ for which the formal power series

$$F(x) = \sum_{i\geq 1} (1 - x^i)^{-a_i}$$

is a rational function of x.

8. [2] Let $F(x) = \sum_{n\geq 0} a_{n+1} x^n \in \mathbb{C}[[x]]$. Show that the following conditions are equivalent.

i. There exists a rational power series $G(x)$ for which $F(x) = G'(x)/G(x)$.

ii. The series $\exp \sum_{n\geq 1} a_n \dfrac{x^n}{n}$ is rational.

iii. There exist nonzero complex numbers (not necessarily distinct) $\alpha_1, \dots, \alpha_j, \beta_1, \dots, \beta_k$ such that for all $n \geq 1$,

$$a_n = \sum \alpha_i^n - \sum \beta_i^n.$$

9. [2+] If $F(x)$ is a rational function over $\mathbb{Q}$ such that $F(n) \in \mathbb{Z}$ for all $n \in \mathbb{Z}$, does it follow that $F(x)$ is a polynomial?

10. [3] Let $f(z)$ be an analytic function in an open set containing the disk $|z| \leq 1$. Suppose that the only singularities of $f(z)$ inside or on the boundary of this disk are poles, and that the Taylor series $\sum a_n z^n$ of $f(z)$ at $z = 0$ has integer coefficients a_n. Show that $f(z)$ is a rational function.

11. Solve the following recurrences.
    a. [2–] $a_0 = 2$, $a_1 = 3$, $a_n = 3a_{n-1} - 2a_{n-2}$ for $n \geq 2$.
    b. [2–] $a_0 = 0$, $a_1 = 2$, $a_n = 4a_{n-1} - 4a_{n-2}$ for $n \geq 2$.
    c. [2] $a_0 = 5$, $a_1 = 12$, $a_n = 4a_{n-1} - 3a_{n-2} - 2^{n-2}$ for $n \geq 2$.
    d. [2+] $a_i = i$ for $0 \leq i \leq 7$, and

    $$a_n = a_{n-1} - a_{n-3} + a_{n-4} - a_{n-5} + a_{n-7} - a_{n-8}, \; n \geq 8.$$

    Rather than an explicit formula for a_n, give a simple description. For instance, compute a_{105} without using the recurrence.

12. [2] Consider the decimal expansion

    $$\frac{1}{9899} = 0.00010203050813213455\cdots.$$

    Why do the Fibonacci numbers $1, 2, 3, 5, 8, 13, 21, 34, 55, \ldots$ appear?

13. [2+] Is it true that for every $n \in \mathbb{P}$ there is a Fibonacci number F_k, $k \geq 1$, divisible by n?

14. [3] Let $a, b \in \mathbb{P}$, and define $f(0) = a$, $f(1) = b$, and $f(n+1) = f(n) + f(n-1)$ for $n \geq 1$. Show that we can choose a, b so that $f(n)$ is composite for all $n \in \mathbb{N}$.

15. [2+] Let I be an order ideal of the poset $\mathbb{N}^m$, and define $f(n) = \#\{(a_1, \ldots, a_m)\} \in I :$ $a_1 + \cdots + a_m = n\}$. In other words, $f(n)$ is the number of $t \in I$ whose rank in $\mathbb{N}^m$ in n. Show that there is a polynomial $P(n)$ such that $f(n) = P(n)$ for n sufficiently large. For instance, if I is finite then $P(n) = 0$.

16. [2+] How many partitions $\lambda = (\lambda_1, \lambda_2, \ldots)$ of n satisfy $\lambda_3 = 2$? Give an exact formula.

17. **a.** [2+]* Let A_k be the set of all permutations $w = a_1 a_2 \cdots a_{2n}$ of the multiset $M_n = \{1^2, 2^2, \ldots, n^2\}$ with the following property: If $r < s < t$ and $a_r = a_t$, then $a_s > a_r$. For instance, $A_2 = \{1122, 1221, 2211\}$. Let B_n be the set of all permutations $w = a_1 a_2 \cdots a_{2n}$ of M_n with the following property: if $r < s$ and $a_r = a_s < a_t$, then $r < t$. For instance, $B_2 = \{1122, 1212, 1221\}$. Let

    $$F_n(x) = \sum_{w \in A_n} x^{\mathrm{des}(w)},$$

    $$G_n(x) = \sum_{w \in B_n} x^{\mathrm{des}(w)},$$

    where $\mathrm{des}(w)$ denotes the number of descents of w. Show that $F_n(x) = G_n(x)$.
    b. [2+] Show that

    $$\sum_{k \geq 0} S(n+k, k)x^k = \frac{x F_n(x)}{(1-x)^{2n+1}},$$

    where $S(n+k, k)$ denotes a Stirling number of the second kind.

18. [2+]* Define polynomials $p_n(u)$ by

    $$\sum_{n \geq 0} p_n(u)x^n = \frac{1}{1 - ux - x^2}.$$

    Use combinatorial reasoning to find $\sum_{n \geq 0} p_n(u) p_n(v) x^n$.

19. [2]* Let $a, b \in \mathbb{R}$. Define a function $f : \mathbb{N} \to \mathbb{R}$ by $f(0) = a$, $f(1) = b$, and

    $$f(n+2) = |f(n+1)| - f(n), \; n \geq 0.$$

Find $F(x) = \sum_{n \geq 0} f(n)x^n$. (If you prefer not to look at a large number of cases, then assume that $0 \leq a \leq b$.)

20. [2+]* Show that the function $f(n)$ of Example 4.1.3 (i.e., the number of words w of length n in the alphabet $\{N, E, W\}$ such that EW and WE are not factors of w) is equal to the number of nonzero coefficients of the polynomial

$$P_n(x) = \prod_{j=1}^{n}(1 + x_j - x_{j+1}).$$

Show moreover that all these coefficients are equal to ± 1. (For a related result, see Exercise 1.35(k).)

21. [3] A *tournament* T on $[n]$ is a directed graph on the vertex set $[n]$ with no loops and with exactly one edge between any two distinct vertices. The *outdegree* of a vertex i is the number of edges $i \to j$. The *degree sequence* of T is the set of outdegrees of its vertices, arranged in decreasing order. (Hence the degree sequence is a partition of $\binom{n}{2}$.) A degree sequence is *unique* if all tournaments with that degree sequence are isomorphic. Let $f(n)$ be the number of unique degree sequences of tournaments on $[n]$. Set $f(0) = 1$. Show that

$$\sum_{n \geq 0} f(n)x^n = \frac{1}{1 - x - x^3 - x^4 - x^5}$$

$$= 1 + x + x^2 + 2x^3 + 4x^4 + 7x^5 + 11x^6 + 18x^7 + 31x^8 + \cdots.$$

22. [2+]* Let $\alpha \in \mathbb{C}$, and define for $n \in \mathbb{N}$,

$$f_\alpha(n) = \sum_{k=0}^{n} \binom{n-k}{k} \alpha^k.$$

Show that $F_\alpha(x) := \sum_{n \geq 0} f_\alpha(n)x^n$ is a rational function, and compute it explicitly. Find an explicit formula for $f_\alpha(n)$. What value of α requires special treatment?

23. a. [2+]* Let S be a finite sequence of positive integers, say 2224211. We can describe this sequence as "three two's, one four, one two, two one's," yielding the *derived sequence* $32141221 = \delta(S)$. Suppose we start with $S = 1$ and form successive derived sequences $\delta(S) = 11$, $\delta^2(S) = 21$, $\delta^3(S) = 1211$, $\delta^4(S) = 111221$, $\delta^5(S) = 312211$, and so on. Show that for all $n \geq 0$, no term of $\delta^n(S)$ exceeds 3.

    b. [3] Beginning with $S = 1$ as in (a), let $f(n)$ be the length (number of terms) of $\delta^n(S)$, and set

$$F(x) = \sum_{n \geq 0} f(n)x^n = 1 + 2x + 2x^2 + 4x^3 + 6x^4 + 6x^5 + \cdots.$$

    Show that $F(x)$ is a rational function, which, when reduced to lowest terms, has denominator $D(x)$ of degree 92. Moreover, the largest reciprocal zero $\lambda = 1.30357726903\cdots$ (which controls the rate of growth of $f(n)$) of $D(x)$ is an algebraic integer of degree 71.

    c. [3] Compute the (integer) polynomial $x^{71} - x^{69} - 2x^{68} - \cdots$ of degree 71 for which λ is a zero.

    d. [3+] What if we start with a sequence other than $S = 1$?

24. a. [3] Let $f(x) = f(x_1,\ldots,x_k) \in \mathbb{F}_q[x_1,\ldots,x_k]$. Show that for each $\alpha \in \mathbb{F}_q - \{0\}$ there exist $\mathbb{Z}$-matrices $A_0, A_1, \ldots, A_{q-1}$ of some square size, and there exist a row vector u and a column vector v with the following property. For any integer $n \geq 1$ let $a_0 + a_1 q + \cdots + a_r q^r$ be its base q expansion, so $0 \leq a_i \leq q - 1$. Let $N_\alpha(n)$ be the number of coefficients of $f(x)^n$ equal to α. Then

$$N_\alpha(n) = u A_{a_0} A_{a_1} \cdots A_{a_r} v.$$

b. [2−]* Deduce that the generating function

$$\sum_{n \geq 1} N_\alpha(1 + q + q^2 + \cdots + q^{n-1}) x^n$$

is rational.

25. Let $k = 1$ in Exercise 4.24. Without loss of generality we may assume $f(0) \neq 0$.

a. [3−] Show that there exist periodic functions $u(m)$ and $v(m)$ depending on $f(x)$ and α, such that

$$N_\alpha(q^m - 1) = u(m) q^m + v(m) \tag{4.51}$$

for all m sufficiently large.

b. [3−] Let d be the least positive integer for which $f(x)$ divides $x^{q^m(q^d-1)} - 1$ for some $m \geq 0$. In other words, d is the degree of the extension field of $\mathbb{F}_q$ obtained by adjoining all zeros of $f(x)$. Then the functions $u(m)$ and $v(m)$ have period d (and possibly smaller periods, necessarily dividing d).

c. [2+] Let μ be the largest multiplicity of any irreducible factor (or any zero) of $f(x)$. Then equation (4.51) holds for all $m \geq \lceil \log_q \mu \rceil$. In particular, if $f(x)$ is squarefree, then (4.51) holds for all $m \geq 0$.

d. If $f(x)$ is primitive over $\mathbb{F}_q$ (i.e., $f(x)$ is irreducible, and any zero ζ of $f(x)$ is a generator of the multiplicative group of the field $\mathbb{F}_q(\zeta)$), then $d = \deg f$ and $u(m) = dq^{d-1}/(q^d - 1)$ (a constant).

e. [3−] Write $[a_0, a_1, \ldots, a_{k-1}]$ for the periodic function $p(m)$ on $\mathbb{Z}$ satisfying $p(m) = a_i$ for $m \equiv i \pmod{k}$. Verify the following examples:
   - If $f(x) = 1 + x \in \mathbb{F}_q^n$ where $q = 2^k$, then $N_1(m) = 2^m$.
   - If $f(x) = 1 + x \in \mathbb{F}_q^n$ where q is odd, then

$$N_1(m) = \frac{1}{2}(q^m + 1), \quad N_{-1}(m) = \frac{1}{2}(q^m - 1).$$

   - If $f(x) = 1 + x + x^2 + x^3 + x^4 \in \mathbb{F}_2[x]$, then $f(x)$ is irreducible but not primitive, and

$$N_1(m) = \frac{1}{5}[8, 12]2^m + \frac{1}{5}[-3, 1, 3, -1].$$

   - If $g(x) = 1 + x^2 + x^5 \in \mathbb{F}_2[x]$, then $g(x)$ is primitive and

$$N_1(m) = \frac{80}{31}2^m + \frac{1}{31}[-49, -67, -41, 11, -9].$$

- If $g(x) = 1 + x + x^3 + x^4 + x^5 \in \mathbb{F}_2[x]$, then $g(x)$ is primitive and

$$N_1(m) = \frac{80}{31} 2^m + \frac{1}{31}[-49, -5, -41, 11, -9].$$

Note the closeness to the previous item.
- Let $g(x) = (1 + x^2 + x^5)^3 \in \mathbb{F}_2[x]$. Then

$$N_1(m) = \begin{cases} 1, & m = 0, \\ 9, & m = 1, \\ \frac{168}{31} 2^m + \frac{1}{31}[297, -243, -393, -507, -177], & m \geq 2. \end{cases}$$

- Let $g(x) = 2 + x + x^2 \in \mathbb{F}_3[x]$. Then $g(x)$ is primitive and

$$N_1(m) = \frac{3}{4} 3^m + \frac{1}{2} - \frac{1}{4}(-1)^m,$$

$$N_2(m) = \frac{3}{4} 3^m - \frac{1}{2} - \frac{1}{4}(-1)^m.$$

- Let $g(x) = 2 + x^2 + x^3 \in \mathbb{F}_3[x]$. Then $g(x)$ is irreducible but not primitive, and

$$N_1(m) = \frac{18}{13} 3^m + \frac{1}{13}[-5, 11, 7],$$

$$N_2(m) = \frac{9}{13} 3^m - \frac{1}{13}[9, 14, 3].$$

26. a. [3+] Let p be a prime, and let $g_n(p)$ denote the number of nonisomorphic groups of order p^n. Write (i, j) for the greatest common divisor of i and j. Show that

$g_1(p) = 1$

$g_2(p) = 2$

$g_3(p) = 5$

$g_4(p) = 15, \quad p \geq 3$

$g_5(p) = 2p + 61 + 2(p - 1, 3) + (p - 1, 4), \quad p \geq 5$

$g_6(p) = 3p^2 + 39p + 344 + 24(p - 1, 3) + 11(p - 1, 4) + 2(p - 1, 5), \quad p \geq 5$

$g_7(p) = 3p^5 + 12p^4 + 44p^3 + 170p^2 + 707p + 2455$

$\quad (4p^2 + 44p + 291)(p - 1, 3) + (p^2 + 19p + 135)(p - 1, 4)$

$\quad + (3p + 31)(p - 1, 5) + 4(p - 1, 7) + 5(p - 1, 8) + (p - 1, 9), \quad p \geq 7.$

b. [3+] Show that for fixed p,

$$g_n(p) = p^{\frac{2}{27} n^3 + O(n^{5/2})}.$$

c. [5] Show that for fixed n, $g_n(p)$ is a quasipolynomial in p for p sufficiently large.

27. Let X be a finite alphabet, and let X^* denote the free monoid generated by X. Let M be the quotient monoid of X^* corresponding to relations $w_1 = w_1', \ldots, w_k = w_k',$

where w_i and w_i' have the same length, $1 \le i \le k$. Thus, if $w \in M$, then we can speak unambiguously of the *length* of w as the length of any word in X^* representing w. Let $f(n)$ be the number of distinct words in M of length n, and let $F(x) = \sum_{n \ge 0} f(n)x^n$.

a. [3–] If $k = 1$, then show that $F(x)$ is rational.

b. [3] Show that in general $F(x)$ need not be rational.

c. [3–] Linearly order the q letters in X, and let M be defined by the relations $acb = cab$ and $bac = bca$ for $a < b < c$, and $aba = baa$ and $bab = bba$ for $a < b$. Compute $F(x)$.

d. [3–] Show that if M is commutative, then $F(x)$ is rational.

28. a. [2+] Let A and B be $n \times n$ matrices (over $\mathbb{C}$, say). Given $\boldsymbol{\alpha} = (\alpha_1, \dots, \alpha_r), \boldsymbol{\beta} = (\beta_1, \dots, \beta_r) \in \mathbb{N}^r$, define

$$t(\boldsymbol{\alpha}, \boldsymbol{\beta}) = \operatorname{tr} A^{\alpha_1} B^{\beta_1} A^{\alpha_2} B^{\beta_2} \cdots A^{\alpha_r} B^{\beta_r}.$$

Show that $T_r(\boldsymbol{x}, \boldsymbol{y}) := \sum_{\boldsymbol{\alpha}, \boldsymbol{\beta} \in \mathbb{N}^r} t(\boldsymbol{\alpha}, \boldsymbol{\beta}) \boldsymbol{x}^{\boldsymbol{\alpha}} \boldsymbol{y}^{\boldsymbol{\beta}}$ is rational. What is the denominator of $T_r(\boldsymbol{x}, \boldsymbol{y})$?

b. Compute $T_1(x, y)$ for $A = \begin{bmatrix} 0 & -1 \\ 1 & 0 \end{bmatrix}$ and $B = \begin{bmatrix} 1 & 1 \\ -1 & 0 \end{bmatrix}$.

29. [2+] Let A, B, C be square matrices of the same size over some field K. *True or false*: for fixed i, j, the generating function

$$\sum_{n \ge 1} (A^n B^n C^n)_{ij} x^n$$

is rational.

30. [2+] Let E be the monoid of $\mathbb{N}$-solutions to the equation $x + y - 2z - w = 0$. Write the generating function

$$E(\boldsymbol{x}) = E(x, y, z, w) = \sum_{\alpha \in E} \boldsymbol{x}^{\alpha}$$

explicitly in the form

$$E(\boldsymbol{x}) = \frac{P(\boldsymbol{x})}{\prod_{\beta \in \mathrm{CF}(E)} (1 - \boldsymbol{x}^{\beta})}.$$

That is, determine explicitly the elements of $\mathrm{CF}(E)$ and the polynomial $P(\boldsymbol{x})$.

31. [3–] Let $f(n)$ denote the number of distinct $\mathbb{Z}/n\mathbb{Z}$-solutions $\boldsymbol{\alpha}$ to equation (4.10) modulo n. For example, if $\Phi = [1 \ -1]$ then $f(n) = n$, the number of solutions $(\alpha, \beta) \in (\mathbb{Z}/n\mathbb{Z})^2$ to $\alpha - \beta = 0 \pmod{n}$. Show that $f(n)$ is a quasipolynomial for n sufficiently large (so in particular $\sum_{n \ge 1} f(n)x^n$ is rational).

32. [2+] Let E^* be the set of all $\mathbb{N}$-solutions to equation (4.10) in *distinct* integers $\alpha_1, \dots, \alpha_m$. Show that the generating function $E^*(\boldsymbol{x}) := \sum_{\alpha \in E^*} \boldsymbol{x}^{\alpha}$ is rational.

33. a. [2]* Let $\Phi = \Phi_n$ be the $1 \times (n+1)$ matrix

$$\Phi = [1, 2, 3, \dots, n, -n].$$

Show that the number of generators of the monoid E_Φ, as a function of n, is superpolynomial (i.e., grows faster than any polynomial in n).

b. [2]* Compute the generating function

$$E_{\Phi_3}(\boldsymbol{x}) = \sum_{\alpha \in E_{\Phi_3}} \boldsymbol{x}^{\alpha}.$$

Express your answer as a rational function reduced to lowest terms.

34. [3] Let $\Phi\alpha = 0$ be a system of r linear equations in m unknowns $x_1, \ldots, x_m$ over $\mathbb{Z}$, as in equation (4.10). Let S be a subset of $[m]$. Suppose that $\Phi\alpha = 0$ has a solution $(\gamma_1, \ldots, \gamma_m) \in \mathbb{Z}^m$ satisfying $\gamma_i > 0$ if $i \in S$ and $\gamma_i < 0$ if $i \notin S$. Let

$$F_S(x) = \sum_\alpha x^\alpha,$$

$$\overline{F}_S(x) = \sum_\beta x^\beta,$$

where α runs over all $\mathbb{N}$-solutions to $\Phi\alpha = 0$ satisfying $\alpha_i > 0$ if $i \in S$, while β runs over all $\mathbb{N}$-solutions to $\Phi\beta = 0$ satisfying $\beta_i > 0$ if $i \notin S$. Show that

$$\overline{F}_S(x) = (-1)^{\operatorname{corank}(\Phi)} F_S(1/x).$$

35. a. [2+] Let Φ be an $r \times m$ $\mathbb{Z}$-matrix, and fix $\beta \in \mathbb{Z}^r$. Let E_β be the set of all $\mathbb{N}$-solutions α to $\Phi\alpha = \beta$. Show that the generating function $E_\beta(x)$ represents a rational function of $x = (x_1, \ldots, x_m)$. Show also that either $E_\beta(x) = 0$ (i.e., $E_\beta = \emptyset$) or else $E_\beta(x)$ has the same least denominator $D(x)$ as $E(x)$ (as given in Theorem 4.5.11).

b. [2+] Assume for the remainder of this exercise that the monoid E is positive and that $E_\beta \neq \emptyset$. We say that the pair (Φ, β) has the *R-property* if $\overline{E}_\beta(x) = (-1)^d E_\beta(1/x)$, where $\overline{E}_\beta$ is the set of $\mathbb{P}$-solutions to $\Phi\alpha = -\beta$, and where d is as in Theorem 4.5.14. (Thus, Theorem 4.5.14 asserts that $(\Phi, 0)$ has the R-property.) For what integers β does the pair $([1\ 1\ -1\ -1], \beta)$ have the R-property?

c. [3] Suppose that there exists a vector $\alpha \in \mathbb{Q}^m$ satisfying $-1 < \alpha_i \leq 0$ $(1 \leq i \leq m)$ and $\Phi\alpha = \beta$. Show that (Φ, β) has the R-property.

d. [3+] Find a "reasonable" necessary and sufficient condition for (Φ, β) to have the R-property.

36. a. [2] Let σ be a d-dimensional simplex in $\mathbb{R}^m$ with integer vertices $v_0, \ldots, v_d$. We say that σ is *primitive* (or *unimodular*) if $v_1 - v_0, v_2 - v_0, \ldots, v_d - v_0$ form part of a $\mathbb{Z}$-basis for $\mathbb{Z}^m$. This condition is equivalent to the statement that the relative volume of σ is equal to $1/d!$, the smallest possible relative volume of an integer d-simplex. Now let $\mathcal{P}$ be an integer polytope in $\mathbb{R}^m$. We say that a triangulation Γ of $\mathcal{P}$ is *primitive* (or *unimodular*) if every simplex $\sigma \in \Gamma$ is primitive. (We are allowed to have vertices of Γ that are not vertices of $\mathcal{P}$. For instance, the line segment $[0,2]$ has a primitive triangulation whose facets are $[0,1]$ and $[1,2]$.) Does every integral polytope have a primitive triangulation?

b. [2+] Let Γ be a primitive triangulation of the integer polytope $\mathcal{P}$. Suppose that Γ has f_i i-dimensional faces. Express the Ehrhart polynomial $i(\mathcal{P}, n)$ in terms of the f_i's.

37. a. [2+]* Let $\mathcal{P}$ be an integer polytope in $\mathbb{R}^d$ with vertex set V. Suppose that $\mathcal{P}$ is defined by inequalities $\alpha_i \cdot x \leq \beta_i$. Given $v \in V$, let the *support cone* at v be the cone $\mathcal{C}_v$ defined by $\alpha_i \cdot x \leq \beta_i$ whenever $\alpha_i \cdot v = \beta_i$. Let

$$F_v(x) = \sum_{\gamma \in \mathcal{C}_v \cap \mathbb{Z}^d} x^\gamma.$$

Show that each $F_v(x)$ is a rational Laurent series.

b. [3] Show that

$$\sum_{v \in V} F_v(x) = \sum_{\gamma \in \mathcal{P} \cap \mathbb{Z}^d} x^\gamma,$$

where the sum on the left is interpreted as a sum of *rational functions* (not formal Laurent series).

Example. Let $\mathcal{P}$ be the interval $[2,5] \subset \mathbb{R}$. Then

$$F_2(x) = \sum_{n \geq 2} x^n = \frac{x^2}{1-x},$$

$$F_5(x) = \sum_{n \leq 5} x^n = \frac{x^5}{1-x^{-1}},$$

and

$$\frac{x^2}{1-x} + \frac{x^5}{1-x^{-1}} = x^2 + x^3 + x^4 + x^5$$

$$= \sum_{n \in [2,5]} x^n.$$

As another example, let $\mathcal{P}$ have vertices $(0,0)$, $(0,2)$, $(2,0)$, and $(4,2)$. Then $\mathcal{C}_{(2,0)}$ is defined by $y \geq 0$ and $x - y \leq 2$, and

$$F_{(2,0)}(x,y) = \sum_{n \geq 0} \sum_{m \leq n+2} x^m y^n$$

$$= \sum_{n \geq 0} y^n \cdot \frac{x^{n+2}}{1-x^{-1}}$$

$$= \frac{x^2}{1-x^{-1}} \cdot \frac{1}{1-xy}$$

$$= -\frac{x^3}{(1-x)(1-xy)}.$$

38. [3] Let Φ be an $r \times m$ matrix whose entries are polynomials in n with integer coefficients. Let β be a column vector of length m whose entries are also polynomials in n with integer coefficients. Suppose that for each fixed $n \in \mathbb{P}$ the number $f(n)$ of solutions $\alpha \in \mathbb{N}^m$ to $\Phi \alpha = \beta$ is finite. Show that $f(n)$ is a quasipolynomial for n sufficiently large.

39. **a.** [4–] Let $P_1, \ldots, P_k \in \mathbb{F}_q[x_1, \ldots, x_m]$. Let $f(n)$ be the number of solutions $\alpha = (\alpha_1, \ldots, \alpha_m) \in \mathbb{F}_{q^n}^m$ to the equations $P_1(\alpha) = \cdots = P_k(\alpha) = 0$. Show that $F(x) := \exp \sum_{n \geq 1} f(n) x^n / n$ is rational. (See Exercise 4.8 for equivalent forms of this condition.)

   b. [4–] Let $P_1, \ldots, P_k \in \mathbb{Z}[x_1, \ldots, x_m]$, and let p be a prime. Let $f(n)$ be the number of solutions $\alpha = (\alpha_1, \ldots, \alpha_m) \in (\mathbb{Z}/p^n\mathbb{Z})^m$ to the congruences

$$P_1(\alpha) \equiv \cdots \equiv P_k(\alpha) \equiv 0 \pmod{p^n}.$$

   Show that $F(x) := \sum_{n \geq 1} f(n) x^n$ is rational.

40. **a.** [2+] Let $X = \{x_1, \ldots, x_n\}$ be an alphabet with n letters, and let $\mathbb{C}\langle\langle X \rangle\rangle$ be the *non-commutative* power series ring (over $\mathbb{C}$) in the variables X; that is, $\mathbb{C}\langle\langle X \rangle\rangle$ consists of all formal expressions $\sum_{w \in X^*} \alpha_w w$, where $\alpha_w \in \mathbb{C}$ and X^* is the free monoid

generated by X. Multiplication in $\mathbb{C}\langle\langle X\rangle\rangle$ is defined in the obvious way, namely,

$$\left(\sum_u \alpha_u u\right)\left(\sum_v \beta_v v\right) = \sum_{u,v} \alpha_u \beta_v uv$$

$$= \sum_w \gamma_w w,$$

where $\gamma_w = \sum_{uv=w} \alpha_u \beta_v$ (a finite sum).

Let L be a set of words such that no proper factor of a word in L belongs to L. (A word $v \in X^*$ is a *factor* of $w \in X^*$ if $w = uvy$ for some $u, y \in X^*$.) Define an *L-cluster* to be a triple $(w, (v_1, \ldots, v_k), (\ell_1, \ldots, \ell_k)) \in X^* \times L^k \times [r]^k$, where r is the length of $w = \sigma_1 \sigma_2 \cdots \sigma_r$ and k is some positive integer, satisfying:

i. For $1 \le j \le k$ we have $w = uv_j y$ for some $u \in X^*_{\ell_j - 1}$ and $y \in X^*$ (i.e., w contains v_j as a factor beginning in position ℓ_j). Henceforth we identify v_j with this factor of w.

ii. For $1 \le j \le k-1$, we have that v_j and v_{j+1} overlap in w, and that v_{j+1} begins to the right of the beginning of v_j (so $0 < \ell_1 < \ell_2 < \cdots < \ell_k < r$).

iii. v_1 contains σ_1, and v_k contains σ_r.

Note that two different L-clusters can have the same first component w. For instance, if $X = \{a\}$ and $L = \{aaa\}$, then $(aaaaa, (aaa, aaa, aaa), (1, 2, 3))$ and $(aaaaa, (aaa, aaa), (1, 3))$ are both L-clusters.

Let $D(L)$ denote the set of L-clusters. For each word $v \in L$ introduce a new variable t_v commuting with the x_i's and with each other. Define the *cluster-generating function*

$$C(x, t) = \sum_{(w, \mu, \nu) \in D(L)} \left(\prod_{v \in L} t_v^{m_v(\mu)}\right) w \in \mathbb{C}[[t_v : v \in L]]\langle\langle X\rangle\rangle,$$

where $m_v(\mu)$ denotes the number of components v_i of $\mu \in L^k$ that are equal to v. Show that in the ring $\mathbb{C}[[t_v : v \in L]]\langle\langle X\rangle\rangle$ we have

$$\sum_{w \in X^*} \left(\prod_{v \in L} t_v^{m_v(w)}\right) w = (1 - x_1 - \cdots - x_n - C(x, t-1))^{-1}, \qquad (4.52)$$

where $m_v(w)$ denotes the number of factors of w equal to v, and where $t - 1$ denotes the substitution of $t_v - 1$ for each t_v.

b. [1+]* Note the following specializations of equation (4.52):

i. If we let the variables x_i in (4.52) commute and set each $t_v = t$, then the coefficient of $t^k x_1^{m_1} \cdots x_n^{m_n}$ is the number of words $w \in X^*$ with m_i x_i's for $1 \le i \le n$, and with exactly k factors belonging to L.

ii. If we set each $x_i = x$ and $t_i = t$ in (4.52), then the coefficient of $t^k x^m$ is the number of words $x \in X^*$ of length m, with exactly k factors belonging to L.

iii. If we set each $x_i = x$ and each $t_v = 0$ in (4.52), then the coefficient of x^m is the number of words $w \in X^*$ with no factors belonging to L.

c. [2] Show that if L is finite and the x_i's commute in (4.52), then (4.52) represents a rational function of $x_1, \ldots, x_n$ and the t_v's.

d. [2] If $w = a_1 a_2 \cdots a_l \in X^*$, then define the *autocorrelation polynomial* $A_w(x) = c_1 + c_2 x + \cdots + c_l x^{l-1}$, where

$$c_i = \begin{cases} 1, & \text{if } a_1 a_2 \cdots a_{l-i+1} = a_i a_{i+1} \cdots a_l, \\ 0, & \text{otherwise.} \end{cases}$$

For instance, if $w = abacaba$, then $A_w(x) = 1 + x^4 + x^6$. Let $f(m)$ be the number of words $w \in X^*$ of length m that don't contain w as a factor. Show that

$$\sum_{m \geq 0} f(m)x^m = \frac{A_w(x)}{(1 - nx)A_w(x) + x^l}. \tag{4.53}$$

41. a. [1+]* Let $B_k(n)$ be the number of ways to place k nonattacking queens on an $n \times n$ chessboard. Show that $B_1(n) = n^2$.

b. [2+] Show that

$$B_2(n) = \frac{1}{6}n(n-1)(n-2)(3n-1).$$

c. [3−] Show that

$$B_3(n) = \begin{cases} \frac{1}{12}n(n-2)^2(2n^3 - 12n^2 + 23n - 10), & n \text{ even,} \\[2mm] \frac{1}{12}(n-1)(n-3)(2n^4 - 12n^3 + 25n^2 - 14n + 1), & n \text{ odd.} \end{cases}$$

d. [2+] Show that for fixed $k \geq 1$,

$$B_k(n) = \frac{1}{k!}n^{2k} - \frac{5}{3 \cdot (k-2)!}n^{2k-1} + O(n^{2k-2}).$$

e. [3−] Show that $\sum_{n \geq 0} B_k(n)x^n$ is a rational power series. In fact, $B_k(n)$ is a quasipolynomial.

42. a. [2+]* Show that the number of ways to place k nonattacking bishops on the white squares of an $(n-1) \times n$ chessboard is the Stirling number $S(n, n-k)$.

b. [3−] Let $A_k(n)$ be the number of ways to place k nonattacking bishops on an $n \times n$ chessboard. Show that $A_k(n)$ is a quasipolynomial with quasiperiod two.

c. [3] Find an explicit formula for $A_k(n)$ in the form of a triple sum.

43. [2+] Let $t(n)$ be the number of noncongruent triangles whose sides have integer length and whose perimeter in n. For instance $t(9) = 3$, corresponding to $3 + 3 + 3, 2 + 3 + 4$, $1 + 4 + 4$. Find $\sum_{n \geq 3} t(n)x^n$.

44. [2+] Let $k, r, n \in \mathbb{P}$. Let $N_{kr}(n)$ be the number of n-tuples $\alpha = (\alpha_1, \ldots, \alpha_n) \in [k]^n$ such that no r consecutive elements of α are equal (e.g., $N_{kr}(r) = k^r - k$.) Let $F_{kr}(x) = \sum_{n \geq 0} N_{kr}(n)x^n$. Find $F_{kr}(x)$ explicitly. (Set $N_{kr}(0) = 1$.)

45. a. [3] Let $m \in \mathbb{P}$ and $k \in \mathbb{Z}$. Define a function $f: \{m, m+1, m+2, \ldots\} \to \mathbb{Z}$ by

$$f(m) = k,$$
$$f(n+1) = \left\lfloor \frac{n+2}{n} f(n) \right\rfloor, \quad n \geq m. \tag{4.54}$$

Show that f is a quasipolynomial on its domain.

b. [5−] What happens when $(n+2)/n$ is replaced by some other rational function $R(n)$?

46. a. [2+] Define $f: \mathbb{N} \to \mathbb{Q}$ by

$$f(n+2) = \frac{6}{5}f(n+1) - f(n), \quad f(0) = 0, \quad f(1) = 1. \tag{4.55}$$

Show that $|f(n)| < \frac{5}{4}$.

b. [2] Suppose that $f : \mathbb{N} \to \mathbb{Z}$ satisfies a linear recurrence (4.2) where each $\alpha_i \in \mathbb{Z}$, and that $f(n)$ is bounded as $n \to \infty$. Show that $f(n)$ is periodic.

c. [3+] Suppose that y is a power series with integer coefficients and radius of convergence one. Show that y is either rational or has the unit circle as a natural boundary.

47. [3] If $\alpha \in \mathbb{N}^m$ and $k > 0$, then let $f_k(\alpha)$ denote the number of partitions of α into k parts belonging to $\mathbb{N}^m$. For example, $f_2(2,2) = 5$, since $(2,2) = (2,2) + (0,0) = (1,0) + (1,2) = (0,1) + (2,1) = (2,0) + (0,2) = (1,1) + (1,1)$. If $\alpha = (\alpha_1,\ldots,\alpha_m)$, then write as usual $x^\alpha = x_1^{\alpha_1} \cdots x_m^{\alpha_m}$. Clearly,

$$\sum_{\alpha \in \mathbb{N}^m} \sum_{k \geq 0} f_k(\alpha) t^k x^\alpha = \prod_{\alpha \in \mathbb{N}^m} \left(1 - t x^\alpha\right)^{-1},$$

the m-dimensional generalization of equation (1.77). Show that

$$\sum_{\alpha \in \mathbb{N}^m} f_k(\alpha) x^\alpha = \left[\sum x_1^{\mathrm{maj}(w_1)} \cdots x_m^{\mathrm{maj}(w_m)}\right] \left[\prod_{i=1}^{m}(1 - x_i)(1 - x_i^2) \cdots (1 - x_i^k)\right]^{-1},$$

where the second sum is over all m-tuples $(w_1,\ldots,w_m) \in \mathfrak{S}_k^m$ satisfying $w_1 w_2 \cdots w_m = 1$. Note that Proposition 1.1.8.6(a) is equivalent to the case $m = 1$.

48. a. [3] Let $\mathcal{P}$ be an integral convex d-polytope with $\mathcal{P}$-Eulerian polynomial $A(\mathcal{P},x)$. Show that the coefficients of $A(\mathcal{P},x)$ are nonnegative.

b. [3+] Let $\mathcal{Q} \subset \mathbb{R}^m$ be a finite union of integral convex d-polytopes, such that the intersection of any two of these polytopes is a common face (possibly empty) of both. Suppose that $\mathcal{Q}$, regarded as a topological space, satisfies

$$H_i(\mathcal{Q}, \mathcal{Q} - p; \mathbb{Q}) = 0 \quad \text{if } i < d, \text{ for all } p \in \mathcal{Q},$$
$$\widetilde{H}_i(\mathcal{Q}; \mathbb{Q}) = 0 \quad \text{if } i < d.$$

Here H_i and $\widetilde{H}_i$ denote relative singular homology and reduced singular homology, respectively. We may define the Ehrhart function $i(\mathcal{Q},n)$ for $n \geq 1$ exactly as for polytopes $\mathcal{P}$, and one easily sees that $i(\mathcal{Q},n)$ is a polynomial of degree d for $n \geq 1$. Define $i(\mathcal{Q},0) = 1$, despite the fact that the value of the polynomial $i(\mathcal{Q},n)$ at $n = 0$ is $\chi(\mathcal{Q})$, the Euler characteristic of $\mathcal{Q}$. Set

$$\sum_{n \geq 0} i(\mathcal{Q},n) x^n = \frac{A(\mathcal{Q},x)}{(1-x)^{d+1}}.$$

Show that the coefficients of the polynomial $A(\mathcal{Q},x)$ are nonnegative.

c. [3] Suppose that $\mathcal{P}$ and $\mathcal{Q}$ are integral convex polytopes (not necessarily of the same dimension) in $\mathbb{R}^m$ with $\mathcal{Q} \subseteq \mathcal{P}$. Show that the polynomial $A(\mathcal{P},x) - A(\mathcal{Q},x)$ has nonnegative coefficients. Note that (a) follows from taking $\mathcal{Q} = \emptyset$.

49. Let $\mathcal{P}$ be an integral convex d-polytope in $\mathbb{R}^m$, and let $A(\mathcal{P},x) = 1 + h_1 x + \cdots + h_d x^d$.

a. [3] Show that

$$h_d + h_{d-1} + \cdots + h_{d-i} \leq h_0 + h_1 + \cdots + h_{i+1}, \tag{4.56}$$

for $1 \leq i \leq \lfloor d/2 \rfloor - 1$.

b. [3] Let $s = \max\{i : h_i \neq 0\}$. Show that

$$h_0 + h_1 + \cdots + h_i \leq h_s + h_{s-1} + \cdots + h_{s-i}, \tag{4.57}$$

for $0 \leq i \leq \lfloor s/2 \rfloor$.

50. [2] Let $\partial \mathcal{P}$ denote the boundary of the d-dimensional integral convex polytope $\mathcal{P}$ in $\mathbb{R}^m$. For $n \in \mathbb{P}$, we can define

$$i(\partial \mathcal{P}, n) = \#(n \cdot \partial \mathcal{P} \cap \mathbb{Z}^m),$$

exactly as was done for $\mathcal{P}$. Set $i(\partial \mathcal{P}, 0) = 1$. Show that

$$\sum_{n \geq 0} i(\partial \mathcal{P}, n) x^n = \frac{h_0 + h_1 x + \cdots + h_d x^d}{(1-x)^d},$$

where $h_i \in \mathbb{Z}$ and $h_i = h_{d-i}$ for $0 \leq i \leq d$.

51. a. [2] Fix $r, s \in \mathbb{P}$. Let $\mathcal{P}$ be the convex polytope in $\mathbb{R}^{r+s}$ defined by

$$x_1 + x_2 + \cdots + x_r \leq 1, \quad y_1 + y_2 + \cdots + y_s \leq 1, \quad x_i \geq 0, \quad y_i \geq 0.$$

Let $i(n) = i(\mathcal{P}, n)$ be the Ehrhart (quasi)polynomial of $\mathcal{P}$. Use Exercise 3.169 to find $F(x) = \sum_{n \geq 0} i(n) x^n$ explicitly; that is, find the denominator of $F(x)$ and the coefficients of the numerator. What is the volume of $\mathcal{P}$? What are the vertices of $\mathcal{P}$?
  b. [2] Find a partially ordered set P_{rs} for which $i(\mathcal{P}, n-1) = \Omega_{P_{rs}}(n)$, the order polynomial of P_{rs}.

52. [3–]* Let σ_d be the d-dimensional simplex in $\mathbb{R}^d$ with vertices $(0,0,0,\ldots,0)$, $(1,0,0,\ldots,0)$, $(1,2,0,\ldots,0)$, $\ldots, (1,2,3,\ldots,d)$. Show that $i(\sigma_d, n) = (n+1)^d$.

53. An *antimagic square* of index n is a $d \times d$ $\mathbb{N}$-matrix $M = (m_{ij})$ such that for every permutation $w \in \mathfrak{S}_d$ we have $\sum_{i=1}^d m_{i,w(i)} = n$. In other words, any set of d entries, no two in the same row or column, sum to n.
  a. [2] For what positive integers d do there exist $d \times d$ antimagic squares whose entries are the distinct integers $1, 2, \ldots, d^2$?
  b. [2+] Let R_i (respectively, C_i) be the $d \times d$ matrix with 1's in the ith row (respectively, ith column) and 0's elsewhere. Show that a $d \times d$ antimagic square has the form

$$M = \sum_{i=1}^n a_i R_i + \sum_{j=1}^n b_j C_j,$$

where $a_i, b_j \in \mathbb{N}$.
  c. [2+] Use (b) to find a simple explicit formula for the number of $d \times d$ antimagic squares of index n.
  d. [2] Let $\mathcal{P}_d$ be the convex polytope in $\mathbb{R}^{d^2}$ of all $d \times d$ matrices $X = (x_{ij})$ satisfying

$$x_{ij} \geq 0, \quad \sum_{i=1}^d x_{i,w(i)} = 1 \quad \text{for all } w \in \mathfrak{S}_d.$$

What are the vertices of $\mathcal{P}_d$? Find the Ehrhart polynomial $i(\mathcal{P}_d, n)$.
  e. [2] Find the $\mathcal{P}_d$-Eulerian polynomial $A(\mathcal{P}_d, x)$ and the relative volume $\nu(\mathcal{P}_d)$.

54. a. [2+] Let

$$H_n(r) = \sum_{i=0}^{(n-1)^2} c(n,i) r^{(n-1)^2 - i},$$

where $H_n(r)$ denotes the number of $n \times n$ $\mathbb{N}$-matrices with line sum r, as in Section 4.6.1. Show that $c(n,1)/c(n,0) = \frac{1}{2} n(n-1)^2$.
  b. [5–] (rather speculative) Fix $k \geq 0$. Then as $n \to \infty$ we have the asymptotic formula

$$\frac{c(n,k)}{c(n,0)} \sim \frac{n^{3k}}{2^k k!}.$$

55. [2+]* Let $f(n)$ denote the number of 2×3 $\mathbb{N}$-matrices such that every row sums to $3n$ and every column to $2n$. Find an explicit formula for $f(n)$ and compute (as a rational function reduced to lowest terms) the generating function $\sum_{n \geq 0} f(n) x^n$.

56. **a.** [2+] Let $P = \{t_1, \ldots, t_p\}$ be a finite poset. Let $\mathcal{C}(P)$ denote the convex polytope in $\mathbb{R}^p$ defined by

$$\mathcal{C}(P) = \{(\varepsilon_1, \ldots, \varepsilon_p) \in \mathbb{R}^p : 0 \leq \varepsilon_{i_1} + \cdots + \varepsilon_{i_k} \leq 1 \text{ whenever } t_{i_1} < \cdots < t_{i_k}\}.$$

Find the vertices of $\mathcal{C}(P)$.

b. [2+] Show that the Ehrhart (quasi)polynomial of $\mathcal{C}(P)$ is given by $i(\mathcal{C}(P), n-1) = \Omega_P(n)$, the order polynomial of P. Thus, we have *two* polytopes associated with P whose Ehrhart polynomial is $\Omega_P(n+1)$, the second given by Example 4.6.17.

c. [2] Given $n, k \geq 1$, let $\mathcal{C}_{n,k}$ be the convex polytope in $\mathbb{R}^n$ defined by $x_i \geq 0$ for $1 \leq i \leq n$ and

$$x_{i+1} + x_{i+2} + \cdots + x_{i+k} \leq 1, \ 0 \leq i \leq n - k.$$

Find the volume $v(\mathcal{C}_{n,2})$. (Note that the volume of $\mathcal{C}_{n,k}$ is the same as the relative volume since $\dim \mathcal{C}_{n,k} = n$.)

d. [5] Find the volume V_n of $\mathcal{C}_{n,3}$. For instance,

$$(1! V_1, 2! V_2, \ldots, 10! V_{12}) = (1, 1, 1, 2, 5, 14, 47, 182, 786, 3774, 19974, 115236).$$

e. [2+] Let $k \leq n \leq 2k$. Show that the volume of $\mathcal{C}_{n,k}$ is $C_{n-k+1}/n!$, where C_{n-k+1} is a Catalan number.

57. [2+] Let P and Q be partial orderings of the same p-element set. Suppose that the incomparability graph $\mathrm{inc}(P)$ of P is a proper (spanning) subgraph of $\mathrm{inc}(Q)$. Use Exercise 4.56 to show that $e(P) < e(Q)$.

58. **a.** [3–] Let P be a finite poset and let $\mathcal{V}(P)$ denote the set of all maps $f : P \to \mathbb{R}$ such that for every order ideal I of P we have

$$0 \leq \sum_{t \in I} f(t) \leq 1.$$

Clearly, $\mathcal{V}(P)$ is a convex polytope in the vector space $\mathbb{R}^P$, called the *valuation polytope* of P. It is linearly equivalent to the polytope of all valuations on $J(P)$ (as defined in Exercise 3.94) with values in the interval $[0,1]$. Show that the vertices of $\mathcal{V}(P)$ consist of all functions f_C, where C is a chain $t_1 < t_2 < \cdots < t_k$ in P, defined by

$$f(t) = \begin{cases} (-1)^{i-1}, & t = t_i \\ 0, & \text{otherwise.} \end{cases}$$

Thus, $\mathcal{V}(P)$ is an integer polytope.

b. [2–]* Show that $\dim \mathcal{V}(P) = \#P$.

c. [2] Compute the Ehrhart polynomial $i(\mathcal{V}(P), n)$ of the valuation polytope of a p-element chain.

d. [2+] Show that

$$A(\mathcal{V}(P + Q), x) = A(\mathcal{V}(P), x) A(\mathcal{V}(Q), x),$$

where $A(\mathcal{P}, x)$ denotes the $\mathcal{P}$-Eulerian polynomial.

e. [2] Show that $i(\mathcal{V}(P), 1)$ is the total number of chains of P (including the empty chain).

f. [2+] Let $p = \#P$, and let m denote the number of minimal elements of P. Show that $\deg A(\mathcal{V}(P), x) = p - m$.

g. [2+] Show that $x^{p-m} A(\mathcal{V}(P), 1/x) = A(\mathcal{V}(P), x)$ if and only if every connected component of P has a unique minimal element.

h. [2+]* Let U_k denote the ordinal sum of k 2-element antichains, as in Exercise 3.139. Let $A(n)$ denote the $n \times n$ real matrix, with rows and columns indexed by $[n+1]$, defined by

$$A(n)_{ij} = \begin{cases} i+j-1, & \text{if } i+j \leq n+2, \\ 2n-i-j+3, & \text{if } i+j \geq n+2. \end{cases}$$

Show that $i(\mathcal{V}(U_k), n)$ is the sum of the entries of the first row of $A(n)^k$. Is there a more explicit formula for $i(\mathcal{V}(U_k), n)$? Is there a nice formula for the volume of $\mathcal{V}(U_k)$? If we write $\mathrm{vol}(\mathcal{V}(U_k)) = u_k/(2k)!$, then

$$(u_1, \ldots, u_6)) = (2, 8, 162, 6128, 372560, 33220512).$$

i. [5−] What more can be said about $\mathcal{V}(P)$ in general? Is there a nice combinatorial interpretation of its volume? Are the coefficients of $i(\mathcal{V}(P), n)$ nonnegative?

59. [3] Let $t \in \mathbb{R}$, and define $v_d(t) = (t, t^2, \ldots, t^d) \in \mathbb{R}^d$. The set of all points $v_d(t), t \in \mathbb{R}$, is called the *moment curve*. Let $n > d$ and $T = \{t_1, \ldots, t_n\}$, where the t_i's are real numbers satisfying $t_1 < \cdots < t_n$. Define the *cyclic polytope* $C_d(T)$ to be the convex hull of the points $v_d(t_1), \ldots, v_d(t_n)$. Suppose that each t_i is an integer, so $C_d(T)$ is an integral polytope. Show that

$$i(C_d(T), m) = \mathrm{vol}(C_d(T)) m^d + i(C_{d-1}(T), m),$$

where we set $i(C_0(T), m) = 1$. In particular, the polynomial $i(C_d(T), m)$ has positive coefficients.

60. [2+] Give an example of a 3-dimensional simplex (tetrahedron) $\mathcal{P}$ with integer vertices such that the Ehrhart polynomial $i(\mathcal{P}, n)$ has a negative coefficient.

61. a. [2+] Let e_j be the jth unit coordinate vector in $\mathbb{R}^d$, and let $\mathcal{P}_d$ be the convex hull of the $2d$ vectors $\pm e_j$. (This polytope is the d-dimensional *cross-polytope*. When $d = 3$ it is an octahedron.) Let $i(\mathcal{P}_d, n)$ denote the Ehrhart polynomial of $\mathcal{P}_d$. Find explicitly the polynomial $P_d(x)$ for which

$$\sum_{n \geq 0} i(\mathcal{P}_d, n) x^n = \frac{P_d(x)}{(1-x)^{d+1}}.$$

b. [3−] Show that every (complex) zero of $i(\mathcal{P}_d, n)$ has real part $-1/2$.

62. Let $1 \leq k \leq n-1$. The *hypersimplex* $\Delta_{k,d}$ is the convex hull of all $(0,1)$-vectors in $\mathbb{R}^d$ with exactly k 1's.
a. [2−]* Show that $\dim \Delta_{k,d} = d-1$.
b. [2+] Show that the relative volume of $\Delta_{k,d}$ is $A(d-1, k)/(d-1)!$, where $A(d-1, k)$ is an Eulerian number (the number of permutations $w \in \mathfrak{S}_{d-1}$ with $k-1$ descents).
c. [2+] Show that

$$i(\Delta_{k,d}, n) = [x^{kn}] \left(\frac{1 - x^{n+1}}{1 - x} \right)^d.$$

d. [2]* Deduce from (c) that

$$i(\Delta_{k,d}, n) = \sum_{j=0}^{\lfloor kn/(n+1) \rfloor} (-1)^j \binom{d}{j} \binom{(k-j)n - j + d - 1}{d - 1}.$$

e. [5–] Are the coefficients of $i(\Delta_{k,d}, n)$ nonnegative?

f. [2]* Let $A(\Delta_{k,d}, x)$ be the $\Delta_{k,d}$-Eulerian polynomial. Show that $A(\Delta_{1,d}, x) = 1$.

g. [2+] Show that

$$A(\Delta_{2,d}, x) = \begin{cases} 1 + \tfrac{1}{2}d(d-3)x + \binom{d}{4}x^2 + \binom{d}{6}x^3 + \cdots + \binom{d}{d}x^{d/2}, \\ \qquad\qquad\qquad\qquad\qquad\qquad\qquad\qquad\qquad\qquad d \text{ even}, \\[2mm] 1 + \tfrac{1}{2}d(d-3)x + \binom{d}{4}x^2 + \binom{d}{6}x^3 + \cdots + \binom{d}{d-1}x^{(d-1)/2}, \\ \qquad\qquad\qquad\qquad\qquad\qquad\qquad\qquad\qquad\qquad d \text{ odd}. \end{cases}$$

h. [5–] Find a combinatorial interpretation of the coefficients of $A(\Delta_{k,d}, x)$.

i. [3] Define the "half-open" hypersimplex $\Delta'_{k,d}$ to be the set of all vectors $(x_1, \ldots, x_d) \in \mathbb{R}^d$ satisfying $0 \le x_i \le 1$ and

$$\begin{aligned} 0 &\le x_1 + \cdots + x_d \le 1, \quad k = 1, \\ k - 1 &< x_1 + \cdots + x_d \le k, \quad 2 \le k \le d. \end{aligned}$$

Thus, the unit cube $[0,1]^d$ is a disjoint union of the $\Delta'_{k,d}$'s. Show that

$$A(\Delta'_{k,d}, x) = \sum_w x^{\mathrm{des}(w)},$$

where w ranges over all permutations in $\mathfrak{S}_d$ with $k - 1$ excedances. For instance, $A(\Delta'_{3,4}, x) = 4x + 6x^2 + x^3$, corresponding to the permutations 2314, 2413, 3412, 1342 (one descent), 2431, 3421, 2143, 3142, 3241, 4312 (two descents), and 4321 (three descents).

63. [3] Let $v_1, \ldots, v_k \in \mathbb{Z}^m$. Let

$$\mathcal{Z} = \{a_1 v_1 + \cdots + a_k v_k : 0 \le a_i \le 1\}.$$

Thus, $\mathcal{Z}$ is a convex polytope with integer vertices. Show that the Ehrhart polynomial of $\mathcal{Z}$ is given by $i(\mathcal{Z}, n) = c_m n^m + \cdots + c_0$, where $c_i = \sum_X f(X)$, the sum being over all linearly independent i-element subsets X of $\{v_1, \ldots, v_k\}$, and where $f(X)$ is the greatest common divisor (always taken to be positive) of the determinants of the $i \times i$ submatrices of the matrix whose rows are the elements of X.

64. a. [3] Let $\mathcal{P}_d$ denote the convex hull in $\mathbb{R}^d$ of the $d!$ points $(w(1), w(2), \ldots, w(d))$, $w \in \mathfrak{S}_d$. The polytope $\mathcal{P}_d$ is called the *permutohedron*. Show that the Ehrhart polynomial of $\mathcal{P}_d$ is given by

$$i(\mathcal{P}_d, n) = \sum_{i=0}^{d-1} f_i n^i,$$

where f_i is the number of forests with i edges on a set of d vertices. For example, $f_0 = 1$, $f_1 = \binom{d}{2}$, $f_{d-1} = d^{d-2}$. In particular, the relative volume of $\mathcal{P}_d$ is d^{d-2}.

b. [3] Generalize (a) as follows. Let G be a finite graph (loops and multiple edges permitted) with vertices $v_1, \ldots, v_d$. An *orientation* o of the edges may be regarded as an assignment of a direction $u \to v$ to every edge e of G, where e is incident to vertices u and v. If in the orientation o there are δ_i edges pointing out of v_i, then call $\delta(o) = (\delta_1, \ldots, \delta_d)$ the *outdegree sequence* of o. Define o to be *acyclic* if there are no directed cycles $u_1 \to u_2 \to \cdots \to u_k \to u_1$, as in Exercise 3.60. Let $\mathcal{P}_G$ denote the convex hull in $\mathbb{R}^d$ of all outdegree sequences $\delta(\mathfrak{o})$ of acyclic orientations of G. Show that

$$i(\mathcal{P}_G, n) = \sum_{i=0}^{d-1} f_i(G) n^i,$$

where $f_i(G)$ is the number of spanning forests of G with i edges. Show also that

$$\mathcal{P}_G \cap \mathbb{Z}^d = \{\delta(\mathfrak{o}) : \mathfrak{o} \text{ is an orientation of } G\},$$

and deduce that the number of distinct $\delta(\mathfrak{o})$ is equal to the number of spanning forests of G. (Note that (a) corresponds to the case $G = K_d$.)

65. [3] An *FHM-graph* is a graph G (allowing multiple edges, but not loops) such that every induced subgraph has at most one connected component that is not bipartite. A *spanning quasiforest* of a graph G is a spanning subgraph H of G for which every connected component is either a tree or has exactly one cycle C, such that C has odd length. Let $c(H)$ denote the number of (odd) cycles of the quasiforest H. If H is a graph with vertices $v_1, \ldots, v_p$ and q edges, then the *extended degree sequence* of H is the sequence $\tilde{d}(H) = (d_1, \ldots, d_p, q) \in \mathbb{R}^{p+1}$, where v_i has degree (number of incident edges) d_i. Let $\widetilde{\mathcal{D}}(G)$ denote the convex hull in $\mathbb{R}^p$ of the extended degree sequence $\tilde{d}(H)$ of all spanning subgraphs H of G. Show that if G is an FHM-graph, then

$$i(\widetilde{\mathcal{D}}(G), n) = a_p n^p + a_{p-1} n^{p-1} + \cdots + a_0, \qquad (4.58)$$

where

$$a_i = \sum_H \max\{1, 2^{c(H)-1}\},$$

the sum being over all spanning quasiforests H of G with i edges.

66. **a.** [3] Let $\mathcal{P}$ be a d-dimensional rational convex polytope in $\mathbb{R}^m$, and let the Ehrhart quasipolynomial of $\mathcal{P}$ be

$$i(\mathcal{P}, n) = c_d(n) n^d + c_{d-1}(n) n^{d-1} + \cdots + c_0(n),$$

where $c_0, \ldots, c_d$ are periodic functions of n. Suppose that for some $j \in [0, d]$, the affine span of every j-dimensional face of $\mathcal{P}$ contains a point with integer coordinates. Show that if $k \geq j$, then $c_k(n)$ is constant (i.e., period one).

   b. [3] Generalize (a) as follows: the (not necessarily least) period of $c_i(n)$ is the least positive integer p such that each i-face of $p\mathcal{P}$ contains an integer vector.

67. [2]* Let M be a diagonalizable $p \times p$ matrix over a field K. Let $\lambda_1, \ldots, \lambda_r$ be the distinct nonzero eigenvalues of M. Fix $(i, j) \in [p] \times [p]$. Show that there exist constants $a_1, \ldots, a_r \in K$ such that for all $n \in \mathbb{P}$,

$$(M^n)_{ij} = a_1 \lambda_1^n + \cdots + a_r \lambda_r^n.$$

68. **a.** [2]* By combinatorial reasoning, find the number $f(r, n)$ of sequences $\emptyset = S_0, S_1, \ldots, S_{2n} = \emptyset$ of subsets of $[r]$ such that for each $1 \leq i \leq 2n$, either $S_{i-1} \subset S_i$ and $|S_i - S_{i-1}| = 1$, or $S_i \subset S_{i-1}$ and $|S_{i-1} - S_i| = 1$.

b. [2]* Let $A(r)$ be the adjacency matrix of the Hasse diagram of the boolean algebra B_r. Thus, the rows and columns of $A(r)$ are indexed by $S \in B_r$, with

$$A(r)_{S,T} = \begin{cases} 1, & \text{if } S \text{ covers } T \text{ or } T \text{ covers } S \text{ in } B_r, \\ 0, & \text{otherwise.} \end{cases}$$

Use (a) to find the eigenvalues of $A(r)$. (It is more customary to use (b) to solve (a).)

69. [2]* Use reasoning similar to the previous exercise to find the eigenvalues of the adjacency matrix of the complete bipartite graph K_{rs}. Thus, first compute the number of closed walks of length ℓ in K_{rs}.

70. **a.** [2+] Let G be a finite graph (allowing loops and multiple edges). Suppose that there is some integer $\ell > 0$ such that the number of walks of length ℓ from any fixed vertex u to any fixed vertex v is independent of u and v. Show that G has the same number k of edges between any two vertices (including k loops at each vertex).

   b. [3–] Again let G be a finite graph (allowing loops and multiple edges). For any vertex v, let d_v be its degree (number of incident edges). Start at any vertex of G and do a random walk as follows: If we are at a vertex v, then walk along an edge incident to v with probability $1/d_v$. Suppose that there is some integer $\ell \geq 1$ such that for any intial vertex u, after we take ℓ steps we are equally likely to be at any vertex. Show that we have the same conclusion as (a) (i.e., G has the same number k of edges between any two vertices).

71. [2+] Let K_p^o denote the complete graph with p vertices, with one loop at each vertex. Let $K_p^o - K_r^o$ denote K_p^o with the edges of K_r^o removed, i.e., choose r vertices of K_p^o, and remove all edges between these vertices (including loops). Thus, $K_p^o - K_r^o$ has $\binom{p+1}{2} - \binom{r+1}{2}$ edges. Find the number $C_G(\ell)$ of closed walks in $G = K_{21}^o - K_{18}^o$ of length $\ell \geq 1$.

72. [3–] Let G be a finite graph on p vertices. Let G' be the graph obtained from G by placing a new edge e_v incident to each vertex v, with the other vertex of e_v being a new vertex v'. Thus, G' has p new edges and p new vertices. The new vertices all have degree one. By combinatorial reasoning, express the eigenvalues of the adjacency matrix $A(G')$ in terms of the eigenvalues of $A(G)$.

73. **a.** [2]* Let $F(n)$ be the number of ways a $2 \times n$ chessboard can be partitioned into

   copies of the two pieces ⊞ and ⊟. (Any rotation or reflection of the pieces is allowed.) For instance, $f(0) = 1$, $f(1) = 1$, $f(2) = 2$, $f(3) = 5$. Find $F(x) = \sum_{n \geq 0} f(n)x^n$.

   b. [2]* Let $g(n)$ be the number of ways if we also allow the piece ⊡. Thus, $g(0) = 1$, $g(1) = 2$, $g(2) = 11$. Find $G(x) = \sum_{n \geq 0} g(n)x^n$.

74. [2+] Suppose that the graph G has 16 vertices and that the number of closed walks of length ℓ in G is $8^\ell + 2 \cdot 3^\ell + 3 \cdot (-1)^\ell + (-6)^\ell + 5$ for all $\ell \geq 1$. Let G' be the graph obtained from G by adding a loop at each vertex (in addition to whatever loops are already there). How many closed walks of length ℓ are there in G'? Give a linear algebraic solution and (more difficult) a combinatorial solution.

75. **a.** [2+] Let $M = (m_{ij})$ be an $n \times n$ circulant matrix with first row $(a_0, \ldots, a_{n-1}) \in \mathbb{C}^n$, that is, $m_{ij} = a_{j-i}$, the subscript $j - i$ being taken modulo n. Let $\zeta = e^{2\pi i/n}$. Show that the eigenvalues of M are given by

$$\omega_r = \sum_{j=0}^{n-1} a_j \zeta^{jr}, \quad 0 \leq r \leq n-1.$$

b. [1] Let $f_k(n)$ be the number of sequences of integers $t_1, t_2, \ldots, t_n$ modulo k (i.e., $t_j \in \mathbb{Z}/k\mathbb{Z}$) such that $t_{j+1} \equiv t_j - 1, t_j$, or $t_j + 1 \pmod{k}$, $1 \leq j \leq n-1$. Find $f_k(n)$ explicitly.

c. [2] Let $g_k(n)$ be the same as $f_k(n)$, except that in addition we require $t_1 \equiv t_n - 1$, t_n, or $t_n + 1 \pmod{k}$. Use the transfer-matrix method to show that

$$g_k(n) = \sum_{r=0}^{k-1} \left(1 + 2\cos\frac{2\pi r}{k} \right)^n.$$

d. [5–] From (c) we get $g_4(n) = 3^n + 2 + (-1)^n$ and $g_6(n) = 3^n + 2^{n+1} + (-1)^n$. Is there a combinatorial proof?

76. a. [2+] Let $A = A(n)$ be the $n \times n$ real matrix given by

$$A_{ij} = \begin{cases} 1, & j = i+1 \ (1 \leq i \leq n-1), \\ 1, & j = i-1 \ (2 \leq i \leq n), \\ 0, & \text{otherwise.} \end{cases}$$

Thus, A is the adjacency matrix of an n-vertex path. Let $V_n(x) = \det(xI - A)$, so $V_0(x) = 1$, $V_1(x) = x$, $V_2(x) = x^2 - 1$, $V_3(x) = x^3 - 2x$. Show that $V_{n+1}(x) = xV_n(x) - V_{n-1}(x)$, $n \geq 1$.

b. [2+]* Show that

$$V_n(2\cos\theta) = \frac{\sin((n+1)\theta)}{\sin(\theta)}.$$

Deduce that the eigenvalues of $A(n)$ are $2\cos(j\pi/(n+1))$, $1 \leq j \leq n$.

c. [2–] Let $u_n(k)$ be the number of sequences of integers $t_1, t_2, \ldots, t_k$, $1 \leq t_i \leq n$, such that $t_{j+1} = t_j - 1$ or $t_j + 1$ for $1 \leq j \leq n-1$, and $t_k = t_1 - 1$ or $t_1 + 1$ (if defined, i.e., 1 can be followed only by 2, and n by $n-1$). Find $u_n(k)$ explicitly.

d. [2+] Find a simple formula for $u_{2n}(2n)$.

77. [2]* Let $f_p(n)$ be as in Example 4.7.5. Give a simple combinatorial proof that $f_p(n-1) + f_p(n) = p(p-1)^{n-1}$, and deduce from this the formula $f_p(n) = (p-1)^n + (p-1)(-1)^n$ (equation (4.36)).

78. a. [2] Let $g_k(n)$ denote the number of $k \times n$ matrices $(a_{ij})_{1 \leq i \leq k, \, 1 \leq j \leq n}$ of integers such that $a_{11} = 1$, the rows and columns are weakly increasing, and adjacent entries differ by at most 1. Thus, $a_{i,j+1} - a_{ij} = 0$ or 1, and $a_{i+1,j} - a_{ij} = 0$ or 1. Show that $g_2(n) = 2 \cdot 3^{n-1}$, $n \geq 1$.

b. [2+] Show that $G_k(x) = \sum_{n \geq 1} g_k(n) x^n$ is a rational function. In particular,

$$G_3(x) = \frac{2x(2-x)}{1 - 5x + 2x^2}.$$

79. a. [2+] Let $G_1, \ldots, G_k$ be finite graphs on the vertex sets $V_1, \ldots, V_k$. Given any graph H, write $m(u, v)$ for the number of edges between vertices u and v. Let $u = (u_1, \ldots, u_k) \in V_1 \times \cdots \times V_k$ and $v = (v_1, \ldots, v_k) \in V_1 \times \cdots \times V_k$. Define the *star product* $G_1 * \cdots * G_k$ of $G_1, \ldots, G_k$ to be the graph on the vertex set $V_1 \times \cdots \times V_k$ with edges defined by

$$m(u, v) = \begin{cases} 0, & \text{if } u, v \text{ differ in at least two coordinates,} \\ \sum_i m(u_i, u_i), & \text{if } u = v, \\ m(u_i, v_i), & \text{if } u, v \text{ differ only in coordinate } i. \end{cases}$$

Find the eigenvalues of the adjacency matrix $A(G_1 * \cdots * G_k)$ in terms of the eigenvalues of $A(G_1), \ldots, A(G_k)$.

b. [2+] Let $V_i = [m_i]$, and regard $B = V_1 \times \cdots \times V_k$ as a k-dimensional chessboard. A rook moves from a vertex u of B to any other vertex v that differs from u in exactly one coordinate. Suppose without loss of generality that $u = (1, 1, \ldots, 1)$ and $v = (1^{k-r}, 2^r)$ (i.e., a vector of $k - r$ 1's followed by r 2's). Find an explicit formula for the number N of ways a rook can move from u to v in exactly n moves.

80. [2+] As in Exercise 4.40, let $X = \{x_1, \ldots, x_n\}$ be an alphabet with n letters. Let N be a finite set of words. Define $f_N(m)$ to be the number of words $w \in X_m^*$ (i.e., of length m) such that w contains no subwords (as defined in Exercise 3.134) belonging to N. Use the transfer-matrix method to show that $F_N(x) := \sum_{m \geq 0} f_N(m) x^m$ is rational.

81. **a.** [2] Fix $k \in \mathbb{P}$, and for $n \in \mathbb{N}$ define $f_k(n)$ to be the number of ways to cover a $k \times n$ chessboard with $\frac{1}{2}kn$ nonoverlapping dominoes (or *dimers*). Thus, $f_k(n) = 0$ if kn is odd, $f_1(2n) = 1$, and $f_2(2) = 2$. Set $F_k(x) = \sum_{n \geq 0} f_k(n) x^n$. Use the transfer-matrix method to show that $F_k(x)$ is rational. Compute $F_k(x)$ for $k = 2, 3, 4$.

b. [3] Use the transfer-matrix method to show that

$$f_k(n) = \prod_{j=1}^{\lfloor k/2 \rfloor} \frac{c_j^{n+1} - \bar{c}_j^{n+1}}{2b_j}, \quad nk \text{ even}, \tag{4.59}$$

where

$$c_j = a_j + \sqrt{1 + a_j^2},$$

$$\bar{c}_j = a_j - \sqrt{1 + a_j^2},$$

$$b_j = \sqrt{1 + a_j^2},$$

$$a_j = \cos \frac{j\pi}{k+1}.$$

c. [3–] Use (b) to deduce that we can write $F_k(x) = P_k(x)/Q_k(x)$, where P_k and Q_k are polynomials with the following properties:

i. Set $\ell = \lfloor k/2 \rfloor$. Let $S \subseteq [\ell]$ and set $\bar{S} = [\ell] - S$. Define

$$c_S = \left(\prod_{j \in S} c_j \right) \left(\prod_{j \in \bar{S}} \bar{c}_j \right).$$

Then

$$Q_k(x) = \begin{cases} \prod_S (1 - c_S x), & k \text{ even}, \\ \prod_S (1 - c_S^2 x^2), & k \text{ odd}, \end{cases}$$

where S ranges over all subsets of $[\ell]$.

ii. $Q_k(x)$ has degree $q_k = 2^{\lfloor (k+1)/2 \rfloor}$.

iii. $P_k(x)$ has degree $p_k = q_k - 2$.

iv. If $k > 1$, then $P_k(x) = -x^{p_k} P_k(1/x)$. If k is odd or divisible by 4, then $Q_k(x) = x^{q_k} Q_k(1/x)$. If $k \equiv 2 \pmod 4$ then $Q_k(x) = -x^{q_k} Q_k(1/x)$. If k is odd, then $P_k(x) = P_k(-x)$ and $Q_k(x) = Q_k(-x)$.

82. For $n \geq 2$, let T_n be the $n \times n$ toroidal graph, that is, the vertex set is $(\mathbb{Z}/n\mathbb{Z})^2$, and (i, j) is connected to its four neighbors $(i - 1, j)$, $(i + 1, j)$, $(i, j - 1)$, $(i, j + 1)$ with entries modulo n. (Thus, T_n has n^2 vertices and $2n^2$ edges.) Let $\chi_n(\lambda)$ denote the chromatic polynomial of T_n, and set $N = n^2$.

a. [1+] Find $\chi_n(2)$.

b. [3+] Use the transfer-matrix method to show that

$$\log \chi_n(3) = \frac{3N}{2} \log(4/3) + o(N).$$

c. [5] Show that

$$\log \chi_n(3) = \frac{3N}{2} \log(4/3) - \frac{\pi}{6} + o(1).$$

d. [5] Find $\lim_{N \to \infty} N^{-1} \log \chi_n(4)$.

e. [3−] Let $\chi_n(\lambda) = \lambda^N - q_1(N)\lambda^{N-1} + q_2(N)\lambda^{N-2} - \cdots$. Show that there are polynomials $Q_i(N)$ such that $q_i(N) = Q_i(N)$ for all N sufficiently large (depending on i). For instance, $Q_1(N) = 2N$, $Q_2(N) = N(2N-1)$, and $Q_3(N) = \frac{1}{3}N(4N^2 - 6N - 1)$.

f. [3] Let $\alpha_i = Q_i(1)$. Show that

$$1 + \sum_{i \geq 1} Q_i(N)x^i = (1 + \alpha_1 x + \alpha_2 x^2 + \cdots)^N$$

$$= (1 + 2x + x^2 - x^3 + x^4 - x^5 + x^6 - 2x^7 + 9x^8 - 38x^9$$

$$+ 130x^{10} - 378x^{11} + 987x^{12} - 2436x^{13} + 5927x^{14}$$

$$- 14438x^{15} + 34359x^{16} - 75058x^{17}$$

$$+ 134146x^{18} + \cdots)^N. \tag{4.60}$$

Equivalently, in the terminology of Exercise 5.37, the sequence $1, 1! \cdot Q_1(N), 2! \cdot Q_2(N), \ldots$ is a sequence of polynomials of binomial type.

g. [5−] Let $L(\lambda) = \lim_{N \to \infty} \chi_n(\lambda)^{1/N}$. Show that for $\lambda \geq 2$, $L(\lambda)$ has the asymptotic expansion

$$L(\lambda) \sim \lambda(1 - \alpha_1 \lambda^{-1} + \alpha_2 \lambda^{-2} + \cdots).$$

Does this infinite series converge?

Solutions to Exercises

1. No. Suppose that $F(x) \in K(x)$ and $G(x) \in L(x)$, where K and L are fields of different characteristics (or even isomorphic fields but with no explicit isomorphism given, such as $\mathbb{C}$ and the algebraic closure of the p-adic field $\mathbb{Q}_p$). Then $F(x) + G(x)$ is undefined.

2. a. Define a formal power series $\sum_{n \geq 0} a_n x^n$ with integer coefficients to be *primitive* if no integer $d > 1$ divides *all* the a_i. One easily shows that the product of primitive series is primitive (a result essentially due to Gauss but first stated explicitly by Hurwitz; this result is equivalent to the statement that $\mathbb{F}_p[[x]]$ is an integral domain, where $\mathbb{F}_p$ is the field of prime order p).

Clearly, we can write $f(x) = P(x)/Q(x)$ for some relatively prime integer polynomials P and Q. Assume that no integer $d > 1$ divides every coefficient of P and Q. Then Q is primitive, for otherwise if $Q/d \in \mathbb{Z}[x]$ for $d > 1$, then

$$\frac{P}{d} = f \frac{Q}{d} \in \mathbb{Z}[x],$$

a contradiction. Since $(P, Q) = 1$ in $\mathbb{Q}[x]$, there is an integer $m > 0$ and polynomials $A, B \in \mathbb{Z}[x]$ such that $AP + BQ = m$. Then $m = Q(Af + B)$. Since Q is primitive, the coefficients of $Af + B$ are divisible by m. (Otherwise, if $d < m$ is the largest integer dividing $Af + B$, then the product of the primitive series Q and $(Af + B)/d$ would be the imprimitive polynomial $m/d > 1$.) Let c be the constant term of $Af + B$. Then $m = Q(0)c$. Since m divides c, we have $Q(0) = \pm 1$.

This result is known as *Fatou's lemma* and was first proved in P. Fatou, *Acta Math.* **30** (1906), 369. The proof given here is due to A. Hurwitz; see G. Pólya, *Math. Ann.* **77** (1916), 510–512.

b. This result, while part of the "folklore" of algebraic geometry and an application of standard techniques of commutative algebra, seems first to be explicitly stated and proved (in an elementary way) by I. M. Gessel, *Utilitas Math.* **19** (1981), 247–251 (Thm. 1).

3. The assertion is true. Without loss of generality we may assume that $f(x)$ is primitive, as defined in the solution to Exercise 4.2. Let $f'(x) = f(x)g(x)$, where $g(x) \in \mathbb{Z}[[x]]$. By Leibniz's rule for differentiating a product, we obtain by induction on n that $f(x)|f^{(n)}(x)$ in $\mathbb{Z}[[x]]$. But also $n!|f^{(n)}(x)$, since if $f(x) = \sum a_i x^i$ then $\frac{1}{n!}f^{(n)}(x) = \sum \binom{i}{n} a_i x^{i-n}$. Write $f(x)h(x) = n!(f^{(n)}(x)/n!)$, where $h(x) \in \mathbb{Z}[[x]]$. Since the product of primitive polynomials is primitive, we obtain just as in the solution to Exercise 4.2 that $n!|h(x)$ in $\mathbb{Z}[[x]]$, so $f(x)|(f^{(n)}(x)/n!)$. In particular, $f(0)|(f^{(n)}(0)/n!)$ in $\mathbb{Z}$, which is the desired conclusion.

NOTE. An alternative proof uses the known fact that $\mathbb{Z}[[x]]$ is a unique factorization domain. Since $f(x)|f^{(n)}(x)$, and since $f(x)$ and $n!$ are relatively prime in $\mathbb{Z}[[x]]$, we get $n!f(x)|f^{(n)}(x)$.

This exercise is due to David Harbater.

4. **a.** This result was first proved by T. A. Skolem, *Oslo Vid. Akad. Skrifter I*, no. 6 (1933), for rational coefficients, then by K. Mahler, *Proc. Akad. Wetensch. Amsterdam* **38** (1935), 50–60, for algebraic coefficients, and finally independently by Mahler, *Proc. Camb. Phil. Soc.* **52** (1956), 39–48, and C. Lech, *Ark. Math.* **2** (1953), 417–421, for complex coefficients (or over any field of characteristic 0) and is known as the *Skolem-Mahler-Lech theorem.* All the proofs use p-adic methods. As pointed out by Lech, the result is false over characteristic p, an example being the series

$$F(x) = \frac{1}{1 - (t+1)x} - \frac{1}{1-x} - \frac{1}{1-tx}$$

over the field $\mathbb{F}_p(t)$. See also J.-P. Serre, *Proc. Konin. Neder. Akad. Weten. (A)* **82** (1979), 469–471.

For an interesting article on the Skolem–Mahler–Lech theorem, see G. Myerson and A. J. van der Poorten, *Amer. Math. Monthly* **102** (1995), 698–705. For a proof, see J. W. S. Cassels, *Local Fields*, Cambridge University Press, Cambridge, 1986. For further information on coefficients of rational generating functions, see A. J. van der Poorten, in Coll. Math. Sci. János Bolyai **34**, *Topics in Classical Number Theory* (G. Hal'asz, ed.), vol. 2, North-Holland, New York, 1984, pp. 1265–1294. (This paper, however, contains many inaccuracies, beginning on page 1276.)

b. Let

$$F(x, y) = \sum_{m,n \geq 0} (m - n^2) x^m y^n$$

$$= \frac{1}{(1-x)^2(1-y)} - \frac{y + y^2}{(1-x)(1-y)^3}.$$

Then

$$\sum_{m,n \geq 0} \chi(m - n^2) x^m y^n = \sum x^{m^2} y^n,$$

which is seen to be nonrational, for example, by setting $y = 1$ and using (a). This problem was suggested by D. A. Klarner.

c. A proof based on the same p-adic methods used to prove (a) is sketched by A. J. van der Poorten, *Bull. Austral. Math. Sòc.* **29** (1984), 109–117.

5. This problem was raised by T. Skolem, *Skand. Mat. Kongr. Stockholm, 1934* (1934), 163–188. For the current status of this problem, see V. Halava, T. Harju, M. Hirvensalo, and J. Karhumäki, Skolem's problem—On the border between decidability and undecidability, preprint.

6. This fundamental result is due to L. Kronecker, *Monatsber. K. Preuss. Akad. Wiss. Berlin* (1881), 535–600. For an exposition, see F. R. Gantmacher, *Matrix Theory*, vol. 2, Chelsea, New York, 1989 (§XV.10).

7. **a.** Write

$$\frac{x F'(x)}{F(x)} = \frac{b_1 x}{1 - x} + G(x). \tag{4.61}$$

where $G(x) = \sum_{n \geq 1} c_n x^n$. By arguing as in Example 1.1.14, we have

$$c_n = \sum_{\substack{(2i-1)|n \\ i \neq 1}} (2i - 1) b_i.$$

If n is a power of 2 then this sum is empty and $c_n = 0$; otherwise, $c_n \neq 0$. By Exercise 4.4, $G(x)$ is not rational. Hence by equation (4.61), $F(x)$ is not rational.

This result is essentially due to J.-P. Serre, *Proc. Konin. Neder. Akad. Weten. (A)* **82** (1979), 469–471.

b. Let $F(x) = 1/(1 - \alpha x)$, where $\alpha \geq 2$. Then by the same reasoning as Example 1.1.14, we have that $a_i \in \mathbb{Z}$ and

$$a_i = \frac{1}{i} \sum_{d|i} \mu(i/d) \alpha^d$$

$$\geq \frac{1}{i} \left(\alpha^i - \sum_{j=1}^{i-1} \alpha^j \right) > 0.$$

It is also possible to interpret a_i combinatorially when $\alpha \geq 2$ is an integer (or a prime power) and thereby see combinatorially that $a_i > 0$. See Exercise 2.7 for the case when α is a prime power.

8. (i)$\Rightarrow$(iii) If $F(x) \in \mathbb{C}[[x]]$ and $F(x) = G'(x)/G(x)$ with $G(x) \in \mathbb{C}((x))$, then $G(0) \neq 0, \infty$. Hence, if $G(x)$ is rational, then we can write

$$G(x) = \frac{c \prod(1 - \beta_i x)}{\prod(1 - \alpha_i x)}$$

for certain nonzero $\alpha_i, \beta_i \in \mathbb{C}$. Direct computation yields

$$\frac{G'(x)}{G(x)} = \sum \frac{\alpha_i}{1 - \alpha_i x} - \sum \frac{\beta_i}{1 - \beta_i x},$$

so $a_n = \sum \alpha_i^n - \sum \beta_i^n$.

(iii)$\Rightarrow$(ii) If $a_n = \sum \alpha_i^n - \sum \beta_i^n$. then

$$\exp \sum_{n \geq 1} a_n \frac{x^n}{n} = \frac{\prod (1 - \beta_i x)}{\prod (1 - \alpha_i x)}$$

by direct computation.

(ii)$\Rightarrow$(i) Set $G(x) = \exp \sum_{n \geq 1} a_n \frac{x^n}{n}$ and compute that

$$F(x) = \frac{d}{dx} \log G(x) = \frac{G'(x)}{G(x)}.$$

9. Yes. Suppose that $F(x) = P(x)/Q(x)$, where $P, Q \in \mathbb{Q}[x]$. By the division algorithm for polynomials we have

$$F(x) = G(x) + \frac{R(x)}{Q(x)},$$

where $\deg R < \deg Q$ (with $\deg 0 = -\infty$, say). If $R(x) \neq 0$, then we can find positive integer p, n for which $pG(n) \in \mathbb{Z}$ and $0 < |R(n)/Q(n)| < 1/p$, a contradiction.

10. This result is due to E. Borel, *Bull. Sci. Math.* **18** (1894), 22–25. It is a useful tool for proving that generating functions are not meromorphic. For instance, let p_n be the nth prime and $f(z) = \sum_{n \geq 1} p_n z^n = 2z + 3z^2 + 5z^3 + \cdots$. It is easy to see that $f(z)$ has radius of convergence 1 and is not rational [why?]. Hence by Borel's theorem, $f(z)$ is not meromorphic.

11. a. *Answer.* $a_n = 2^n + 1$. A standard way to solve this recurrence that does not involve guessing the answer in advance is to observe that the denominator of the rational function $\sum_{n \geq 0} a_n x^n$ is $1 - 3x + 2x^2 = (1 - x)(1 - 2x)$. Hence, $a_n = \alpha 2^n + \beta 1^n = \alpha 2^n + \beta$. The initial conditions give $\alpha + \beta = 2$, $2\alpha + \beta = 3$, whence $\alpha = \beta = 1$.
    b. *Answer.* $a_n = n2^n$.
    c. *Answer.* $a_n = 3^{n+1} + 2^n + 1$.
    d. The polynomial $x^8 - x^7 + x^5 - x^4 + x^3 - x + 1$ is just the 15th cyclotomic polynomial (i.e., its zeros are the primitive 15th roots of unity). It follows that the sequence $a_0, a_1, \ldots$ is periodic with period 15. Thus, we need only compute $a_8 = 4$, $a_9 = 0$, $a_{10} = -5$, $a_{11} = -7$, $a_{12} = -9$, $a_{13} = -7$, $a_{14} = -4$ to determine the entire sequence. In particular, since $105 \equiv 0 \pmod{15}$, we have $a_{105} = a_0 = 0$. To solve this problem without recognizing that $x^8 - x^7 + x^5 - x^4 + x^3 - x + 1$ is a cyclotomic polynomial, simply compute a_n for $8 \leq n \leq 22$. Since $a_n = a_{n+15}$ for $0 \leq n \leq 7$, it follows that $a_n = a_{n+15}$ for all $n \geq 0$.

12. Note that

$$\frac{10000}{9899} = \frac{1}{1 - \frac{1}{100} - \frac{1}{100^2}}$$

and

$$\frac{1}{1 - x - x^2} = \sum_{n \geq 0} F_{n+1} x^n.$$

13. Because the Fibonacci recurrence can be run in reverse to compute F_i from F_{i+1} and F_{i+2}, and because there are only finitely many pairs $(a, b) \in \mathbb{Z}/n\mathbb{Z} \times \mathbb{Z}/n\mathbb{Z}$, it follows that the sequence $(F_i)_{i \in \mathbb{Z}}$ is periodic modulo n for all $n \in \mathbb{Z}$. Since $F_0 = 0$, it follows that some F_k for $k \geq 1$ must be divisible by n. Although there is an extensive literature on Fibonacci numbers modulo n (e.g., D. D. Wall, *Amer. Math. Monthly* **67** (1960), 525–532, and S. Gupta, P. Rockstroh, and F. E. Su, Splitting fields and periods of Fibonacci

sequences modulo primes, `arXiv:0909.0362`), it is not clear who first came up with the preceding elegant argument. Problem A3 from the 67th Putnam Mathematical Competition (2006) involves a similar idea. Note that the result of the present exercise fails for the Lucas numbers when $n = 5$. The foregoing proof breaks down because $L_i \neq 0$ for all $i \in \mathbb{Z}$.

14. The first such sequence was obtained by R. L. Graham, *Math. Mag.* **37** (1964), 322–324. At present the smallest pair (a,b) is due to J. W. Nicol, *Electronic J. Combinatorics* **6** (1999), #R44, namely,

$$a = 62638280004239857 = 127 \cdot 2521 \cdot 195642524071,$$

$$b = 49463435743205655 = 3 \cdot 5 \cdot 83 \cdot 89 \cdot 239 \cdot 1867785589.$$

15. Two solutions appear in R. Stanley, *Amer. Math. Monthly* **83** (1976), 813–814. The crucial lemma in the elementary solution given in this reference is that every antichain of $\mathbb{N}^m$ is finite.

16. Denote the answer by $f(n)$. The Young diagram of λ contains a 2×3 rectangle in the upper-left-hand corner. To the right of this rectangle is the diagram of a partition with at most two parts. Below the rectangle is the diagram of a partition with parts 1 and 2. Hence,

$$\sum_{n \geq 0} f(n)x^n = \frac{x^6}{(1-x)^2(1-x^2)^2}$$

$$= 1 + \frac{1}{4(1-x)^4} - \frac{5}{4(1-x)^3} + \frac{39}{16(1-x)^2} - \frac{9}{4(1-x)}$$

$$+ \frac{1}{16(1+x)^2} - \frac{1}{4(1+x)}.$$

It follows that for $n \geq 1$,

$$f(n) = \frac{1}{4}\binom{n+3}{3} - \frac{5}{4}\binom{n+2}{2} + \frac{39}{16}(n+1) - \frac{9}{4} + \frac{1}{16}(-1)^n(n+1) - \frac{1}{4}(-1)^n$$

$$= \frac{1}{48}\left(2n^3 - 18n^2 + 49n - 39\right) + \frac{1}{16}(-1)^n(n-3).$$

This problem was suggested by A. Postnikov, private communication, 2007.

17. **b.** See Exercise 3.62 and I. M. Gessel and R. Stanley, *J. Combinatorial Theory, Ser. A* **24** (1978), 24–33. The polynomial $F_n(x)$ is called a *Stirling polynomial*.

21. See P. Tetali, *J. Combinatorial Theory Ser. B* **72** (1998), 157–159. For a connection with radar tracking, see T. Khovanova, Unique tournaments and radar tracking, `arXiv:0712.1621`.

23. **b.** This remarkable result is due to J. H. Conway, *Eureka* **46** (1986), 5–18, and §5.11 in *Open Problems in Communication and Computation* (T. M. Cover and B. Gopinath, eds.), Springer-Verlag, New York, 1987 (pp. 173–188). For additional references, see item A005150 of *The On-Line Encyclopedia of Integer Sequences*.

c. $F(x) = x^{71} - x^{69} - 2x^{68} - x^{67} + 2x^{66} + x^{64} - x^{63} - x^{62} - x^{61} - x^{60} - x^{59} + 2x^{58} + 5x^{57} + 3x^{56} - 2x^{55} - 10x^{54} - 3x^{53} - 2x^{52} + 6x^{51} + 6x^{50} + x^{49} + 9x^{48} - 3x^{47} - 7x^{46} - 8x^{45} - 8x^{44} + 10x^{43} + 6x^{42} + 8x^{41} - 5x^{40} - 12x^{39} + 7x^{38} - 7x^{37} + 7x^{36} + x^{35} - 3x^{34} + 10x^{33} + x^{32} - 6x^{31} - 2x^{30} - 10x^{29} - 3x^{28} + 2x^{27} + 9x^{26} - 3x^{25} +$

$14x^{24} - 8x^{23} - 7x^{21} + x^{20} - 3x^{19} - 4x^{18} - 10x^{17} - 7x^{16} + 12x^{15} + 7x^{14} + 2x^{13} - 12x^{12} - 4x^{11} - 2x^{10} - 5x^9 + x^7 - 7x^6 + 7x^5 - 4x^4 + 12x^3 - 6x^2 + 3x - 6.$

d. For any initial sequence the generating function $F(x)$ is still rational, though now the behavior is more complicated and more difficult to analyze. See the references in (b).

24. **a.** Suppose that $a_0, a_1, \ldots$ is an infinite sequence of integers satisfying $0 \le a_i \le q - 1$. Let $\mathcal{P}'$ be the Newton polytope of f, i.e., the convex hull in $\mathbb{R}^k$ of the exponent vectors of monomials appearing in f, and let $\mathcal{P}$ be the convex hull of $\mathcal{P}'$ and the origin. If $c > 0$, then write $c\mathcal{P} = \{cv : v \in \mathcal{P}\}$. Set $S = (q-1)\mathcal{P} \cap \mathbb{N}^k$ and $r_m = \sum_{i=0}^m a_i q^i$.

Suppose that $f(x)^{r_m} = \sum_\gamma c_{m,\gamma} x^\gamma$. We set $f(x)^{r-1} = 1$. Let $\mathbb{F}_q^S$ be the set of all functions $F: S \to \mathbb{F}_q$. We will index our matrices and vectors by elements of $\mathbb{F}_q^S$ (in some order). Set

$$R_m = \{0, 1, \ldots, q^{m+1} - 1\}^k.$$

For $m \ge -1$, define a column vector ψ_m by letting $\psi_m(F)$ (the coordinate of ψ_m indexed by $F \in \mathbb{F}_q^S$) be the number of vectors $\gamma \in R_m$ such that for all $\delta \in S$ we have $c_{m, \gamma + q^{m+1}\delta} = F(\delta)$. Note that by the definition of S we have $c_{m, \gamma + q^{m+1}\delta} = 0$ if $\delta \notin S$. (This is the crucial finiteness condition that allows our matrices and vectors to have a fixed finite size.) Note also that given m, every point η in $\mathbb{N}^k$ can be written uniquely as $\eta = \gamma + q^{m+1}\delta$ for $\gamma \in R_{m+1}$ and δ in $\mathbb{N}^k$.

For $0 \le i \le q - 1$ define a matrix Φ_i with rows and columns indexed by $\mathbb{F}_q^S$ as follows. Let $F, G \in \mathbb{F}_q^S$. Set

$$g(x) = f(x)^i \sum_{\beta \in S} G(\beta) x^\beta$$

$$= \sum_\gamma d_\gamma x^\gamma \in \mathbb{F}_q[x].$$

Define the (F, G)-entry $(\Phi_i)_{FG}$ of Φ_i to be the number of vectors $\gamma \in R_0 = \{0, 1, \ldots, q-1\}^k$ such that for all $\delta \in S$ we have $d_{\gamma + q\delta} = F(\delta)$. A straightforward computation shows that

$$\Phi_{a_m} \psi_{m-1} = \psi_m, \quad m \ge 0. \tag{4.62}$$

Let $u = u_\alpha$ be the row vector for which $u(F)$ is the number of values of F equal to α, and let $n = a_0 + a_1 q + \cdots + a_r q^r$ as in the statement of the theorem. Then it follows from equation (4.62) that

$$N_\alpha(n) = u \Phi_{a_r} \Phi_{a_{r-1}} \cdots \Phi_{a_0} \psi_{-1},$$

completing the proof.

This proof is an adaptation of an argument of Y. Moshe, *Discrete Math.* **297** (2005), 91–103 (Theorem 1) and appears in T. Amdeberhan and R. Stanley, Polynomial coefficient enumeration, preprint, dated 3 February 2008;

⟨http://math.mit.edu/~rstan/papers/coef.pdf⟩ (Theorem 2.1).

25. See T. Amdeberhan and R. Stanley, ibid. (Theorem 2.8), where the result is given in the slightly more general context of the polynomial $f(x)^{q^n - c}$ for $c \in \mathbb{P}$.

26. **a.** The case $n = 5$ was obtained by G. Bagnera, *Ann. di Mat. pura e applicata* (3) **1** (1898), 137–228. The case $n = 6$ is due to M. F. Newman, E. A. O'Brien, and M. R. Vaughan-Lee, *J. Algebra* **278** (2004), 283–401. The case $n = 7$ is due to E. A. O'Brien and M. R. Vaughan-Lee, *J. Algebra* **292** (2005), 243–258. An interesting book on enumerating groups of order n is S. R. Blackburn, P. M. Neumann,

and G. Venkataraman, *Enumeration of Finite Groups*, Cambridge University Press, Cambridge, 2007.

b. The lower bound $g_n(p) \geq p^{\frac{2}{27}n^2(n-6)}$ is due to G. Higman, *Proc. London Math. Soc. (3)* **10** (1960), 24–30. The upper bound with the error term $O(n^{8/3})$ in the exponent is due to C. C. Sims, *Proc. London Math. Soc. (3)* **15** (1965), 151–166. The improved error term $O(n^{5/2})$ is due to M. F. Newman and C. Seeley, appearing in Blackburn, et al., ibid. (Chapter 5).

c. This is a conjecture of G. Higman, ibid. (page 24). See also Higman, *Proc. London Math. Soc.* **10** (1960), 566–582.

27. a. Follows from J. Backelin, *C. R. Acad. Sc. Paris* **287**(A) (1978), 843–846.

b. The first example was given by J. B. Shearer, *J. Algebra* **62** (1980), 228–231. A nice survey of this subject is given by J.-E. Roos, in *18th Scandanavian Congress of Mathematicians* (E. Balslev, ed.), *Progress in Math.*, vol. 11, Birkhäuser, Boston, 1981, pp. 441–468.

c. Using Theorems 4 and 6 of D. E. Knuth, *Pacific J. Math.* **34** (1970), 709–727, one can give a bijection between words in M of length n and symmetric $q \times q$ $\mathbb{N}$-matrices whose entries sum to n. It follows that

$$F(x) = \frac{1}{(1-x)^q (1-x^2)^{\binom{q}{2}}}.$$

See Corollary 7.13.6 and Section A1.1 (Appendix 1) of Chapter 7.

d. This result is a direct consequence of a result of Hilbert-Serre on the rationality of the Hilbert series of commutative finitely generated graded algebras. See, for example, M. F. Atiyah and I. G. Macdonald, *Introduction to Commutative Algebra*, Addison-Wesley, Reading, Mass., 1969 (Theorem 11.1); W. Bruns and J. Herzog, *Cohen–Macaulay Rings*, Cambridge University Press, Cambridge, 1993 (Chapter 4); and D. Eisenbud, *Commutative Algebra with a View Toward Algebraic Geometry*, Springer-Verlag, New York, 1995 (Exercise 10.12).

28. a. We have

$$T_r(x,y) = \text{tr} \sum A^{\alpha_1} B^{\beta_1} \cdots A^{\alpha_r} B^{\beta_r} x^{\alpha} y^{\beta}$$

$$= \text{tr} \left(\sum A^{\alpha_1} x_1^{\alpha_1} \right) \left(\sum B^{\beta_1} y_1^{\beta_1} \right) \cdots \left(\sum A^{\alpha_r} x_r^{\alpha_r} \right) \left(\sum B^{\beta_r} y_r^{\beta_r} \right)$$

$$= \text{tr}(1 - Ax_1)^{-1}(1 - By_1)^{-1} \cdots (1 - Ax_r)^{-1}(1 - By_r)^{-1}.$$

Now for any invertible matrix M, the entries of M^{-1} are rational functions (with denominator $\det M$) of the entries of M. Hence, the entries of $(1 - Ax_1)^{-1} \cdots (1 - By_r)^{-1}$ are rational functions of x and y with coefficients in $\mathbb{C}$, so the trace has the same property. The denominator of $T_r(x,y)$ can be taken to be

$$\det(1 - Ax_1)(1 - By_1) \cdots (1 - Ax_r)(1 - By_r)$$

$$= \prod_{i=1}^{r} \det(1 - Ax_i) \cdot \prod_{j=1}^{r} \det(1 - By_j).$$

b.

$$(1 - Ax)(1 - By) = \begin{bmatrix} 1 - y + xy & x - y \\ -x + y + xy & 1 + xy \end{bmatrix}$$

$$\Rightarrow T_1(x,y) = \frac{2 - y + xy}{(1 + x^2)(1 - y + y^2)}.$$

29. True. The generating function $\sum_{n\geq 0}(A^n)_{rs}\lambda^n$ is rational by Theorem 4.7.2. Hence, $(A^n)_{rs}$ has the form of Theorem 4.1.1(iii), namely,

$$(A^n)_{rs} = \sum_{m=1}^{k} P_m(n)\gamma_m^n, \quad n \gg 0.$$

The same is true of B and C and hence of $(A^n B^n C^n)_{ij}$ by the definition of matrix multiplication, so the proof follows.

30. *Answer.*

$$\frac{1 + xyz - x^2yzw - xy^2zw}{(1 - x^2z)(1 - xw)(1 - y^2z)(1 - yw)}.$$

31. Let $S = \{\beta \in \mathbb{Z}^m : \text{ there exist } \alpha \in \mathbb{N}^m \text{ and } n \in \mathbb{P} \text{ such that } \Phi\alpha = n\beta \text{ and } 0 \leq \alpha_i < n\}$. Clearly, S is finite. For each $\beta \in S$, define $F_\beta = \sum y^\alpha x^n$, summed over all solutions $\alpha \in \mathbb{N}^m$ and $n \in \mathbb{P}$ to $\Phi\alpha = n\beta$ and $\alpha_i < n$. Now $\sum_{n\geq 0} f(n)x^n = \sum_{\beta\in S} F_\beta(\mathbf{1}, x)$ (where $\mathbf{1} = (1,\dots,1) \in \mathbb{N}^m$), and the proof follows from Theorem 4.5.11.

32. For $S \subseteq \binom{[m]}{2}$, let E_S denote the set of $\mathbb{N}$-solutions α to equation (4.10) that also satisfy $\alpha_i = \alpha_j$ if $\{i, j\} \in S$. By Theorem 4.5.11, the generating function $E_S(x)$ is rational, whereas by the Principle of Inclusion-Exclusion

$$E^*(x) = \sum_{S}(-1)^{\#S} E_S(x), \tag{4.63}$$

and the proof follows.

NOTE. For practical computation, one should replace $S \subseteq \binom{[m]}{2}$ by $\pi \in \Pi_m$ and should replace equation (4.63) by Möbius inversion on Π_m.

34. See Proposition 8.3 of Stanley [4.53].

35. **a.** Given $\beta \in \mathbb{Z}^r$, let $S = \{i : \beta_i < 0\}$. Now if $\gamma = (\gamma_1,\dots,\gamma_r)$, then define $\gamma^S = (\gamma_1',\dots,\gamma_r')$ where $\gamma_i' = \gamma_i$ if $i \notin S$ and $\gamma_i' = -\gamma_i$ if $i \in S$. Let F^S be the monoid of all $\mathbb{N}$-solutions (α,γ) to $\Phi\alpha = \gamma^S$. By Theorem 4.5.11, the generating function

$$F^S(x,y) = \sum_{(\alpha,\gamma)\in F^S} x^\alpha y^\gamma$$

is rational. Let $\beta^S = (\beta_1',\dots,\beta_r')$. Then

$$E_\beta(x) = \frac{1}{\beta_1'!\cdots\beta_r'!} \frac{\partial^{\beta_1'}}{\partial y_1^{\beta_1'}} \cdots \frac{\partial^{\beta_r'}}{\partial y_r^{\beta_r'}} F^S(x,y)\bigg|_{y=0}, \tag{4.64}$$

so $E_\beta(x)$ is rational. Moreover, if $\alpha \in \mathrm{CF}(E)$ then $(\alpha, \mathbf{0}) \in \mathrm{CF}(F^S)$. The factors $1 - x^\alpha$ in the denominator of $F^S(x,y)$ are unaffected by the partial differentiation in equation (4.64), whereas all other factor disappear upon setting $y = 0$. Hence, $D(x)$ is a denominator of $E_\beta(x)$. To see that it is the *least* denominator (provided $E_\beta \neq \emptyset$), argue as in the proof of Theorem 4.5.11.
 b. *Answer.* $\beta = 0, \pm 1$.
 c. Let $\alpha_i = p_i/q_i$ for integers $p_i \geq 0$ and $q_i > 0$. Let ℓ be the least common multiple of $q_1, q_2, \dots, q_m$. Let $\Phi = [\gamma_1, \dots, \gamma_m]$ where γ_i is a column vector of length r, and define $\gamma_i' = (\ell/q_i)\gamma_i$. Let $\Phi' = [\gamma_1', \dots, \gamma_m']$. For any vector $v = (v_1, \dots, v_m) \in \mathbb{Z}^m$

satisfying $0 \le \nu_i < q_i$, let $E'_{(\nu)}$ be the set of all $\mathbb{N}$-solutions δ to $\Phi'(\delta) = \mathbf{0}$ such that $\delta_i \equiv \nu_i \pmod{q_i}$. If E' denotes the set of *all* $\mathbb{N}$-solutions δ to $\Phi'(\delta) = \mathbf{0}$, then it follows that $E' = \bigcup_\nu E'_{(\nu)}$ (disjoint union). Hence by Theorem 4.5.14,

$$\overline{E}'(x) = \pm E'(1/x) = \pm \sum_\nu E'_{(\nu)}(1/x). \tag{4.65}$$

Now any monomial x^ε appearing in the expansion of $E'_{(\nu)}(1/x)$ about the origin satisfies $\varepsilon_i \equiv -\nu_i \pmod{q'_i}$. It follows from equation (4.65) that $E'_{(\nu)}(1/x) = \pm E'_{(\bar\nu)}(x)$, where $\bar\nu_i = q_i - \nu_i$ for $\nu_i \ne 0$ and $\bar\nu_i = \nu_i$ for $\nu_i = 0$, and where $\overline{E}'_{(\mu)} = E'_{(\mu)} \cap \overline{E}'$. Now let σ_i be the least nonnegative residue of p_i modulo q_i, and let $\sigma = (\sigma_1, \ldots, \sigma_m)$. Define an affine transformation $\phi : \mathbb{R}^m \to \mathbb{R}^m$ by the condition

$$\phi(\delta) = (\delta_1/q_1, \ldots, \delta_m/q_m) + \alpha.$$

One can check that ϕ defines a bijection between $E'_{(\sigma)}$ and E_β and between $\overline{E}'_{(\sigma)}$ and $\overline{E}_\beta$, from which the proof follows.

This proof is patterned after Theorem 3.5 of R. Stanley, in *Proc. Symp. Pure Math.* (D. K. Ray-Chaudhuri, ed.), vol. 34, American Math. Society, Providence, RI, 1979, pp. 345–355. This result can also be deduced from Theorem 10.2 of [4.53], as can many other results concerning inhomogeneoous linear equations. A further proof is implicit in [4.57] (see Theorem 3.2 and Corollary 4.3).

  d. See Corollary 4.3 of [4.57].

36. **a.** No. The simplest example is the simplex σ in $\mathbb{R}^3$ with vertices $(0,0,0)$, $(1,1,0)$, $(1,0,1)$, and $(0,1,1)$. Note that σ has no additional integer points, so σ itself is the only integer triangulation of σ. But σ is not primitive, for example, since

$$\det \begin{bmatrix} 0 & 1 & 1 \\ 1 & 0 & 1 \\ 1 & 1 & 0 \end{bmatrix} = 2 > 1.$$

  b. If σ is a primitive d-simplex, then it is not hard to see that $i(\sigma, n) = \binom{n+d}{d}$, so $\bar{i}(\sigma, n) = (-1)^d \binom{-n+d}{d} = \binom{n-1}{d}$. Since $\mathcal{P}$ is the disjoint union of the interior of the faces of Γ, we get

$$i(\mathcal{P}, n) = \sum_{j \ge 0} f_j \binom{n-1}{j}.$$

An elegant way to state this formula is $\Delta^j i(\mathcal{P}, 1) = f_j$, where Δ denotes the first difference operator.

37. **b.** This remarkable result is due to M. Brion, *Ann. Sci. École Norm. Sup.* (4) **21** (1988), 653–663. Many subsequent proofs and expositions have been given, such as M. Beck, C. Haase, and F. Sottile, *Math. Intell.* **31** (2009), 9–17.

38. This result is due to S. Chen, N. Li, and S. V. Sam, Generalized Ehrhart polynomials, arXiv:1002.3658. Their result generalizes the conjecture of Exercise 4.12 of the first edition of this book. A conjectured multivariate generalization is due to Ehrhart [4.14, p. 139].

39. **a.** This result was conjectured by A. Weil as part of his famous "Weil conjectures." It was first proved by B. M. Dwork, *Amer. J. Math.* **82** (1960), 631–648, and a highly

readable exposition appears in Chapter V of N. Koblitz, *p-adic Numbers, p-adic Analysis, and Zeta-Functions*, 2nd ed., Springer-Verlag, New York, 1984. The entire Weil conjectures were subsequently proved by P. R. Deligne (in two different ways) and later by G. Laumon and K. S. Kedlaya (independently).

b. This exercise is a result of J.-I. Igusa, *J. Reine Angew. Math.* **278/279** (1975), 307–321, for the case $k = 1$. A simpler proof was later given by Igusa in *Amer. J. Math.* **99** (1977), 393–417 (appendix). A proof for general k was given by D. Meuser, *Math. Ann.* **256** (1981), 303–310, by adapting Igusa's methods. For another proof, see J. Denef, *Lectures on Forms of Higher Degree*, Springer-Verlag, Berlin/Heidelberg/New York, 1978.

40. a. Let D_w denote the set of all factors of w belonging to L. (We consider two factors u and v different if they start or end at different positions in w, even if $u = v$ as elements of X^*.) Clearly for fixed w,

$$\sum_{T \subseteq D_w} \left(\prod_{v \in T} s_v \right) = \prod_{v \in L} (1 + s_v)^{m_v(w)}.$$

Hence, if we set each $s_v = t_v - 1$ in equation (4.52), we obtain the equivalent formula

$$\sum_{u \in X^*} \sum_{T \subseteq D_w} \left(\prod_{v \in T} s_v \right) = (1 - x_1 - \cdots - x_n - C(\boldsymbol{x}, \boldsymbol{s}))^{-1}. \tag{4.66}$$

Now given $w \in X^*$ and $T \subseteq D_w$, there is a *unique* factorization $w = v_1 \cdots v_k$ such that either

i. $v_i \in X$ and v_i does not belong to one of the factors in T, or

ii. v_i is the first component of some L-cluster (v_i, μ, v) where the components of μ consist of all factors of v_i contained in D_w.

Moreover,

$$\prod_{v \in T} s_v = \prod_i \prod_{v \in L} s_v^{m_v(\mu)},$$

where i ranges over all v_i satisfying (ii), and where μ is then given by (ii). It follows that when the right-hand side of equation (4.66) is expanded as an element of $\mathbb{C}[[t_v : v \in L]]\langle\langle X \rangle\rangle$, it coincides with the left-hand side of (4.66).

This result is due to I. P. Goulden and D. M. Jackson, *J. London Math. Soc. (2)* **20** (1979), 567–576, and also appears in [3.32, Ch. 2.8]. A special case was proved by D. Zeilberger, *Discrete Math.* **34** (1981), 89–91. (The precise hypotheses used in this paper are not clearly stated.)

c. Let $C_v(\boldsymbol{x}, \boldsymbol{t})$ consist of those terms of $C(\boldsymbol{x}, \boldsymbol{t})$ corresponding to a cluster (w, μ, v) such that the last component of μ is v. Hence, $C(\boldsymbol{x}, \boldsymbol{t}) = \sum_{v \in L} C_v(\boldsymbol{x}, \boldsymbol{t})$. By equation (4.52) or (4.66), it suffices to show that each C_v is rational. An easy combinatorial argument expresses C_v as a linear combination of the C_u's and 1 with coefficients equal to polynomials in the x_i's and t_v's. Solving this system of linear equations by Cramer's rule (it being easily seen on combinatorial grounds that a unique solution exists) expresses C_v as a rational function. (Another solution can be given using the transfer-matrix method.) An explicit expression for $C(\boldsymbol{x}, \boldsymbol{t})$ obtained in this way appears in Goulden and Jackson, ibid., Prop. 3.2, and in [3.32, Lem. 2.8.10]. See also L. J. Guibas and A. M. Odlyzko, *J. Combinatorial Theory Ser. A* **30** (1981), 193–208.

d. The right-hand side of equation (4.53) is equal to $(1 - nx + x^\ell A_w(x)^{-1})^{-1}$. The proof follows from analyzing the precise linear equation obtained in the proof of (c). This result appears in L. J. Guibas and A. M. Odlyzko, ibid.

41. b. E. Lucas, *Théorie des nombres*, Gauthier-Villars, Paris, 1891.

c. E. Landau, *Naturwissenschaftliche Wochenschrift*, **11** (1896), 367–371.

d. This result and many more on this topic, such as the determination of $B_4(n)$ and similar results for other chess pieces, can be found in V. Kotěšovec, Non-attacking chess pieces, 2nd ed., 2010,

$$\langle\texttt{http://web.telecom.cz/vaclav.kotesovec/math.htm}\rangle.$$

e. Identify an $n \times n$ chessboard with the set $[0, n-1]^2$. Then $k!B_k(n)$ is equal to the number of vectors $v = (\alpha_1, \ldots, \alpha_k, \beta_1, \ldots, \beta_k, \gamma) \in \mathbb{Z}^{2k+1}$ satisfying

$$\gamma = n - 1, \tag{4.67}$$

$$0 \le \alpha_i \le \gamma, \ 0 \le \beta_i \le \gamma, \tag{4.68}$$

$$i \ne j \Rightarrow [(\alpha_i \ne \alpha_j) \& (\beta_i \ne \beta_j) \& (\alpha_i - \beta_i \ne \alpha_j - \beta_j) \&$$
$$\alpha_i + \beta_i \ne \alpha_j + \beta_j)]. \tag{4.69}$$

Label the $r = 4\binom{k}{2}$ inequalities of (4.69), say $I_1, \ldots, I_r$. Let $\bar{I}_i$ denote the negation of I_i, that is, the equality obtained from I_i by changing $\ne$ to $=$. Given $S \subseteq [r]$, let $f_S(n)$ denote the number of vectors v satisfying (4.67), (4.68), and I_i for $i \in S$. By the Principle of Inclusion-Exclusion,

$$k!B_k(n) = \sum_S (-1)^{\#S} f_S(n). \tag{4.70}$$

Now by Theorem 4.5.11 the generating functions $F_S = \sum x_1^{\alpha_1} \cdots x_k^{\alpha_k} y_1^{\beta_1} \cdots y_k^{\beta_k} x^{\gamma}$ are rational, where the sum is over all vectors v satisfying (4.68) and $\bar{I}_i$ for $i \in S$. But $\sum f_S(n) x^{n-1}$ is obtained from F_S by setting each $x_i = y_j = 1$, so $\sum f_S(n) x^n$ is rational. It then follows from (4.70) that $\sum B_k(n) x^n$ is rational.

Note that the basic idea of the proof is same as in Exercise 4.32; namely, replace non equalities by equalities and use Inclusion-Exclusion. For many more results of this nature, see S. Chaiken, C. R. H. Hanusa, and T. Zaslavsky, Mathematical analysis of a q-queens problem, preprint, dated 19 May 2011.

42. The explicit formula is due to C. E. Arshon, Reshenie odnoĭ kombinatoromoĭ zadachi, *Math. Proveschchenie* **8** (1936), 24–29, from which it is clear that $A_k(n)$ has the properties stated in (b). A polytopal approach to nonattacking bishops was developed by S. Chaiken, C. R. H. Hanusa, and T. Zaslavsky, op cit. Proofs can also be given using the theory of Ferrers boards of Section 2.4.

43. We want to count triples $(a, b, c) \in \mathbb{P}^3$ satisfying $a \le b \le c, a + b > c$, and $a + b + c = n$. Every such triple can be written uniquely in the form

$$(a, b, c) = \alpha(0, 1, 1) + \beta(1, 1, 1) + \gamma(1, 1, 2) + (1, 1, 1),$$

where $\alpha, \beta, \gamma \in \mathbb{N}$; namely,

$$\alpha = b - a, \quad \beta = a + b - c - 1, \quad \gamma = c - b.$$

Moreover, $n - 3 = 2\alpha + 3\beta + 4\gamma$. Conversely, any triple $(\alpha, \beta, \gamma) \in \mathbb{N}^3$ yields a valid triple (a, b, c). Hence, $t(n)$ is equal to the number of triples $(\alpha, \beta, \gamma) \in \mathbb{N}^3$ satisfying $2\alpha + 3\beta + 4\gamma = n - 3$, so

$$\sum_{n \ge 3} t(n) x^n = \frac{x^3}{(1 - x^2)(1 - x^3)(1 - x^4)}.$$

From the viewpoint of Section 4.5, we obtained such a simple answer because the monoid E of $\mathbb{N}$-solutions (a,b,c) to $a \leq b \leq c$ and $a + b \geq c$ is a *free* (commutative) monoid (with generators $(0,1,1)$, $(1,1,1)$, and $(1,1,2)$).

Equivalent results (with more complicated proofs) are given by J. H. Jordan, R. Walch, and R. J. Wisner, *Notices Amer. Math. Soc.* **24** (1977), A-450, and G. E. Andrews, *Amer. Math. Monthly* **86** (1979), 477–478. For some generalizations, see G. E. Andrews, *Ann. Combinatorics* **4** (2000), 327–338, and M. Beck, I. M. Gessel, S. Lee, and C. D. Savage, *Ramanujan J.* **23** (2010), 355–369.

44. A simple combinatorial argument shows that

$$N_{kr}(n+1) = kN_{kr}(n) - (k-1)N_{kr}(n-r+1), \quad n \geq r. \tag{4.71}$$

It follows from Theorem 4.1.1 and Proposition 4.2.2(ii) that $F_{kr}(x) = P_{kr}(x)/(1 - kx + (k-1)x^r)$, where $P_{kr}(x)$ is a polynomial of degree r (since the recurrence (4.71) fails for $n = r - 1$). In order to satisfy the initial conditions $N_{kr}(0) = 1$, $N_{kr}(n) = k^r$ if $1 \leq n \leq r - 1$, $N_{kr}(r) = k^r - k$, we must have $P_{kr}(x) = 1 - x^r$. Hence,

$$F_{kr}(x) = \frac{1 - x^r}{1 - kx + (k-1)x^r}.$$

If we reduce $F_{kr}(x)$ to lowest terms, then we obtain

$$F_{kr}(x) = \frac{1 + x + \cdots + x^{r-1}}{1 - (k-1)x - (k-1)x^2 - \cdots - (k-1)x^{r-1}}.$$

This formula can be obtained by proving directly that

$$N_{kr}(n+1) = (k-1)[N_{kr}(n) + N_{kr}(n-1) + \cdots + N_{kr}(n-r+2)],$$

but then it is somewhat more difficult to obtain the correct numerator.

45. a. (I. M. Gessel and R. A. Indik, A recurrence associated with extremal problems, preprint, 1989) Let $q \in \mathbb{P}$, $p \in \mathbb{Z}$ with $(p,q) = 1$, and $i \in \mathbb{N}$. First one shows that the two classes of functions

$$f(n) = i + \sum_{j=1}^{n} \left\lceil \frac{pj}{q} \right\rceil, \quad \text{where } n \geq 2iq + q,$$

$$f(n) = -i + 1 + \sum_{j=1}^{n} \left\lceil \frac{pj+1}{q} \right\rceil, \quad \text{where } n \geq 2iq + 2q,$$

satisfy the recurrence (4.54). Then one shows that for any $m \in \mathbb{P}$ and $k \in \mathbb{Z}$, one of the preceding functions satisfies $f(m) = k$.

b. The most interesting case is when $R(n) = P(n)/Q(n)$, where

$$P(n) = x^d + a_{d-1}x^{d-1} + a_{d-2}x^{d-2} + \cdots + a_0$$

$$Q(n) = x^d + b_{d-2}x^{d-2} + \cdots + b_0,$$

where the coefficients are integers and $a_{d-1} > 0$. (Of course, we should assume that $Q(n) \neq 0$ for any integer $n \geq m$.) In this case, $f(n) = O(n^a)$, where $a = a_{d-1}$, and we can ask whether $f(n)$ is a quasipolynomial. Experimental evidence suggests that the answer is negative in general, although in many particular instances the answer is afirmative. Gessel has shown that in all cases the function $\Delta^a f(n)$ is bounded. A further reference is Z. Füredi and A. Kündgen, *J. Graph Theory* **40** (2002), 195–225 (Theorem 7).

46. a. A simple computation shows that

$$f(n) = \frac{5i}{8}(\alpha^n - \beta^n),$$

where $\alpha = \frac{1}{5}(3 - 4i)$ and $\beta = \frac{1}{5}(3 + 4i)$. Since $|\alpha| = |\beta| = 1$, we have

$$|f(n)| \le \frac{5}{8}(|\alpha|^n + |\beta|^n) = \frac{5}{4}.$$

The easiest way to show $f(n) \ne \pm 5/4$ is to observe that the recurrence (4.55) implies that the denominator of $f(n)$ is a power of 5.

b. Since f is integer-valued and bounded, there are only many finitely many different sequences $f(n+1), f(n+2), \ldots, f(n+d)$. Thus for some $r < s$, we have $f(r+i) = f(s+i)$ for $1 \le i \le d$; and it follows that f has period $s - r$.

c. This result was conjectured by G. Pólya in 1916 and proved by F. Carlson in 1921. Subsequent proofs and generalizations were given by Pólya and are surveyed in *Jahrber. Deutsch. Math. Verein.* **31** (1922), 107–115; reprinted in *George Pólya: Collected Papers*, vol. 1 (G. Pólya and R. P. Boas, eds.), M.I.T. Press, Cambridge, Mass., 1974, pp. 192–198. For more recent work in this area, see the commentary on pp. 779–780 of the *Collected Papers*. •

47. See A. M. Garsia and I. M. Gessel, *Advances in Math.* **31** (1979), 288–305 (Remark 22). There is now a large literature on the subject of *vector partitions*. See for example B. Sturmfels, *J. Combin. Theory Ser. A* **72** (1995), 302–309; M. Brion and M. Vergne, *J. Amer. Math. Soc.* **1** (1997), 797–833; A. Szenes and M. Vergne, *Advances in Appl. Math.* **3)** (2003), 295–342; W. Baldoni and M. Vergne, *Transformation Groups* **13** (2009), 447–469.

48. a. Several proofs of this result are known. One [4.56, Thm. 2.1] uses the result (H. Bruggesser and P. Mani, *Math. Scand.* **29** (1971), 197–205) that the boundary complex of a convex polytope is shellable. The second proof (an immediate generalization of [4.54, Prop. 4.2]) shows that a certain commutative ring $R_{\mathcal{P}}$ associated with $\mathcal{P}$ is Cohen–Macaulay. A geometric proof was given by U. Betke and P. McMullen, *Monatshefte für Math.* **99** (1985), 253–265 (a consequence of Theorem 1, Theorem 2, and the remark at the bottom of page 257 that $h(K, t)$ has nonnegative coefficients). Further references include M. Beck and F. Sottile, *Europ. J. Combinatorics* **28** (2007), 403–409 (reproduced in M. Beck and S. Robins [4.4, Thm. 3.12]), and A. Stapledon, Ph.D. thesis, University of Michigan, 2009.

b. See R. Stanley, in *Commutative Algebra and Combinatorics* (M. Nagata and H. Matsumura, eds.), Advanced Studies in Pure Mathematics **11**, Kinokuniya, Tokyo, and North-Holland, Amsterdam/New York, 1987, pp. 187–213 (Theorem 4.4). The methods discussed in U. Betke, *Ann. Discrete Math.* **20** (1984), 61–64, are also applicable.

c. This result was originally proved using commutative algebra by R. Stanley, *Europ. J. Combinatorics* **14** (1993), 251–258. A geometric proof was given by A. Stapledon, Ph.D. thesis, University of Michigan, 2009, and `arXiv:0807.3542`.

49. Equation (4.56) is due to T. Hibi, *Discrete Math.* **83** (1990), 119–121, while (4.57) is a result of R. Stanley, *Europ. J. Combinatorics* **14** (1993), 251–258. Both proofs were based on commutative algebra. Subsequently geometric proofs were given by A. Stapledon, *Trans. Amer. Math. Soc.* **361** (2009), 5615–5626. Stapledon gives a small improvement of Hibi's inequality and some additional inequalities.

50. Let

$$F(x) = \sum_{n \ge 0} i(\mathcal{P}, n) x^n = \frac{\sum_{j=0}^{d} a_i x^i}{(1-x)^{d+1}}.$$

By the reciprocity theorem for Ehrhart polynomials (Theorem 4.6.9), we have

$$\sum_{n \geq 0} i(\partial \mathcal{P}, n) x^n = F(x) - (-1)^{d+1} F(1/x)$$

$$= \frac{\sum_{j=0}^{d+1} a_j (x^j - x^{d+1-j})}{(1-x)^{d+1}},$$

from which the proof follows easily. For further information, including a reference to a proof that $h_i \geq 0$, see Exercise 4.48(b).

51. a. We have that $i(n)$ is equal to the number of $\mathbb{N}$-solutions to $x_1 + \cdots + x_r \leq n$, $y_1 + \cdots + y_s \leq n$. There are $\binom{n+r}{r}$ ways to choose the x_i's and $\binom{n+s}{s}$ ways to choose the y_i's, so $i(n) = \binom{n+r}{r}\binom{n+s}{s} = \left(\!\binom{n+1}{r}\!\right)\left(\!\binom{n+1}{s}\!\right)$. Hence by Exercise 3.169(b) we get

$$F(x) = \sum_{n \geq 1} \left(\!\binom{n}{r}\!\right)\left(\!\binom{n}{s}\!\right) x^{n-1}$$

$$= \frac{\sum_{k=0}^{r} \binom{r}{k}\binom{s}{k} x^k}{(1-x)^{r+s+1}}.$$

The volume of $\mathcal{P}$ is by Proposition 4.6.13

$$V(\mathcal{P}) = \frac{1}{(r+s)!} \sum_{k=0}^{r+s} \binom{r}{k}\binom{s}{k} = \frac{1}{r! s!}.$$

There are $(r+1)(s+1)$ vertices—all vectors $(x_1, \ldots, x_r, y_1, \ldots, y_s) \in \mathbb{N}^{r+s}$ such that $x_1 + \cdots + x_r \leq 1$ and $y_1 + \cdots + y_s \leq 1$.

b. $P_{rs} = r + s$. See Exercise 4.56 for a generalization to any finite poset P.

53. a. For any d, the matrix

$$\begin{bmatrix} 1 & 2 & \cdots & d \\ d+1 & d+2 & \cdots & 2d \\ & & \vdots & \\ d^2-d+1 & d^2-d+2 & \cdots & d^2 \end{bmatrix}$$

is an antimagic square.

b. Let $M = (m_{ij})$ be antimagic. Row and column permutations do not affect the antimagic property, so assume that m_{11} is the minimal entry of M. Define $a_i = m_{i1} - m_{11} \in \mathbb{N}$ and $b_j = m_{1j} \in \mathbb{N}$. The antimagic properties implies $m_{ij} = m_{i1} + m_{1j} - m_{11} = a_i + b_j$.

c. To get an antimagic square M of index n, choose a_i and b_j in (b) so that $\sum a_i + \sum b_j = n$. This can be done in $\left(\!\binom{2d+n-1}{2d-1}\!\right)$ ways. Since the only linear relations holding among the R_i's and C_j's are scalar multiples of $\sum R_i = \sum C_j$, it follows that we get each M exactly once if we subtract from $\left(\!\binom{2d+n-1}{2d-1}\!\right)$ the number of solutions to $\sum a_i + \sum b_j = n$ with $a_i \in \mathbb{P}$ and $b_j \in \mathbb{N}$. It follows that the desired answer is $\left(\!\binom{2d+n-1}{2d-1}\!\right) - \left(\!\binom{d+n-1}{2d-1}\!\right)$. (Note the similarity to Exercise 2.15(b).)

d. The vertices are the $2d$ matrices R_i and C_j; this result is essentially a restatement of (b). An integer point in $n\mathcal{P}_d$ is just a $d \times d$ antimagic square of index n. Hence by (c),

$$i(\mathcal{P}_d, n) = \binom{2d + n - 1}{2d - 1} - \binom{d + n - 1}{2d - 1}.$$

e. By (d) we have

$$\sum_{n \geq 0} i(\mathcal{P}_d, n) x^n = \frac{1}{(1-x)^{2d}} - \frac{x^d}{(1-x)^{2d}}$$

$$= \frac{1 + x + \cdots + x^{d-1}}{(1-x)^{2d-1}},$$

whence $A(\mathcal{P}_d, x) = 1 + x + \cdots + x^{d-1}$ and $v(\mathcal{P}_d) = d/(2d-2)!$.

54. a. It follows from equation (4.29) that the average of the zeros of $H_n(r)$ is $-n/2$. Since $\deg H_n(r) = (n-1)^2$, we get that the sum of the zeros is $-\frac{1}{2}n(n-1)^2/2$, and the proof follows. This result was observed empirically by R. Stanley and proved by B. Osserman and F. Liu (private communication, dated 16 November 2010).

56. a. The vertices are the characteristic vectors χ_A of antichains A of P; that is, $\chi_A = (\varepsilon_1, \ldots, \varepsilon_p)$, where

$$\varepsilon_i = \begin{cases} 1, & \text{if } x_i \in A, \\ 0, & \text{if } x_i \notin A. \end{cases}$$

b. Let $\mathcal{O}(P)$ be the order polytope of Example 4.6.17. Define a map $f : \mathcal{O}(P) \to \mathcal{C}(P)$ by $f(\varepsilon_1, \ldots, \varepsilon_p) = (\delta_1, \ldots, \delta_p)$, where

$$\delta_i = \min\{\varepsilon_i - \varepsilon_j : x_i \text{ covers } x_j \text{ in } P\}.$$

Then f is a bijection (and is continuous and piecewise-linear) with inverse

$$\varepsilon_i = \max\{\delta_{j_1} + \cdots + \delta_{j_k} : t_{j_1} < \cdots < t_{j_k} = t_i\}.$$

Moreover, the image of $\mathcal{O}(P) \cap (\frac{1}{n}\mathbb{Z})^p$ under f is $\mathcal{C}(P) \cap (\frac{1}{n}\mathbb{Z})^p$, and the proof follows from Example 4.6.17.

NOTE. Essentially the same bijection f is given in the solution to Exercise 3.143(a). Indeed, it is clear that $\mathcal{C}(P)$ depends only on $\text{Com}(P)$, so any property of $\mathcal{C}(P)$ (such as its Ehrhart polynomial) depends only on $\text{Com}(P)$.
The polytope $\mathcal{C}(P)$ is called the *chain polytope* of P. For more information on chain polytopes, order polytopes, and their connections, see R. Stanley, *Discrete Comput. Geom.* **1** (1986), 9–23. For a generalization, see F. Ardila, T. Bliem, and D. Salazar, Gelfand-Tsetlin polytopes and Feigin-Fourier-Littelmann polytopes as marked poset polytopes, arXiv:1008.2365.

c. Choose P to be the zigzag poset Z_n of Exercise 3.66. Then $\mathcal{C}(Z_n) = \mathcal{C}_{n,2}$. Hence by (b) and Proposition 4.6.13, $v(\mathcal{C}_n)$ is the leading coefficient of $\Omega_{Z_n}(m)$. Then by Section 3.12 we have $v(\mathcal{C}_{n,2}) = e(Z_n)/n!$. But $e(Z_n)$ is the number E_n of alternating permutations in $\mathfrak{S}_n$ (see Exercise 3.66(c)), so

$$\sum_{n \geq 0} e(Z_n) \frac{x^n}{n!} = \tan x + \sec x.$$

A more *ad hoc* determination of $v(\mathcal{C}_{n,2})$ is given by I. G. Macdonald and R. B. Nelsen (independently), *Amer. Math. Monthly* **86** (1979), 396 (problem proposed by R. Stanley), and R. Stanley, *SIAM Review* **27** (1985), 579–580 (problem proposed by E. E. Doberkat). For an application to tridiagonal matrices, see P. Diaconis and P. M. Wood, Random doubly stochastic tridiagonal matrices, preprint.

d. Using the integration method of Macdonald and Nelsen, ibid., the following result can be proved. Define polynomials $f_n(a,b)$ by

$$f_0(a,b) = 1, \quad f_n(0,b) = 0 \text{ for } n > 0,$$

$$\frac{\partial}{\partial a} f_n(a,b) = f_{n-1}(b-a, 1-a).$$

For instance,

$$f_1(a,b) = a,$$

$$f_2(a,b) = \frac{1}{2}(2ab - a^2),$$

$$f_3(a,b) = \frac{1}{6}(a^3 - 3a^2 - 3ab^2 + 6ab).$$

Then $v(\mathcal{C}_{n,3}) = f_n(1,1)$. A "nice" formula or generating function is not known for $f_n(a,b)$ or V_n. Similar results hold for $v(\mathcal{C}_{n,k})$ for $k > 3$.

e. Let P be the poset with elements $t_1,\ldots,t_n$ satisfying $t_1 < t_2 < \cdots < t_k$, $t_{k+1} < t_{k+2} < \cdots < t_n$, and $t_{k+i} < t_{i+1}$ for $1 \le i \le n-k$, except that when $n = 2k$ we omit the relation $t_{2k} < t_{k+1}$. The equations defining $\mathcal{C}(P)$ are exactly the same as those defining $\mathcal{C}_{n,k}$, so $v(\mathcal{C}_{n,k}) = e(P)/n!$. If we add a $\hat{0}$ to P and remove successively $2k - n - 1$ $\hat{1}$'s (where when $n = 2k$ we add a $\hat{1}$), we don't affect $e(P)$ and we convert P to $\mathbf{2} \times (\mathbf{n-k+1})$. It is easy to see that $e(\mathbf{2} \times (\mathbf{n-k+1})) = C_{n-k+1}$ (see Exercise 6.19(aaa), Vol. II), and the proof follows.

57. By Exercise 4.56(a), the set of vertices of the polytope $\mathcal{C}(P)$ is a proper subset of the set of vertices of $\mathcal{C}(Q)$, so $\mathrm{vol}(\mathcal{C}(P)) < \mathrm{vol}(\mathcal{C}(Q))$. By Exercise 4.56(b) we have $\mathrm{vol}(\mathcal{C}(P)) = e(P)/p!$ and similarly for $\mathrm{vol}(\mathcal{C}(Q))$, so the proof follows. No other proof of this "obvious" inequality (communicated by P. Winkler) is known.

58. a. This result was conjectured by L. D. Geissinger, in *Proc. Third Caribbean Conference on Combinatorics and Computing*, University of the West Indies, Cave Hill, Barbados, pp. 125–133, and proved by H. Dobbertin, *Order* **2** (1985), 193–198.

c. To compute $i(\mathcal{V}(p),n)$ choose $f(1)$, $f(2)$, $\ldots, f(p)$ in turn so that $0 \le f(1) + f(2) + \cdots + f(j) \le n$. There are exactly $n+1$ choices for each $f(j)$, so $i(\mathcal{V}(P),n) = (n+1)^p$.

d. Let $0 \le k \le n$. There are $i(\mathcal{V}(P),k) - i(\mathcal{V}(P),k-1)$ maps $f\colon P \to \mathbb{Z}$ for which every order ideal sum is nonnegative and the maximum such sum is exactly k. Given such an f, there are then $i(\mathcal{V}(Q),n-k)$ choices for $g\colon Q \to \mathbb{Z}$ for which every order ideal sum is nonnegative and at most $n-k$. It follows that

$$i(\mathcal{V}(P+Q),n) = \sum_{k=0}^{n}(i(\mathcal{V}(P),k) - i(\mathcal{V}(P),k-1))i(\mathcal{V}(Q),n-k).$$

Hence,

$$\left((1-x)\sum_{n\ge 0}i(\mathcal{V}(P),n)x^n\right)\left(\sum_{n\ge 0}i(\mathcal{V}(Q),n)x^n\right) = \sum_{n\ge 0}i(\mathcal{V}(P+Q),n)x^n$$

$$= \frac{A(\mathcal{V}(P+Q),x)}{(1-x)^{p+q+1}},$$

where $p = \#P$ and $q = \#Q$. The result now follows from $\sum_{n \geq 0} i(\mathcal{V}(P), n)x^n = A(\mathcal{V}(P), x)/(1-x)^{p+1}$ and $\sum_{n \geq 0} i(\mathcal{V}(Q), n)x^n = A(\mathcal{V}(Q), x)/(1-x)^{q+1}$.

e. In view of (a), we need to show that the only integer points of $\mathcal{V}(P)$ are the vertices. This is a straightforward argument.

f. It follows from Corollary 4.2.4(ii) and the reciprocity theorem for order polynomials (Theorem 4.6.9) that the quantity $p + 1 - \deg A(\mathcal{V}(P), x)$ is equal to the least $d > 0$ for which there is a map $f \colon P \to \mathbb{Q}$ such that every order ideal sum lies in the open interval $(0, 1)$ and $df(t) \in \mathbb{Z}$ for all $t \in P$. Since every subset of the set of m minimal elements $t_1, \ldots, t_m$ is an order ideal, we have $f(t_i) > 0$ and $f(t_1) + \cdots + f(t_m) < 1$. Hence the minimal d for which $df(t_i) \in \mathbb{Z}$ is $d = m + 1$, obtained by taking each $f(t_i) = 1/(m+1)$. We can extend f to all of P by defining $f(t) = 0$ if t is not minimal, so the proof follows.

g. Let $f(t) = 1/(m+1)$ for each minimal $t \in P$. By the proof of (f) and by Corollary 4.2.4(iii), we have that $x^{p-m} A(\mathcal{V}(P), 1/x) = A(\mathcal{V}(P), x)$ if and only if there is a unique extension of f to P for which each order ideal sum lies in $(0, 1)$ and for which $(m + 1)f(t) \in \mathbb{Z}$ for all $t \in P$. It is not difficult to show that this condition holds if and only if every connected component of P has a unique minimal element (in which case $f(t) = 0$ for all nonminimal $t \in P$).

59. This result was conjectured by M. Beck, J. A. De Loera, M. Develin, J. Pfeifle, and R. Stanley, *Contemp. Math.* **374** (2005), 15–36 (Conjecture 1.5) and proved by F. Liu, *J. Combinatorial Theory Ser. A* **111** (2005), 111–127. Liu subsequently greatly generalized this result, culminating in the paper Higher integrality conditions, volumes and Ehrhart polynomials, *Advances in Math.* **226** (2011), 3467–3494.

60. The tetrahedron with vertices $(0,0,0)$, $(1,0,0)$, $(0,1,0)$, $(1,1,r)$, $r \geq 1$, has four integer points and volume going to ∞, so for sufficiently large r the Ehrhart polynomial must have a negative coefficient. In fact, $r = 13$ yields $\frac{13}{6}n^3 + n^2 - \frac{1}{6}n + 1$.

61. **a.** The facets of $\mathcal{P}_d$ are given by the 2^d inequalities

$$\pm x_1 \pm x_2 \pm \cdots \pm x_d \leq 1.$$

Hence, $i(\mathcal{P}_d, n)$ is the number of integer solutions to

$$|x_1| + |x_2| + \cdots + |x_d| \leq n,$$

or, after introducing a slack variable y,

$$|x_1| + |x_2| + \cdots |x_d| + y = n.$$

Equivalently,

$$i(\mathcal{P}, n) = \sum f(a_1)f(a_2) \cdots f(a_d),$$

summed over all weak compositions $a_1 + \cdots + a_d + b = n$ of n into d parts, where

$$f(a) = \begin{cases} 1, & a = 0, \\ 2, & a > 0. \end{cases}$$

Now $\sum_{a \geq 0} f(a)x^a = (1+x)/(1-x)$, so

$$\sum_{n \geq 0} i(\mathcal{P}, n)x^n = \left(\frac{1+x}{1-x}\right)^d \cdot \frac{1}{1-x} = \frac{(1+x)^d}{(1-x)^{d+1}}.$$

Hence, $P_d(x) = (1+x)^d$.

b. Follows from F. R. Rodriguez-Villegas, *Proc. Amer. Math. Soc.* **130** (2002), 2251–2254. It is also a consequence of Theorem 3.2 of R. Stanley, *Europ. J. Combinatorics* **32** (2011), 937–943. For some related results, see T. Hibi, A. Higashitani, and H. Ohsugi, *Proc. Amer. Math. Soc.* **139** (2011), 3707–3717.

62. b. The projection of $\Delta_{k,d}$ to the first $d-1$ coordinates gives a linear bijection φ with the polytope $\mathcal{R}_{d-1,k}$ of Exercise 1.51. Moreover, φ takes aff$(\Delta_{k,d}) \cap \mathbb{Z}^d$ to $\mathcal{R}_{d-1,k} \cap \mathbb{Z}^{d-1}$, where aff denotes affine span. Since $\Delta_{k,d}$ and $\mathcal{R}_{d-1,k}$ are both integer polytopes, it follows that they have the same Ehrhart polynomial and therefore the same relative volume.

c. The polytope $\Delta_{k,d}$ is defined by

$$0 \le x_i \le 1, \quad x_1 + \cdots + x_d = k.$$

Hence, $i(\Delta_{k,d}, n)$ is equal to the number of integer solutions to the equation $x_1 + \cdots + x_d = kn$ such that $0 \le x_i \le n$, so

$$i(\Delta_{k,d}, n) = [x^{kn}](1 + x + \cdots + x^n)^d$$

$$= [x^{kn}]\left(\frac{1 - x^{n+1}}{1 - x}\right)^d.$$

g. This result is equivalent to Theorem 13.2 of T. Lam and A. E. Postnikov, *Discrete & Comput. Geom.* **38** (2007), 453–478, after verifying that the triangulation of $\Delta_{2,d}$ appearing there is primitive and appealing to Exercise 4.36(b). It can also be proved directly from part (c) or (d) of the present exercise.

(i) This result was conjectured by R. Stanley and proved by N. Li (December, 2010).

63. This result follows from the techniques in §5 of G. C. Shephard, *Canad. J. Math.* **26** (1974), 302–321, and was first stated by R. Stanley [4.56, Ex. 3.1], with a proof due to G. M. Ziegler appearing in *Applied Geometry and Discrete Combinatorics*, DIMACS Series in Discrete Mathematics, vol. 4, American Mathematical Society, Providence, RI, 1991, pp. 555–570 (Theorem 2.2). The polytope $\mathcal{Z}$ is by definition a *zonotope*, and the basic idea of the proof is to decompose $\mathcal{Z}$ into simpler zonotopes (namely, parallelopipeds, the zonotopal analogue of simplices), each of which can be handled individually. There is also a proof based on the theory of mixed volumes.

64. b. The crucial fact is that the polytope $\mathcal{P}_G$ is a zonotope and so can be handled by the techniques of Exercise 4.63. See [4.56, Ex. 3.1]. A purely combinatorial proof that the number of $\delta(o)$'s equals the number of spanning forests of G is given by D. J. Kleitman and K. J. Winston, *Combinatorica* **1** (1981), 49–54. The polytope $\mathcal{P}_G$ was introduced by T. K. Zaslavsky (unpublished) and called by him an *acyclotope*. For a vast generalization of the permutohedron, see A. Postnikov, *Int. Math. Res. Notices* **2009** (2009), 1026–1106.

65. The crucial fact is that for a loopless graph G, the integer points in the polytope $\tilde{\mathcal{D}}(G)$ are the extended degree sequences of spanning subgraphs of G if and only if G is an FHM-graph. This result is due to D. R. Fulkerson, A. J. Hoffman, and M. H. McAndrew, *Canad. J. Math.* **17** (1965), 166–177, whence the term "FHM-graph." Equation (4.58) is due to R. Stanley, in *Applied Geometry and Discrete Combinatorics*, DIMACS Series in Discrete Mathematics, vol. 4, American Mathematical Society, Providence, RI, 1991, pp. 555–570 (§5). For the case $G = K_k$, see Exercise 5.16.

66. a. This result was conjectured by Ehrhart [4.14, p. 53] and solved independently by R. Stanley [4.56, Thm. 2.8] and P. McMullen, *Arch. Math. (Basel)* **31** (1978/79), 509–516. A polytope $\mathcal{Q}$ whose affine span contains an integer point is called *reticular* by Ehrhart [4.14, p. 47], and the least j for which every j-face of $\mathcal{P}$ is reticular is called the *grade* of $\mathcal{P}$ [4.14, p. 12].

b. See McMullen, ibid., §4.

70. a. Let A denote the adjacency matrix of G. We have $A^\ell = pJ$ for some $p \geq 0$, where J is the all 1's matrix. The matrix pJ has one nonzero eigenvalue, so the same is true for A. Since A is symmetric, it therefore has rank one. Since $[1, 1, \ldots, 1]$ is a left eigenvector and $[1, 1, \ldots, 1]^t$ is a right eigenvector for pJ, the same is true for A. These conditions suffice to show that all entries of A are equal.

The analogous problem for directed graphs is much more complicated; see Exercise 5.74(f).

b. Let V be the vertex set of G, with $p = \#V$. Note that G must be connected so $d_v > 0$ for all $v \in V$. Let D be the diagonal matrix with rows and columns indexed by V, such that $D_{vv} = 1/d_v$. Let $M = DA$. Note that M_{uv} is the probability of stepping to v from u. The hypothesis on G is therefore equivalent to $M^\ell = \frac{1}{p}J$. Let E be the diagonal matrix with $E_{vv} = 1/\sqrt{d_v}$. Then $E^{-1}ME = EAE$, a symmetric matrix. Thus, M is conjugate to a symmetric matrix and hence diagonalizable. The proof is now parallel to that of (a).

71. The adjacency matrix A of G has only two distinct rows, so rank $A = 2$. Thus, there are two nonzero eigenvalues (since A is symmetric), say x and y. We have

$$\operatorname{tr}(A) = x + y = \text{number of loops} = 3.$$

Furthermore, $\operatorname{tr}(A^2) = C_G(2)$, which is twice the number of nonloop edges plus the number of loops [why?]. Thus,

$$\operatorname{tr}(A^2) = x^2 + y^2 = 2\left(\binom{21}{2} - \binom{18}{2}\right) + 3 = 117.$$

(There are other ways to compute $\operatorname{tr}(A^2)$.) The solutions to the equations $x + y = 3$ and $x^2 + y^2 = 117$ are $(x, y) = (9, -6)$ and $(-6, 9)$. Hence, $C_G(\ell) = 9^\ell + (-6)^\ell$.

72. Let us count the number $c_{G'}(n)$ of closed walks of length n in G'. We can do a closed walk W in G of length $n - 2k$, and then between any two steps of the walk (including before the first step and after the last) insert "detours" of length two along an edge e_v and back. There are $\binom{n-k}{k}$ ways to insert the detours [why?]. Thus, the number of closed walks of G' that start at a vertex of G is

$$c_G(n) + \binom{n-1}{1}c_G(n-2) + \binom{n-2}{2}c_G(n-4) + \binom{n-3}{3}c_G(n-6) + \cdots.$$

On the other hand, we can start at a vertex v'. In this case, after one step we are at v and can take $n - 2$ steps as in the previous case, ending at v, and then step to v'. Thus, the number of closed walks of G' that start at a vertex v' is

$$c_G(n-2) + \binom{n-3}{1}c_G(n-4) + \binom{n-4}{2}c_G(n-6) + \binom{n-5}{3}c_G(n-8) + \cdots.$$

Therefore

$$c_{G'}(n) = c_G(n) + \left(\binom{n-1}{1} + 1\right)c_G(n-2)$$

$$+ \left(\binom{n-2}{2} + \binom{n-3}{1}\right)c_G(n-4)$$

$$+ \left(\binom{n-3}{3} + \binom{n-4}{2}\right)c_G(n-6) + \cdots.$$

The following formula can be proved in various ways and is closely related to Exercise 4.22: If $\lambda^2 \neq -4$, then

$$\lambda^n + \left(\binom{n-1}{1} + 1\right)\lambda^{n-2} + \left(\binom{n-2}{2} + \binom{n-3}{1}\right)\lambda^{n-4}$$

$$+ \left(\binom{n-3}{3} + \binom{n-4}{2}\right)\lambda^{n-6} + \cdots = \alpha^n + \overline{\alpha}^n,$$

where

$$\alpha = \frac{\lambda + \sqrt{\lambda^2 + 4}}{2}, \quad \overline{\alpha} = \frac{\lambda - \sqrt{\lambda^2 + 4}}{2}.$$

Since $c_G(n) = \sum \lambda_i^n$, where the λ_i's are the eigenvalues of $A(G)$, and similarly for $c_{G'}(n)$, we get that the eigenvalues of $A(G')$ are $(\lambda_i \pm \sqrt{\lambda_i^2 + 4})/2$. (We don't have to worry about the special situation $\lambda_i^2 = -4$ since the λ_i's are real.)

For a slight generalization and a proof using linear algebra, see Theorem 2.13 on page 60 of D. M. Cvetković, M. Doob, and H. Sachs [4.10].

74. *Answer.* $9^\ell + 2 \cdot 4^\ell + (-5)^\ell + 5 \cdot 2^\ell + 4$. Why is there a term $+4$?

75. **a.** The column vector $(1, \zeta^r, \zeta^{2r}, \ldots, \zeta^{(n-1)r})^t$ (t denotes transpose) is an eigenvector for M with eigenvalue ω_r. The attempt to generalize this result from cyclic groups and other finite abelian groups to arbitrary finite groups led Frobenius to the discovery of group representation theory; see T. Hawkins, *Arch. History Exact Sci.* **7** (1970/71), 142–170; **8** (1971/72), 243–287; **12** (1974), 217–243.

b. $f_k(n) = k \cdot 3^{n-1}$

c. Let $\Gamma = \Gamma_k$ be the directed graph on the vertex set $\mathbb{Z}/k\mathbb{Z}$ such that there is an edge from i to $i+1 \pmod{k}$. Then $g_k(n)$ is the number of closed walks in Γ of length n. If for $(i, j) \in (\mathbb{Z}/k\mathbb{Z})^2$, we define

$$M_{ij} = \begin{cases} 1, & \text{if } j \equiv i-1, i, i+1 \pmod{k}, \\ 0, & \text{otherwise,} \end{cases}$$

then the transfer-matrix method shows that $g_k(n) = \text{tr} \, M^n$, where $M = (M_{ij})$. By (a), the eigenvalues of M are $1 + \zeta^r + \zeta^{-r} = 1 + 2\cos(2\pi r/k)$, where $\zeta = e^{2\pi i/k}$, and the proof follows.

76. **a.** Expand $\det(xI - A)$ by the first row. We get $V_n(x) = xV_{n-1}(x) + \det(xI - A : 1, 2)$. Subtract the first column of the matrix $(xI - A : 1, 2)$ from the second. The determinant is then clearly $-V_{n-2}(x)$, and the result follows. NOTE. If $U_n(x)$ is the *Chebyshev polynomial of the first kind*, then $V_n(2x) = U_n(x)$.

c. *Answer.* $\sum_{j=1}^{n} \left(2 \cos \frac{j\pi}{n+1}\right)^k$.

d. *Answer.* $(2n+1)\binom{2n}{n} - 4^n$.

78. **a.** There are two choices for the first column. Once a column has been chosen, there are always exactly three choices for the next column (to the right).

b. Let Γ_n be the graph whose vertex set is $\{0, 1\}^{n-1}$, with m edges from $(a_1, \ldots, a_{n-1})$ to $(b_1, \ldots, b_{n-1})$ if there are m ways to choose the next column to be of the form $[d, d + b_1, d + b_1 + b_2, \ldots, d + b_1 + \cdots + b_{n-1}]^t$ when the current column has the form $[c, c + a_1, c + a_1 + a_2, \ldots, c + a_1 + \cdots + a_{n-1}]^t$. In particular, there are two loops at each vertex; otherwise, $m = 0$ or 1. Then $g_k(n)$ is the total number of walks of length $n - 1$ in Γ_n, so by the transfer-matrix method (Theorem 4.7.2) $G_k(x)$ is rational.

For the case $k = 3$, we get a 4×4 matrix A with $\det(I - xA) = (1 - x)(2 - x)(1 - 5x + 2x^2)$, but the factor $(1 - x)(2 - x)$ is cancelled by the numerator. Thus, we are led to the question: What is the degree of the denominator of $G_k(x)$ when this rational function is reduced to lowest terms?

This exercise is due to L. Levine (private communication, 2009).

79. a. Write $c_i(n)$ for the number of closed walks of length n in G_i, and similarly $c(n)$ for the number of closed walks of length n in $G = G_1 * \cdots * G_k$. Then

$$c(n) = \sum_{\substack{i_1 + \cdots + i_k = n \\ i_j \geq 0}} \binom{n}{i_1, \ldots, i_k} c_1(i_1) \cdots c_k(i_k).$$

It follows that if $F_i(x) = \sum_{n \geq 0} c_i(n) \frac{x^n}{n!}$ and $F(x) = \sum_{n \geq 0} c(n) \frac{x^n}{n!}$, then $F(x) = F_1(x) \cdots F_k(x)$. If the eigenvalues of a graph G are $\lambda_1, \ldots, \lambda_p$, then

$$\sum_{n \geq 0} (\lambda_1^n + \cdots + \lambda_p^n) \frac{x^n}{n!} = e^{\lambda_1 x} + \cdots + e^{\lambda_p x}.$$

It now follows from equation (4.35) that the eigenvalues of G are the numbers $\mu_1 + \cdots + \mu_k$, where μ_i is an eigenvalue of G_i.

The star product is usually called the *sum* and is denoted $G_1 + \cdots + G_k$, but this notation conflicts with our notation for disjoint union. The result of this exercise is a special case of a more general result of D. M. Cvetković, Grafovi i njihovi spektri (thesis), *Univ. Beograd Publ. Elektrotehn. Fak.*, Ser. Mat. Fiz., no. 354–356 (1971), 1–50, and also appears in [4.10, Thm. 2.23].

b. We are asking for the number of walks of length n in the star product $K_{m_1} * \cdots * K_{m_k}$, from $(1, 1, \ldots, 1)$ to $(1^{n-r}, 2^r)$. Write $f_m(n)$ for the number of closed walks of length n in K_m from some specified vertex i. Write $g_m(n)$ for the number of walks of length n in K_m from some specified vertex i to a specified different vertex j. Then the number N we seek is given by

$$N = \sum_{\substack{i_1 + \cdots + i_k = n \\ i_j \geq 0}} \binom{n}{i_1, \ldots, i_k} f_{m_1}(i_1) \cdots f_{m_{k-r}}(i_{k-r}) g_{m_{k-r+1}}(i_{k-r+1}) \cdots g_{m_k}(i_k).$$

$$(4.72)$$

By Example 4.7.5 we have

$$f_m(n) = \frac{1}{m} \left((m-1)^n + (m-1)(-1)^n \right),$$

$$g_m(n) = \frac{1}{m} \left((m-1)^n - (-1)^n \right).$$

Substituting into equation (4.72), expanding the product, and arguing as in (a) gives

$$N = \frac{1}{m_1 \cdots m_k} \sum_{S \subseteq [k]} (-1)^{\#([r+1,k]-S)} \left(\prod_{i \in [r+1,k] \cap S} (m_i - 1) \right) \left(\sum_{j \in S} m_j - k \right)^n.$$

For instance, if $B = [a] \times [b]$, then the number of walks from $(1, 1)$ to $(1, 1)$ in n steps is

$$N = \frac{1}{ab} \left((a+b-2)^n + (b-1)(a-2)^n + (a-1)(b-2)^n + (a-1)(b-1)(-2)^n \right).$$

The number of walks from $(1,1)$ to $(1,2)$ in n steps is

$$N = \frac{1}{ab}\left((a+b-2)^n - (a-2)^n + (a-1)(b-2)^n - (a-1)(-2)^n\right).$$

The number of walks from $(1,1)$ to $(2,2)$ in n steps is

$$N = \frac{1}{ab}\left((a+b-2)^n - (a-2)^n - (b-2)^n + (-2)^n\right).$$

This problem can also be solved by explicitly diagonalizing the adjacency matrix of G and using Corollary 4.7.4.

80. Let $N = \{w_1, w_2, \ldots, w_r\}$. Define a digraph $D = (V, E)$ as follows: V consists of all $(r+1)$-tuples $(v_1, v_2, \ldots, v_r, y)$ where each v_i is a left factor of w_i and $v_i \neq w_i$ (so $w_i = v_i u_i$ where $\ell(u_i) \geq 1$) and where $y \in X$. Draw a directed edge from $(v_1, v_2, \ldots, v_r, y)$ to $(v_1', v_2', \ldots, v_r', y')$ if $v_i y' \notin N$ for $1 \leq i \leq r$, and if

$$v_i' = \begin{cases} v_i y', & \text{if } v_i' y \text{ is a left factor of } w_i \\ v_i, & \text{otherwise.} \end{cases}$$

A walk beginning with some $(1, 1, \ldots, 1, y_1)$ (where 1 denotes the empty word) and whose vertices have last coordinates $y_1, y_2, \ldots, y_m$ corresponds precisely to the word $w = y_1 y_2 \cdots y_m$ having no subword in N. Hence by the transfer-matrix method, $F_N(x)$ is rational.

81. **a.** Let D be the digraph with vertex set $V = \{0, 1\}^n$. Think of $(\varepsilon_1, \ldots, \varepsilon_k) \in V$ as corresponding to a column of a $k \times n$ chessboard covered with dimers, where $\varepsilon_i = 1$ if and only if the dimer in row i extends into the next column to the right. There is a directed edge $u \to v$ if it is possible for column u to be immediately followed by column v. For instance, there is an edge $01000 \to 10100$, corresponding to Figure 4.31. Then $f_k(n)$ is equal to the number of walks in D of length $n-1$ with certain allowed initial and final vertices, so by Theorem 4.7.2 $F_k(x)$ is rational. (There are several tricks to reduce the number of vertices which will not be pursued here.)

Example. Let $k = 2$. The digraph D is shown if Figure 4.32. The paths must start at 00 or 11 and end at 00. Hence, if

$$A = \begin{bmatrix} 1 & 1 \\ 1 & 0 \end{bmatrix},$$

then

$$F_2(x) = \frac{-\det(I - xA : 1, 2) + \det(I - xA : 2, 2)}{\det(I - xA)}$$

$$= \frac{x + (1-x)}{1 - x - x^2} = \frac{1}{1 - x - x^2},$$

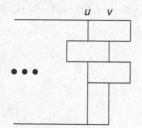

$u \quad v$ Figure 4.31 Dimers in columns u and v.

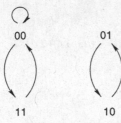

Figure 4.32 The digraph for dimer coverings of a $2 \times n$ board.

the generating function for Fibonacci numbers.

This result can also be easily be obtained by direct reasoning. We also have (see J. L. Hock and R. B. McQuistan, *Discrete Applied Math.* **8** (1984), 101–104; D. A. Klarner and J. Pollack, *Discrete Math.* **32** (1980), 45–52; R. C. Read, *Aequationes Math.* **24** (1982), 47–65):

$$F_3(x) = \frac{1 - x^2}{1 - 4x^2 + x^4},$$

$$F_4(x) = \frac{1 - x^2}{1 - x - 5x^2 - x^3 + x^4},$$

$$F_5(x) = \frac{1 - 7x^2 + 7x^4 - x^6}{1 - 15x^2 + 32x^4 - 15x^6 + x^8},$$

$$F_6(x) = \frac{1 - 8x^2 - 2x^3 + 8x^4 - x^6}{1 - x - 20x^2 - 10x^3 + 38x^4 + 10x^5 - 20x^6 + x^7 + x^8}.$$

b. Equation (4.59) was first obtained by P. W. Kastelyn, *Physica* **27** (1961), 1209–1225. It was proved *via* the transfer-matrix method by E. H. Lieb, *J. Math. Phys.* **8** (1967), 2339–2341. Further references to this and related results appear in the solution to Exercise 3.82(b). See also Section 8.3 of Cvetković, Doob, and Sachs [4.10].

c. See R. Stanley, *Discrete Applied Math.* **12** (1985), 81–87.

82. a. $\chi_n(2) = \begin{cases} 2, & n \text{ even}, \\ 0, & n \text{ odd}. \end{cases}$

b. This is equivalent to a result of E. H. Lieb, *Phys. Rev.* **162** (1967), 162–172. More detailed proofs appear in Percus [4.40, pp. 143–159] (this exposition has many minor inaccuracies), E. H. Lieb and F. Y. Wu, in *Phase Transitions and Critical Phenomena* (C. Domb and M. S. Green, eds.), vol. 1, Academic Press, London/New York, 1972, pp. 331–490, and Baxter [4.2] (see eq. (8.8.20) and p. 178).

c. The constant $-\pi/6$ has been empirically verified to eight decimal places.

e,f. See N. L. Biggs, *Interaction Models*, Cambridge University Press, Cambridge, 1977; Biggs, *Bull. London Math. Soc.* **9** (1977), 54–56; D. Kim and I. G. Enting, *J. Combinatorial Theory Ser. B* **26** (1979), 327–336. In particular, the expansion (4.60) is equivalent to equation (16) of the last reference. Part (f) is due to J. Schneider, 2011.

Appendix

Graph Theory Terminology

The number of systems of terminology presently used in graph theory is equal, to a close approximation, to the number of graph theorists. Here we describe the particular terminology that we have chose to use throughout this book, though we make no claims about its superiority to any alternate choice of terminology.

A *finite graph* is a triple $G = (V, E, \phi)$, where V is a finite set of *vertices*, E is a finite set of *edges*, and ϕ is a function that assigns to each edge e a 2-element multiset of vertices. Thus, $\phi \colon E \to \left(\binom{V}{2} \right)$. If $\phi(e) = \{u, v\}$, then we think of e as *joining* the vertices u and v. We say that u and v are *adjacent* and that u and e, as well as v and e, are *incident*. If $u = v$, then e is called a *loop*. If ϕ is injective (one-to-one) and has no loops, then G is called *simple*. In this case, we may identify e with the set $\phi(e) = \{u, v\}$, sometimes written $e = uv$. In general, the function ϕ is rarely explicitly mentioned in dealing with graphs, and virtually never mentioned in the case of simple graphs.

A *walk* (called by some authors a path) of length n from vertex u to vertex v is a sequence $v_0 e_1 v_1 e_2 v_2 \cdots e_n v_n$ such that $v_i \in V$, $e_i \in E$, $v_0 = u$, $v_n = v$, and any two consecutive terms are incident. If G is simple then the sequence $v_0 v_1 \cdots v_n$ of vertices suffices to determine the walk. A walk is *closed* if $v_0 = v_n$, a *trail* if the e_i's are distinct, and a *path* if the v_i's (and hence the e_i's) are distinct. If $n \geq 1$ and all the v_i's are distinct except for $v_0 = v_n$, then the walk is called a *cycle*.

A graph is *connected* if it is nonempty and any two distinct vertices are joined by a path (or walk). A graph without cycles is called a *free forest* or simply a *forest*. A connected graph without cycles is called a *free tree* (called by many authors simply a tree).

A *digraph* or *directed graph* is defined analogously to a graph, except now $\phi \colon E \to V \times V$; that is, an edge consists of an *ordered* pair (u, v) of vertices (possibly equal). The notions of walk, path, trail, cycle, and so on, carry over in a natural way to digraphs; see the beginning of Section 4.7.1 for further details.

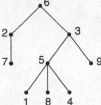

Figure A.1 A tree.

We come next to the concept of a tree. It may be defined recursively as follows. A *tree* (or *rooted tree*) T is a finite set of vertices such that:

a. One specially designated vertex is called the *root* of T, and
b. The remaining vertices (excluding the root) are partitioned into $m \geq 0$ disjoint nonempty sets $T_1, \ldots, T_m$, each of which is a tree. The trees $T_1, \ldots, T_m$ are called *subtrees* of the root.

Rather than formally defining certain terms associated with a tree, we will illustrate these terms with an example, trusting that this will make the formal definitions clear. Suppose that $T = [9]$, with root 6 and subtrees T_1, T_2. The subtree T_1 has vertices $\{2, 7\}$ and root 2, while T_2 has vertices $\{1, 3, 4, 5, 8, 9\}$, root 3, and subtrees T_3, T_4. The subtree T_3 has vertices $\{1, 4, 5, 8\}$, root 5, and subtrees T_5, T_6, T_7 consisting of one vertex each, while T_4 consists of the single vertex 9. The tree T is depicted in an obvious way in Figure A.1. Note that we are drawing a tree with its root at the *top*. This is the most prevalent convention among computer scientists and combinatorialists, though many graph theorists (as well as Nature herself) would put the root at the bottom. In Figure A.1, we call vertices 2 and 3 the *children* or *successors* of vertex 6. Similarly 7 is the child of 2, 5 and 9 are the children of 3, and 1, 4, and 8 are the children of 5. We also call 2 the *parent* or *predecessor* of 7, 5 the parent of 1,4, and 8, and so on. Every vertex except the root has a unique parent. Those vertices without children are called *leaves* or *endpoints*; in Figure A.1 they are 1, 4, 7, 8, and 9.

If we take the diagram of a tree as in Figure A.1 and ignore the designation of the root (i.e., consider only the vertices and edges), then we obtain the diagram of a free tree. Conversely, given a free tree G, if we designate one of its vertices as a root, then this defines the structure of a tree T on the vertices of G. Hence a "rooted tree," meaning a free tree together with a root vertex, is conceptually identical to a tree.

A tree may also be regarded in a natural way as a poset; simply consider its diagram to be a Hasse diagram. Thus, a tree T, regarded as a poset, has a unique maximal element, namely, the root of the tree. Sometimes it is convenient to consider the dual partial ordering of T. We therefore define a *dual tree P* to be a poset such that the Hasse diagram of the dual poset P^* is the diagram of a tree.

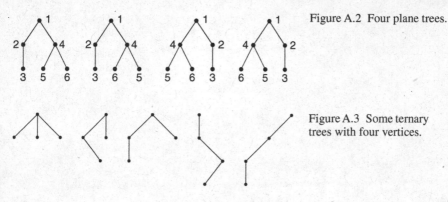

Figure A.2 Four plane trees.

Figure A.3 Some ternary trees with four vertices.

Some important variations of trees are obtained by modifying the recursive definition. A *plane tree* or *ordered tree* is obtained by replacing (b) in the definition of a tree with:

b′. The remaining vertices (excluding the root) are put into an *ordered* partition $(T_1, \ldots, T_m)$ of $m \geq 0$ pairwise disjoint, nonempty sets $T_1, \ldots, T_m$, each of which is a plane tree.

To constrast the distinction between (rooted) trees and plane trees, an ordinary (rooted) tree may be referred to as an *unordered tree*. Figure A.2 shows four *different* plane trees, each of which has the same "underlying (unordered) tree." The ordering $T_1, \ldots, T_m$ of the subtrees is depicted by drawing them from left-to-right in that order.

Now let $m \geq 2$. An *m-ary tree* T is obtained by replacing (a) and (b) with:

a″. Either T is empty, or else one specially designated vertex is called the *root* of T, and
b″. The remaining vertices (excluding the root) are put into a (weak) ordered partition $(T_1, \ldots, T_m)$ of exactly m disjoint (*possibly empty*) sets $T_1, \ldots, T_m$, each of which is an *m*-ary tree.

A 2-ary tree is called a *binary tree*. When drawing an *m*-ary tree for m small, the edges joining a vertex v to the roots of its subtrees $T_1, \ldots, T_m$ are drawn at equal angles symmetric with respect to a vertical axis. Thus, an empty subtree T_i is inferred by the absence of the ith edge from v. Figure A.3 depicts 5 of the 55 nonisomorphic ternary trees with four vertices. We say that an *m*-ary tree is *complete* if every vertex not an endpoint has m children. In Figure A.3, only the first tree is complete.

The *length* $\ell(T)$ of a tree T is equal to its length as a poset; that is, $\ell(T)$ is the largest number ℓ for which there is a sequence $v_0, v_1, \ldots, v_\ell$ of vertices such that v_i is a child of v_{i-1} for $1 \leq v_i \leq \ell$ (so v_0 is necessarily the root of T). The *complete m-ary tree of length ℓ* is the unique (up to isomorphism) complete *m*-ary tree with *every* maximal chain of length ℓ; it has a total of $1 + m + m^2 + \cdots + m^\ell$ vertices.

First Edition Numbering

We give the numbers of theorems and the like from the original (first) edition of Volume 1, together with their numbers in this Second edition. References in Volume 2 to Volume 1 refer to the First edition, so the following list can be used to find the referents in the Second edition of Volume 1.

First edition	Second edition	First edition	Second edition
Example 1.1.1	Example 1.1.1	Proposition 1.3.12	Proposition 1.4.3
Example 1.1.2	Example 1.1.2	Example 1.3.13	Example 1.5.2
Example 1.1.3	Example 1.1.3	Proposition 1.3.14	Proposition 1.5.3
Example 1.1.4	Example 1.1.4	Example 1.3.15	Example 1.5.4
Example 1.1.5	Example 1.1.5	Proposition 1.3.16	Proposition 1.5.5
Example 1.1.6	Example 1.1.6	Proposition 1.3.17	Proposition 1.7.1
Example 1.1.7	Example 1.1.7	Proposition 1.3.18	Proposition 1.7.2
Proposition 1.1.8	Proposition 1.1.8	Proposition 1.3.19	Proposition 1.7.3
Proposition 1.1.9	Proposition 1.1.9	Proposition 1.4.4	Proposition 1.8.1
Example 1.1.10	Example 1.1.10	Corollary 1.4.5	Corollary 1.8.2
Example 1.1.11	Example 1.1.11	Proposition 1.4.1	Proposition 1.9.1
Example 1.1.12	Example 1.1.12	Proposition 1.4.2	Proposition 1.9.2
Example 1.1.13	Example 1.1.13	Corollary 1.4.3	Corollary 1.9.3
Example 1.1.14	Example 1.1.14	Theorem 2.1.1	Theorem 2.1.1
Example 1.1.15	Example 1.1.15	Example 2.2.1	Example 2.2.1
Example 1.1.16	Example 1.1.16	Proposition 2.2.2	Proposition 2.2.2
Example 1.1.17	Example 1.1.17	Example 2.2.3	Example 2.2.3
Proposition 1.3.1	Proposition 1.3.1	Example 2.2.4	Example 2.2.4
Proposition 1.3.2	Proposition 1.3.2	Example 2.2.5	Example 2.2.5
Lemma 1.3.3	Lemma 1.3.6	Proposition 2.2.6	Proposition 2.2.6
Proposition 1.3.4	Proposition 1.3.7	Theorem 2.3.1	Theorem 2.3.1
Example 1.3.5	Example 1.3.8	Example 2.3.2	Example 2.3.2
Example 1.3.6	Example 1.3.9	Example 2.3.3	Example 2.3.3
Proposition 1.3.7	Proposition 1.3.10	Lemma 2.3.4	Lemma 2.3.4
Corollary 1.3.8	Corollary 1.3.11	Corollary 2.3.5	Corollary 2.3.5
Proposition 1.3.9	Proposition 1.3.12	Theorem 2.4.1	Theorem 2.4.1
Corollary 1.3.10	Corollary 1.3.13	Corollary 2.4.2	Corollary 2.4.2
Proposition 1.3.11	Proposition 1.4.1	Corollary 2.4.3	Corollary 2.4.3

First edition	Second edition	First edition	Second edition
Theorem 2.4.4	Theorem 2.4.4	Corollary 3.12.2	Corollary 3.13.2
Corollary 2.4.5	Corollary 2.4.5	Theorem 3.12.3	Theorem 3.13.3
Proposition 2.5.1	Proposition 2.5.1	Definition 3.13.1	Definition 3.14.1
Proposition 2.5.2	Proposition 2.5.2	Theorem 3.13.2	Theorem 3.14.2
Corollary 2.5.3	Corollary 2.5.3	Example 3.13.3	Example 3.14.3
Example 2.6.1	Example 2.6.1	Example 3.13.4	Example 3.14.4
Theorem 2.7.1	Theorem 2.7.1	Example 3.13.5	Example 3.14.5
Example 2.7.2	Example 2.7.2	Proposition 3.14.1	Proposition 3.16.1
Example 3.1.1	Example 3.1.1	Proposition 3.14.2	Proposition 3.16.2
Proposition 3.3.1	Proposition 3.3.1	Lemma 3.14.3	Lemma 3.16.3
Proposition 3.3.2	Proposition 3.3.2	Lemma 3.14.4	Lemma 3.16.4
Proposition 3.3.3	Proposition 3.3.3	Proposition 3.14.5	Proposition 3.16.5
Theorem 3.4.1	Theorem 3.4.1	Corollary 3.14.6	Corollary 3.16.6
Proposition 3.4.2	Proposition 3.4.2	Example 3.14.7	Example 3.16.7
Proposition 3.4.3	Proposition 3.4.3	Example 3.14.8	Example 3.16.8
Proposition 3.4.4	Proposition 3.4.5	Theorem 3.14.9	Theorem 3.16.9
Proposition 3.5.1	Proposition 3.5.1	Example 3.15.1	Example 3.18.1
Proposition 3.5.2	Proposition 3.5.2	Definition 3.15.2	Definition 3.18.2
Example 3.5.3	Example 3.5.3	Example 3.15.3	Example 3.18.3
Example 3.5.4	Example 3.5.4	Theorem 3.15.4	Theorem 3.18.4
Example 3.5.5	Example 3.5.5	Proposition 3.15.5	Proposition 3.18.5
Definition 3.6.1	Definition 3.6.1	Example 3.15.6	Example 3.18.6
Proposition 3.6.2	Proposition 3.6.2	Example 3.15.7	Example 3.18.7
Proposition 3.7.1	Proposition 3.7.1	Example 3.15.8	Example 3.18.8
Proposition 3.7.2	Proposition 3.7.2	Example 3.15.9	Example 3.18.9
Example 3.8.1	Example 3.8.1	Example 3.15.10	Example 3.18.10
Proposition 3.8.2	Proposition 3.8.2	Example 3.15.11	Example 3.18.11
Example 3.8.3	Example 3.8.3	Lemma 3.16.1	Lemma 3.19.1
Example 3.8.4	Example 3.8.4	Lemma 3.16.2	Lemma 3.19.2
Proposition 3.8.5	Proposition 3.8.5	Corollary 3.16.3	Corollary 3.19.3
Proposition 3.8.6	Proposition 3.8.6	Proposition 3.16.4	Proposition 3.19.4
Example 3.8.7	Example 3.8.7	Theorem 4.1.1	Theorem 4.4.1.1
Proposition 3.8.8	Proposition 3.8.8	Example 4.1.2	Example 4.1.3
Proposition 3.8.9	Proposition 3.8.9	Corollary 4.2.1	Corollary 4.2.1
Example 3.8.10	Example 3.8.10	Proposition 4.2.2	Proposition 4.2.2
Proposition 3.8.11	Proposition 3.8.11	Proposition 4.2.3	Proposition 4.2.3
Definition 3.9.1	Definition 3.9.1	Corollary 4.2.4	Corollary 4.2.4
Theorem 3.9.2	Theorem 3.9.2	Proposition 4.2.5	Proposition 4.2.5
Corollary 3.9.3	Corollary 3.9.3	Corollary 4.3.1	Corollary 4.3.1
Corollary 3.9.4	Corollary 3.9.4	Proposition 4.4.1	Proposition 4.4.1
Corollary 3.9.5	Corollary 3.9.5	Example 4.4.2	Example 4.4.2
Example 3.9.6	Example 3.9.6	Lemma 4.5.1	Definition 1.4.10
Proposition 3.10.1	Proposition 3.10.1	Lemma 4.5.2	Lemma 3.15.4
Example 3.10.2	Example 3.10.2	Lemma 4.5.3	Lemma 3.15.3
Example 3.10.3	Example 3.10.3	Theorem 4.5.4	Theorem 3.15.5
Example 3.10.4	Example 3.10.4	Example 4.5.5	Example 3.15.6
Proposition 3.11.1	Proposition 3.12.1	Lemma 4.5.6	Lemma 3.15.9
Example 3.11.2	Example 3.12.2	Theorem 4.5.7	Theorem 3.15.10
Theorem 3.12.1	Theorem 3.13.1	Theorem 4.5.8	Theorem 3.15.7

First edition	Second edition	First edition	Second edition
Corollary 4.5.9	Proposition 1.4.6	Proposition 4.7.6	Proposition 4.7.8
Corollary 4.5.10	Corollary 3.15.13	Example 4.7.7	Example 4.7.9
Example 4.5.11	Example 3.15.14	Proposition 4.7.8	Proposition 4.7.10
Lemma 4.5.12	Lemma 3.15.15	Proposition 4.7.9	Proposition 4.7.11
Theorem 4.5.13	Theorem 3.15.16	Proposition 4.7.10	Proposition 4.7.12
Theorem 4.5.14	Theorem 3.15.8	Proposition 4.7.11	Proposition 4.7.13
Corollary 4.5.15	Corollary 3.15.12	Lemma 4.7.12	Lemma 4.7.14
Lemma 4.5.16	Lemma 3.15.17	Example 4.7.13	Example 4.7.15
Corollary 4.5.17	Corollary 3.15.18	Example 4.7.14	Example 4.7.16
Example 4.5.18	Example 3.15.19	Example 4.7.15	Example 4.7.17
Lemma 4.6.1	Lemma 4.5.1	Example 4.7.16	Example 4.7.18
Lemma 4.6.2	Lemma 4.5.2	Example 4.7.17	Example 4.7.19
Example 4.6.3	Example 4.5.3	Proposition 4.7.19	Proposition 4.7.20
Lemma 4.6.4	Lemma 4.5.4	Exercise 1.1	Exercise 1.2
Example 4.6.5	Example 4.5.5	Exercise 1.2(a–d)	Exercise 1.3(a–d)
Lemma 4.6.6	Lemma 4.5.6	Supplementary Exercise 1.19	Exercise 1.3(e)
Lemma 4.6.7	Lemma 4.5.7	Supplementary Exercise 1.20	Exercise 1.3(g)
Corollary 4.6.8	Corollary 4.5.8	Supplementary Exercise 1.21	Exercise 1.3(h)
Example 4.6.9	Example 4.5.9	Supplementary Exercise 1.14	Exercise 1.4
Proposition 4.6.10	Proposition 4.5.10	Supplementary Exercise 1.5	Exercise 1.5
Theorem 4.6.11	Theorem 4.5.11	Supplementary Exercise 1.28	Exercise 1.6
Lemma 4.6.12	Lemma 4.5.12	Exercise 1.3	incorporated into the text
Lemma 4.6.13	Lemma 4.5.13	Exercise 1.4	Exercise 1.8
Theorem 4.6.14	Theorem 4.5.14	Exercise 1.5	Exercise 1.9
Example 4.6.15	Example 4.5.15	Supplementary Exercise 1.25	Exercise 1.11
Corollary 4.6.16	Corollary 4.5.16	Supplementary Exercise 1.10	Exercise 1.12
Lemma 4.6.17	Lemma 4.5.17	Exercise 1.6	Exercise 1.14
Lemma 4.6.18	Lemma 4.6.1	Exercise 1.7	Exercise 1.17
Proposition 4.6.19	Proposition 4.6.2	Exercise 1.8	Exercise 1.19
Lemma 4.6.20	Lemma 4.6.3	Supplementary Exercise 1.27	Exercise 1.20
Proposition 4.6.21	Proposition 4.6.4	Supplementary Exercise 1.2	Exercise 1.21
Example 4.6.22	Example 4.6.5	Exercise 1.9	Exercise 1.22
Lemma 4.6.23	Lemma 4.6.6	Exercise 1.10	Exercise 1.24
Lemma 4.6.24	Lemma 4.6.7	Supplementary Exercise 1.1	Exercise 1.26
Theorem 4.6.25	Theorem 4.6.8		
Theorem 4.6.26	Theorem 4.6.9		
Example 4.6.27	Example 4.6.10		
Corollary 4.6.28	Corollary 4.6.11		
Example 4.6.29	Example 4.6.12		
Proposition 4.6.30	Proposition 4.6.13		
Corollary 4.6.31	Corollary 4.6.14		
Example 4.6.32	Example 4.6.15		
Example 4.6.33	Example 4.6.16		
Example 4.6.34	Example 4.6.17		
Theorem 4.7.1	Theorem 4.7.1		
Theorem 4.7.2	Theorem 4.7.2		
Corollary 4.7.3	Corollary 4.7.3		
Example 4.7.4	Example 4.7.6		
Example 4.7.5	Example 4.7.7		

First edition	Second edition	First edition	Second edition
Supplementary Exercise 1.3	Exercise 1.28	Supplementary Exercise 1.24	Exercise 1.126
Supplementary Exercise 1.8	Exercise 1.29	Exercise 1.31	Exercise 1.127
Exercise 1.42	Exercise 1.30	Exercise 1.32	Exercise 1.128
Exercise 1.11	Exercise 1.31	Supplementary Exercise 1.15	Exercise 1.131
Exercise 1.12	Exercise 1.32		
Exercise 1.13	Exercise 1.34	Exercise 1.33	Exercise 1.133
Exercise 1.14	Exercise 1.35	Exercise 1.34	Exercise 1.136
Supplementary Exercise 1.6	Exercise 1.18	Exercise 1.35	Exercise 1.137
		Exercise 1.36(a)	Exercise 1.154(a)
Exercise 1.15	Exercise 1.36	Exercise 1.36(b)	Exercise 1.154(b)
Exercise 1.16	Exercise 1.45	Exercise 1.37	Exercise 1.155
Exercise 1.17	Exercise 1.46	Exercise 1.38	Exercise 1.156
Supplementary Exercise 1.19(a,b)	Exercise 1.47(a,b)	Exercise 1.39	Exercise 1.157
		Exercise 1.40	Exercise 1.158
		Exercise 1.41(a)	Exercise 1.168(c)
Exercise 1.18	Exercise 1.48(c)	Exercise 1.41(b)	Exercise 1.168(d)
Supplementary Exercise 1.22	Exercise 1.51	Exercise 1.41(c)	Exercise 1.168(i)
		Exercise 1.41(d)	Exercise 1.168(j)
Supplementary Exercise 1.4	Exercise 1.54	Exercise 1.43(a–c)	Exercise 1.170(a–c)
Exercise 1.19	Exercise 1.62	Exercise 1.44	Exercise 1.173
Exercise 1.20	Exercise 1.63	Exercise 1.45	Exercise 1.203
Supplementary Exercise 1.11	Exercise 1.71	Exercise 2.1	Exercise 2.3
		Exercise 2.2	Exercise 2.4
Exercise 1.21	Exercise 1.74	Exercise 2.3	Exercise 2.5
Exercise 1.22	Exercise 1.75	Exercise 2.4	Exercise 2.8 (modified)
Exercise 1.23	Exercise 1.76(d)		
Exercise 1.24	Exercise 1.78	Exercise 2.5	Exercise 2.9
Exercise 1.25	Exercise 1.79	Supplementary Exercise 2.4	Exercise 2.14
Exercise 1.26	Exercise 1.80		
Supplementary Exercise 1.18	Exercise 1.81	Exercise 2.6(a,b)	Exercise 2.15
		Exercise 2.7	Exercise 2.16
Supplementary Exercise 1.29	Exercise 1.96	Exercise 2.8	Exercise 2.17
		Exercise 2.9	Exercise 2.18
Supplementary Exercise 1.12	Exercise 1.97	Exercise 2.10	Exercise 2.20
Supplementary Exercise 1.26	Exercise 1.102	Supplementary Exercise 2.1	Exercise 2.25
		Supplementary Exercise 2.2	Exercise 2.26
Exercise 1.27	Exercise 1.105		
Exercise 1.28	Exercise 1.106	Exercise 2.11(a,b)	Exercise 2.27
Exercise 1.29	Exercise 1.108		
Supplementary Exercise 1.9	Exercise 1.113	Exercise 2.12	Exercise 2.28
		Exercise 2.13	Exercise 2.30
Supplementary Exercise 1.16	Exercise 1.121	Exercise 2.14	Exercise 2.21(e)
		Exercise 2.15	Exercise 2.33
Exercise 1.30	Exercise 1.124	Exercise 2.16	Exercise 2.34

First edition	Second edition	First edition	Second edition
Exercise 2.17	Exercise 2.36	Supplementary	Exercise 3.91
Supplementary	Exercise 2.37	Exercise 3.5	
Exercise 2.3		Exercise 3.33	Exercise 3.92
Exercise 3.1	Exercise 3.2	Exercise 3.34	Exercise 3.93
Exercise 3.2	Exercise 3.3	Exercise 3.35	Exercise 3.94
Exercise 3.3	Exercise 3.5	Exercise 3.36	Exercise 3.95
Exercise 3.4	Exercise 3.6	Exercise 3.37	Exercise 3.96
Exercise 3.5	Exercise 3.7	Exercise 3.38	Exercise 3.97
Exercise 3.6	Exercise 3.8	Supplementary	Exercise 3.98
Exercise 3.7	Exercise 3.10	Exercise 3.8	
Exercise 3.8	Exercise 3.11	Supplementary	Exercise 3.99
Exercise 3.9	Exercise 3.19	Exercise 3.12	
Exercise 3.10	Exercise 3.20	Exercise 3.38.5	Exercise 3.101
Exercise 3.11	Exercise 3.25	Exercise 3.39	Exercise 3.102
Exercise 3.12	Exercise 3.27	Exercise 3.40	Exercise 3.103
Exercise 3.13	Exercise 3.28	Exercise 3.41	Exercise 3.104
Exercise 3.14	Exercise 3.29	Exercise 3.42	Exercise 3.105
Exercise 3.15	Exercise 3.35	Exercise 3.43	Exercise 3.106
Exercise 3.16	Exercise 3.36	Exercise 3.44	Exercise 3.108(b)
Exercise 3.17	Exercise 3.38	Exercise 3.56	Exercise 3.118
Exercise 3.18	Exercise 3.40	Exercise 3.45	Exercise 3.119
Supplementary	Exercise 3.45	Exercise 3.46	Exercise 3.121
Exercise 3.18		Exercise 3.47	Exercise 3.122
Exercise 3.19	Exercise 3.47	Exercise 3.48	Exercise 3.123
Exercise 3.20	Exercise 3.48	Exercise 3.49	Exercise 3.124
Exercise 3.21	Exercise 3.50	Exercise 3.49.5	Exercise 3.126
Exercise 3.22	Exercise 3.51	Supplementary	Exercise 3.127
Supplementary	Exercise 3.56	Exercise 3.3	
Exercise 3.14		Supplementary	Exercise 3.128
Supplementary	Exercise 3.57	Exercise 3.11	
Exercise 3.1		Supplementary	Exercise 3.129
Supplementary	Exercise 3.58	Exercise 3.4	
Exercise 3.15		Exercise 3.50	Exercise 3.130
Supplementary	Exercise 3.62	Exercise 3.51	Exercise 3.131
Exercise 3.2		Exercise 3.52	Exercise 3.132
Exercise 3.23	Exercise 3.66	Exercise 3.53	Exercise 3.133
Supplementary	Exercise 3.70	Supplementary	Exercise 3.134
Exercise 3.6		Exercise 3.16	
Exercise 3.24	Exercise 3.71	Exercise 3.54	Exercise 3.135
Exercise 3.25	Exercise 3.72	Exercise 3.55	Exercise 3.136
Exercise 3.26	Exercise 3.73	Exercise 3.57	Exercise 3.137
Exercise 3.28	Exercise 3.74	Exercise 3.58	Exercise 3.138
Exercise 3.29	Exercise 3.76	Exercise 3.59	Exercise 3.141
Exercise 3.30	Exercise 3.84	Exercise 3.60	Exercise 3.143
Supplementary	Exercise 3.85	Supplementary	Exercise 3.144
Exercise 3.13		Exercise 3.9	
Exercise 3.31	Exercise 3.86	Exercise 3.61(a)	Exercise 3.145
Exercise 3.32	Exercise 3.87	Exercise 3.61(b)	Exercise 3.146

First edition	Second edition	First edition	Second edition
Exercise 3.62	Exercise 3.147	Supplementary Exercise 4.5	Exercise 3.17
Exercise 3.63	Exercise 3.149		
Exercise 4.21	Exercise 3.150	Supplementary Exercise 4.6	Exercise 3.18
Exercise 3.64	Exercise 3.154		
Exercise 3.65	Exercise 3.155	Supplementary Exercise 4.7	Exercise 3.19
Exercise 3.66	Exercise 3.156		
Exercise 3.67	Exercise 3.157	Supplementary Exercise 4.9	Exercise 3.22
Exercise 3.68	Exercise 3.158		
Exercise 4.22	Exercise 3.163	Supplementary Exercise 4.8	Exercise 3.23
Exercise 4.23	Exercise 3.166		
Exercise 4.26	Exercise 3.169	Exercise 4.7	Exercise 4.27
Exercise 3.27	Exercise 3.170	Exercise 4.8	Exercise 4.28
Exercise 4.24	Exercise 3.171	Exercise 4.9	Exercise 4.31
Exercise 4.25	Exercise 3.172	Exercise 4.10	Exercise 4.32
Exercise 3.69(a,b,c)	Exercise 3.174	Supplementary Exercise 4.12	Exercise 3.33
Exercise 3.69(d)	Exercise 3.175	Exercise 4.11	Exercise 4.35
Exercise 3.70	Exercise 3.176	Exercise 4.12	Exercise 4.38
Exercise 3.71	Exercise 3.177	Exercise 4.13	Exercise 4.39
Supplementary Exercise 3.7	Exercise 3.178	Exercise 4.14	Exercise 4.40
		Exercise 4.15	Exercise 4.41
Exercise 3.72	Exercise 3.179	Exercise 4.16	Exercise 4.43
Exercise 3.73	Exercise 3.180	Exercise 4.17	Exercise 4.44
Exercise 3.74	Exercise 3.182	Exercise 4.18	Exercise 4.45
Exercise 3.75(a)	Exercise 3.183(a)	Exercise 4.19	Exercise 4.46
Exercise 3.75(b,c,d)	Exercise 3.185(c,d,e)	Exercise 4.20	Exercise 4.47
		Exercise 4.27	Exercise 4.51
Exercise 3.76	Exercise 3.188	Exercise 4.28	Exercise 4.48
Exercise 3.78	Exercise 3.196	Exercise 4.29	Exercise 4.53
Exercise 3.79	Exercise 3.201	Supplementary Exercise 4.13	Exercise 4.55
Exercise 3.80	Exercise 3.203		
Exercise 3.81	Exercise 3.204	Exercise 4.30	Exercise 4.56
Exercise 4.1	Exercise 4.2	Exercise 4.31	Exercise 4.63
Exercise 4.2	Exercise 4.3	Exercise 4.32	Exercise 4.64
Exercise 4.3	Exercise 4.4	Exercise 4.33	Exercise 4.66
Supplementary Exercise 4.10	Exercise 3.6	Exercise 4.34	Exercise 4.75
		Supplementary Exercise 4.11	Exercise 4.77
Exercise 4.4	Exercise 4.7		
Exercise 4.5	Exercise 4.8	Exercise 4.35	Exercise 4.80
Exercise 4.6	Exercise 4.15	Exercise 4.36	Exercise 4.81
Supplementary Exercise 4.1	Exercise 3.67	Exercise 4.37	Exercise 4.82
Supplementary Exercise 4.4	Exercise 3.73		

List of Notation (Partial)

$\mathbb{C}$	complex numbers
$\mathbb{N}$	nonnegative integers
$\mathbb{P}$	positive integers
$\mathbb{Q}$	rational numbers
$\mathbb{R}$	real numbers
$\mathbb{R}_+$	nonnegative real numbers
$\mathbb{Z}$	integers
$\mathbb{C}^*$	$\mathbb{C} - \{0\}$
$[n]$	the set $\{1, 2, \ldots, n\}$ for $n \in \mathbb{N}$ (so $[0] = \emptyset$)
$[i, j]$	for integers $i \le j$, the set $\{i, i + 1, \ldots, j\}$ (when the context is clear, it can also be the set $\{x \in \mathbb{R} : i \le x \le j\}$)
δ_{ij}	the Kronecker delta, equal to 1 if $i = j$ and 0 otherwise
$:=$	equals by definition
$\mathaccent"0016U$	disjoint union
$S \subseteq T$	S is a subset of T
$S \subset T$	S is a subset of T and $S \ne T$
$\lfloor x \rfloor$	greatest integer $\le x$
$\lceil x \rceil$	least integer $\ge x$
card X, $\#X$, $\lvert X \rvert$	all used for the number of elements of the finite set X
$\{a_1, \ldots, a_k\}_<$	the set $\{a_1, \ldots, a_k\} \subset \mathbb{R}$, where $a_1 < \cdots < a_k$
2^S	the set of subsets of S
X^N	the set of all functions $f : N \to X$ (a vector space if X is a field)
$\binom{S}{k}$	the set of k-element subsets of S
$\left(\!\binom{S}{k}\!\right)$	the set of k-element multisets on S
$\binom{n}{a_1, a_2, \ldots, a_k}$	multinomial coefficient
$\binom{n}{k}$	q-binomial coefficient (in the variable q)
$\binom{n}{k}_x$	q-binomial coefficient in the variable x
(i)	$1 + q + \cdots + q^{i-1}$, for $i \in \mathbb{P}$

$(n)!$	$(1)(2)\cdots(n)$, for $n \in \mathbb{N}$
$[i]$	$1 - x^i$, for $i \in \mathbb{P}$
$[n]!$	$[1][2]\cdots[n]$, for $n \in \mathbb{N}$
$\mathfrak{S}_M$	the set of permutations of the multiset M
$s(n,k)$	Stirling number of the first kind
$c(n,k)$	signless Stirling number of the first kind
$S(n,k)$	Stirling number of the second kind
$B(n)$	Bell number
$A(n,k)$	Eulerian number
$A_d(x)$	Eulerian polynomial
E_n	Euler number
C_n	Catalan number
$\mathcal{A}(w)$	set of functions $f : [n] \to \mathbb{N}$ that are compatible with $w \in \mathfrak{S}_n$
$\mathcal{A}_m(w)$	set of functions $f : [n] \to [m]$ that are compatible with $w \in \mathfrak{S}_n$
Δ	first difference operator (i.e., $\Delta f(n) = f(n+1) - f(n)$)
$\lambda \vdash n$	λ is a partition of the integer $n \geq 0$
$\mathrm{Par}(n)$	the set of all partitions of the integer $n \geq 0$
$\mathrm{Comp}(n)$	the set of all compositions of the integer $n \geq 0$
$p(n)$	number of partitions of n
$p_k(n)$	number of partitions of n into k parts
$\mathfrak{S}_n$	set (or group) of all permutations of $[n]$
$\mathrm{inv}(w)$	number of inversions of the permutation (or sequence) w
$\mathrm{maj}(w)$	major index of the permutation (or sequence) w
$\mathrm{des}(w)$	number of descents of the permutation (or sequence) w
$D(w)$	descent set of the permutation (or sequence) w
$\mathrm{exc}(w)$	number of excedances of the permutation $w \in \mathfrak{S}_n$
$\mathrm{Exc}(w)$	excedance set of the permutation $w \in \mathfrak{S}_n$
$c(w)$	number of cycles of the permutation w
$\alpha_n(S), \alpha(S)$	$\#\{w \in \mathfrak{S}_n : D(w) \subseteq S\}$
$\beta_n(S), \beta(S)$	$\#\{w \in \mathfrak{S}_n : D(w) = S\}$
$\mathbb{F}_q$	a finite field (unique up to isomorphism) with q elements
$\mathbb{F}_q^*$	$\mathbb{F}_q - \{0\}$
$\mathrm{GL}(n,q)$	group of invertible linear transformations $\mathbb{F}_q^n \to \mathbb{F}_q^n$
$\mathrm{Mat}(n,q)$	algebra of $n \times n$ matrices over $\mathbb{F}_q$
$\gamma_n = \gamma_n(q)$	$\#\mathrm{GL}(n,q)$
$\mathcal{I}(q)$	set of all nonconstant monic irreducible polynomials over $\mathbb{F}_q$
$\mathrm{im}\, A$	image of the linear transformation (or function) A
$\ker A$	kernel of the linear transformation A
$\mathrm{tr}\, A$	trace of the linear transformation A

$\det(B : j,i)$	determinant of the matrix obtained from B by removing the jth row and ith column
$\text{aff}(S)$	affine span of the subset S of a vector space V (i.e., all linear combinations of elements of S whose coefficients sum to 0)
$R[x]$	ring of polynomials in the indeterminate x with coefficients in the integral domain R
$R(x)$	ring of rational functions in x with coefficients in R, so $R(x)$ is the quotient field of $R[x]$ when R is a field
$R[[x]]$	ring of formal power series $\sum_{n\geq 0} a_n x^n$ in x with coefficients a_n in R
$R((x))$	ring of formal Laurent series $\sum_{n\geq n_0} a_n x^n$, for some $n_0 \in \mathbb{Z}$, in x with coefficients a_n in R, so $R((x))$ is the quotient field of $R[[x]]$ when R is a field
$\Psi_{k,j} F(x)$	(k, j) multisection of the power series $F(x)$
x^α	$x_1^{\alpha_1} \cdots x_k^{\alpha_k}$, where $\alpha = (\alpha_1,\dots,\alpha_k)$
$[x^n]F(x)$	coefficient of x^n in the series $F(x) = \sum a_n x^n$
$F(x)^{\langle -1\rangle}$	compositional inverse of the power series $F(x) = a_1 x + a_2 x^2 + \cdots, a_1 \neq 0$
$f(n) \sim g(n)$	$f(n)$ and $g(n)$ are asymptotic as $n \to \infty$ (i.e., $\lim_{n\to\infty} f(n)/g(n) = 1$)
$s \parallel t$	s and t are incomparable (in a poset P)
$s \lessdot t$	t covers s (in a poset P)
P^*	dual of the poset P
$\widehat{P}$	the poset P with a $\hat{0}$ and $\hat{1}$ adjoined
$P + Q$	disjoint union of the posets P and Q
$P \times Q$	cartesian (or direct) product of the posets P and Q
$P \oplus Q$	ordinal sum of the posets P and Q
$P \otimes Q$	ordinal product of the posets P and Q
Λ_t	$\{s \in P : s \leq t\}$, where P is a poset
V_t	$\{s \in P : s \geq t\}$, where P is a poset
$\text{Int}(P)$	the poset of (nonempty) intervals of the poset P
$J(P)$	lattice of order ideals of the poset P
$J_f(P)$	lattice of *finite* order ideals of the poset P
$e(P)$	number of linear extensions of the poset P
ρ	rank function of the graded poset P
$\ell(s,t)$	length of the longest chain of the interval $[s,t]$
P_S	the S-rank-selected subposet of the graded poset P (i.e., $P_S = \{t \in P : \rho(t) \in S\}$)
$\alpha_P(S)$	number of maximal chains of the rank-selected subposet P_S
$\beta_P(S)$	$\sum_{T \subseteq S}(-1)^{\#(S-T)}\alpha_P(T)$

$\mathcal{L}(P,\omega)$	set of linear extensions (regarded as permutations of the labels) of the labeled poset (P,ω) on $[p]$
$|\sigma|$	$\sum_{t\in X}\sigma(t)$, for a function $\sigma: X \to \mathbb{Z}$, where X is a finite set
$\mathcal{L}(P)$	the set $\mathcal{L}(P,\omega)$ when ω is natural
$\Omega_{P,\omega}(m)$	order polynomial of the labeled poset P,ω
$\Omega_P(m)$	$\Omega_{P,\omega}(m)$ when ω is natural
$|\Delta|$	geometric realization of the simplicial complex Δ
$\partial\Gamma$	boundary of a triangulation Γ
Γ°	interior of a triangulation Γ
$r(\mathcal{A})$	number of regions of the arrangement $\mathcal{A}$
$b(\mathcal{A})$	number of bounded regions of the arrangement $\mathcal{A}$

Index

Printed in the United States
By Bookmasters